ANNUAL REVIEW OF MICROBIOLOGY

ANNUAL REVIEW OF MICROBIOLOGY

CHARLES E. CLIFTON, *Editor*
Stanford University

SIDNEY RAFFEL, *Associate Editor*
Stanford University

MORTIMER P. STARR, *Associate Editor*
University of California

VOLUME 25

1971

ANNUAL REVIEWS INC.
PALO ALTO, CALIFORNIA, USA 94306

ANNUAL REVIEWS INC.
PALO ALTO, CALIFORNIA, USA

Standard Book Number 8243-1125-6
Library of Congress Catalog Card Number: 49-432

FOREIGN AGENCY

Maruzen Company, Limited
6, Tori-Nichome Nihonbashi
Tokyo

PRINTED AND BOUND IN THE UNITED STATES OF AMERICA BY
GEORGE BANTA COMPANY, INC.

PREFACE

After twenty-four prefaces which have fit comfortably within the stereotype, the Editor is finally asserting his prerogative of freedom. In this twenty-fifth volume he is going to say only thank you to all of the thousands of researchers whose work our several scores of writers have reviewed for these pages, to the writers themselves, and all others who make these volumes possible

THE EDITOR

CONTENTS

REPRINTS

The conspicuous number (1560 to 1585) aligned in the margin with the title of each review in this volume is a key for use in the ordering of reprints.

Reprints of most articles published in the *Annual Reviews of Biochemistry* and *Psychology* from 1961, and the *Annual Reviews of Microbiology* and *Physiology* from 1968 are now maintained in inventory. Beginning in July 1970, this reprint policy was extended to all other *Annual Reviews* volumes.

Available reprints are priced at the uniform rate of $1 each, postpaid. Payment must accompany orders of less than $10. The following discounts will be given for large orders: $5–9, 10%; $10–24, 20%; $25 and over, 30%. All remittances are to be made payable to Annual Reviews Inc. in U. S. dollars. California orders are subject to sales tax. One-day service is given on items in stock.

For orders of 100 or more, any *Annual Reviews* article will be specially printed and shipped within 6 weeks. Reprints that are out of stock may also be purchased from the Institute for Scientific Information, 325 Chestnut Street, Philadelphia, Pa. 19106. Direct inquiries to Annual Reviews, Inc., Reprint Department.

The sale of reprints of articles published in the *Reviews* has been expanded in the belief that reprints as individual copies, as sets covering stated topics, and in quantity for classroom use will have a special appeal to students and teachers.

André Lwoff

FROM PROTOZOA TO BACTERIA AND VIRUSES 1560

Fifty Years With Microbes

André Lwoff

Institut de Recherches Scientifiques sur le Cancer
B. P. 8, 94—Villejuif (France)

To Marguerite

Contents

Introduction

My father, born in Russia, in Sinferopol, came to France in 1880 at the age of 21. As a student, he had been involved in political activities and was sentenced to a few months imprisonment in the Peter and Paul prison in Saint-Petersburg, later Petrograd, then Leningrad. My father then left Russia, came to Paris, where he studied medicine and became a psychiatrist. My mother, too, was Russian. She had studied sculpture in Saint-Petersburg with Antokolski and, as a young girl, came to Paris as so many artists did.

My father soon became head physician in a psychiatric hospital and I was born in the small village of Ainay-le-Château (Allier) in the center of France, on May the 8th, 1902. Then we went to Neuilly-sur-Marne, close to Paris, to another hospital. The psychiatrist had a house just at the outskirts of the hospital. At home, the servants, the cook, the chambermaid, and the gardener were inmates of the hospital. The cook sometimes answered voices which were speaking to her and which I could not hear. The chambermaid sometimes behaved strangely, and also the gardener. This was perfectly normal, of course, and things went all right. Maybe I should mention that one person was not an inmate of the hospital, namely, "Fraulein." She taught me German and at the time I became bilingual. However, Fraulein used to speak to herself . . . Have I been influenced in one way or another by this unusual environment? I am the last to be able to answer this question.

The house was surrounded by a garden with beautiful trees and flowers. There were also fruit trees and a vegetable garden. In the vicinity was a tennis court, and not far away a river where we used to swim and row. Moreover, I had been presented with a small rifle and used to kill rats and sparrows with bullets and even, once—a great event—a partridge in the field behind the garden. From time to time my father interrupted the games and I was summoned to his office to read Plato or Kant but, on the whole, life was marvelous.

Guests came to visit us. Relatives from Russia, of course, and from Italy and Austria, because the family had been scattered by emigration. There were also Russian socialists just out of jail or from Siberia, freed in one way or another. Also members of illustrious aristocratic Russian families living in Paris who had taken the wrong path and needed the care of a psychiatrist.

My father had decided that the study of mental illnesses should be part of my education, therefore at a tender age, I was taken to the wards. I do not know if this exposure has been useful but I am under the impression that the contacts, perhaps somewhat premature, with the manifestations of mental disorder considerably reinforced my inclination for scientific disciplines.

I was twelve when the first World War started. The German troops had reached a point twenty miles from our place. Later on during the battles, I could hear the roaring of the cannons from the battle line. From time to time at night German planes were trying to bomb Paris. The antiaircraft guns were close and the splitters were whistling and drumming on the roof. I listened to the strange music with curiosity, perfectly unaware of the danger. Later on, the psychiatric hospital was turned into a military hospital but I was not mature enough to realize the depth of the war tragedy.

I went to school, read an incredible number of books, and suddenly I decided to study biology in order to do research. How this idea germinated in my head, I do not know. Perhaps because of Elie Metchnikoff who was a friend of my father. When a patient had died of a disease which interested

Metchnikoff, he went to the hospital for the autopsy, and then came home for lunch. Cotton-plugged tubes showed outside the pocket of his coat and the cotton was soiled with blood. My poor mother was horrified.

It was Elie Metchnikoff who showed me a microbe for the first time. It was in 1915 when I was thirteen. My father had taken me to the Pasteur Institute. Metchnikoff asked, "have you ever seen a microbe?" I had not. So, Metchnikoff took a glass slide, put it under the microscope and said "it is the typhoid bacillus. Look." I was very excited. What happened, I remember very well. I looked into the microscope and saw nothing. I was very impressed. Such was my first contact with bacteria.

Anyhow, at seventeen, I decided to study biology and to do research. My father explained to me that research is not always successful and as I would have to earn my living, I had better study medicine. At the time, before entering the medical school the student had to spend a year in the *Faculté des Sciences* studying biology, chemistry, and physics. This was not enough. As I was very enthusiastic I had spent two months in the Marine Biological Laboratory at Roscoff in Brittany during the summers of 1919 and 1920, and during my first years of medicine I took the botany and zoology examinations at the Sorbonne and again in 1921 went to Roscoff where a determining event took place.

Impressed by my youthful ardour, the assistant recommended me to Edouard Chatton. Thus, I had the privilege of becoming the pupil and the collaborator, later on the disciple and friend, of the most brilliant representative of the brilliant French school of protozoology and, in my judgment, the greatest protozoologist of all time. Could I note for the readers of the *Annual Review of Microbiology* that it is Chatton who, in 1928, separated the Eukaryotes from the Prokaryotes, and coined the two names. From 1921 on, we worked together two or three months each year in the marine laboratories at Roscoff, Banyuls, Villefranche-sur-Mer, Wimereux or Sète. Our collaboration was interrupted by the war and ended only with the death of Edouard Chatton in 1947.

In 1921, Chatton was professor in Strasburg. He recommended me to Felix Mesnil who was head of the Department of Protozoology at the Pasteur Institute. He had been the secretary of Louis Pasteur and the collaborator of Alphonse Laveran. In October 1921, I received a fellowship and worked part time whilst studying medicine. I earned 350 francs a month, that is, 70 U.S. dollars—and felt rich. I had, of course, a tendency to sacrifice my medical studies to my passion for ciliates. But still I learned to examine patients; I swallowed anatomy, histology, embryology, bacteriology, parasitology, physiology, and biochemistry, etc., and succeeded in passing the examinations for pathology, obstetrics, therapeutics, and the rest.

The Eyes of Copepods

So, I passed the examinations. But in order to become an M.D. one had to defend a thesis. Whilst in the marine station in Roscoff I had worked on parasitic ciliates which evolved in copepods engulfed by coelenterates. Often they were red and soon it was clear that the eye of the copepod was the origin of the pigment. So I started working on copepod eyes and in 1927 I presented my thesis at the Medical School. The title was "Le cycle du pigment carotinoide chez *Idya furcata* Baird, Copépode Harpacticide." For the first time, a carotinoid pigment was described in an eye. The eyes of copepods are normally dark red. I raised copepods on a carotinoid-poor diet. The red layers became less important and a beautiful blue structure appeared which is normally masked. I showed, among other things, that the blue pigment is a combination of a carotinoid with a protid and proposed the hypothesis that it plays a role in the vision. The president of the "Jury" was the well-known histologist, André Prenant, but a gynecologist was a member of the examination board. He shrugged his shoulders in disgust. Who knows, perhaps he was not really interested in copepods. Anyhow, as a result of the work, my name is now on the list of world copepodologists!

However, there was a byproduct: whilst working at the Kaiser Wilhelm Institut in Heidelberg in 1932, I called to the attention of Richard Kuhn and Edgar Lederer the fact that carotinoids were present in crustaceans. They bought a few lobsters and discovered astacin.

Morphology and Morphogenesis of Ciliates

However beautiful a copepod with blue eyes might be, I was in love with ciliates. During the summer of 1921 in Roscoff I had been asked by Edouard Chatton to explore the gills of acephals. The first mollusc examined, *Dosinia exoleta,* showed a strange organism devoid of cilia which reproduces by ciliated buds. This was the type of a new genus, *Sphenophrya,* and of a new family, the *Sphenophryidae,* to which we later added two other genera, *Pelecyophrya* and *Gargarius.* The systematic investigation of the ciliates of the gills brought the discovery of many new organisms and the organization of new families into a new suborder, the *Thigmotricha.* I should make clear that the work on ciliates, with a few exceptions, was performed with Edouard Chatton and, from 1925 on, with Marguerite Lwoff.

The study of the morphology and morphogenesis of ciliates was greatly facilitated by the introduction of a new technique of silver staining. The Klein method, silver staining of dried specimens, was not applicable to marine ciliates owing to the formation of precipitates of silver chloride. But we noticed the absence of a precipitate and nice staining of encysted ciliates. So, the idea germinated that cysts could be replaced by gelatin. After proper fixation the ciliates were embedded in salted gelatin, then impregnated with silver nitrate which was reduced by ultraviolet light. The shape of the specimens was preserved and the kinetosomes were beautifully stained as also were the trichocysts and many fibers. Moreover, the whole organism was stained. So the ciliary system of many ciliates was described accurately for the first time, new structures were discovered as well as the intimate process of stomatogenesis. In *Tetrahymena piriformis,* before division, the kinetosomes of the stomatogenic kinety n° 1 start dividing toward their left. It is in or around the newly formed field of kinetosomes that the membranelles are organized and that the morphogenesis of the mouth takes place (with E. Chatton, M. Lwoff, and J. Monod).

One of the groups studied was *Apostomata.* Numerous new species were discovered and studied. We described the life cycle in the two hosts and the ciliary metamorphosis which take place during the cycle.

Some general features of the morphology of ciliates were disclosed: 1. the genetic continuity of kinetosomes and the concept of the "infraciliature"; 2. the origin of trichocysts; 3. the kinetodesma; and 4. the law of desmodexy. We shall examine them separately.

1. When one looks at a silver-stained ciliate one sees that new kinetosomes always appear in the immediate vicinity of a persisting one and that the two kinetosomes are united by a "desmose." Therefore, it was concluded that kinetosomes are endowed with genetic continuity and reproduce by division. We know today that only a double-stranded nucleic acid made of two complementary parts can "divide" and reproduce its kind. No other molecule is reproduced by division. So, *sensu stricto,* a kinetosome cannot divide. How, then, does a kinetosome reproduce? The simplest approach is to assume that a kinetosome is located in a specific morphogenetic field or territory. The molecules of this territory can increase in number. A new kinetosome can be organized only within the vicinity of a pre-existing one. The reproduction of a kinetosome is the result of the division of a specific morphogenetic territory.

Let us now consider a ciliate as a whole. Ciliates reproduce by binary fission or division. Before division, two new ciliates are organized within the framework of the parent. Division separates two new ciliates, one anterior, the other posterior, for which we proposed the name "proter" and "opisthe," respectively. Whatever the case might be, if the reproduction of a ciliate is called division, then the reproduction of a kinetosome is also a division.

Some adult ciliates are deprived of cilia but are reproduced by ciliated buds. This is the case for most suctorians. The study of the suctorian *Podophrya* revealed that kinetosomes devoid of cilia are present in the adult and that they are the origin of the kinetosome of the bud. The ensemble of kinetosomes was called an "infraciliature."

2. Let us consider the origin of trichocysts. In the Apostomatous ciliates, trichocysts are formed only during a short and unique definite period of the life cycle. One then sees all the kinetosomes "dividing" to their left and, from the daughter kinetosomes, one sees trichocysts growing. Needless to say, the electron microscopists studying apostomous ciliates were, up to now, unable to see this phenomenon which is very clearly visible with an ordinary microscope when one examines silver stained ciliates *at the right phase of the cycle.*

It is currently said that cilia are produced by kinetosomes. The kinetosome as such cannot synthesize the proteins of the cilium, but they have the property to organize and to orient the ciliary proteins, and also the proteins of the trichocysts and of the fibers. A kinetosome and its territory represent a remarkable morphogenetic machine.

3. The study of ciliates revealed the existence of a fiber which we called the "kinetodesma" and which always runs parallel to the rows of kinetosomes on their right. This is the law of desmodexy. The kinety is thus asymmetrical—one can recognize an anterior and a posterior end, a right and a left. With Chatton, we also saw that the kinetodesma was formed by fibers connected to the kinetosomes. This was rediscovered much later by the electron microscopists.

In some ciliates at a given phase of the life cycle, the kinetodesma of a given kinety disappears. As a result, the corresponding kinetosomes are scattered in disorderly fashion. This exemplifies the role of the cortex in the maintenance of the kinetodesma and the role of the kinetodesma in the organization of the kinetosomal pattern.

Thigmotricha and Apostomata were our main concern. But, when working in a marine laboratory, one comes across strange organisms which nobody yet has seen. We thus described a curious peritrichous ciliate, *Ellobiophrya,* which is appended to the gills of the acephalus *Donax vittatus* like an earring and reproduces by budding a strange and beautiful "embryo." We also described the *Conidophrys pilisuctor* a ciliate which, as embryo, empales itself on the secretory hairs of the amphipod *Corophium,* loses its cilia, encysts, and reproduces by budding.

Finally, I discovered an amazing ciliate whose nuclear apparatus is a protocaryon: *Stephanopogon mesnili* is homokaryotic. It was not possible to state any longer that ciliates are characterized by the differentiation of the nuclear apparatus into a macro- and a micronucleus. Maybe I should add that *Stephanopogon* is a true ciliate which possesses mouth characteristics both of Gymnostomes and of Hymenostomes, and is a most devilish animal. We may note here that Opalinids, which are homokaryotic, were generally and are sometimes still considered as ciliates despite the fact that Chatton and Brachon, a long time ago, showed that their division is longitudinal; it cuts between two sets of longitudinal kineties, whereas in ciliates the kineties are cut transversally by division.

I started playing with ciliates in 1921. The milestones of the work were monographs on *Stephanopogon, Conidophrys, Ellobiophrya, Apostomata,* and the two monographs on *Thigmotricha.* In 1948, I was invited to Harvard to deliver the Dunham Lectures. The theme was the morphogenesis of ciliates and a book was published, *Problems of Morphogenesis in Ciliates,* which I will try to summarize.

The study of the development of ciliates shows that the activity of the kinetosomes, their organization into specific patterns, their division, and the expression of their potentialities, depend on their position and on the phase of the life cycle. As already said, before division of *Tetrahymena piriformis* occurs, the kinetosomes located around the middle of the "stomatogenic kinety" n° 1 start dividing actively toward their left; a dense field of kinetosomes is formed which is the base for the organization of the membranelles and the mouth. It is clear that kinetosomes are induced to reproduce by the properties of the underlying cortex. Hence, the conclusion that "if kinetosomes are necessary for morphogenesis, they seem not to command but to obey some mysterious force . . ." In certain ciliates, trichocysts, for example, are formed only at one phase of the cycle. The analysis of the movements of the cortical structures in various ciliates led to the hypothesis that the cortex differs according to its location and that cortical structures command morphogenesis. Morphogenesis is in part the result of the response of an apparently homogeneous population of kinetosomes to their environment, the cortex. An orderly and organized asymmetry like that of an egg or of a ciliate may be only the reflection of cortical properties. A constantly flowing or potentially flowing endoplasm cannot be asymmetrical. The organelles may be asymmetrical. But when the ciliate is considered as an organism the conclusion is reached that organized asymmetry, or simply organization, can belong only to a more or less rigid, a more or less permanent system, that is to say, the cortex. Finally, one finds that, in well-defined evolutionary series of ciliates the structure of the daughter ciliates just before division corresponds to that of the primitive type. Ontogeny repeats phylogeny.

It came as a surprise when Beisson and Sonneborn, in a paper which appeared in the *Proceedings of The National Academy of Science* in 1965, made me responsible for the statement that: ". . . kinetosomes . . . are directive or instrumental in morphogenesis" (this sentence is the only reference to my book). Impressed by this statement, Nanney wrote later: "Persistent reports of DNA associated with the ciliate cortex . . . lend plausibility to the suggestion (see Lwoff 1950) that gene-like elements imbedded in the cortex represent peripheral "nucleic" reservoirs responsible in some way for cortical characteristics." This, supposedly a view of mine, is stated by Nanney to be opposed to Sonneborn's ideas concerning the role of cortical structures in morphogenesis!

Of course, kinetosomes have a morphogenetic role in the sense that they are responsible for the production of cilia, trichocysts, and fibers. However, I have never thought nor said that kinetosomes were controlling the cortex or were directives in the morphogenesis. As a matter of fact, my conclusions as can be judged by the above quotations point to the opposite concept, namely, that the presence, location, organization, and activity of cortical organelles—including kinetosomes—are not determined by the kinetosomes but by the cortical structures and their "internal" environment as determined by the phase of development. Beisson and Sonneborn have reached the same conclusion. I am, however, unable to understand why my colleagues made me say *exactly the contrary* of what I clearly said. Owing to the authority of Sonneborn, his distorted interpretation of my concepts is now spreading. I regret to have been obliged to put things straight myself.

Nutrition of Free-Living Protozoa

The work on morphology and morphogenesis of ciliates was performed each year at several marine biological laboratories during the months of July, August, and September and sometimes during the Easter holidays. At the Pasteur Institute I first took the course of microbiology. Later on, I became responsible for the classes in protozoology and parasitology and had the responsibility of maintaining pure cultures of trypanosomides, trichomonas, and amoebas. So, I gained some experience in handling protozoa and decided to work on the nutrition of ciliates. This was mad. So far as I can remember, I had read a review of an article dealing with the alleged pure culture of a ciliate in a medium in which ammonium sulfate was the nitrogen source and dextrose the sole organic substance. The author of the review said that these results shattered the doctrinal corpus concerning the power of synthesis of protozoa. The literature was searched for data concerning this power of synthesis. Nothing was found except concepts such as "heterotrophy" defined as the need for organic substances, and "autotrophy," the absence of need for organic substances which was considered as bound to photosynthesis. I could not discover anything about the nutritional requirements of protozoa. This is why I decided to investigate the nutrition of ciliates. In order to do so, "pure," that is bacteria-free, cultures were needed. From a wild culture of *Tetrahymena piriformis,* I obtained in 1923 a bacteria-free pure culture. A ciliate was put in a drop of sterile medium and transferred from drop to drop with a micropipette under a dissecting microscope. The ciliate grew in peptone solutions but refused to multiply in media with an ammonium salt and with dextrose as sole organic source. So it needed other "organic substances" which I tried bravely to identify. Knowing today the large number of amino acids and of vitamins necessary for this organism, it is retrospectively not surprising that my work failed, but not entirely as will be seen later. However, the first pure culture proved to be a useful tool for the solution of biological and morphological problems.

Flagellates seemed more promising. A number of groups are interesting because they comprise chlorophyll-bearing organisms as well as their chlorophyl-less counterparts. The nutrition of the "green" and "white" species was systematically investigated with the collaboration of Marguerite Lwoff first and, later on, Hisatake Dusi and Luigi Provasoli.

Some species existed in culture collections and I isolated a few strains by washing single individuals. The green flagellates utilize nitrates as nitrogen source. Most white organisms thrive in the presence of an ammonium salt but not with a nitrate. However, it was found that one species, *Polytoma ocellatum,* utilizes nitrates. This was the first example of a protozoan able to reduce nitrates.

It was known that *Polytoma uvella* utilizes acetic acid as a carbon source. The systematic study of white Chlamydomonadinas, Phytomonadinas, Cryptomonadinas, and Euglenidas revealed that some of them can thrive on acetic acid only, whereas others are satisfied with a number of lower fatty acids and some also lactic and pyruvic acids. The obvious idea was to study the green counterpart of the white flagellates in the absence of photosynthesis. *Chlamydomonas* grows beautifully in the dark if provided with acetic acid, as does *Haematococcus* also. It turned out that in the presence of acetic acid only, *Haematococcus* manufactures large amounts of carotenoids in the dark as well as in the light. Finally, we investigated a few Euglenidas. For *Astasia* (with H. Dusi) the fatty acids were the right carbon sources and this gave the key to the culture in the dark of green Euglenidas. They had been grown in the dark, but the cultures were always poor except when acetic acid was added. Acetic acid was supposed to act by lowering the pH, but we showed that, in fact, acetic acid acted as a carbon source. The green *Euglena* in the absence of photosynthesis behave as do the other flagellates (with H. Dusi).

All the investigated flagellates manufacture starch or paramylon. None of them, however—as already said—utilize glucose or any other sugar as an extrinsic carbon source. However, *Polytomella caeca* contains a phosophorylase which can transform starch into glucose-1-phosphate in the presence of phosphate (with Hélène Ionesco). It could be that the synthesis and utilization of starch in these organisms takes place without glucose as intermediate except in the phosphorylated form.

Hematin as a Growth Factor

In 1932, I received a Rockefeller fellowship to work with Otto Meyerhof in Heidelberg at the "Kaiser Wilhelm Institut für Medizinische Forschung." I stayed there fifteen months, until the end of the year 1933. It was a very interesting year, not only from the scientific point of view. The government of Germany changed in the spring of 1933 and a tragic era began. It was clear that war would come sooner or later. The laboratory was very active and everyone behaved apparently as if science only was important. However, the numerous foreign workers attracted by the fame of Meyerhof did not fail to observe and to comment on the situation without any illusion for the future. Later on, Meyerhof, as a Jew, had to leave Germany with his family. His relatives who stayed disappeared in extermination camps together with millions of other human beings.

In the Pasteur Institute, Marguerite Lwoff had shown that hematin could replace blood for the growth of *Crithidia fasciculata*—at this time called *Strigomonas fasciculata*. I decided to investigate the role of hematin as a growth factor. The field of growth factors was rather confused. Microbiologists were convinced that a "trace" of a growth factor would induce an unlimited development of microorganisms. Growth factors were supposed to act on multiplication as catalysts. At the time, a few essential amino acids had been shown to be necessary for the development of bacteria, but no "vitamin" had been identified. There was only one exception, namely hematin. So, hematin as a growth factor for *Crithidia fasciculata* was investigated. It was first shown that blood and hematin acted quantitatively; the number of flagellates which developed was, within certain limits, proportional to the amount of hematin. Each flagellate needed 520,000 molecules of hematin. What was its function?

The respiration of flagellates grown in a medium in which the hematin concentration is the limiting factor is lower than the respiration of "normal" flagellates. When hematin is added the respiration increases, and the increase is proportional to the amount of hematin added. It was then easy to calculate the amount of hematin needed by one flagellate in order for its respiration to be normal: it was 720,000 molecules, in good agreement with the number 520,000 found by measuring the growth as a function of the growth factor. Thus, it was clear that hematin was not acting as a "catalyst" either on multiplication or on respiration. It entered into the constitution of the catalytic respiratory system and its action was quantitative.

Why is hematin necessary? Blood was known to be necessary for the growth of the bacteria *Hemophilus influenzae*. As it was found that hematin works, but that hematoporphyrin is inactive, the action of hematin had been ascribed to its iron atom. A number of "active" iron preparations had been proposed as a substitute for hematin and were supposed to work. However, hematoporphyrin does not differ from protohematin only in the absence of iron. Therefore, the specificity of the hematin molecule was investigated.

Protoporphyrin proved to be active. Thus, iron—the catalytically active atom of hematin—was not the growth factor. Moreover, a number of hematins were investigated, all of which were inactive. The activity of hematin was bound to the structure: tetramethyl 1,3,5,8-divinyl 2,4-dipropionic 6,7-porphyrin. This was the first study of the specificity of a growth factor. The need for hematin was the consequence of the inability to synthesize it.

The properties of hematin as a growth factor for the flagellate *Crithidia fasciculata* turned out to be the general feature of growth factors. Growth factors act quantitatively and not catalytically on growth. They enter into the constitution of catalytic systems. Their activity is bound to a specific structure.

Growth factors were for the first time defined as specific substances which the organism is unable to synthesize and which are necessary for its growth and multiplication. Trypanosomids are parasitic flagellates and the need for hematin is found only in parasitic organisms. It was concluded that the need for hematin was the result of the loss of the power to perform its synthesis. This concept was to be extended to all growth factors.

There is today a general consensus about growth factors. Let me tell what happened to me at the 2nd International Congress of Microbiology held in London in 1936. I had to open a session on growth factors. The title of the paper was "study of lost functions." I discussed especially the results concerning hematin. When I had finished, an eminent microbiologist and biochemist, the head of a brilliant school, stood and said, "I do not like substances which produce miracles." A beautiful execution—which I survived. The judge-executioner also survived. The miracle soon became the daily bread provided by text books but I still feel the rope around my neck, and how I was thrown into the emptiness.

Moraxella

The results concerning *Crithidia* were extended to *Hemophilus influenzae*. Moreover, with Ignacio Pirosky we showed that hematin was a growth factor for *Hemophilus ducreyi*. So it was decided to investigate systematically all the members of the *Hemophilus* group. Among them, in *Bergey's Manual,* the Morax bacillus was included. At the time, the group was defined essentially by the "need for body fluids." Officially, the Morax bacillus was unable to grow in the absence of serum. So I secured a culture of the bacillus and realized that it was widely different from *Hemophilus influenzae,* that it could belong neither to the genus nor to the family, and that it could not fit in any known bacterial genus. So I proposed the new name *Moraxella* given in the honor of Victor Morax who "invented" the organism known today as *Moraxella lacunata.* It grows in broth only if serum is added, and the question of the nature of the substance involved was posed. It turned out that the serum acts by neutralizing the toxic action of fatty acids present in broth. *Moraxella lacunata* grows in broth provided it is diluted with distilled water.

The systematic study of the various species of *Moraxella* was started and Alice Audureau discovered a new species (*Moraxella lwoffii*). We tried to grow it in synthetic media with an ammonium salt and glucose. It multiplied only when peptone was added. So I tried to identify the responsible growth factors. Thiamin was active, but was the organism really unable to synthesize thiamin from pyrimidine and thiazole? An astonishing phenomenon was observed. *Moraxella* grew in the presence of pyrimidine as well as thiazole. It was then realized that thiamine, pyrimidine, and thiazole were dissolved in ethanol. The addition of ethanol permitted growth. It was found that *Moraxella lwoffii* is unable to utilize any sugar as carbon and energy source but utilizes ethanol. It does not need any growth factor.

The wild strain of *Moraxella lwoffii* is unable to utilize malic acid as carbon source. I found a mutant able to do so and which possesses an enzyme converting malic acid directly into pyruvic acid without oxaloacetic acid as an intermediary step. The enzyme requires K^+ which could be replaced by rubidium or caesium but not by sodium.

Growth Factors for Free-Living Protozoa

Since 1932 the situation concerning vitamins had undergone considerable changes. The first vitamin, vitamin C, had been identified. This was followed by the identification of vitamin B_2 and B_1. During investigations on the carbon sources for flagellates it had been noticed that some flagellates would not grow in a medium containing acetic acid as sole organic substance. A "trace" of something was necessary. The growth factors for some of the flagellates were identified with the thiazole or pyrimidine moiety of thiamine.

Thiazole (methyl-2, β-hydroxyethyl-5, thiazol) is the only growth factor for *Polytoma obtusum, P. ocellatum,* and for *Chilomonas paramoecium.* Both thiazole and pyrimidine (methyl-2, amino-4, aminomethyl-6 pyrimidine) are necessary for *Polytomella caeca* (with Hisataka Dusi). The investigation was extended to an amoeba, *Acanthamoeba castellanii,* which was available in pure culture. In addition to numerous growth factors, it needed pyrimidine. All the investigated organisms are thus able to manufacture thiamine from pyrimidine and thiazole. This is not possible for the ciliate *Tetrahymena piriformis* which needs the complete molecule of thiamine (with M. Lwoff). Thiazole, pyrimidine, and thiamine were the first growth factors identified for free-living protozoa. The specificity of thiazole and pyrimidine was investigated.

The substitution of the -hydroxyethyl in the molecule of thiazole by one hydrogen or by methyl leads to an inactivation of the molecule, but the replacement of -hydroxyethyl by an acetoxyethyl is compatible with utilization as a growth factor by the flagellates (with H. Dusi).

The replacement of the aminomethyl in position 5 in the pyrimidine by thioformylaminomethyl or by a hydroxymethyl is compatible with the utilization. The substitution by a methyl inactivates the molecule as a growth factor. The substitution of the NH_2 in position 4 by hydroxyl or by $-OCH_3$ leads to inactivation. The transfer of the methyl from position 2 to position 6 also leads to inactivation.

The suppression of the hydroxyethyl of thiazole, the transfer of the methyl group of pyrimidine from position 2 to position 6, and the presence of a supplementary bond -C-N=C- between pyrimidine and thiazole inactivates the molecule of thiamine as growth factor for the ciliate *Tetrahymena.*

Growth Factor V

In 1936, the Rockefeller Foundation gave me a second fellowship to work in Cambridge with David Keilin, then Director of the Molteno Institute.

Hemophilus influenzae can be grown in broth only if blood is added. It was known that blood provides two factors: one, the factor X, is hematin, the other one being the growth factor "V." "V" is often interpreted as a Roman figure for five, whereas it is a V as in victory. In fact, it stands for "vitamin-like" because it is known to be destroyed by heat; sensitivity to heat being long considered as a characteristic property of vitamins. At any rate, the growth factor V is not destroyed by "heat" in an acid medium. Nothing was known about its nature.

With Marguerite Lwoff, we decided to try to identify it. Very valuable help was received from David Ezra Green, who was then working in the Department of Biochemistry, and from Tadeusz Mann, David Keilin's collaborator. Thirty pounds of yeast were extracted and fractionated and the fractions tested for their "V" activity. The active substance was finally identified with coenzymes I or II, later on, phosphopyridino-nucleotides. Bacteria grown with factor V as limiting factor had a very low respiration rate and were unable to reduce methylene blue. The addition of coenzyme I or II restored the movement of hydrogen within 60 seconds. The need for growth factor V was found to be due to the lack of power to synthesize phosphopyridino-nucleotides. A new growth factor had been identified and its physiological role determined.

Nicotinamide

After Paul Fildes had discovered that nicotinamide was the only growth factor for *Proteus vulgaris,* we investigated, together with Andriès Querido, the specificity and established the effects of various substitutions on its activity. Finally, a quantitative test for the estimation of nicotinamide was devised. We estimated nicotinamide in various organs and the first value for blood was found to be 0.75 mg per 100 ml. Vitamin PP had not yet been detected in milk and the hypothesis was put forward that the newborn synthesizes nicotinamide. With Madeleine Morel, we showed that nicotinamide was present in colostrum of human milk and that the nicotinamide content decreases for two to nine days and then increases up to a value of 15 to 34 mg/100 ml. The administration of nicotinamide was followed by a rapid but limited (0.5 mg/100 ml) increase. We also estimated vitamin PP of various tissues in various pathological conditions.

A System of Nutrition

In the "good old days," autotrophy was defined as the ability to grow in the absence of any organic substance, and was considered to be correlated either with photosynthesis or with chemosynthesis. Heterotrophy was defined as the need for organic substances.

It was clear that organic substances could represent either energy and carbon sources or growth factors. Moreover, it was shown by Hisatake Dusi that some photosynthetic *Euglenas* need one or many growth factors not yet identified (later shown to be vitamin B_{12}). Thus, it was proposed in 1932 to consider separately the energy and carbon sources and the growth factors.

The problem was considered anew in 1946 in Cold Spring Harbor with C. B. van Niel, F. J. Ryan, and E. L. Tatum. We took into account the latest developments in microbial physiology. A nomenclature of nutritional types was proposed which was based upon energy source on the one hand, and the ability to synthesize essential metabolites on the other. Phototrophy, of course, corresponded to energy provided by photochemical reactions, of which there are two types: photolithotrophy and photo-organotrophy, depending on whether the exogenous hydrogen donor was inorganic or organic. Chemotrophy corresponded to the energy provided by dark chemical reactions with two types, chemolithotrophy and chemo-organotrophy, depending on whether growth depended on inorganic or organic substances. Autotrophy corresponded to the synthesis of essential metabolites—the term, prototrophy, is now commonly used; heterotrophy, to the need for growth factors.

For the definition of categories one considers separately the energy source, the hydrogen donor, and the power to synthesize essential metabolites. This principle is now widely accepted and the terms proposed are commonly used.

Lysogeny

In 1949, I started working on lysogeny. At a time when genetic material had not been identified, a few bacteriologists had understood the strangeness of lysogenic bacteria, but knowledge concerning viruses and their reproduction was too cloudy. Moreover, a number of papers were obscured by useless polemics. Some textbooks of microbiology contained a paragraph about lysogeny but their reading did not bring much light; no adequate review or discussion was available.

My work had led to the following conclusions: (*a*) In lysogenic bacteria the phage is perpetuated in a noninfectious form which was called the prophage. (*b*) Bacteriophage is not secreted but is liberated by the lysis of a lysogenic bacterium. (*c*) The production of bacteriophage is the result of an "induction." Irradiation with ultraviolet light induces the quasi totality of the lysogenic *Bacillus megaterium* to produce bacteriophage (with Louis Siminovitch and Niels Kjelgaard). (*d*) Hydrogen peroxide is an inducer but only in organic media. Organic peroxides are inducers; the development of bacteria in an organic medium containing copper ends with the oxidation of thio compounds, the formation of hydrogen peroxide, and of organic peroxides which account in part for the "spontaneous" production of bacteriophage. (*e*) In *Salmonella typhimurium,* the fate of a bacterium infected with a temperate bacteriophage is decided within seven minutes (with Evelyne Ritz). (*f*) Lysogeny was defined as the perpetuation of the power to produce bacteriophage in the absence of infection.

Finally, the hypothesis was proposed in 1953 that the potential power of a cell to become malignant may be perpetuated in the form of a genelike structure—the genetic material of the oncogenic virus—and that carcinogenic agents induce the expression of the potentiality of this genetic material, which would culminate in the formation of virions. The history of lysogeny, together with the development and state of the new concepts, were discussed in the review "Lysogeny" which appeared in 1953.

In 1966, a collection of essays was dedicated to Max Delbrück on the occasion of his sixtieth birthday. The stories had been written by his friends, colleagues, and disciples. Max Delbrück had paradoxically played a role in the development of lysogeny. I say paradoxically because the founder of the "phage church" did not believe in the existence of lysogeny. Falling from the lips of Max Delbrück, the death sentence, "I do not believe" had been often heard by many of us. It was an excellent catalyst.

I contributed a paper entitled "The Prophage and I" which is the story of my own contribution—and I am not going to repeat myself. However, I would like to say that the lysogeny period had been something quite apart in my scientific life. In 1949, the "occupation" and its sequels were just over. After a long tragic period of isolation the outside world had flowed in with all its marvelous news, an awakening in flourish after a long sleep. Around us, groups were forming again, Louis Rapkine and Jacques Monod joined the Pasteur Institute. Young men, freed from the war, were starting their career, foreign scientists were visiting.

The 1946 Cold Spring Harbor Symposium had marked the rediscovery of freedom and the beginning of the new era. It was in this rather exceptional atmosphere that the experiments on lysogenic bacteria were started, that the work developed and that unexpected results cropped up.

The Poliovirus

In July 1953, I attended a Cold Spring Harbor Symposium on Viruses. It was sponsored by the National Foundation for Infantile Paralysis which gave considerable support to fundamental work on viruses, including bacteriophage, wisely considered as a good model.

After the meeting, I visited Harry Weaver, head of the Research Department of the Foundation. He invited me to lunch at the Banker's Club. The National Foundation for Infantile Paralysis was on Broadway, close to Wall Street. The Banker's Club was located on the top floor of a skyscraper. The weather was perfectly clear and the view of New York Bay extended for miles and miles. When I returned to earth, I suddenly became conscious of what had happened. The Foundation had invited me to spend a few months in the States, to visit a few laboratories in order to learn tissue culture and make contact with animal viruses. Thereafter, it would provide my laboratory with the necessary equipment and support my research for a few years. I would work on the poliovirus. I was abashed.

I have always been extremely sensitive to the charm, beauty, and personality of soul-inhabited cities, even if the soul is of stainless steel. Each one exerts its own specific influence and induces a given mood. The standing city was obviously not conducive to dreams but to action. I gave some thoughts to the wealth, power, and efficiency of American foundations. Of course, there was still time for reflection, but in a way it was too late; squashed on the bottom of the black Wall Street Canyon I was muddled by vertigo. Antivertigo would be more correct; and more fashionable too.

It was of course foolish to abandon the still quiet—although not for long—field of lysogeny in order to intrude into the jungle of animal virology.

To enter a new field at the age of fifty-two is unwise anyhow; but unwise decisions debouch on the unexpected which is the salt of research. Beforehand, the unexpected is necessarily entirely hypothetical. The aposteriori nature of these remarks will not escape the perspicacity of the reader.

I forgot, or tried to forget, that I was worried about my future as an animal virologist and in March 1954 we, that is, my wife and myself, embarked courageously for a long trip in the United States. It started in Bethesda where we spent two weeks in the National Institutes of Health with Wilton Earle. Later on we visited, in succession, Joseph Melnick in New Haven, John Enders in Boston, Raymond Parker in Toronto, Jonas Salk in Pittsburg, Gerome Syverton in Minneapolis. I suppose there are a number of recipes for the rejuvenation of aging scientists. One of them is to become a student.

After a fascinating and beautiful drive we finally reached Pasadena. At the California Institute of Technology, we worked with Renato Dulbecco and Marguerite Vogt from July to December. How was the poliovirus released from the cell? Throughout the vegetative phase or at its end? In order to solve the problem it was necessary to study single cells, and a number of technical difficulties had to be overcome. Finally, the question was answered. Infectious particles are liberated all at the same time by the burst of the infected cell. On the way back I stopped in New York and delivered a Harvey Lecture, "Control and interrelations of metabolic and viral diseases of bacteria." The year had been busy.

Back in Paris, something unexpected happened. A colleague from the Pasteur Institute tried to persuade the Foundation that he, and not I, should receive a grant: he was defeated. He also made efforts that I should be forbidden by the Director to work on poliovirus but without success. I did what I had decided to do. Bad or good, I have always done.

It took some time to start the experiments. The Service de Physiologie Microbienne was really crowded. Jacques Monod had not yet completed the organization of his new laboratory and the density of scientists per square foot—I wish the United States had adopted the metric system—was high. As a matter of fact, the research on the poliovirus started only in the fall of 1955.

I had no idea, not the slightest idea, in what direction things would go. I knew only that I was expected to meet with success. A grant is more or less an investment. Of course, foundations are aware of the hazard of research and the National Foundation for Infantile Paralysis was very kind. It was only after two years when research was developing that Theodor Boyd, who had succeeded Harry Weaver, told me that the Foundation had been for a time worried because it felt responsible for throwing me into the adventure. The work had developed slowly. I had started to play with the virus, to study its multiplication under various conditions, and it was necessary to know the optimal temperature.

One-step growth cycles were performed at various temperatures. The curves of viral multiplication as a function of temperature showed that different strains exhibited different patterns. Beforehand, and even now, the sensitivity of development to temperature was expressed by a^+ or a^-. We proposed to express the sensitivity by the temperature at which the viral development is decreased by 90 percent. This was the *rt* (*r* for reproduction, *t* for temperature). To determine *rt* is a rather long procedure. First, one needs a series of water baths at different temperatures which have to be rigidly controlled. For certain critical values, a difference of 1° may modify the yield by a factor of 2. One needs to have a growth curve at each temperature, that is to say a number of estimations. It is probably the reason why the ++, +, ± and – are still popular.

The multiplication of the virulent strains was less sensitive to "high" temperatures—between 37 and 41°C—than the multiplication of the nonvirulent strain including the vaccine strains.

The vaccine strains of Hilary Koprowski and of Albert Sabin have a *rt* of 37.8. By growing the type I vaccine at 41°C a strain of *rt* 40°8 was obtained. An injection of 600,000 virions of the vaccine strain in the spinal cord of the monkey—the most sensitive route—does not produce lesions. With Albert Sabin, we injected the "hot" virions into the brain, the less sensitive route; three particles killed the monkey. A similar type of experiment was performed with the MEF strain of type II. The hot strain was much more virulent for the mouse than the normal one. The LD_{50} was correlated with the *rt*.

This type of experiment was extended to the virus of encephalomyocarditis. One particle of the wild strain kills the animal; the virulence is maximum. By growing the virus at lower temperature, strains of lower *rt* and of a higher LD_{50} were obtained (M. Lwoff, Y. Perol-Vauchez, and P. Tournier). This, a relation was established between virulence and the sensitivity to high temperatures. What does this mean?

Fever and the Fight of the Organism Against a Primary Viral Infection

It had been known that by growing the poliovirus at low temperatures (23°C) strains devoid of virulence were obtained. That a strain, unable to grow, or growing poorly, at the temperature of the animal, would be devoid of virulence was not in the least surprising. The existence of a correlation between the ability to multiply at temperatures above the normal temperature of the animal and virulence posed a problem.

It had been known that an elevation of temperature can decrease the severity of a viral disease. Experiments showed that the value of the LD_{50} is increased when fever is induced within an animal. Suddenly, everything became clear. Fever is one of the mechanisms by which the organism fights against the primary viral infection. A virus is virulent when it can multiply despite fever. Fever is a byproduct of the inflammatory reaction. Moreover, in an inflammatory zone the pH can drop to values below 6.00 and the poliovirus is unable to multiply below 6.8. It thus appeared that the inflammatory reaction played an important role during a primary infection. Anyhow, prior to the production of antibodies only nonspecific reactions could be responsible for the fight of the organism against the virus.

During the past ten years many examples have been given of the relation between virulence and the resistance of viral development to temperature. The importance of fever in viral infection was recognized very slowly. Text books of virology are now often written by molecular virologists who are not interested in infectious diseases, and the books are strangely lacking in discussions concerning the fight of the organism against the viral infection. Is it so strange after all?

Viruses: Definition, Terminology, Classification

A virologist is necessarily bound to ask questions. One of them is, what is a virus? The question was asked and has been answered. He is also faced with problems of terminology; new terms were proposed. Finally, when one enters the field of virology rather late one can experience difficulties in recognizing the place of each virus, hence the need for a classification.

Definition.—Words should have a meaning. The "Concept of Virus" was discussed in a Marjory Stephenson Lecture delivered in 1957 before the Society for General Microbiology in London. It was proposed to define viruses as infectious particles possessing only one type of nucleic acid and which reproduced from their sole genetic material. A few other characteristics were sifted out: inability to grow and to divide, absence of metabolism, absence of the information for the enzymes of energy metabolism. We added later on, the absence of transfer RNA and of ribosomes and also of the corresponding information. Thus, by the virtue of a few discriminating traits, viruses were separated from nonviruses: the category, virus, was at last defined. An infectious particle could no longer belong to the group of viruses by the sole virtue of its size. A number of "small" bacteria were thus excluded from viruses and reinstated where they belong.

The concept of virus as it was proposed is now of universal acceptance. For historical purposes I should note that the concept had already been proposed in 1953 in the review "Lysogeny." At the time, however, nobody paid any attention to the proposals which were a few years in advance on the viral calendar.

Terminology.—Virologists interested in the structure of the infectious particle came across the inadequacy of terminology. With Thomas Anderson and François Jacob, we proposed three terms: *virion, capsid,* and *capsomere.* Later on, in Cold Spring Harbor, a group of virologists added *nucleocapsid.* All these terms are now part of the virological vocabulary.

Classification.—A synoptic table of viruses is certainly useful. Now, either you like order or not. If you like it, your love can be either active or platonic. If it is active then you are thrown into systematics. To classify is an amusing game—one tries to select characters and to define categories. However, there are drawbacks. First, categories do not exist in nature. They are creations of our mind; a category is the result of an arbitrary grouping. This does not matter as long as you are aware of the arbitrariness. If you were alone there would be no problem. But if you are not . . . The hell, according to Sartre, the hell, it is the others.

Moreover, when one builds categories, one has to provide them with names, hence nomenclature. A nomenclature has to be international. No wonder that there are conflicting views on nomenclature as well as on classification, hence discussions and even polemics. Before the war, I had been a member of the Judicial Committee of Bacterial Nomenclature. One day, I complained to one of my colleagues, an eminent biochemist, about the total lack of interest in the sessions. He said that if one would leave nomenclature to people interested in nomenclature the result would be a catastrophy.

This being said, with Robert Horne, Paul Tournier, and Peter Wildy, we discussed the problem of classification and finally succeeded in producing a system with the use of four characters. Once the work was brought to an end, Peter Wildy decided that he could not sign the paper because of hierarchy, the hierarchy of viral characteristics, of course. I have always regretted this decision. So the proposed system had to be issued without him. It became the L.H.T. system. It suffered the fate of all classifications, adopted by some, villipended by others. This system, however has a few advantages: 1. it exists, 2. it is the only one to exist, and 3. it allows us to classify viruses. If I were not a co-father I would be inclined to say that it is not such a bad system after all.

We had made use of four discriminating characteristics and I am still convinced that discriminating characteristics should be the basis for a classification. It is clear, for example, that the category, virus, can be defined only by the use of discriminating characteristics. What could be the use of nondiscriminating ones?

However, a number of virologists have not yet understood that the principles which apply to the category virus necessarily apply to categories of lower hierachial rank. They have escaped the difficulty by forgetting to provide their own definition of viruses. So the selective use of discriminating characteristics and the LHT system are not universally accepted and battles are raging. A taxonomical war, because it deals with categories which do not exist in nature and with opinions, is the equivalent of a religious war. There is, however, a difference. The heretics, that is the others, not being burned, the war cannot come to an end.

The Cancer Institute

In October 1966, the director of the Centre National de la Recherche Scientifique asked me if I would within two years consider taking the directorship of the Institut de Recherches Scientifiques sur le Cancer in Villejuif (one mile south of Paris). This could only be a full time job. Acceptance would therefore mean abandoning the Pasteur Institute where I had worked for forty-five years and where I felt quite at home, and also the Sorbonne (Faculty of Sciences) where I had taught microbiology since 1959.

The decision was postponed. In 1967, I spent seven months with Renato Dulbecco at the Salk Institute and we worked together on the biology of the Simian virus 40. Back in Paris, I considered the situation.

The "Délégation Générale à la Recherche Scientifique et Technique" had, in 1961, offered the Pasteur Institute the funds necessary to build an Institute of Molecular Biology. However, the director and the board of trustees ruled that molecular biology held no interest whatsoever for the Pasteur Institute.

In 1965, the obstruction ceased. A new director took up the matter again and after a number of vicissitudes, the question was settled. But the edification would start in the spring of 1969 and the building be ready in the fall of 1971—only a year before my retirement. No opportunity was offered to me anywhere else. The attic still harboured various residues and I had no chance to develop what was since the adventure of lysogeny, my principal and enduring interest, namely, the cancer problem.

Good groups were at work at Villejuif. Why not spend a few years helping them to develop research on cancer? So I decided to move and on February the first, 1968, started my last scientific—or maybe parascientific—endeavour.

Remembrances

During the first part of my career at the Pasteur Institute, I had the fatherly and efficient support of Felix Mesnil who, just before his death in 1938, had obtained for me the creation of the Service de Physiologie Microbienne. So I organized an attic into a laboratory. Later on, I experienced some difficulties. For example, a director told me once that my work was devoid of any interest for the Pasteur Institute and I should throw out a few workers in order to save money! I paid no attention whatsoever to this preposterous command for I had decided long ago and once forever that the scientists transcended the director and the board of trustees, and that everything good for science was good for the Pasteur Institute.

In the Institute, salaries were low, promotions almost nonexistent, and the budget of the laboratories poor, but freedom, holy freedom, was provided with unlimited generosity. In passing, freedom, if provided without discernment, can be very costly. Yet, freedom is not enough. If the work could be pursued and developed it was, thanks to the help of the Centre National de la Recherche Scientifique, of the Institut National de la Santé et de la Recherche Médicale, of the Délégation Générale à la Recherche Scientifique et Technique, of the National Foundation for Infantile Paralysis, and of the National Institutes of Health. Thus, the Service de Physiologie Microbienne was amply provided with technicians, equipment, and a budget for daily life in such a way that money has never been the limiting factor for the work. May I be allowed here to express my deep gratitude to all those who have given a testimony of their confidence by generously supporting our work.

A scientist should be aware of the existence of competition and not be obnubilated by "the others," but I am now more and more conscious of the intensity of competition and of the pace of scientific development. Of course, in the past few decennaries, competition and pace have increased markedly. So, the alteration I observe in my mind might be a sign of the time as well as an evidence of maturation, or a symptom of aging, who knows? Whatever the case might be, science has always been competitive and it has certainly never been wise to enter, necessarily unprepared, widely different new fields of research. I have never given any thought to this aspect of scientific endeavour, and it is why, with perfect unconsciousness I have worked in succession on the morphology and morphogenesis of protists, on growth factors, and on various aspects of cell physiology, on lysogeny, on the virulence of viruses, and on the role and mechanism of nonspecific factors in the fight of the organism against viral infection.

In fact, during many years, research was performed simultaneously in different disciplines: the work on ciliates had started in 1921 and ended with the monograph on Apostomes in 1935, and with "Problems of Morphogenesis in Ciliates" in 1950. The work on nutrition of protozoa started in 1923 and was discussed in "Recherches biochimiques sur la Nutrition des Protozoaires."

The milestones of the work on growth factors were the papers on hematin (1933–34), on growth factor V (1936), and the book "La vitamine PP et les avitaminoses nicotiniques" (1942), l'Evolution physiologique, the editing of *Biochemistry and Physiology of Protozoa* (1951). The work on lysogeny extended from 1949 to 1953 and ended—almost—with the review "Lysogeny" (1953). The "Concept of Viruses" was published in 1957, the "System of Viruses" in 1962. In the meantime, "Biological Order" had seen the light (1962) and also the new concepts concerning the virulence of viruses and the fight of the organism against a primary viral infection (1959), and finally the mechanism of the action of fever on viral development (1969).

Retrospect

Biology in its widest sense has, since 1921, undergone extraordinary development: the structure and functions of vitamins, of growth factors and coenzymes, the steps of anabolism and catabolism, the activation and movements of oxygen and hydrogen, the cytochromes, the storing and utilization of energy, the antibiotics, the antimetabolites, the nature and structure of genetic material, one gene–one enzyme, the messenger, the operator, the repressor, the transcription and translation, the code, colinearity, the structure of proteins, allostery, the nature of mutations, the sexuality of bacteria, and, more widely speaking, molecular biology and also molecular virology.

It happens that I have been associated with, or known many, if not most, of the scientists responsible for these revolutions in our knowledge and in our thinking. In one way or another, by their achievements or their personality, they have influenced what I may describe as my evolution. A few men, however, played an especially important role in my scientific life. Edouard Chatton, my master, with whom I collaborated intensively for sixteen years; Otto Meyerhof who accepted me in the Kaiser Wilhelm Institut in Heidelberg; David Keilin who provided to me a kind hospitality in the Molteno Institute in Cambridge; Louis Rapkine, the friend too soon disappeared and, finally, the members of the Service de Physiologie Microbienne. I had been fortunate enough to attract a few exceptionally gifted scientists to the Pasteur Institute and to be able to provide everyone with everything needed for research. My collaborators have certainly influenced me at least as much as I might have influenced them. Their names and achievements are well known. Thanks to them, the attic has been for many years the theater of remarkable successes. The work was pleasurable despite its intensity, and the atmosphere festive. As research pertains to ludic[1] activity I should perhaps have described the attic as a playground. Anyhow, it has been for me a constant ravishment to see important problems solved, great discoveries blooming, and new concepts piling up day after day. I sometimes said to my friends that I never felt jealous . . . and that this was meritorious.

[1] From the latin ludus, meaning game.

BIBLIOGRAPHY

The prefatory chapter is not a review. The bibliography is therefore not organized according to the rules of *Annual Reviews*. It is not an exhaustive list of the author's publications but represents a selection of papers considered to be characteristic of the various scientific periods.

Sur une nouvelle famille d'Acinétiens, les Sphénophryidés, adaptés aux branchies des mollusques acéphales (avec E. Chatton).
C. R. Acad. Sci., 1921, *173*, 1495.
Sur la nutrition des Infusoires.
C. R. Acad. Sci., 1923, *176*, 928.
Reproduction d'um Hydraire gymnoblaste par poussées répétées de propagules.
Bull. Soc. Zool. France, 1925, *50*, 405.
Pottsia infusoriorum, n.g., n.sp., Acinétien parasite des Folliculines et des Cothurnies (avec E. Chatton).
Bull. Inst. océan. Monaco, 1927, *489*, 1–12.
Le cycle du pigment carotinoïde chez *Idya furcata* Baird, Copépode Harpacti-vide. Nature, origine, évolution du pigment et des réserves ovulaires au cours de la segmentation. Structure de l'oeil chez les Copépodes.
Bull. biol. France Belgique, 1927, *61*, 193–240.
Les infraciliatures et la continuité des systèmes ciliaires récessifs (avec E. Chatton et M. Lwoff).
C. R. Acad. Sci., 1929, *190*, 1190.
L'infraciliature et la continuité génétique des blépharoplastes chez l'Acinétien Podophrya fixa (O. F. Muller), (avec E. Chatton, M. Lwoff et L. Tellier).
C. R. Soc. Biol., 1929, *100*, 1191.
Contribution à l'étude de l'adaptation d'*Ellobiophrya donacis* CH. et LW., Péritriche vivant sur les branchies de l'Acéphale *Donax vittatus* da Costa (avec E. Chatton).
Bull. Biol. France Belgique, 1929, *63*, 321–349.
Imprégnation par diffusion argentique de l'infraciliature des Ciliés marins et d'eau douce après fixation cytologique et sans dessication (avec E. Chatton).
C. R. Soc. Biol., 1930, *104*, 834.
Détermination expérimentale de la synthèse massive de pigment carotinoïde par le Flagellé *Haematococcus pluvialis* Flot. (avec M. Lwoff).
C. R. Soc. Biol., 1930, *105*, 454.
L'apparition de groupements -SH avant la division chez les Foettingeriidae (Ciliés). (avec E. Chatton et L. Rapkine).
C. R. Soc. Biol. 1931, *106*, 626.
La formation de l'ébauche postérieure buccale chez les Ciliés en division et ses relations de continuité topographique et génétique avec la bouche antérieure (avec E. Chatton, M. Lwoff et J. Monod).
C. R. Soc. Biol., 1931, *107*, 540.
Recherches morphologiques sur *Leptomonas ctenocephali* Fanth. Remarques sur l'appareil parabasal (avec M. Lwoff).
Bull. Biol. France Belgique, 1931, *65*, 170–215.
Recherches morphologiques sur *Leptomonas oncopelti* Noguchi et Tilden, et *Leptomonas fasciculata* Novy, Mac Neal et Torrey.
Arch. Zool. exp. et gén. (Protistologica), 1931, *71*, 21–37.
Bartonelloses et infections mixtes (avec M. Vaucel).
Ann. Inst. Pasteur, 1931, *46*, 258.
Recherches biochimiques sur la nutrition des Protozoaires. Thèse de Doctorat ès-Sciences. Collections des Monographies de l'Institut Pasteur, Masson éd., Paris 1932.
Die Bedeutung des Blutfarbstoffes für die parasitschen Flagellaten.
Zbl. Bakt. I. Orig., 1934, *130*, 497–518.
L'appareil parabasal des Flagellés (avec M. Lwoff).
Arch. Zool. exp. et gén., 1934, *76*, 56.
Le pouvoir pathogène de *Trichomonas foetus* pour le système nerveux central (avec S. Nicolau).
Bull. Soc. Path. exot., 1935, *28*, 277.
Les Ciliés Apostomes. I. Aperçu historique et général; étude monographique des genres et des espèces (avec E. Chatton).
Arch. Zool. exp. et gén., 1935, *77*, 1–453.
Le cycle nucléaire de *Stephanopogon mesneli Lw. (Cilié homocaryote)*.
Arch. Zool. exp. et gén., 1936, *78*, 117.
Etude sur les fonctions perdues. Rapport du 2e Congrès international de Microbiologie, Londres.
Ann. Fermentations, 1936, *2*, 419.
Les *Pilisuctoridae* CH. et LW. Ciliés parasites des poils sécréteurs des Crustacés Edriophthalmes. Polarité, orientation et desmodexie ches les Infusoires (avec E. Chatton).

Bull. Biol. France Belgique, 1936, *70,* 86.

Les remaniements et la continuité du cinétome au cours de la scission chez les Thigmotriches Ancistrumidés (avec E. Chatton).
Arch. Zool. exp. et gén., 1936, *78,* 84.

La pyrimidine et le thiazol, facteurs de croissance pour le Flagellé *Polytomella coeca* (avec H. Dusi).
C. R. Acad. Sci., 1937, *205,* 630.

Le thiazol, facteur de croissance pour *Polytoma ocellatum* (Chlamydomonadiné). Importance des constituants de l'aneurine pour les Flagellés leucophytes (avec H. Dusi).
C. R. Acad. Sci., 1937, *205,* 882.

Le thiazol, facteur de croissance pour les Flagellés Polytoma caudatum et *Chilomonas paramaecium* (avec H. Dusi).
C. R. Acad. Sci., 1937, *205,* 756.

Caractères physiologiques du Flagellé *Polytoma obtusum.* (avec L. Provasoli).
C. R. Soc. Biol., 1937, *126,* 279.

Détermination du facteur de croissance pour *Haemophilus ducreyi* (avec I. Pirosky).
C. R. Soc. Biol., 1937, *126,* 1169.

Studies on codehydrogenases. I. Nature of growth factor "V" (with M. Lwoff).
Proc. Roy. Soc. London, Series B, 1937, *122,* 352.

Studies on codehydrogenases. II. Physiological function of growth factor "V" (with M. Lwoff).
Proc. Roy. Soc. London, Series B, 1937, *122,* 360.

Rôle physiologique de l'hématine pour *Haemophilus influenza* Pfeiffer. (avec M. Lwoff).
Ann. Inst. Pasteur, 1937, *59,* 129.

L'aneurine, facteur de croissance pour le Cilié *Glaucoma piriformis* (avec M. Lwoff).
C. R. Soc. Biol., 1937, *126,* 644.

La spécificité de l'aneurine, facteur de croissance pour le Cilié *Glaucoma piriformis* (avec M. Lwoff).
C. R. Soc. Biol., 1938, *127,* 1170.

Influence de diverses substitutions sur l'activité du thiazol considéré comme facteur de croissance pour quelques Flagellés leucophytes (avec H. Dusi).
C. R. Soc. Biol., 1938, *127,* 238.

La synthèse de l'aneurine par le Protozoaire *Acanthamoeba castellanii.*
C. R. Soc. Biol., 1938, *128,* 455.

L'activité de diverses pyrimidines considérées comme facteur de croissance pour les Flagellés Polytoma coeca et *Chilomonas paramaecium* (avec H. Dusi).
C. R. Soc. Biol., 1938, *127,* 1408.

Dosage de l'amide de l'acide nicotinique au moyen du test *Proteus;* principe de la méthode (avec A. Quérido).
C. R. Soc. Biol., 1938, *129,* 1039.

Révision et démembrement des *Hemophilae.* Le genre *Moraxella n. g.*
Ann. Inst. Pasteur, 1939, *62,* 168.

La nutrition carbonée de *Moraxella Lwoffi* (avec A. Audureau).
Ann. Ins. Pasteur, 1941, *66,* 417.

Recherches sur le sulfamide et les antisulfamides. I. Action du sulfamide sur le Flagellé *Polytomella coeca.* II. Action antisulfamide de l'acide aminobenzoïque en fonction du pH. (avec F. Nitti, Mme J. Tréfouël et Mlle V. Hamon).
Ann. Inst. Pasteur, 1941, *67,* 9.

La nicotinamide dans les tissus du foetus humain (avec M. Morel et L. Digonnet).
C. R. Acad. Sci., 1941, *213,* 1030.

Enrichissement du lait de la femme en **vitamine PP** après injection de nicotinamide (avec L. Digonnet et H. Dusi).
C. R. Acad. Sci., 1942, *214,* 39.

L'évolution de la teneur en nicotinamide du lait de la femme et le besoin du nourrisson (avec M. Morel et M. Bilhaud).
C. R. Acad. Sci., 1942, *214,* 244.

Conditions et mécanisme de l'action bactéricide de la vitamine C. Rôle de l'eau oxygénée (avec M. Morel).
Ann. Inst. Pasteur, 1942, *68,* 323.

L'évolution de la teneur du lait de la femme en nicotinamide (avec M. Morel).
C. R. Soc. Biol., 1942, *136,* 187.

Vitamine antipellagreuse et avitaminoses nicotiniques (avec L. Justin-Besançon).
1 volume in 8° de 284 pages. Masson édit. Paris, 1942.

L'agglutination réversible des Moraxella par les cations bi ou polyvalents (avec A. Audureau).
Ann. Inst. Pasteur, 1944, *70,* 144.

L'évolution physiologique. Etude des pertes de fonctions chez les microorganismes.
Actualités scientifiques. Collection de microbiologie, Hermann éd. Paris, 1944, vol. in 8° de 308 p.

Un nouveau réactif biologique de l'acide p-aminobenzoïque le Trypanosomide *Strigomonas oncopelti* (avec M. Lwoff).
Ann. Inst. Pasteur, 1945, *71,* 206.

Nomenclature of nutritional types of microorganisms (with C. B. van Niel, F. Ryan and E. L. Tatum).
Cold Spring Harbor Symp., 1946, *11,* 302–303.

Essai d'analyse du rôle de l'anhydride carbonique dans la croissance microbienne (avec J. Monod).
Ann. Inst. Pasteur, 1947, *73,* 323–347.

Production bactérienne directe d'acide pyruvique aux dépens de l'acide malique (avec R. Cailleau).
C. R. Acad. Sci., 1947, *224*, 678–679.
Nécessité de l'ion potassium pour la décarboxylation oxydative bactérienne de l'acide malique en acide pyruvique (avec H. Ionesco).
C. R. Acad. Sci., 1947, *224*, 1664–1666.
Sur le rôle du sérum dans de développement de *Moraxella lacunata* et de *Neisseria gonnorrhae*.
Ann. Inst. Pasteur, 1947, *73*, 735.
Nécessité de l'ion Mg pour la décarboxylation oxydative de l'acide malique et la croissance de la bactérie *Moraxella Lwoffi* (avec H. Ionesco).
Ann. Inst. Pasteur, 1948, *74*, 433.
Culture du Flagellé opalinide *Cepedea dimidiata* (avec S. Valentini).
Ann. Inst. Pasteur, 1948, *75*, 1.
Recherches sur les Ciliés Thigmotriches (avec E. Chatton).
Arch. Zool. exp., 1949, *86*, 169–253.
Recherches sur les Ciliés Thigmotriches. II. (avec E. Chatton).
Arch. Zool. exp., 1950, *86*, 393–485.
Induction de la lyse bactériophagique de la totalité d'une population microbienne lysogène (avec L. Siminovitch, et N. Kjeldgaard).
C. R. Acad. Sci., 1950, *231*, 190–191.
Problems of morphogenesis in ciliates. The kinetosomes in development, reproduction and evolution.
John Wiley & Sons Inc., New York, 1950.
Introduction to biochemistry of Protozoa.
In Biochemistry of Protozoa. Academic Press, New York, 1951, 1–26.
Conditions de l'efficacité inductrice du rayonnement ultra-violet chez une bactérie lysogène.
Ann. Inst. Pasteur, 1951, *81*, 370–388.
Induction de la production de bactériophages et d'une colicine par les peroxydes, les éthylèneimines et les halogénoalcoylamines (avec F. Jacob).
C. R. Acad. Sci., 1952, *234*, 2308.
L'induction du développement du prophage par les substances réductrices (avec L. Siminovitch).
Ann. Inst. Pasteur, 1952, *82*, 676–690.
Définition de quelques termes relatifs à la lysogénie (avec F. Jacob, L. Siminovitch et E. L. Wollman).
Ann. Inst. Pasteur, 1953, *84*, 222.
L'induction.
Ann. Inst. Pasteur, 1953, *84*, 225.
Lysogeny.
Bact. Rev., 1953, ***17***, 269–337.
Recherches sur la lysogénisation de Salmonella typhi-murium (avec A. S. Kaplan et E. Ritz).
Ann. Inst. Pasteur, 1954, *86*, 127.
Kinetics of the release of poliomyelitis virus from single cells (with R. Dulbecco, M. Vogt and M. Lwoff).
Virology, 1955, *1*, 128–139.
Control and interrelations of metabolic and viral diseases of bacteria.
The Harvey Lectures, series L (1954–55), 92–111, Academic Press, New York.
The concept of virus.
J. Gen. Microb., 1957, *17*, 239–253.
The Mammalian Cell as an Independent Organism.
Spec. Pub. New York Acad. Sci., 1957, *V*, 300–302.
L'espèce bactérienne.
Ann. Inst. Pasteur, 1958, *94*, 137–140.
Factors influencing the evolution of viral diseases at the cellular level and in the organism.
Bact. Rev., 1959, *23*, 109–124.
Remarques sur les caractéristiques de la particule virale infectieuse (avec T. F. Anderson et F. Jacob).
Ann. Inst. Pasteur, 1959, *97*, 281–289.
Sur les facteurs du développement viral et leur rôle dans l'évolution de l'infection (avec M. Lwoff).
Ann. Inst. Pasteur, 1960, *98*, 173–203.
Tumor, viruses and the cancer problem: a summation of the conference.
Cancer Research, 1960, *20*, 820–829.
Les événements cycliques du cycle viral. I. Effets de la température (avec M. Lwoff).
Ann. Inst. Pasteur, 1961, *101*, 469–477.
Les événements cycliques du cycle viral. II. Les effets de l'eau lourde (avec M. Lwoff).
Ann. Inst. Pasteur, 1961, *101*, 478–489.
Les événements cycliques du cycle viral. III. Discussion (avec M. Lwoff).
Ann. Inst. Pasteur, 1961, *101*, 490–504.
Mutations affecting neurovirulence.

In "Poliomyelitis." 5e Conférence Internationale sur la Poliomyélite. Lippincott éd., Philadelphia, 1961, 13–20.

Biological Order (Karl Taylor Compton Lectures).
M.I.T. Press, Massachusetts Institute of Technology, Cambridge, Mass., 1962.

The thermosensitive critical event of the viral cycle.
Cold Spring Harb. Symp. Quant. Biol., 1962, *27*, 159–174.

Proposals (with D. L. D. Caspar, R. Dulbecco, A. Klug, M. S. Stoker, P. Tournier and P. Wildy).
Cold Spring Harb. Symp. Quant. Biol., 1962, *27*, 49–50.

A system of viruses (with R. W. Horne and P. Tournier).
Cold Spring Harb. Symp. Quant. Biol., 1962, *27*, 51–55.

Un mutant du poliovirus insensible aux effets de la deutération (avec M. Lwoff).
C. R. Acad. Sci., 1964, *258*, 2702–2704.

The specific effectors of viral development. (The first Keilin Memorial Lecture).
Biochem. J., 1965, *96*, 289–301.

La synthèse du RNA chez le poliovirus. Effet de la guanidine (avec C. Burstein et E. Batchelder).
Ann. Inst. Pasteur, 1966, *111*, 1–13.

The classification of viruses (with P. Tournier).
Ann. Rev. Microb., 1966, *20*, 45–74.

Les effecteurs de l'infection virale primaire. (Conférence prononcée au Congrès de Microbiologie de Moscou, le 24 juillet 1966).
Extrait du Maroc-Médical, n° 500–47–67.

The Prophage and I.
In "Phage and the Origins of Molecular Biology." Edited by J. Cairns, G. S. Stent, J. D. Watson. Cold Spring Harbor Lab. of Quant. Biol., publisher, 1966.

Le rôle de la biologie moderne dans la médecine.
In "Scientia valemus," published by CIBA, Basel, 1967.

Death and Transfiguration of a Problem.
Bact. Rev., 1969, *33*, 390–403.

BIOLOGY OF THE LARGE AMOEBAE

CICILY CHAPMAN-ANDRESEN
Physiological Department, Carlsberg Laboratory,
Copenhagen, Denmark

CONTENTS

INTRODUCTION

According to the latest taxonomic system proposed for the phylum Protozoa by a committee of the Society of Protozoologists (58), the large amoebae are placed in the subphylum Sarcomastigophora, superclass III, Sarcodina, class I, Rhizopodea. Within this class are found many interesting amoeboid organisms which are indeed very large and as such might fall within the scope of this review. The subclass Mycetozoa contains species which may attain a plasmodial length of up to 30 cm or more, and through the streaming channels of which numerous granules flow. Some species of slime moulds, such as *Physarum polycephalum* (29) and *Dictyostelium discoideum* (104) can now be maintained in axenic culture in defined media, thus opening up many new possibilities for their study. In the subclass 3 Granuloreticulosia, the marine foraminiferan *Allogromia laticollaris* Arnold, with a test greater than 450 μ in diameter and a pseudopodial reticulum which may cover several square millimeters (4), may also be designated a large amoeboid organism. This species has a complex and fascinating life cycle (5), and can now be kept in sterile culture with one bacterial species as food (67).

However, the limitations of space and of special knowledge necessitate a rather strict limit to be placed on the number of species, as well as on the number of topics which can usefully be included in this review. "The Amoebae" will be restricted to the genera *Amoeba, Chaos,* and *Pelomyxa,* and the topics will be limited to considerations of their suitability for different types of studies and to selected aspects in which fine structure is related to physiological processes.

The three genera of amoebae, *Amoeba, Chaos,* and *Pelomyxa* have increased in popularity as research organisms during the last decade and their suitability for studies of many kinds is indicated by the number of papers (ca 600) which have appeared since the last published volume devoted entirely to the large free-living fresh-water amoebae appeared in 1959 (107). Much effort has been invested in elucidation of the fine structural organization of cytoplasmic organelles, endocytic processes, nuclear-cytoplasmic interactions, genetics of heterokaryons, and growth and renewal problems. In fact, so many new data have been accumulated that a new publication *The Biology of Amoebae* is at present being prepared under the editorship of Dr. K. W. Jeon (60), and is due to appear in 1972. This new volume will present in detail many aspects of amoeba biology, which are omitted or only briefly mentioned in the present review.

The Large Amoebae as Experimental Organisms

Taxonomy.—Although this review is not a suitable place for extensive taxonomic discussion, brief mention must be made here concerning the names used for the species to be included. The vexing question of the taxonomy of the large fresh-water amoebae has not so far been clarified. The species which causes the most trouble was discovered, or re-discovered by Wilson in 1900 (109), and has been found only with certainty in North America; the names *Chaos chaos* and *Pelomyxa carolinensis* are still equally used for this species, and are fiercely defended by their protagonists (1-3, 33, 64). This usage causes confusion among those not familiar with taxonomic problems, and constant irritation to those who work with this species. As I hope will become apparent later in this review, the most recent observations on the fine structure and physiology of this species, as compared with that of *Pelomyxa palustris,* the specific designation of which has never been questioned, indicate that the differences between *P. palustris* and the large North American amoeba are so great that their inclusion within one genus seems unjustifiable. Corliss (28) recognizes the genus *Chaos,* and includes it, together with *Amoeba* and *Pelomyxa* within the order Amoebida. For these reasons, Wilson's amoeba will be consistently termed *Chaos chaos* in this review. A full discussion of the present taxonomic and phylogenetic position of these amoebae by Bovee & Jahn (6) is in preparation, and I agree with the nomenclature proposed.

In connection with taxonomic problems, it should perhaps be mentioned that characteristics, other than morphological (in the case of the large amoebae, size, number of nuclei, typical pseudopodial type, etc.) have been put forward as possible aids to a satisfactory classification. The content of hydrolytic enzymes (3) in *Chaos chaos, Amoeba proteus,* and *Pelomyxa palustris,* the free amino acid levels, protein amino acid composition, and content of metallic trace ions (39–42) in *C. chaos, A. proteus,* and *Amoeba dubia,* have been discussed separately and in combination as possible "quantities" characteristic for the different species. While biochemical characteristics may be of considerable use in classification of species which can be grown on defined media under axenic conditions, it seems doubtful whether such studies on amoebae can yield useful and applicable information until the amoebae can be grown under more controlled conditions than those at present available. The complications which may arise in such studies owing to bacterial symbionts and contaminants in many strains of these amoebae will be discussed more fully in a later section.

Common features.—The genera *Amoeba, Chaos,* and *Pelomyxa* are all fresh-water forms, characterized by a simple life cycle consisting of nuclear division, followed by cytoplasmic division into two or more daughter cells. There is no well-documented evidence for any more complex type of life cycle than fission and formation of cysts.

The size range of these amoebae is large, a factor of 10^4 separates the volume of the smallest from the largest; the small *Amoeba* spp. having a volume of ca 0.002 μl or less, while some specimens of *P. palustris* may have a volume of 20 μl or more.

Large size and multinuclearity are common to *Pelomyxa* and *Chaos,* and these characteristics have been used as an indication for placing these amoebae in one genus (64). However, the type of nucleus, feeding habits and food vacuolar structure, and mitochondria are but a few of the features, which are dissimilar for these two genera, while these organelles and characteristics are very similar in *Amoeba* and *Chaos,* apart from the structure of the nuclear envelope. *Amoeba* and *Chaos* are carnivorous and their digestive processes are similar while *P. palustris* does not capture active prey, but feeds on algae, plant remains, i. e., nonmotile food, and ingests all that adheres to its surface, including mineral grains.

Culture.—The main disadvantage of the genera *Amoeba, Chaos,* and *Pelomyxa* as experimental organisms is the regrettable fact that so far none of these amoebae have been cultured axenically. Recent progress in this direction will be discussed in the chapter on "Culture; large yields, maintenance and problems of approaching axenic cultures" by Griffin (47). Some earlier work on *A. proteus* (88) indicated that amoebae, freed of extracellular bacteria by passage across sterile agar (80) or treatment by antibiotics (79) could not grow or reproduce on a diet of washed *Tetrahymena,* even when liver extract or other complex nutrients were added, in the absence of bacteria. It may prove that the amoebae have a requirement for particles in the medium, as has been shown for other protozoa, such as *Allogromia* (67), *Acanthamoeba* (50), and *Tetrahymena* (91). Particle-free media may lack a factor essential for the membrane turnover process, designated as "permanent pinocytosis" by Wohlfarth-Bottermann & Stockem (112). Recent studies on the growth of *Tetrahymena* in the presence of protein solutions (Rasmussen, personal communication) have shown clearly that growth is promoted by the addition to proteose-peptone media of sterile-filtered, heat-denatured 0.1 percent egg albumin, while the same concentration of sterile-filtered, undenatured protein is not effective. It may be that the media tested for the axenic culture of amoebae lack such a particulate factor or an inducing factor, which might act as a stimulant for pinocytic uptake of the medium.

The culture methods most generally used at present for *Chaos* and *Amoeba* are derived from the method introduced by Prescott & James (89). These methods consist in a regular and controlled food supply, using one defined species of ciliate as food, and have resulted in improved yield and uniformity of the cultures. In the hands of skillful investigators, yields of *Chaos* of up to 40 g or ca 4×10^5 cells (12), and of *A. proteus* of 20 ml or 4×10^6 cells (82) could be harvested weekly. However, although good and reproducible growth conditions can be obtained by the daily feeding with *Tetrahymena* (46, 87, 97) or with *Paramecium aurelia* (95, 98), intra- and extracellular bacterial contamination is still a problem with these cultures.

These methods are time-consuming, but the cells are readily collected for experiments, in contrast to the older culture type (1) in which a variety of food organisms, ciliates, flagellates, and bacteria, multiplied together with the amoebae and a species of *Saprolegnia,* from which it was difficult to detach the amoebae. The latter type of cultures is still useful as stock cultures, as they require little attention, and last for several months.

So far there has been no report of the establishment of permanent cultures of *P. palustris,* although the amoebae may survive for several months, using the culture method of Kudo (65) with Carrel flasks, filled with pond water and chopped filamentous algae as food. The saprobiotic conditions under which this organism is found indicate that a low oxygen tension and decaying vegetable matter are the natural habitat for this species. Recent studies have shown (Chapman-Andresen and Holter, unpublished findings) that the oxygen tension in a small stagnant pond from which large numbers of these amoebae have been collected ranges from ca 20–40 mm Hg. Attempts at culture under different partial pressures of oxygen, using naturally occurring and artificially produced low oxygen tension in the medium showed clearly that the amoebae survive for longer periods under low oxygen tensions (ca 20–40 mm Hg) and at low temperature (4–8°C), but that there is a marked seasonal variation in their capacity for withstanding unfavorable conditions.

Clone cultures, and also heterokaryon clones have been established for the *Amoeba* spp. (71). Three nuclear clones of *C. chaos,* obtained by cutting off a single pseudopod containing only one nucleus, were established by Prescott (87). After feeding on *Tetrahymena,* these fragments attained the polynuclear state and size of normal specimens of *Chaos,* and one of these clones is still surviving after sixteen years. Apart from their application to genetic studies too little use has been made of such standardized, but readily available, experimental material, and strain designations are still sometimes omitted from publications.

Intracellular organisms.—Many strains of the large amoebae contain numerous endosymbionts or contaminants, some of which are clearly bacteria, and may easily be observed by light microscopy in living organisms. It is so far unknown whether any of these organisms are symbionts to the extent that their presence is essential for the metabolic processes of the amoebae, but the self-replicating nature of some types has been demonstrated by Wolstenholme (114).

In electron microscopic studies, Roth & Daniels (94) found spiral and flagellar bacteria in vacuoles in *A. proteus.* Attempts to eliminate the contaminating bacteria by starvation and by penicillin treatment were unsuccessful, although the infection was reduced. Apparently different strains of *A. proteus* differ in their capacity to resist and to harbor bacterial infections. In studies on two strains, "Bristol" (strains 1a, 1b, 15; 40, 42) strain which contains large numbers of rod-shaped contaminants, and "Adam" (strain 4, 15) strain which is without light microscopically visible contaminants, Chapman-Andresen & Hayward (22) confirmed that starvation does not eliminate the contaminants. The number of bacteria present in one amoeba of Bristol strain is of the order of 4000 (22), an order of magnitude similar to the number of mitochondria. Attempts to infect the Adam strain with the bacteria of the Bristol strain were not successful. A certain strain specificity is indicated by the constant presence of short rods in Bristol strain and lack of bacteria in Adam strain during culture under the same conditions during the past ten years. While these contaminants of *A. proteus* do not affect the growth of the amoebae, Jeon & Lorch (61) observed bacterial contaminants of *Amoeba discoides* which behaved as parasites, affecting growth and vitality. No reports could be found in the literature on contaminants in *Chaos* and none have been seen in strains used at the Carlsberg Laboratories (unpublished observations). So far, no extensive survey and identification of contaminants or symbionts in amoebae has been made which is in any way comparable with that published by Sonneborn (102) for *Paramecium aurelia.* There is, however, good indication (114) that the so-called DNA and RNA bodies in the cytoplasm of amoebae may be endosymbionts, and that these may influence hereditary processes in the cell (48).

While the bacterial symbionts or contaminants of the genus *Amoeba* are found as widely divergent types of bacteria in different strains (22, 94, 114), and may be facultative symbionts or sometimes parasites (36, 61), the bacterial symbionts of *P. palustris* appear to be well defined and constant cell components, and may participate in cellular metabolism, perhaps as a substitute for mitochondria, which are absent from this species (2, 33, 68). From light microscopic observations, at least two types of bacteria appear to be constant cell constituents in strains of *Pelomyxa* found in many different countries (2). The most striking type is large, measuring ca 0.8 by 3 μ, often found in short chains, and classified by Gould-Veley (45) as *Cladothrix pelomyxae,* i.e., among the iron bacteria; in addition, slender rods have also been found. The available electron microscopic observations indicate that these bacteria are similar in *Pelomyxa* from North America (33, 34), Germany (68), and Denmark (2). However, recent studies by Murray and Birch-Andersen (in preparation) on a Danish strain of *Pelomyxa* have demonstrated that three types of bacteria are constantly present: two types of slender rods, one Gram-positive with pointed ends, the other Gram-negative, with rounded ends, and a large type with a very characteristic shape, especially at the ends, and a large central "mesosome" apparently always present, regardless of the division stage of the cell. The three types of bacteria are illustrated in Figure 1[1].

Although fine structural studies show that the morphology of the large and small bacteria are quite distinct, Leiner and co-workers (68, 69) consider that the large bacteria probably originate from fusion of the slender rods. It should be noted that under optimal light microscopic conditions the mesosome of the large bacteria may be visible, giving the impression that this organism is longitudinally split.

→

FIGURE 1.[1] Endosymbionts of *Pelomyxa palustris,* seen within the amoeba. 1. Gram-negative bacterium, longitudinal section showing rounded end, and typical complex surface layers. 2. Gram-positive bacterium, longitudinal section through recently divided cell, which has separated into two vacuoles; the cells show pointed ends and typical smooth simple membranes. 3. Large bacterium, usually Gram-negative, transverse section, showing complex surface layers and mesosome. Magnification 90,000 ×. (Murray and Birch-Andersen).

[1] The author wishes to express her best thanks to Drs. R. G. E. Murray and A. Birch-Andersen for permission to include these unpublished micrographs, made at the State Serum Institute, Copenhagen.

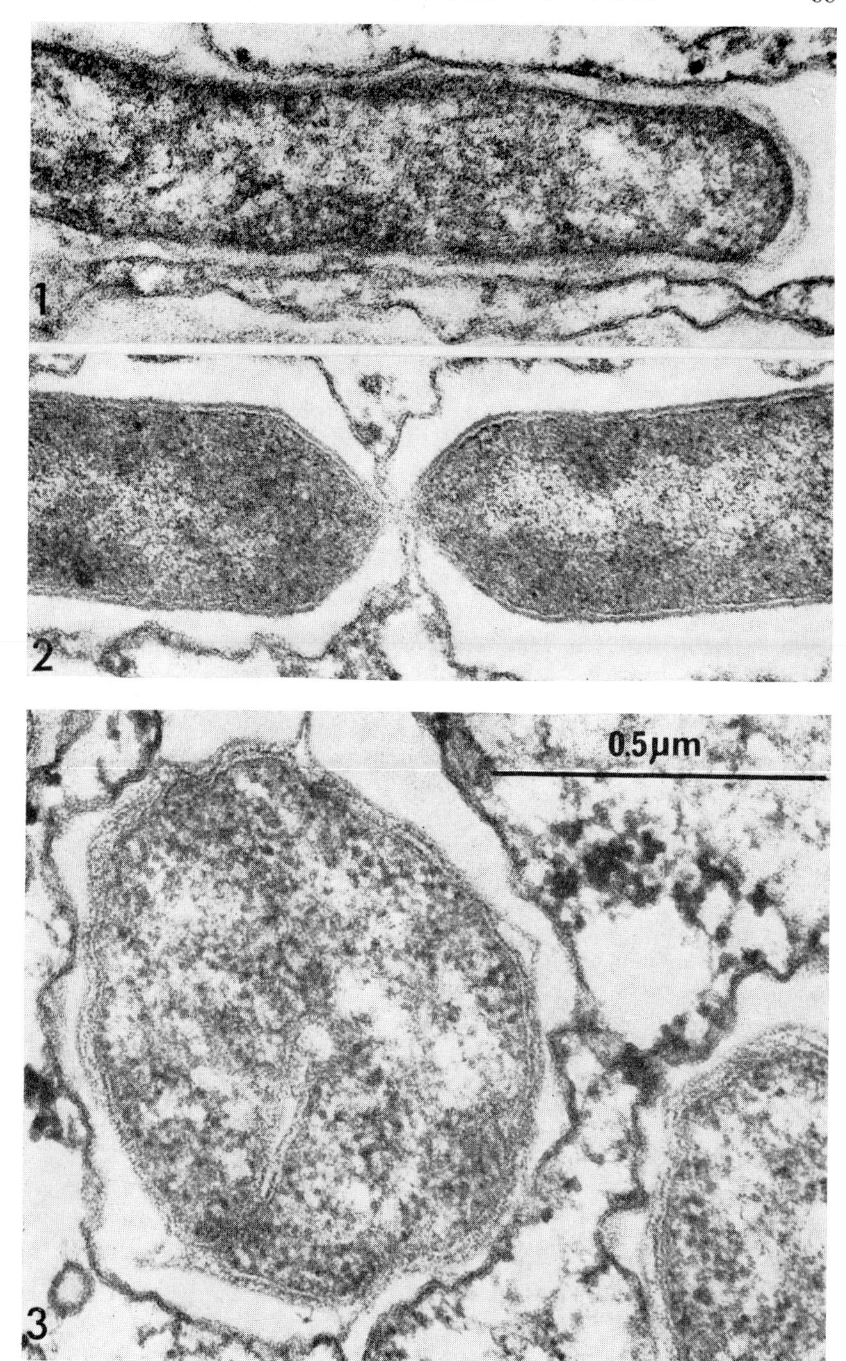
1
2
0.5µm
3

TABLE 1. ***Pelomyxa palustris.*** **Distribution of Bacteria in Different Types of Cells**[a]

Cell type	Percentage of 3 different bacteria in homogenates		
	Large	Gram positive rods	Gram negative rods
Light 40 cells mean ± S.D.	24.1 ± 1.1	68.3 ± 1.0	7.6 ± 0.7
Grey 40 cells mean ± S.D.	42.5 ± 2.2	45.5 ± 2.1	12.0 ± 1.3
Black 40 cells mean ± S.D.	47.6 ± 0.7	37.5 ± 0.9	14.9 ± 0.7

[a] Data from Chapman-Andresen and Holter (unpublished).
Gram-stained homogenates, 1000 bacteria counted in each cell.

Bacterial counts on Gram-stained preparations of homogenized *Pelomyxa* show that the proportion of these three different bacteria varies according to cell type, and according to the oxygen tension of the environment. See Table 1 for cell types and bacterial counts.

Three types of *Pelomyxa* may be distinguished in the Danish material from one pond, and these types can be recognized from other material. The "light" cells, characterized by lively movement, generally well orientated with a pronounced uroid, contain few quartz grains, many glycogen bodies, and a high proportion of small Gram-positive rods; the "grey" type, generally more sluggish in movement, less well orientated, with many quartz grains, and few or small glycogen bodies, is characterized by a high number of the large bacteria. The small "black" cells which are remarkably uniform in size (ca 0.1 μl) have similar bacterial composition to the grey cells, many small quartz grains, and small structural vacuoles. Intermediate types are also encountered, and it is suggested that these types represent stages in normal life cycle: the light cells feed intensely and store large quantities of glycogen, the grey type generally contains less food, and is extremely variable in size, occasionally reaching a volume of 20 μl or more; these may be the log phase cells which divide to form the small black cells, which in turn form cysts.

Studies in which the light cells were cut into pieces with a glass needle and placed under different conditions of oxygen tension but in the same medium, showed that the large bacteria multiplied more rapidly under low oxygen tension (ca 20–40 mm Hg), while at high oxygen tension (equilibration with atmospheric air, ca 155 mm Hg), the small Gram-positive bacteria grew most rapidly (see Table 2).

Cysts.—While the conditions for encystment are known and well defined

TABLE 2. ***Pelomyxa palustris.*** **Influence of Oxygen Tension in Medium on Distribution of Bacteria in Halved Cells**[a]

Experimental conditions	Percentage of 3 different bacteria in homogenates		
	Large	Gram-positive rods	Gram-negative rods
High O_2 tension 12 half-cells mean value ± S.D.	27.3 ± 1.0	70.1 ± 1.1	2.6 ± 0.3
Low O_2 tension 12 half-cells mean value ± S.D.	48.0 ± 1.2	46.8 ± 1.4	5.2 ± 0.3

[a] Data from Chapman-Andresen and Holter (unpublished). Gram-stained homogenates. 1000 bacteria counted for each cell half. Cells cut into 2, and starved for 6 days, one-half at low, the other at high oxygen tension.

for some small species of amoebae such as *Acanthamoeba castellani* (81), and fine structural and biochemical changes which accompany the initial stages have been studied (7), encystment in the large amoebae has not so far been achieved by environmental changes. Encystment has been sporadically observed in culture and was described in detail by Kudo (63) for the large amoeba *P. illinoisensis,* and a note on the fine structure of these cysts was published by McClellan (72). The Carlsberg strains of *C. chaos,* originally obtained from The Biological Supply Houses, Chicago, have formed cysts at irregular intervals, after being in culture for a period of up to ten or more years. Electron microscopic studies on these cysts are in progress (Chapman-Andresen and Nilsson, unpublished observations). In a Danish strain of *P. palustris,* cysts have appeared in culture, especially during the autumn, but the conditions for cyst formation have not been defined so far, nor has excystment been observed. Fine structural studies on these cysts, of known age (one day to two years), by Chapman-Andresen and Nilsson (unpublished findings) indicate that the formation of cysts is similar to that found by Bowers & Korn (7) for *Acanthamoeba castellani.* There are two cyst walls, the endo- and exocyst walls have been observed and large numbers of autolytic vacuoles in the central zone, and food debris extruded into the space between the inner and outer cyst walls. The three types of symbiotic bacteria were identified within the cyst, the small Gram-negative bacteria in colonies in the inter-wall space, the large and small Gram-positive bacteria lined up as "palisades" just within the inner cyst wall in mature cysts. There is some indication of sporulation of the large bacteria (68). No reports of encystment of *Amoeba* species could be found.

Experimental techniques.—The large size and relative ease of culture has made the amoebae of the genera *Amoeba* and *Chaos* very suitable organisms for many types of experimental studies. Their capacity for withstanding environmental conditions of varying osmolarity and pH (15, 20) and their adaptability to various microsurgical procedures are high (59). Conditions which are lethal to mammalian cells can be tolerated by amoebae, and in spite of their simple life cycle, the technique of nuclear transplantation has opened up wide possibilities for the study of inheritance (62). Centrifugation in vivo facilitates the identification of sites of enzyme distribution (24, 52–55) and the function of various organelles in the digestive cycle (19, 25).

However, progress with certain types of studies is severely limited by the lack of axenic culture technique, and by the sporadic or permanent occurrence of endosymbionts in some strains. Thus, determinations of biochemical cell components include the contribution of compounds found in varying contaminants, and within old food vacuoles. The proteins, amino acids, and trace metals of the symbionts contribute to the values found for the amoebae, and comparison of species with and without these contaminants may give misleading results. Furthermore, the usual procedure of starving the cells for a few days before the experiments does not in fact eliminate food residues. Studies with natural (ciliates containing symbiotic algae) and unnatural tracers (Alcian blue, ferritin), show (16, 19, 25) that starved cells retain residues from a test meal for considerably longer time (six to eight days) than do cells which are fed daily after the test meal. However, determinations on the protein amino acid composition of amoebae starved for three to twelve days (42), indicate that these residues are not protein, and that the protein from food organisms is degraded within three days.

An extreme case of the influence of food residues on experimental results was found recently by Chapman-Andresen and Holter (unpublished observation) during studies on the oxygen uptake of homogenates of *Pelomyxa palustris*. Large cells were divided into two, and one-half exposed to low, the other half to high oxygen tension during a starvation period of six days. Earlier experiments on starved cells had shown that the half exposed to low oxygen tension consumes several times as much oxygen as the corresponding half maintained at high oxygen tension. During the mild autumn of 1970 when algae were more plentiful than usual in the natural environment of the amoebae, a progressive change was found in the ratio of oxygen consumption in the two halved cells; finally, the cell half kept at high oxygen tension showed the highest oxygen consumption. The cause of this high oxygen consumption was finally traced to increasing numbers of green algae surviving in the food vacuoles of the amoebae during the starvation period, the high oxygen tension apparently favoring the survival of the algae, which contributed significantly to the oxygen consumption measured for the amoebae.

In measurements on the content of trace metals, the contribution of food organisms and media to the "basic" metal content of the cells may be high, as shown by studies on *C. chaos* with iron (19). The "true basic iron content" being ca 0.1 ng per cell, while the value found in amoebae kept in an iron-containing inorganic medium in which precipitated iron salts were irregularly taken up by the cells, ranged from 0.2 to 5 ng per cell. These factors may be of considerable importance in comparison of data from different sources, and from different strains.

Amoebae may be treated individually for certain types of studies, such as in the classical methods of microrespirometry (51), measurement of reduced weight (115), centrifugation in vivo (52), and nuclear transfer (59, 71, 84), or may be used in bulk, thanks to the cultural methods which enable several grams of amoebae to be harvested weekly (12, 82) for the study of the composition of certain cellular components.

The simple technique of enucleation by cutting *A. proteus* into two pieces has also yielded valuable information on the influence of the nucleus on a variety of vital processes, such as survival (83), movement (43), fine structure of organelles (37), contractile vacuole formation (38), and pinocytosis (15, 25a).

Some Aspects of Fine Structure Related to Physiological Processes

The cytology of the large amoebae, as revealed by light microscopy, is well documented for *Chaos chaos* by Andresen (1) and by Torch (106), for the closely related species *Pelomyxa illinoisensis* by Kudo (63), and the classicial descriptions of *Amoeba proteus* by Mast & Doyle (73) are still valid. Following the detailed light microscopic observations of Gould-Veley (44, 45) on *Pelomyxa palustris,* several additional cytological studies have been published (2, 65, 70).

At first sight, electron micrographs of amoebae appear as a chaotic confusion of large and small vacuoles, vesicles, granules and membranes; however, fine structure studies have resulted in identification of some of the organelles involved in at least three important cellular functions—membrane renewal, the function of the Golgi apparatus, and the digestive cycle.

The cell surface.—The plasmalemma, a normal unit membrane, is protected on its outer surface by a mucous layer. The appearance of the mucous coat in the electron microscope depends to a large extent on the method of fixation used (76); slightly alkaline osmium tetroxide as used by Pappas (85) shows the coat to be composed of filaments about 60 Å thick and up to 1700 Å long. However, when uranyl acetate is added to the osmium tetroxide fixative (76) the individual filaments in *Chaos* appear to consist of a light core surrounded by an electron opaque layer. While the length of the filament was observed to be the same as that found by Pappas, the width was appreciably greater, ca 120 Å. However, the diameter of the light core was approximately the same as that of the entire diameter measured by Pappas. As an explanation for this difference Nachmias points out the evidence suggesting the presence of anionic sites which may be bound by uranyl ions, these sites are not fixed by alkaline osmium alone. Numerous observations on adsorption of pinocytic inducers to the mucous coat (8, 10, 12, 15, 19, 77, 96, 103, 110) under conditions in which the effective compound is predominantly cationic (15) support this suggestion. The presence of the fibrillar layer on the outer surface of the plasmalemma, and its capacity for adsorption of large amounts of pinocytic inducers has been utilized for the identification of ingested material and the study of its pathways through the cell (10, 19, 25, 77). The significance of the mucous coat for binding of pinocytic inducers will be discussed in the section on endocytosis.

The mucous coats appear to be characteristic for different species of amoebae. While those of *C. chaos* and *A. proteus* are thick, 1100 to 1700 Å (85), observations on *Amoeba dubia* (100a) indicate that the mucous coat of this species is less pronounced; this may be correlated with the less intense response of *A. dubia* to inducers of pinocytosis (15). In *P. palustris* only sparse threads have been observed on the plasmalemma (2).

Golgi systems.—The Golgi bodies in the large fresh-water amoebae are small, with a diameter of ca 0.5–1 μ. Attempts to identify these bodies in light microscopy by impregnation with silver or osmium were unsuccessful (106). These bodies were first demonstrated in *A. proteus* and *C. chaos* by Pappas (85), as typically stacked parallel membranes with vesicles of different types orientated at the two poles.

The first observations connecting Golgi vesicles with the plasmalemma were made by Daniels (31), who observed vesicles of about 0.3 μ in diameter, lying on the concave side of the Golgi body and noted that these vesicles on their inner aspect were fringed as the plasmalemma. These vesicles were particularly abundant shortly after starved amoebae had been fed or induced to pinocytosis. These observations led Daniels to the conclusion that the fringed vesicles passed from the plasmalemma to the Golgi body. Two recent studies on the fine structure of the Golgi system (103, 110) present convincing evidence that the flow of fringed vesicles is directed from, and not toward the Golgi apparatus. Wise & Flickinger (110) used electron-opaque tracers to induce pinocytosis in *A. proteus* and *C. chaos* and showed that none of the tracers appeared in the Golgi apparatus, but after increasing intervals the tracers were found in cytoplasmic vesicles and food vacuoles. Cytochemical tests for glycoproteins gave positive results for the mucous coat, the fringed vesicles and for the cisternae on the concave side of the Golgi apparatus. These results, indicating that the fringed vesicles on the concave side of the Golgi apparatus are not derived from the cell surface, agree with the studies of Stockem (103), who employed a different fixative, and different methods for demonstration of mucopolysaccharide.

Stockem made careful measurements on the components and membranes contained within the Golgi body and suggested that the fringed vesicles enlarge as they approach the plasmalemma, finally fusing with it to renew both membrane and mucous coat; the final decisive evidence, observation of fusion between vesicles and plasmalemma is still lacking. Insertion of vesicles into the plasmalemma was also suggested by Nachmias (77) in her study on the formation of new surface by *C. chaos,* using the ingenious method of counting the clumps of dye-mucous complex formed on the plasmalemma after treatment with the inducer Alcian blue. Confirmation that these vesicles should be considered as exocytic is also given by observations that the vesicles do not contain acid phosphatase (24), or pinocytically ingested dyes (16, 25). The exocytic vesicles correspond to the alpha granules described by Mast (73) and are found in the small granule zone of centrifuged amoebae (25, 30).

No characteristic Golgi bodies have been observed in *Pelomyxa palustris* (2, 32); the cycle of plasmalemma renewal has not yet been studied in this species, but labeling experiments with Alcian blue (Chapman-Andresen, unpublished observations) indicated that large, optically empty vacuoles were formed at the uroid.

In addition to its function in production of mucopolysaccharides and plasmalemma, the Golgi apparatus is also concerned in production of, or perhaps packaging of hydrolytic enzymes into the primary lysosomes. Acid phosphatase activity is present in the small granule zone of centrifuged *Chaos,* which also contains the Golgi bodies (24). Recent electron microscopic studies on the Golgi apparatus of *A. proteus* (111) showed that the cisternae on the convex pole contain acid phosphatase and thiamine pyrophosphatase activity. These enzymes were uniformly found in all Golgi complexes observed and were limited to the three to five cisternae at the convex pole.

Lysosomes.—The digestive cycle of the fresh-water amoebae includes many lysosomal components (75), and a recent review by Müller (74) gives a comprehensive survey of the digestive enzymes found in protozoa, including the amoebae. Of the 37 enzymes which have so far been found in lysosomes (105), only a few have been demonstrated in amoebae. Acid phosphatase, acid protease, and amylase were shown to be associated with the formed organelles of *Chaos chaos* by Holter and co-workers (52–55). In addition, alpha-glucosidase, aryl sulfatase (21), nonspecific esterase (57), and thiamine pyrophosphatase (111) activities have been found in *Chaos.* With regard to the cytochemical localization of hydrolytic enzymes, acid phosphatase activity is found in food vacuoles soon after their formation (24, 75). The association of acid phosphatase activity with other cytoplasmic organelles is limited to those cell constituents which are concentrated in the small granule zone of cells centrifuged in vivo (24). In fractionated homogenates of *Chaos,* Rothschild (95) found most of the acid phosphomonoesterase and acid protease activities to be associated with particles with sucrose-equilibrium densities greater than 1.25 g/ml. Of the organelles in this zone the Golgi bodies have recently been shown by Wise & Flickinger (111) to contain acid phosphatase activity, in addition to endocytic vesicles, the *a*-vesicles (24, 25). These *a*-vesicles may be considered as secondary lysosomes since they contain acid phosphatase activity and accumulate around newly formed food vacuoles and may also be labeled with tracers ingested by endocytosis. The *a*-vesicles may be considered as a product of coalescence between pinocytic vesicles and primary lysosomes, probably formed from vesicles produced from the acid phosphatase-containing cisternae at the convex pole of the Golgi body.

The relationship of the secondary lysosomes (*a*-vesicles) to the digestive cycle has been studied by the use of labeled inducers of pinocytosis. Using Alcian blue, which is a stable and easily identified marker for both light and electron microscopy, it has been found (25) that the labeled lysosomes (*a*-vesicles) accumulated around newly formed food vacuoles at the same sites and in the same sequence as the acid phosphatase-positive granules (24), within half an hour of their formation, but disappeared from this site within 2 hours. The label, together with acid phosphatase activity, was found during this period within the food vacuoles. That the lysosome cycle is not dependent on ingestion of nutritive material is indicated by observations on accumulation of labeled lysosomes around large polysterene spheres (Chapman-Andresen and Hendil, unpublished findings) and acid phosphatase activity associated with small polysterene spheres, 3 hours after ingestion (24).

The rate of disappearance of labeled lysosomes from the small granule zone of cells centrifuged in vivo is dependent on the state of nutrition of the amoebae following the labeling (19, 25). When cells are fed daily after labeling, the label is completely eliminated after three feedings, presumably by coalescence of the lysosomes with the new food vacuoles. When cells are starved after the labeling, the label is retained for periods up to eight days. Quantitative data on the turnover of lysosomes have been obtained by the use of ferritin as a tracer and by measurements of ferritin content by determination of the iron content of this protein by atomic absorption (19). This method is sufficiently sensitive to permit measurements on ten *Chaos* halves. Cells centrifuged in vivo were cut in two parts to separate the small granule zone from the heavier portion containing food vacuoles; the content of ferritin in these two portions was followed in a series of cells fed daily after heavy ferritin labeling, and in similar cells starved for eight days after the labeling. The data showed that the turnover of lysosomes is promoted by feeding, while starvation conserves the supply.

It may be interesting to attempt some correlation between this turnover and observations on the activity of the Golgi system. Daniels (31) observed that Golgi bodies were more abundant in amoebae shortly after a period of active endocytosis, when presumably hydrolytic enzymes would be required for renewal of the secondary lysosomes, in addition to a requirement for the fringed vesicles to renew the plasmalemma utilized during endocytosis. The observation of Flickinger (37) on the disappearance of Golgi bodies after enucleation of *A. proteus* may be correlated with the inability of enucleates to pinocytose when tested two days after enucleation (15), while freshly enucleate halves pinocytose more actively than the corresponding nucleate halves (15, 25a).

Some very approximate numerical data on the utilization of secondary lysosomes during feeding (19) indicate that approximately one half of the total number present in *Chaos* is used in one feeding. Hence, the renewal rate would be expected to be high but how much of this renewal is synthesis, i.e., formation of new primary lysosomes from the Golgi complex, and how much is re-use, that is, transfer from food vacuoles, is problematic. Data on the total acid phosphatase content of starving amoebae (21, 66) showed that while the specific activity of the enzyme does not decrease during starvation for sixteen days, the content per cell is reduced to about 50 percent of the initial value by the end of starvation. Autolytic vacuoles are formed about five days after the last feeding and increase in number and size during prolonged starvation (18). Loss of enzyme is presumably reduced by the low rate of defecation of autolytic vacuoles, and Lagunoff (66) could detect only small amounts of acid phosphatase activity in the culture fluid of starving amoebae. In *Pelomyxa palustris,* lysosomes have not yet been defined, but small amounts of catheptic proteinase and peptidase have been demonstrated (3).

Endocytosis.—The present definition of the term, endocytosis, implies inclusion of all types of uptake processes which involve the transformation of part of the plasmalemma into the lining of a cytoplasmic vesicle or vacuole regardless whether the material ingested is solid or fluid, natural or unnatural. Two main types of normal physiological endocytosis may be distinguished in the fresh-water amoebae: (*a*) the normal feeding cycle in which the intake of membrane is discontinuous, and during which normal locomotion is interrupted during the initial stages of digestion; and (*b*) a slow, but continuous uptake of the membrane at the uroid, with a correspondingly low membrane renewal rate (113), compatible with and, according to Wohlfarth-Bottermann & Stockem (112), an essential part of the normal locomotory process.

The normal digestive cycle.—The normal digestive cycle is a dynamic process, involving a series of coalescences and subdivisions of vacuoles and vesicles, during which water and soluble products of digestion are removed from, and hydrolytic enzymes added to, the original food vacuole. Although many reports on the fine structure of food vacuoles of amoebae are available (25, 26, 30, 85, 93) in only a few of these studies is the age of the food vacuole sufficiently well defined to permit the sequence of events around the cytoplasmic aspect of the vacuole to be followed. Christiansen & Marshall (26) have given a detailed description of the earliest phases of food vacuole formation in *C. chaos,* describing a reorganization zone at the vacuole membrane and initiation of the vacuolar dehydration process. Using *Stentor polymorphus* as a tracer meal at known time, later phases of the digestive process were studied (19, 25) during the period from one-half hour to three days after ingestion. The fusion of secondary lysosomes with the food vacuoles has been described in the section on lysosomes. The subdivision of the original food vacuole commencing about 3 hours after ingestion, coincided with the appearance of a band of fibrils around the cytoplasmic aspect of the vacuolar membrane. These fibrils were similar to those observed by Nachmias (78) in the ground cytoplasm of *Chaos,* and are presumably involved in the vacuolar division process. Older food vacuoles are surrounded by a cap of small dense vesicles (93) which remain adjacent to the food vacuole until defecation. However, if amoebae are starved until autolytic vacuoles are formed (18), the large debris vacuoles which appear later are not surrounded by these vesicles.

Induced pinocytosis.—This process also involves a discontinuous uptake of membrane. Under optimal conditions of induction (15) *A. proteus* ingests about 50 percent of the plasmalemma during a full cycle of pinocytosis lasting about 30 minutes (14). That the duration of the cycle is limited by the availability of membrane was indicated by observatons that a rest period of about 4 hours was necessary before the amoeba could again pass through a full cycle (14). This indication has recently been fully confirmed by Sanders & Bell (100) in studies using puromycin to inhibit membrane synthesis.

Many compounds, from simple inorganic salts to complex macromolecules, act as inducers of pinocytosis; they show no great specificity, the presence of some cationic charges, however, appears necessary (15). However, pinocytosis is not an all-or-none process, the response of the amoebae depends on the concentration and pH of the inducing solution and on temperature as well as on the diluent added to the inducer (15). The number of pinocytic channels formed in an inducing solution may vary from one to two up to a hundred or more during a cycle, i. e., the response is graded, and when it is slight channels are formed only at the uroid as previously reported (96, 113) in culture medium. These observations indicate that while the permanent pinocytosis (112) may well be physiological, there is no evidence that its continuance does not depend on the presence of traces of proteins or salts in the medium. The present methods of amoeba culture make it impossible to test pinocytic activity in the complete absence of contaminations. In this connection it may be noted that attempts to culture amoebae in sterile medium were unsuccessful (79, 80, 88).

It may be of interest here to review the conditions under which amoebae respond most actively to pinocytosis, induced under conditions optimal for the chosen compound. A short starvation period (one to three days) and the temperature, ca 20°C, at which the amoebae are cultured, give a maximal response. Enucleate halves of *Amoeba proteus* respond more intensely than do the corresponding nucleate halves within the first few hours of enucleation; however, starved enucleate cells give a lower response than their corresponding nucleate halves (15, 25a). Recently, polylysine has been shown to stimulate pinocytosis of globulin, while treatment with the polycation alone results in the formation of numerous, very fine, atypical channels which are not suppressed by metabolic inhibitors (100a). Physiological conditions which reduce the response of the amoebae to an inducing solution include recent intense endocytosis or prolonged starvation (14, 15) and the period immediately after cell division (100). In addition, the centrifugal half of *A. proteus* centrifuged in vivo and cut into two responds poorly to an inducer, while the centripetal half exhibits a maximal response (99). Certain inhibitors of respiration (8, 17, 35) and glycolysis (17) rapidly depress the formation of pinocytic channels in *A. proteus,* while inhibitors of oxydative phosphorylation (8, 17) act more slowly. Treatment with puromycin (100) to inhibit membrane synthesis limits both normal locomotion and pinocytosis, confirming that the other conditions under which pinocytic response is low may be also ascribed to lack of available membrane.

The study of pinocytosis induced by inorganic ions has given useful information on the influence of different cations (12, 13, 27, 49), especially the effect of calcium on interaction of ions (27, 49), on plasmalemma structure (9), and in demonstrating the ion exchange properties of the mucous coat (12, 49). Protein inducers have been useful for studying the fate of pinocytosed material. The pattern of loss with time of different labeled proteins from the cell (11, 19, 23, 56) suggests that while bovine albumin and gamma globulin are degraded, ferritin is excreted unchanged, following a loss curve similar to that of the presumably indigestible Alcian blue (16). More direct evidence of protein degradation is given by the finding of labeled nonprotein material in both cells and medium (11).

Contractile vacuoles.—Light microscopic studies on the contractile vacuole of *Chaos* and *Amoeba* (1, 73, 106) show that the membrane of the vacuole appears as a layer about 0.5 μ thick surrounded by an outer layer of granules, suggested by Andresen (1) to be mitochondria. Fine structural studies (31, 86) showed that this wall was a typical unit membrane ca 70 Å thick, surrounded by a layer of closely packed small vesicles again surrounded by a layer of mitochondria. The increased activity of the contractile vacuole observed shortly after feeding, together with the reduction in size of the initial food vacuoles, and the complex reorganization zone (26) found on the cytoplasmic aspect of newly formed food vacuoles, support the suggestion that dehydration of the food vacuoles is accomplished by water transport in vesicles.

The inhibition of contractile vacuolar activity observed after a full cycle of pinocytosis (15) is rapidly reversed when the amoeba is transferred to a noninducing solution. This inhibition may be in part due to the greatly reduced cytoplasmic streaming, which occurs during pinocytosis. There is some evidence that the contractile vacuole may excrete substances other than water (20). Direct measurements on the freezing point depression of the cytoplasm and the contractile vacuole fluid, removed by micropuncture of the vacuole (101) showed that the fluid is hypo-osmotic to the cytoplasm, suggesting that salt is subsequently reabsorbed from the vacuole. The results were confirmed by Riddick (92).

Wigg, Bovee & Jahn (108) consider that the contractile vacuole should be designated "water expulsion vesicle" as there is no fine structural evidence of fibrils surrounding the organelle which might promote its contraction. That the force for systole is generated by the vacuole itself and is not dependent on motive forces in the surrounding cytoplasm, has also been shown by Prusch & Dunham (90), who isolated contractile vacuoles from *A. proteus* and observed immediate contraction of the vacuoles after application of adenosine triphosphate and magnesium ions. It would be relevant to know the thickness of these isolated vacuole walls, as it seems unlikely that contractile proteins could be present in a unit membrane.

The presence of the nucleus appears to be necessary for the formation of new contractile vacuoles in *A. proteus* (38). The contractile vacuole was maintained as long as the cell lived, regardless whether it was present in the nucleate or in the enucleate half. However, if the enucleate half lacked a contractile vacuole, no vacuole was regenerated, while nucleate portions without a contractile vacuole regenerated a new contractile vacuole within 1 hour. These studies raise speculations on whether the source of membrane for the new contractile vacuole may be the Golgi system, which disappears from enucleate cells (38).

Concluding Remarks

In this brief survey of a few selected topics of amoeba biology, emphasis has been placed on certain physiological aspects of fine structure, and this has necessarily resulted in the omission of many important and interesting papers, even within this narrowly defined field. However, the author hopes that it has given some impression that the large amoebae are model cells for many types of studies, as well as being interesting and unique species of protozoa. Lack of establishment of axenic cultures for the large amoebae is still a limiting factor, and development of such cultures would promote and extend the possibilities for studies on biochemical characterization of cell constituents and on the part played by an uncontaminated cell surface in uptake processes.

LITERATURE CITED

1. Andresen, N. 1956. *Compt. Rend. Trav. Lab. Carlsberg Sér. Chim.* 29:435–555
2. Andresen, N., Chapman-Andresen, C., Nilsson, J. R. 1968. *Compt. Rend. Trav. Lab. Carlsberg* 36: 285–317
3. Andresen, N., Holter, H. 1949. *Science* 110:114–15
4. Arnold, Z. M. 1948. *Trans. Am. Microsc. Soc.* 67:231–35
5. Arnold, Z. M. 1955. *Univ. Calif. Publ. Zool.* 61:167–252
6. Bovee, E. C., Jahn, T. L. See Ref. 60
7. Bowers, B., Korn, E. D. 1969. *J. Cell Biol.* 41:786–805
8. Brandt, P. W. 1958. *Exp. Cell Res.* 15:300–13
9. Brandt, P. W., Freeman, A. R. 1967. *Science* 155:582–85
10. Brandt, P. W., Pappas, G. D. 1960. *J. Biophys. Biochem. Cytol.* 8:675–87
11. Brownstone, Y. S., Chapman-Andresen, C. In preparation
12. Bruce, D. L., Marshall, J. M., Jr. 1965. *J. Gen. Physiol.* 49:151–78
13. Chapman-Andresen, C. 1958. *Compt. Rend. Trav. Lab. Carlsberg Sér. Chim.* 31:77–92
14. Chapman-Andresen, C. 1961. Progress in Protozoology. *Proc. Int. Conf. Protozool., 1st, Prague,* 267–70
15. Chapman-Andresen, C. 1962. *Compt. Rend. Trav. Lab. Carlsberg* 33: 73–264
16. Chapman-Andresen, C. 1967. *Compt. Rend. Trav. Lab. Carlsberg* 36: 161–87
17. Chapman-Andresen, C. 1967. *Protoplasma* 63:103–5
18. Chapman-Andresen, C. 1969. Progress in Protozoology. *Proc. Int. Conf. Protozool., 3rd., Leningrad,* 23
19. Chapman-Andresen, C., Christensen, S. 1970. *Compt. Rend. Trav. Lab. Carlsberg* 38:19–57
20. Chapman-Andresen, C., Dick, D. A. T. 1962. *Compt. Rend. Trav. Lab. Carlsberg* 32:445–68
21. Chapman-Andresen, C., Hansen, T. B., Müller, M. 1969. *Exp. Cell Res.* 58:452
22. Chapman-Andresen, C., Hayward, A. F. 1963. *Bacterial Complexes in* Amoeba proteus. Presented at Brit.-Dutch-Scand. Meeting Soc. Exp. Biol., Oxford
23. Chapman-Andresen, C., Holter, H. 1955. *Exp. Cell Res. Suppl.* 3:52–63
24. Chapman-Andresen, C., Lagunoff, D. 1966. *Compt. Rend. Trav. Lab. Carlsberg* 35:419–36
25. Chapman-Andresen, C., Nilsson, J. R. 1967. *Compt. Rend. Trav. Lab. Carlsberg* 36:189–207

25a. Chapman-Andresen, C., Prescott, D. M. 1956. *Compt. Rend. Trav. Lab. Carlsberg Sér. Chim.* 30:57–78

26. Christiansen, R. G., Marshall, J. M., Jr. 1965. *J. Cell Biol.* 25:443–57
27. Cooper, B. A. 1968. *Compt. Rend. Trav. Lab. Carlsberg* 36:385–403
28. Corliss, J. O., 1967. *Chem. Zool.* 1: 1–20
29. Daniel, J. W., Baldwin, H. H. 1964. *Methods Cell Physiol.* 1:9–41
30. Daniels, E. W. 1964. *J. Protozool.* 11:281–90
31. Daniels, E. W. 1964. *Z. Zellforsch. Mikrosk. Anat.* 64:38–51
32. Daniels, E. W., Breyer, E. P. 1967. *J. Protozool.* 14:167–79
33. Daniels, E. W., Breyer, E. P., Kudo, R. R. 1965. *Am. Zool.* 5:734–37
34. Daniels, E. W., Breyer, E. P., Kudo, R. R. 1966. *Z. Zellforsch. Mikrosk. Anat.* 73:367–83
35. De Terra, N., Rustad, R. C. 1959. *Exp. Cell Res.* 17:191–95
36. Drozanski, W. 1963. *Acta Microbiol. Polon.* 12:9–24
37. Flickinger, C. J. 1969. *J. Cell Biol.* 43:250–62
38. Flickinger, C. J., Coss, R. A. 1970. *Exp. Cell Res.* 62:326–30
39. Friz, C. T. 1963. *J. Protozool. Suppl.* 10:23
40. Friz, C. T. 1968. *J. Protozool.* 15: 149–52
41. Friz, C. T. 1970. *J. Protozool. Suppl.* 17:21
42. Friz, C. T. 1970. *J. Protozool.* 17: 235–39
43. Goldstein, L., Jelinek, W. 1966. *Exp. Cell Res.* 43:51–55
44. Gould-Veley, L. J. 1894. *Quart. J. Micros. Sci.* 36:295–306
45. Gould-Veley, L. J. 1905. *J. Linn. Soc.* 29:374–95
46. Griffin, J. L. 1960. *Exp. Cell Res.* 21:170–78
47. Griffin, J. L. See Ref. 60

48. Hawkins, S. E., Wolstenholme, D. R. 1967. *Nature (London)* 214:928–29
49. Hendil, K. 1971. *Compt. Rend. Trav. Lab. Carlsberg* 38:187–211
50. Holst-Sørensen, H., Rasmussen, L. 1971. *Compt. Rend. Trav. Lab. Carlsberg* 38:163–70
51. Holter, H. 1943. *Compt. Rend. Trav. Lab. Carlsberg Sér. Chim.* 24:399–478
52. Holter, H. 1954. *Proc. Roy. Soc. Ser. B* 142:140–46
53. Holter, H., Doyle, W. L. 1938. *Compt. Rend. Trav. Lab. Carlsberg Sér. Chim.* 22:219–25
54. Holter, H., Lowy, B. A. 1959. *Compt. Rend. Trav. Lab. Carlsberg Sér. Chim.* 31:105–27
55. Holter, H., Løvtrup, S. 1949. *Compt. Rend. Trav. Lab. Carlsberg Sér. Chim.* 27:22–62
56. Holter, H., Marshall, J. M., Jr. 1954. *Compt. Rend. Trav. Lab. Carlsberg Sér. Chim.* 29:7–37
57. Holter, H., Thompson, A. In preparation
58. Honigberg, B. M. et al. 1964. *J. Protozool.* 11:7–20
59. Jeon, K. W. 1970. *Methods Cell Physiol.* 4:179–94
60. Jeon, K. W. Ed. 1972. *The Biology of Amoebae.* New York: Academic
61. Jeon, K. W., Lorch, I. J. 1967. *Exp. Cell Res.* 48:236–40
62. Jeon, K. W., Lorch, I. J., Danielli, J. F. 1970. *Science* 167:1626–27
63. Kudo, R. R. 1951. *J. Morphol.* 88:145–84
64. Kudo, R. R. 1952. *Trans. Am. Microsc. Soc.* 71:108–13
65. Kudo, R. R. 1957. *J. Protozool.* 4:154–64
66. Lagunoff, D. 1964. *Compt. Rend. Trav. Lab. Carlsberg* 34:433–50
67. Lee, J. J., Pierce, S., 1963. *J. Protozool.* 10:404–11
68. Leiner, M., Bohwmick, D. K. 1967. *Z. Mikrosk. Anat. Forsch.* 77:529–52
69. Leiner, M., Wohlfeil, M. 1953. *Arch. Protistenk.* 98:227–86
70. Leiner, M., Wohlfeil, M., Schmidt, D. 1954. *Ann. Sci. Nat. Zool. Biol.* 16:537–94
71. Lorch, I. J., Danielli, J. F. 1953. *Quart. J. Microsc. Sci.* 94:445–60, 461–80
72. McClellan, J. F. 1958. *J. Protozool. Suppl.* 5:10
73. Mast, S. O., Doyle, W. L. 1935. *Arch. Protistenk.* 86:155–80, 278–306
74. Müller, M. See Ref. 28, 351–80
75. Müller, M. Röhlich, P., Tóth, J., Törö, I. 1963. *Ciba Found. Symp. Lysosomes* 201–16
76. Nachmias, V. T. 1965. *Exp. Cell Res.* 38:128–32
77. Nachmias, V. T. 1966. *Exp. Cell Res.* 43:583–601
78. Nachmias, V. T. 1968. *J. Cell Biol.* 38:40–50
79. Nardone, R. M. 1959. *J. Protozool. Suppl.* 6:9
80. Neff, R. J. 1958. *J. Protozool.* 5:226–31
81. Neff, R. J., Ray, S. A., Benton, W. F., Wilborn, W. See Ref. 29, 55–84
82. O'Neill, C. H. 1964. *Exp. Cell Res.* 35:477–96
83. Ord, M. J. 1968. *J. Cell Sci.* 3:81–88
84. Ord, M. J., Bell, L. G. E. 1968. *Nature (London)* 218:384
85. Pappas, G. D. 1959. The Biology of the Amoeba. *Ann. NY Acad. Sci.* 78:448–73
86. Pappas, G. D., Brandt, P. W. 1958. *J. Biophys. Biochem. Cytol.* 4:485–88
87. Prescott, D. M. 1956. *Compt. Rend. Trav. Lab. Carlsberg Sér. Chim.* 30:1–12
88. Prescott, D. M., Carrier, R. F. 1964. See Ref. 29, 85–95
89. Prescott, D. M., James, T. W. 1955. *Exp. Cell Res.* 8:255–58
90. Prusch, R. D., Dunham, P. B. 1970. *J. Cell Biol.* 46:431–34
91. Rasmussen, L., Kludt, T. A. 1970. *Exp. Cell Res.* 59:457–63
92. Riddick, D. H. 1968. *Am. J. Physiol.* 215:736–40
93. Roth, L. E. 1960. *J. Protozool.* 7:176–85
94. Roth, L. E., Daniels, E. W. 1961. *J. Biophys. Biochem. Cytol.* 9:652–53
95. Rothschild, J. 1967. *Compt. Rend. Trav. Lab. Carlsberg* 35:457–500
96. Rustad, R. C. 1961. *Sci. Am.* April: 121–30
97. Salt, G. W. 1961. *Exp. Cell Res.* 24:618–20
98. Salt, G. W. 1968. *J. Protozool.* 15:275–80
99. Sanders, E. J. 1970. *Exp. Cell Res.* 61:461–65
100. Sanders, E. J., Bell, L. G. 1970. *Exp. Cell Res.* 63:379–84

100a. Sanders, E. J., Bell, L. G. 1970. *J. Cell Sci.* 7:739–53
101. Schmidt-Nielsen, B., Schrauger, C. R. 1963. *Science* 139:606–7
102. Sonneborn, T. M. See Ref. 59, 241–339
103. Stockem, W. 1969. *Histochem.* 18: 217–40
104. Sussman, R., Sussman, M. 1967. *Biochem. Biophys. Res. Commun.* 29:53–55
105. Tappel, A. L. 1969. *Lysosomes in Biology and Pathology,* 2:208–44, ed. J. T. Dingle, H. B. Fell. Amsterdam: North Holland
106. Torch, R. 1959. *Ann. NY Acad. Sci.* 78:407–20
107. Whitelock, O. V. St., Ed. 1959. The Biology of the Amoeba. *Ann. NY Acad. Sci.* 78:401–704
108. Wigg, D., Bovee, E. C., Jahn, T. L. 1967. *J. Protozool.* 14:104–8
109. Wilson, H. V. 1900. *Am. Nat.* 34: 535–50
110. Wise, G. E., Flickinger, C. J. 1970. *Exp. Cell Res.* 61:13–23
111. Wise, G. E., Flickinger, C. J. 1970. *J. Cell Biol.* 46:620–26
112. Wohlfarth-Bottermann, K. E., Stockem, W. 1966. *Z. Zellforsch. Mikrosk. Anat.* 73:444–74
113. Wolpert, L., O'Neill, C. H. 1962. *Nature* (*London*) 196:1261–66
114. Wolstenholme, D. R. 1966. *Nature* 211:652–53
115. Zeuthen, E. 1948. *Compt. Rend. Trav. Lab. Carlsberg Sér. Chim.* 26:243–76

THERMOPHILIC ENTERIC YEASTS[1]

1562

L. R. R. G. TRAVASSOS AND A. CURY

Departmento de Microbiologia Geral, Instituto de Microbiologia, Universidade Federal do Rio de Janeiro, Brasil

CONTENTS

INTRODUCTION

The rather obscure thermophilic enteric yeasts have so fascinating an ecology—presumably reflecting odd metabolic specializations—that a panoramic description seems useful. They likely have practical value: some grow so well at body temperature—37°C or even higher—in simple defined media that they may be valuable for assaying choline and carnitine. *Saccharomycopsis guttulata* offers extreme nutritional-ecological specialization including an elevated CO_2 requirement.

Thermophilic enteric yeasts live in domestic and some wild animals. Thermophilic here denotes a minimal growth temperature above room temperature (159). Maximal temperatures are not, however, as high as those for thermophilic bacteria and certain fungi (39), therefore they have been called psychrophobic (25).

This review updates van Uden's (159).

[1] This work was supported by the Conselho Nacional de Pesquisas, Brasil.

Systematics

Yeasts described here are psychrophobic species of *Saccharomycopsis, Saccharomyces, Candida,* and *Torulopsis.* The only species accepted as a *Saccharomycopsis* is *S. guttulata* (Robin) Schiönning. This species includes budding yeasts with endospores having two membranes and germinating by budding (120). In malt extract (acidified to pH 2.5 with HCl) inoculated with stomach contents or feces of rabbits, growth of *S. guttulata* appeared after two days at 37°C; cells were long-oval to cylindrical 4–7 × 11–21 μ (75), in pairs, or in larger groups. Older cells had drop-like vacuoles, giving the yeast its name. Spores are oval, 1–4 per ascus, 0.9–1.6 × 1–2.8 μ (127). Sporulation was observed on Gorodkowa agar and turnip plugs at 18°C. No apparent conjugation was observed before ascus formation (99). Parle studied the fermentation reactions of *S. guttulata* in stomach-extract broth. Since the medium was rather acid (pH 3.5) gas production served to detect fermentation. Glucose and sucrose were fermented; maltose, galactose, lactose, and raffinose were not; fermentable sugars served as sole carbon sources. Slow assimilation of glucose, sucrose, raffinose, and citrate was obtained in a yeast autolysate basal medium (127). Nitrate utilization was unclear because of the complexity of the medium. *S. guttulata* is identified by its habitat (discussed later), fastidious nutritional requirements, and narrow temperature range (35–40°C). Shifrine & Phaff (128) found also that *S. guttulata* has an unusual cell wall: high in protein and low in mannan compared with *Saccharomyces cerevisiae.*

Candida slooffii, like *S. guttulata,* sediments on growing at 37°C in liquid medium; no ring or pellicle forms. Cells of *C. slooffii* are oval, single, or in pairs, but mostly in groups, 4–7 × 6–9 μ (160). Bigger roundish cells may form. Pseudomycelium is evident in Dalmau plates at 37°C (161) with corn meal agar. Characteristic is the formation of well-developed, verticillated, branched chains of oval or roundish blastospores. No ascospores are formed. *C. slooffii* ferments glucose but not galactose, sucrose, maltose, lactose, or raffinose. Only glucose is assimilated. It does not grow at room temperature and require six vitamins (33, 161).

Van Uden and co-workers (163, 167) stressed temperature relations as a taxonomic criterion and applied it in differentiating several *Candida* species. Maximal growth temperature was a more convenient systematic character than optimal temperature. Growth of *C. slooffii* is sharply inhibited at temperatures higher than 44°C; that of *T. pintolopesii,* more than 40°C. Habitat and nutritional imbalances also help define *C. slooffii* (33, 82).

The description of *Torulopsis pintolopesii* (158, 169) specifies oval cells, 2.5–3 × 3–5 μ and, in older cultures, roundish cells, single or in pairs. At 35°C a white sediment forms in liquid media and no ring or pellicle is formed. *T. pintolopesii* ferments glucose but not galactose, sucrose, maltose, or lactose. Only glucose is assimilated. Van Uden (158) could not grow *T.*

pintolopesii with peptone unless yeast extract (0.04 percent) was added. Yeast extract masked the assimilation (if any) of nitrate, ammonium, asparagine or urea. In a synthetic medium supplemented with 7 vitamins (33, 155) to be described later, 12 strains of *T. pintolopesii,* including the original strain of van Uden, nevertheless grew with asparagine or ammonium as sole N sources. Failure to sporulate or to form pseudomycelium, and habitat and temperature range also help define *T. pintolopesii.* Cultures of *T. pintolopesii* are short-lived in malt agar, perhaps related to the respiratory deficiency exhibited by this yeast as with *T. lactis-condensi* (67). Among the obligate saprophytes besides *T. pintolopesii, C slooffii* is also respiratory-deficient whereas *S. telluris* and *T. bovina* are not. Respiratory-deficient mutants arose spontaneously from *S. telluris* and *T. bovina* (23). The Custers' effect—O_2-stimulated fermentation or negative Pasteur effect—useful in classifying *Brettanomyces* (181), was lacking in *C. slooffii, T. bovina, T. pintolopesii,* and *S. telluris* (119).

Candida bovina (162), later moved to *Torulopsis* (169), forms in liquid medium with oval to long-oval cells 4–6 × 6–9 μ, single, in pairs but mostly in groups, again forming sediment. On solid medium cells appeared round, oval, sometimes irregular. On corn meal agar pseudomycelium formed (162). Van Uden et al (166) noted as distinctions between *C. slooffii* on one side and *C. bovina* and *S. telluris* on the other, the capacity of *C. slooffii* to form a well-developed pseudomycelium and the others only a "primitive" one. In the "standard" description of *T. bovina* (169), "No pseudomycelium is formed, but dense ramified chains of ovoid and elongate cells may be present." Only glucose is fermented and assimilated by *Torulopsis bovina.*

Saccharomyces tellustris, later corrected to *S. telluris* (174), isolated from soil (172) and from carious teeth in South Africa (173), formed in malt extract at 25°C round to short-oval cells, 3.5–6.5 × 4.5–8.5 μ, single, in pairs, short chains or small clusters. No pseudomycelium was observed in slide cultures but a few elongated cells sometimes formed. The spores on Gorodkowa agar were large, round with small oil droplets, usually one and seldom two per ascus. Only glucose was fermented and assimilated. Growth was reported on ethanol (172) but not confirmed (33). Kreger-van-Rij (66) studied the morphology, temperature relations, and vitamin requirements of *S. telluris* and concluded that it represented the perfect stage of *C. bovina.*

Ecology and Nutrition

Obligate saprophytes.—Yeasts in the digestive tract of warm-blooded animals may be obligate saprophytes, facultative saprophytes, or transients (159). Only obligate saprophytes are considered here; Carmo-Sousa (25) so grouped: *Saccharomycopsis guttulata, Torulopsis pintolopesii, Candida slooffii, Saccharomyces telluris (Torulopsis bovina), Torulopsis glabrata,*

Candida stellatoidea and *Candida albicans.* These yeasts have not usually been found outside the animal body. However, isolations from other sources indicate that these yeasts may survive a while outside animals. *T. pintolopesii* was reported from soil (9). *T. glabrata* was first isolated from nonhuman sources among yeasts from shrimp (101). Evidence has since accumulated that this yeast is isolable from soil, seawater, and from creek and pond waters and their sediments (26). The position, then, of *T. glabrata* as an obligate saprophyte is doubtful (166). *C. albicans,* commonly found in man, monkeys, cats, dogs, donkeys, rats, hedgehogs, chickens, turkeys, and pigeons, is occasionally isolated from soil and vegetable substrates (86, 95, 111, 168). If a yeast can form ascospores, survival outside the host is understandable. *S. telluris,* the perfect form of *T. bovina* (66), has been isolated from South African soil (172). *S. guttulata* does not grow at less than 35°C, yet sporulates at 18–24°C (127).

What may preclude vegetative life of these yeasts outside the animal body is the high minimum temperature exhibited by some species. *C. slooffii* (161) will not grow at less than 28°C. The minimum temperature for *T. pintolopesii* is approximately 24°C; and *T. bovina* grows slowly at 20–22°C (33). Nevertheless, additional important factors, mainly nutritional, must determine whether a microorganism is an obligate saprophyte. In fact, *T. glabrata,* isolated by Phaff et al (101), grew at 5°C, and several other strains from human sources at 15–17°C (3). The same applies to *C. albicans,* a true obligate saprophyte of warm-blooded animals, which does not have a high minimum growth temperature. Although usually saprophyte, *T. glabrata* (180) and *C. albicans,* under certain conditions of the host, cause disease. Van Uden et al (166) postulated that "the higher the suitability of a given species to be a host for *C. albicans,* the more individuals of this species will actually harbor *C. albicans* and the higher will be the incidence of moniliasis." Wickerham (180) noted the increasing importance of *T. glabrata* as human pathogen. Other species of the ecological group of obligate saprophytes are not reported to cause disease naturally or experimentally (25, 126). *C. slooffii* was found nevertheless in the hyperplasic mucosal lesions of the pars oesophagea of 35 otherwise healthy piglets (134).

Oddly enough, although *T. pintolopesii, T. bovina,* and *C. slooffii* are common in domestic animals, they were first recognized by van Uden and co-workers in the 1950–60 decade, probably because of previously inadequate procedures, especially media and incubation temperature. Since species highly adapted to warm-blooded animals may not grow at room temperature in classical Sabouraud medium, surveys of the intestinal yeast flora of these animals generally overlooked them (159).

Isolation of yeasts from the bovine cecum (162) exemplifies suitable methodology: fecal material was collected so as to avoid external contamination and was immediately inoculated in broth with the following composition (percent): glucose 2 g, yeast extract 0.5 g, peptone 1 g, penicillin (Na) approximately 3 mg, and streptomycin 10 mg. To discourage transients not thriving at body temperature, tubes were inoculated at 37°C. After three–four days and before the filamentous fungi, which commonly contaminated the cultures, sporulated, a loopful of the broth was streaked on plates of the following medium (g percent): glucose, 2; yeast extract, 0.5; peptone, 1; and agar, 2. To suppress bacterial contamination the pH was adjusted to 3.5 with lactic acid, then the plates were incubated at 37°C. This procedure, although not highly specific, favors psychrophobic yeasts.

Such yeasts usually can grow even at pH 1 (160, 169). They also grow abundantly at 37°C. For counting yeasts in the intestinal contents direct microscopic examination is sometimes best, e.g., for *S. guttulata* which grows poorly in most media suitable for the others. For yeasts that grow in solid enriched media, surface plate counts are preferred (165).

Host specificity and nutritional requirements.—The obligate saprophytes exhibit different specificities; their abundance is influenced by the feed. A correlation may emerge between suitability of host and nutritional requirements of the yeasts. Van Uden (159) supposed that the fussier a yeast's nutritional requirements the harder it would be to find it in more than one animal host.

Host specificity of *S. guttulata* is extreme: it occurs commonly in the rabbit's digestive tract, also in the stomachs of chinchillas (115, 127). Isolation of *S. guttulata* from rabbits is obtainable regularly if they are fed an uncooked vegetarian diet and not at all if kept on a milk diet. A cycle of *S. guttulata* in nature was proposed: vegetative cells propagate by budding in rabbit stomach; in the intestine, growth stops or is much reduced because of the alkaline pH. Yeasts are excreted in feces and, due to the coprophagy of rabbits, night feces are ingested, bringing the yeast back into the stomach of the rabbit (127). Early cell granulation of *S. guttulata* appeared to be a function of H^+ concentration of the medium. Cells grown at pH 6 in vitro, did not become as rapidly and as highly granulated as those at lower pHs. However, since *S. guttulata* grows in the rabbit stomach at pH 1.5 to 2 in nongranulated condition, it seems that other environmental factors are important: continuous gassing with an optimal gas phase (15 percent CO_2, 2 percent O_2 in nitrogen) increases growth rate and granulated cells do not appear in the log phase (22). Cultivation of *S. guttulata* in pure culture was only recently successful although this yeast has been known since 1845. The medium proposed for growing *S. guttulata* contained an autoclaved extract of stomach contents of rabbits supplemented with glucose and yeast extract (99). Later on, growth was also obtained with certain peptones and yeast autolysates (129). Growth on solid media required a high CO_2 atmosphere (109). No defined medium is yet completely satisfactory for *S. guttulata*. Its temperature range is very narrow: 35–40°C.

Host specificity of *C. slooffii* is less strict. Originally isolated from 6 out of 252 horses (161), it was then found commonly (48.4 percent) in pigs in Lisbon. Van Uden suggested that the pig rather than the horse might be the commoner host for *C. slooffii* (165). A study of six sites of the digestive tract of 57 swine showed the most cells in the cecum and rectum. The cells were most abundant in feces, suggesting that an optimal constellation of growth factors resulted from contributions of the microflora superadded to mucous secretions (165). Important in determining the number of cells in positive animals (approximately 50 percent) was diet: pigs kept on a green food diet did not harbor cells of *C. slooffii;* pigs with few *C. slooffii* had been fed a mixed diet, including kitchen refuse; pigs with high counts had been on a grain diet, chiefly maize. Apparently feed rich in cellulose and protein lower the yeast count; starchy food raises it. *C. slooffii* was also found in African bush pigs (*Potamochoerus choeropotamus*) but not in the wart hog (*Phacochoerus aethiopicus*) or hippopotamus (164). It was not found in humans, baboons, African pied crows (*Corvus albus*), cattle, goats, or sheep. Nutritionally, this yeast is not as exacting as *S. guttulata;* however, its unexpected sensitivity to the usual concentrations of glutamic acid, choline, and metals in yeast media made its cultivation in defined media tricky at first (33). Reliable growth of *C. slooffii* requires appropriately balanced media; presumably, this may be correlated with the observed different densities in population along the intestinal tract of pigs, where amino acids, lipids, and metallic ions are progressively absorbed and intestinal secretions and sloughed-off cells added.

Torulopsis pintolopesii, first isolated from the peritoneum of mice inoculated with the fungus *Acladium castellani* (158), has since been isolated from the digestive tract of rats, mice, other small rodents, and pigeons (10, 40, 100, 159). Mackinnon (78–79) observed it in the stomach of all mice examined and therefore termed this species acidophilic; it is only transient in the intestine. *T. pintolopesii* from mice grew at pH 2 from 24 to 40°C (153); minimum pH growth at 37°C was 0.9–1.1 (169). Unlike *S. guttulata,* it does not require unknown growth factors and, unlike *C. slooffii,* it is rather insensitive to nutritional imbalances. Some strains require choline (33) which is not being added currently to synthetic media for yeasts (75, 179).

Torulopsis bovina, first isolated from the cecal contents of a cow (162), has also been found in swine, rodents, and fowls (102, 159). The 14 percent incidence in swine suggests that these animals are natural hosts for *T. bovina.* As pointed out by van Uden (159), *T. bovina* is "... intermediate ... between the thermophilic, fastidious and highly host-specific obligate saprophytes and the other obligate saprophytes such as *C. albicans* which grows easily at room temperature on ordinary culture media and can be isolated from a wide range of hosts." *T. bovina* grows from 20 to 43°C (33). Growth on synthetic media is easily obtained; the nutritional requirements are much like that of *T. pintolopesii;* some strains likewise require choline.

Synthetic media and temperature factors.—As stated by Shifrine & Phaff (129) "nearly all yeasts isolated from natural substrates can be grown in relatively simple defined media, containing a carbon source, minerals, and an inorganic nitrogen source." B vitamins are needed by some strains. Thermophilic enteric yeasts display a varied picture in their nutritional requirements. The most exacting species is *S. guttulata.* A mineral basal medium supplemented with glucose, asparagine, casein hydrolysate, and small amounts of yeast autolysate, permitted growth at 37°C but all factors required have not been identified (31). Apparently indispensable are: tryptophan, thiamine, inositol, biotin, and oleic acid (as Tween 80). Conditions of growth included narrow ranges of pH (3 to 4) and temperature (35–41°C). B_6 vitamins were toxic; pyridoxal more so than pyridoxamine and pyridoxine; growth inhibition was overcome by thiamine. Inhibition by thymine and uracil was antagonized by thymidine (31). Shifrine & Phaff (129) reported cultivation of *S. guttulata* in a medium for *Bacillus subtilis* (36) containing 21 amino acids, glutamine, asparagine, salts, trace elements, vitamins, two purines, and a pyrimidine. The growth rate was slower than in natural media. Eight consecutive transfers were made in the defined medium. These authors (129) observed that Difco yeast carbon base supplemented with adenine, guanine, uracil, 1 percent additional glucose, and the 21 amino acids of the medium of Demain & Hendlin (36) did not support growth of *S. guttulata.* This was attributed to imbalance arising upon removal of certain vitamins—pantetheine, pyridoxamine, pyridoxal, lipoic acid, folinic acid, and B_{12}—from the medium. Growth took place with the further omission of adenine, uracil, and guanine. Use of a slightly modified medium brought evidence of the essentiality of nicotinic acid, inositol, pantothenic acid, and thiamine. Which amino acids were essential was undetermined. Since some casein hydrolysates, Bacto peptone, and yeast extract did not support growth of *S. guttulata,* but a fully autolysed bakers' yeast, Proteose Peptone and Trypticase did, Shifrine & Phaff (129) suggested that the amino acid balance was probably critical. The additional requirement of a high CO_2 atmosphere for growing the yeast on solid media (109) complicates the entire picture. Buecher & Phaff (22) found that two strains of *S. guttulata* grew at a maximal rate under continuous gassing with high CO_2 concentrations. Growth rates at 15 percent CO_2 were almost the same between O_2 concentrations of 0.25 and 20 percent. Growth increased at 2 percent O_2 in direct proportion to the CO_2 concentration up to 10–15 percent. Some of the discrepancies in reports on *S. guttulata* may well be due to varying purity and preparation of nutrients and variations in inocula. A thorough nutritional study is needed.

C. slooffii is not as fastidious as *S. guttulata.* It requires biotin, pantothenate, thiamine, pyridoxine, nicotinic acid, and inositol (33, 161). Under certain conditions *C. slooffii* is inhibited by choline. At 37°C as little as 30 μg/liter of choline inhibited whatever was the inositol concentration (152). The inhibition of choline was overcome with time or by adding to the basal medium glutamic acid plus cytosine (or orotic acid) or a mixture of purines (152, 153). Thymine was inactive, and uracil as well as guanine inhibited. Although stimulatory, methionine did not overcome the inhibition; other amino acids were likewise ineffective. Dimethylethanolamine inhibited in concentrations 100 times greater than choline. Ethanolamine and *N*-methylethanolamine were inactive. Choline was noninhibitory at 40°C (152); at 43°C, choline proved essential (114), protecting *C. slooffii* against thermal death (113). Inhibition by choline in a medium containing asparagine, glucose, salts, and the required vitamins held for four new strains received from Carmo-Sousa (155). In all instances, glutamic acid and purines reversed the inhibition. In a richer medium with several stimulatory amino acids at pH 3.0 (153), inhibition by choline was not observed in most strains of *C. slooffii*. Inhibition by choline was reported also in the mesophiles *Saccharomyces carlsbergensis* and *S. cerevisiae* (44, 147). This hinders the use of these yeasts as test organisms for inositol assay because of the variable concentration of choline in biological materials. Inositol-requiring, choline-insensitive yeasts—*Kloeckera brevis* and *Schizosaccharomyces pombe*—are now available for the assays (35, 97).

Since inositol does not influence choline inhibition in *C. slooffii*, competition for the synthesis of specific phosphatides seems not to apply (2). As *C. slooffii* cultivated in the absence of choline synthesizes this base by the phospholipid pathway (152), presumably this pathway was operative at 37°C. At 43°C the choline requirement might be reflecting an increased need for phospholipids at higher temperatures or as a block in de novo synthesis. The influence of temperature in utilization of free choline by the Kennedy pathway for biosynthesis of lecithin is unreported. Cytosine protects against choline inhibition. Perhaps cytosine and choline form, in this yeast, a complex (CDP-choline) with low turnover which results in lowered cytosine triphosphate concentration (28). Since free cytosine is deaminated to uracil (152) and orotic acid is also active in counteracting inhibition by choline, the stimulation of pyrimidine nucleotide synthesis seems important to annul choline inhibition.

Kreger-van-Rij (66) found that *C. slooffii* required pantothenate at 37°C but not at 30°C. This could not be confirmed for three strains grown in a different synthetic medium (33). Utilization of ammonium by *C. slooffii* was irregular. Ammonium citrate, not used by three strains (33), served as a nitrogen source for four other strains investigated later (155) although growth was slow. Glutamic acid is consistently not utilized as a sole N source. Furthermore, it is inhibitory in certain concentrations (33, 153). High concentrations (0.4–0.6 percent) of asparagine (33) or aspartic acid (153), otherwise suitable N sources, inhibit growth of *C. slooffii* which is overcome with time or by adding small amounts (0.005–0.01 percent) of glutamic acid. Similar interactions of glutamic acid and aspartic acid were reported in *Lactobacillus dextranicum* (105, 106): glutamic acid inhibited competitively utilization of aspartic acid for biosynthesis of threonine; conversely, aspartic acid inhibited utilization of glutamic acid for the synthesis of proline, citrulline, and glutamine. In the absence of these amino acids, growth would only occur—as for *C. slooffii*—at a definite ratio of glutamic acid:aspartic acid (153). Seven strains of *C. slooffii* exhibited these peculiarities in glutamic-aspartic utilization (33, 155), a feature perhaps of taxonomic interest. *C. slooffii* was highly sensitive to certain metal mixtures. When the metal mixture used by Hutner et al (59) for *Ochromonas malhamensis* above 35°C, was added to a citrated medium at pH 5.0, full inhibition of yeast growth occurred (33, 155). This was not observed in an enriched medium at pH 3.0. At this pH and at 43°C, Fe, Mn, and V, together with choline and leucine, were temperature factors (114). On what the emergence of trace elements depends on direct competition by H^+ was not determined. Requirements for trace elements do rise steeply with temperature for several microorganisms including *Saccharomyces* (58). Although different salts of V were active in *C. slooffii*, further purification would confirm the role of vanadium as a temperature factor. Contamination of salts used with other trace elements and the increasing number of reports on the essentiality for unfamiliar growth factors such as Ni, Cr, As, I (16, 123, 142) and stimulation of red seaweed *Porphyra tenera* (63) by Br, Li, Rb, and Sr, are hints which deserve attention.

Many synthetic media include citrate, aspartic (or asparagine) and glutamic acids, choline, metals, vitamins of the B group and, for more exacting organisms, additional amino acids, purines, and pyrimidines. These substances are considered "physiological" and normally stimulate yeast growth. Nutritional imbalances as observed in *S. guttulata* and *C. slooffii* may, however, impede attempts to cultivate them in defined media. Once controlled, these imbalances may, on the other hand, be remarkably reproducible, as with *C. slooffii*. We therefore propose that nutritional imbalances, if proven specific, may be an additional taxonomic criterion for species determination. The behavior of seven strains of *C. slooffii* was as uniform concerning nutritional imbalances as it was for vitamin requirements (33, 155). Other yeasts—*Cryptococcus neoformans, Torulopsis glabrata, Trichosporon cutaneum* (166), *Candida albicans, C. tropicalis, C. guilliermondi,* and *C. pelliculosa* (38)—also show fixed deficiency patterns.

Van Uden et al (166) and Kreger-van-Rij (66) observed variations in vitamin requirements in *S. telluris, T. bovina,* and *T. pintolopesii.* This was confirmed for 23 strains of these species (155) with a simple glucose-asparagine-salts medium and small inocula. All strains required biotin. Requirements for pantothenate, thiamine, nicotinic acid, pyridoxine, inositol, and choline showed three different patterns: (*a*) essential requirement for different vitamins on short incubation; (*b*) stimulatory or no action of single vitamins; and (*c*) interchangeable requirement of vitamins, e.g., pantothenate or thiamine; inositol or choline. This last behavior suggests that the rate of synthesis of a cofactor may be influenced indirectly by supplying vitamins which spare the cell's biosynthetic resources (182).

Interestingly, choline was required by strains of *T. pintolopesii, T. bo-*

vina, and *S. telluris*. For 7 out of 23 strains (from the Centraalbureau voor Schimmelcultures and the Instituto Gulbenkian de Ciência, Portugal) choline was an absolute requirement or highly stimulatory. This requirement may not emerge in Wickerham's medium (179) because of its methionine (20 mg/liter) which partly substitutes for choline as for *N. crassa* (56) and *T. pintolopesii* (156). Although the choline-requiring strains grow slowly with methionine, heavy inocula and prolonged incubation (up to three weeks) assures good growth.

Choline as an essential growth factor for microorganisms is rather uncommon. In bacteria this requirement is rare. Choline-less *Pneumococcus* were described; growth was also possible with ethanolamine (12). Among yeasts, choline is usually stimulatory (92). Known choline-requiring yeasts include strains of *T. pintolopesii, T. bovina, S. telluris,* (33, 155), the choline-less shoyu yeast (118), and mutants employed in genetic studies by C. C. Lindegren (34). Horowitz & Beadle (56) introduced choline-less mutants of *N. crassa* for microbiological assay of choline. Only a few protozoa need choline. The apparent choline requirement of *Leishmania tarentolae* was eliminated by pyridoxal (151). *Trichomonas gallinae* was reported to require choline (73).

For choline-requiring strains of *T. pintolopesii* *N,N*-dimethylethanolamine substituted for choline (156) as in *N. crassa* (64). *N*-methylethanolamine was poorly utilized on short incubation and ethanolamine was inactive alone. Formaldehyde (0.01–0.02 mg/ml) promoted growth on prolonged incubation especially in conjunction with ethanolamine; methionine was active on short incubation and stimulated growth in the presence of *N*-methylethanolamine. Amino acid mixtures without methionine did not permit growth of *T. pintolopesii*. The following were inactive in replacing choline: lecithin, creatine, methanol, acetone, sarcosine, dimethylamine, trimethylamine, tetramethylammonium, carnitine, γ-butyrobetaine, *O*-acetylcarnitine, folic and folinic acids, vitamin B_{12}, betaine, and purines. Esters of choline—acetylcholine, succinylcholine, phosphorylcholine—were active but at higher levels than needed for choline or *N,N*-dimethyletanolamine (156).

The nutritional replacement of choline by carnitine in the insect *Phormia regina* (53) inspired a search for carnitine-less mutants starting with a choline-less *T. bovina* (157). Replica plating revealed yeasts growing on choline or on choline or carnitine. The latter on subculture in media containing only carnitine finally yielded clones that required carnitine on short incubation; choline was stimulatory. Only *O*-acetylcarnitine, substituted for carnitine with equal efficiency. The following were inactive: γ-aminobutyric, α-aminobutyric and α-amino-β-hydroxybutyric acids, γ-butyrobetaine, betaine, creatine, methionine, acetylcholine, purines, putrescine, and cadaverine. Glutamic acid and ornithine were active but at much higher concentrations than those for carnitine. Citrulline and arginine promoted scanty growth; other amino acids were inactive (157).

Emergence of carnitine as a yeast growth factor, hitherto important only in the nutrition of certain insects (47, 53) and of bacteria in soy sauce brewing, notably *Pediococcus soyae* (117), and the frequent requirement for choline in the yeasts studied, suggest that carnitine and choline be kept in mind in the cultivation of exacting microbes.

Table 1 details synthetic media for thermophilic enteric yeasts.

Physiology

Choline- and carnitine-requiring yeasts.—Utilization of choline by choline-less strains of *T. pintolopesii* was studied in growing cells and cells starved in phosphate buffer (104). Choline was assayed in the supernatants and in the cells by the *Torulopsis* assay (32). Since total choline was determined, it summed exogenous choline incorporated and choline synthesized by the de novo pathway at the lecithin level. If feedback inhibition stops de novo synthesis, then choline assayed in the cells would represent choline incorporated less choline degraded. Cell choline increased when the yeast was grown with exogenous concentrations greater than 1 mg/liter. Growth was maximal at 0.8–1 mg/liter. Cells grown on 6 mg/l accumulated the most choline—1 mg/g dry cells. This amount is the same as in *C. slooffii* grown without choline (152). In *T. pintolopesii* 80–90 percent of incorporated choline was degraded and was undetected by microbiological assay after cell hydrolysis. In starving cells, increased exogenous choline favored cellular accumulation but less than in growing cells. Degradation of choline under this condition was favored. These results showed that metabolism of choline in this yeast is not simply directed toward synthesis of lecithin or acetylcholine and the like.

Suspensions of *T. pintolopesii* oxidizing glucose in manometric experiments were inactive with choline, dimethylethanolamine or betaine (29). Since betaine never substituted for choline as a growth factor (156), there is as yet no evidence of the oxidation of choline to betaine as initial step in utilization of choline by yeast cells. Although responding to methionine as does *T. pintolopesii* (156), choline-requiring mutants of *N. crassa* utilize choline at least in part by oxidizing it to betaine (55). Richardson & Speed (108) showed that betaine aldehyde could replace choline as a growth factor, suggesting that choline in *N. crassa* in part serves as a methyl donor. The inability of betaine to promote growth was explained by the action of betaine aldehyde dehydrogenase apparently being irreversible (116).

No evidence was obtained in *T. pintolopesii* for formation of trimethylamine. The use of choline-Me-^{14}C and choline-Me-^{3}H demonstrated that radioactivity initially taken up by growing cells in the first 24–48 hours of incubation was excreted in the next hours up to 85–95 percent. The similar excretion of choline-Me-^{14}C and choline-Me-^{3}H seemed to indicate that the methyl group as a unit was transferred to the excreted compound. Formation and excretion of substances usually related to choline, either chemically or metabolically, was ruled out by paper chromatography. Yet, the reaction leading to the synthesis of the excreted compound involves choline specifi-

TABLE 1. Synthetic Media Proposed for Growing Thermophilic Enteric Yeasts (in g/l)

	Saccharomycopsis guttulata	*Candida slooffii*		*T. bovina* *T. pintolopesii* *S. telluris*
	(129)	(153)[a]	(33,152)	(33,155)
Glucose	20.0	20.0	20.0	20.0
Citric acid·H_2O		2.5		
Succinic acid		2.5		
$(NH_4)_2SO_4$	5.0	0.5		0.5
NaCl	0.1	0.1	0.1	0.1
KH_2PO_4	1.0	0.4	0.4	0.5
$MgSO_4 \cdot 7H_2O$	0.5		0.5	0.5
$CaCl_2 \cdot 2H_2O$	0.1			
H_3BO_3	0.0005	0.0002		
$CuSO_4 \cdot 5H_2O$	0.00004			
$CuSO_4$		0.001		
KI	0.0001			
$FeCl_3$	0.0002			
$MnSO_4 \cdot H_2O$	0.0004	0.003		
$Na_2MoO_4 \cdot 2H_2O$	0.0002	0.0025		
$ZnSO_4 \cdot 7H_2O$	0.0004	0.003		
$Fe(NH_4)_2(SO_4)_2 \cdot 6H_2O$		0.0002		
$CaCO_3$		0.025		
$MgCO_3$		0.1		
NH_4HCO_3		0.5		
$NaVO_3$		0.00001		
$CoSO_4 \cdot 7H_2O$		0.00005		
Biotin		0.00002	0.00002	0.00002
Ca pantothenate	0.0004	0.002	0.002	0.002
Thiamine HCl	0.0004	0.002	0.002	0.002
Pyridoxine HCl		0.002	0.002	0.002
Nicotinic acid	0.0004	0.002	0.002	0.002
Inositol	0.002	0.01	0.003	0.01
Choline Cl				0.001
Carnitine Cl				0.001
L-Asparagine·H_2O	0.2		0.5	1.0
DL-Aspartic acid	0.6	0.5		
L-Glutamine	0.2			
L-Glutamic acid	2.6	0.3[b]	0.1[b]	
DL-Alanine	0.19			
L-Arginine HCl	0.19	0.1		
L-Cystine	0.02			

[a] This medium permits growth at 37°C; at 43°C higher amounts of Fe, Mn, and V are required along with choline (114).

[b] Higher amounts of glutamic acid inhibit.

TABLE 1. *(Continued)*

	Saccharo-mycopsis guttulata		*Candida slooffii*	*T. bovina* *T. pintolopesii* *S. telluris*
Glycine	0.025			
L-Histidine HCl	0.12			
L-Hydroxyproline	0.01			
DL-Isoleucine	0.48	0.16		
DL-Leucine	0.48	0.16		
L-Lysine HCl	0.32	0.05		
DL-Methionine	0.33	0.02		
DL-Phenylalanine	0.39	0.04		
L-Proline	0.44			
DL-Serine	0.48			
DL-Threonine	0.4			
DL-Tryptophan	0.12	0.12		
L-Tyrosine	0.33	0.04		
DL-Valine	0.79	0.1		
Adenine			0.03[c]	0.01[d]
Adenylic acid			0.03[c]	
Hypoxanthine			0.03[c]	0.01[d]
Cytosine			0.03[c]	
Orotic acid			0.03[c]	
pH	4.5	3.0	5.0	5.0

[c] Highly stimulatory; reverses toxicity by glutamic acid.
[d] Optional; usually stimulate growth.

cally because the radioactivity of methionine-Me-^{14}C previously incorporated by growing cells was not excreted in an appreciable amount (29). An equilibrium between excretion and assimilation of tracer may, however, have occurred. The simultaneous addition of increasing amounts of "cold" choline did not alter incorporation of methionine-Me-^{14}C. Likewise, "cold" methionine did not interfere with uptake and excretion of radioactivity from choline-Me-^{14}C. Betaine-Me-^{14}C was not incorporated regardless of the final pH (3.0 or 5.0). *T. pintolopesii,* like other yeasts, can form *S*-adenosylmethionine (SAM) even when growing without exogenous methionine but with choline instead. Cells of *T. pintolopesii* harvested when maximal incorporation of radioactivity from choline-Me-^{14}C was obtained, although forming SAM, this sulfonucleoside was not radioactive. Direct participation of choline in the synthesis of the excreted substance is also suggested by the amount of radioactivity excreted. Were betaine an intermediate, and if the reaction leading to the synthesis of the excreted substance involved a simple transmethylation, then the radioactivity excreted would be much less. Since the main transmethylation reaction involving betaine is synthesis of methionine (24), radioactivity incorporated in the cells should, on the contrary, be higher because methyl groups from methionine or from dimethylglycine and sarcosine when oxidized should give rise to SAM or activated 1-carbon units used in several reactions in a growing cell.

The nature of the excreted substance in *T. pintolopesii* has not yet been elucidated. It may comprise a heteroside containing mannitol as the polyhydroxylated moiety (29). An interesting situation lies in correlation of excretion of a substance which carries most of the radioactivity out of choline-Me-^{14}C with the requirement of choline in *T. pintolopesii.* Cells of this yeast, precipitated with TCA and acetone and extracted for lipids in three different solvent systems with heating, still disclosed the presence of choline by microbiological assay (6). A suggestion was made for participation of choline in a structure similar to that described in choline-less *Pneumococcus* (94, 149, 150) in which a novel polymer-bound choline (polysaccharide or teichoic acid) was identified. Substitution of choline by ethanolamine, *N*-methylethanolamine, or dimethylethanolamine in the sites of the macromolecule usually occupied by choline changed several characteristics of *Pneumococcus* including the capacity for genetic transformation. The choline component of the teichoic acid of *Pneumococcus* also plays a key role in determining the sensitivity to autolytic enzymes: cell walls with choline replaced by ethanolamine were totally resistant to the autolytic enzyme. The occurrence of enzymes able to utilize choline in *T. pintolopesii* for synthesis of a polyhydroxylated nonusable product, would hinder covalent-binding of choline to surface polymers, thus impairing function (29); growth would depend on successful competition for exogenous choline.

The requirement of carnitine by *T. bovina* (157) probably involved a quite different mechanism. Seeking a model system for carnitine studies, Miranda et al (90) re-examined the main characteristics of the carnitine-requiring strain of *T. bovina* and found that with large inocula growth seemed to be independent of carnitine. The carnitine requirement could only be demonstrated with dilute inocula and only during initial incubation. They suggested that during lag phase a growth-promoting substance might have accumulated in the medium. Carnitine itself seemed to be the substance, indicating that its biosynthesis in the requiring yeasts remained unaltered, and that there was leakage of free carnitine (90). Growth of *T. bovina* required added carnitine or prolonged incubation during which the yeast leaked carnitine until a critical external concentration was reached. Cultures of *T. bovina* constantly dialyzed against basal medium without carnitine did not grow even on prolonged incubation. No evidence was obtained for incorporation of carnitine in phospholipids but, since the specific activity of the isotope was somewhat low, this must be interpreted with caution.

Sterol synthesis: temperature relationships.—How temperature affects biosynthesis of ergosterol in yeasts has been studied (138). Maximal yield in a mesophile like *Saccharomyces cerevisiae* was obtained when incubation was at optimal growth temperature. At 35°C sterol synthesis was depressed and inhibited at 40°C or above. Lysis of cells was prominent at 45–50°C. *S. cerevisiae* synthesizes ergosterol in aerated cultures but needs ergosterol plus fatty acids for anaerobic growth (4, 5, 72, 103). Death of *S. cerevisiae* at 40°C is preventable by supplementing the medium with ergosterol plus oleic acid (124, 138). The similarity between nutritional requirements in anaerobiosis and at high temperatures is noteworthy. Sherman (125) observed induction of petites as an important temperature effect.

Incorporation of $^{14}CH_3$-labeled methionine was used to monitor synthesis of ergosterol in *C. slooffii, T. pintolopesii, T. bovina, S. telluris, T. glabrata,* and *S. cerevisiae* (74). Sterol synthesis in thermophilic yeasts was determined at the extremes of the growth range—25–28°C and 40–43°C—as well as at 37°C. The highest sterol synthesis was in aerated suspensions with cells previously grown in stationary cultures rather than with forced aeration. With few exceptions, thermophilic strains grew very poorly in strict anaerobiosis in a complex medium. Sterol synthesis as a rule was higher at the lower temperatures tested, decreasing as the temperature was raised, regardless of the optimal for each strain (74). These results were rather unexpected since thermophilic yeasts grow abundantly at 37°C and above, suggesting that a speeding of the whole metabolism, including sterol synthesis and building up cell membranes, should take place. More ergosterol was synthesized by *S. cerevisiae* than by *Candida* and *Torulopsis* species. *Torulopsis glabrata* whose minimal temperature is low (3) nevertheless synthesized ergosterol at a rate comparable to that of *S. cerevisiae.* Enhanced synthesis of ergosterol at low temperatures may, on the other hand, indicate the need of a special composition of lipids for maintaining membrane integrity as suggested for *Tetrahymena* (107). These results with thermophilic yeasts parallel those of Adams & Parks (1) in that not all sterol is recovered by conventional saponification of whole yeast cells. Best yields were generally obtained with cold acid treatment (HCl 0.1 *N* at 30°C, 1 hr) before saponification (KOH 20 percent at 100°C, 6 hr). Another form (1) of ergosterol (which is acid-labile) could not be detected in the thermophiles. Only ergosterol was identified by thin-layer chromatography and ultraviolet spectroscopy (74). Differences in extraction procedures might correlate with the stratified surface structure in yeasts in which the glucan component is insoluble in alkali. The insolubility was explained tentatively by the presence of an outer insoluble membrane containing chitin among other components. Extractability of glucan and glycogen can be increased by ultrasonics or mild acid (11). Lipid extraction in yeasts in relation to dissolution of surface polymers should make an interesting study.

Microbiological Assays

Choline.—The value of hardy organisms for microbiological determination of key metabolites is emphasized by Baker & Frank (13, 14); they did not, however, discuss choline. Since metabolism of choline as such or as a component of lecithin depends on availability of other metabolites—folic acid, vitamin B_6, vitamin B_{12}, and amino acids (methionine, serine, glycine) —a need arises for a convenient specific determination of choline in animal materials to further understanding of metabolic distortions represented by cirrhosis and like diseases.

Chemical estimations of free and bound choline was discussed (41); since then new methods have been introduced but no present chemical method is quite satisfactory for biological materials. Chemical procedures demand extensive purification of extracts to remove interfering substances. The usual chemical method is colorimetry of the reineckate; for greater sensitivity one measures the ultraviolet absorption of the acetone solution of the reineckate. A more sensitive method, although not much more specific, was proposed by Appleton et al (7), then improved in ease and specificity (68): extraction of choline into acetone, evaporation of acetone, removal of interfering substances by butanol and isobutanol. Choline was precipitated as the enneaiodide which was dissolved in 1,2-dichloroethane and its absorption measured at 365 nm. This method was employed for determining free choline in plasma, but with urine and homogenized tissues much noncholine material reacted like choline.

The best known, much employed microbiological method depends on *N. crassa* mutants (56). It is rather specific: only choline, lecithin, and methionine support growth. Precursors of choline, *N,N*-dimethylethanolamine and *N*-methylethanolamine, are active but are not usually accumulated in biological materials. Ethanolamine was inactive. Choline is detectable at 0.02 mg/liter. The method has been used to determine free choline in plasma, urine, animal tissues, milk products, pharmaceutical biologicals and many other materials (15, 54, 61, 76, 77). Bioautography for identification of choline and derivatives with *N. crassa* mutants detects as little as 0.03 μg (71). Choline assay with *Torulopsis pintolopesii* might prove preferable. Growth from dilute inocula is measurable turbidimetrically after two days which is certainly simpler than measuring growth by weighing the dry mycelium after three to five days incubation which is the procedure with *N. crassa*. The two choline-requiring strains of *T. pintolopesii* respond to similarly low concentrations of choline, 0.02–0.05 mg/liter, and are more specific than *N. crassa* since they do not respond to lecithin (156). The *T. pintolopesii* method was applied to determination of total choline in biological materials (32).

Materials to be assayed were hydrolyzed in sulfuric acid 3 percent (v/v) for 2 hours at 121°C (56); hydrolysis in HCl was inadequate (30). The acid hydrolysates were neutralized with saturated barium hydroxide (to pH 5–6), then passed through Permutit. This step eliminates methionine which substitutes for choline but only in much higher concentrations. The Permutit columns were washed with 0.3 percent NaCl and eluted with 5 percent NaCl. The eluates were used directly for the assays or after dilution. Tubes inoculated with *T. pintolopesii* were incubated at 37°C for 48 hours. Results with several biological materials agreed with those obtained simultaneously with *N. crassa.* The average recovery of added choline was approximately 97 percent (32); the standard curve for choline ranges from 0.05 to 1 mg/liter. The hydrolysates, after passage through Permutit, contained few if any stimulating substances other than choline. The basal medium can therefore be simple: glucose, asparagine, six vitamins, and salts.

Carnitine.—The ubiquity of carnitine (45), its indispensability for the transport of fatty acids into mitochondria (20, 51, 148), and its role as a vitamin for some insects motivated study of its determination in biological materials. Carnitine assay methods are chemical or biological (46, 143). Enzymatic determination is possible (50, 80). Chemical methods present some of the limitations noted for choline. The periodide method for carnitine requires the preliminary esterification of the carboxyl (110). The previous biological test for carnitine used larvae of the mealworm *Tenebrio molitor* which is a cumbersome method (48, 49). The occurrence of a carnitine-less mutant of *Torulopsis bovina* (157) may yield a better assay; only 1 μg/liter initiated growth, with saturation at 15–20 μg. Carnitine was replaceable, as noted earlier, by high concentrations of glutamic acid, arginine, or ornithine. Sensitivity to carnitine was such that in practice simple dilution of the material to be assayed eliminates interference by amino acids. The carnitine mutant of *T. bovina* has so far been used only to detect carnitine bioautographically (90). Since the yeast responds also to some carnitine esters, it was used to detect ^{32}P-phosphorylcarnitine to confirm the identity of the synthetic compound (91).

Antimetabolites

Ethionine.—The striking transformation of ethionine—ostensibly merely the *S*-ethyl homolog of methionine—into a powerful hepatic carcinogen and metabolic inhibitor (42), raises the question of how well information on its metabolism, especially mechanisms of resistance and protection against toxicity, contribute to the carcinogenesis theory.

Inhibition of *C. slooffii* by ethionine is counteracted by methionine, adenine plus uracil, and by aromatic amino acids (154). Protection by tyrosine and tryptophan (and perhaps by adenine) appears to favor the theory that carcinogens, as distinguished from precursors, are electrophiles (88) reacting with nucleophilic sites, and that free tyrosine and tryptophan act as nucleophilic scavengers of electrophilic reactants (89). In a rich synthetic medium at pH 3.0, ethionine (0.1 g/liter) plus methionine (0.002 g/liter), just failed to restore growth of *C. slooffii;* addition of phenylalanine, tryptophan, tyrosine, and adenine, and less by isoleucine and 7-methylguanine, restored growth (152). Singly, 5-methylcytosine, 6-methyluracil, uracil, and thymine were inactive. With choline inhibiting growth synergistically with ethionine, 6-mercaptopurine annulled the overall toxicity. Similar results were obtained with heavy inocula of log phase cells in the absence of methionine (83).

Incorporation of ethionine-ethyl-1-^{14}C by resting cells of *C. slooffii* was inhibited by phenylalanine plus tryptophan plus tyrosine, or by methionine (83, 84). Ethionine, reciprocally, inhibited incorporation of phenylalanine-^{3}H. This reciprocal interference seemed to apply only for amino acids since ethionine did not affect incorporation of adenine-^{3}H. Although protein synthesis was not blocked in these experiments, the suggestion was made that the transport systems for amino acids had low specificity as in *S. cerevisiae* (144). After incorporation of adenine-^{3}H by *C. slooffii,* incubated with methionine or ethionine, the cold TCA-extractable fraction was tested for radioactivity in the sulfonucleosides: half of the total radioactivity of the extract was in *S*-adenosylmethionine or *S*-adenosylethionine, indicating that adenine, which does not stimulate sulfonucleoside synthesis in *S. cerevisiae* (121), is, for *C. slooffii,* an important precursor. Unlike the protective effect of adenine on ethionine toxicity in *C. slooffii,* which accords with results in rats (43, 130), and reinforcing the hypothesis for an ethionine-induced, ATP-trapping effect (139, 170, 171), Smith & Salmon (135) found in *S. typhimurium* that adenine enhanced the growth inhibition induced by ethionine.

Several transfers of *C. slooffii* in a medium containing ethionine plus adenine and/or aromatic amino acids, induced the formation of ethionine-resistant mutants tolerating up to 0.5 g/liter ethionine in a medium devoid of protectants (84). The mechanism of resistance did not center around a derepressed methionine synthesis as in *T. utilis* and *N. crassa* (65). Nor did the resistant mutants appear less permeable to ethionine as in *S. cerevisiae* (81, 136, 137). A possible sensitivity to ethylated analogs that might originate from ethionine was not studied further; triethylcholine, formed in ethionine-treated rats (141), was not toxic for *C. slooffii* (152). Ethionine-resistant cells of this yeast, although incorporating ethionine faster than sensitive cells, did not excrete 5-ethylthioadenosine as in resistant cells of *S. cerevisiae* (137). Levels of *S*-adenosylmethionine in resistant and sensitive cells of *C. slooffii* were similar, and the slightly higher accumulation of *S*-adenosylethionine in sensitive cells incubated with ethionine in Stekol's mineral solution (140) could not account for the resistance. Utilization of ethionine sulfur for synthesis of methionine and cystine, as in resistant strains of *S. cerevisiae* and *T. utilis* (81), tested with ethionine-^{35}S, if holding true for *C. slooffii,* would be consistent with the utilization of ethionine as a S source for growth; *C. slooffii* did not do this. Most of the radioactivity from ethionine-ethyl-1-^{14}C previously incorporated by resting cells of *C. slooffii* was gradually excreted; no similar excretion was observed with methionine-Me-^{14}C (83). Apparently, ethionine resistance depends on induction of degradative mechanism yielding a substance chromatographically similar to the one excreted in *N. crassa* (17). Radioactivity from ethionine-ethyl-1-^{14}C was distributed in proteins, nucleic acids, lipids, and in the soluble metabolic pool of sensitive cells of *C. slooffii* (83). About 60 percent was recovered in the cold TCA fraction mainly as *S*-adenosylethionine; 10 percent was found in the lipid fraction, and 20 percent in the hot TCA-insoluble residue, mainly protein. The lipid-free, hot TCA-soluble fraction contained ca 5 percent of the total radioactivity incorporated. To determine roughly the distribution of radioactivity in RNA and DNA, fractionation in $HClO_4$ (122) was used: the fraction corresponding to RNA carried half the radioactivity of the DNA fraction. This yeast may therefore be useful in studying the physiology of biochemically ethylated DNA. That ethionine is mutagenic via *S*-adenosylethionine is inferrable from the fact that no resistant mutants appeared when subculturings were made in presence of methionine (84). Ethionine-resistant mutants of *C. slooffii* were stable: no back-mutation was observed after six serial subcultures in a nonselective complex medium. They were also thermosensitive at 43°C, contrary to the wild type which thrives at this temperature (85). Resistant mutants of *N. crassa* (87) behaved similarly.

Urethane.—This drug, introduced as a sedative, is used in multiple myeloma, myelogenous leukemia, and other cancers. It can induce pulmonary tumors in mice (96). For carcinogenesis, urethane must presumably be converted into *N*-hydroxyurethane (18, 19). If this reaction is present in yeasts, protection against urethane toxicity could be expected (as in ethionine inhibition) by nucleophilic reactants (89). Urethane inhibition in *C. slooffii* was reversed by the combination adenine plus cytosine (or orotic acid), or adenine plus aromatic amino acids plus glutamic acid, or adenine plus ammonium carbonate (or bicarbonate). Ureidosuccinic acid was also active which may denote an interference with carbamyl compounds (93). These results parallel others in bacteria and in animals. Pentose nucleotides counteract carcinogenesis by urethane (27). Growth inhibition of *Escherichia coli* by urethane was reversed by 2,6-diaminopurine (132), phenylalanine plus glutamic acid (177), or by adenine (133). Aminopterin-potentiated carcinogenesis by urethane in mice was antagonized by ureidosuccinic, dihydroorotic, and cytidylic acids or thymine (112). Thymine also relieved urethane toxicity in *Poteriochromonas stipitata* (62).

The anomalous behavior of urethane as an inhibitor of yeast respiration (69) was attributed to its surfactant properties. Whether this is relevant to *C. slooffii* remains to be seen; however, butyl carbamate is much more toxic to it than urethane. Toxicity by butyl carbamate was partly reversed by adenine plus aromatic amino acids or adenine plus glutamic acid. Among compounds tested as protectants against urethane toxicity for *C. slooffii* the following were inactive: thymine, uracil, 7-methylguanine, 5-methylcytosine, 6-methyluracil, xanthine, amino acids (with the exception of aromatics and glutamic acid), uridine, xanthopterin, folic acid (70), PABA, and some inorganic salts. Hypoxanthine and 6-mercaptopurine were as active as adenine (93).

It may not be fortuitous that purines, certain pyrimidines, glutamic acid, and aromatic amino acids—all active against ethionine, urethane, or choline—tend to reverse these inhibitions for *C. slooffii* (93, 152). Choline and derivatives of the others are electrophilic reactants; if not removed metabolically they may bind to sites causing damage unless somehow sequestered as by nucleophilic or other targets.

Aminopterin.—Swendseid & Nyc (146) observed that a choline-less *N. crassa* was rather insensitive to aminopterin compared with the wild type. Microorganisms resistant to folic acid antagonists usually require substances whose biosynthesis depends on 1-carbon units. An amethopterin-resistant strain of *S. faecalis* required purines even when supplemented with folic acid (57). Okada et al (98) found that aminopterin-resistant *E. coli* cannot synthesize thymidylate and is therefore dependent on thymine for growth. Requirement for thymine or thymine plus adenine appeared in *Lactobacillus casei* resistant to pyrimethamine (131) and the aminopterin-resistant mutants of *N. crassa* needed choline. In a medium supplemented with choline, choline-less strains of *T. pintolopesii* resisted up to 0.1 g/liter aminopterin on prolonged incubation (96 hours). With short incubation, inhibition by aminopterin (0.03 g/liter) was annulled by aspartic acid and partly by purines but not by alanine, glycine, cystine, or pyrimidines (156). In *Saccharomyces cerevisiae,* growth stimulation by biotin and aspartate was abolished by aminopterin. This could be overcome with thymine and adenine (178). Toxicity by aminopterin in the wild type of *N. crassa,* and in *E. coli, B. megaterium,* and *Torula cremoris* (21, 146, 175) was not reversed by N^5-formyltetrahydrofolic acid. In *T. pintolopesii* this compound only weakly protected against aminopterin toxicity. The possibility of degradation of citrovorum factor (8, 37) was not studied; enhanced synthesis and excretion of this substance in yeast cells would parallel human leukemic cells (52, 145). Degradative reactions of folic acid and folic acid analogs (176), conspicuous in microorganisms, may explain resistance to antifolics (60). Resistance to aminopterin, if denoting altered metabolism of folic acid in *T. pintolopesii,* may be connected with the choline growth requirement.

Prospects

Conditions favoring survival of obligate saprophytes and determinants of host specificity exhibited by some yeasts are far from understood. Some highly exacting yeasts are also highly exacting in respect to habitat. Some species, however, that thrive on ordinary culture media at room temperature nevertheless are only sporadically found outside the animal host. Obvious limitations in monosaccharide and vitamin content of certain natural habitats would preclude vegetative growth but this applies to most yeasts, not to a particular species. If the lower temperature limit for growth is high then an important constraint on development outside the warm-blooded animal is manifest. It should be worthwhile to plot survival curves of obligate saprophytes and other yeasts in soil either under natural conditions or after enrichment with appropriate growth factors. Long generation times and respiratory deficiency in certain yeasts would lessen the probability of a successful biological competition in natural habitats unless a constant supply of growth factors is available as in the animal body. The possibility of a higher susceptibility to antibiotics and some physical conditions are other features which may impose obligate saprophytism.

Implantation experiments aiming to install yeasts in the "wrong" hosts should be tried in relation to diet, also the population equilibrium achieved in the host after drinking suspensions of obligately saprophytic yeasts. Germ-free animals and synthetic diets would indubitably be of great aid. A comparative nutritional study of the animal hosts may prove to be important in fixing absorption rates of critical nutrients and the nutritional value of secretions along the intestinal tract.

The extreme fastidiousness of one species, *S. guttulata,* challenges the microbiologist: amino acid balance as well as responses of the yeast to different peptones and hydrolysates may reflect nonspecific permeability interferences inducting nutritional imbalance, or a need for peptides. Incorporation of CO_2 should be investigated to elucidate the requirement for high CO_2 for growth on solid media.

Nutritional imbalances, when reproducible as in *Candida slooffii,* may join the array of taxonomic tools rather than being considered a nuisance. Choline and carnitine ought to have a place in synthetic media for yeasts generally; slow-growing yeasts on the standard synthetic media for determinations of vitamin requirements may suggest the need for additional growth factors, e.g., choline and carnitine.

Relationships among obligate or near-obligate saprophytes should be studied immunologically and by DNA homology determination. Also, a comparative study of the structure of surface layers and extracellular polysaccharides from different species is worth doing, especially since the composition of the cell wall in *S. guttulata* is peculiar and *T. pintolopesii* may bind choline to surface polymers.

The need for more strains for physiological studies should stimulate ecological surveys. Isolation and study of new obligate saprophytes should broaden present concepts of host-microbe interactions: conversion from a transient yeast to an obligate saprophyte and pathogen must depend on a constellation of still uncontrollable factors.

ACKNOWLEDGEMENTS

We are indebted to Dr. S. H. Hutner for help with the manuscript. We thank also the persons who provided us reprints, papers in press, and yeast cultures, especially Dr. L. do Carmo-Sousa and Dr. N. van Uden.

LITERATURE CITED

1. Adams, B. G., Parks, L. W. 1967. *J. Cell. Comp. Physiol.* 70:161–68
2. Agranoff, B. W., Fox, M. R. S. 1959. *Nature (London)* 183:1259–60
3. Altman, P. L., Dittmer, D. S. 1966. *Environmental Biology*. Bethesda: Fed. Am. Soc. Exp. Biol.
4. Andreasen, A. A., Stier, T. J. B. 1953. *J. Cell. Comp. Physiol.* 41:23–36
5. Andreasen, A. A., Stier, T. J. B. 1954. *J. Cell. Comp. Physiol.* 43:271–82
6. Angluster, J., Travassos, L. R. In preparation
7. Appleton, H. D., LaDu, B. N., Jr., Levy, B. B., Steele, J. M., Brodie, B. B. 1953. *J. Biol. Chem.* 205: 803–13
8. Aravindakshan, I., Braganca, B. M. 1958. *Biochim. Biophys. Acta* 27: 345–54
9. Artagaveytia-Allende, R. C. 1953. *Arch. Soc. Biol. Montevideo* 20: 37–43
10. Aschner, M., Halevy, S., Awram, D. 1954. *Bull. Res. Counc. Isr.* 4, No. 2
11. Bacon, J. S. D., Farmer, V. C., Jones, D., Taylor, I. F. 1969. *Biochem. J.* 114:557–67
12. Badger, E. 1944. *J. Biol. Chem.* 153: 183–91
13. Baker, H., Frank, O. 1968. *Clinical Vitaminology*. New York: Intersci. Pub.
14. Baker, H., Frank, O. 1968. *World Rev. Nutr. Diet.* 9:124–60
15. Bandelin, F. J., Tuschhoff, J. V. 1951. *J. Am. Pharm. Assoc.* 40:245–48
16. Bartha, R., Ordal, E. J. 1965. *J. Bacteriol.* 89:1015–19
17. Beeman, E. A., Smith, R. C. 1969. *Can. J. Microbiol.* 15:445–49
18. Berwald, Y., Sachs, L. 1963. *Nature (London)* 200:1182–84
19. Boyland, E., Nery, R. 1965. *Biochem. J.* 94:198–208
20. Bressler, R., Friedberg, S. J. 1964. *J. Biol. Chem.* 239:1364–68
21. Broquist, H. P. 1957. *Arch. Biochem. Biophys.* 70:210–16
22. Buecher, E. J., Phaff, H. J. 1970. *J. Bacteriol.* 104:133–37
23. Bulder, C. J. E. A. 1963. *On respiratory deficiency in yeasts.* PhD Thesis, Univ. of Delft.
24. Cantoni, G. L. 1965. *Transmethylation and methionine biosynthesis,* ed. S. K. Shapiro, F. Schlenk. Chicago: The Univ. of Chicago Press
25. Carmo-Sousa, L. do. 1969. *The Yeasts,* vol I, ed. A. H. Rose, J. S. Harrison. London, New York: Academic
26. Cooke, W. B. 1962. Personal Communication to H. F. Hasenclever, W. O. Mitchell, 1962. *Sabouraudia,* 2:87–95
27. Cowen, P. N. 1949. *Brit. J. Cancer* 3:94–97
28. Cruz, F. S., Cury, A., Roitman, I., Travassos, L. R. 1963. *Ciência Cult. (Sao Paulo)* 15:289
29. Cruz, F. S., Travassos, L. R. 1970. *Arch. Mikrobiol.* 73:111–20
30. Cury, A., Barreiros, M. P. 1956. *An. Acad. Bras. Cienc.* 28:489–99
31. Cury, A., Hutner, S. H. 1958. *An. Acad. Bras. Cienc.* 30:34–35
32. Cury, A., Oliveira, E. N. S., Cruz, F. S., Travassos, L. R. 1965. *An. Microbiol.* 13:11–24
33. Cury, A., Suassuna, E. N., Travassos, L. R. 1960. *An. Microbiol.* 8:13–64
34. Cury, A., Travassos, L. R., Suassuna, E. N. 1961. *An. Microbiol.* 9:441–64
35. Darbre, A., Norris, F. W. 1956. *Biochem. J.* 64:441–46
36. Demain, A. L., Hendlin, D. 1958. *J. Bacteriol.* 75:46–50
37. Dinning, J. S., Sime, J. T., Work, P. S., Allen, B., Day, P. L. 1957. *Arch. Biochem. Biophys.* 66:114–19
38. Drouhet, E., Vieu, M. 1957. *Ann. Inst. Pasteur* 92:825–31
39. Emerson, R. 1968. *The Fungi,* vol. III, ed. G. C. Ainsworth, A. S. Sussman. New York, London: Academic
40. Emmonds, C. W. 1955. Unpublished data
41. Engel, R. W., Salmon, W. D., Ackerman, C. J. 1954. *Methods of Biochemical Analysis,* vol. 1, ed. D. Glick. New York: Intersci. Pub.
42. Farber, E. 1963. *Advan. Cancer Res.* 7:383–474
43. Farber, E., Shull, K. H., Villa-Trevino, S., Lombardi, B., Thomas, M. 1964. *Nature (London)* 203:34–40
44. Folch, J., Baron, F. N. 1956. *Can. J. Biochem. Physiol.* 34:305–19
45. Fraenkel, G., 1954. *Arch. Biochem. Biophys.* 50:486–95
46. Fraenkel, G. 1957. *Methods Enzymol.* 3:662–67
47. Fraenkel, G., Blewett, M. 1947. *Biochem. J.* 41:469–75
48. Fraenkel, G., Friedman, S. 1957.

Vitam. Horm. (New York) 15: 73–118
49. Fraenkel, G., Leclercq, J. 1956. *Arch. Intern. Physiol.* 64:601–22
50. Fritz, I. B., Schultz, S. S., Srere, P. A. 1963. *J. Biol. Chem.* 238: 2509–17
51. Fritz, I. B., Yue, K. T. N. 1963. *J. Lipid Res.* 4:279–88
52. Girdwood, R. H. 1953. *Brit. Med. J.* 2:741–45
53. Hodgson, E., Cheldelin, V. H., Newburgh, R. W. 1956. *Can. J. Zool.* 34:527–32
54. Hodson, A. Z. 1945. *J. Biol. Chem.* 157:383–85
55. Hogg, J. A., Richardson, M. 1968. *Arch. Mikrobiol.* 62:153–56
56. Horowitz, N. H., Beadle, G. W. 1943. *J. Biol. Chem.* 150:325–33
57. Hutchison, D. J. 1954. *Ann. N. Y. Acad. Sci.* 60:212–19
58. Hutner, S. H. et al. 1958. *Trace Elements,* ed. C. A. Lamb, O. G. Bentley, J. M. Beatie. New York: Academic
59. Hutner, S. H., Baker, H., Aaronson, S., Nathan, H. A., Rodriguez, E. et al. 1957. *J. Protozool.* 4:259–69
60. Hutner, S. H., Nathan, H., Baker, H. 1959. *Vitam. Horm.* 17:1–52
61. Ikawa, M., Borowski, D. T., Chakravarti, A. 1968. *Appl. Microbiol.* 16:620–23
62. Isenberg, H. D., Seifter, E., Berkman, J. I., Mueller, A., Henson, E. 1962. *J. Protozool.* 9:262–64
63. Iwasaki, H. 1967. *J. Phycol.* 3:30–34
64. Jukes, T. H., Dornbush, A. C. 1945. *Proc. Soc. Exp. Biol. Med.* 58: 142–43
65. Kappy, M. S., Metzenberg, R. L. 1965. *Biochim. Biophys. Acta* 107:425–33
66. Kreger-van-Rij, N. J. W. 1958. *Antonie van Leeuwenhoek; J. Microbiol. Serol.* 24:137–44
67. Kreger-van-Rij, N. J. W. 1969. *The Yeasts,* vol. 1, ed. A. H. Rose, J. S. Harrison. London, New York: Academic
68. Kushner, D. J. 1956. *Biochim. Biophys. Acta* 20:554–55
69. Lamanna, C., Campbell, J. J. R. 1963. *J. Bacteriol.* 65:596–600
70. Lee, K. Y., Shubik, P. 1965. *Nature (London)* 206:1051–52
71. Lewin, L. M., Marcus, N. 1965. *Anal. Biochem.* 10:96–100
72. Light, R. J., Lennarz, W. J., Bloch, K. 1962. *J. Biol. Chem.* 237:1793–1800
73. Lilly, D. M. 1967. *Chemical Zoology,* vol. 1, ed. M. Florkin, B. T. Scheer; *Protozoa,* vol. 1, ed. G. W. Kidder. New York: Academic
74. Lima, M. E., Angluster, J., Travassos, L. R. 1970. *Rev. Microbiol. (Brasil)* 1:61–69
75. Lodder, J., Kreger-van-Rij, N. J. W. 1952. *The Yeasts.* Amsterdam: North Holland Publ. Co.
76. Luecke, R. W., Pearson, P. B. 1944. *J. Biol. Chem.* 153:259–63
77. Luecke, R. W., Pearson, P. B. 1944. *J. Biol. Chem.* 155:507–12
78. Mackinnon, J. E. 1959. *Arch. Soc. Biol. Montevideo* 24:43–48
79. Mackinnon, J. E. 1959. *Mycopathol. Mycol. Appl.* 10:207–8
80. Marquis, N. R., Fritz, I. B. 1964. *J. Lipid Res.* 5:184–87
81. Maw, G. A. 1966. *Arch. Biochem. Biophys.* 115:291–301
82. Mehnert, B., Koch, U. 1963. *Zentralbl. Bakteriol. Parasitenk. Infektionskr. Abt. I* 188:103–19
83. Mendonça, L. C. S. 1970. *Metabolismo de etionina em Candida slooffii.* DSc thesis, Universidade Federal do Rio de Janeiro
84. Mendonça, L. C. S., Travassos, L. R. 1969. *An. Acad. Bras. Cienc.* 41: 650R–1R
85. Mendonça, L. C. S., Travassos, L. R. 1970. *2° Congr. Brasileiro de Microbiol., 2nd, Sao Paulo* Proceedings in press
86. Menna, M. E. Di 1955. *J. Gen. Microbiol.* 12:54–62
87. Metzenberg, R. L., Kappy, M. S., Parson, J. W. 1964. *Science* 145: 1434–35
88. Miller, E. C., Miller, J. A. 1969. *Ann. N. Y. Acad. Sci.* 163:731–50
89. Miller, J. A., Miller, E. C. 1969. *Symp. Quantum Chem., Biochem.,* vol. 1, ed. E. D. Bergman, B. Pullman. Jerusalem; Israeli Acad. Sci. and Human.
90. Miranda, M., Picard, I., Cruz, W. B., Ribeiro, A. F. F. 1967. *An. Acad. Bras. Cienc.* 39:429–32
91. Miranda, M., Billek, G., Rocha, H. 1967. *An. Acad. Bras. Cienc.* 39: 433–35
92. Mitchell, H. K., Williams, R. J. 1940. *Biochem. J.* 34:1532–36
93. Moreira, M. C. B., Cruz, F. S., Travassos, L. R. 1965. *An. Microbiol.* 13:35–52
94. Mosser, J. L., Tomasz, A. 1970. *J. Biol. Chem.* 245:287–98
95. Negroni, P., Fischer, I. 1941. *Rev. Inst. Bacteriol. Malbran* 10:334–42

96. Nettleship, A., Henshaw, P. S., Meyer, H. L. 1943. *J. Nat. Cancer Inst.* 4:309–19
97. Northam, B. E., Norris, F. W. 1952. *J. Gen. Microbiol.* 7:245–56
98. Okada, T., Yanagisawa, K., Ryan, F. J. 1960. *Nature (London)* 188: 340–41
99. Parle, J. N. 1956. *Antonie van Leeuwenhoek; J. Microbiol. Serol.* 22: 237–42
100. Parle, J. N. 1957. *J. Gen. Microbiol.* 17:363–67
101. Phaff, H. J., Mrak, E. M., Williams, O. B. 1952. *Mycologia* 44:431–51
102. Prophet, K. 1963. Inaug. Diss. Tierarztliche Hochschule Hannover, 35, 1277
103. Proudlock, J. W., Wheeldon, L. W., Jollow, D. J., Linnane, A. W. 1968. *Biochim. Biophys. Acta* 152:434–37
104. Rabinovitch, L., Cruz, F. S., Travassos, L. R., Cury, A. 1965. *An. Microbiol.* 13:25–33
105. Ravel, J. M., Felsing, B., Shive, W. 1954. *J. Biol. Chem.* 206:791–96
106. Ravel, J. M., Reger, J. L., Shive, W. 1955. *Arch. Biochem. Biophys.* 57: 312–22
107. Reid, R., Cow, D., Baker, H., Frank, O. 1969. *J. Protozool.* 16:231–35
108. Richardson, M., Speed, D. J., 1969. *Arch. Mikrobiol.* 66:195–98
109. Richle, R., Scholer, H. J. 1961. *Pathol. Microbiol.* 24:783–93
110. Rocha, H., Miranda, M. 1970. Personal communication
111. Rogers, A. L., Beneke, E. S. 1964. *Mycopathol. Mycol. Appl.* 22:15–20
112. Rogers, S. 1957. *J. Exp. Med.* 105: 279–306
113. Roitman, I. 1967. *Fatores de temperatura em* Candida slooffi. DSc thesis, Universidade Federal do Rio de Janeiro
114. Roitman, I., Travassos, L. R., Azevedo, H. P., Cury, A. 1969. *Sabouraudia* 7:15–19
115. Rolle, M., Mehnert, B. 1957. *Zentralbl. Bakteriol. Parasitenk. Infektionskr., Abt. I* 168:268–77
116. Rothschild, H. A., Barron, E. S. G. 1954. *J. Biol. Chem.* 209:511–23
117. Sakaguchi, K. 1962. *Agr. Biol. Chem.* 26:72–74
118. Sato, M., Saito, K. 1957. *J. Agr. Soc. Japan* 31:675–85
119. Scheffers, W. A., Wiken, T. O. 1969. *Antonie van Leeuwenhoek; J. Microbiol. Serol.* 35:A31–32; Proc. Int. Symp. on Yeasts, 3rd; Delft-The Hague
120. Schionning, H. 1903. *Compt. Rend. Trav. Lab. Carlsberg.* 6:103; quoted from Lodder, J., Kreger-van-Rij, N. J. W., Ref. 75
121. Schlenk, F., Zidek, C. R., Ehninger, J. D., Dainko, J. L. 1965. *Enzymologia* 29:283–98
122. Schneider, W. C. 1957. *Methods Enzymol.* 3:680–84
123. Schoeder, H. A. 1968. *Am. J. Clin. Nutr.* 21:230–44
124. Sherman, F. 1959. *J. Cell. Comp. Physiol.* 54:29–35
125. Sherman, F. 1959. *J. Cell. Comp. Physiol.* 54:37–52
126. Shiefer, B., Mehnert, B. 1967. *Zentralbl. Bakteriol. Parasitenk. Infektionskr. Abt. I* 202:233–47
127. Shifrine, M., Phaff, H. J. 1958. *Antonie van Leeuwenhoek; J. Microbiol. Serol.* 24:193–209
128. Shifrine, M., Phaff, H. J. 1958. *Antonie van Leeuwenhoek; J. Microbiol. Serol.* 24:274–80
129. Shifrine, M., Phaff, H. J. 1959. *Mycologia* 51:318–28
130. Shull, K. H., McConomy, J., Vogt, M., Castillo, A., Farber, E. 1966. *J. Biol. Chem.* 241:5060–70
131. Singer, S., Elion, G. B., Hitchings, G. H. 1966. *J. Gen. Microbiol.* 42: 185–96
132. Skipper, H. E., Schabel, F. M., Jr. 1952. *Arch. Biochem. Biophys.* 40: 476–78
133. Skipper, H. E., Schabel, F. M., Jr., Binns, V., Thomson, J. R., Wheeler, G. P. 1955. *Cancer Res.* 15:143–46
134. Smith, J. M. B. 1967. *Sabouraudia* 5:220–25
135. Smith, R. C., Salmon, W. D. 1965. *J. Bacteriol.* 89:1494–98
136. Sorsoli, W. A., Spence, K. D., Parks, L. W. 1964. *J. Bacteriol.* 88:20–24
137. Spence, K. D., Parks, L. W., Shapiro, S. K. 1967. *J. Bacteriol.* 94:1531–37
138. Starr, P. R., Parks, L. W. 1962. *J. Cell. Comp. Physiol.* 59:107–10
139. Stekol, J. A. 1963. *Advan. Enzymol.* 25:369–93
140. Stekol, J. A. 1963. *Methods Enzymol.* 6:566–77
141. Stekol, J. A., Weiss, K. 1950. *J. Biol. Chem.* 185:577–83
142. Stosch, H. A. von. 1963. *Proc. Int. Seaweed Symp., 4th.* Oxford: Pergamon Press
143. Strack, E., Kunz, W. 1963. *Z. Physiol. Chem.* 333:46–56

144. Surdin, Y., Sly, W., Sire, J., Bordes, A. M., Robichon-Szulmajster, H. 1965. *Biochim. Biophys. Acta* 107: 546–66
145. Swendseid, M. E., Bethell, F. H., Bird, O. D. 1951. *Cancer Res.* 11: 864–67
146. Swendseid, M. E., Nyc, J. F. 1958. *J. Bacteriol.* 75:654–59
147. Taylor, W. E., McKibbin, J. M. 1952. *Proc. Soc. Exp. Biol. Med.* 79:95–96
148. Thomitzek, W. D. 1970. *Reviews of Physiology,* 62. Berlin, Heidelberg, New York: Springer-Verlag
149. Tomasz, A. 1967. *Science* 157:694–97
150. Tomasz, A. 1968. *Proc. Nat. Acad. Sci. (Wash.)* 59:86–93
151. Trager, W. 1957. *J. Protozool.* 4: 269–76
152. Travassos, L. R. 1967. *Ação antimetabólica de etionina e colina em Candida slooffii.* DSc thesis, Universidade Federal do Rio de Janeiro
153. Travassos, L. R., Cury, A. 1966/67. *An. Microbiol.* 14:11–35
154. Travassos, L. R., Cury, A., Hutner, S. H. 1964. *Bacteriol. Proc.* 35
155. Travassos, L. R., Leon, W., Mendonça, L. 1970. In press
156. Travassos, L. R., Suassuna, E. N., Cury, A. 1963. *An. Microbiol.* 11 (B):13–76
157. Travassos, L. R., Suassuna, E. N., Cury, A., Hausmann, R. L., Miranda, M. 1961. *An. Microbiol.* 9: 465–89
158. Uden, N. van. 1952. *Arch. Mikrobiol.* 17:199–208
159. Uden, N. van. 1960. *Ann. N. Y. Acad. Sci.* 89:59–68
160. Uden, N. van, Buckley, H. 1970. *The Yeasts,* ed. J. Lodder. Amsterdam: North Holland Publ. Co.
161. Uden, N. van, Carmo-Sousa, L. do. 1957. *Port. Acta Biol., Ser. A* 5: 7–17
162. Uden, N. van, Carmo-Sousa, L. do. 1957. *J. Gen. Microbiol.* 16:385–95
163. Uden, V. van, Carmo-Sousa, L. do. 1959. *Port. Acta Biol., Ser. B* 6: 239–56
164. Uden, N. van, Carmo-Sousa, L. do. 1962. *Antonie van Leeuwenhoek; J. Microbiol. Serol.* 28:73–77
165. Uden, N. van, Carmo-Sousa, L. do. 1962. *J. Gen. Microbiol.* 27:35–40
166. Uden, N. van, Carmo-Sousa, L. do, Farinha, M. 1958. *J. Gen. Microbiol.* 19:435–45
167. Uden, N. van, Farinha, M. 1958. *Port. Acta Biol. Ser. B* 6:161–78
168. Uden, N. van, Matos-Faia, M., Assis-Lopes, L. 1956. *J. Gen. Microbiol.* 15:151–53
169. Uden, N. van, Vidal-Leiria, M. 1970. *The Yeasts,* ed. J. Lodder. Amsterdam: North Holland Publ. Co.
170. Villa-Trevino, S., Farber, E. 1962. *Biochim. Biophys. Acta* 61:649–51
171. Villa-Trevino, S., Shull, K. H., Farber, E. 1963. *J. Biol. Chem.* 238: 1757–63
172. Walt, J. P. van der. 1957. *Antonie van Leeuwenhoek, J. Microbiol. Serol.* 23:23–29
173. Walt, J. P. van der. 1957. *J. Dent. Assoc. S. Afr.* 12:1–3
174. Walt, J. P. van der. 1970. *The Yeasts,* ed. J. Lodder. Amsterdam: North Holland Publ. Co.
175. Webb, M. 1954. *Chem. Biol. Pteridines, Ciba Found. Symp.*
176. Webb, M. 1955. *Biochim. Biophys. Acta* 17:212–25
177. Wheeler, G. P., Grammer, M. G. 1960. *Biochem. Pharmacol.* 3:316–27
178. Whitaker, J. M., Umbreit, W. W. 1961. *J. Bacteriol.* 81:730–32
179. Wickerham, L. J. 1951. *Taxonomy of Yeasts, U. S. Dep. Agr. Tech. Bull. No. 1029,* Washington, D.C.
180. Wickerham, L. J. 1957. *J. Am. Med. Assoc.* 165:47–48
181. Wiken, T., Scheffers, W. A., Verhaar, A. J. M. 1961. *Antonie van Leeuwenhoek; J. Microbiol. Serol.* 27:401–33
182. Williams, R. J., Eakin, R. E., Snell, E. E. 1940. *J. Am. Chem. Soc.* 62: 1204–7

AGGREGATION AND DIFFERENTIATION IN THE CELLULAR SLIME MOLDS

1563

John Tyler Bonner

Department of Biology, Princeton University, Princeton, N.J.

Contents

The front line of developmental biology today consists of attempts to identify the key biochemical mechanisms that control specific developmental processes. Our starting point continues to be the gene and from the striking results of many studies of viruses and bacteria it is possible, to an extraordinary degree, to follow the specific chain of events from gene to action, that is, to some ultimate particular of differentiation. By contrast, in eukaryotes the distance between the gene and its end result seems so enormous, either because the chemical steps are so very numerous or so very complex, that this has become the crucial problem of modern developmental biology.

The main difference between eukaryotes and bacteria is one of size. The individual cell is much larger and furthermore eukaryotes compound this difference by developing into multicellular organisms. From the point of view of genetics this means a vast increase in the amount of DNA per cell, and of special current interest is the role of the large quantities of redundant, or genetically identical DNA. From the point of view of development, this size difference means a great increase in the number of biochemical steps and therefore the time required for development.

Many multicellular eukaryotes develop from a fertilized egg, and therefore much of the recent thought and experimental work on the biochemistry of development concerns DNA, RNA, and enzyme changes in oögenesis and post-fertilization stages, cleavage, gastrulation, etc. in animal embryos. The microbiological models are much in evidence, and the eukaryotes show differences of great interest. But it has long been recognized that the process of forming a multicellular organism has been achieved, in nature, in a number of different ways; not all forms come from large eggs. And this must mean that there are many variations in the method of going from gene to differentiation, yet we all feel that there must be some basic aspects of the process that are common to all higher organisms, just as all of them have genes and chromosomes.

The cellular slime molds (especially the species *Dictyostelium discoideum,* Raper) provide excellent material for such a comparative approach. They differ from most other eukaryotes in that they undergo a period of growth—feeding and cell division—as separate, independent cells. When the food supply (bacteria) is gone, they aggregate into cell masses (Fig. 1). Eventually the cells at the anterior end of these masses turn into stalk cells, while the cells at the posterior and turn into spores, any one of which can start a new generation. These organisms are an important component of the soil fauna. From an adaptive point of view one assumes that their feeding site is not the optimal one for spore dispersal, and as a result one has the multicellular phase which permits the mass of amoebae to move away (for the mass moves towards light and heat) and up into the air by differentiating a delicate stalk which supports the apical spore mass.

The basic notion I would like to emphasize in this review is the necessity, in any study of the biochemistry of development, to do the biochemistry and the biology hand-in-hand. Any other procedure, it seems to me, will be bound by severe limitations. If one concentrates merely on the biology, one ends by knowing nothing of the key chemical events. If one is concerned merely with the biochemistry one soon loses sight of what the substances themselves could possibly be doing in the developing organism. This presents less difficulty in virus or even bacterial developmental studies for the morphological changes are relatively simple, linear, and straightforward. But the very opposite is true of multicellular eukaryotes; the biological steps are subtle and intricately interwoven, making their biochemical interpretation that much more difficult, and at the same time particularly interesting. To illustrate my point I will review two specific aspects of slime mold development: the chemotaxis of aggregation and the differentiation of spores and stalk cells and try to link what we know of the gross changes in morphology with what we know of the molecular changes.

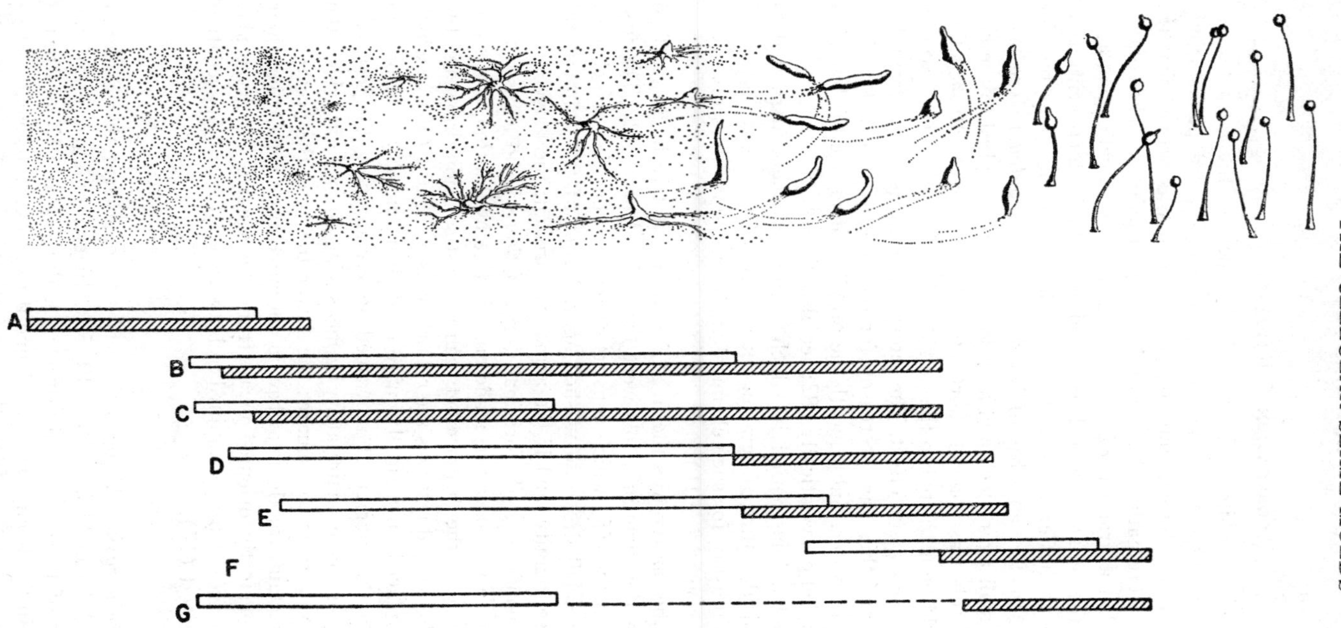

FIGURE 1. *Above:* The development of *Dictyostelium discoideum* from the end of feeding through aggregation, migration, and the final culmination stage. (Drawing by Patricia Collins from the *Scientific American.*) *Below:* The periods susceptible to RNA synthesis inhibitors (open bars) and protein synthesis inhibitors (black bars) for various enzymes. A, acetyl glucosaminidase; B, trehalose-phosphate synthetase; C, threonine dehydrase; D, UDP-glucose phosphorylase; E, UDP-galactose polysaccharide transferase; F, β-glucosidase; C, alkaline phosphatase. The time span for each enzyme is drawn in a nonlinear fashion to correspond to the drawing above. [This figure is adapted from Loomis (63) who cites the original sources.]

Aggregation Chemotaxis

Early work.—Using quite indirect methods, such as the attraction of amoebae by centers across semipermeable membranes (78) and the disturbing of aggregation patterns by currents of water (6), it was argued that the separate amoebae were gathered to central collection points by chemotaxis. It was assumed that a gradient of a substance which was called "acrasin" somehow orients the cells; the amoebae moved toward a high concentration. A few years later, Shaffer (81) was able to devise an ingenious test for acrasin in which sensitive amoebae of *Dictyostelium discoideum* were sandwiched under a small agar block and acrasin water was applied to the outside meniscus. If this was done at short intervals (e.g., every 10 sec) then the amoebae would be attracted to the outside edge; if the intervals were a minute or over there was no effect. The interpretation was that acrasin disappeared rapidly and this was the reason for the need of repeated applications of the acrasin water. Subsequently he provided evidence to show that acrasin was degraded enzymatically (82; see also 95).

There followed a long period in which various workers attempted to identify the chemical nature of acrasin, but because of the shortcomings of the Shaffer acrasin assay for such chemical analysis, the results were unsatisfactory. (For a review, see 11)

The crux of the matter was the need for a quantitative bioassay. Two such assays using the same principle were devised independently. The first was that of Konijn (54, 55) which consists of putting sensitive amoebae of *D. discoideum* in small drops on washed agar. Drops of test solution are placed in the close vicinity, and if the acrasin is present, the amoebae will burst out of the confines of the original drop; if there is no acrasin they will remain inside. The test can be made quantitative by dilution of the test solutions or by varying the distance between the test solution drops and the amoeba drops. The second test was the cellophane square test (19) in which the amoebae are placed on a small square of dialysis membrane which, in turn, is placed on the surface of unwashed agar containing the test substance. The assay is made quantitative by measuring the rate at which the amoebae move away from the square; the faster the rate, the greater the power of the attractant, for the amoebae do not move more rapidly but their paths are more perfectly oriented away from the square. We have also shown that chemotaxis can be examined in both vegetative and aggregating amoebae with this test (17).

Cyclic AMP as acrasin.—Using both tests it became clear that bacterial extracts attract slime mold amoebae (19, 56). In examining some of the properties of one of these bacterial attractants, Barkley and Konijn tested cyclic 3′,5′ AMP in the Konijn test and found it to be extraordinarily active with *D. discoideum* amoebae. (60). Subsequently this was found to be the component secreted by *Escherichia coli* which affects aggregating amoebae (61).

The cells of *D. discoideum* also secrete large quantities of a specific phosphodiesterase which converts cyclic AMP to 5′ AMP (24), and in a number of other respects it is somewhat similar, but not identical to mammalian phosphodiesterase (23, 30). Using a variety of different techniques to prevent this phosphodiesterase from immediately converting the slime mold's own cyclic AMP, it was possible to show that *D. discoideum* does synthesize its own cyclic AMP (4, 59). From this we assume that this species uses cyclic AMP as an acrasin and that its phosphodiesterase is the acrasinase of Shaffer.

There is one more key bit of descriptive information concerning these substances and aggregation. We have followed the course of their production through time and shown that, during the vegetative stage, cyclic AMP is either not produced at all, or below levels we can measure on our bioassays, but at the onset of aggregation there is a sudden surge of secretion, reaching a maximum of 100-fold above the first detectable concentration (13). During migration this secretion falls to a lower level, but that may be due to the fact that the cells are now packaged in masses in a slime sheath. The secretion of phosphodiesterase has also been followed through time, but unfortunately only in liquid culture. This has been done independently by Riedel & Gerisch (110) and by ourselves (unpublished findings, 76) using different methods and obtaining different results, although both groups (but especially clearly shown by Reidel & Gerisch) have evidence of a phosphodiesterase inhibitor. Also, Konijn (57) has shown that *D. discoideum* amoebae grown in liquid culture in the presence of added cyclic AMP secrete an increased amount of the enzyme. It is now imperative to find out the relation between substrate, enzyme, and inhibitor under normal conditions at an air-water interface on a solid substratum. Presumably these substances do control aggregation and we must find their exact relations during normal aggregation, a program we have already begun.

Spatial and temporal patterns of aggregation.—Already at this point in our brief description of aggregation in *D. discoideum* it is obvious that to understand development in biochemical terms means more than identifying the substances and their interactions and tracking their changes in concentration through time: their distribution in space is another major consideration. Generally this is beyond the ken of classical biochemistry and recently the biomathematicians have attempted to bring together the spatial, temporal, and chemical events into unifying schemes. Their models provide useful hypothetical bridges between the biology and the chemistry, and supply stepping stones for future experiments.

One of the prime spatial-temporal problems is initiation. In *Polysphondylium,* Shaffer (86) showed that certain key cells (founder cells) are responsible for beginning the process, but no such cell has yet been observed in *D. discoideum.* One of the useful predictions of the Keller-Segel (52, 53) equilibrium instability theory is that one does not necessarily need such special cells to explain aggregation patterns; the acrasin-acrasinase conditions are sufficient.

This possibly applies also to the fact that under certain conditions the size of the aggregation territories remains constant regardless of the cell density (15). Even though theoretically no special inhibitor or "spacing" substance might be necessary, we gave some evidence that there was such a substance (18; see also 50), as Shaffer (87) did for founder cells inhibiting the formation of other founder cells in *Polysphondylium* (see also 51). More recently, Feit (31) has presented some preliminary evidence that in *D. discoideum* this substance might be ammonia.

Francis (33) and Kahn (51) have made the important point that acrasin itself could cause the inhibition of centers forming in the close neighborhood of well-established centers; in fact, Francis (33) specifically suggested such inhibition might be the result of a gradient of acrasin. It has been possible to confirm this possibility by applying artificial gradients of cyclic AMP. The fact that gradients are essential is clear for high, even concentrations of cyclic AMP do not inhibit centers but cause a large number to be formed, each with a minute territory. (13, 56, 58). Needless to say, the question of how this occurs on a biochemical level is crucial, but unfortunately we have no inkling. Recently Mason, Rasmussen, and DiBella (personal communication) have shown that high external concentrations of cyclic AMP inhibit its secretion by amoebae, a significant experiment which may lead to a better understanding of these problems.

Pulses and chemotaxis.—Another ancient observation (Arndt, 1) has attracted the attention of many investigators: the aggregating cells move inward in a series of pulses which start in the center and move outward. Shaffer (84) made the hypothesis that this was a relay system where small acrasin gradients were passed outward from the center on a principle analagous to the passage of an impulse down a nerve. More recently, Goodwin & Cohen (43) have devised a general oscillator model to explain the spatial and temporal organization in all developing systems, and Cohen & Robertson (27, 28; Robertson, 77) have developed this specifically for *D. discoideum* and suggest that all the important aspects of aggregation chemotaxis can be thought of in terms of these oscillating pulsations, including territory formation, an idea already put forward in some detail by Gerisch (38, 40). Their hypothesis is an extension of Shaffer's relay hypothesis, and one of their specific suggestions brings us to a key problem.

This is the question of how cyclic AMP orients an amoeba. All the early work assumed that a stable gradient is necessary and that one end of each cell is being constantly bombarded by more molecules of the attractant than the other. Cohen & Robertson (27, 28) suggest rather that the orientation of cells is not dependent upon a permanent concentration gradient (which may or may not exist), but that a puff of acrasin (above a specific threshold) hits one end of the cell first, and as Shaffer (84; see also 38) suggested, once the cell responds it goes into a refractory period for some time before any new acrasin puff can affect it. One must never forget the extremely hypothetical nature of these stimulating ideas, but they open up, in this case, a totally new way of looking at amoeba chemotaxis, providing an attractive possibility which is sound theoretically and therefore should provide increased sophistication for future experimental efforts.

But whether we think of the cell as being oriented by a steady-state gradient or by a puff hitting one end first (a momentary gradient), we still have the hard question of how the specific molecule, cyclic AMP, achieves this feat. This question is of special interest because of the great variety of other physiological accomplishments of cyclic AMP, from bacteria to mammals. Recently, Chi & Francis (25) have shown that additions of $10^{-4}M$ cyclic AMP to *D. discoideum* amoebae near or at the aggregation stage caused a large increase in the calcium efflux from the cells, while no such effect was evident with the sodium ions. They suggest that this effect on calcium is directly related to the stimulation of amoeboid movement in one direction. In this regard it is also interesting that Mason, Rasmussen and DiBella (unpublished observations) have shown that a reduction of calcium in the surrounding medium causes an increase in cyclic AMP secretion, but a failure to aggregate. That there is a relation between calcium and cyclic AMP in this system (as has been demonstrated for various other mammalian systems (Rasmussen & Tenenhouse, 75) is indeed likely.

Acrasinase and chemotaxis.—Another interesting question is the role of the acrasinase (phosphodiesterase) in chemotaxis. Some years ago Shaffer (83) argued the enzyme was needed to maintain gradients, but this, of course, would only apply if gradients are necessary, a proposition that is no longer obvious considering Cohen & Robertson's (27, 28) theoretical speculations.

It should be added that the problem here is in many ways comparable to chemical signaling in insects and the analysis of Bossert & Wilson (review: 100) may have a direct application to slime molds. In particular they show the relation between the emission rate of the key molecules and the sensitivity threshold of the receiving organism, and how this relation is affected by molecular size and other parameters. They stress the importance of decreasing the "fade out" time of the attractant and how this can be enhanced by a mechanism such as enzymatic deactivation. Also, Bossert (22) shows that theoretically much information can be passed on to the receiver if the attractant is given off in pulses.

The presence of the enzyme is possible mandatory for the successful operation of the two bioassays for acrasin (the Konijn test and the cellophane square test). The assays both involve a concentrated group of amoebae, and the measurement of the spreading of the amoebae from the group. Since the acrasin is widespread in the agar, the only way to achieve a gradient is by having the attractant consumed in the vicinity of the cells, so that the cells will move outward to the higher concentrations of the gradient (13). There is also some evidence for a separate repellant which causes the cells to move away from one another (80; Bonner, unpublished data), but this is the same in the controls as in the experimentals and does not influence the tests as acrasin tests.

In *D. discoideum* especially obvious, expanding rings appear when a drop of amoebae is placed on agar containing cyclic AMP (13, 58). The hypothesis is that the ring, which forms back from the farthest edge of advancing amoebae, is the region of the steepest part of the outward acrasin gradient, caused by the phosphodiesterase and that the cells here tend to clump together. The ring itself moves outward. This suggests the interesting possibility that cells can be oriented both by gradients and directional acrasin puffs and that the movement of the ring is evidence for the former.

Cell adhesiveness.—The clumping or adhesiveness induced in the ring is of special interest. The first to note such an acrasin effect was Shaffer (84, 85), and now we have clear evidence that it is not the mere presence of acrasin that will produce this effect, but a gradient of acrasin (13). Vegetative cells can be induced to clump, although cells ready to aggregate exhibit the phenomenon to a much greater degree, and one suspects without the need of a gradient.

It has long been thought that this increased adhesiveness characteristic of normal development was due to a change of the cell surface, and Sonneborn, Sussman & Levine (91) showed that new surface antigens (which they suggested might be lipoproteins) appeared at the aggregation stage. This matter has been investigated in depth by Gerisch and his co-workers (review: 40). The first step was to demonstrate that the aggregation cells were resistant to EDTA; they clumped even in its presence, which was not true of vegetative cells (35, 39, 40). Next, they analyzed the chemical changes in the cell surface and showed that it was possible to extract increasingly larger amounts of a polysaccharide antigen as development proceeded into the aggregation stage. They also revealed species differences in the surface polysaccharides between *Dictyostelium* and *Polysphondylium;* the sugars involved were identical but the relative amounts differed (41).

Of special interest are Gerisch's (37, 40) experiments with 2,4-dinitrophenol. He showed that the changes in adhesive properties of aggregating cells were unchanged by this inhibitor of metabolism, but that the cells rounded up and tended to form cell clumps of flat sheets at the aggregation stage while the clumps were spherical for vegetative amoebae. From this he suggests that possibly an adhesion factor exists on the cells in some sort of equatorial ring.

Recently, Gerisch and his co-workers (108, 109) had shown that they can completely block adhesiveness with a univalent antibody prepared from antisera of aggregation cells. Yet, even though the cells fail to adhere to one another, they have not in any way lost their ability to respond to acrasin and will stream toward centers as isolated cells. This important experiment shows that chemotaxis and adhesion are totally separate phenomena.

Other examples of chemotaxis in the slime molds.—This discussion has centered largely around the aggregation chemotaxis, and some immediately related problems. To emphasize the narrowness of our discussion, let me list some of the other processes, many of them chemotactic, which are also related and known to exist during slime mold development, yet they have been barely studied. Among the other chemotactic factors there is evidence for a substance given off by bacteria which specifically attracts vegetative amoebae, which, in turn, make an enzyme (not a phosphodiesterase) that destroys the attractants (17, and unpublished data). There is evidence for a substance which causes vegetative amoebae to move away from one another (80; Bonner, unpublished observations), and evidence for a volatile repellant in fruiting cell masses (16). Among the nonchemotactic factors there are substances which affect the rate of movement of cells (17, 40, 42, 80), and the spacing substance or substances which were discussed previously (15, 18, 31). Finally, we have not mentioned all the work on the effect of environmental changes (temperature, light, humidity) and how they affect aggregation.

The complexity of the interactions between cells in these "simple" organisms is extraordinarily impressive. Not only are there many substances and many reactions, but there is the further matter of changing patterns of these chemical events through time. And here we are considering one isolated developmental process, aggregation, and considering it only at the level of the actual event, omitting any mention of its genetic control, of which virtually nothing is known.

Differentiation

Sex.—The study of differentiation in the cellular slime molds has the advantage of allowing us to consider the control mechanisms on the gene level as well as the product level. One difficulty in the past has been the lack of any clear-cut sexuality for breeding experiments, but recently there is some interesting evidence that these organisms might have a parasexual cycle similar to that described by Pontecorvo (71) for fungi. The first suggestion for this came from Sussman (94), and recently Sinha & Ashworth (89) have presented evidence for a parasexual system, a case that is supported by evidence from other workers (64, 79, 105, 106). Hopefully, this may ultimately lead to a better understanding of the genetics of *Dictyostelium.*

Chromosome repair.—One approach to matters concerning the activities of genes is the use of irradiation. Deering et al (29) have shown that gamma rays will cause, in a particular mutant, the complete cessation of DNA duplication and cell proliferation, yet such irradiated cells will aggregate and differentiate even though the spores they produce are incapable of further growth. They give evidence that the DNA repair system is defective in this mutant, and that once the DNA is fractured by the irradiation, it cannot come together again and therefore cannot replicate. This means that either the DNA fragments can manufacture normal messenger RNA or all the messages have been prefabricated before the irradiation. The former possibility seems more promising, although Hirschberg et al (48) suggest, because of the greater sensitivity of dividing cells to RNA synthesis inhibitors as compared to cells in later stages of development, that all the messages could conceivably be manufactured before the beginning of the aggregation stage.

The timing of gene messages.—The support for the possibility that the messages are made at different times during development comes from Sussman and his associates (reviews: 2, 34, 96). Their evidence does have the well-known weakness of depending upon the action of inhibitors, but they have found, for numerous enzymes, that the period of susceptibility to enzyme synthesis inhibition by actinomycin D is characteristic for each one, and invariably this period precedes, to different degrees, the period of susceptibility to cycloheximide. They interpret this to mean that there are varying periods of delay between transcription and translation and that the most compelling argument in favor of this view is that this relation has been consistent for all enzymes studied and the exact timing characteristic of each enzyme (Fig. 1).

It should be noted, however, that recently Watts & Ashworth (99) have isolated a mutant (Ax-2) that grows without bacteria on a simple medium containing yeast extract, peptone, and glucose. Ashworth (Ashworth & Watts, 3, and personal communication) has made the most interesting discovery that the pattern of synthesis of enzymes varies radically depending upon the nutrient conditions during the growth phase. With this new important advance it should be possible to find what are the parts of the enzyme activity patterns that are essential to development, and what parts are due to the vagaries imposed by the growth conditions.

"Control" of development.—There has been some vigorous discussion of the question of how development in the cellular slime molds is controlled, and the reader is urged to consult the excellent review of Francis (34) who stresses the different kinds of interaction that might exist between gene, product, and all the intermediate reactions. As we have seen, Sussman and his group have demonstrated definite time steps for translation and transcription but, as Francis in 1969 pointed out, the time of appearance of the final product is extremely variable and exhibits none of this regularity. This facet of the matter has been central to Wright's (102–104) thesis that the control of development is exercised by substrate and product concentrations involving feedback inhibition or stimulation. Surely, the answer must be that both ideas are correct and perhaps the real problem lies simply in our

tendency to forget that development does not have one key controlling step but many. What we are really looking for is the order or sequence of developmental control mechanisms. The confusion has partly been the result of the unique position of DNA in developing systems, and although it is a master controller, it is nothing without all the other associated mechanisms. And to add to the difficulty, this is less obvious in viral and bacterial development than it is in eukaryotic, multicellular development.

It is helpful to look at the question of developmental control mechanisms from the point of view of evolution. In order for natural selection to operate it is necessary to have stable units of inheritance, that is, genes. But as the size and complexity of the organism increases, so does its development, and this means a larger number of secondary, tertiary (and beyond), reactions over and above the ones immediately controlled by the gene. Furthermore, these other substances, the products and substrates of the reaction, are also involved in control, as we have just stressed. The difficulty is apparently overcome by having genes and all the substances they affect in a series of reactions in some sort of a unit that has a degree of independence of neighboring units (Bonner, 10). Gene changes which are subject to selection, carry these entire units of interrelated products with them, the reason they are units is just so they might be separated or moved in some way from the other units. Their very existence through selection depends upon this ability.

During the course of evolution one imagines that first there existed a rigid system in which a series of genes are read off in a sequence along the chromosome. As flexible responses to environmental changes began to be fixed by selection the need for alternate developmental pathways became imperative, and for this any fixed reading of a linear set of instructions is impossible. It is at this point that one sees the complex control mechanisms appearing in large numbers, and there is no reason, from either the point of view of selection or the mechanics of development, why these sensitive controlling steps should occur only on the genome; it is just as likely that it could be some other part of the entire gene-controlled chain of steps that would affect the successful operation of the whole unit. Furthermore, there is always the excellent possibility that a change in one such unit will affect the operation of neighboring units; for instance, this is probably the case in the majority of lethal mutants.

There are innumerable examples of ways in which the cellular slime molds respond to environmental changes (e.g., 11, 73, 74, 90). Temperature, humidity, and light changes in particular have a marked effect. For instance, it has been known for a long time that one could completely circumvent the migration stage by lowering the atmospheric humidity, or increasing the osmotic pressure of the substratum. Recently, Newell & Sussman (69) have used this fact and made an exceedingly interesting observation on the formation of UDP-glucose pyrophosphorylase. The specific activity of this enzyme (which is presumably involved in cell wall synthesis) rises slowly during migration, but completes its synthesis rapidly in cell masses that fruit immediately. By examining the susceptibility of the enzyme to actinomycin D, they show that the presumed DNA-dependent RNA synthesis is correspondingly depressed in the migrating slug. From this they suggest that the amount of enzyme is controlled at the transcription level in response to high humidity and low osmotic pressure. This, of course, does not say anything about how these external factors affect the RNA synthesis; it would be interesting to know the chemical steps that connect the two. But as is clear from Wright's work and Francis' (34) discussion, there is no reason why environmental effects need only reach the transcriptive level; they could just as well affect the level of translation or that of the substrate and product.

One example of a small molecule having a profound effect on differentiation is the observation that cyclic AMP stimulates the differentiation of stalk cells in the unaggregated cells of *D. discoideum* (12). From previous work it was known that in the cell mass, acrasin is produced solely in the region of stalk formation during fruiting (7) and that stalk formation involves high levels of catabolism (47). This is of special interest because of the recent work on *Escherichia coli* in which it has been shown that cyclic AMP is involved in regulating the synthesis of many inducible enzymes both at the transcriptional and translational level (70). The fact is, at least in *D. discoideum,* one substance can be involved in both aggregation and differentiation, and the fact that the attributes of this substance are of such active concern in other systems, gives one hope that this will provide a fertile approach.

The biochemical methods we have discussed so far have all involved the details of known substances. There is also some information on unknown substances which affect differentiation, a fact that widens the horizon for future work. For instance, Yanagisawa et al (107) have shown that a substance is given off by wild-type *D. discoideum* which inhibits spore (but not stalk) differentiation. This has no effect on the wild type, but they found a susceptible mutant which forms fruiting bodies normal in external appearance, but has no spore differentiation. Nothing is known of the nature of this substance or what role it might play in normal development. They have some evidence that it is produced throughout development and suggest, therefore, that it might be a normal metabolic product.

Microanatomy of differentiation.—Another approach to the problem of differentiation is through electron microscopy. Special emphasis has been placed on the prespore cells and two different kinds of vacuoles have been discovered that do not exist in the prestalk cells. The PV vacuoles have been discovered by Hohl & Hamamoto (49) and independently by Maeda & Takeuchi (66). Hohl & Hamamoto show that these vacuoles contribute to the formation of the spore wall, and suggest that they present the nonstarch polysaccharides known from previous histochemical studies (14), and the spore antigens known from previous immunological studies (98). More recently, Gregg & Badman (46) have found a new vesicle in the spore which they call SV which appears after the PV vacuoles and suggest that this might be the storage site of trehalose which Clegg & Filosa (26) find in mature spores. Furthermore, Gregg & Badman (46) were able to inhibit the formation of both vacuoles using actinomycin D in early aggregates, indicating a dependence on RNA synthesis. They were also able to prevent spore differentiation by mechanical disturbance, yet this did not interfere with the formation of both types of vacuoles. Clearly, these ultrastructure studies will play an increasingly important role together with the biochemical studies, for this is one way to bring together spatial or structural considerations with chemical ones.

Proportions of stalk and spore cells.—We have already pointed out in our discussion of aggregation the need for introducing spatial relations of molecules in developing systems. Here it is necessary to remind the reader that we have two quite different kinds of differentiation going on at the same time (Bonner, 11): One is *temporal* during which one follows a sequence of chemical steps leading in time to a specific kind of differentiation. This is precisely what we have done here in our examination of spore differentiation. The other is *spatial* in which we have a simultaneous distribution of differentiation properties in different parts of the developing organism. This is clearly illustrated by the fact that the slug of *D. discoideum,* regardless of its size, will show a very strict proportional relationship in the relative number of spore cells and stalk cells (review: Bonner, 11). Furthermore, Raper (72) showed that this was true even for fractions of the slug that were cut free. Each fraction, even if it was originally destined to be all stalk or all spore, given enough time would "regulate" so as to produce ultimately the normal ratio of stalk and spore cells. There has been a beginning at following these regulatory changes histochemically (14, 45), and on the level of ultrastructure (46). Perhaps most interesting of all, in our attempt at a biochemical understanding of regulations, are the fluorescent antibody studies of Gregg (44) in which macromolecular changes can be followed in both fractions: the isolated anterior (prestalk) fraction resynthesizes antigens normally found in vegetative cells, while the isolated posterior (prespore) fraction resynthesizes antigens characteristic of the prespore region of intact migrating slugs.

The fact that the number of prestalk and the number of prespore cells bear a constant allometric relation to each other, and the fact that this relation can readjust itself through regulation is a matter of great importance for it is a fundamental property of all multicellular differentiation. As I have pointed out previously (8), there must be some method of communication between parts for the posterior cells "know" how many anterior cells there are, and vice versa. My suggestion was that this communication was by a few messenger cells which move at a faster and slower rate than the entire mass, and these cells carry some critical chemical information. A small number of fast and slow cells do exist (8), but what function, if any, they might perform is unkown. An entirely different method of communication between parts has been suggested by Goodwin & Gohen (43) in the form of two periodic events of different wavelength moving down a multicellular mass, and positional information being theoretically possible through phase shifts. This model has been applied to *Dictysotelium* slugs by Robertson (77) who suggests that the known pulses in migrating slugs (5) pass down the entire slug and might be one of the propagating waves involved. Another positional theory of Wolpert (101) suggests gradients and thresholds along the axis of the slug. Finally, there has been a recent suggestion by Loomis (unpublished findings) and Ashworth, (2) that variation in the slime sheath along the axis of the cell mass is responsible for directing the spore-stalk proportional relations. These speculations are not all mutually exclusive and hopefully this important matter will be pursued further.

In addition to the fact that temperature is known to affect proportions of spore and stalk cells (21), there are some attempts to examine the question biochemically. Kostellow (62) mentions briefly that argenine occasionally produces an excess of spores. Mitchell (67) showed that the cells which aggregated in the presence of ethionine produced fruiting bodies with disporportionally large stalks, the proportions had been shifted so that more stalk cells were produced. This is especially interesting in that previously Filosa (32) reported that ethionine caused the phenotypic reversion of wild type of a mutant with an abnormal fruiting body. In a recent preliminary note, Maeda (65) reports that the ionic conditions of the medium have a clear effect on stalk and spore differentiation. Lithium ions promote stalk cells and inhibit spore formation (in the presence of calcium and magnesium). Also, a high concentration of calcium stimulated stalk cell formation, while fluoride stimulated the formation of spores.

It must be stressed here that there is considerable evidence that normal spore and stalk differentiation are not mutually dependent. Mutants showing complete cell differentiation, but with cells irregularly placed, have been described (Sonneborn, White & Sussman, 92) as well as similar effects being produced by raising normal cells in the presence of various chemical agents (EDTA, ethylurethane, 36, cyclic AMP, 12). Mutants are also known which have complete morphological differentiation, but lack final spore differentiation when grown in crowded conditions (107).

Sorting out.—To return to the matter of the proportion of stalk and spore cells in normal development, there is now overwhelming evidence that at the end of aggregation the cells of the slug go through a period of total redistribution within the cell mass. The first indication was of nonrandom distribution of marked cells in the slug, and clear evidence that in one strain of *D. discoideum* the larger cells ended up in the anterior end (9). Next, Takeuchi (97) showed that the preaggregation cells contained varying amounts of spore antigens and while these were randomly distributed in aggregating masses, after aggregation all the cells containing spore antigens were in the posterior end. More recently, he (98) has marked the cells with ^{3}H-thymidine and confirmed the evidence for sorting out in detail. Especially interesting is his demonstration that if preaggregation cells are centrifuged in dextrin equal to the mean specific gravity of the cells, he can separate the heavy and light cells, and if he mixes a heavy and light fraction, one of them labeled with ^{3}H-thymidine and the other not, the heavy cells clearly predominate in the anterior end. We have confirmed this for different strains of *D. discoideum* using cell size as a marker (20).

One obvious and immediate problem is the mechanism of sorting out. It is tempting to apply the differential adhesion hypotheses of Steinberg (review, 93), and indeed the anterior cells appear to possess great mutual adhesion for, as Maeda & Takeuchi (66) demonstrate, the cells are in far closer contact in the anterior, prestalk region. The argument would be that the cells with greater mutual adhesion are anterior rather than central in a sphere because all the cells have a polarity and are moving in the same direction, which would change the configuration from three dimensions to essentially one. There is also the interesting suggestion made by Gerisch (37, 40) that only the equatorial zone of each cell is adhesive [a fact which fits in with the ideas of Shaffer (88) on cell movement within the slug], and this could also play a role. Considering the increasing importance attached to sorting out in animal development it is not surprising to find it plays a significant role in the cellular slime molds. But the next key problem here, as it is for animal embryos (e.g. 68), is to understand the chemical basis of adhesion changes; it would, among other things, be interesting to know if the cell specific gravity is in any way chemically related to cell adhesion in the cellular slime molds.

There is one apparent difficulty involving the idea that cells sort out; it is the matter of regulation in fragments of a slug. The only possible conclusion is that the changes in the preaggregation cells must be revisible; for that reason I have in the past referred to the states of preaggregation cells as showing stalk and spore "tendencies." One can assume that the cells form a gradient of such tendencies and any fraction of the slug would possess a portion of this gradient and can accordingly regulate (9).

Finally, it should be pointed out that there is no conflict between sorting out and the idea that transcription and translation of different enzymes occur in a sequence (Fig. 1), even if the transcription begins early before aggregation. In that case, it must begin in some of the cells and not others while the cells are still separate. We presume that those cell differences are somatic and arise perhaps as a result of unequal cell division, and at all times their condition is potentially reversible (11). Transcription and translation occur over a considerable time span (Fig. 1) and there is no reason why this could not begin in separate scattered cells which eventually sort out in one region. But because of the reversibility, clearly, any change of the internal environment in the slug due to cutting must be able to reverse or inhibit the formation of some enzymes, and stimulate the synthesis of others. Therefore position and differentiation certainly can be related even though in some circumstances as we have seen, one can achieve one without the other and vice versa.

CONCLUSION

In the development of multicellular eukaryotes, such as cellular slime molds, one not only contends with genes, RNA and protein synthesis, substrates and products, and all the switches which produce their complex interrelations, but also with the position of the substances, both within the cells, between the cells, and then the position of the cells themselves, and the change of their positions with time. These processes involve, in addition to chemical reaction kinetics, diffusion, permeability and other factors which affect molecular movement, adhesion, cell motility, and reactivity to environmental cues. The ordering and simplification of this maze is the great task of developmental biology.

ACKNOWLEDGMENTS

I am indebted to, and wish to thank a number of friends who gave me helpful criticisms of an early draft of the manuscript: Drs. J. M. Ashworth, D. S. Barkley, D. W. Francis, G. Gerisch, J. H. Gregg, T. M. Konijn, A. Newton, J. L. Rifkin, A. D. J. Robertson, and L. A. Segel.

All the work done in our laboratory and referred to in this review was supported by funds from research grant BG-3332 of the National Science Foundation.

LITERATURE CITED

1. Arndt, A. 1937. *Roux' Arch. Entwicklungsmech. Organ.* 136:681–747
2. Ashworth, J. M. 1971. *Symp. Soc. Exp. Biol.* In press
3. Ashworth, J. M., Watts, D. J. 1970. *Biochem. J.* 119:175–82
4. Barkley, D. S. 1969. *Science* 165: 1133–34
5. Bonner, J. T. 1944. *Am. J. Bot.* 31: 175–82
6. Bonner, J. T. 1947. *J. Exp. Zool.* 106:1–26
7. Bonner, J. T. 1949. *J. Exp. Zool.* 110:259–71
8. Bonner, J. T. 1957. *Quart. Rev. Biol.* 32:232–46
9. Bonner, J. T. 1959. *Proc. Nat. Acad. Sci. USA* 45:379–84
10. Bonner, J. T. 1965. *Size and Cycle.* Princeton Univ. Press. 219 pp.
11. Bonner, J. T. 1967. *The Cellular Slime Molds,* 2nd ed. Princeton Univ. Press. 205 pp.
12. Bonner, J. T. 1970. *Proc. Nat. Acad. Sci. USA* 65:110–13
13. Bonner, J. T. et al. 1969. *Develop. Biol.* 20:72–87
14. Bonner, J. T., Chiquoine, A. D., Kolderie, M. Q. 1955. *J. Exp. Zool.* 130:133–58
15. Bonner, J. T., Dodd, M. R. 1962a. *Biol. Bull.* 122:13–24
16. Bonner, J. T., Dodd, M. R. 1962b. *Develop. Biol.* 5:344–61
17. Bonner, J. T., Hall, E. M., Sachsenmaier, W., Walker, B. K. 1970. *J. Bacteriol.* 102:682–87
18. Bonner, J. T., Hoffman, M. E. 1963. *J. Embryol. Exp. Morphol.* 11: 571–89
19. Bonner, J. T., Kelso, A. P., Gillmor, R. G. 1966. *Biol. Bull.* 130:28–42
20. Bonner, J. T., Sieja, T. W., Hall, E. M. 1971. *J. Embryol. Exp. Morphol.* In press
21. Bonner, J. T., Slifkin, M. K. 1949. *Am. J. Bot.* 36:727–34
22. Bossert, W. H. 1968. *J. Theoret. Biol.* 18:157–70
23. Butcher, R. W., Sutherland, E. W. 1962. *J. Biol. Chem.* 237:1244–50
24. Chang, Y. Y. 1968. *Science* 161:57–59
25. Chi, Y.-Y, Francis, D. W. 1971. *J. Cell. Physiol.* In press
26. Clegg, J. S., Filosa, M. F. 1961. *Nature (London)* 192:1077–78
27. Cohen, M. H., Robertson, A. D. J. 1971a. *J. Theoret. Biol.* In press
28. Cohen, M. H., Robertson, A. D. J. 1971b. *J. Theoret. Biol.* In press

29. Deering, R. A., Smith, M. S., Thompson, B. K., Adolf, A. C. 1971. *Radiation Res.* In press
30. Drummond, G. I., Perrott-Yee, S. 1961. *J. Biol. Chem.* 236:1126–29
31. Feit, I. 1969. *Evidence for the Regulation of Aggregate Density by the Production of Ammonia in the Cellular Slime Molds.* PhD thesis. Princeton University
32. Filosa, M. F. 1960. *Anat. Rec.* 138: 348
33. Francis, D. W. 1965. *Develop. Biol.* 12:329–46
34. Francis, D. W. 1969. *Quart. Rev. Biol.* 44:277–90
35. Gerisch, G. 1960. *Roux' Arch. Entwicklungsmech. Organ.* 152:632–54
36. Gerisch, G. 1961. *Roux' Arch. Entwicklungsmech. Organ.* 153:158–67
37. Gerisch, G. 1962. *Exp. Cell Res.* 26: 462–84
38. Gerisch, G. 1965a. *Roux' Arch. Entwicklungsmech. Organ.* 156:127–44
39. Gerisch, G. 1965b. *Umschau* 13:392–95
40. Gerisch, G. 1968a. *Curr. Topics Develop. Biol.* 3:157–97
41. Gerisch, G. 1968b. *Excerpta Medica Int. Congr. Ser. No. 166,* Abstract No. 24
42. Gerisch, G., Norman, I., Beug, H. 1966. *Naturwissenschaften* 53:618
43. Goodwin, B. C., Cohen, M. H. 1969. *J. Theoret. Biol.* 25:49–107
44. Gregg, J. H. 1965. *Develop. Biol.* 12:377–93
45. Gregg, J. H. 1968. *Exp. Cell Res.* 51:633–42
46. Gregg, J. H., Badman, N. S. 1970. *Develop. Biol.* 22:96–111
47. Gregg, J. H., Hackney, A. L., Krivanek, J. O. 1954. *Biol. Bull.* 107:226–35
48. Hirschberg, E., Ceccarini, C., Osnos, M., Carchman, R. 1968. *Proc. Nat. Acad. Sci. USA* 61:316–23
49. Hohl, H. R., Hamamoto, S. T. 1969. *J. Ultrastruct. Res.* 26:442–53
50. Kahn, A. J. 1964. *Biol. Bull.* 127: 25–96
51. Kahn, A. J. 1968. *Develop. Biol.* 18: 149–62
52. Keller, E. F., Segel, L. A. 1970a. *Nature (London)* 227:1365–66
53. Keller, E. F., Segel, L. A. 1970b. *J. Theoret. Biol.* 26:399–415
54. Konijn, T. M. 1961. *Cell aggregation in* Nictyostelium discoideum. PhD thesis. Univ. of Wisconsin, Madison
55. Konijn, T. M. 1965. *Develop. Biol.* 12:487–97
56. Konijn, T. M. 1969. *J. Bacteriol.* 99:503–9
57. Konijn, T. M. 1970. Symposium *Int. Congr. Microbiol., Mexico City, 10th* In press
58. Konijn, T. M., Barkley, D. S., Chang, Y. Y., Bonner, J. T. 1968. *Am. Nat.* 102:225–33
59. Konijn, T. M., Chang, Y. Y., Bonner, J. T. 1969. *Nature (London)* 224:1211–12
60. Konijn, T. M., van de Meene, J. G. C., Bonner, J. T., Barkley, D. S., 1967. *Proc. Nat. Acad. Sci. USA* 58:1152–54
61. Konijn, T. M., van de Meene, J. G. C., Chang, Y. Y., Barkley, D. S., Bonner, J. T. 1969. *J. Bacteriol.* 99:510–12
62. Kostellow, A. B. 1956. *Developmental responses of* Dictyostelium discoideum *to some amino acids and their analogues.* PhD thesis. Columbia University, New York
63. Loomis, W. F., Jr. 1969. *J. Bacteriol.* 100:417–22
64. Loomis, W. F., Jr., Ashworth, J. M. 1969. *J. Gen. Microbiol.* 53:181–96
65. Maeda, Y. 1970. (summary in *Cell. Slime Mold Bull.* No. 7, 4)
66. Maeda, Y., Takeuchi, I. 1969. *Develop. Growth Differentiation* 11: 232–45
67. Mitchell, J. L. A. 1966. *The Effects of Certain Amino Acid Analogs on the Morphogenesis of* Dictyostelium discoideum. Senior thesis, Oberlin College, Ohio
68. Moscona, A. 1968. *Develop. Biol.* 18:250–77
69. Newell, P. C., Sussman, M. 1970. *J. Mol. Biol.* 49:627–37
70. Pastan, I., Perlman, R. L. 1970. *Science* 169:339–44
71. Pontecorvo, G. 1958. *Trends in Genetic Analysis.* Columbia Univ. Press
72. Raper, K. B. 1940. *J. Elisha Mitchell Sci. Soc.* 56:241–82
73. Raper, K. B. 1956. *Mycolgia* 48:169–205
74. Raper, K. B. 1963. *Harvey Lect.* 57: 111–41
75. Rasmussen, H., Tenenhouse, A. 1968. *Proc. Nat. Acad. Sci. USA* 59: 1364–70
76. Roberts, A. B. 1970. *Enzymatic secretions of the cellular slime mold amoebae,* Dictyostelium discoideum

and Polysphondylium violaceum. Senior thesis, Princeton University
77. Robertson, A. D. J. 1971. *Proc. Am. Math. Soc.* In press
78. Runyon, E. H. 1942. *Collecting Net* 17:88
79. Sackin, M. J., Ashworth, J. M. 1969. *J. Gen. Microbiol.* 59:275–84
80. Samuel, E. W. 1961. *Develop. Biol.* 3:317–35
81. Shaffer, B. M. 1953. *Nature (London)* 171:975
82. Shaffer, B. M. 1956a. *Science* 123: 1172–73
83. Shaffer, B. M. 1956b. *J. Exp. Biol.* 33:645–57
84. Shaffer, B. M. 1957a. *Am. Nat.* 91: 19–35
85. Shaffer, B. M. 1957b. *Quart. J. Microsc. Sci.* 98:377–92
86. Shaffer, B. M. 1961. *J. Exp. Biol.* 38:833–49
87. Shaffer, B. M. 1963. *Exp. Cell Res.* 31:432–35
88. Shaffer, B. M. 1964. In *Primitive Motile Systems in Cell Biology,* ed. R. D. Allen, N. Kamiya, 387–405. New York: Academic Press
89. Sinha, U., Ashworth, J. M. 1968. *Proc. Roy. Soc. London Ser. B* 173:531–40
90. Slifkin, M. K., Bonner, J. T. 1952. *Biol. Bull.* 102:273–77
91. Sonneborn, D. R., Sussman, M., Levine, L. 1964. *J. Bacteriol.* 87: 1321–29
92. Sonneborn, D. R., White, G. J., Sussman, M. 1963. *Develop. Biol.* 7:79–93
93. Steinberg, M. S. 1970. *J. Exp. Zool.* 173:395–434
94. Sussman, M. 1961. In *Growth in Living Systems,* ed. M. X. Zarrow. New York: Basic Books, Inc.
95. Sussman, M., Lee, F., Kerr, N. S. 1956. *Science* 123:1171–72
96. Sussman, M., Sussman, R. R. 1969. *Symp. Soc. Gen. Microbiol.* 19: 403–35
97. Takeuchi, I. 1963. *Develop. Biol.* 8: 1–26
98. Takeuchi, I. 1969. In *Nucleic Acid Metabolism, Cell Differentiation, and Cancer Growth,* ed. E. V. Cowdry, S. Seno, 297–304. New York: Pergamon Press
99. Watts, D. J., Ashworth, J. M. 1970. *Biochem. J.* 119:171–74
100. Wilson, E. O. 1970. In *Chemical Ecology.* ed. E. Sondheimer, J. B. Simeone, 133–55. New York: Academic Press
101. Wolpert, L. 1969. *J. Theoret. Biol.* 25:1–47
102. Wright, B. E. 1966a. *Science* 153: 830–37
103. Wright, B. E. 1966b. In *Development and Metabolic Control and Neoplasia,* ed. D. N. Ward, 296–316. Baltimore: The Williams & Wilkins Co.
104. Wright, B. E. 1968. In *Systems Theory in Biology,* ed. M. D. Mesarovic. Springer Verlag
105. Yamada, T., Yanagisawa, K., Ono, H. 1969. *Bot. Mag.* (Tokyo) 82: 171–79 (Transl. in *Cell. Slime Mold Bull.* No. 3, 6, 1970)
106. Yanagisawa, K., Yamada, T., Hashimoto, Y. 1970. In press. (Summary in *Cell. Slime Mold Bull.* No. 2, 3, 1969)
107. Yanagisawa, K., Yamada, T., Ono, H. 1969. *Z. Mag.* 78:277–86 (Transl. in *Cell. Slime Mold Bull.* No. 4, 9. 1970)
108. Beug, H., Gerisch, G. 1970. *Naturwissenschaften* 56:374
109. Beug, H., Gerisch, G., Kempff, S., Riedel, V., Cremer, G. 1970. *Exp. Cell. Res.* 63:147–58
110. Riedel, V., Gerisch, G. 1971. *Biochem. Biophys. Res. Commun.* 42:119–23

PROSTHECATE BACTERIA

1564

JEAN M. SCHMIDT

Department of Botany and Microbiology, Arizona State University, Tempe, Arizona

CONTENTS

INTRODUCTION

While most bacteria with rigid cell walls fall into precise morphological categories of spheres, rods, filaments, or spirals, there exist some with notable complications and variations of these relatively simple topographies. Staley (123) has put forward a serviceable definition for one kind of morphological complication, the prostheca: "a semi-rigid appendage extending from a procaryotic cell, with a diameter which is always smaller than that of the mature cell, and which is bounded by the cell wall." Included among the better known prosthecate microorganisms are bacteria of the genus *Caulobacter,* with their cellular stalks, and the budding bacteria *Hyphomicrobium* and *Rhodomicrobium,* which possess hyphae or filaments (126). There are several other genera which meet the prosthecate criteria, some of which have been characterized only recently. Also to be considered here are some marginal cases whose appearance may suggest the presence of prosthecae, but for which definitive evidence is lacking, and several examples of nonprosthecate bacterial appendages.

In attempting to categorize prosthecate bacteria from among the procaryons, two slight qualifications of Staley's definition (123) of prosthecae will be used here: we will assume the structure should contain murein (peptidoglycan) in its enclosing layers, as implied by the terms "semirigid" and "bounded by the cell wall"; secondly, the prostheca should be a structure originating from a unicellular procaryon, as opposed to the branching and occasionally septate filaments (hyphae) of the Actinomycetales (15, 75). Most of the branching hyphal structures and germination tubes of the actinomycetes are eliminated as prosthecae, since prosthecate structures always have a diameter less than that of the mature cell from which they originate. However, some sporophores of Actinoplanaceae appear to be thinner than the main mycelial threads (130, 138), and the second condition purports to exclude any such mycelial structures from further consideration here.

A lumping of bacteria that possess prosthecate appendages gives rise to a heterogeneous collection with respect to physiology, morphology, mode of reproduction, and genetic composition. Lest taxonomists take offense at this collection, it should be mentioned that no taxonomic significance is implied; it is intended as a convenient handle to compare examples of unusual morphological development for those interested in procaryotic differentiation and diversity.

Bacteria With Cellular Stalks

The biology of *Caulobacter* and *Asticcacaulis* has been treated extensively by Poindexter (99). Ultrastructural studies have demonstrated the prosthecate nature of the stalk (98, 99), the site of stalk growth and development (112), and the occurrence of a polar membranous organelle, distinct from a mesosome, which occurs in the developmental region at the base of the stalk (22, 98). The prosthecae of caulobacters differ from those of *Hyphomicrobium* in that the *Caulobacter* stalks have no physical role in the generation of daughter cells (128) and they appear to contain no deoxyribonucleic acid (DNA) (Schmidt, unpublished data). Initiation of stalk development in the *Caulobacter* swarmer necessarily occurs prior to cell division so that only stalked bacteria divide (99, 128). The morphogenesis of the caulobacters provides promising material for the study of procaryotic cellular differentiation and morphogenetic regulation (99, 112, 117), and mutants with defects in their stalk development have been reported (109, 110). An interesting feature of the prosthecae of the caulobacters (and also *Hyphomicrobium*) is that their lengths can be greatly increased by starving the bacteria for inorganic phosphate (112). The biochemical basis for this striking morphological alteration has yet to be elucidated.

Prosthecae of caulobacters contain, interspersed at various intervals, bulkheads (92), also called "Querbalken" (58), and crossbands (14, 26, 98, 99). The bulkheads, which probably have an annular structure, extend across the stalks and appear to be attached to the inner layer of the outer unit membrane (92, 99). They appear to be very electron-dense in thin sections; their chemical composition has not been determined although it is possible they are composed of murein, for they appear as very rigid structures (92).

The genus *Asticcacaulis* was proposed by Poindexter (99) to include stalked bacteria whose stalk and flagellum originate from an excentral position rather than from the center of the pole of the rod-shaped cell. The adhesive holdfast of these bacteria is secreted at or near the center of the pole so its position does not coincide with the location of the stalk. Thus, the prostheca of *Asticcacaulis,* unlike that of *Caulobacter,* does not have adhesive properties, and Pate & Ordal (92) preferred the term "pseudostalk" to distinguish it from the adhesive stalks of *Caulobacter.* Ultrastructurally, the prosthecae of these two genera appear to have similar, if not identical, organization (92, 98).

Another striking feature of *Asticcacaulis excentricus* is the inequality in size of the daughter cells; the motile cell (swarmer) is smaller than the sister stalked cell (99). The daughter cells arise by asymmetric binary fission and without the intermediate tube formation between dividing cells that characterizes most budding bacteria (139).

The DNA composition of *Asticcacaulis excentricus* distinguishes it from *Caulobacter* spp.; *A. excentricus* has a mole percent G + C content of 55, lower than bacteria of the *Caulobacter* genus which show 62 to 67 (99).

Both DNA and RNA bacteriophages have been isolated for *Caulobacter* (64, 111, 129). There appears to be an absolute generic host range specificity among the DNA phages of *Caulobacter* and of *Asticcacaulis* (99, 111). The RNA phages of *Caulobacter* adsorb to polar pili of the swarmer bacteria, and stalked bacteria are usually devoid of pili (113). This phenomenon has become the basis for a differentiation assay, utilizing RNA phage adsorption (117). The occurrence of RNA phage-specific pili suggested that conjugation should be sought (and such a possibility was also suggested by Nemec & Bystrický, 82), but results of conjugation studies have been disappointing. Neither has transduction proved a very useful genetic transfer method in *C. crescentus* (28, 107, 141), although some temperate phages have been found (29).

BUDDING BACTERIA

Of the bacteria that reproduce by budding, not all are obviously prosthecate, and it appears that even definitions of budding in bacteria are not agreed upon, for there are organisms which do not fit either those based on general or specialized characteristics. Buds are not considered to be prosthecae (123). A few of the budding bacteria, *Hyphomicrobium* and *Rhodomicrobium* (23), *Pedomicrobium* (52), and *Planctomyces* (*Blastocaulis*) (49, 55, 119) produce structures that are readily recognized as prosthecae. Others, including *Pasteuria* (*Blastobacter*) (49, 143), *Rhodopseudomonas palustris* and *R. viridis* (139), *R. acidophila* (97), *Nitrobacter winogradskii* (143), and *Arthrobacter atrocyaneus* (125), bud without production of distinctive prosthecae although a "tube formation" may be involved in their division process. *Metallogenium* (143) and *Seliberia* (5) have been described as budding bacteria, but are incompletely characterized. Pfennig (95) has suggested that the manner of branching of the green sulfur bacterium *Pelodictyon* suggests budding. *Geodermatophilus* (59) and *Ancalomicrobium* (123) are prosthecate budding bacteria. Budding bacteria have been reviewed by Zavarzin (143) and Starr & Skerman (126).

The term, budding, was applied by Starr & Skerman (126) to unequal cell division in which two derivative cells with obvious inequality rather than morphological equivalence, are produced. However, there are some bacteria, notably *Caulobacter* and *Asticcacaulis* (99), which divide by transverse binary fission, but the fission is asymmetrical, where one daughter cell inherits the stalk of the parent, and the other is nonstalked, motile, and in *Asticcacaulis,* smaller that its sister stalked cell. The general criterion of inequivalence in morphology of daughter cells did not appear to describe budding adequately since it applied to caulobacters also, and Whittenbury & McLee (139) pointed out features which budding bacteria, characterized at the time, possessed "specialized systems of polar growth (in which) the new cell produces a tube of variable dimension, depending on the species but always narrower in width than the original cell, at the end of which can be seen the developing daughter cell." A complex membranous structure of parallel lamellae occurs along the periphery of the cell, probably arising by an infolding of the plasma membrane. The budding process is seen as a mechanism that avoids complex reorganization of the procaryotic cellular structure, particularly the membranous system. Controlled growth conditions, specifically the light intensities employed, have been shown to influence the development of the lamellate system of *Rhodomicrobium vannielii* (131).

Budding bacteria possessing the complex membranous structure include *Hyphomicrobium vulgare* and *Rhodomicrobium vannielii* (23), *Rhodopseudomonas palustris* and *R. viridis* (139), and *Nitrobacter winogradskii* (*N. agilis*) (139). A recently characterized budding purple nonsulfur bacterium, *Rhodopseudomonas acidophila,* also possesses the parallel lamellate system (97). In a few exceptional budding bacteria, *Ancalomicrobium,* and *Geodermatophilus* (2, 59), the lamellae are not present. Whittenbury & McLee (139) have suggested that there is a lack of evidence to indicate a fundamental difference between budding and binary fission. In budding, morphological inequivalence and tube formation between mother and daughter cell characterize most examples.

Hyphomicrobium and Rhodomicrobium.—In *Hyphomicrobium vulgare* and *Rhodomicrobium vannielii* (23) the buds are produced at the tips of hyphae or filaments. Ultrastructural studies showed the cytoplasm, cytoplasmic membrane, and cell wall to be continuous between the hyphae and bacterial cells. Cell wall analyses of *Hyphomicrobium* (61) showed a composition typical of Gram-negative bacteria; it is not known if the hyphal wall composition differs from that of the cell wall. Relative to the occurrence of their extensive intracellular membrane systems, phospholipid analyses of *Hyphomicrobium* (41, 46) revealed the presence of lecithin, which is rarely found in bacterial membranes, other than in photosynthetic bacteria. Lecithin is also present in *Rhodomicrobium* (91).

Consistent with the involvement of hyphae in reproduction, some DNA is found to be located in buds and hyphal tips, but not ordinarily in the hyphal bases (54). The hyphae usually grow out from their bases, suggesting that at least one genome is present in the distal tip of the hypha as it is pushed away from the mother cell. Since the same hypha of a mother cell can repeatedly produce buds at its tip, DNA replication must take place in the hyphal tip or developing bud, or, much less probably, DNA might travel down the hypha to the bud formation site during subsequently generations. Hirsch and co-workers (50, 51, 53, 55) have done extensive studies on the isolation, nutrition, and ecology of many of the budding bacteria, particularly *Hyphomicrobium* and *Rhodomicrobium.*

An anaerobic photoheterotroph similar to *Rhodomicrobium vannielii,* but with an unusual feature, was reported by Gorlenko (44). In addition to normal ovoid cells, there were refractile, heat-resistant, angular cells present. Their refractility was not due to the presence of poly β hydroxybutyrate which occurs commonly in the normal ovoid cells. The angular cells were referred to as "spores," although they are not analogous to bacterial endospores. Whether this organism is identical to *R. vannielii* (32) and its "spores" are abnormal growth forms, remains to be determined.

Pedomicrobium.—This budding, hyphal organism which deposits iron or manganese is closely related to bacteria of the genus *Hyphomicrobium.* Its establishment into a separate genus, *Pedomicrobium,* has gained some support (52). *Pedomicrobium* was described by Aristovskaya (4) as a new group of iron- and manganese-oxidizing bacteria, similar in morphology to *Rhodomicrobium vannielii.* Hirsch (52) isolated several strains of iron-depositing "hyphomicrobia," tentatively designated *Pedomicrobium,* which develop hyphae from several sites on the rod-shaped cell. This growth pattern was not observed in a large number of *Hyphomicrobium* strains. Ferric hydroxide was deposited at localized sites along the hyphae and on mother cells. The deposits were much heavier on older cells. Tyler & Marshall (133–135) have concluded that *Hyphomicrobium* strains can oxidize manganese and exhibit the pleomorphy that is supposed to set the genus *Pedomicrobium* apart; they have suggested that *Pedomicrobium* may not be a valid genus.

Planctomyces/Blastocaulis and Pasteuria/Blastobacter.—As of the present, no ultrastructural studies on these budding bacteria are available. Observations using phase contrast microscopy have revealed the presence of stalks. While the stalks of some other rare microorganisms, *Nevskia* and *Siderophacus,* are thought to be secreted material (gum in the case of *Nevskia* and ferric hydroxide in *Siderophacus*) (53, 119, 123), stalks observed in *Planctomyces/Blastocaulis* and in *Pasteuria* appear to be cellular, and so these latter two microorganisms probably can be considered prosthecate. The microorganism named *Blastocaulis sphaerica* by Henrici & Johnson (49) had long slender stalks with a holdfast at the distal end, usually attached to a substrate. It bears some resemblance to *Hyphomicrobium* (126). The same organism had been described earlier by Gimesi in 1924, under the name *Planctomyces* (123).

In the genus *Pasteuria* as described by Henrici & Johnson (49, 119), the cells are spherical or more often pear-shaped, multiplication taking place by budding from the pole opposite the holdfast, or by longitudinal fission. Stalks are very short or absent, and motile forms are present. Zavarzin (143) did not concur with the findings of Henrici & Johnson (49) that their *Pasteuria* was the same as the organism originally described as *P. ramosa* in Metchnikov's study of parasites of Daphnia; he proposed another scientific name, *Blastobacter henricii,* for the organism called *Pasteuria ramosa* by Henrici & Johnson. The nomenclature of these groups is presently somewhat confused, and not much is known about their prosthecate nature. Recently, Staley has undertaken the isolation of microorganisms of the *Pasteuria-Blastocaulis-Planctomyces* groups (124) so perhaps the biology of these budding bacteria may become more approachable.

Recently Recognized Prosthecate Bacteria

Prosthecomicrobium and Ancalomicrobium.—In a recent study of freshwater bacteria, Staley (123) focused particular attention on forms found in open water rather than "attaching" bacteria, with the discovery of some previously unrecognized procaryons with numerous cellular appendages. Nine morphologically distinct forms, most possessing apparent prosthecae, were found in direct electron microscopic preparations of concentrated water samples, and several strains were isolated. Two new genera, *Prosthecomicrobium* and *Ancalomicrobium,* were proposed to accommodate the isolated bacteria. Some of the forms (#3 and #8) observed by Staley (123) had also been found in electron microscopic examination of aqueous soil extracts and lake samples by Nitikin and co-workers (84–86, 127); they did not isolate the bacteria in question, and regarded them as protozoa. A procaryotic nature has been demonstrated for organisms of both the prosthecate genera proposed by Staley (123). Their multiple distinctive appendages are cellular. Two species have been proposed for the *Prosthecomicrobium* genus: *P. pneumaticum* divides by binary fission, is nonmotile, and contains numerous gas vacuoles in the form of spindle-shaped vesicles (100 × 300 nm). The prosthecae of *P. pneumaticum* are approximately as long as the body of the cell, on the order of 0.75 to 1.0 μm but are sometimes longer.

They taper in width from 0.25 μm at the base and end in a blunt tip. Many prosthecae emanate from diverse locations on the cell surface, giving it an irregular spiked outline. The occurrence of multiple prosthecae in *Prosthecomicrobium* sets it distinctly apart from *Caulobacter.* The second species, *P. enhydrum,* is motile by a single polar flagellum, does not contain gas vacuoles, and its prosthecae are shorter than those of *P. pneumaticum.* Its DNA content is 65.8 mole percent G+C, as compared to 69.4 and 69.9 for *P. pneumaticum* strains (123).

Ancalomicrobium adetum, the only species in the genus, is a nonmotile budding bacterium, although lacking a complex lamellate membrane system (123). The bud is formed on the mother cell and prosthecae differentiate from the bud before division is completed, with two to eight prosthecae per cell. The bud attains about the same size as the mother cell before division ensues. Prosthecae of *A. adetum* reach a length of 3 μm or about three times the length of the cell. No holdfasts are present, and bulkheads are not observed. Gas vacuoles are often present, predominantly in the cell body proper, although sometimes in the proximal region of the prosthecae. The physiological characteristics and isolation procedures for bacteria of these two prosthecate genera have been described by Staley (123).

Prosthecochloris.—This obligate photoautotroph was isolated from hydrogen sulfide-containing mud by Gorlenko (42, 45). *Prosthecochloris aestuarii* is a green sulfur bacterium, strictly anaerobic, and in its physiological characteristics has much in common with organisms of the genus *Chlorobium.* Its major pigment is bacteriochlorophyll c, and it can utilize sodium sulfide as an electron donor. The most characteristic feature, according to Gorlenko (42), is its cell morphology. There are approximately 20 appendages (prosthecae) per spherical cell surface. The cells are approximately 0.6 by 1.0 μm, and the diameter of the prosthecae, from 0.1 to 0.17 μm. The prosthecae are seldom longer than the diameter of the cell and are usually about 0.25 μm long (42, 45). In morphology, it is most similar to the nonphotosynthetic heterotroph, *Prosthecomicrobium.* Reproduction takes place by binary fission in several different planes. Sometimes cells do not part but form a chain branching in many different directions, due to the multiplicity of planes of division. Gorlenko & Zhilina (43) suggested from their ultrastructural studies that the prosthecae could possibly be formed as the organism divided. The constriction which occurs during cell division seems to bring about the formation of one prostheca on each daughter cell. The prosthecae of *P. aestuarii* appear to be quite different in structure and formation from the branches of another green sulfur bacterium, *Pelodictyon clathratiforme* (96).

Ultrastructural study of *P. aestuarii* (43) has demonstrated that the photosynthetic vesicles take up most of the volume of the prosthecae. The vesicles are ellipsoidal, 100 × 30 nm, enclosed by a membrane 30 Å in thickness, and are apparently connected to the cytoplasmic membrane on one side. There are usually three to five vesicles per prostheca, as seen in transverse section.

Geodermatophilus.—One genus among the actinomycetales, *Geodermatophilus,* cannot be overlooked in a survey of prosthecate bacteria, notwithstanding the limitations placed here in defining prosthecae. Luedemann (76) first characterized this new genus of Dermatophilaceae. A second study of a *Geodermatophilus* sp. (59) concentrated on ultrastructural and morphogenetic aspects, and defined two morphological forms: the C-form, coccoid cells occurring in irregular aggregates and dividing by fission, and the R-form, a motile rod that multiplies by budding. Both the C-form, during morphogenesis or generation of an R-form, and the R-form, during budding, produced cellular extensions termed stalks. These stalks are prosthecae. The stalk occurs between the C-form and the bud that becomes the R-form, and when the mature daughter R-form detaches after septum formation, the stalk is left behind with the parent cell. The stalks are also found between buds and mother R-cells, either as polar or subpolar outgrowths. *Geodermatophilus* is unusual as a budding bacterium in that it lacks the lamellate membrane system (2, 59) found in most budding bacteria (139).

Other prosthecate bacteria.—The following organism may not be a very impressive example of a prosthecate bacterium, since its prosthecae are of miniature size. *Helicoidal polyspheroides* is a rod 0.5 μm wide, and one to several μm long, covered with rounded protuberances 0.1 μm in diameter and arranged in helical rows. The cell wall and plasma membrane appear to be continuous around the cell's numerous appendages (21, 89, 90). A very similar organism has been discovered by Nitikin, Vasileva & Lokmacheva (86), in electron microscopic studies of soil microorganisms, and upon isolation and characterization, it was called *Agrobacterium polyspheroidum* (83). The organism is Gram-negative and aerobic.

Another intriguing creature, probably prosthecate, among Nitikin's soil microorganisms was designated *Tuberoidobacter* (83, 86). It is a Gram-negative rod with many "tubercular appendages, about 0.1 μ wide and of varying lengths, extending in all directions from the cell."

Other microorganisms which might possibly be prosthecate include the budding *Kuznetsovia polymorpha,* and *Caulococcus manganifer* (94). There is relatively little information available on these rare manganese- and iron-depositing soil organisms; they have been observed only in peloscopes.

Nonprosthecate Bacteria of Interest

The bifid form of *Lactobacillus bifidus,* which occurs when calcium ions are not available to the organism (65, 66), is an interesting example of bacterial pleomorphism. This is not prostheca formation, but a conditional aberration which occurs during severe physiological stress. Under conditions of calcium limitation, crosswalls in the bifid forms are not found (67), and the branches which are formed bear a superficial resemblance to prosthecae. When calcium ions are once again available, crosswall formation is reinitiated.

In *Leucothrix mucor,* bulb formation occurs from fusion of knots which form in the cellular filaments (18) or from enlargements of individual cells of the filaments (121). In one instance, a bulb was caught in the act of putting forth a prostheca-like extension, which was termed a germ tube (121). Its diameter was approximately 0.2 μm, as compared to a diameter of 2 μm for the bulb. Whether the germ tube has some significance in the morphogenesis of *L. mucor* is not known. In some procaryons with well-studied germination tube morphogenesis, the tubes develop into filaments or hyphae with the same diameter as mature hyphae in the culture. Accordingly, germ tubes cannot be considered prosthecae. A well-documented example of germ tube morphogenesis is available for *Bacterionema matruchotii,* an oral filamentous Gram-positive bacterium. Gilmour (38) showed with time-lapse photography that germ tubes arose from bacillus-like bodies or from branching filaments, and after some growth became identical to other filaments in the culture. The germ tubes of spores of certain actinomycetes undergo a similar morphogenesis (75, 138).

Gallionella.—The controversy concerning the nature of iron-containing ribbons of *Gallionella ferruginea* was discussed by Starr & Skerman (126). On the one hand, there is the widely held view that "the twisted ribbons are noncellular secretions of the bean-shaped cells" (126; see also 73, 100, 136, 137). However, more recent investigations (7, 9, 47) have tended to support Van Iterson's thesis (136) that the microfibrils or strands which make up the twisted ribbons are not inert secretions but possess an organic matrix and some reproductive capacity. The ribbons (also referred to as stalks) are not analogous to the prostheca of *Caulobacter* or *Hyphomicrobium.*

A method of obtaining pure cultures of *G. ferruginea* was discovered by Nunley & Kreig (87, 88) which makes use of the selective survival of *Gallionella* in the presence of 0.5 percent formalin. The availability of axenic cultures should be a great help in further characterization studies. Previous investigations were done with enrichment cultures (136, 137) and monocultures (6, 7) (in which only one type of *Gallionella* was present, but which were also inhabited by noniron-oxidizing bacteria). Hanert (47) reported studies of "pure" cultures of *G. ferruginea* in which at least chemoheterotrophic contaminants were absent.

Electron microscopic investigations of *Gallionella* ribbons have shown them to be composed of many thin ferruginous strands or microfibrils, with a diameter of approximately 0.1 μm (6–8, 47, 136, 137). Van Iterson (136), and later Balashova (and others, reviewed by Balashova, 7), observed small budding cells on the side of the ribbon fibrils, and sporangia-like membrane sacs associated with the ribbons. Hanert (47) was unable to find any terminal bacterial cells associated with the twisted fibrils in his electron microscopic studies of cultures free of pseudomonad contaminants. However, individual microcolonies from these cultures contained 60,000 to 80,000 reproductive units, leading to his conclusion that reproductive capacity must lie in some structure other than the bacterial cells, most likely the fibrils.

The only ultrastructural study of a *Gallionella* so far available has shed some light on the aforesaid problems. Balashova & Cherni (9) examined monocultures of *G. filamenta,* which has only six microfibrils per ribbon as compared to 40 or more for *G. ferruginea.* Micrographs of several transverse sections of *G. filamenta* ribbons demonstrated that the six microfibrils are tubular (80 to 110 nm in diameter) with an electron-dense core surrounded by an electron-transparent layer, and a dense boundary layer. Whether this outer dense layer had a membrane structure was not readily apparent, due to residual iron deposits, but it was suggested. The fibrils were seen to give rise to rounded bud-like structures (250 nm in diameter) which had a dense interior and a boundary layer in common with the fibril. The round structure was compared to elementary bodies of the mycoplasmas. Finally, in a thin section of a bacterial cell of the monoculture, an unusual cytoplasmic organelle with structure analogous to cross-sections of the six tubular fibrils was found, evidence that the ribbon fibrils do originate or are secreted from a bacterial cell. It is a strong possibility that both secretions of organic tubular fibrils from bacterial cells, and elementary body production on the fibrils may occur in *Gallionella.* Their microfibrils do not appear to be prosthecae; information on their chemical composition would be desirable, however.

Seliberia.—A soil microorganism, *Seliberia stellata,* has been compared to *Gallionella* and *Blastocaulis* (5) but the spiral filaments making up its rosettes appear to be large bacteria-like cells, 2.5–3 μm wide and several μm long. Rounded germinative cells are produced at the peripheral tips. The spiral filaments probably represent a cellular stage in the complex life cycle of *Seliberia,* rather than being comparable to the ribbons of *Gallionella* or prosthecae akin to those of *Hyphomicrobium.*

Metallogenium *and its probable relationship to the mycoplasmas.*—Initial descriptions of the manganese-oxidizing microorganisms *Metallogenium* (93, 94, 116, 144, 147, 148) suggested that it is related to budding bacteria of the Hyphomicrobiales. Growth forms of *Metallogenium personatum* were studied mainly in peloscopes (rectangular capillary tube-soil cultures) (94); the microcolonies were predominantly of the trichospherical form with numerus fine, bent radial branches. Individual cells were coccoidal, arranged in chains, and connected by fine filaments. The organism reproduced by budding (94). Another species, *Metallogenium symbioticum,* could be grown only in symbiotic culture with an unidentified fungus (143–145). Morphologically and in mode of reproduction it resembled *M. personatum.* Motility of the spherical cells was demonstrated. Zavarzin (146, 147) proposed the term *arais* for the extremely fine threads or filaments of *M. symbioticum;* the minimum diameter of the arais was 10 to 20 nm.

More recently, the morphological resemblance of *Metallogenium* to the mycoplasmas has been noted and substantiated (31, 119). Pure cultures were obtained, using culture media enriched with horse serum (30). Using negative contrast, electron microscopic methods, Dubinina (31) studied the morphology of *Metallogenium* grown under a variety of cultural conditions. It was found to be highly pleomorphic in the size and proportion of filaments (arais) and spherical cells. In older cultures, chains of spherical cells and membrane sacs containing many smaller cells were found, similar to phenomena observed in *Mycoplasma* (36, 63), and also in *Gallionella* (136).

The mycoplasmas are not considered to be prosthecate because they lack a cell wall. Their pliable cell envelope is a tripartite membrane structure and does not contain muramic acid or diaminopimelic acid, characteristic components of bacterial cell walls (35, 104, 118, 120). Thus, the filaments and filamentous appendages which are frequently observed in these organisms are not prosthecae. Most species of *Mycoplasma* are pleomorphic and produce filaments, polar or terminal blebs, filaments between dividing cells, and tubular membrane-bound extensions of the surface of apparently nondividing cells (3, 10–12, 27, 77, 78, 80, 81). In certain cases, the tubular extensions may be artifacts, occurring without fixation of cells due to hypotonic effects of the negative contrast, electron microscopic preparative technique (105). However, the occurrence of filaments as a normal stage in the life cycle has been established for many different *Mycoplasma* species (62, 103). Several reviews on morphology and reproduction of the mycoplasmas are available (36, 48, 104).

There are definite similarities in morphology and reproduction between *Metallogenium* and the mycoplasmas (31). More needs to be known concerning the ultrastructure, wall composition, and growth requirements of *Metallogenium* but, judging from the apparent plasticity and size variability of *Metallogenium,* a prosthecate nature for this organism is considered doubtful. Relatively few *Mycoplasma* species are saprophytic: only *M. laidlawii* and the recently described *Thermomycoplasma acidophila* (24). Neither of these species has been reported to oxidize manganese.

Krassilnikovae.—A new class of marine "microorganisms," Krassilnikovae, described as filaments with botryoid heads (71, 72) might have deserved some attention as possible prostheca-producers, but Sorokin (122) has cast serious doubt that these entities are microorganisms at all. Rather, they are thought to be parts of dismembered invertebrate marine animals "colloblasts, sticky cells of the raptorial tentacles of ctenophores." The unprotected glass collecting slides apparently came into destructive contact with the invertebrates as they were lowered and raised through sea zones rich in marine life.

Nonprosthecate Bacterial Appendages

Bacterial flagella and pili (fimbriae) are recognized examples of nonprosthecate appendages (16, 17, 34, 60, 74). However, there are a few unusual types of flagella and pili whose appearances deserves some comment here, because they bear some superficial resemblance to prosthecae.

Most pili (16) or fimbriae (33, 34) have very small diameters ranging from 40 to 85 Å. A few examples of pili with wider diameters, up to 250 Å, have been described (17). Moll & Ahrens (79) described a new type of fimbrium of *Agrobacterium,* 410 to 640 Å in width, occurring as peritrichous appendages, 1 to 3 μm long. These unusually wide fimbriae are flat ribbons or tubes, built up of a spirally twisted fiber, 100 Å in diameter. Their chemical composition has not been determined, but from available electron micrographs they do not appear to contain an extension of the cell wall as an enclosing layer.

Bacterial flagella would not be discussed in a survey of prosthecate appendages, were it not for the occurrence of sheathed flagella. Flagella are not prosthecae (123), and an examination of available ultrastructural studies on the sheathed flagella supports this view. Sheathed flagella have been reported in a Gram-positive bacterium, *Bacillus brevis* (25), (although, in a more recent study of another strain, no sheath was present, 1), and in several Gram-negative species, as summarized by Fuerst & Hayward (37). Some examples in which electron microscopic observation of thin sections has demonstrated or suggested a continuity between the outer cell envelope layer and the flagellum sheath include *Vibrio metchnikovii* (39), *Bdellovibrio bacteriovorus* (19, 20, 114, 115), and *Pseudomonas stizolobii* (37). In thin sections which traverse the site of flagellum attachment to the cell, the sheath of the flagellum is seen to originate from the outer component of the cell envelope. This is a tripartite structure of two electron-dense layers separated by a less dense layer, with a total diameter of about 75 Å (39, 114, 115). There is no ultrastructural evidence that a murein layer, which ordinarily lies under the outer envelope component and close to the cytoplasmic membrane in most Gram-negative bacteria (40), is associated with the flagellum sheath. Although biochemical analyses for muramic acid in these sheathed flagella have not been reported, it is very unlikely that a semirigid murein layer is present, since it would impair the flagellar motion due to its rigidity, and these organisms with sheathed flagella are actively motile.

Another filamentous bacterial process which is not prosthecate is the "degenerated" flagellum of *Azotobacter chroococcum* (70) which occurs as a function of nutrient depletion prior to encystment (13) or, in another case, possibly due to interaction with capsular polysaccharide (68).

Finally, there are the spore appendages of several types of *Clostridium* endospores (56, 69, 101, 102, 108, 142), and of some *Streptomyces* spore coats (72, 132, 138, 140). Little is known of the chemical composition of the endospore appendages (106), but in ultrastructural appearance they differ considerably from prosthecae.

Theories on Prostheca Function

While the presence of prosthecae produces some striking bacterial morphologies which can be readily recognized, the physiological and ecological contributions of prosthecae are somewhat more difficult to ascertain. The hyphae or tubes of the budding bacteria are involved in their reproduction process (126, 139), but prosthecae of nonbudding bacteria do not share this reproductive involvement. In the case of the manganese-depositing bacteria, *Hyphomicrobium* or *Pedomicrobium*, or both, Tyler & Marshall (134) have theorized that the mechanism of reproduction provides a means of temporary "escape" from the encasing manganese oxide precipitate and thus a perpetuation of the colony. Both older manganese-covered cells and prosthecae, and manganese-free daughter cells could be found in the microcolonies.

The presence of prosthecate cellular extensions greatly increases surface area, including that of the plasma membrane. This might enhance membrane-associated activities such as respiration (for the aerobic prosthecate bacteria) (92) and nutrient uptake (53). Prosthecate bacteria are frequently isolated from environments of very dilute nutrient concentrations. However, investigations in our laboratory, involving autoradiography and kinetic studies of nutrient uptake, have failed to demonstrate that *Caulobacter crescentus* stalked cells can take up nutrients more efficiently than sister populations of swarmers. The increase in surface area associated with the stalked (and often dividing) *Caulobacter* cells does correlate with an enhancement in oxygen uptake (141).

In *Prosthecochloris*, Gorlenko & Zhilina (43) have reported that the number of photosynthetic vesicles is greater by 30 to 60 per cell than could be contained in a cell without prosthecae. In this bacterium the vesicles occur only in the prosthecae which, in fact, make up most of the cell surface.

Poindexter (99) has pointed out the suspensory function of the caulobacter stalk. The stalked bacterium's tendency to float had been demonstrated in the centrifugal cell separation procedure (128). In nature, caulobacters, which are obligate aerobes, frequently inhabit an air-water interface, and Poindexter favored the theory that the stalk, as a flotation organelle, maintained the cells close to the surface.

The idea of the prostheca as an attachment organelle is no longer very popular (92, 99). An adhesive holdfast is present on the tip of the *Caulobacter* stalk, but not in *Asticcacaulis*, and holdfasts are not present at all in several free-floating prosthecate bacteria (123). The adhesive stalk of *Caulobacter* does not contribute to ectocommensalism since the holdfast is formed by the swarmers, and they attach while still in the swarmer cell stage (99).

Houwink (57, 58) suggested that the *Caulobacter* stalk is used to parasitize other microorganisms, since caulobacters are frequently observed attached to microorganisms in nature. However, Poindexter (99) showed that *Caulobacter* gains no advantage by their attachment, nor are the organisms to which they attach harmed by the association.

Nemec & Bystrický (82) proposed that rosette formation in *C. vibrioides* represents a conjugation phenomenon, with the stalks involved in the process. This was based on morphological observations only, without genetic evidence.

Since speculation on prostheca functions tends toward teleological considerations and there is no single outstanding function, this area of discussion must be left in an unresolved condition as a challenge to further investigation. It is, of course, possible that prosthecae may serve multiple functions, or perhaps none at all.

LITERATURE CITED

1. Abram, D., Vatter, A. E., Koffler, H. 1966. *J. Bacteriol.* 91:2045–68
2. Ahrens, R., Moll, G. 1971. *Arch. Mikrobiol.* 70:243–65
3. Anderson, D., Barile, M. F. 1965. *J. Bacteriol.* 90:180–92
4. Aristovskaya, T. V. 1961. *Dokl. Acad. Sci. USSR* (Engl. transl.) 136:111–14
5. Aristovskaya, T. V. 1964. *Microbiology, USSR* (Engl. transl.) 33: 823–28
6. Balashova, V. V. 1967. *Microbiology, USSR* (Engl. transl.) 36:541–44
7. Balashova, V. V. 1968. *Microbiology, USSR* (Engl. transl.) 37:590–98
8. Balashova, V. V. 1967. *Microbiology, USSR* (Engl. transl.) 36:879–81
9. Balashova, V. V., Cherni, N. E. 1970. *Microbiology, USSR* (Engl. transl.) 39:298–302
10. Bernstein-Ziv, R. 1969. *Can. J. Microbiol.* 15:1125–28
11. Biberfield, G., Biberfield, P. 1970. *J. Bacteriol.* 102:855–61
12. Boatmaan, E. S., Kenny, G. E. 1970. *J. Bacteriol.* 101:262–77
13. Boltyanskaya, E. V. 1967. *Microbiology, USSR* (Engl. transl.) 36: 82–86
14. Bowers, L. E., Weaver, R. H., Grula, E. A., Edwards, O. F. 1954. *J. Bacteriol.* 68:194–200
15. Bradley, S. G., Rizzi, D. 1968. *J. Bacteriol.* 95:2358–64
16. Brinton, C. C. 1965. *Trans. N.Y. Acad. Sci. Ser. II.* 27:1003–53
17. Brinton, C. C. 1967. Contributions of Pili to the Specificity of the Bacterial Surface. In *The Specificity of Cell Surfaces,* ed. B. D. Davis, L. Warren, 37–70. New Jersey: Prentice-Hall. 290 pp.
18. Brook, T. D. 1964. *Science* 144: 870–72
19. Burger, A., Drews, G., Ladwig, R. 1968. *Arch. Mikrobiol.* 61:261–79
20. Burnham, J. C., Hashimoto, T., Conti, S. F. 1968. *J. Bacteriol.* 96:1366–81
21. Bystricky, V. 1970. *Proc. Int. Congr. Electron Microscopy, 7th, Grenoble,* 379
22. Cohen-Bazire, G., Kunisawa, R., Poindexter, J. S. 1966. *J. Gen. Microbiol.* 42:301–8
23. Conti, S. F., Hirsch, P. 1965. *J. Bacteriol.* 89:503–12
24. Darland, G., Brock, T. D., Samsonoff, W., Conti, S. F. 1970. *Science* 170:1416–18
25. De Robertis, E., Franchi, C. M. 1951. *Exp. Cell Res.* 2:295–98
26. Desjardins, P. R. 1968. *Trans. Am. Microsc. Soc.* 87:392–94
27. Domermuth, C. H., Nielsen, M., Freundt, E. A., Birch-Andersen, A. 1964. *J. Bacteriol.* 88:1428–32
28. Driggers, L. J. 1970. *Bacteriophages and Genetic Transfer in* Caulobacter. MS thesis. Ariz. State Univ., Tempe. 41 pp.
29. Driggers, L. J., Schmidt, J. M. 1970. *J. Gen. Virol.* 6:421–27
30. Dubinina, G. A. 1969. *Dokl. Akad. Sci. USSR* 184:1433–37
31. Dubinina, G. A. 1970. *Z. Allg. Mikrobiol.* 10:309–20
32. Duchow, E., Douglas, H. C. 1949. *J. Bacteriol.* 58:409–16
33. Duguid, J. P., Smith, I. W., Dempster, G., Edmunds, P. N. 1955. *J. Pathol. Bacteriol.* 70:335–48
34. Duguid, J. P. 1968. *Arch. Immunol. Ther. Exp.* 16:173–80
35. Eaton, M. D. 1965. *Ann. Rev. Microbiol.* 19:379–406
36. Freundt, E. A. 1969. Cellular Morphology and Mode of Replication of the Mycoplasmas. In *The Mycoplasmatales and the L-Phase of Bacteria,* ed. L. Hayflick, New York: Appleton-Century-Crofts. 731 pp.
37. Fuerst, J. A., Hayward, A. C. 1969. *J. Gen. Microbiol.* 58:239–45
38. Gilmour, M. 1961. *Bacteriol Rev.* 25:142–51
39. Glauert, A. M., Kerridge, D., Horne, R. W. 1963. *J. Cell Biol.* 18: 327–36
40. Glauert, A. M., Thornley, M. J. 1969. *Ann. Rev. Microbiol.* 23: 159–98
41. Goldfine, H., Hagen, P.-O. 1968. *J. Bacteriol.* 95:367–75
42. Gorlenko, V. M. 1968. *Dokl. Akad. Sci. USSR* (Engl. transl.) 179: 195–98
43. Gorlenko, V. M., Zhilina, T. N. 1968. *Microbiology, USSR* (Engl. transl.) 37:892–97
44. Gorlenko, V. M. 1969. *Microbiology, USSR* (Engl. transl.) 38:106–11
45. Gorlenko, V. M. 1970. *Z. Allg. Mikrobiol.* 10:147–49

46. Hagen, P.-O., Goldfine, H., Williams, P. J. Le B. 1966. *Science* 151:1543–44
47. Hanert, H. 1968. *Arch. Mikrobiol.* 60:348–76
48. Hayflick, L., Chanock, R. M. 1965. *Bacteriol. Rev.* 29:185–221
49. Henrici, A. T., Johnson, D. E. 1935. *J. Bacteriol.* 30:61–93
50. Hirsch, P., Conti, S. F. 1964. *Arch. Mikrobiol.* 48:339–57
51. Hirsch, P., Conti, S. F. 1964. *Arch. Mikrobiol.* 48:358–67
52. Hirsch, P. 1968. *Arch. Mikrobiol.* 60:201–16
53. Hirsch, P. 1968. *Mitt. Int. Verein. Limmol.* 14:52–63
54. Hirsch, P., Jones, H. E. 1968. *Bacteriol. Proc.* 44
55. Hirsch, P., Rheinheimer, G. 1968. *Arch. Mikrobiol.* 62:289–306
56. Hodgkiss, W., Ordal, Z. J. 1966. *J. Bacteriol.* 91 :2031–36
57. Houwink, A. L. 1951. *Nature (London)* 168:654
58. Houwink, A. L. 1955. *Antonie van Leeuwenhoek J. Microbiol. Serol.* 21:49–64
59. Ishiguro, E. E., Wolfe, R. S. 1970. *J. Bacteriol.* 104:566–80
60. Jahn, T. L., Bovee, E. C. 1965. *Ann. Rev. Microbiol.* 19:21–58
61. Jones, H. E., Hirsch, P. 1968. *J. Bacteriol.* 96:1037–41
62. Kammer, G. M., Pollack, J. D., Klainer, A. S. 1970. *J. Bacteriol.* 104:499–502
63. Kang, K. S., Casida, L. E., Jr. 1967. *J. Bacteriol.* 93:1137–42
64. Khavina, E. S., Rautenstein, Y. I. 1963. *Dokl. Acad. Sci. USSR* (Engl. transl.) 153:1569–71
65. Kojima, M., Suda, S., Hotta, S., Hamada, K. 1968. *J. Bacteriol.* 95:710–11
66. Ibid 1970. 102:217–20
67. Kojima, M., Suda, S., Hotta, S., Hamada, K., Suganuma, A. 1970. *J. Bacteriol.* 104:1010–13
68. Konvicka, J. L., Wyss, O., Lucke, L. 1971. *Bacteriol. Proc.* 1
69. Krasil'nikov, N. A., Duda, V. I., Sokolov, A. A. 1964. *Microbiology USSR* (Engl. transl.) 33:404–10
70. Krasil'nikov, N. A., Boltyanskaya, E. V., Sokolov, A. A. 1966. *Microbiology USSR* (Engl. transl.) 35:252–56
71. Kriss, A. E., Mitzkevich, I. N. 1959. *J. Gen. Microbiol.* 20:1–12
72. Kriss, A. E. 1963. *Marine Microbiology (Deep Sea)* New York: Interscience. 536 pp.
73. Kucera, S., Wolfe, R. S. 1957. *J. Bacteriol.* 74:344–49
74. Kushner, D. J. 1969. *Bacteriol. Rev.* 33:302–45
75. Lechevalier, H. A., Lechevalier, M. P. 1967. *Ann. Rev. Microbiol.* 21: 71–100
76. Luedemann, G. M. 1968. *J. Bacteriol.* 96:1848–58
77. Maniloff, J., Morowitz, H. J., Barrnett, R. J. 1965. *J. Bacteriol.* 90: 193–204
78. Maniloff, J. 1969. *J. Bacteriol.* 100: 1402–8
79. Moll, G., Ahrens, R. 1970. *Arch. Mikrobiol.* 70:361–68
80. Morowitz, H. J., Maniloff, J. 1966. *J. Bacteriol.* 91:1638–44
81. Nelson, J. B., Lyons, M. J. 1965. *J. Bacteriol.* 90:1750–63
82. Nemec, P., Bystrický, V. 1963. *Biol. Zentralbl.* 82:595–600
83. Nitikin, D. I. 1970. *Proc. Int. Congr. Microbiol., 9th, Mexico City,* 23
84. Nitikin, D. I. 1964. *Sov. Soil Sci.* No. 6, June:636–41
85. Nitikin, D. I., Kuznetsov, S. I. 1967. *Microbiology, USSR* (Engl. transl.) 36:789–94
86. Nitikin, D. I., Vasileva, L. V., Lokmacheva, R. A. 1966 *New and Rare Forms of Soil Microorganisms.* Moscow: Science Publ. House (in Russian) 70 pp.
87. Nunley, J. W., Krieg, N. R. 1968. *Can. J. Microbiol.* 14:385–89
88. Nunley, J., Krieg, N. R. 1967. *Bacteriol Proc.* 47
89. Orenski, S. W., Bystrický, V., Maramorosch, K. 1966. *Nature (London)* 210:221
90. Orenski, S. W., Bystrický, V., Maramorosch, K. 1966. *Can. J. Microbiol.* 12:1291–92
91. Park, C.-E., Berger, L. R. 1967. *J. Bacteriol.* 93:221–29
92. Pate, J. L., Ordal, E. J. 1965. *J. Cell. Biol.* 27:133–50
93. Perfil'yev, B. V., Gabe, D. R. 1961. *Capillary Methods of Studying the Microorganisms.* Moscow: Science Publ. House (in Russian) 534 pp.
94. Perfil'yev, B. V., Gabe, D. R. 1965. The Use of the Microbial Landscape Method to Investigate Bacteria which concentrate Manganese and Iron. In *Applied Capillary Microscopy.* 9–54. Transl. F.

Sinclair. New York: Consultants Bureau. 122 pp.
95. Pfennig, N. 1967. *Ann. Rev. Microbiol.* 21:285–324
96. Pfennig, N., Cohen-Bazire, G. 1967. *Arch. Mikrobiol.* 59:226–36
97. Pfennig, N. 1969. *J. Bacteriol.* 99: 597–602
98. Poindexter, J. S., Cohen-Bazire, G. 1964. *J. Cell Biol.* 23:587–607
99. Poindexter, J. S. 1964. *Bacteriol Rev.* 28:231–95
100. Poindexter, J. S., Lewis, R. F. 1966. *Int. J. Sys. Bacteriol.* 16:377–82
101. Pope, L., Yolton, D. P., Rode, L. J. 1967. *J. Bacteriol.* 94:1206–15
102. Pope, L., Rode, L. J. 1969. *J. Bacteriol.* 100:994–1001
103. Razin, S., Cosenza, B. J. 1966. *J. Bacteriol.* 91:858–69
104. Razin, S. 1969. *Ann. Rev. Microbiol.* 23:317–56
105. Reuss, K. 1967. *J. Bacteriol.* 93: 490–92
106. Rode, L. J., Crawford, M. A., Williams, M. G. 1967. *J. Bacteriol.* 93:1160–73
107. Ruby, C. L. 1967. *Genetic Exchange in* Caulobacter crescentus. MS thesis. Indiana University, Bloomington, 39 pp.
108. Samsonoff, W. A., Hashimoto, T., Conti, S. F. 1970. *J. Bacteriol.* 101:1038–45
109. Schmidt, J. M. 1968. *J. Gen. Microbiol.* 53:291–98
110. Schmidt, J. M. 1969. *J. Bacteriol.* 98:816–17
111. Schmidt, J. M., Stanier, R. Y. 1965. *J. Gen. Microbiol.* 39:95–107
112. Schmidt, J. M., Stanier, R. Y. 1966. *J. Cell Biol.* 28:423–36
113. Schmidt, J. M. 1966. *J. Gen. Microbiol.* 45:347–53
114. Seidler, R. J., Starr, M. P. 1967. *Bacteriol Proc.* 42
115. Seidler, R. J., Starr, M. P. 1968. *J. Bacteriol.* 95:1952–55
116. Shapiro, N. I. 1965. The Chemical Composition of Deposits Formed by *Metallogenium* and *Siderococcus*. In *Applied Capillary Microscopy*. 82–87. Trans. F. Sinclair. New York: Consultants Bureau. 122 pp.
117. Shapiro, L., Agabian-Keshishian, N. 1970. *Proc. Nat. Acad. Sci. USA* 67:200–3
118. Sharp, J. T. 1963. *J. Bacteriol.* 86: 692–701
119. Skerman, V. B. D. 1967. *A Guide to the Identification of the Genera of Bacteria*. Baltimore: Williams & Wilkins. 2nd ed. 303 pp.
120. Smith, P. F. 1964. *Bacteriol. Rev.* 28:97–125
121. Snellen, J. E., Raj, H. D. 1970. *J. Bacteriol.* 101:240–49
122. Sorokin, Y. I. 1963. *Microbiology, USSR* (Engl. transl.) 32:362–71
123. Staley, J. T. 1968. *J. Bacteriol.* 95: 1921–42
124. Staley, J. T. 1970. *Proc. Int. Congr. Microbiol., 9th, Mexico City,* 6
125. Starr, M. P., Kuhn, D. A. 1962. *Arch. Mikrobiol.* 42:289–98
126. Starr, M. P., Skerman, V. B. D. 1965. *Ann. Rev. Microbiol.* 19: 407–54
127. Stefanov, S. B., Nitikin, D. I. 1965. *Microbiology, USSR* (Engl. transl.) 34:261–64
128. Stove, J. L., Stanier, R. Y. 1962. *Nature (London)* 196:1189–92
129. Szeyko, G. H., Gerencser, V. F. 1967. *Bacteriol. Proc.* 27
130. Thiemann, J. E., Beretta, G. 1968. *Arch. Mikrobiol.* 62:157–66
131. Trentini, W. C., Starr, M. P. 1967. *J. Bacteriol.* 93:1699–701
132 Tresner, H. D., Davies, M. C., Backus, E. J. 1961, *J. Bacteriol.* 81: 70–80
133. Tyler, P. A., Marshall, K. C. 1967. *J. Bacteriol.* 93:1132–36
134. Tyler, P. A., Marshall, K. C. 1967. *Arch. Mikrobiol.* 56:344–53
135. Tyler, P. A., Marshall, K. C. 1967. *Antonie van Leeuwenhoek J. Microbiol. Serol.* 23:171–83
136. Van Iterson, W. 1958. Gallionella ferruginea *Ehrenberg in a Different Light*. Amsterdam: Academisch Proefschrift, N. V., Noord-Hollandsche Uitgevers Maatschappij. 185 pp.
137. Vatter, A. E., Wolfe, R. S. 1956. *J. Bacteriol.* 72:248–52
138. Waksman, S. A. 1967. *The Actinomyces: A Summary of Current Knowledge*. New York: Ronald Press. 280 pp.
139. Whittenbury, R., McLee, A. G. 1967. *Arch. Mikrobiol.* 59:324–34
140. Wildermuth, A. 1970. *J. Bacteriol.* 101:318–22
141. Wong, H. 1971. *Survey for Genetic Transfer Systems and Studies on Stalk Function in* Caulobacter crescentus. M S thesis. Arizona State University, Tempe. 80 pp.

142. Yolton, D. P., Pope, L., Williams, M. G., Rode, L. J. 1968. *J. Bacteriol.* 95:231–38
143. Zavarzin, G. A. 1961. *Microbiology, USSR* (Engl. transl.) 30:774–91
144. Zavarzin, G. A. 1961. *Microbiology, USSR* (Engl. transl.) 30:343–45
145. Zavarzin, G. A., Epikhina, V. V. 1963. *Dokl. Acad. Sci. USSR* (Engl. transl.) 148:151–52
146. Zavarzin, G. A. 1963. *Microbiology, USSR* (Engl. transl.) 32:864–67
147. Zavarzin, G. A. 1964. *Z. Allg. Mikrobiol.* 4:390–95
148. Zavarzin, G. A. 1968. Bacteria in Relationship to Manganese Metabolism. In *The Ecology of Soil Bacteria.* Ed. T. R. G. Gray, D. Parkinson, 612–23. Toronto: U. of Toronto Press. 681 pp.

EFFECT OF HYPERBARIC OXYGEN ON MICROORGANISMS[1]

1565

Sheldon F. Gottlieb

Biological Sciences Division, Purdue University, Fort Wayne, Indiana

Contents

[1] The following abbreviations will be used: Ata (atmospheres absolute); AUP (air under pressure); ETP (effective therapeutic pressure of O_2); HO (hyperbaric oxygenation); INH (isonicotinic acid hydrazide); NAD (nicotinamide adenine dinucleotide); NADH (reduced NAD); OHP (oxygen under high pressure); OHPT (oxygen under high pressure therapy); ORS (oxygen-resistant strains); PAS (*p*-aminosalicylic acid).

INTRODUCTION

Oxygen, a drug, acts rather indiscriminately. Molecular O_2 is required for energy production by most forms of life and, especially among eukaryotes, vital synthesis, yet all forms of life are susceptible to O_2 toxicity. Partial pressures of oxygen (Po_2) initiating metabolic derangements in many instances are not much greater than the Po_2 at which the organism usually grows. Like other organisms, microbes vary in sensitivity to O_2; within a genus, species may differ in response to O_2 (27, 107, 111, 112, 159, 180, 188, 189, 224). That certain bacteria grow at the Po_2 of air at one atmosphere pressure (1 Ata) and the inhibition of growth by the same Po_2 has been known since Leeuwenhoek and has become a cardinal character in microbial taxonomy. Responsitivity to high oxygen tensions may differentiate species (106, 107, 112).

Since hyperbaric oxygenation as it pertains to microbes receives impetus from clinical applications, not surprisingly, most hyperbaric studies deal with in vitro and in vivo toxic effects of oxygen on microorganisms. This review focuses on recent developments of high O_2 tensions on microbes, and on pitfalls and areas of research of likely heuristic and practical value. The entire literature on the effects of O_2 on microbial metabolism has not been culled. Wimpenny (*277*) extensively reviewed O_2 as a regulator of growth and metabolism. Theories underlying the use of hyperbaric oxygenation and physiological, pathological, and biochemical aspects of O_2 toxicity, have been reviewed (103, 122).

Nomenclature

"Hyperbaric oxygen"—a common expression left in the title of this article—erroneously implies a special type of O_2. The issue is the physiological and biochemical effects of *increased* or *high* O_2 tensions; hyperbaric or pressure chambers are only means for attaining partial pressures of $O_2 > 1$ Ata (760 mm Hg). A $Po_2 > 1$ Ata usually denotes hyperbaric oxygenation (HO) or some other suitable synonym—O_2 under high pressure (OHP), high pressure O_2 (HPO), high O_2 drenching (HOD), high O_2 pressure (HOP), etc. Strictly, any gas mixture with a $Po_2 >$ air (159 mm Hg; 0.209 Ata) at 1 Ata may induce "hyperbaric oxygenation." (More strictly, any gas mixture with a Po_2 greater than that to which the organism is adapted may be considered hyperbaric oxygenation.) Hyperbaric oxygenation usually implies the use of increased atmospheric pressure, i.e., pressure > 1 Ata and is a confusing term as are its synonyms since investigators working with 2 Ata O_2 use this term with the same justification as those working with pressures > 3 Ata O_2. Hyperbaric oxygenation should be accompanied by the value of the specific pressure as well as exposure time. The physiological, pathological, and clinical sequelae may vary markedly both with the pressure and duration of exposure (103).

IN VITRO

General Considerations

Oxygen under high pressure, used in treating infections associated with anaerobic organisms, has had fair success. Can this success be matched with diseases associated with aerobic bacteria?—which directs much attention to in vitro and in vivo effects of OHP on growth and disease processes associated with organisms common in wounds or burns.

Factors Affecting Response to Oxygen Under High Pressure

Partial pressures of oxygen.—Since O_2, as noted, is a drug, its use can be viewed pharmacologically to study dose-response relationships in vitro and in vivo. Oxygen toxicity is time-pressure-dependent; dose-response versus time is very critical. Oxygen tolerance decreases with increasing pressure (103).

Gottlieb et al (112) introduced intermittent exposure, later used by others (28, 107, 111, 141, 228), to simulate in vitro a possible clinical regimen and to help define the time oxygen pressure relationships. A twice daily, 3-hr exposure to 2.87 Ata O_2 for 5 days markedly retarded the growth of *Mycobacterium tuberculosis,* whereas a twice daily 2-hr exposure to 1.87 Ata O_2 for 5 days slightly inhibited growth (112), which suggested a critical relationship between Po_2 and exposure time before damage by O_2 became manifest. The critical relationship between Po_2 and duration of exposure has been confirmed for other bacteria (28, 107, 111, 141).

Most in vitro studies involved growing microorganisms in nonagitated broth cultures; under such conditions oxygen concentration is poorly controlled. Depending upon surface:volume ratio, i.e., diffusion of O_2 from the overlying gas phase into the medium is the limiting factor in maintaining a constant O_2 tension throughout the vessel or tissue. The diffusion barrier makes it unlikely that full Po_2 ever prevails in the medium. Growth could initiate at the bottom of the tube where the Po_2 would be least and, by consumption of O_2 (whose replenishment by diffusion is slow) in the region around the growing cells, reduce the Po_2 thereby creating a favorable environment for further growth (106, 108, 137, 140, 160, 209, 254). Here, size of inoculum is important (106, 107, 112, 159, 224): a large inoculum could rapidly lower the Po_2 and thereby lessen or mask inhibition by O_2. Also, complex infusion broths can consume O_2 (158).

One way around diffusional limitation is the use of a fritted dispersion tube for delivering O_2 in small bubbles while agitating (106, 181, 182, 251). Also, diffusion can be speeded by increasing the surface:volume ratio by use of broader culture flasks (108, 200, 219), laying culture tubes nearly horizontal (111, 140), or using surface cultures (27, 106, 112, 189). Unlike non-agitated broth cultures, in surface cultures the organisms are exposed to full ambient Po_2. Such variations in technique of exposing organisms to O_2 helps account for different results with the same organism (8, 132, 151, 199, 200).

Increasing the overlying Po_2 enhances the growth of *Escherichia coli, Staphylococcus aureus,* and *Pseudomonas aeruginosa* (140, 198, 200, 209). Above 1.3 Ata, O_2 inhibits (198). Knowledge of the actual Po_2 in the medium during the growth cycle is essential in understanding how O_2 affects physiology. ZoBell & Hittle (290) suggested that "it is best to determine and report the dissolved O_2 in the medium rather than merely to report its partial or absolute pressure in the atmosphere." They reported initial dissolved O_2 concentration in the medium.

Continuous monitoring of the Po_2 in liquid media during growth in the presence of OHP has not been done although techniques are available for measuring oxygen during aerobic growth (80, 162, 176, 190). Polarographic techniques are readily adaptable for such purposes and have been used for microbial metabolic studies (26, 172) as well as for measuring diffusion rates of O_2 into sterile broth under hyperbaric conditions (141). Continuous culture could be used to study the effects of constant but elevated Po_2 on microbial growth and metabolism (277), especially since it is possible to control the dissolved O_2 concentration automatically (146, 176).

Oxidation-reduction potential.—In hyperbaric studies no attention has been given to the oxidation-reduction potential (E_h) of the medium, probably reflecting the lack of interest in it during the last 30 years. Redox potentials, the resultant of many factors, depend, among other things, on the composition and pH of the medium as well as degree of oxygenation. The redox potential may reflect the reducing intensities of substances liberated by microbial cells rather than any intrinsic characteristic of the cells. However, Wimpenny (277) states, "it seems likely, though not proven, that the extracellular components of the culture which generate the E_h figure are in equilibrium with the metabolic pool inside the cell, and that changes in external E_h are reflections of, or are reflected by, the redox state of this pool." Should the equilibrium relationship between components of the intracellular metabolic pool and extracellular medium with respect to E_h prove valid, then E_h measurements joined with simultaneous continuous culture studies involving nutrition (106), drug interactions (111, 112), and biophysical measurements (39, 190) may provide insights into mechanisms of oxygen toxicity, metabolic regulation, drug-O_2 synergisms (vide infra) and, by extension, chemotherapy.

Carbon dioxide.—The need for exogenous CO_2 for the growth of probably all bacteria is long known (99, 225, 277). Unfortunately, many of the in vitro OHP studies were done in the presence of pure O_2 at 1 or higher Ata pressures and thereby suffer from the absence of CO_2 (1, 8, 24, 25, 27, 28, 36, 88, 90–92, 132, 137, 141, 148, 149, 151, 163, 180, 188, 189, 196, 198–200, 219, 233, 255, 263, 279, 285, 286). Results with oxygen under high pressure are usually (in some cases wrongly) compared to those obtained in air at 1 Ata. Air at 1 Ata contains 0.03 Ata CO_2 (0.22 mm Hg) and 0.79 Ata N_2. Whether the absence of CO_2 affected the results is unknown. Some investigators, although omitting CO_2 from the gas mixture, claim that its absence may not have influenced their results (55, 224, 279, 285, 286); others (16, 111, 112, 159, 202, 274) have incorporated CO_2 into their experiments. That delays in the onset of growth, metabolic, and permeability changes, drug sensitivity changes, or other inhibitory effects, etc., may be due to the absence of CO_2, or the combined effects of CO_2 absence and an increased Po_2, may be inferred from the following: growth of an *Achromobacter* was markedly inhibited when pure oxygen was bubbled through at 1 Ata; inhibition was markedly lessened by the addition of 0.05 Ata CO_2 (106). Knox et al (159) showed that absence of CO_2 from the overlying O_2 atmosphere inhibited the growth of *M. tuberculosis,* thereby supporting Novy & Soule's (195) criticisms of Adams (1). Choe & Bertani (42) showed that in the absence of CO_2 *Serratia marcescens* and several strains of *E. coli* had increased osmotic fragility. Lack of a CO_2 effect in OHP experiments from which CO_2 was omitted may be explicable in part to local accumulation of CO_2 in nonagitated cultures or the CO_2-sparing by nutrients in complex media (4, 175, 184, 273) or both. There has been no systematic study of the effects of OHP on CO_2-sparing by specific nutrients.

Pressure.—Direct and indirect effects of pressure (volume changes, increased solution of cytoplasmic components, particle acceleration, pH changes, enzyme denaturation, etc.) must not be overlooked in studying effects of OHP. A detailed review of the effects of pressure exceeds the scope of this review. For detailed discussions on the general biological and microbiological effects of pressure see Johnson et al (144), Hedén (124), ZoBell (289), and Zimmerman (288). Because of the practical limitations imposed by the clinical applications, the upper pressure ranges in OHP, as it has developed to date, rarely exceeds 4 Ata—a pressure much smaller than the hundreds and thousands of atmospheres encountered in marine microbiology, sterilization, and experiments on survival under extreme conditions. Biological limitations do not preclude experiments using $P_{O_2} > 4$ Ata; such experiments have been done. At 1–4 Ata, pressure per se did not affect the growth of several species of protista (16, 24, 25, 27, 36, 88, 111, 112, 137, 140, 141, 202, 209, 224, 255, 263). Air under pressure (AUP) is the most commonly used pressure control to help differentiate between effects of oxygen and pressure per se. Air under pressure has one limitation as a pressure control: at 2, 3, or 4 Ata, etc., the corresponding P_{O_2} will be, respectively, 2, 3, or 4 times that in 1 Ata air. There also will be a corresponding increase in P_{CO_2}, P_{N_2}, etc. Since some organisms are inhibited and growth of others is stimulated (140, 198, 209) by these O_2 tensions, and since the increased P_{O_2} found in AUP may alter microbial responsitivity to antibiotics (28), the question of a proper pressure control is raised. Hydrostatic pressure would serve as the ideal control since it avoids all the problems related to gases. However, technically, it is more involved than using AUP. AUP can be used as a pressure control only if it can be shown that the phenomenon under study does not differ from the 1 Ata air control. The next most practical control would be to increase the total pressure by adding an "inert'" atmospheric gas to the 1 Ata air in the chamber. Under such conditions the P_{O_2} remains at 0.2 Ata although the total pressure is raised. Because of its relative biological (103) and microbiological (78) inertness, helium would be the best gas.

ZoBell & Hittle (290) have shown that the effects of oxygen on bacterial growth are enhanced by increased hydrostatic pressure and that in a closed system *E. coli, Bacillus subtilis,* and *Bacillus megaterium* grew as well at 100 Ata as at 1 Ata in media in which the initial dissolved O_2 content was 7.0 μg/ml (0.2 Ata O_2). These same 3 organisms also grow well in the presence of 35 μg O_2/ml (1.0 Ata O_2). When as little as 4 Ata hydrostatic pressure was superimposed on the media containing 35 μg O_2/ml, all 3 species were killed. Similar results were also obtained for several Gram-negative, facultative aerobes from marine environments.

Air under pressure, theoretically, can be used for studying the independent effects of P_{O_2} and its interactions with increased pressure, assuming the increased P_{N_2} does not exert adverse effects. For example, 5 Ata air has the same P_{O_2} as 1 Ata O_2; any organism sensitive to 1 Ata O_2 should show identical sensitivity at 5 Ata air. Differences in observed sensitivity to O_2 at 5 Ata air from 1 Ata O_2 may be due to the superimposed effects of pressure, the inert gas or, less likely, absence of N_2 from the 1 Ata O_2.

Temperature.—Thaysen (259) showed that increasing the temperature of incubation above the normal optimum in the presence of 10 Ata O_2 was lethal; all vegetative cells (*S. aureus, Bacillus mesentericus, E. coli, Streptococcus lactis*) were destroyed within 5 hr. The spores of *B. mesentericus* were not killed. Inhibition by temperature and oxygen on the growth of bacteria and protozoa was confirmed (45, 101, 107, 173, 280) despite the decreased solubility of O_2 at the higher temperatures.

Hydrogen ion potential.—Oxygen inhibition of growth of an *Achromobacter* was not influenced by pH when 1-amino-2-propanol was used as a substrate (106). Nutritional mitigation of O_2 toxicity in *Vibrio cholera* was independent of pH (111). Oxygen inhibition of pyruvate utilization by *S. aureus* was less sensitive to changes in pH than was the methylene blue inhibition of the same reaction (282). There has been no other systematic study of the effects of pH on O_2 inhibition of growth of protista with a wide range of substrates. Various acidic (110) and alkaline (213) solutions protected animals against oxygen-induced convulsions.

Genetics, adaptations.—Fenn et al (77) demonstrated that a 16-hr exposure to 6–10 Ata O_2 doubled the mutation rate of a streptomycin-dependent *E. coli* while pyruvate prevented this mutagenicity. Mutagenicity of OHP for *E. coli* was confirmed (90). Oxygen is teratogenic (103).

Mechanisms for mutagenicity and teratogenicity are unknown. Chalkley & Voegtlin (37) demonstrated decreased nuclear growth and fission of *Amoeba proteus* on increasing the P_{O_2} for 24 hr and attributed the adverse effects of O_2 to the nuclear redox potential being upset by interference with sulfhydryls. Chromosomal breakage in *Tradecantia paludosa* increased with increasing P_{O_2} (49). Oxygen under high pressure increased the viscosity of DNA even in the presence of reducing compounds (96).

Irvin et al (141) could not demonstrate differences in mannitol fermentation or pigment production of *S. aureus* exposed to 4 Ata O_2 for 120 hr as compared to air controls, nor could they isolate oxygen-resistant strains of *S. aureus.* Using adaptation techniques, I was unable to isolate an oxygen-resistant strain of *Achromobacter* P6. In a separate study of *Achromobacter* P6, oxygen-grown cells had a higher respiration rate under oxygen than air-grown cells (106). In contrast to the unsuccessful attempts with bacteria, Wagner & Welch (267) isolated an ORS of *Chlorella sorokiniana.* This adaptation to 1 Ata O_2 was irreversible and independent of photosynthesis. However, frequency of cell division was markedly different between ORS and O_2-sensitive strains. Tolerance to oxygen was not absolute (222) since growth of the ORS *C. sorokiniana* decreased linearly as the Po_2 increased from 1 to 2 Ata. Two Ata of oxygen were not algicidal. The nature of the adaptation process remains obscure. Oxygen-resistant strains of any organism (uni- or multicellular) should provide a tool for elucidating mechanisms of O_2 toxicity. Similar advantages would accrue from comparative studies of the effect of O_2 on O_2-sensitive strains versus normal or resistant strains of the same species. An O_2-sensitive strain of *P. aeruginosa* is known (27, 28).

Nutrition.—Responsitivity of an organism to oxygen depends in part on the growth medium (22, 59, 82, 88, 92, 106, 107, 111, 163, 173, 255, 271, 281); nutrients may mitigate the toxicity of O_2, varying in effect from one organism to another. As a nutritional approach to the mechanism of O_2 toxicity, Gottlieb (106) found oxygen inhibition of growth of *Achromobacter* P6 when 1-amino-2-propanol, acetate, lactate, citrate, or glucose served as sole substrate; inhibition was absent with succinate, fumarate, malate, or glutamate. The succinate data parallel the protective effects of this compound in mammals (229).

Yeasts survived exposure to 10 Ata O_2 better with ethanol as the carbon source rather than glucose (92); yet, Stuart et al (255) reported that glucose decreased potassium leakage from bakers' yeast exposed to 10–143 Ata O_2. Glucose, fructose, or sucrose mitigated O_2 inhibition of growth of *V. cholera* (111). Glucose as well as high phosphate concentrations protected *Euglena gracilis* against 0.95 Ata O_2 (22). Pyruvate, but not glucose, lactate, or ethanol increased the survival time of *Paramecium caudatum* exposed to 9 Ata O_2 (281). Monosaccharides and thiourea protected lyophilized *E. coli* against the lethal effects of oxygen (169). Glycerol and thiourea enhanced viability of aerosolized *Serratia marcescens* (127). The addition of nutrient broth, replaceable by a vitamin-free casein hydrolysate or an amino acid mixture, decreased O_2 inhibition of growth of *Achromobacter* P6 (106). The oxygen-protective effects of the nutrients for bacteria are not due to reducing substances (65, 106, 173), some of which may be toxic even in the presence of O_2 (106, 169). Sulfhydryl compounds protected *P. caudatum* against O_2 toxicity (281). Fletcher & Plastridge (82) found that growth on a yeast-extract agar doubled the tolerance of microaerophilic vibrios to oxygen as compared to growth on a defined medium. The addition of yeast extract or brain heart infusion to the growth medium of aerobic vibrios in the presence of oxygen tensions that are inhibitory to these organisms (2.87 Ata) permitted growth equaling that in 1 Ata air (111). Fresh complex biological media are never in equilibrium with the atmosphere since they keep absorbing O_2 at a rate proportional to their reducing potentials (158).

Cations may profoundly affect responsitivity to oxygen (59, 88, 127, 163, 169, 271, 281). Dedic & Koch (59) grew *Clostridium tetani* aerobically by adding cobalt to Difco "micro-inoculum broth." Gerschman et al (88) also reported that Co^{++} as well as Mn^{++} increased the survival times of *P. caudatum* exposed to 9 Ata O_2 but decreased them at 2 Ata O_2. Wittner (281) confirmed protection by Co^{++} and Mn^{++} for *P. caudatum* at 9 Ata O_2 and extended the observations to include Mg^{++}; Fe^{++}, Zn^{++}, and Ni^{++} were not protective. Manganese^{++} and Co^{++} enhanced the viability of aerosolized *S. marcescens* (127). Neither Mn^{++}, Co^{++}, Mg^{++}, nor Ca^{++} protected *Micromonospora vulgaris* against the 1 Ata O_2 inhibition of respiration (271). *Pseudomonas fluorescens* grown at a high partial oxygen pressure (but < 1 Ata) had an increased Mo^{++} requirement compared to cells grown at a lower Po_2 (163). Sodium chloride, NaBr, and $NaNO_3$ were more protective for lyophilized *E. coli* than the corresponding K^+ salts (169); LiCl and $MgCl_2$ were protective; $CaCl_2$, NH_4Cl, RbCl were less so. Anions markedly affected the results: protection by Mg^{++} and Na^+ disappeared when they were given as sulfates or phosphates. Sodium iodide and KI were highly protective as were $NaNO_2$, KNO_2, NaSCN, and KSCN.

Oxygen as an Antimicrobial Agent

Because of the peculiarities of the many different organisms used, it may be profitable to examine oxygen inhibitions of specific microbes, then review interactions of O_2 with other antimicrobial agents and, finally, examine in vitro mechanisms of O_2 toxicity. Table 1 lists some organisms used in experiments involving O_2; it is not meant to be exhaustive as to organisms or references.

Mycobacterium.—Gottlieb (105), in reviewing early in vitro studies of the effects of oxygen on the growth of mycobacteria, proposed OHP for treating tuberculosis and Hansen's disease. Subsequent detailed study (112) indicated that intermittent exposures of the tubercle bacillus to 2.87 Ata O_2 retarded growth. The lag phase of drug-resistant strains of *M. tuberculosis* as well as Battey-type and scotochromogenic organisms was markedly prolonged by 76 hr continuous exposure to 2.87 Ata O_2. Continuous exposure to 1 Ata O_2 killed *M. tuberculosis;* growth inhibition by oxygen was clearer with the slower growing strains, the order of susceptibility being BCG > *M. tuberculosis* > chromogenic acid-fast bacilli > saprophytic mycobacteria (159).

TABLE 1. Organisms and Oxygen

	References
Achromobacter	
sp.	72, 106, 166
fischeri	123
Actinomyces	
sp.	1
israeli	177
Aerobacter	
aerogenes	27, 152, 173, 190, 215, 274, 279
Alkaligenes	
faecalis	18
Amoebae	
Pleomyxa carolinensis	101
Amoeba proteus	37
Aphanomyces	
euteiches	55
Aspergillus	
fumigatus	180
niger	36
ochraceus	173
wentii	36
Azotobacter	
sp.	66
chroococcum	54
vinelandii	55, 64–66
Bacillus	
anthracis	189, 274
cereus	27
megaterium	189, 274, 290
mesentericus	173, 259
mycoides	274
niger	274
stearothermophilus	68, 69
subtilis	66, 91, 127, 148, 164, 173, 274, 279, 290
Bacteroides	
sp.	196
Candida	
albicans	180, 189
utilis	92, 150, 217
Chilomonas	
paramecium	202
Chlorella	
pyrenoidosa	267
sorokiniana	222
Chromatium	
sp.	134
Chromobacterium	
orangium	219
violaceum	219
Clostridium	
sp.	145, 147
botulinum	84
butyricum	85
cryptocercus	45
histolyticum	85
novyi	133
perfringens	5, 11, 23, 30, 31, 33, 34, 35, 48, 83, 85, 98, 100, 121, 128, 130, 133, 136, 138, 151, 153, 154, 156, 179, 180, 185, 192, 194, 196, 203, 207, 210, 238, 248, 249, 252, 253, 262, 265, 268
septicum	58
sporogenes	85
tetani	3, 85, 86, 153, 170, 171, 187, 208, 220, 223, 271, 278
Corynebacterium	
diphtheriae	189, 279
Cylindrocarpon	
radicicola	266
Diplococcus	
pneumoniae	16, 148, 151, 152, 189, 228
Escherichia	
coli	27, 42, 75, 77, 81, 90, 91, 125, 140, 152, 157, 167, 168, 169, 173, 174, 180, 188, 189, 199, 200, 209, 254, 257–259, 274, 279, 290

TABLE 1. *(Continued)*

	References		References
Eremothecium		*Pasteurella*	
ashbyii	17	*pestis*	1, 189
Euglena		*tularensis*	235
gracilis	22	*Penicillium*	
Flavobacterium		*cyclopium*	36
arborescens	219	*digitatum*	36
Fusarium		*expansum*	173
culmorum	266	*Peptococcus*	11
solani	224	*Periconia*	
Gaffkya		sp.	266
tetragena	27	*Phoma*	
Haemophilus		*violacea*	36
parainfluenzae	245	*Phycomycetes*	5, 217, 224, 266
Klebsiella			
pneumoniae	71–73, 148	*Proteus*	
Lactobacillus		sp.	132, 180, 209, 274
bulgaricus	227		
casei	173, 174	*mirabilis*	151, 152
Malleomyces		*vulgaris*	148, 173, 189, 250, 274, 279
mallei	189		
Merulius		*Protozoa*	
lacrymans	36	sp.	43, 44
Micrococcus		*Pseudomonas*	
dentrificans	214	sp.	199, 287
flavus	173	*aeruginosa*	9, 18, 25, 26, 28, 115, 132, 137, 149, 152, 180, 188, 189, 219, 274, 279
niger	85		
Micromonospora			
vulgaris	271		
Mycetoma	1		
Mycobacterium		*fluorescens*	163, 173, 189, 274
"atypical"	112		
bovis	40	*perfectomarinus*	290
phlei	159, 173	*saccharophila*	285, 286
photochromogenic	112	*Rhizobium*	
scotochromogenic	112	*trifolii*	66
smegmatis	159	*Ristella*	
stercoris	159	*fragilia*	85
tuberculosis	1, 12, 112, 159, 160, 188, 189, 195	*Saccharomyces*	
		capsularis	17
		cerevisiae	2, 66, 82, 173
Neisseria		*Salmonella*	
catarrhalis	104, 107	sp.	111
meningitidis	103, 104, 107	*enteritidis*	18, 189, 274
Papularia		*paratyphi*	18
arundines	266	*typhimurium*	151, 152, 188, 270, 276
Paramecium			
sp.	280	*typhosa*	18, 111, 148, 189
caudatum	88, 89, 281		

TABLE 1. *(Continued)*

	References		References
Sarcina		*epidermidis*	143, 279
sp.	173	*Streblomastix*	
aurantiaca	189	*strix*	64
lutea	27, 189	*Streptococcus*	
Schizosaccharomyces		*faecalis*	78, 180, 256
pombe	263	*hemolyticus*	5, 148
Serratia		*lactis*	173
fuchsina	219	*lactis acidi*	259
kiliensis	219	*viridans*	180
marcescens	17, 27, 42, 61, 62, 127, 173, 189, 193, 219, 274	*Streptomyces*	
		sp.	173
		Streptothrix	
		sp.	1
Shigella		*Tetrahymena*	
sp.	111	*pyriformis*	202
dysenteriae	18, 111, 188, 189	*Trichoderma*	
Staphylococcus		*viride*	266
sp.	132, 199	*Trichomomas*	
albus	1, 27, 148, 188, 189, 274	*termopsidis*	64
		Trichonympha	
aureus	1, 8, 14, 18, 24, 74, 81, 113, 115, 119, 125, 137, 139, 141, 148, 151, 152, 165, 180, 189, 209. 211, 233, 234, 236, 246, 249, 258, 259, 274, 279, 282	*campanula*	64
		Veillonella	
		sp.	85
		Vibrio	
		sp.	82, 111
		cholera	18, 188, 189
		Viruses	
		Coxsackie	201
		Friend virus	165
		influenza	230
citreus	1, 188, 189	mongirous	97

Gram-Positive Bacteria

Recorded effects of oxygen on Gram-positive organisms are few and do not yet fall into a clear pattern; most of the studies were yes-no as to inhibition by O_2. Detailed studies on responses of these organisms to O_2 are desirable.

Staphylococcus.—The observations of Moore & Williams (188, 189) on O_2 inhibition of growth of *S. aureus* have been confirmed (132, 148, 180, 199, 236, 259, 274); they reported that *S. aureus* was more susceptible to growth inhibition by O_2 than was *Staphylococcus albus* and that both were more susceptible than *Staphylococcus citreus*. Adams (1) did not confirm this relative O_2 susceptibility of these three species. Karsner (148) was unable to demonstrate an oxygen effect on pigment production by *S. aureus.* Hopkinson & Towers (132) found that 4 Ata O_2 did not kill *S. aureus.* One strain of *S. aureus,* after 6 exposures at 4 Ata O_2, grew well when incubated at 1 Ata air; but the colonies were smaller, less pigmented, and tended more toward Gram-negativity than controls. Unlike the wild type, L-forms of staphylococci did not grow after exposure to 4 Ata O_2 for 18 hr; there was no growth on reincubation or subculture nor tendency to revert to normal forms. Hemolysis by *S. aureus* was abolished after exposure to OHP (180). Growth and β-hemolysin activity increased in the range of 1–4 Ata O_2 compared to controls (8).

Pneumococcus.—Karsner et al (148) reported that 0.8–1.0 Ata O_2 did not inhibit surface growth. Bean (16) found that 3.5 Ata O_2 were bacteriostatic and 4.8 Ata O_2 bactericidal for surface cultures. Kaye (151) reported that *Diplococcus pneumoniae* type 1 grew in 3.0 Ata O_2 in broth (vide supra Po_2). In contrast, Ross & McAllister (228) found that a type III pneumococcus was inhibited by 2 Ata O_2 when administered either continuously or intermittently (3 periods of 2 hr with 2-hr intervals of air exposure in between) to surface cultures on horse blood agar.

Gram-Negative Bacteria

Azotobacter.—Parker & Scott (206) reviewed early work on O_2 inhibitions of N_2 fixation. They showed that 0.1–0.2 Ata O_2 competitively inhibited N_2 fixation in *A. vinelandii;* marked subculture variation was noted, and at $Po_2 > 0.3$ Ata marked inhibition of respiration and growth. Sensitivity of 3 N_2-fixing *Azotobacter* species to 0.2 Ata O_2 was confirmed (54, 55); in contrast to the marked oxygen sensitivity seen with these organisms in nitrogen-fixing situation, cells growing in the presence of fixed N_2 (NH_4^+) were not sensitive to 0.2 Ata O_2; growth inhibition was seen at 0.6 Ata O_2. Other effects of oxygen on *Azotobacter* are discussed under mechanisms of O_2 toxicity. Nitrogen-fixing cells of blue-green algae are O_2-sensitive (76).

Escherichia.—Early reports (1, 148, 189) stated that growth of *E. coli* was not inhibited by 1 Ata O_2; later reports described O_2 inhibition at partial pressures > 1.3 Ata (173, 174, 198–200, 259). Hopkinson & Towers (132), using surface cultures, showed that an 18-hr exposure to 2–4 Ata O_2 inhibited growth of wild-type and L-forms of *E. coli.* Two strains of *E. coli* were not recovered on subsequent reincubation in air. Unlike staphylococci, oxygen-exposed L-forms of *E. coli* subcultured into nutrient broth reverted to normal structure.

Proteus.—Neither Moore & Williams (189) nor Adams (1) demonstrated O_2 inhibition of growth of *P. vulgaris* at a $Po_2 \leq 1$ Ata. Karsner et al (148) found 1 Ata O_2 bactericidal to surface cultures of *P. vulgaris.* Oxygen inhibition of growth of *P. vulgaris* at 1 Ata was confirmed (132, 180, 274). *P. mirabilis* grew in broth culture exposed continuously for 24 hr to 3 Ata O_2 (151). Whether these observations reflect true species difference in response to O_2 or merely growth conditions (surface versus broth cultures; vide supra Po_2) is undetermined.

Pseudomonas.—Growth of neither *P. aeruginosa* nor *P. fluorescens* is inhibited by 1 Ata O_2 (1, 148, 189). Higher O_2 tensions are bacteriostatic (132, 173, 180, 274).

Salmonella-Shigella-Vibrio.—Growth of *Salmonella typhosa, Shigella dysenteriae,* and *Salmonella typhimurium* was not inhibited by 1 Ata O_2 (1, 111, 148, 189, 276); *S. typhimurium* under 1 Ata O_2 grew 5 times more than unaerated controls (276) and grew in broth cultures under 3 Ata O_2 (151). On surface culture, 0.87 Ata O_2 inhibited the growth of *Shigella flexneri* but not *S. typhosa* (111). Growth of the latter was inhibited by 1.87 Ata O_2. A 24-hr exposure to 2.87 Ata O_2 was bacteriostatic to representatives of the *Salmonella-Shigella* groups. Vibrios seem more sensitive to oxygen than other Gram-negative, aerobic enteric organisms (111).

Fungi

Studies on fungi have been mainly exploratory. Karsner & Saphir (149) reported that growth of common air molds was not inhibited by 1 Ata O_2; animal parasitic molds were inhibited at partial pressures > 0.75 Ata; inhibition increased with increasing Po_2; of the plant parasitic molds studied, only three were inhibited by O_2. Growth inhibitions by O_2 bore no fixed relationship to parasitism; there was no differential inhibition of pigment production. Oxygen was not cytocidal since organisms grew normally when reincubated in air. Oxygen inhibition of growth of fungi has been confirmed (239, 266, 275). Robb (224) studied growth patterns of 103 species of fungi (not listed in Table 1) at 10 Ata O_2. There was no correlation between taxonomy and physiological response to O_2; species variation was observed. The impairment of colony formation of bakers' yeast exposed to 10–143 Ata O_2 may relate to post-exposure K^+ leakage (255). One Ata O_2 suppresses aerial mycelial growth of *Micromonospora vulgaris* while bottom growth is either unaffected or stimulated (271).

Protozoa

Cleveland (43, 44) studied the time needed to kill free-living ciliates and flagellates as well as the intestinal protozoa of a wide range of hosts (*vide infra* in vivo).

Chilomonas paramecium grew well in 6.6×10^{-4} Ata O_2 with optimal growth at 0.1 Ata O_2; Po_2's > 0.66 Ata were cytocidal (202). In contrast, *Tetrahymena pyriformis* ("*geleii*") did not grow at 6.6×10^{-4} Ata O_2 but did in the range 0.013–1.0 Ata O_2 (202). This experiment is not a valid comparison between plant- and animal-like protozoans since *C. paramecium* was grown on a minimal, chemically defined medium, whereas *T. pyriformis* was grown in 2.0 percent proteose peptone. The peptone could be protecting *T. pyriformis* against oxygen by several mechanisms. Browning et al (29) found the optimal Po_2 range for reproduction of *T. pyriformis* in 1.0 percent proteose peptone to be 0.05–0.2 Ata. Up to 9 Ata O_2, the lethal time for *Paramecium caudatum* is inversely related to Po_2 (280) and can be mitigated nutritionally (281). Toxic effects of O_2 on *Paramecium* and *Tetrahymena* have been confirmed (95, 243).

Interaction of Oxygen With Other Antimicrobial Agents

Studies on mechanisms of O_2 toxicity may yield insight into selective enhancement of O_2 toxicity and so enhance its effectiveness for therapy of infections caused by aerobes (106). Studies were therefore undertaken to detect synergism between O_2 and several common drugs. Synergism was found between 2.87 Ata O_2 and either *p*-aminosalicylic acid (PAS), isonicotinic acid hydrazide (INH), or streptomycin (ST) as measured by the effects of each, alone or together with O_2, on in vitro growth of *M. tuberculosis;* synergism between O_2 and these drugs held even for drug-resistant strains and even with intermittent exposure to OHP (3 Ata, 3 hr, 2×/day) (112). Synergistic growth inhibition between ST and O_2 was also reported for *P. aeruginosa* (28) but others have not confirmed this (279). Schreiner (236) found that ST activity against *S. aureus* was enhanced by 3 Ata O_2 but analysis with penicillin suggested that the effects were additive not synergistic. The minimal inhibitory concentration (MIC) of ST was lowered as the time of exposure of *S. aureus* to 3 Ata O_2 increased (24).

Simultaneous in vitro exposure of *P. aeruginosa* to polymyxin B and O_2 decreased the MIC as O_2 exposure was lengthened (25, 26). Argamaso & Wiseman (9) reported that polymyxin B plus oxygen under high pressure (2 Ata for 2 hr) prolonged survival of mice infected intraperitoneally with *P. aeruginosa,* yet these investigators (279) and others (28), unlike Bornside (25, 26), failed to show synergism of oxygen with polymyxin B in the presence of 3 Ata O_2 in treating burns of rats infected with *P. aeruginosa.* The tetracycline antibiotics enhanced growth inhibition of *S. aureus* under 3 Ata O_2 but not 1 or 2 Ata. Increased activity of oxytetracycline by O_2 tensions $>$ 1 Ata have been confirmed (198). Enhanced growth inhibition by kanamycin by O_2 (or vice versa) against *S. aureus* and *P. aeruginosa* has been reported (28, 236). Sulfisoxazole activity against *V. comma* was potentiated 100 times by 2.87 Ata O_2 (111). These inhibitory and lethal effects of sulfisoxazole were potentiated even where O_2 inhibition of growth was obviated nutritionally. Oxygen did not synergize the activity of sulfisoxazole or trimethoprim against *Neisseria catarrhalis* but 1.87 Ata O_2 did increase by 4- to 25-fold the effectiveness of trimethoprim plus sulfisoxazole (104, 107). Oxygen at 1.87 Ata increased sulfisoxazole activity against *P. aeruginosa* two-fold; synergism was absent with trimethoprim (204) even at high concentrations in a synthetic medium lacking end products of folate metabolism (114). In vitro studies indicated that OHP (2–3 Ata) enhances the activity of mafenide (26, 204) against *P. aeruginosa,* but did not appear to enhance the response to mafenide in experimental burns of rats infected with *P. aeruginosa* (26). Oxygen synergized growth inhibition of *Streptococcus faecalis* by the narcotic gases Ar, Xe, and N_2O (78).

Synergism was not found between O_2 and antimicrobial agents, among them penicillin, chloramphenicol, viomycin, nitrofurantoin, erythromycin, and neomycin (136, 236, 271), although Bornside (24) reports decreased minimal inhibitory concentrations to *S. aureus* of some of these by 3 Ata O_2.

Antibiotic inhibition of oxytetracycline was reversed when the partial pressure of oxygen was <1 Ata (198), and for mafenide by 3 Ata O_2 (26). Approximately 0.2 Ata O_2 induced sensitivity to the respiratory inhibitor picricidin A in *Candida utilis* (150). More study is needed of the interaction of O_2 and antibiotics on a diversity of microbes. The significance of the drug-O_2 studies to mechanisms of O_2 toxicity will be discussed below.

Miscellaneous Effects of Oxygen

Light production by *Achromobacter fischeri* was reversibly decreased in direct proportion to increase in Po_2, reaching 66 percent reduction at 10 Ata O_2. Decreased production was not due to pressure per se since 28 Ata N_2 was inert (123). A growing literature describes the lethal effects of O_2 on lyophilized and aerosolized bacteria (17, 61, 62, 127, 167–169, 193, 227). Growth of *B. subtilis* and *E. coli* became synchronous after exposure to 10 Ata O_2 for 18 hr (91). L-Asparaginase production by *S. marcescens* is inhibited by $<$ 1.0 Ata O_2; prodigiosin production is enhanced by 0.2 Ata O_2 (126) and inhibited by 2–3 Ata O_2 (219). Oxygen acts as a substrate in late stages of heme synthesis in *Staphylococcus epidermidis* (143). It induces sterol synthesis (2, 155), cytochrome formation (41), and mitochondrial DNA synthesis (218) in yeasts. Induction of nitrate and nitrite reductase systems and their terminal electron acceptors are markedly affected by O_2 (68, 69, 166, 215, 245). Although exposure to 0.95 Ata O_2 for 6 hr did not affect motility or viability of *Chromatium* D, it did stop synthesis of bacteriochlorophyll (134). Carotenoid and bacteriochlorophyll synthesis, are markedly inhibited by 0.2 Ata O_2 but not growth and protein synthesis of *Rhodospirillum molischianum* (247). Two to 3 Ata O_2 reversibly inhibit growth and pigment production of chromogenic bacteria in deep cultures but kill them in media in shallow layers (219). One Ata O_2 did not inhibit sporulation of *Bacillus subtilis* (164). Algal photosynthesis is inhibited by 0.95 Ata O_2 (267, 269).

Mechanism(s) of Microbial Oxygen Toxicity

Search for sites of O_2 action continues. The varied effects of O_2 on cellular and subcellular organization suggests that it acts at multiple sites (103, 122). The question is open as to whether there is single or multiple-site action.

Growth.—Oxygen, in the range of 1–4 Ata, exerts bacteriostatic and bactericidal effects on growth. The few growth curves obtained for mycobacteria (106), *Achromobacter* P6 (112), *P. aeruginosa* (25), *Aerobacter aerogenes* (53), and *E. coli* (53) indicate a prolonged lag phase; the log growth phase of *S. aureus* was decreased 60 percent by OHP (24). *Clostridium perfringens* and *Bacteroides* sp. were most susceptible to O_2 in log phase growth than in stationary phase (196). In fungi, length of lag varied with species and inoculum; there is generally a direct relationship between length of lag and length of O_2 exposure (224). Hydrogen peroxide lengthened the lag of *Salmonella typhimurium* in direct proportion to its concentration but did not affect log growth (270).

How O_2 inhibits growth can be looked at from several physiological and biochemical aspects. Each will be treated only to the extent that information on microbes is available. For more extensive reviews of mechanisms of O_2 toxicity see references 103 and 122.

Permeability.—Stuart et al (255) reported that bakers' yeast exposed to 2–8 Ata O_2 for 2–40 hr showed decreased K^+ leakage, whereas exposure to 10–143 Ata O_2 for 0.5–22 hr increased leakage. Young (285, 286) reported that 1 Ata O_2 inhibited the uptake of leucine and sucrose by *Pseudomonas saccharophilia.* That O_2 may be acting primarily on cell membrane transport rather than secondarily through a decrease in available energy may be surmised from Vaughan & Wilbur (263) who showed that 1 Ata O_2 noncompetitively inhibited the transport of glycine, L-leucine, and uracil by the yeast *Schizosaccharomyces pombe;* this O_2 exposure slowed growth and protein synthesis by 44 percent without affecting O_2 consumption.

Respiration.—Inhibition of growth of microorganisms by oxygen may in part reflect inhibition of respiration with resultant unavailability of energy. Dilworth & Parker (66) reviewed early work on the effects of oxygen on *Azotobacter* and showed that above 0.3 Ata, O_2 inhibition of respiration of *A. vinelandii* is linearly related to the Po_2 in the presence of sucrose, glucose, pyruvate, acetate, succinate, and malate but not glycerol. Neither cysteine nor glutathione protected against O_2. Later, Dilworth (64) found O_2 inhibition of respiration in *A. vinelandii* was reversible and could be mimicked by methylene blue and arsenite; in cell-free extracts, pyruvic oxidase and alpha-ketoglutamic oxidase were O_2-sensitive. In contrast to the earlier report (66), he found cell-free extracts insensitive to O_2 when succinate was the substrate (64). He did not discuss the reason for the discrepancy; it may be due to whole cells versus cell extracts; the possibility of a permease involvement has not been explored. Inactivation of phosphate transacetylase in cell-free extracts of *A. vinelandii* could account for the decrease in acetate oxidation in whole cells if it occurs in vivo.

Dilworth & Parker (66) thought the O_2 inhibition of respiration was peculiar to *A. vinelandii* since they were unable to find it in *B. subtilis, Sac-*

charomyces cerevisiae, and *Rhizobium trifolii* even with 1.0 Ata O_2. Gottlieb (106) found that 0.8 Ata O_2 inhibited growth and 0.94 Ata O_2 markedly inhibited respiration of *Achromobacter* P6 with 1-amino-2-propanol but not with succinate as substrate. In line with Dilworth's data (66), cysteine was not protective (106).

Wolin et al (282) showed that 1 Ata O_2 reversibly inhibits oxidation of pyruvate by *S. aureus,* an inhibition reversed by thiamine and Mg^{++} over the pH range 6.0–7.9; cocarboxylase could replace thiamine. Oxidation of lactic dehydrogenase and cytochrome b_2 of bakers' yeast by H_2O_2 and by Cu^{++} under aerobic conditions results in dissociation of the flavine prosthetic group and loss of dehydrogenase activity (6, 7).

Inhibition of respiration by oxygen may vary with the stage in the growth cycle (271). Oxygen stimulated the O_2 uptake of developing mycelium of *M. vulgaris* but irreversibly inhibited O_2 uptake by an older mycelium preparation in which O_2 inhibition was not due to substrate exhaustion.

Respiration of *S. pombe* was unaffected during a 2-hr exposure to 2.5 Ata O_2 (263). The respiration of mycelia of *Myrathecium verrucaria* was not inhibited by 1 Ata O_2 (56).

Intermediary metabolism.—Nutritional reversal of inhibition of growth by oxygen (vide supra) is not understood and awaits identification of the active components. Its occurrence implies that these nutrients by-pass O_2-inhibited enzymes of intermediary metabolism. Nutritional mitigation of O_2 inhibition may also be due to nutrients providing intermediates for regenerating reduced NAD (38). Young (286) found 1 Ata O_2 to inhibit lipid synthesis in *P. saccharophilia* but not leucine incorporation into protein (285); also O_2 stimulated polysaccharide formation from sucrose (286). The mechanism of cation protection (59, 88, 127, 163, 169, 271, 281) is unknown.

"Experimentation on the biochemical effects of gaseous environments may add depth to knowledge of intermediary metabolism with all its attendant ramifications" (106). Interaction of O_2 with antibiotics may provide insight into reactions that may be sites of O_2 action. Oxygen synergism with PAS (112) and sulfisoxazole (111) suggested that pathways involving PABA may be O_2-labile. Studies with trimethoprim (107) suggest that O_2 may affect the dihydrofolate reductase system, but this view is questioned (204). Oxygen either affects the enzyme directly or keeps tetrahydrofolic acid oxidized, thereby altering the kinetics of all the reactions in which folates are involved. Tetrahydrofolic acid, extremely O_2-labile, is readily oxidized by increased O_2 tensions; its solutions are stored under N_2. The interaction between O_2 and drugs helps elucidate not only O_2 toxicity but also provides leads to the synthesis of new drugs (107).

Oxidation-reduction potential.—Limiting growth of *Bacillus megatherium* in vacuum ($\leq$ 10 mm Hg) is the O_2 content and not E_h (158). Barnes &

Ingram (13) demonstrated that toxic effects of O_2 on *C. perfringens* may be less a direct effect of molecular O_2 than an effect on the redox potential. *C. perfringens* grew even under 1.0 percent O_2 at low E_h's. They pointed out the direct relationship of E_h to the length of lag phase: as E_h increased from −45 mV (no lag) to +230 mV, the log growth, once initiated, was similar to that at lower potentials. Hanke & Bailey (120) studied the redox requirements of three clostridia over a wide pH range.

Enzymes.—Other than for respiratory enzymes, relatively few studies have been done with other enzymes. Lück (173) found that 10 Ata O_2 inhibited dehydrogenase activity irreversibly and alkaline phosphatase of *E. coli* even in the presence of substrate; lactate and CO_2 production also decreased; proteolytic enzymes, lipase, and catalase were unaffected. The accumulation of keto acids in respirometric experiments suggested that 1 Ata O_2 was inactivating the pyruvic oxidase of *M. vulgaris* (271). The pyruvic oxidase and α-ketoglutaric oxidase of *A. vinelandii* are O_2-sensitive (64). Pyruvic oxidase from several sources is O_2-sensitive (103, 122). Phosphate transacetylase, acetate kinase, malate synthase, and isocitrate lyase in cell-free extracts of *A. vinelandii* were inactivated by 1.0 Ata O_2; transacetylase was most sensitive, losing 94 percent of its activity within 6 hr (65).

Sulfhydryl (SH) groups.—Inhibition of growth and of respiratory and other enzymes by oxygen may be partly explicable by oxidation of SH since many tricarboxylic acid cycle and cytochrome enzymes are SH-dependent. One prominent theory of O_2 toxicity is based on oxidation of SH. A requirement for metazoan and protistan cell division is high SH concentration (15, 216). In bacteria, reducing activity and SH content are maximal during lag phase (161, 191), which implies that substances interfering with accumulation of SH should inhibit growth. Such interfering substances might be most effective during late lag phase since that metabolism would then be shifting toward rapid production and accumulation of SH. As mentioned, oxygen prolongs the lag phase of growth of aerobic bacteria (25, 53, 106, 112), and, by increasing E_h, prolongs the lag phase for growth of anaerobes (13). Reducing substances may not reverse O_2 inhibition of growth in bacteria (64, 106, 173). Wittner (281) found that SH reversed the inhibition of *Paramecium* by O_2, yet Chalkley & Voegtlin (37) did not find that 10^{-5} *M*-reduced glutathione reversed O_2 inhibition of fission and nuclear growth in *Amoeba proteus.* Reducing agents may be toxic even in the presence of O_2 or may even enhance the toxic effects of O_2 (106, 169). The efficacy of reducing agents in protecting lyophilized bacteria against O_2 seems to be in question: dithionite, cysteine, and reduced glutathione enhanced the lethal effects of O_2 on *E. coli* (169); on the other hand, a combination of thiodipropionate and morpholinohexose reductone protected *S. marcescens* (17). Webley (271) and Wittner (281) presented indirect evidence that O_2 toxicity is connected with inactivation of thiol enzymes. Armstrong et al (10) demonstrated that O_2 inactivation of cytochrome b_2 is due to dissociation of riboflavin phosphate from protein and aggregation of the protein due to oxidation of SH and formation of intermolecular disulfide bonds. Any SH enzyme or coenzyme is suspect as a site of O_2 action.

Peroxide.—A theory of O_2 toxicity especially pertinent to anaerobes is the formation and subsequent inhibitory action of H_2O_2. This glib explanation has little weight unless we know what H_2O_2 is oxidizing. Schubert et al. (237) showed that H_2O_2 reacts with carbonyl compounds to form compounds more toxic to *S. typhimurium* than the parent compounds. Because of its reactivity, H_2O_2 may be involved in oxidizing SH groups and flavins and/or reacting with lipids to yield peroxides deranging the structure and function of lipoproteins, membrane constituents, etc. Dilworth (64) did not hold H_2O_2 responsible for O_2 inhibition of pyruvate oxidation by *A. vinelandii.* Indeed, Dickens (63) had suggested that it is unlikely H_2O_2 is involved in mechanisms of O_2 toxicity in mammalian tissues (vide infra).

Free radicals.—The underlying event(s) that may account for inactivation of various enzymes, coenzymes, etc., is free radical formation. As with H_2O_2 free radical formation means little without understanding of how and where free radicals are formed, the systems affected, and why these systems are vulnerable. Little is known about free radicals in microbes. Hedén & Malmberg (125) discuss free radical involvement in O_2 toxicity but without hard evidence. For a detailed discussion of the role of free radicals in O_2 toxicity, see references 94, 122, 186.

Mechanism of Resistance to Oxygen Toxicity

The appearance of O_2 in the atmosphere presented both advantage and challenge—i.e., more efficient energy production along with the hazard of O_2 toxicity. The challenge was met by the development of various antioxidant mechanisms (57, 87, 94). A ubiquitous antioxidant mechanism is the heme enzyme, catalase, which decomposes H_2O_2. Since catalase is present in virtually all tissues, Dickens (63) suggested that H_2O_2 is unlikely to be involved in O_2 toxicity, and showed that susceptibility of mammalian tissues to O_2 toxicity is unrelated to catalase content. In microbes the picture is somewhat different. As Dickens showed for mammalian tissue, Pritchard & Hudson (217) showed in fungi and plants no apparent correlation between initial catalase activity and susceptibility to O_2 toxicity. They suggested that it is changes in catalase activity in response to OHP rather than its preexposure concentration that matters, an idea which appears to have been substantiated (91, 92, 174). Lück (174) found *E. coli* grown under OHP to have higher catalase activity than the 1 Ata controls. Watson & Schubert (270) reported that H_2O_2-resistant strains of *S. typhimurium* had more catalase. Gifford (91) found that in O_2-induced synchronous cultures of *B. subtilis,* maximum resistance to 10 Ata O_2 coincided with time of maximum catalase activity. In synchronous cultures of *S. cerevisiae* the cell cycle has two points in which the cells are most resistant to 10 Ata O_2: just after cell division and just before budding (92). These periods coincide with maximum catalase production: the prebudding period of catalase synthesis is shorter than the post-division period. Yeasts in the stationary phase are more resistant to 10 Ata O_2 than cells from other phases of the growth cycle presumably because of faster catalase synthesis (92). Anaerobes can be protected against O_2 by adding catalase to the growth medium (128).

The consumption of O_2 during bioluminescence suggested that bioluminescence was an effective antioxidation mechanism (183).

Dalton & Postgate (55) present data to suggest that O_2-sensitive nitrogenase can be protected from excess O_2 in two ways: increase in respiratory rate and structural rearrangement of components of the nitrogenase complex. An increased respiratory rate as an antioxidant mechanism has support (66, 106, 206).

IN VIVO

Inhibition of bacterial growth by oxygen provided the initial impetus for its use as an antimicrobial agent. It was reasoned that O_2 applied to sites of inflammation might destroy or inhibit invading microbes. This was the basis for earlier O_2 and H_2O_2 treatments of anaerobic infections; it bolsters the theory underlying the present use of OHP in treating anaerobic infections and presents hope that oxygen under high pressure will aid in treating diseases caused by aerobes.

Many in vivo data come from studies on man. In part, this situation arose because some human diseases cannot be duplicated in animals; in part as a "last resort"; and in part, the ability of some clinicians to convince patients to risk exposure to an unknown and unproved therapeutic procedure. Since clinicians are in an unenviable position in that moral restraints, together with insufficient clinical material in certain diseases, make carefully controlled clinical studies of OHP difficult, data obtained often are statistically uninterpretable. In medical research, evaluation of clinical experiences is often important. Unfortunately, conclusions drawn from some clinical studies with OHP were not critically analyzed. Undoubtedly, more information is stored in the files of the hyperbaric centers than has been published. This makes comment by a laboratory scientist on the rather few published clinical studies difficult and tentative.

Factors Affecting Response to Oxygen Under High Pressure

Partial pressure of oxygen.—Limitations are associated with periodic injections of gaseous O_2 to maintain a barrier around the phlegmon: the danger of O_2 embolism, difficulty of creation of an O_2 barrier in some regions of the body, and the area of diffusion from this barrier is small. Some limitations associated with creating a localized O_2 emphysema are overcome by OHP.

Anatomical, physiological, and biochemical considerations determine cellular Po_2. The only practical method of elevating cellular Po_2 significantly is by increasing the Po_2 of the arterial blood (Pao_2). The Pao_2 is primarily determined by the Po_2 in the alveoli ($P_{A_{O_2}}$) and the ventilation/perfusion relationship of the entire lung. Assuming a normal lung, the $P_{A_{O_2}}$, and thereby the Po_2 of the blood in contact with the functioning alveoli, can be elevated only by altering the composition of the inspired air. Compared to air breathing, the inhalation of O_2 at 1 Ata results in an increase of the Po_2 of the inspired air ($P_{I_{O_2}}$) and alveolar air by factors of 4.75 and 6.6, respectively. If 100 percent oxygen is inhaled at 3 Ata there is a 14.8-fold increase in the $P_{I_{O_2}}$ and 21.5-fold elevation of the $P_{A_{O_2}}$ as compared to breathing air at 1 Ata. The theoretical consequences of breathing 100 percent O_2 at 3 Ata are that there is enough O_2 physically dissolved in the plasma capable of meeting the normal O_2 requirements of most tissues, and that this volume of O_2 is being delivered at a high pressure (theoretically 2193 mm Hg compared to 100 mm Hg breathing air at 1 Ata), thereby increasing the rate of diffusion of O_2 to the tissues. That O_2 is delivered via the circulatory system implies that the area for diffusion is maximal and O_2 is readily accessible to all parts of the body.

Therapeutic partial pressure of oxygen.—In using a drug one seeks an exploitable difference in sensitivity between host and parasite. Generally, one deals with differences in concentration: the concentration required to inhibit microbes being one or more orders of magnitude less than LD_{50} for man. With oxygen it is different. The Po_2 required for mammals in order to raise tissue Po_2 to a level such that it will have adverse effects on growth or metabolism (including toxin production), or both, of invading microbes is in the same range as that which gives rise to pulmonary and CNS manifestations of O_2 toxicity in mammals (103). The exploitable difference in sensitivity one looks for in OHPT is the time factor. Because O_2 toxicity is a time/pressure-dependent phenomenon (actually a dose-response relationship with time very critical), one hopes that the exposure to OHP will be long enough to affect the microbe and give the body's defenses time to recuperate and prevail, but not so long as to result in convulsions or lung damage. These considerations are extremely important when using OHP to treat infections by aerobes. The clinical significance of more O_2-drug synergism studies (vide supra) becomes obvious.

Furthermore, unless duration of exposure and Po_2 relationships as they affect microorganisms and associated disease processes are understood in perspective, confusion will result (vide infra, gas gangrene). Many in vivo investigations (9, 29, 100, 137, 147, 154, 156, 157, 200, 238) are marked by the absence of comparative studies of different O_2 tensions. Comparison between 2 or 3 Ata O_2 with 1 Ata air may seem a valid experimental design. I submit that on at least three counts that it is incomplete: (*a*) such experiments are, in essence, dose-response curves, and more than two points on the Po_2 curve should be required; (*b*) the change from a gaseous milieu of essentially 20 percent O_2 and 80 percent N_2 (air) to one of 100 percent O_2 at increased pressure raises the question as to whether 100 percent oxygen at 1 Ata may not be as effective as OHP; (*c*) a pressure control is required, i.e., built into the total experimental design should be a series of animals exposed to increased ambient pressure; ideally, the Po_2 should equal that of 1 Ata air. This can be accomplished practically by raising the pressure in the air-containing chamber by means of a compressor or compressed air cylinders. This technique permits chambers to be continually flushed to prevent CO_2 accumulation, hypoxia, and build-up of heat. The latter depends on the number of animals in a chamber, heat conductivity of materials from which a chamber is constructed, thickness of chamber wall, and trans-chamber temperature differential.

At 3 Ata air, PI_{O_2} is 3 times and PA_{O_2} 4.1 times $>$ 1 Ata air. If, under such conditions, there still is an effect greater than 1 Ata air, an alternate pressure control must be used. Another method consists of raising pressure in a chamber to desired levels by adding an inert gas, N_2 or He, to the 1 Ata air in the chamber. This design has the merit of maintaining the Po_2 the same as that at 1 Ata air. With this technique it is difficult to flush the chamber without an accurate gas-mixing device or unless gas mixtures are employed. To control CO_2 build-up and avoid hypoxia (due to consumption of the limited O_2 supply), one can incorporate a CO_2 absorber (which should be included even in the pure O_2 experiments) and a solid chemical O_2 supply in conjunction with a small, explosion-proof, nonsparking (if it is to be used in pure O_2 environments) fan. Potassium superoxide (KO_2) can be used to absorb CO_2 and simultaneously release O_2 to maintain a sufficient Po_2 to avoid hypoxia.

Heat in the chamber can be controlled by immersing it in water or by wrapping it in copper tubing and insulation and running H_2O through the tubing; alternatively, success was had with lining the inside of a chamber with tubing attached to connectors on the chamber wall and running water through this system. The latter methods have been integrated with a constant temperature bath and water pump to maintain precise internal temperature conditions (109, 112).

Unless explosive decompression is employed, decompression sickness is generally not a problem since small animals are more resistant than man. Depending on the nature of the disease under study, aeroembolism may be a problem.

"Anaerobic" Infections: Clostridia

Clostridium perfringens

The anaerobic nature of clostridia suggested the use of O_2 or H_2O_2 in therapy. Successful treatment of gas gangrene with O_2 or H_2O_2 injections has been described (129, 131, 272). Present clinical use of OHP in treating clostridial gas gangrene was pioneered in Holland by Boerema and Brummelkamp (23, 30, 35). Successful clinical use of OHP for this disease was an accidental discovery since they were unaware of the discouraging results with OHP obtained by DeAlmeida & Pacheco (58) 20 years before in Brazil in treating experimental gas gangrene in guinea pigs. DeAlmeida & Pacheco's (58) work was based on findings that cultures of *C. perfringens* exposed to 3 Ata O_2 were almost completely killed within three days, even though the spores survived (203). The initial studies of Brummelkamp et al (35) in treating experimental gas gangrene were successful: later experiments were unsuccessful—the results were like DeAlmeida's (58). In contrast to animal experiments, clinical use of OHP in treating *C. perfringens* infections in man was dramatically successful. The clinical experience of Brummelkamp and colleagues has been summarized (30, 31, 33, 34).

Their clinical regimen for treating *C. perfringens* infections, as it evolved since 1960, essentially consists of 7 sessions of OHP in a 3-day period; 3 exposures to O_2 on the first day and 2 sessions each for the next 2 days. Each treatment consists of exposure to 3 Ata O_2 for 2 hr. Neither antibiotics nor antigas gangrene serum is used routinely. Antibiotics are used to prevent secondary infections. They prefer to perform any surgery for removal of necrotic tissue after OHPT when the patient is nontoxemic and improved. Their experience is that OHP facilitates demarcation of the necrotic tissue and radical surgery through healthy tissue becomes unnecessary.

The successful use of OHP to treat gas gangrene has been confirmed (5, 11, 48, 67, 98, 121, 145, 179, 192, 207, 210, 226, 248, 252, 262, 265, 268). Agreement is general that it is an extremely important addition to the armamentarium for treating gas gangrene. There is some question concerning the relative importance of OHP and surgery in this therapy: Brummelkamp prefers surgery, if required, after OHP; Smith et al (138, 253) concluded from their experience that OHP "is an aid to, but does not replace, adequate surgery"; they do not mention treatment of gas gangrene without prior surgery.

Initial and weak support for Smith's position came from McSwain et al (185) who described their 10-year experience with clostridial infections of the abdominal wall. Their conclusions, based on 10 patients, only 2 of whom received OHP (1 died) at 3 Ata after (not before) debridement, is that "surgical debridement is the keystone of the treatment and that OHP is an adjunct to therapy."

The question is not whether O_2 is beneficial but whether it should be given before or after debridement. It is important to understand the reason for these somewhat divergent attitudes. Smith et al (253) use a regimen different from that of Brummelkamp; 2 Ata O_2 for 2 hours versus 3 Ata O_2 for 2 hours. A marked difference exists between OHP at 2 versus 3 Ata. Diffusion of O_2 is greatly enhanced by the extra Ata O_2. Increased pressure will further decrease volume and diameter of gas pockets (Boyle's Law), thereby presenting a smaller diffusion barrier to O_2 than with 2 Ata; diminished size of the gas pockets would probably result in improvement of perfusion of diseased areas. Later information indicated that patients of Smith et al were not receiving theoretical amounts of oxygen; Pa_{O_2} varied in the range of 600–1000 mm Hg (147). One reviewer (103) therefore pointed out that "it would seem that before a group comes to a conclusion about a certain technique or mode of therapy, different from that of those who divised and used it successfully, it should be incumbent upon them to employ the same conditions used by the innovators; otherwise unnecessary confusion results."

Need for adequate O_2 pressure to ensure sufficient oxygenation of infected areas is apparent from theoretical and experimental evidence. Clostridial myositis, an inflammatory and necrotizing process, includes circulatory impairment. Theoretically, one must assure a sufficient volume of O_2 at an effective therapeutic Po_2 (ETP) to perform the dual function of inhibiting organismal growth or its toxin production or both, and supplying ample O_2 at sufficient pressure to affected surrounding healthy tissue to maintain metabolism at a rate adequate to combat infectious processes. The ETP must be administered for a therapeutically beneficial length of time but yet within safe time limits to avoid O_2 toxicity to patients. To increase the total time of O_2 exposure one may increase number of exposures per day.

Need for early and adequate oxygenation (clinicians have limited, if any control over early treatment) can be surmised from experimental in vivo studies (100, 133, 154). Kelley & Page (154) produced a standard trauma in albino mice and injected the wound with a culture of *C. perfringens,* American Type A, along with sterilized dirt as a foreign body. Animals were exposed to 3 Ata O_2 for 30 min every 12 hr for 4 days. Simultaneously the effects of delay (12, 18, 24, and 30 hr) of OHPT on the outcome of the disease were studied. Ninety-one percent of mice unexposed to O_2 died compared to a 4 percent mortality for the OHP group. Delay of OHPT for 12 hr resulted in a rise in mortality to 42 percent. Further delay of OHPT increased mortality. Increased mortality resulting from delay has been confirmed (100, 128, 156).

Glover & Mendelson (100) used rabbits for studying OHP effects in clostridial myonecrosis. Two Ata O_2 for 1 hr every 6 hr (total of 12 times in 3 days) instituted immediately after injection, had definite but limited prophylactic effects.

If partial pressures of oxygen < 3 Ata are to be used, it would appear that either the number of exposures per day must be increased, exposure time lengthened, or both (128, 133, 147, 156). Studies of Karasevich et al (147) tend to confirm the hypothesis that at 2 Ata more exposures should be employed in therapy. Using hepatic artery ligation of dogs, these investigators found a therapeutic regimen of 2 Ata, 2 hr, once per day for 2 days to be ineffective. An exposure regimen of 2 Ata, 3 hours, 3 times per day for 2 days resulted in decreasing mortality from 74 percent to 43 percent. Klopper (156), using the ligated hepatic artery technique (rabbits instead of dogs), found clostridial infection to be prevented in over 70 percent of the animals if they were exposed to 3 Ata O_2 within 2 hr after surgery. Two Ata O_2 comparisons were not made.

Sevcik & Vymola (238) injected rabbits with *C. perfringens* and exposed them to 2 Ata O_2 for 2 hr then followed with a 3-hr exposure to 2 Ata O_2 after being at 1 Ata air for 1 hr. The untreated controls died whereas 1 animal in the O_2-treated group died. If antigas gangrene serum was used in conjunction with OHP all animals survived. They did not report on the effectiveness of antigas gangrene serum alone. They concluded that O_2 was an auxiliary method of treatment. Neither 1 nor 3 Ata O_2 comparisons were made.

Hunt et al (133) failed to modify experimental clostridial infection in rabbits as measured by bacterial counts, duration of infection, or mortality with 2 Ata O_2: preliminary experiments with 3 Ata O_2 immediately after initiation of infection resulted in survivors.

It may be concluded that the data point to the greater effectiveness of 3 Ata O_2 on *C. perfringens* infections in the various model systems, and support the conclusions of Brummelkamp et al (30, 31, 34) as to the relative roles of OHP and surgery in treatment of *C. perfringens* infections. Recently, Hitchcock changed from supporting Smith's position (11) to supporting that of Brummelkamp (130).

Mechanism of action of oxygen on C. perfringens *infections.*—In general, mechanisms of action of O_2 on any disease process fall into one or more of three categories: effect on (*a*) organisms, (*b*) toxins elaborated by organisms, and (*c*) host defense mechanisms. This discussion will be confined to *C. perfringens.*

Effects on organisms: in vitro.—Inactivation of *C. perfringens* is time-dependent (83, 196, 203, 264). Van Unnik (264) did not find growth of *C. perfringens* to be arrested by a 1.5-hr exposure to 3 Ata O_2. Pacheco & Costa (203) found the organism to be killed after continuous 3-day exposure to 1 Ata air (0.2 Ata O_2) or 3 Ata O_2; however, the spores survived. Three Ata O_2 inactivated *C. perfringens* in log growth faster than in stationary phase (196). The lethality of O_2 is reduced by blood (196); specifically by its catalase (128).

Data are few on the Po_2 in broth in relation to the effects of O_2 on growth. Brummelkamp (31) reported that the in vitro lethal Po_2 for *C. perfringens* was 90 mm Hg; Van Unnik (as cited by 30) found that 250 mm Hg only inhibited growth. Different results are probably due to differences in the duration of exposure and growth media. In neither of the two experiments were the organisms exposed to the full ambient Po_2 since the broths were not agitated to facilitate solution and diffusion of O_2. One and a half hours is insufficient time for broth to equilibrate with the overlying gas phase. Diffusional limitations could also be inferred from growth data (30, 264).

Effects on organisms: in vivo.—Brummelkamp (30, 31) isolated viable *C. perfringens* cells from patients treated with 3 Ata O_2. Viable cells were recovered from oxygen-treated patients not only because of time-pressure inadequacies but probably because of catalase in infected tissues (128).

Experiments such as that of Van Unnik (264), besides giving insight into the effect of O_2 on growth and toxin production (vide infra), have the virtue of stressing the importance of diffusion barriers encountered in many disease processes (edema, by increasing the distance of diffusion as well as affecting solubility of tissues for O_2, imposes a potent barrier to diffusion) and the need for higher partial pressures of oxygen to provide sufficient driving force for the diffusion of O_2 into the infected area. Data of van Unnik coupled with the above theoretical discussions bolster arguments for therapeutic use of 3 Ata O_2. Further support stems from Po_2 measurements in a clostridia-induced inflammation (30): Po_2 while breathing 1 Ata air, 100 percent O_2 at 1, 2, and 3 Ata, was 50 mm Hg, 110 mm Hg, 250 mm Hg, and 330 mm Hg, respectively. Since the partial pressure of oxygen of 90 mm Hg is supposedly lethal for *C. perfringens* in vitro, one may deduce that breathing 100 percent O_2 at 1 Ata should suffice for treating such infections, which is apparently not true. These data underscore the need for studying the effects of O_2 at more than just two points on a dose-response curve, and for understanding the limitations of in vivo polarographic O_2 measurements.

Effects on toxin.—Van Unnik (264) found that 1.5 hr at 3 Ata O_2 reversibly inhibited alpha toxin production by *C. perfringens;* 2 Ata O_2 was not inhibitory; theta toxin production was not affected. From Po_2 studies (31) it can be deduced that 250 mm Hg is required to inhibit alpha toxin production. The site(s) of O_2 inhibition of this O_2-labile synthetic sequence is not known. Fredette (83) and Nora et al (194) were unable to demonstrate in vitro direct adverse effects of O_2 on alpha toxin.

Extrapolation of the above in vitro and in vivo data to clinical situations may tell something about the mechanism of beneficial effects of OHPT. Perhaps, during OHPT, there is no new production of alpha toxin; toxin present before OHPT becomes rapidly "fixed" in vivo and so disappears (30). Necrotic muscle tissue may release intracellular proteases which may proteolyze the toxin. Thus, the lethal activity of *C. perfringens* ceases.

Host defense mechanisms.—The above explanation does not clarify why, after one O_2 exposure, toxin production is not resumed, as in the vitro studies of van Unnik (264), especially since viable *C. perfringens* can be isolated from tissues after the use of OHP. One conjectures that OHP causes an "environmental" change lasting beyond periods of OHPT. The nature of this change is unknown. There are other complexities involved in working with *C. perfringens:* alpha toxin is not its only cause of pathogenicity, but beta hemolysin and kappa toxin also are lethal; also many infections are mixed infections. There is little knowledge on the effects of O_2 on these toxins and various mixed infections.

Clostridium tetani

The rationale for the use of OHPT in tetanus is identical with that for clostridial gas gangrene: both are caused by anaerobes; organismal growth and metabolism should be inhibited by O_2. Oxygen inhibits *C. tetani* toxin production (86), is bactericidal (151) but, contrary to a previous report (283), does not inactivate tetanus toxin (32, 86). The use of OHP to treat tetanus—a disease caused by exotoxins of *C. tetani*—has not met with the same success as for *C. perfringens* infections (3, 32, 154, 170, 171, 208, 220, 238, 278). Some clinicians think that OHPT may help (3, 170, 208, 278) but this impression is based largely on subjective observations (170, 208, 278) with little quantitative support (3, 32, 238). In experimental models, the initial positive effects of OHP (3, 32, 238) which are disputed (154), may be ascribed to its administration soon after infection and can temporarily inhibit growth and toxin production. Recent studies of the effects of 3 Ata O_2 on experimental tetanus in mice with tetanus toxin-induced disease which permits a mimicking of clinical conditions more accurately than spore-induced or organism-induced experimental designs, indicated no improvement in symptoms or survival time. In fact, a slight detrimental effect was noted, thus confirming various clinical and experimental observations (83, 187, 220). In one comparable experiment of toxin-induced disease, OHP was given before tetanus symptoms were evident (223) and, in a sense, meaninglessly.

Since *C. perfringens* and *C. tetani* are anaerobes belonging to the same genus, it is interesting that their disease processes differ so markedly in response to O_2. Dissimilarity in responsiveness to O_2 is probably due to differences in pathogenesis. The alpha toxin of *C. perfringens* is rapidly fixed in vivo and is no longer available for further tissue destruction: a continuous supply of alpha toxin is required for progressive damage; proteases released upon tissue destruction probably inactivate the toxin. Tetanus toxin attaches to skeletal muscle and nervous tissue, especially to the anterior horn cells and motor cranial nerves. Once fixed there, and in the absence of overt tissue destruction, this toxin may not be as readily proteolyzed as the alpha toxin; it continues to paralyze.

Other Anaerobic Bacterial Infections

Growth and toxin production of *Clostridium botulinum* (84) are inhibited by oxygen. *Bacteroides* spp., impregnated in agar discs, are rapidly inactivated in vitro and in vivo by 3 Ata O_2 given for 2 hr twice daily; log phase cultures are more sensitive than those in stationary phase (196); controls receiving O_2 at 1 or 3 Ata or 3 Ata of air were not described. The anaerobic nature of *Bacteroides* and its relationship to puerperal infections, abortal and postabortal infections, vaginal and pelvic infections may point to another role for OHP. The use of OHP in a *Bacteroides* infection resulted in the patient becoming afebrile but eventually dying of staphylococcal septicemia (268). Adams (1) reported strains of actinomycetes to be killed by $<$ 1 Ata O_2. Manheim et al (177) successfully used OHP to treat a perirectal actinomycosis, thereby confirming a prediction (105, 112).

Oxygen should alter the intestinal flora. Hopkinson & Towers (132) were unable to recover anaerobes from fecal samples of rats exposed to 2 Ata O_2 for 2 hr. Detailed studies are needed on time-pressure relationships on specific intestinal anaerobic and aerobic species as well as on the time required to re-establish a normal flora. As with the gut, there has been no systematic study made of the oral flora as affected by O_2. Such studies may have implications in caries formation. Fusiform bacilli and spirochetes in Vincent's disease should be susceptible to O_2 inhibition.

The idea of using OHP in treating syphilis as proposed by Cunningham (cited in 103) is sound: *Treponema pallidum* is anaerobic. Conceivably, O_2 could be used in treating diseases associated with other treponemes, i.e., bejel, yaws, and pinta. Therapy of these diseases by OHP is probably of theoretical interest only—at least in temperate zones—since they are rare and conventional antibiotic therapy appears highly effective.

"Aerobic" Infections

Tuberculosis.—The effects of O_2 on tubercle bacilli in vitro were confirmed in part by the vivo studies of Chandler et al (40). They studied the effects of OHP (3 Ata O_2 for three hr/day, 22–28 days) on a miliary infection of rabbits caused by *Mycobacterium bovis*. They found marked changes in the course of the infection: less disease in the lungs of the treated animals as evidenced by fewer and smaller tubercles, less caseat, and containing fewer tubercle bacilli: also, the average weight of the unexposed controls was significantly heavier than the OHP-treated animals, presumably because of more extensive disease in the controls. Tuberculosis of the axillary, inguinal, cervical, tracheobronchial, and mesenteric lymph nodes was less in the treated animals. No difference was found in the liver and kidney. Disease in the heart and spleen of the O_2-treated animals was more extensive. Neither 1 Ata nor 2 Ata O_2 nor 3 Ata air experiments were done.

Barach (12) did not demonstrate retardation of tuberculosis in rabbits inhaling 0.6 Ata O_2. Moore (Cited by 1), administering 0.8 Ata O_2 1 hr daily to 3 tuberculous patients for several weeks, observed no benefit. There are reports (cited in 105) on the beneficial effects of O_2 introduced directly into local tuberculous lesions. J. Vance and F. Beerel, Millard Fillmore Hospital, Buffalo, N. Y., (personal communication) treated several drug-resistant cases of tuberculosis at 2.5 Ata O_2, once/day for 4–6 weeks and obtained encouraging results: 2 patients for the first time in many years had negative smears although cultures were still positive; also, organisms from one patient had increased sensitivity to PAS. In light of in vitro data and the study by Chandler et al, the time-pressure relationships were inadequate to affect the disease process. In Barach's study and the recent clinical trial, the $P_{I_{O_2}}$ was too low and the duration of exposure too short. Although data from clostridial studies indicate difficulty in extrapolating the effects of oxygen on disease processes associated with closely related species, much is to be gained from comparative physiology. On the basis of O_2 inhibition of *M. tuberculosis,* it was postulated (105, 112) that O_2 may be useful in treating other mycobacterial infections, i.e., Hansen's disease. S. Jakubowski (Bethlehem Corp., Bethlehem, Pa.; personal communication) received favorable reports on OHPT of Hansen's disease at a special meeting in Spain in 1970. No publications have yet issued from this meeting and he has yet to receive answers to his inquiries to the participants. His experience parallels the lack of response to letters almost a decade ago (1963–1964) which I addressed to various leprosaria and medical directors of leprosy missions concerning the possible use of OHP in therapy.

Staphylococcus.—Grogan (115–117) reported that up to 3 hr of continuous exposure to 2 Ata O_2 did not affect the course of *S. aureus* infection in mice. At 3 Ata O_2, mortality depended on the duration of exposure: mice exposed for 1 hr died at the same rate as controls; 2- or 3-hr exposures increased the death rate; after 3 hr even uninfected animals had high mortality because of O_2 toxicity. It is unlikely that O_2 increased mouse susceptibility to infection since there was no relationship between mortality and duration of OHP exposure for mice similarly infected with *P. aeruginosa;* differences in mechanism of pathogenesis do not preclude this possibility. Most in vitro data (8, 137, 141) do not support the idea of increased production or enhanced potency of toxins produced by *S. aureus* or enhanced pathogenicity.

Using a different model, Irwin et al (137) showed that 2 Ata O_2 depressed growth of *S. aureus* in a standard skin wound; the inhibition was reversed on exposure to air. They avoided host O_2 toxicity complications by permitting the guinea pigs to breathe air at 2 Ata while passing warmed, humidified O_2 over the wound.

Barnwell et al (14) reported that mice infected with *S. aureus* and exposed to low P_{O_2} (0.06–0.098 Ata), outlived animals exposed to 0.2 Ata O_2 (air), 2 or 3 Ata O_2. The OHP experiments unfortunately consisted of continuous as opposed to intermittent exposure and therefore control and infected mice died of O_2 toxicity. The paradoxical beneficial effects of hypoxia, in contrast to those of Schmidt et al (234) with a different model, may be due to its effects as a stressor causing secondary effects, or it could be due to secondary effects resulting from cardiovascular changes as suggested by Rhoden et al (221).

The effectiveness of OHP in treating *S. aureus* osteomyelitis has been discussed (60, 113, 211, 246, 250). Since osteomyelitis in part is characterized by local hypoxia caused by ischemia, and since the OHP time-pressure relationships (2–2.5 Ata O_2, 1.5–2 hours) are not of the magnitude to raise wound Po_2 to where it would inhibit staphylococcal growth, an effect on the host (direct O_2 effect on osteogenesis, 50, 212) may account for the observed benefit. This view is supported by the findings of Hamblen (119) that OHP did not decrease the *S. aureus* bone-infection rate in rats but did speed healing of established osteomyelitis.

Pneumococcus.—In vivo studies revealed that the primary effect of O_2 on pneumococcal septicemia in mice is to prolong survival time rather than mortality. Oxygen prolongation of survival depends on inoculum size (228).

Escherichia.—Ollodart & Blair (199), studying effects of 3 Ata O_2 on fecal peritonitis in dogs, showed than when O_2 is introduced via the circulatory system, peritoneal cavity $Po_2 = 0.1$ Ata. Such animals developed immediate septicemia and died within 5 hr. Similar results were obtained by Filler et al (79) in 3-Ata O_2 treatment of *E. coli* peritonitis in rats, and by Klotz (157) using 2-Ata O_2 in *E. coli*-induced pyelonephritis in rats. These results are explicable on the basis of in vitro data relating growth to Po_2 (vide supra). In contrast to O_2 delivery via the circulation, Ollodart & Blair (199) found that surgically exposing the abdomen to the full 3-Ata O_2 cleared the initial septicemia within 2.5–3.5 hr. Septicemia eventually returned and the infected dogs died after 8 hr; the overall effect was 3-hr increased survival. These experiments, reminiscent of those of Irvin et al (137, vide infra) on direct exposure of surface infections to O_2, underscore the importance of using sufficient Po_2 at the infection site.

Pseudomonas.—Results from in vivo studies appear unencouraging as to the therapeutic role of O_2 alone in *Pseudomonas* infections. Grogan (115, 116) did not find 3 hr of 2 or 3 Ata O_2 to be effective in altering the survival time or mortality of mice having acute *P. aeruginosa* infections.

Zaroff et al (287), using either flap incisions or burns on rats infected with *P. aeruginosa,* found that 3 Ata O_2 given 2 hr daily until death had no effect on mortality rate or survival time. Argamaso & Wiseman (9) had similar results with an intraperitoneal infection in mice; however, 2 Ata O_2 combined with polymyxin B significantly enhanced survival. In view of the fact that O_2 is primarily bacteriostatic, it would appear that the exposure times in the in vivo experiments were too low; also these investigators (9, 115, 116, 287) attempted to supply the O_2 via the circulatory system to areas, especially burns, where the vasculature was destroyed. It is doubtful if they had a sufficient Po_2 prolonged enough to reach the infected areas and inhibit the organism or bolster local defenses, or both.

Support for this analysis comes from Irvin et al (137): in studies on

Pseudomonas skin infections in guinea pigs treated by continuous application of 2 Ata O_2 for 72 hr by passing the humidified gas over the wound while the animal was breathing 2 Ata air, there was within 72 hr a significant decreased bacterial count in O_2-treated wounds compared to air-treated wounds. After decompression, the number of organisms increased until they equaled that of controls. This model has much to offer since it approximates clinical situations.

Miscellaneous Bacterial Infections

Septic shock is a complex syndrome resulting from different types of infection, i.e., bacteremia, endotoxins, exotoxins, or invasion of a vital organ, resulting in inadequate tissue perfusion. Approximately two-thirds of the cases of septic shock are caused by Gram-negative bacteria. There are conflicting reports on the value of OHP in endotoxin or bacteremic shock. Evans et al (75) found that OHP lengthened the survival of dogs subjected to *E. coli* endotoxin shock, whereas Blair et al (21) found no beneficial effects of O_2 on bacteremic shock (instillation of saline suspension of feces) in dogs. Bacteriostasis induced by OHP benefits early bacteremic shock (200); in later stages its primary benefit, in the absence of data indicating direct O_2 inactivation of endotoxin, is relieving tissue hypoxia. The discrepancy between Evans et al and Blair et al may be partly explained by the role of the Schwartzman reaction in septic shock. Oxygen enhances the Schwartzman reaction (257). Lack of benefit from OHP in the studies reported by Blair et al could be due to balancing of the beneficial hypoxia-relieving role of OHP by the adverse stimulating effects of O_2 on the Schwartzman reaction, whereas the Schwartzman reaction may not have been as important in the experimental model of Evans et al.

The prediction (103) that OHP might help control *Neisseria* infections seems to be borne out by in vitro (104, 107) and clinical (178) experience. Unfortunately, in the clinical cases (178) the causative organisms were not isolated nor identified.

Oxygen under high pressure has been used successfully in treating nonclostridial gangrene (70), infected pressure sores, and skin ulcers (81, 102).

Fungal and Protozoan Infections

To my knowledge there are no clinical or animal experiments with OHP on fungal infections. Cleveland's (43) work with protozoa deserves special mention because it gives some insight into what may be accomplished in conjunction with antimicrobial agents for man and animals. He found that after exposing the termite, *Termopsis nevadensis,* to 1 Ata O_2 for 24 hr, the flagellate *Trichomonas termopsidis* was completely absent from its intestine; other protozoa, *Trichonympha campanula, Leidyopsis sphaerica,* and *Streblomastix strix* were removed after the termite has been exposed to O_2 for 72 hr. Termites did not suffer direct ill effects from O_2 but died of starvation within 3–4 weeks after defaunation. In a more detailed study, Cleveland (44) reported on the time-pressure relationships required to kill free-living ciliates and flagellates as well as the intestinal protozoa of cockroaches, frogs, fish, salamanders, and rats. For the free-living ciliates, *Paramecium, Chilodon, Diophrys,* and *Holostrica,* the time (in hours) required to kill all individuals at 3.5 Ata O_2 was 5, 4, 60, and 50 respectively; for the free-living flagellates *Euglena* and *Heteronema,* the time required for death was 65 and 50 hr, respectively. Cleveland's studies were confirmed (45, 284). Because of their greater resistance to O_2 toxicity, it is impossible to kill the parasites of rats and man by O_2 alone at 3.5 Ata without killing the host. There is need for detailed studies on the effects of O_2 alone and in combination with drugs on the intestinal parasites of man and animals. Intrarectal oxygen has been used to treat acute ascariasis (142).

Viruses

No systematic studies have been done on direct in vitro and in vivo effects of O_2 on viruses. Panov & Remezov (205) reported on the Po_2 duration of exposure relationship in affording protection to white mice against experimental lymphocytic choriomeningitis and acute human disseminated encephalomyelitis. At 2 Ata O_2, 5–10 hr was maximal time of exposure to protect mice against both neurotropic virus infections.

Sawicki et al (230) infected mice with either Lee strain B or WS strain A influenza virus; mice were subsequently exposed to 0.5 Ata O_2. The O_2-treated animals showed earlier pneumonia and increased final mortality, results attributed to a decrease in the antiviral activity of interferon. Gifford (93) reported that interferon activity was not significantly affected by marked changes in Po_2.

Libet & Siegel (165) demonstrated that high Po_2 suppresses the development of leukemia symptoms induced by Freund virus in white mice. Once symptoms appear, O_2 can cause symptoms to regress; the beneficial effects of O_2 were not permanent as shown by a recurrence of symptoms after oxygen was discontinued. Three Ata O_2 enhances Coxsackie virus infection in mice (201).

Host Defense Mechanisms

Although in vitro data suggest a direct effect of O_2 on the microbe, interpretation of in vivo experiments should include consideration of possible O_2-induced physiological and biochemical alterations in the host which may make the host (or specific tissues) more or less resistant to pathogens. The effect of OHP on host resistance mechanisms is virtually unexplored. Yet, that host resistance is affected by O_2 can be surmised from data on clostridial and osteomyelitis infections. Further studies are required on relationships between various infectious diseases and host susceptibility to O_2 toxicity; there are some clinical indications but no quantitative studies indicating an increased host O_2 sensitivity. It is not known whether this is due to increased body temperature or to the presence of specific pathogens or their metabolic products.

Immune mechanisms.—No systematic study of the effects of oxygen on immune mechanisms has been undertaken. Depending on the partial pressure of oxygen, barometric pressure and specific immune system, oxygen may depress or potentiate. Oxygen depressed the size and weight of thymus and spleens of mice (51, 241) perhaps because of a decrease in oxidative enzymes (52). These observations suggested that alone, or in combination with drugs, oxygen might suppress the immune response. Yet, it stimulates the Schwartzman reaction (257) and reticuloendothelial system (197). Oxygen under high pressure did not affect circulating antibody titers to sheep red blood cells (SRBC) in mice (242), and 100 percent O_2 at 0.33 to 0.5 Ata total pressure decreased antibody titers to heat-killed *Brucella abortus* tube antigen in mice (51). Oxygen potentiates the immunosuppressive effects of cytoxan in mice sensitized to bovine serum albumin in complete Freund's adjuvant (244) yet, in conjunction with cytoxan, it did not affect circulating antibody titers to SRBC in mice (242). Antigenicity of culture-grown cells did not appear to be diminished by 4 Ata oxygen (118). Effects of altitude on immune response in rabbits and guinea pigs have been studied (261).

"Environment."—Extensive discussion is not now in order. But a growing literature describes the effects of sealed environmental systems, i.e., low-pressure (space-cabin) and high-pressure (underwater) environments, on the susceptibility of hosts to bacterial and viral infections (15, 19, 20, 46, 47, 71–74, 97, 231, 232, 234, 235, 240, 260).

Warnings

I will not belabor problems of O_2 toxicity, decompression sickness, aeroembolism, fire hazards, or similar factors related to oxygen or pressure, or both, since they have been reviewed (103).

The administration of O_2 to infected ischemic areas entails danger. Up to 1.3 or more Ata, O_2 stimulated the in vitro growth of many organisms. In a wound dependent on diffusion for oxygenation, the Po_2 attained by OHP might well stimulate rather than inhibit bacterial growth. After reading many clinical reports one wonders whether the secondary infections encountered were not due to O_2 stimulation of bacterial growth. Support for this suspicion comes from such observations as O_2 intensifying the severity of chronic pyelonephritis in rats infected with *E. coli*, and the increasing incidence of superinfection with *Proteus* (157). Thurston (260) noticed an increased incidence of chest infections in patients treated with OHP for acute myocardial infarction. Although OHP appeared beneficial in experimental miliary tuberculosis in rabbits (40), it must be recalled that OHP exacerbated the disease in heart and spleen. In view of the enhancement of virus infection in mice (201, 230) by oxygen, perhaps caution should be exercised in treating viremic patients. Detailed experiments and carefully analyzed statistics from clinical studies are required to assess the risk of secondary or superinfection after OHPT.

Oxygen effects on virulence.—There are only limited data on whether virulence of microbes is altered by oxygen, especially since it is mutagenic. Most data do not indicate an increase in production or potency of toxins or pathogenicity of *S. aureus* (8, 137, 141, 180) although there is one report of increased β-hemolysin production in the range of 1–4 Ata O_2 (8). In contrast to the above biochemical indicators, Wilson (276), in addition to reviewing earlier literature, found by animal experimentation that *S. typhimurium* did not alter its virulence when grown anaerobically; in the range of 0.01–0.21 Ata O_2 virulence declined; with increasing Po_2—0.4–1.0 Ata—virulence increased. In the range 0.4–0.75 Ata O_2 but not < 0.4 or > 0.75 Ata O_2, growth of *S. typhimurium* was in the nonmotile O form. Detailed animal studies are needed to assess O_2 effects on the virulence of various pathogens.

CONCLUDING REMARKS

There are grounds for enthusiasm among microbiologists advocating the use of O_2 as a research and clinical tool. That it may be toxic to microorganisms in vitro and in vivo is established, but the molecular mechanisms are not understood. Microbial responsiveness to high O_2 tensions may be a useful character for differentiating microbial species.

Studies on microbial O_2 toxicity may promote understanding of O_2 toxicity in higher life forms as well as provide information contributing to understanding convulsive states. Data on SH compounds in protecting microbes against O_2 toxicity suggest that protozoa may be more profitable to study than bacteria since SH-stabilizing agents, e.g., cysteine, glutathione, or 2,3-mercaptopropanol, are not protective in bacterial systems but do protect protozoa and mammals (vide supra)—a suggestion in line with Hutner's musings on protozoan origins of metazoan responsitivities (135). The data on O_2 synergism with PABA or folic acid antagonists suggest that such antimetabolites used in conjunction with O_2 may find wider clinical usefulness. A disease which may yield to combined O_2-antifolate therapy is malaria.

As implied, moral constraints intensify the need for model system studies. Few microbiologists study in vivo effects of O_2 on microbial growth. Work is needed on in vivo and in vitro model systems based on pure cultures, also on models involving mixed infections. Gnotobiotic animals may be useful in these studies. More study is needed on the effects of O_2, alone and in combination with drugs, on model diseases involving drug-sensitive and resistant pathogens.

ZoBell's (290) finding that pressure may enhance the bactericidal effects of O_2 raises an interesting possibility. Would increased inert gas pressure enhance the toxic effects of O_2 in vivo and thus become a therapeutic technique?—ignoring, for the present, problems of decompression. What are the conjoint effects of O_2, pressure, and antimicrobial agents?

Elucidation of interrelationships between O_2 and the structure-activity relationships of drugs should help delineate drug structures requisite for synergism. Knowledge gained from nutritional or other analytical studies may pinpoint O_2-sensitive enzyme systems, thereby opening a rational approach to new chemotherapeutic agents; it may then be possible to substitute such new drugs for O_2, thus retaining its synergistic action without having to use expensive, cumbersome, and manpower-, time-, and space-consuming gadgetry.

Note added in proof:

Recently, Demello et al (291) confirmed the observation (203) that spores of *C. perfringens* are more resistant to the effects of OHP. Their conclusions that 2 or 3 Ata were bacteriostatic rather than bactericidal probably were due to insufficient time of exposure and diffusional limitations and therefore do not contradict previous reports (58; vide supra "Anaerobic Infections: Clostridia"). Hill & Osterhout (292) showed that *Clostridium bifermentans* and *Clostridium septicum* are more resistant to 3 Ata O_2 than *C. perfringens, C. tetani, C. novyi,* or *C. histolyticum: C. perfringens* was the most sensitive to OHP. These data lend further support to the hypothesis that responsitivity to high oxygen tensions may differentiate species (vide supra, "Introduction").

LITERATURE CITED

1. Adams, A. 1912. *Biochem. J.* 6:297–314
2. Adams, B. G., Parks, L. W. 1969. *J. Bacteriol.* 100:370–76
3. Allgood, M. A., Holmes, D. D. 1967. *Aerospace Med.* 38:169–70
4. Ajl, S. J., Werkman, C. H. 1949. *Proc. Soc. Exp. Biol. Med.* 70:522–24
5. Andrews, E. C., Rockwood, C. A., Cruz, A. B., Jr. 1969. *Texas Med.* 65:44–49
6. Appleby, C. A., Morton, R. K. 1954. *Nature (London)* 173:749–52
7. Appleby, C. A., Morton, R. K. 1959. *Biochem. J.* 73:539–50
8. Argamaso, R. V., Wiseman, G. M., Penn, I. 1966. *Can. J. Microbiol.* 12:863–64
9. Argamaso, R. V., Wiseman, G. M. 1967. *Plastics, Reconstr. Surg.* 40:81–85
10. Armstrong, J. McD., Coates, J. H., Morton, R. K. 1960. *Nature (London)* 186:1033–34
11. Arnar, O., Bitter, J. E., Haglin, J. J., Hitchcock, C. R. 1966. *Proc. Int. Conf. Hyperbaric Medicine, 3rd,* ed. I. W. Brown, Jr., B. G. Cox. 508–73. *Nat. Acad. Sci.-Nat. Res. Counc. Pub. 1404.* Washington, D.C.
12. Barach, A. L. 1926. *Am. Rev. Tuberc.* 13:293–316
13. Barnes, E. M., Ingram, M. 1956. *J. Appl. Bacteriol.* 19:117–28
14. Barnwell, P., Sopher, S., Fleckinger, R. R., II, Rhoden, C. H., Smith, I. M. 1966. *Am. Rev. Resp. Dis.* 94:756–76
15. Barron, E. S. G. 1953. *Texas Rep. Biol. Med.* 11:653–70
16. Bean, J. W. 1941. *J. Cell. Comp. Physiol.* 17:277–84
17. Benedict, R. G. et al. 1961. *Appl. Microbiol.* 9:256–62
18. Berghaus, S. 1907. *Arch. Hyg.* 62:172–99
19. Berry, L. J., Mitchell, R. B. 1953. *Texas Rep. Biol. Med.* 11:379–401
20. Berry, L. J., Mitchell, R. B., Rubenstein, D. 1955. *Proc. Soc. Exp. Biol. Med.* 88:543–48
21. Blair, E. et al. 1964. *J. Traumatol.* 4:652–63
22. Blum, J. J., Begin-Heick, N. 1967. *Biochem. J.* 105:821–29
23. Boerema, I., Brummelkamp, W. H. 1960. *Ned. Tijdsch. Geneesk.* 104:2548–50
24. Bornside, G. H. 1967. *Appl. Microbiol.* 15:1020–24
25. Bornside, G. H. 1967. *Proc. Soc. Exp. Biol. Med.* 125:1152–56
26. Bornside, G. H., Nance, F. C. 1969. *Antimicrob. Ag. Chemother. 1968:* 497–500
27. Brown, O. R., Huggett, D. O. 1968. *Appl. Microbiol.* 16:476–79
28. Brown, O. R., Silverberg, R. G., Huggett, D. O. 1968. *Appl. Microbiol.* 16:260–62
29. Browning, F., Bergendahl, J. C., Brittain, M. S. 1952. *Texas Rep. Biol. Med.* 10:790–93
30. Brummelkamp, W. H., 1965. *Ann. N.Y. Acad. Sci.* 117:688–99
31. Brummelkamp, W. H. 1964. *Clinical Application of Hyperbaric Oxygen,* ed. Boerema, I., Brummelkamp, W. H., Meijne, N. G., 20–30. Amsterdam: Elsevier
32. Brummelkamp, W. H. See Ref. 31, pp. 63–67
33. Brummelkamp, W. H. See Ref. 11, pp. 492–500
34. Brummelkamp, W. H., Boerema, I., Hoogendijk, J. L. 1963. *Lancet* 2:235–38
35. Brummelkamp, W. H., Hoogendijk, J. L., Boerema, I. 1961. *Surgery* 49:299–302
36. Caldwell, J. 1963. *Nature (London)* 197:772–74
37. Chalkley, H. W., Voegtlin, C. 1940. *J. Nat. Cancer Inst.* 1:63–75
38. Chance, B., Jamieson, D., Coles, H. 1965. *Nature (London)* 206:257–63
39. Chance, B., Jamieson, D., Williamson, J. R. See Ref. 11, pp. 15–41
40. Chandler, P. J., Allison, M. J., Margolis, G., Gerszten, E. 1965. *Am. Rev. Resp. Dis.* 91:855–60
41. Chen, W. L., Charalampous, F. C. 1969. *J. Biol. Chem.* 244:2767–76
42. Choe, B. K., Bertani, G. 1968. *J. Gen. Microbiol.* 54:59–66
43. Cleveland, L. R. 1925. *Biol. Bull.* 48:309–27
44. Ibid. 1925. 48:455–68
45. Cleveland, L. R., Burke, A. W., Jr. 1956. *J. Protozool.* 3:74–77
46. Cobet, A. B., Dimmick, R. L. 1970. *Aerosp. Med.* 41:617–20
47. Cobet, A. B., Wright, D. N., Warren, P. I. 1970. *Aerosp. Med.* 41:611–16
48. Colwill, M. R., Maudsley, R. H. 1968. *J. Bone Joint Surg.* 50B:732–42
49. Conger, A. D., Fairchild, L. M.

1952. *Proc. Nat. Acad. Sci. USA* 38:289–99
50. Coulson, D. B., Ferguson, A. B., Jr., Diehl, R. C., Jr. 1966. *Surg. Forum* 17:449–50
51. Coyne, R. V., Ackerman, G. A. 1969. *Aerosp. Med.* 40:1219–23
52. Coyne, R. V., Ackerman, G. A. 1969. *Aerosp. Med.* 40:1224–31
53. Dagley, S., Dawes, E. A., Morrison, G. A. 1951. *J. Bacteriol.* 61:433–41
54. Dalton, H., Postgate, J. R. 1967. *J. Gen. Microbiol.* 48:v
55. Dalton, H., Postgate, J. R. 1969. *J. Gen. Microbiol.* 54:463–73
56. Darby, R. T., Goddard, J. R. 1950. *Am. J. Bot.* 37:379–87
57. Davies, H. C., Davies, R. E. 1965. *Handbook of Physiology,* Sect. 3, Vol. 2, ed. W. O. Fenn, H. Rahn, 1047–58. Am. Physiol. Soc., Washington, D.C.
58. DeAlmeida, A. O., Pacheco, G. 1941. *Rev. Brasil. Biol.* 1:1–10
59. Dedic, G. A., Koch, O. G. 1956. *J. Bacteriol.* 71:126
60. DePalacios y Carrajal, J., Cué, A. A. 1969. *Int. Soc. Orthoped. Surg. Traumatol.* E 57
61. Dewald, R. R. 1966. *Appl. Microbiol.* 14:561–67
62. Ibid. 1966. 14:568–72
63. Dickens, F. 1946. *Biochem. J.* 40: 145–71
64. Dilworth, M. J. 1962. *Biochim. Biophys. Acta* 56:127–38
65. Dilworth, M. J., Kennedy, I. R. 1963. *Biochim. Biophys. Acta* 67: 240–43
66. Dilworth, M. J., Parker, C. A. 1961. *Nature (London)* 191:520–21
67. Dossa, J., Lefebvre, F., Serre, L., Joyeux, R., du Cailar, J. 1970. *Sem. Hop.* 46:125–26
68. Downey, R. J., Kiszkiss, D. F. 1969. *Microbios* 2:145–53
69. Downey, R. J., Kiszkiss, D. F., Nuner, J. H. 1969. *J. Bacteriol.* 98:1056–62
70. Dunlop, J., Trapp, W. G., Granshorn, J. A. 1968. *Can. Med. Assoc. J.* 99:1138–40
71. Ehrlich, R., Mieszkuc, B. J. 1969. *Aerosp. Med.* 40:176–79
72. Ehrlich, R., Mieszkuc, B. J. 1964. *Develop. Ind. Microbiol.* 5:207–15
73. Ehrlich, R., Mieszkuc, B. J. 1962. *J. Infect. Dis.* 110:278–81
74. Ehrlich, R., Mieszkuc, B. J. 1965. *Nature (London)* 207:1109
75. Evans, W. E., Darin, J. C., End, E., Ellison, E. H. 1964. *Surgery* 56: 184–92
76. Fay, P., Cox, R. M. 1967. *Biochim. Biophys. Acta* 143:562–69
77. Fenn, W. O., Gerschman, R., Gilbert, D. L., Terwilliger, D. E., Cothran, F. V. 1957. *Proc. Nat. Acad. Sci. USA* 43:1027–32
78. Fenn, W. O., Marquis, R. E. 1968. *J. Gen. Physiol.* 52:810–24
79. Filler, R. M., Reeves, E., Sleeman, H. K. See Ref. 11, pp. 593–600
80. Finn, R. K. 1954. *Bacteriol. Rev.* 18:254–74
81. Fischer, B. H. 1969. *Lancet* 2:405–9
82. Fletcher, R. D., Plastridge, W. N. 1964. *J. Bacteriol.* 87:352–55
83. Fredette, V. 1965. *Ann. N.Y. Acad. Sci.* 117:700–05
84. Fredette, V. See Ref. 11, pp. 555–62
85. Fredette, V., Plante, C., Roy, A. 1967. *J. Bacteriol.* 93:2012–17
86. Fredette, V. 1964. *Rev. Can. Biol.* 23:241–46
87. Gerschman, R. 1964. *Oxygen in the Animal Organism,* ed. F. Dickens, E. Neil, 475–92. New York: Macmillan
88. Gerschman, R., Gilbert, D. L., Frost, J. N. 1958. *Am. J. Physiol.* 192: 572–76
89. Gerschman, R., Gilbert, D. L., Nye, S. W., Fenn, W. O. 1956. *Fed. Proc.* 15:72
90. Gifford, G. D. 1968. *Biochem. Biophys. Res. Commun.* 33:294–98
91. Gifford, G. D. 1968. *J. Gen. Microbiol.* 52:375–79
92. Gifford, G. D. 1969. *J. Gen. Microbiol.* 56:143–49
93. Gifford, G. E. 1963. *Proc. Soc. Exp. Biol. Med.* 114:644–49
94. Gilbert, D. L. 1963. *Radiat. Res. Suppl.* 3:44–53
95. Gilbert, D. L., Frost, J., Caccaminc, D., Gerschman, R. 1957. *Fed. Proc.* 16:46
96. Gilbert, D. L., Gerschman, R., Cohen, J., Sherwood, W. 1957. *J. Am. Chem. Soc.* 79:5677–80
97. Giron, D. J., Pindak, F. F., Schmidt, J. P. 1967. *Aerosp. Med.* 38:832–34
98. Glad, R. M., Bouhoutsos, D. C., Douglass, F. M. 1965. *Am. J. Surg.* 109:230–32
99. Gladstone, G. P., Fildes, P., Richardson, G. M., 1935. *Brit. J. Exp. Pathol.* 16:335–48
100. Glover, J. L., Mendelson, J. 1964. *J. Trauma* 4:624–51
101. Gold, A. J., Silver, E. C., Hance,

H. E. 1964. *Aerosp. Med.* 35:563–77
102. Gorecki, Z. 1964. *J. Am. Geriat. Soc.* 12:1147–48
103. Gottlieb, S. F. 1965. *Advan. Clin. Chem.* 8:69–139
104. Gottlieb, S. F. 1969. *Bacteriol. Proc.* 80
105. Gottlieb, S. F. 1963. *Dis. Chest* 44: 215–17
106. Gottlieb, S. F. 1966. *J. Bacteriol.* 92:1021–27
107. Gottlieb, S. F. 1970. *Proc., Int. Congr. on Hyperbaric Medicine, 4th,* ed. J. Wada, T. Iwa. 288–96. Tokyo: Igaku Shoin Ltd.
108. Gottlieb, S. F. Unpublished results
109. Gottlieb, S. F., Cymerman, A. 1970. *Aerosp. Med.* 41:661–65
110. Gottlieb, S. F., Cymerman, A. See Ref. 11, pp. 97–104
111. Gottlieb, S. F., Pakman, L. M. 1968. *J. Bacteriol.* 95:1003–10
112. Gottlieb, S. F., Rose, N. R., Maurizi, J., Lanphier, E. H. 1964. *J. Bacteriol.* 87:838–43
113. Goulen, M., Rapin, M., Letournel, E., Barois, A. See Ref. 11, pp. 584–91
114. Gresley, J., Gottlieb, S. F. Unpublished results
115. Grogan, J. B. 1966. *Arch. Surg.* 92: 740–42
116. Grogan, J. B. 1966. *Bacteriol. Proc.* 65
117. Grogan, J. B. 1965. *Surg. Forum* 16:81–82
118. Halasz, N. A., Stier, H. A., Seiffert, L. N., Orloff, M. J. 1966. *Proc. Soc. Exp. Biol. Med.* 122:220–23
119. Hamblen, D. L. 1968. *J. Bone Joint Surg.,* 50:1129–41
120. Hanke, M. E., Bailey, J. H. 1943. *Proc. Soc. Exp. Biol. Med.* 59: 163–66
121. Hanson, G. C., Chew, H. E. R., Slack, W. K., Thomas, D. A. 1966. *Postgrad. Med. J.* 42:499–505
122. Haugaard, N. 1968. *Physiol. Rev.* 48: 311–73
123. Haywood, C., Hardenberg, H. C., Jr., Harvey, E. N. 1956. *J. Cell. Comp. Physiol.* 47:289–93
124. Hedén, C.-G. 1964. *Bacteriol. Rev.* 28:14–29
125. Hedén, C.-G., Malmborg, A. S. 1961. *Sci. Rep. Super. Sanita., 1st* 1: 213–21
126. Heinemann, B., Howard, A. J., Palocz, H. J. 1970. *Appl. Microbiol.* 19:800–4
127. Hess, G. E. 1965. *Appl. Microbiol.* 13:781–87
128. Hill, G. B., Osterhout, S. S. See Ref. 11, pp. 538–43
129. Hinton, D. 1947. *Am. J. Surg.* 73: 223–32
130. Hitchcock, C. R. See Ref. 107, p. 294
131. Hoge, S. F. 1932. *J. Arkansas Med. Soc.* 28:4–9
132. Hopkinson, W. I., Towers, A. G. 1963. *Lancet* 2:1361–63
133. Hunt, T. K., Ledingham, I. McA., Hutchison, J. G. P. See Ref. 11, pp. 572–77
134. Hurlbert, R. E. 1967. *J. Bacteriol.* 93:1346–52
135. Hutner, S. H. 1961. *Symp. Microbiol Reaction to Environment,* No. 12 Soc. Gen. Microbiol. 1–18
136. Irvin, T. T., Moir, E. R. S., Smith, G. 1968. *Surg., Obstet. Gynecol.* 127:1058–66
137. Irvin, T. T., Norman, J. N., Suwanagul, A., Smith, G. 1966. *Lancet* 1:392–94
138. Irvin, T. T., Smith, G. 1968. *Surgery* 63:363–76
139. Irvin, T. T., Suwanagul, A., Norman, J. N., Smith, G. 1967. *Brit. J. Surg.* 54:595–97
140. Irvin, T. T., Suwanagul, A., Norman, J. N., Smith, G. 1966. *Lancet* 2: 1222–24
141. Irvin, T. T., Suwanagul, A., Norman, J. N., Smith, G. 1967. *Surg. Obstet. Gynecol.* 125:1217–31
142. Islam, N., Chowdhury, R. I. 1963. *J. Trop. Med. Hyg.* 66:285–86
143. Jacobs, N. J., Jacobs, J. M., Sheng, G. S. 1969. *J. Bacteriol.* 99:37–41
144. Johnson, F. H., Eyring, H., Polissar, M. J. 1954. *The Kinetic Basis of Molecular Biology,* 286–368. New York: Wiley
145. Johnson, J. T., Gillespie, T. E. Cole, J. R., Markowitz, H. A. 1969. *Am. J. Surg.* 118:839–43
146. Johnson, M. J. 1958. *Int. Congr. Microbiol., 7th Stockholm,* 398–402. Springfield, Illinois: Thomas
147. Karasevich, E. G. et al. See Ref. 31, pp. 36–40
148. Karsner, H. T., Brittingham, H. H., Richardson, M. L. 1923. *J. Med. Res.* 44:83–88
149. Karsner, H. T., Saphir, O. 1926. *J. Infect. Dis.* 39:231–36
150. Katz, R. 1970. *Fed. Proc.* 29:892
151. Kaye, D. 1964. *Clin. Res.* 12:454
152. Kaye, D. 1967. *Proc. Soc. Exp. Biol. Med.* 124:360–64
153. Kaye, D. 1967. *Proc. Soc. Exp. Biol. Med.* 124:1090–93
154. Kelley, H. G., Page, W. G. 1963.

Surg. Forum 14:46–47
155. Klein, H. P. 1955. *J. Bacteriol.* 69: 620–27
156. Klopper, P. J. See Ref. 31, pp. 31–35
157. Klotz, P. G. 1968. *Invest. Urol.* 6:1–8
158. Knaysi, G., Dutky, S. R. 1934. *J. Bacteriol.* 27:109–19
159. Knox, R., Thomas, C. G. A., Lister, A. J., Saxby, C. 1961. *Guys Hosp. Rep.* 110:174–80
160. Knox, R., Lister, A. J., Thomas, C. G. A. 1957. *Soc. Gen. Microbiol. Proc.* 17 :IX–X
161. Kopper, P. H. 1952. *J. Bacteriol.* 63:639–45
162. Lemp, J. 1960. *J. Biochem. Microbiol. Tech. Eng.* 2:215–25
163. Lenhoff, H. M., Nicholas, D. J. D., Kaplan, N. O. 1956. *J. Biol. Chem.* 220:983–95
164. Levine, P. P. 1936. *J. Bacteriol.* 31:151–60
165. Libet, B., Siegel, B. V. 1962. *Cancer Res.* 22:737–42
166. Lindeberg, G., Lode, A., Somme, R. 1963. *Acta Chem. Scand.* 17:232–38
167. Lion, M. B., Avi-Dor, Y. 1963. *Isr. J. Chem.* 1:374–78
168. Lion, M. B., Bergmann, E. D. 1961. *J. Gen. Microbiol.* 24:191–200
169. Ibid. 1961. 25:291–96
170. Lippincott, C. L., Harter, W. L. 1963. *J. Am. Vet. Med. Assoc.* 142:872–74
171. Lockwood, W. R., Langston, L. L., Jr. 1970. *Curr. Therap. Res., Clin. Exp.* 12:311–16
172. Longmuir, I. S. 1954. *Biochem. J.* 57:81–87
173. Lück, H. 1953. *Schweiz. Z. Allg. Pathol. Bakteriol.* 16:728–39
174. Lück, H. 1954. *Schweiz. Z. Allg. Pathol. Bakteriol.* 17:106–17
175. Lwoff, A., Monod, J. 1947. *Ann. Inst. Pasteur* 73:325–47
176. MacLennan, D. G., Pirt, S. J. 1966. *J. Gen. Microbiol.* 45:289–302
177. Manheim, S. D., Voleti, C., Luddig, A., Jacobson, J. H. 1969. *J. Am. Med. Assoc.* 210:552–53
178. Martin C., Bildstein, G., Meyrand, M. S., Fagart, D., Laigle, J. C. 1969. *Ann. Pediat.* 16:113–16
179. Marty, A. T., Filler, R. M. 1969. *Lancet* 2:79–81
180. McAllister, T. A., Stark, J. M., Norman, J. N., Ross, R. M. 1963. *Lancet* 2:1040–42
181. McDaniel L. E., Bailey, E. G., Zimmerli, A. 1965. *Appl. Microbiol.* 13:109–14
182. Ibid. 1965. 13:115–19
183. McElroy, W. D., Selinger, H. H. 1962. *Horizons in Biochemistry*, ed. M. Kasha, B. Pullman, 91. New York: Academic Press
184. McLean, D. J., Pardie, E. F. 1952. *J. Biol. Chem.* 197:539–45
185. McSwain, B., Sawyers, J. L., Lawler, M. R., Jr. 1966. *Ann. Surg.* 163:859–65
186. Menzel, D. B. 1970. *Ann. Rev. Pharmacol.* 10:739–94
187. Milledge, J. S. 1968. *J. Am. Med. Assoc.* 203:875–76
188. Moore, B., Williams, R. S. 1909. *Biochem. J.* 4:177–90
189. Moore, B., Williams, R. S. 1911. *Biochem. J.* 5:181–87
190. Moss, F. 1956. *Aust. J. Exp. Biol. Med.* 34:395–405
191. Mortenson, L. E., Beinert, H. 1953. *J. Bacteriol.* 66:101–4
192. Muller, D. A. 1964. *S. Afr. Med. J.* 38:539–41
193. Naylor, H. B., Smith, P. A. 1946. *J. Bacteriol.* 52:565–73
194. Nora, P. F., Mousavipour, M., Laufman, H. See Ref. 11, pp. 544–57
195. Novy, F. G., Soule, M. H. 1925. *J. Infect. Dis.* 36:168–232
196. Nuckolls, J. G., Osterhout, S. S. 1964. *Clin. Res.* 12:244
197. Okada, T. 1968. *Nagoya J. Med. Sci.* 31:299–324
198. Ollodart, R. M. See Ref. 11, pp. 565–71
199. Ollodart, R. M., Blair, E. 1964. *J. Am. Med. Assoc.* 188 :450
200. Ollodart, R. M., Blair, E. 1965. *J. Am. Med. Assoc.* 191:736–39
201. Orsi, E. V., Mancini, R., Barriso, J. 1970. *Aerosp. Med.* 41:1169–72
202. Pace, D. M., Ireland, R. L. 1945. *J. Gen. Physiol.* 28:547–57
203. Pacheco, G., Costa, G. A. 1941. *Rev. Brasil. Biol.* 1:145–53
204. Pakman, L. M. 1970. *Bacteriol. Proc.* 99
205. Panov, A. G., Remezov, P. I. 1960. *Virology* 5:290–96
206. Parker, C. A., Scott, P. B. 1960. *Biochim. Biophys. Acta* 38:230–38
207. Parker, M. T. 1969. *Brit. Med. J.* 3:671–76
208. Pascale, L. R., Wallyn, R. J., Goldfein, S., Gumbinar, S. H. 1964. *J. Am. Med. Assoc.* 189:408–10
209. Pennock, C. A. 1966. *Lancet* 1:1348–50
210. Perrin, L. E., Ostergard, D. R., Mishell, D. R., Jr. 1970. *Am. J. Obstet. Gynecol.* 106:666–68

211. Perrins, D. J. D., Maudsley, R. H., Colwill, M. R., Slack, W. K., Thomas, D. A. See Ref. 11, pp. 578–84
212. Persson, B. M. 1968. *Acta Orthop. Scand. Suppl.* 117:1–99
213. Pfeiffer, C. C., Gersh, I. 1944. *Naval Med. Res. Inst. (Bethesda) Rep. No. 2.*
214. Pichinoty, F., D'Ornano, L. 1961. *Biochim. Biophys. Acta* 52:386–89
215. Pichinoty, F., D'Ornano, L. 1961. *Nature (London)* 191:879–81
216. Pine, L. 1954. *J. Bacteriol.* 68:671–79
217. Pritchard, G. G., Hudson, M. A. 1967. *Nature (London)* 214:345–46
218. Rabinowitz, M., Getz, G. S., Casey, J., Swift, H. 1969. *J. Mol. Biol.* 41:381–400
219. Reynolds L. L., King, J. W. 1967. *Cleveland Clin. Quart.* 34:171–77
220. Rhea, J. W., Graham, A. W., Jr., Akhoukh, F., Parthew, C. T. 1967. *J. Pediat.* 71:33–38
221. Rhoden, C. H., Lowry, L., Rabinovich, S., Smith, I. M. 1969. *Am. Rev. Resp. Dis.* 100:699–705
222. Richardson, B., Wagner, F. W., Welch, B. E. 1969. *Appl. Microbiol.* 17:135–38
223. Rö, J. 1965. *Acta Pathol. Microbiol. Scand.* 65:421–25
224. Robb, S. M. 1966. *J. Gen. Microbiol.* 45:17–29
225. Rockwell, G. E., Highberger, J. H. 1927. *J. Infect. Dis.* 40:438–46
226. Roesel, R. W., O'Sullivan, D. D., Baffes, T. G. 1969. *Illinois Med. J.* 136:580–83
227. Rogers, L. A. 1914. *J. Infect. Dis.* 14:100–23
228. Ross, R. M., McAllister, T. A. 1965. *Lancet* 1:579–81
229. Sanders, A. P., Hall, I. H., Woodhall, B. 1965. *Science* 150:1830–31
230. Sawicki, L., Barron, S., Isaacs, A. 1961. *Lancet* 2:680–82
231. Schmidt, J. P. 1969. *Fed. Proc.* 28:1099–1103
232. Schmidt, J. P., Ball, R. J. 1970. *Aerosp. Med.* 41:1238–39
233. Schmidt, J. P., Ball, R. J. 1967. *Appl. Microbiol.* 15:757–58
234. Schmidt, J. P., Cordaro, J. T., Ball, R. J. 1967. *Appl. Microbiol.* 15:1465–67
235 Schmidt, J. P., Cordaro, J. T., Busch, L. F., Jr., Ball, R. J. 1967. *US School Aerosp. Med.* TR–67–9
236. Schreiner, H. R. 1965. *Hyperbaric Oxygenation, Proc. Int. Conf., 2nd,* ed. I. McA. Ledingham, 267–74. Edinburgh: E. S. Livingstone, Ltd.
237. Schubert, J., Watson, J. A., White, E. R. 1967. *Int. J. Radiat. Biol.* 13:485–89
238. Sevcik, V., Vymola, F. 1964. *J. Hyg. Epidemiol. Microbiol. Immunol.* 8:313–17
239. Sherwood, R. J., Hagedorn, D. J. 1961. *Phytopathology* 51:492–93
240. Shilov, V. 1970. *Aerosp. Med.* 41:1353
241. Siegel, B. V. 1967. *Aerosp. Med.* 38:399–401
242. Siegel, B. V. 1967. *Life Sci.* 6:–1009–12
243. Siegel, S. M., Halpern, L., Davis, G., Giumarro, C. 1963. *Aerosp. Med.* 34:1034–37
244. Siegel, B. V., Morton, J. I. 1967. *Experientia* 23:758–59
245. Sinclair, P. R., White, D. C. 1970. *J. Bacteriol.* 101:365–72
246. Sippel, H. W., Nyberg, C. D., Alvis, H. J. 1969. *J. Oral Surg.* 27:739–41
247. Sistrom, W. R. 1965. *J. Bacteriol.* 89:403–8
248. Slack, W. K., Hanson, G. C., Chew, H. E. R. 1969. *Brit. J. Surg.* 56:505–10
249. Slack, W. K. et al. See Ref. 11, pp. 521–25
250. Slack, W. K., Thomas, D. A., Perrins, D. 1965. *Lancet* 1:1093–94
251. Smith, C. G., Johnson, M. J. 1954. *J. Bacteriol.* 68:346–50
252. Smith, G., McDowall, D. G. See Ref. 11, pp. 36–40
253. Smith, G. et al. 1962. *Lancet* 2:756–57
254. Stark, J. M., Orr, J. S. 1966. *Lancet* 2:108–9
255. Stuart, B., Gerschman, R., Stannard, J. N. 1962. *J. Gen. Physiol.* 45:1019–30
256. Swienciki, J. F., Hartman, R. E. 1967. *Can. J. Microbiol.* 13:1445–50
257. Szilagy, T., Toth, S., Miltenyi, L., Jona, G. 1968. *Acta Microbiol. Acad. Sci. Hung.* 15:5–9
258. Takahashi, H. 1968. *Jap. J. Dermatol.* 78:178
259. Thaysen, A. C. 1934. *Biochem. J.* 28:1330–35
260. Thurston, J. G. B. 1969. *A Controlled Trial of Hyperbaric Oxygen in Acute Myocardial Infarction-Preliminary Results.* Pre-

sented at Am. Heart. Assoc. Meeting, Dallas, Tex.
261. Trapani, I. L. 1966. *Fed. Proc.* 25: 1254–63
262. Van Elk, J., Trippel, O., Ruggie, A., Staley, C. See Ref. 11, pp. 526–37
263. Vaughan, G. L., Wilbur, K. M. 1970. *Fed. Proc.* 29:845
264. Van Unnik, A. J. M. 1965. *Antonie van Leeuwenhoek J. Microbiol. Serol.* 31:181–86
265. Van Zyl, J. J. W., Maartens, P. R., DuToit, F. D. See Ref. 11, pp. 515–20
266. Waid, J. S. 1962. *Trans. Brit. Mycol. Soc.* 45:479–87
267. Wagner, F. W., Welch, B. E. 1969. *Appl. Microbiol.* 17:139–44
268. Wallyn, R. J., Gumbinar, S. H., Goldfein, S., Pascale, L. R. 1964. *Surg. Clin. N. Am.* 44:107–12
269. Warburg, O. 1920. *Biochem. Z.* 103: 188–91
270. Watson, J. A., Schubert, J. 1969. *J. Gen. Microbiol.* 57:25–34
271. Webley, D. M. 1954. *J. Gen. Microbiol.* 11:114–22
272. White, J. H. 1928. *J. Okla. State Med. Assoc.* 21:59–60
273. Wiame, J. M., Bourgeois, S. 1955. *Biochim. Biophys. Acta* 18:269–78
274. Williams, J. W. 1939. *Growth* 3:21–33
275. Williams, J. W. 1938. *J. Lab. Clin. Med.* 24:39–43
276. Wilson, G. S. 1930. *J. Hyg.* 30:433–67
277. Wimpenny, J. W. T. 1969. *Symp. Soc. Gen. Microbiol. No.* 19:161–97
278. Winkel, C. A., Kroon, T. A. J. See Ref. 31, pp. 52–62
279. Wiseman, G. M., Violago, F. C., Roberts, E., Penn, I. 1966. *Can. J. Microbiol.* 12:521–29
280. Wittner, M. 1957. *J. Protozool.* 4: 20–23
281. Wittner, M. 1957. *J. Protozool.* 4: 24–29
282. Wolin, M. J., Evans, J. B., Niven, C. F., Jr. 1955. *Arch. Biochem. Biophys.* 58:356–64
283. Wright, G. P. 1955. *Pharmacol. Rev.* 7:413–65
284. Yamasaki, M. 1931. *Mem. Coll. Sci. Kyoto Imp. Univ. Ser. B* 7:179–88
285. Young, H. L. 1968. *Nature (London)* 219:1068–69
286. Young, H. L. 1969. *J. Bacteriol.* 97:1498–99
287. Zaroff, L., II, Walker, H. L., Lowenstein, E., Evans B. W., Kroos, S. 1965. *Arch. Surg.* 91:586–88
288. Zimmerman, A. M. 1970. *High Pressure Effects on Cellular Processes.* New York: Academic Press, 324 pp.
289. ZoBell, C. E. 1964. *Recent Researches in the field of Hydrosphere, Atmosphere, and Nuclear Geochemistry,* ed. M. Miyake, T. Koyama, 83–116. Tokyo, Japan: Mauruzen Co., Ltd.
290. ZoBell, C. E., Hittle, L. L. 1967. *Can. J. Microbiol.* 13:1311–1319
291. Demello, F. J., Hashimoto, T., Hitchcock, C. R., Haglin, J. J. 1970. *Proc. Int. Congr. on Hyperbaric Medicine, 4th.* Ed. J. Wada, T. Iwa, 276–81. Toyko: Igaku Shoin, Ltd.
292. Hill, G. B. Osterhout, S. 1970. See Ref. 291, 282–87

DNA RESTRICTION AND MODIFICATION MECHANISMS IN BACTERIA[1]

1566

Herbert W. Boyer

Department of Microbiology, University of California, San Francisco, San Francisco, California

Contents

Introduction

The restriction and modification of DNA as currently understood (10, 15, 16, 100) was first recognized as a unique biological mechanism because of the effect it has on the ability of a bacteriophage to propagate on related bacterial strains (1, 24, 85, 94). Similar observations were made prior to

[1] Work in the laboratory of the author has been supported by grants from the United States Public Health Service (GM 14378 and Training Grant AI 00299).

these reports, but they were never clearly defined or pursued on an experimental basis. For example, restriction and modification is part of the basis for phage typing of pathogenic bacteria established in the late 1930's (3, 41), and many of d'Herelle's observations (43) appear to be descriptions of the restriction and modification of phage.

Interest in the mechanism of restriction and modification was not stimulated until the early 1950's (1, 24, 85, 94). These experiments led to Luria's definition of the mechanism which clearly distinguished host-controlled modification of phage as a nonhereditary and phenotypic extension of host range rather than a permanent genotypic adaption (86). Thus, a bacterial strain A is considered to have a host modification and restriction mechanism if the restricted growth of a phage on host A as a result of one cycle of growth on host B is released upon one cycle of growth on host A. The host specificity of strain A, or the ability of strain A to discern the identity of the phage's previous host, is unique. This operational definition is still the most convenient way of demonstrating the presence of a functioning restriction and modification mechanism, but since not all phages are subject to a given restriction and modification mechanism, its usefulness is limited.

The elegant experiments of Arber & Dussoix published in 1962 redefined the restriction and modification of phage in molecular terms (7, 46). They proved that the modification of the phage imposed by the host was associated with the phage DNA, and that when modified DNA replicated in a nonmodifying host, the modification was maintained only in those progeny particles with one or both parental strands of DNA (7). Moreover, they found that unmodified phage DNA was rapidly degraded to nucleotides and nucleosides when injected into a restricting host (46). These observations led to the generalization that not only viral DNA but plasmid, episomal, and bacterial DNA were subject to restriction and modification and that the enzymes involved in this process were capable of making subtle distinctions between DNA molecules. Thus, the concept of host specificity could be explained on the basis of a specific sequence of base pairs being restricted or modified.

The restriction and modification of phage λ by several host specificities found in *E. coli* has been extensively exploited and has led to the general concept that two enzymes, an endonuclease and a methylase, are responsible for the restriction and modification of DNA. The restriction endonuclease makes a double-strand scission at a specific sequence of base pairs if it is unmodified, and the modification methylase modifies this sequence of base pairs to render it insensitive to the restriction endonuclease. This mechanism (*hs*) appears to be quite general in bacteria, but it should be pointed out that some restriction and modification system (e.g., the restriction and modification of T-even phages, 87) could have a different molecular basis.

The plan of this review is to treat the restriction and modification of DNA as a general bacterial mechanism. Emphasis will be placed on the two types of *hs* mechanisms associated with *E. coli,* and two models of protein-polynucleotide interaction are presented to account for the two types of restriction and modification mechanisms.

Restriction and Modification Terminology

Those mechanisms of restriction and modification based on endonuclease and methylase enzymes will be referred to as *hs* mechanisms. The following terminology has been adopted for the *hs* type of restriction and modification, and changes have been instituted only when dictated by recent experimental results (16). The restriction and modification of T-even phages will be referred to as the T* mechanism.

Host specificity.—The unique specificity of a restriction and modification mechanism is genetically controlled either by chromosomal alleles or by extrachromosomal elements. The designation for host specificities are either extracted from the strain or plasmid designation or arbitrarily enumerated. For example, the host specificities of *E. coli* K and B are designated K and B while the R-factor host specificities will be designated fi^+ and fi^-. The symbol "O" designates no known host specificity.

Phage modification.—The modification of a phage is designated by the host specificity symbol following the phage symbol, e.g., $\lambda \cdot K$ designates that the phage λ was grown on an *E. coli* $r^+_K m^+_K$ strain; fd · B designates that the phage fd was grown on an *E. coli* $r^+_B m^+_B$ strain, etc.

Phenotypes.—Functional restriction and modification mechanisms of the *hs* type are designated as r^+m^+, and the host specificity can be designated as a subscript, e.g., $r^+_K m^+_K$. Mutant phenotypes are denoted as $r^-_K m^+_K$, $r^-_K m^-_K$, $r^-_B m^+_B$ etc.

Enzymology of the *hs* Modification and Restriction Mechanisms

Much of the rationale for the experimental approach used to identify the enzymes involved in the modification and restriction mechanism relied on the information provided by the experiments of Arber and Dussoix (7, 8, 46). Other investigations provided additional guidelines which led to the identification of the restriction enzyme as an endonuclease and the modification enzyme as a DNA methylase in the *hs* mechanism. The *hs* enzymes involved in restriction and modification of DNA will be referred to as restriction endonucleases and modification methylases.

Since unmodified λ DNA was found to be rapidly degraded after infection of a restricting host, it was clear that a DNA nuclease (or nucleases) was involved in the restriction process (8, 46). Several other observations also indicated that an endonuclease must be involved in the restriction process. First of all, it was known (44, 46, 79) that genetic markers of unmodified λ could be rescued in a restricting host by recombining with a modified genome. Subsequently, it was shown that in bacterial crosses with a restricting female recipient the linkage of unselected markers to selected markers was drastically reduced (28, 40, 93). It could also be rationalized that only DNA endonucleases would have the versality necessary for the recognition of unmodified DNA.

The best bet for the chemical basis of host-controlled modification of DNA was the methylation of adenine and cytosine bases in DNA. The investigations of two laboratories had demonstrated that in bacteria there were two DNA methylases which resulted in the methylation (0.01 to 0.20 mole percent) of adenine to 6-methylaninopurine and cytosine to 5-methylcytosine (52, 58). And, more importantly, the DNA of different bacterial strains had unique methylation patterns, so that the DNA of one strain which would not act as a substrate for the homologous methylases would serve as a substrate for the methylases of another strain (58). Arber provided experimental support for this idea by obtaining partially unmodified λ phage from a Met⁻ *E. coli* K-12 lysogen whose DNA was synthesized while the cell was deprived of L-methionine (11). No loss of modification was found when L-proline or L-arginine were withdrawn during the lytic cycle of the induced λ. L-methione is a direct precursor of S-adenosyl-methionine (SAM), the methyl donor for the methylation reaction, thereby relating DNA methylation to host-controlled modification of DNA. An unpublished observation by Wood, quoted by Arber (11), that *E. coli* K-12 was nonrestricting when deprived of L-methionine explained how the partially unmodified λ escaped the host restriction. The effect of methionine on restriction has been confirmed for the B and K mechanism (61). The observation also suggested that SAM might be involved with the restriction process. In addition, nonreplicating and nonfunctioning λ DNA was found to be restricted and modified in vivo and DNA modification persisted through phenol extractions (7, 47, 80).

The first detection of in vitro restriction was reported in 1966 by Takano, Watanabe & Fukasawa (116) who showed that unmodified infectious λ DNA was inactivated more than modified infectious λ DNA by a crude extract of *E. coli* carrying an *fi*⁻ R-factor. Later work by Takano et al (117) demonstrated by zone centrifugation that crude extracts of the R-factor degraded unmodified λ DNA but not R-modified λ DNA. Only Mg^{++} was necessary for this activity.

Meselson & Yuan were the first to purify and characterize a restriction endonuclease in detail (88). They achieved a 5000-fold purification of the *E. coli* K restriction endonuclease and, in contrast to the R-factor restriction endonuclease, the K-12 restriction endonuclease was found to require ATP, SAM, and Mg^{++} for activity on unmodified DNA.

With additional purification and characterization of the *fi*⁻ R-factor restriction endonuclease (107), the *E. coli* B restriction endonuclease (84, 106), and a restriction endonuclease of *Hemophilus influenzae* (113), it is now clear that there are at least two types of restriction endonucleases which are responsible for the *hs* system of restriction and modification. The enzyme(s) involved in the restriction of T-even phage have not been detected in vitro but may represent a third type of restriction endonuclease (100).

The E. coli *K and B restriction endonuclease.*—The *E. coli* K-12 and B restriction endonucleases are representatives of the first class (Type I) of restriction endonucleases. These enzymes have stringent requirements for ATP, SAM, and Mg^{++} for endonuclease activity on unmodified DNA (84, 88, 106). The substrate specificity has been determined with modified and unmodified λ DNA and modified and unmodified fd RF DNA in zonal velocity centrifugation experiments (88, 106) and infectious biological assays (84, 88). With purified preparations of these enzymes only unmodified DNA is degraded or inactivated. They produce a limited number of double-strand breaks in unmodified DNA with no single-strand breaks in the product fragments (88).

Both the B and K endonucleases produce double-strand breaks by a two-step mechanism (88, 106). By following the kinetics of the production of nicked circles and linear molecules from supercoiled DNA by the restriction endonuclease, it has been shown that a phosphodiester bond cleavage is made in one strand and followed by another cleavage nearby in the opposite strand. This temporal dissociation of the two scissions has been shown to be the result of a physical dissociation of the enzyme-substrate complex after the first cleavage occurs (30). This was demonstrated by showing that the production of linear molecules from unmodified, supercoiled fd RF DNA with one substrate site for the B restriction endonuclease was inhibited by the addition of cold, unmodified *E. coli* DNA shortly after the initiation of the reaction. The assumption was made that all possible enzyme-substrate complexes occurred immediately after the addition of the enzyme; this is supported by binding studies (30, 126). At the time of the addition of competitive DNA (30 sec), 23 percent of the supercoiled DNA (RFI) was converted to nicked circular DNA (RFII), but no linear DNA (RFIII) had been formed. Without competitive DNA, an additional 90 sec of incubation resulted in a distribution of 42 percent RFII and 38 percent RFIII, while 10 min of incubation yielded 5 percent RFII and 95 percent RFIII. Competitive DNA at a mass ratio of 160 (μg unmodified *E. coli* DNA/μg unmodified fd RFI DNA) added at 30 sec after the reaction was initiated resulted in a 50 percent inhibition of the appearance of RFIII after an additional 90 sec of incubation. Complete inhibition was reached at a mass ratio of about 500. The appearance of RFII was not significantly altered by the competitive DNA, and 90–95 percent conversion of RFI to RFIII occurred after 15 min additional incubation. These results predict that the RFII generated by the restriction endonuclease would also serve as a substrate for the restriction endonuclease and be converted to RFIII. This was confirmed by showing that RFII generated by the B restriction endonuclease and purified by zone centrifugation was completely converted to RFIII by B restriction endonuclease.

It has been demonstrated clearly that the B restriction endonuclease does not attack unmodified single-strand DNA since no in vitro restriction of unmodified fd virion DNA is observed under conditions which give complete inactivation of unmodified fd RF DNA in an infectious assay (84). Also, λ DNA consisting of one modified and one unmodified polynucleotide strand is completely resistant to attack by the K restriction endonuclease (88). Thus, a newly replicated restriction endonuclease substrate site would be resistant to cellular restriction endonuclease molecules for an additional round of replication, more than sufficient time for the modification methylase fully to modify the hybrid site. This would normally be sufficient leeway since it is known that methylation of DNA occurs soon after a substrate site is replicated (26, 76). It has been shown that a restricting cell will degrade its own DNA if methylation can be suppressed for one round of replication and if the restriction endonuclease does not require SAM (77, 78).

The K and B restriction endonucleases sediment at 11S-12S in zonal centrifugation experiments and their estimated molecular weights are 250,000 daltons (88, 107). On the basis of in vivo complementation experiments, they consist of more than one type of subunit (29, 68). These two enzymes appear to be very similar except for substrate specificity. Indeed, the genetic loci of the K and B restriction endonucleases are allelic and complementation experiments have shown that their subunits are interchangeable (28, 29, 68, 123). This suggests a common genetic origin for these enzymes, and it will be of interest to examine the restriction endonuclease of *E. coli* 15 which appears to be a member of this family (17, 78, 115).

The role of ATP and SAM in the endonucleolytic reaction is not clear. Only dATP can replace ATP in the reaction, although less effectively, and no substitute for SAM is known (84, 88, 106). The requirement for SAM is especially interesting, considering the possible structural relationship between the restriction endonuclease and the modification methylase. In vivo complementation experiments suggest that the restriction endonuclease is the modification methylase molecule with a third type of subunit (68). In the case of the modification methylase, SAM is a substrate, presumably as it is in other methylase reactions (75), but in the restriction mechanism it is not altered to any measurable extent (127).

However, during the course of the K and B restriction endonuclease reactions, an enormous amount of ATP is hydrolyzed to ADP + P_i, far in excess (5000-fold) of the amount of phosphodiester bonds cleaved (64, 127). No ATP hydrolysis occurs with modified DNA or single-strand DNA in lieu of unmodified native DNA.

The kinetics of ATP hydrolysis as a function of SAM concentration suggest that SAM is an allosteric effector for this reaction (64). In the absence of SAM some ATP hydrolysis occurs (about 1/20 of the normal rate), perhaps because of the structural similarity between substrate and effector. The relationship between ATP hydrolysis and ATP concentration is linear for a given concentration of SAM. Thus, the type I restriction endonucleases are complex multisubstrate enzymes with apparently heterotropic ligand effects (73). It will be of interest to compare the complexity of these enzymes and their substrates *vis a vis* the less complex restriction endonucleases (see below) and their substrates (124).

The filter technique used to measure repressor-operator DNA interaction (102) has been used to study the interaction of the K and B restriction endonucleases with unmodified DNA (30, 126). Detectable binding requires SAM, but up to 50 percent of the normal binding takes place in the absence of ATP. It is possible that SAM effects the conformational state of the restriction endonuclease required for stabilization of the enzyme-DNA complex, and ATP hydrolysis occurs during or after, or both, phosphodiester bond cleavage takes place. It should be remembered that no detectable phosphodiester bond cleavages occur in the absence of SAM or ATP so ATP, but not necessarily ATP hydrolysis, is necessary for endonuclease activity. Unmodified DNA exhaustively treated with restriction endonuclease does not stimulate the hydrolysis of ATP, and the addition of pancreatic DNase during the course of the reaction immediately inhibits ATP hydrolysis (64).

What is the role of ATP in the reaction? In the absence of SAM and unmodified DNA the enzyme is inactivated but no ATP is hydrolyzed, so it is not involved in direct regulation of the enzyme activity (64, 88, 106). Nor is it likely to be involved in the process by which the enzyme screens DNA for unmodified substrate sites, since no hydrolysis occurs with modified DNA (64, 127). There is a good deal of similarity between the ATP-dependent exonuclease of *E. coli* (22, 33, 90) and the ATP-dependent restriction endonucleases, which suggests a possible common role for ATP in these cases. The exonuclease hydrolyzes ATP to ADP and P_i, and it is structurally complex with an estimated molecular weight of 350,000 daltons (90) and at least two subunits controlled by the *recB* and *recC* genes (22). It is also enzymatically complex, having at least two nuclease activities associated with it (59). Either ATP hydrolysis is involved in phosphodiester bond cleavage or it occurs after cleavage for some other purpose such as the regulation of enzymatic activity.

Hemophilus and fi$^-$ *restriction endonucleases.*—The second type of restriction endonuclease (Type II) is represented by the *Hemophilus* and *fi*$^-$ R-factor enzymes. Both of these enzymes have been purified extensively and have been found to require only Mg^{++} for activity on unmodified DNA (107, 113, 117). The R-factor enzyme has an estimated molecular weight of 72,000 daltons on the basis of zonal velocity centrifugation experiments (107), and on the basis of gel filtration the *Hemophilus* enzyme has a molecular weight of 67,000 daltons (113). The third feature which distinguishes these enzymes from the Type I enzyme is that they produce more double-strand breaks per given length of unmodified genome. Both the *He-*

mophilus and R-factor restriction endonuclease make approximately one break per 1000 base pairs (71, 107), while the B restriction endonuclease makes 1–2 breaks per 5000 base pairs (30).

The *Hemophilus* restriction endonuclease appears to make double-strand breaks by a two-step procedure similar to Type I restriction endonucleases. No data are available for the R-factor restriction endonuclease. Neither enzyme uses unmodified single-strand DNA as a substrate. The similarities between the *Hemophilus* and R-factor restriction endonucleases, even though they are found in organisms of different genera, suggest that these enzymes are representatives of a second major class of restriction endonucleases.

Modification methylase.—At the present time only one modification enzyme, from *E. coli* B has been purified and characterized (74, 75). The modification activity was detected by using an fd RF DNA biological assay (23). The modification activity imparts an increased ability of unmodified fd RF DNA to produce a lytic response in restricting *E. coli* B spheroplasts. The only requirement for this activity is SAM. The reaction has optimum activity at pH 6.0, 30°C, in contrast to a pH optimum of 7.6, 30°C, for the B restriction endonuclease. Using an ^3H label in the donor methyl moiety of SAM, it has been demonstrated that about two methyl groups are incorporated into unmodified fd RF DNA per substrate site, and the radioactivity is associated only with 6-methylaminopurine (75). This observation fulfills the prediction that there are two modified elements (methylated bases) per substrate site, and presumably there is one in each polynucleotide strand (7, 70, 88). No incorporation is detected with B-modified fd RF DNA or with fd RF DNA made from mutants resistant to the *E. coli* B restriction endonuclease. The latter observation is in keeping with the idea that the restriction and modification methylase have a common subunit involved in the recognition of the same substrate site. Because of the results of recent in vivo experiments (77, 78), it is fairly safe to predict that the Type II modification enzymes will be methylases.

Thus, one biological role of methylated bases in DNA is that of protecting certain regions of DNA from endonucleolytic attack. However, the methylation of bacterial and phage DNA involved in host-controlled modification represents only a small part of the total methylated bases in a given genome, since isogenic strains of bacteria differing only in modification function by virtue of mutation or Pl prophage had no detectable differences of 5-MC or 6-MAP in bacterial DNA or λ DNA (60). It has also been shown that isogenic strains differing only in K and B restriction and modification alleles had the same quantitative patterns of methylation, and strains isogenic except for the 5-methylcytosine methylase gene had no change in restriction and modification specificity (87). Thus, the biological role of the majority of methylated bases in bacterial DNA and phage DNA remains obscure.

The restriction endonuclease and modification methylase substrate.—The substrate for the *hs* restriction endonuclease and modification methylase is postulated to be a specific sequence of base pairs. Genetic experiments demonstrating marker rescue (28, 44, 46, 93) or function rescue (118) and experiments with restriction endonuclease-treated unmodified λ DNA and fd RF DNA indicate a limited number of substrate sites per genome (30, 88, 106). This has been confirmed by the isolation of mutants of fd phage resistant to the B restriction mechanism (14). Ordinarily, the efficiency of plating (eop) of fd·0 on *E. coli* r^+_B m^+_B is 10^{-4}, but the partially resistant mutants plate with an eop of 10^{-2}. Fully resistant mutants were obtained after a second selection procedure. A related phage, M13, is restricted by *E. coli* B with an eop of 10^{-2} and a one-step mutational event leads to complete resistance to B restriction. Arber & Kühnlein postulated that fd has two restriction sites per genome and M13 has one per genome, and loss of the restriction site occurs by mutation to yield phages partially or fully resistant to the restriction mechanism (14). This hypothesis has been confirmed in the following way (30). Unmodified radioactively labeled RFI DNA was prepared from W.T. fd phage and a partially resistant fd phage, treated with the B restriction endonuclease, and subjected to zonal velocity centrifugation. The mutant RFI DNA was degraded to a form with the properties of a linear fd RF molecule and the WT RFI DNA was degraded to fragments about ⅔ and ⅓ of the size of the linear molecule. Protein-DNA binding studies showed that the WT RF DNA had twice as many B restriction endonuclease binding sites as the RF DNA prepared with the partially resistant mutant (30). When RF DNA of the fully resistant fd phage was treated in vitro with the B restriction endonuclease, no double- or single-strand scissions were observed. However, significant binding of the restriction endonuclease to RF DNA made from one fully resistant mutant was found; this suggests that binding and recognition of the substrate site by the restriction endonuclease is independent of phosphodiester bond cleavage.

The B modification methylase does not use the mutant fd RF as a substrate (75), which confirms the idea that the restriction endonuclease and modification methylase recognize the same substrate by virtue of having a common recognition subunit (16, 29). Even though these two enzymes interact with the same unmodified substrate, the restriction endonuclease does not interact with a partially modified substrate (7, 70, 88). The modification methylase must interact with the partially modified substrate to make it fully modified, and interact with the unmodified substrate as well.

Since no protein-DNA complex is detectable between restriction endonuclease and modified DNA (30, 126), the presence of a methyl group at the 6 amino position of adenine must have a pronounced effect on the ability of the restriction endonuclease to bind to the recognition site. Mutational changes of the substrate site lead to complete loss of binding capacity or to residual binding capacity (30). Thus, some base pair change of the sequence has less effect on the recognition process than the presence of a methyl group in the sequence.

The nature of the substrate site.—At the present time only the sequence of base pairs recognized by the *Hemophilus* (R) restriction endonuclease is known. Kelly & Smith (71) labeled the 5′-phosphoryl groups of the 5′-termini made by the R restriction endonuclease with polynucleotide kinase (101), and analyzed the sequence or bases at the 5′- and 3′-termini by specific nuclease digestion and chromatographic procedures. On the basis of these data they were able to construct the following sequence:

```
                ↓
5′ · · · G T Py Pu A C · · · 3′
          p p  p  p p
3′ · · · C A Pu Py T G · · · 5′
          p p  p  p p
               ↑
```

The enzyme cleaves the phosphodiester bond where the arrows indicate to generate 5′-phosphoryl, 3′-hydroxyl termini. This sequence should be generated at random at a frequency of 10^{-3}, which agrees with experimental observations (71).

The most interesting feature of this sequence is the twofold rotational symmetry perpendicular to the helix axis. In other words, the sequence of bases in each strand is the same when oriented in the same direction if the partial variability of the internal base pairs is ignored. As pointed out by Kelly & Smith, the symmetry of the site probably reflects some underlying symmetry in the *Hemophilus* restriction endonuclease and suggests that this will be a general feature of the restriction and modification substrate. It will be of great interest to determine the sequences of other restriction endonuclease sites [see Arber & Linn (16) for a comprehensive discussion of the possible types of sites and site symmetry].

Models of restriction endonuclease and modification methylase interaction with a symmetrical substrate.—Simplified models for the Type I and II restriction endonuclease and modification methylase reactions can accommodate most of the experimental observations made with these enzymes. The major assumption is that the substrate for both classes of enzymes is a symmetrical sequence of base pairs similar to the one found for the *Hemophilus* restriction endonuclease. The number of base pairs—6, 8, or 10—comprising the site is not important for the development of these models.

Figure la depicts a possible interaction of the Type I modification methylase with a hypothetical substrate. The principle feature of this model is that the two recognition subunits, related by a twofold axis of symmetry, recognize the same sequence of bases on opposite polynucleotide strands, and that methylation takes place on the strand not directly interacting with the recognition subunits. In this diagram, both recognition subunits are placed in the same region of the major groove, although they could just as well be depicted as clamping over the minor groove, i.e., straddling the two phospho-deoxyribose chains, and interacting with the symmetrical sequence. The position of the modified base in this model is arbitrarily placed and could be placed at any of the three possible positions. SAM, bound to the modification subunit serves as a methyl donor for the methylation of adenine in the 6 amino position. Either a fully unmodified or partially unmodified (i.e., one of the adenines is not methylated) sequence serves as a substrate for the methylase (75). It is not known if both methylation events occur in one enzyme-substrate complex in the former case, although the partially modified substrate appears to be the normal substrate for the methylase.

In Figure 1b, the Type I restriction endonuclease is depicted interacting with the same site. Hubacek & Glover (68) recently presented evidence that the restriction endonuclease is the modification methylase plus a third subunit. The third subunit is therefore responsible for phosphodiester bond cleavage. The observation that the restriction endonuclease is not capable of introducing nicks on either strand if one or the other strand is modified, can be accounted for by the modification subunit interacting with either methylated base which rejects the protein by virtue of an allosteric inhibition (73). It is not yet known if the enzyme binds to the partially modified substrate, but it certainly does not bind to the fully modified substrate (30, 126). Binding to the unmodified site requires SAM bound to the ligand site on the modification subunit. Once bound, the restriction subunit, in the presence of ATP and Mg^{++}, would cleave the phosphodiester bonds at the same point on the opposite strands. There is no experimental evidence to support this position for the breaks of the Type I restriction endonuclease, but it is assumed on the basis of the general similarity to the R restriction endonuclease (71). Presumably, the ATP is bound to the restriction subunit and ATP hydrolysis would either take place during or after phosphodiester bond cleavage. The reason for the dissociation of the enzyme-substrate complex after one phosphodiester bond cleavage (30) is not made clear by the model.

Models for the Type II restriction endonuclease interacting with a symmetrical site have been suggested (71) and are similar to the ones presented here. The uncertain relationship between the modification enzyme and restriction endonuclease makes it a little harder to consider a detailed model for this mechanism. This class of restriction endonucleases is not as structurally complex as Type I enzymes, and the genetics suggest that the modification and restriction functions do not share a common factor (see below). The small molecular weight of the Type II restriction endonucleases (70,000 daltons) limits the possible number of subunits to two, and in order

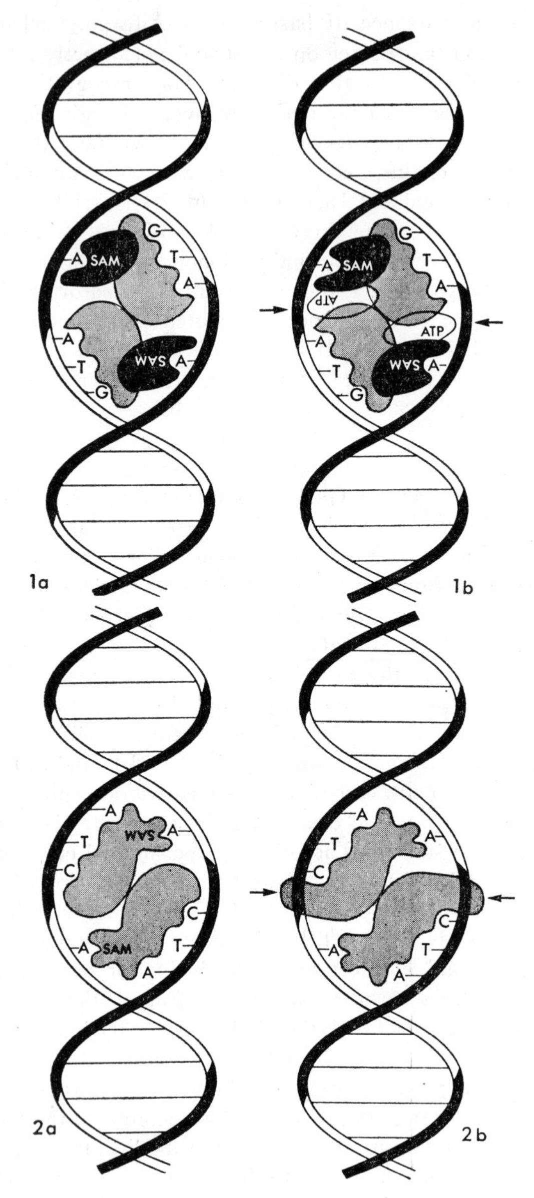

FIGURES 1 & 2. Schematic representations of Type I (1a & 1b) and Type II (2a & 2b) restriction endonucleases and modification methylases interacting with a hypothetical symmetrical substrate. See text for discussion.

to retain the utility of the twofold symmetry of the site, the Type II restriction endonuclease and modification methylase enzymes must have two recognition subunits per molecule. The Type II restriction and modification enzymes could be constructed of polypeptide chains which are very similar and have a common polypeptide sequence capable of recognizing the same substrate. The modification methylase containing two identical subunits could interact with the substrate site as depicted in Figure 2a and modify the adenine residues in the strands opposite the recognition sequence. The restriction endonuclease containing two identical subunits similar to the methylase subunits could interact with the same substrate as depicted in Figure 2b and introduce double-strand scissions at the indicated point. The restriction endonuclease subunit must have some remnant of the methylase portion of the polypeptide which can still interact with the modified base to account for its inability to restrict modified DNA. This model predicts one gene for the restriction endonuclease and one for the modification methylase, but assumes that part of both genes are identical and evolved by gene duplication.

The models presented above explain why these enzymes require double-strand DNA for activity. If both recognition subunits must interact with their substrates, then both polynucleotide strands must be present. This would also explain the partial loss of restriction activity in cells containing recognition subunits with different host specificities, which are capable of interacting (29, 68). The recognition subunit may also require a rigid structure furnished by the proper stacking of bases in the helical structure.

It is of interest to consider some of the consequences of these models with respect to the evolution of different host specificities and the possible number of different host specificities that might exist in bacteria. The number of possible sites and therefore possible host specificities that could be generated for a site similar to the one recognized by the *Hemophilus* restriction endonuclease would be 16 if one base pair must remain fixed for methylation, and if recognition of the internal base is unambiguous (16, 125). Thus, one family of restriction and modification enzymes with at most 16 different host specificities could evolve for this particular set of sequences. It is conceivable that another family of enzymes could evolve for a similar sequence, but with a fixed modifiable base at another position. This would result in three families of related enzymes with a potential for 48 host specificities. If four base pairs composed the subsite, there would be four possible families with 256 potential host specificities.

The versatility of the Type I restriction and modification mechanism is in the fact that alteration of one gene, the gene controlling the recognition subunit, could alter the specificity of the restriction and modification enzymes at the same time. However, in the case of the Type II mechanism, the generation of host specificities becomes more complicated. For the Type II mechanism, all new specificities would have to evolve through an intermediate state in which the enzymes have ambiguous recognition of one or more bases. One can imagine that a Type II modification methylase would develop ambiguous recognition at one base without serious consequence. By virtue of recombination with the gene for the restriction endonuclease, or by a second mutation occurring in this gene, both enzymes could acquire the ambiguity. Additional mutation or recombination, or both, could lead to a new host specificity without ambiguity.

Genetic Control of Restriction and Modification Mechanisms

Restriction and modification functions are controlled by chromosomal genes (28, 36, 39, 82, 123), phage genes (53), and extrachromosomal elements such as R-factors (12, 18–20, 62, 120, 121), and defective P1-like plasmids (78). There are several selective procedures available for isolating mutants defective in restriction function, and these are based on the recovery of transductants, recombinants, or lysogens from a restricting host (37, 38, 53, 57, 123). The most striking feature of the r^- mutants is that in most cases about 50 percent of these are m^- (21, 38, 53, 104, 123). However, 35 r^- mutants of an fi^- R-factor (R15), isolated in three different backgrounds ($r^+_B m^+_B$, $r^+_K m^+_K$, and *E. coli* C) were all m^+ (125). This is in conflict with another report that 5 out of 10 r^- mutants of a similar fi^- R-factor were m^- (21). Since drug resistance markers can be lost easily by virtue of deletions (49), the latter observation might be misleading. Unfortunately, no genetic analysis of the *Hemophilus* system or other potential Type II restriction and modification mechanisms has been reported.

The *E. coli* K, B, and A (strain 15T⁻) host specificities are controlled by alleles mapping close to *serB* (17, 28, 56, 82, 115, 123) while the *hs* complex of *Salmonella typhimurium* maps near *proC* (39, 128). The genetic control of restriction and modification in *Pseudomonas aeruginosa* strains appears to be a more complex situation (66, 67). Two different host specificities have been identified with R-factors (18-21, 120, 121) in *E. coli* and apparently the same two have been associated with R-factors in *Salmonella panama* (62, 63). Many independent isolates of fi^- R-factors have a restriction and modification mechanism which restricts unmodified λ at an eop of 10^{-2}, while only one fi^+ R-factor has been found in *E. coli* which restricts unmodified λ at an eop of 10^{-4} (21). The fi^+ restriction and modification is found more frequently in *Salmonella* strains (62). The two host specificites are mutually exclusive.

The prophage P1 and a plasmid related to phage P1 genetically control mutually exclusive host specificities (53, 78).

Mutant phenotypes other than those described above have been found in several systems, but these have not been analyzed extensively in vitro. These include $r^{\pm}m^+$, and $r^-m^{\pm}$ mutants as well as temperature-sensitive mutants. A clone restricting unmodified λ with an eop of 10^{-2} is referred to as $r^{\pm}$, while a clone which modifies λ so that it plates with an eop of 10^{-2} on the parental host strain is said to have an $m^{\pm}$ phenotype. These mutants may have partially defective restriction endonucleases and modification methylases. The $m^{\pm}$ phenotype can be defined with λ phage, but when fd phage is grown on the $r^-m^{\pm}$ mutant, it is not modified (32). This may be explained by the different modes of DNA replication, genome size, and the inability of the single strand fd genome to be modified. In vitro degradation of fd RF prepared from the $m^{\pm}$ strain is complete while λ $m^{\pm}$ DNA is slightly less degraded than unmodified DNA (32). Perhaps the λ DNA has two sequences which B restriction and modification enzymes recognize, and the $r^-m^{\pm}$ mutant retains the capacity to modify one of the sites.

The r^+m^- phenotype has not been observed and two possible explanations for this failure have been discussed (29, 68).

Complementation studies.—Complementation analyses of the various restriction and modification mutants of *E. coli* K and B have revealed much of what is currently known about the relationship between the restriction and modification enzymes of the Type I mechanism. Analysis of the restriction and modification properties of permanent diploids with various arrangements of mutant and wild-type alleles and of different host specificities has demonstrated that at least three cistrons are involved in the K and B restriction and modification mechanisms (30, 55, 68, 84). These cistrons have been designated *hss, hsr,* and *hsm* (84). Most mutents with r^-m^- phenotypes in the haploid state can be complemented by the cistronic products of the r^-m^+ mutant to restore the r^+m^+ phenotype to the diploid. Moreover, if r^-m^- and r^-m^+ mutants derived from parents with different host specificities are established in a diploid arrangement, e.g., $r^-_Km^+_K/r^-_Bm^-_B$, the restored restriction function has only the host specificity associated with the r^-m^+ mutant. Thus, the defective cistron resulting in the r^-m^- phenotype, *hss,* makes an inactive recognition factor associated with the restriction and modification function. The r^-m^+ mutant has an intact modification function and therefore must have a defective cistron, *hsr,* controlling another factor associated with the restriction function. These observations are interpreted on the basis of the restriction endonuclease and modification methylase having a common protein subunit, the *hss* cistronic product, which is responsible for these two enzymes recognizing the same substrate (29).

The third cistron, *hsm,* has been defined in two ways (29, 68). Some modification mutants derived from r^-m^+ (*hsr*$^-$) strains and which are phenotypically r^-m^- can be complemented by r^-m^- (*hss*$^-$) strains so that the diploid is r^+m^+ with the parental host specificity of the double mutant (29). Thus, the double mutant must have an intact *hss* cistron and the mutation responsible for the m^- phenotype must be in a third cistron, *rsm,* associated with the modification function. Recently, the $r^-m^{\pm}$ mutant phenotype has been demonstrated to be the result of an altered *hsm* cistron (68). The cistronic products of an $r^-_Bm^+_B$ (*hsr*$^-$) mutant complement the $r^-_Km^{\pm}_K$ cistronic products to restore the B and K restriction and modification specificities to the diploid. Therefore, the $r^-m^\pm$ mutant must have an intact *hsr* cistron and in order to explain these results Hubacek & Glover (68) postulated that the restriction endonuclease contains all three cistronic products. Thus, some *hsm*⁻ mutations might result in a loss of the ability of the subunits to interact normally so that the *hsr* product would not be able to form a stable enzyme in the presence of the mutant *hsm* product but this would result in a partially active modification methylase. This interpretation is incorporated in the model presented above for the interaction of the Type I restriction endonuclease and modification methylase with a common substrate. The K and B restriction endonucleases are certainly large enough (250,000 daltons) to accommodate 4–6 protein subunits.

When diploids contain WT alleles of the K and B host specificity, they restrict λ·C, λ·K, and λ·B phage but with efficiencies less than the respective haploids. This has been interpreted by assuming two *hss* subunits per restriction endonuclease molecule, so that in such a diploid half of the restriction endonuclease molecules would be hybrid with respect to recognition specificities and would probably be nonfunctional (29). This possibility is one of the features of the general model presented above which relates the restriction endonuclease structure to the possible twofold dimensional symmetry of restriction and modification substrate sites (71).

Clearly, the products of the *hsr* and *hsm* cistrons are interchangeable (29, 55, 68), and therefore it is not surprising that the K and B restriction endonucleases have the same enzymological features. The K and B host specificities must be the result of alterations in the *hss* cistron, and therefore they have a common origin. The nice feature of this model is that it provides an easy way for the generation of new host specificities. A mutation of the *hss* cistron resulting in a protein with the ability to recognize a different substrate simultaneously changes the substrate recognition capacity of the restriction endonuclease and modification methylase (29). No complementation occurs between the Pl restriction and modification proteins and the K or B restriction and modification proteins. There is some question as to whether or not the Pl restriction and modification mechanism is Type I or Type II (61, 78, 88), or perhaps a third type.

Some r^- m^- mutants isolated as one-step mutants are trans-dominant to the wild type and r^- m^+ alleles (29, 55). These mutants could possibly represent mutations in a fourth gene, responsible for gene regulation, or they could represent alterations of the *hsm* cistron so that the mutant protein interacting with the wild-type proteins inactivates the complex enzyme structures. The trans-dominant mutants are of two types: one class is dominant to the restriction function only and one is dominant to restriction and modification (29). These are interpretable on the basis of the model of Hubacek & Glover (68).

Recent experiments of Lark (77) and Lark & Arber (78) have shown that the various *E. coli* restriction and modification mechanisms could be divided into two classes. Lark observed that a culture of a Met⁻ *E. coli* strain grown in the presence of L-methionine analogs, such as L-ethionine, started to die and degrade DNA after one round of replication (77). This was subsequently shown to be a result of the restriction and modification mechanism of the cell, since r^- mutants of the strain remained viable and did not degrade its DNA under the same condition (78). However, this effect was not seen with all restriction and modification specificities. The presence of K, B, and A restriction and modification mechanisms do not result in cell death and DNA degradation when the cells are grown in the presence of ethionine. The presence of the fi^- R-factor, the 15 plasmid, and the P1 phage restriction and modification mechanisms result in cell death and DNA degradation. The restriction endonucleases of K and B (and A by inference) require SAM for endonuclease activity, so in the presence of ethionine a cell is depleted of SAM, and even though unmodified substrate sites become available after one round of DNA replication, no self-degradation can occur. On the other hand, the restriction endonuclease of the fi^- R-factor does not require SAM for endonuclease activity and could initiate degradation of the unmodified chromosome after one round of replication in the absence of SAM. By inference, the P1 and 15 restriction endonucleases would be expected not to require SAM for activity.

Although no genetic analysis of any other Type II system is available, the failure to recover one-step r^- m^- mutants of the fi^- R-factor, and the characteristics of the fi^- R-factor and *Hemophilus* restriction endonucleases suggest that a distinct difference exists between this mechanism and the more complex Type I mechanism.

Biological Effects of Restriction and Modification

Restriction of the phage lytic cycle.—The most dramatic effect of restriction and modification is observed with the low eop of unmodified phage stocks on restricting hosts (12). In some cases in which the restricting host has three unrelated restriction and modification mechanisms, the eop can be 10^{-7} to 10^{-8}. Determination of the eop of phage stocks on related strains represents the easiest and most efficient procedure for establishing the presence of a restriction and modification mechanism and its host specificity. The reduced eop observed when unmodified phage are plated on a restricting host is presumably the result of the restriction endonuclease introducing double-strand scissions into the phage genome (88) within a few minutes after infection (110, 122). With the genetic continuity of the replicon interrupted, the fragments would be degraded further by cellular exonucleases (83). However, some functions of a restricted phage genome persist in the host (118) probably because of the positions of the restriction endonuclease substrate sites relative to the phage genes in question (51).

E. coli diploid either for the K or B restriction and modification genes restrict with eop's a little lower (10- to 50-fold) than the respective haploids (29). A strain with Type I and Type II restriction and modification mechanisms dramatically decrease (1000-fold) the eop of an unmodified phage (12). Thus, the intracellular concentration of restriction endonuclease is apparently not the crucial limiting factor in the eop values but rather a combination of the number of substrate sites per genome and the concentration of active enzymes. Several other factors also determine the ability of a cell to restrict an unmodified phage.

Since the unmodified DNA is a substrate for the restriction endonuclease and modification methylase, one can imagine a race between the methylase which must partially modify all sites and the restriction endonuclease which only has to make one double-strand scission in the genome to be successful. Thus, the eop of an unmodified phage must also be related to the number of susceptible sites per genome as demonstrated with the phage fd (25). Other parameters influence the restricting ability of a cell as indicated by the following observation. The eop of $\lambda \cdot O$ on an *E. coli* $r^{+}{}_{fi^-}$-$m^{+}{}_{fi^-}$ (*fi*$^-$ R-factor) is 10^{-2} to 10^{-3} as compared to 10^{-4} on an *E. coli* $r_B{}^{+}m_B{}^{+}$, but in vitro the R-factor endonuclease degrades $\lambda \cdot O$ more extensively than does the B endonuclease (107).

Physiological perturbations of the restricting host cell such as amino acid starvation (11, 61), heating in low salt (81, 109, 110, 119), by growth at 43°C (65), age of the culture (109), and ultraviolet irradiation (24) also result in less efficient restriction. High multiplicities of infection also reduce the restricting efficiency of a cell (67, 92, 119).

A strain known to restrict one phage is not always capable of restricting other phages for which it is a host. No restriction or modification of RNA phages has been observed, and the autonomous virulent phages (T-even) are resistant to restriction in general (48) by *E. coli.* The dependent virulent phages T1, T3, and T7 are restricted only by the P1 restriction mechanisms, while the temperate coliphages as a group are the most susceptible to restriction by the B, K, and P1 mechanisms.

All of the F-specific (fd, f1, f12, Ec9, and HR) and I-specific (If1 and If2) bacteriophages are restricted and modified by the B restriction mechanism, while none of the F-specific phages are sensitive to K, A, or 15 restriction (13, 25). The I-specific phages are sensitive to P1, 15, and A restriction but not K restriction. All of the F-specific phages except F12 are restricted partially (eop of 0.3) by P1.

It should be pointed out that even though a few cases of phage restriction have been associated with restriction endonucleases and specific double-strand scissions, other mechanisms for restriction are possible. For example, a restriction endonuclease or a similar protein might possibly be able to interact with a specific sequence of base pairs without making phosphodiester bond cleavages and remain tightly bound to the DNA in a manner analogous to repressor-operator interactions. This may produce an abortive attempt at DNA replication with subsequent degradation in vivo. No endonuclease activity would be detected in vitro with a mechanism of this type. It has been demonstrated that the nonglucosylated T-even phages are restricted by P1, and that this restriction is genetically controlled by the same locus responsible for the restriction of λ (72, 99). However, purified P1 restriction endonuclease which degrades unmodified λ DNA does not degrade nonglucosylated T-even DNA (quoted in 100). In this context, another observation is relevant. Hybrids of λ and ϕ80 can be produced which are only partially sensitive to K restriction (10^{-2} eop) and from which one-step mutants, totally resistant to K restriction, can be derived. The partially resistant mutant is not degraded by purified preparations of K restriction endonuclease (89). Thus, it is very possible that mechanisms exist for restriction of phage which do not involve phosphodiester bond cleavages.

Phage typing, in part, relies on the restriction and modification of phages (1–5), although this aspect of phage typing has not been explored fully. Anderson has called attention to the value of restriction and modification mechanisms to typing procedures for several *Salmonella* species. Other phage typing schemes obviously depend in part on restriction and modification although this is difficult to verify on the basis of available information.

Bacterial conjugation.—The first report of the effect of restriction and modification on conjugation was that few recombinants were obtained from a cross between an Hfr *E. coli* K-12 donor and an F^- *E. coli* K-12 recipient lysogenic for P1, in contrast to a cross wherein the donor was lysogenic for P1 (6). This observation was subsequently exploited to show that in addition to the reduced number of recoverable recombinants there was also a dramatic reduction in the linkage of one genetic marker to another when the female recipient had a restriction and modification specificity not present in the donor (12, 27, 28, 37, 38, 40, 82, 91, 93, 103, 105, 123). Also, such episomes as *F-lac, F-gal,* the *E. coli* sex factor, and R-factors are restricted during conjugation and manifested in the reduction (about 10^{-2}) of the number of ex-conjugants carrying the episome (12, 28, 53, 78). The successful transfer and establishment of episomes in such cases are in part the result of r^- mutants in the recipient population (53).

The extent of the restriction effect on conjugation is dependent on several parameters. Recombinants are recovered more frequently from large fragments of donor chromosomes than from smaller fragments, and the extent of marker-unlinking depends on the region of the chromosome and the distance between the markers (28, 40, 105). The proximity of the restricted site to the genetic marker probably influences the recovery of recombinants as well. The greater efficiency in recovering recombinants from large chromosome fragments is probably related to the breakdown of restriction observed at high multiplicities of infection by unmodified phage (67, 92, 119) and the impaired restriction of phage by conjugal zygotes (54). One can imagine that with a limited number of restriction endonuclease molecules per cell, substrate saturation would enhance the modification of a genome. The effects of a restricting recipient on recombination are correlated with limited endonuclease degradation, since it has been demonstrated that unmodified bacterial DNA from various enteric genera is degraded in vitro by purified restriction endonuclease of *E. coli* B (50).

Transduction and transfection.—Restriction of specialized and generalized transducing phages are known to occur in *E. coli.* Transduction of a restricting *E. coli* Gal^- strain to Gal^+ with unmodified λ *dg* is reduced a thousand-fold (9, 47). Transduction of bacterial markers by an $r^- m^-$ mutant of Pl to an $r^+ m^+$ PL lysogen is also affected (2- to 10-fold reduction), but the extent is variable depending on the marker examined (69). Restriction of Pl transduction of the Thr Ara Leu markers between isogenic *E. coli* strains with K or B host specificity has also been observed (31). Leu^+ transduction was depressed 20-fold while Ara^+ and Thr^+ transductions were reduced about tenfold. Significant loss of linkage was observed between Leu^+ and Ara^+.

Restriction is known to depress specific infectivities of infectious λ DNA (47, 88) and fd RF and virion DNA (23). Competent *Hemophilus influenzae* cell preparations degrade foreign DNA, including phage P22 DNA, but not *H. influenzae* DNA (113). One would expect that any restricting cell capable of being transformed would degrade unmodified transforming DNA provided the DNA could achieve a double-strand configuration after uptake, and that the state of competency did not alter the restricting capacity of the cell (23). Several observations with inter- and intra-species transformation of *Bacillus subtilis* (45, 114) and transformations between a pneumococcal and streptococcal strains (34, 98) are superficially similar to what one would expect as a result of restriction, but no experimental conformation is available.

Speciation.—It is obvious from the above discussion that restriction and modification of DNA serve as a protective barrier to phage infection. Are there other functions of this mechanism? The $r^- m^+$ and $r^- m^-$ mutants have no apparent shortcomings for survival in the laboratory, so one is left with the following type of speculation. In higher organisms, the geographic isolation of genetic material is considered to be one of the factors in the development of a new species. In microorganisms, genetic isolation in an ecological niche cannot be accomplished in ordinary ways, but restriction and modification of DNA could at least partially isolate the genetic material of a newly developing strain. Whether or not restriction and modification mechanisms play a role in speciation is debatable, but as Crick has recently pointed out, nature probably has evolved special processes to speed the rate of evolution (42).

Survey of Restriction and Modification in Bacteria

Documented cases of restriction and modification in eight genera of bacteria, including Gram-negative and Gram-positive representatives, insure the generality of this mechanism in the microbial world. The observations described below satisfy Luria's definition (86) and therefore are considered to be bona fide restriction and modification mechanisms. Restriction and modification mechanism probably can be found in other genera and strains if a systematic approach is taken.

Gram-negative bacteria.—In *E. coli* strains there are at least six and possibly seven different host specificities known (12, 21, 78). Three of the specificities are genetically controlled by allelic gene loci on the bacterial chromosome and appear to have Type I mechanisms (17, 78). The other four, P1, 15, fi^+, and fi^-, are genetically controlled by the phage P1, the plasmid 15 (found originally in *E. coli* 15), and the fi^+ and fi^- R-factors (17, 21, 78). The P1, 15, and fi^- specificities appear to have Type II mechanisms (78) while the mechanism associated with the fi^+ specificity has not been examined.

Several species of *Salmonella* are known to restrict and modify phage. *S. anatum* and *S. butantan* restrict and modify the temperate phage ε^{15} with mutually exclusive host specificities (119). Restriction mutants of *S. typhimurium* identify the presence of a mechanism in that strain (37, 38). The presence of fi^- and fi^+ R-factors confer two different host specificities on *S. panama.* However, these probably are the same R-factor host specificities found in *E. coli* (62). At least one instance of restriction and modification is known in *S. typhi* and probably there are others as well (3).

Restriction and modification of DNA in *P. aeruginosa* has been examined in detail by Holloway and his collaborators. Three different host specificities have been found in *P. aeruginosa* strains with phage B3 (67). Many aspects of the *P. aeruginosa* mechanism are similar to the *E. coli* mechanism, but the genetics appear to be unusual. No occurrences have been reported in other species.

One instance of restriction (and probably modification) is known to occur in *H. influenzae* (113). Although Luria's criterion has not been satisfied in this case, the purification of an endonuclease from this strain which degrades "foreign" DNA to fragments about 1000 base pairs in length, but does not degrade *H. influenzae* DNA, is sufficient evidence.

Two species of *Rhizobium, R. legumenosarum* and *R. trifolii,* have mutually exclusive restriction and modification mechanisms for phage ϕL1 and ϕL5 (112). Three strains of *R. legumenosarum,* L1, L4, and L7 have the same host specificity, while strains L2 and L25 do not restrict or modify phages ϕL1 and ϕL5.

Gram-positive bacteria.—One instance of restriction and modification has been reported for *Staphylococcus aureus* and has been extensively investigated (94–97). Two mutually exclusive host specificities were demonstrated in other staphylococcal strains (108). This mechanism of restriction and modification shares many of the *E. coli* features.

Three strains of *Streptococcus cremoris* (M1, M2, and M3) isolated from three different commercial cultures used in making cottage cheese were shown to restrict and modify three strains of bacteriophages isolated from slowly fermenting vats of cottage cheese inoculated with the above cultures (35). Strains M1 and M3 have mutually exclusive host specificities on the basis of eop values with phages M1 and M2 (isolated from strains M1 and M2). The M2 strain does not restrict or modify phages M1 and M2. However, phage M3 which plaques only on strains M2 and M3, is modified and restricted by M2 but not M3.

Concluding Remarks

It is becoming evident that restriction and modification of DNA in bacterial organisms is not limited to several strains of *E. coli* but rather is a common mechanism prevalent in microorganisms. The available enzymological and genetic information although limited, clearly demonstrates that there are at least two types of restriction and modification mechanisms. Endonucleases and methylases are the enzymological bases of both mechanisms. Answers to some important questions about the generality of the symmetrical substrate, the distribution of complex and simple restriction endonucleases and modification methylases, etc., are expected soon.

The various restriction endonucleases with exclusive recognition capacities offer a unique probe for dissecting small genomes such as λ, fd, polyoma, and SV40. This work is now being undertaken in several laboratories including the author's.

Literature Cited

1. Anderson, E. S., Felix, A. 1952. *Nature (London)* 170:492–94
2. Anderson, E. S., Fraser, A., 1956. *J. Gen. Microbiol.* 15:255–39
3. Anderson, E. S. 1962. *Brit. Med. Bull.* 18:64–68
4. Anderson, E. S. 1966. *Nature (London)* 212:795–99
5. Anderson, E. S., Pitton, J., Mayhew, J. 1968. *Nature* 219:640–41
6. Arber, W. 1962. *Pathol. Microbiol.* 25:668–81
7. Arber, W., Dussoix, D. 1962. *J. Mol. Biol.* 5:18–36
8. Arber, W., Hattman, S., Dussoix, D. 1963. *Virology* 21:30–35
9. Arber, W. 1964. *Virology* 23:173–82
10. Arber, W. 1965. *Ann. Rev. Microbiol.* 19:365–78
11. Arber, W. 1965. *J. Mol. Biol.* 11:247–56
12. Arber, W., Morse, M. L. 1965. *Genetics* 51:137–48
13. Arber, W. 1966. *J. Mol. Biol.* 20:483–96
14. Arber, W., Kühnlein, U. 1967. *Pathol. Microbiol.* 30:946–52
15. Arber, W. 1968. *Symp. Soc. Gen. Microbiol., 18th* 295–314
16. Arber, W., Linn, S. 1969. *Ann. Rev. Biochem.* 38:467–500
17. Arber, W., Wauters-Willems, D. 1971. *Mol. Gen. Genet.* In press
18. Bannister, D., Glover, S. W. 1968. *Biochem. Biophys. Res. Commun.* 30:735–38
19. Bannister, D. 1970. *J. Gen. Microbiol.* 61:273–81
20. Bannister, D. 1970. *J. Gen. Microbiol.* 61:283–87
21. Bannister, D., Glover, S. W. 1970. *J. Gen. Microbiol.* 61:63–71
22. Barbour, S. D., Clark, A. J. 1970. *Proc. Nat. Acad. Sci. USA* 65:955–61
23. Benzinger, R. 1968. *Proc. Nat. Acad. Sci. USA* 59:1294–99
24. Bertani, G., Weigle, J. J. 1953. *J. Bacteriol.* 65:113–21
25. Bickle, T., Arber, W. 1969. *Virology* 39:605–7

26. Billen, D. 1968. *J. Mol. Biol.* 31: 477–86
27. Boice, L. B., Luria, S. E. 1963. *Virology* 20:147–57
28. Boyer, H. W. 1964. *J. Bacteriol.* 88: 1652–60
29. Boyer, H. W., Roulland-Dussoix, D. 1969. *J. Mol. Biol.* 41:459–72
30. Boyer, H. W., Hedgpeth, J., Scibienski, L., Roulland-Dussoix, D. 1970. In preparation
31. Boyer, H. W. Unpublished observation
32. Bulkacz, J., Boyer, H. W. Unpublished observations
33. Buttin, G., Wright, M. 1968. *Cold Spring Harbor Symp. Quant. Biol.* 33:259–69
34. Chen, K., 1966. *J. Mol. Biol.* 22: 109–34
35. Collins, E. B. 1956. *Virology* 2:261–71
36. Colson, C., Glover, S. W., Symonds, N., Stacey, K. 1965. *Genetics* 52: 1043–50
37. Colson, C., Colson, A. M. 1967. *Biochem. Biophys. Res. Commun.* 29:692–95
38. Colson, A. M., Colson, C., Van Pel, A. 1969. *J. Gen. Microbiol.* 58: 57–64
39. Colson, C., Colson, A., Van Pel, A. 1970. *J. Gen. Microbiol.* 60 :265–71
40. Copeland, J. C., Bryson, V. 1966. *Genetics* 54:441–52
41. Craigie, J., Yen, C. H. 1938. *Can. J. Public Health* 29:448–63
42. Crick, F. 1970. *Nature (London)* 228:613–19
43. d'Herelle, F. 1926. *The Bacteriophage and Its Behaviour.* Baltimore: Williams & Wilkins. 609 pp.
44. Drexler, H., Christensen, J. R. 1961. *Virology* 13:31–39
45. Dubnau, D., Smith, I., Morell, P., Marmur, J. 1965. *Proc. Nat. Acad. Sci. USA* 54:491–98
46. Dussoix, D., Arber, W. 1962. *J. Mol. Biol.* 5:37–49
47. Dussoix, D., Arber, W. 1965. *J. Mol. Biol.* 11:238–46
48. Eskridge, R. W., Weinfeld, H., Paigen, K. 1967. *J. Bacteriol.* 93: 835–44
49. Falkow, S., Citarella, R. V., Wohlhieter, J. A., Watanabe, T. 1966. *J. Mol. Biol.* 17:102–16
50. Ford, E., Boyer, H. W. 1970. *J. Bacteriol.* 104:594–95
51. Franklin, N., Dove, W. 1969. *Genet. Res.* 14:151–57
52. Fujiimoto, D., Srinivasan, P., Borek, E. 1965. *Biochemistry* 4: 2849–55
53. Glover, S. W., Schell, J., Symonds, N. Stacey, K. A. 1963. *Genet. Res.* 4:480–82
54. Glover, S. W., Colson, C. 1965. *Genet. Res.* 6:153–55
55. Glover, S. W. 1968. *J. Gen. Microbiol.* 53:i-ii
56. Glover, S. W., Colson, C. 1969. *Genet. Res.* 13:227–40
57. Glover, S. W. 1970. *Genet. Res.* 15: 237–50
58. Gold, M., Hurwitz, J. 1963. *Cold Spring Harbor Symp. Quant. Biol.* 28:149–56
59. Goldmark, P., Linn, S. 1970. *Proc. Nat. Acad. Sci. USA* 67:434–41
60. Gough, M., Lederberg, S. 1966. *J. Bacteriol.* 91:1460–68
61. Grasso, R., Paigen, K. 1968. *J. Virol.* 2:1368–73
62. Guinée, P., Willens, H. 1967. *Antonie van Leeuwenhock J. Microbiol. Serol.* 33:397–406
63. *Ibid.* 33:407–12
64. Hedgpeth, J., Boyer, H. W. Unpublished observations
65. Holloway, B. W. 1965. *Virology* 25: 634–42
66. Holloway, B. W., Rolfe, B. 1964. *Virology* 23:595–602
67. Holloway, B. W. 1969. *Bacteriol. Rev.* 33:419–43
68. Hubacek, J., Glover, S. W. 1970. *J. Mol. Biol.* 50:111–27
69. Inselburg, J. 1966. *Virology* 30: 257–65
70. Kellenberger, G., Symonds, N., Arber, W. 1966. *Z. Verebungslehre* 98:247–56
71. Kelly, T. J., Jr., Smith, H. O. 1970. *J. Mol. Biol.* 51:393–409
72. Klein, A. 1965. *Z. Verebungslehre* 96:324–45
73. Koshland, D. E., Neet, K. E. 1968. *Ann. Rev. Biochem.* 37:359–410
74. Kühnlein, U., Linn, S., Arber, W. 1969. *Proc. Nat. Acad. Sci. USA* 63:556–62
75. Kühnlein, U. 1970. *The B-Specific DNA Modification Enzyme of* Escherichia coli *B.* Doctoral thesis Université de Genève. 69 pp.
76. Lark, C. 1968. *J. Mol. Biol.* 31: 389–400
77. Lark, C. 1968. *J. Mol. Biol.* 31:401–14
78. Lark, C., Arber, W. 1970. *J. Mol. Biol.* 52:337–48

79. Lederberg, S. 1957. *Virology* 3:496–513
80. Lederberg, S., Meselson, M. 1964. *J. Mol. Biol.* 8:623–28
81. Lederberg, S. 1965. *Virology* 27:378–87
82. Lederberg, S. 1966. *J. Bacteriol.* 91:1029–36
83. Lehman, I. R. 1963. In *Progress in Nucleic Acid Research,* ed. J. N. Davidson, W. E. Cohn, 2:84–118. New York: Academic
84. Linn, S., Arber, W. 1968. *Proc. Nat. Acad. Sci. USA* 59:1300–6
85. Luria, S. E., Human, M. L. 1952. *J. Bacteriol.* 64:557–69
86. Luria, S. E. 1953. *Cold Spring Harbor. Symp. Quant. Biol.* 18:237–44
87. Mamelak, L., Boyer, H. W. 1970. *J. Bacteriol.* 104:57–62
88. Meselson, M., Yuan, R. 1968. *Nature (London)* 217:1110–14
89. Murray, K., Murray, N. Unpublished observation
90. Oishi, M. 1969. *Proc. Nat. Acad. Sci. USA* 64:1292–99
91. Okada, M., Watanabe, T., Miyake, T. 1968. *J. Gen. Microbiol.* 50:241–52
92. Paigen, K., Weinfeld, H. 1963. *Virology* 19:565–72
93. Pittard, J. 1964. *J. Bacteriol.* 87:1256–57
94. Ralston, D. J., Krueger, M. P. 1952. *Proc. Soc. Exp. Biol. Med.* 80:217–20
95. Ralston, D. J., Baer, B. S. 1964. *J. Gen. Microbiol.* 36:1–16
96. Ralston, D. J., Baer, B. S., 1964. *J. Gen. Microbiol.* 36:17–24
97. Ralston, D. J., Baer, B. S., 1964. *J. Gen. Microbiol.* 36:25–36
98. Ravin, A., Chen, K. 1967. *Genetics* 57:851–64
99. Revel, H. R., Georgopoulos, C. P. 1969. *Virology* 39:1–17
100. Revel, H. R., Luria, S. E. 1970. *Ann. Rev. Genet.* 4:177–92
101. Richardson, C. C. 1965. *Proc. Nat. Acad. Sci. USA* 54:158–65
102. Riggs, A. D., Suzuki, H., Bourgeois, S. 1970. *J. Mol. Biol.* 48:67–83
103. Rolfe, B., Holloway, B. W. 1966. *J. Bacteriol.* 92:43–48
104. Rolfe, B., Holloway, B. 1968. *Genet. Res.* 12:94–102
105. Rolfe, B., Holloway, B. W. 1969. *Genetics* 61:341–49
106. Roulland-Dussoix, D., Boyer, H. 1969. *Biochim. Biophys. Acta* 195:219–29
107. Roulland-Dussoix, D., Yoshimori, R., Boyer, H. W. MS. in preparation
108. Rountree, P. M. 1956. *J. Gen. Microbiol.* 15:266–79
109. Schell, J., Glover, S. W. 1966. *Genet. Res.* 7:273–76
110. Schell, J., Glover, S. 1966. *J. Gen. Microbiol.* 45:61–72
111. Schnegg, B., Hofschneider, P. H. 1969. *J. Bacteriol.* 3:541–42
112. Schwinghamer, E. A. 1965. *Aust. J. Biol. Sci.* 18:333–43
113. Smith, H. O., Wilcox, K. W. 1970. *J. Mol. Biol.* 51:379–91
114. Šrogl, M. 1966. *Folia Microbiol.* 11:39–42
115. Stacy, K. A. 1965. *Brit. Med. Bull.* 21:211–16
116. Takano, T., Watanabe, T., Fukasawa, T. 1966. *Biochem. Biophys. Res. Commun.* 25:192–98
117. Takano, T., Watanabe, T., Fukasawa, T. 1968. *Virology* 34:290–302
118. Terzi, M. 1968. *J. Mol. Biol.* 34:165–73
119. Uetake, H., Toyama, S., Hagiwara, S. 1964. *Virology* 22:202–13
120. Watanabe, T., Nishida, H., Ogata, C., Arai, T., Sato, S. 1964. *J. Bacteriol.* 88:716–26
121. Watanabe, T., Takano, T., Arai, T., Nishida, H., Sato, S. 1966. *J. Bacteriol.* 92:477–86
122. Weinfeld, H., Paigen, K. 1964. *Virology* 24:71–83
123. Wood, W. B. 1966. *J. Mol. Biol.* 16:118–33
124. Yarus, M. 1969. *Ann. Rev. Biochem.* 38:841–79
125. Yoshimori, R., Roulland-Dussoix, D. MS. in preparation
126. Yuan, R., Meselson, M. 1970. *Proc. Nat. Acad. Sci. USA* 65:357–62
127. Yuan, R. Personal communication
128. Zinder, N. D. 1960. *Science* 131:813–15

AUTOTROPHY: CONCEPTS OF LITHOTROPHIC BACTERIA AND THEIR ORGANIC METABOLISM

1567

DONOVAN P. KELLY
Microbiology Department, Queen Elizabeth College,
Campden Hill, London, W.8, England

CONTENTS

THE LITHOTROPHIC BACTERIA

For the purposes of this review the lithotrophic bacteria are defined as those organisms that can produce metabolically useful energy by processes that depend on the oxidation of inorganic compounds. Three classes of lithotrophic bacteria are recognized in this discussion. These are the chemolithotrophic bacteria, the photolithotrophic bacteria, and the photosynthetic blue-green bacteria (blue-green algae). A fourth group, the methylotrophic bacteria are also considered briefly because of their marked metabolic similarities to some lithotrophic species.

Chemolithotrophic bacteria can obtain all the energy necessary for growth and carbon assimilation from the oxidation, in the absence of light, of reduced inorganic sulfur compounds (98, 113, 202), ammonium ion (5, 181, 253), nitrite ion (5, 181, 253), iron metal or ferrous iron (181, 209), and hydrogen (181). Photolithotrophic bacteria obtain energy from light but their photosynthesis uses the oxidation of inorganic reduced sulfur compounds or hydrogen (243, 244). The principal photolithotrophs are the Thiorhodaceae and Chlorobacteriaceae, but some normally photoorganotrophic Athiorhodaceae (246) (*Rhodospirillum* and *Rhodopseudomonas*) can oxidize hydrogen and sulfur compounds, thus exhibiting some photolithotrophic capability. The blue-green bacteria (Cyanophyta) use light energy and oxidize water most probably by mechanisms exactly like those for higher plants and algae. Apart from the affinity of these organisms for the eubacteria through both being prokaryotes, certain aspects of their carbon metabolism suggest metabolic parallels between them and other lithotrophs, and seem to support further the view that they are best regarded as photolithotrophic blue-green bacteria (33, 51). The methylotrophs (85, 173, 174, 196) obtain energy from oxidizing one-carbon organic compounds such as methane. Some, like *Methylococcus capsulatus* seem obligatory in their requirement for one-carbon substrates and thus exhibit a metabolic fastidiousness comparable with that of the obligate chemolithotrophs like *Thiobacillus neapolitanus* and photolithotrophs like *Chlorobium*. Moreover, ultrastructural studies on all the lithotrophs indicate that internal membrane systems are common (122, 157, 187, 194, 214, 261). The photolithotrophic green, purple, and blue-green bacteria have highly developed photosynthetic membranes. Notable similarities exist in the extent and complexity of the membranes of blue-green bacteria and those of the nitrifying bacteria (166, 187, 253) and of *Methylococcus* (44, 189, 196). Similarities of lipid composition have also been shown between blue-green bacteria and nitrifiers (21).

In recent excellent reviews, Rittenberg (197), Peck (181), and Aleem (5) have discussed problems of the energy metabolism and some of the roles of organic matter in the physiology of chemolithotrophic bacteria, and Pfennig (183) and Holm-Hansen (92) have surveyed the physiology of photosynthetic bacteria and blue-green algae. Since these reviews were published, advances have been made regarding several of the topics whose importance was stressed by these reviewers. In this survey I am concerned primarily with those lithotrophs that can grow wholly autotrophically and particularly with considerations relevant to understanding the small number among those that are regarded as obligate lithotrophs. My use of these terms derives from the clear summary of the meanings presented by Rittenberg (197) when considering chemolithotrophy. Obligate chemolithotrophs seem to have no means of obtaining metabolically available energy from their environment other than from specific inorganic oxidations. Obligate photolithotrophs are similarly dependent on light as well as specific inorganic oxidations. "Autotrophy" is used to describe the ability of some lithotrophs to develop in wholly inorganic media, using dissolved carbon dioxide as their only source of carbon. "Mixotrophy" describes the situation in which concurrent use is made of organic and inorganic sources of carbon or energy, or both. "Heterotrophy" implies that all energy is from organic oxidations (chemoorganotrophy), while organic compounds also provide most of the cell carbon, with carbon dioxide as a minor source, assimilated by established anaplerotic mechanisms.

The sources of carbon and energy available to any lithotrophic bacterium allow a comparative classification of them to be constructed as shown in Figure 1. Both the chemo- and photolithotrophs vary widely in the physi-

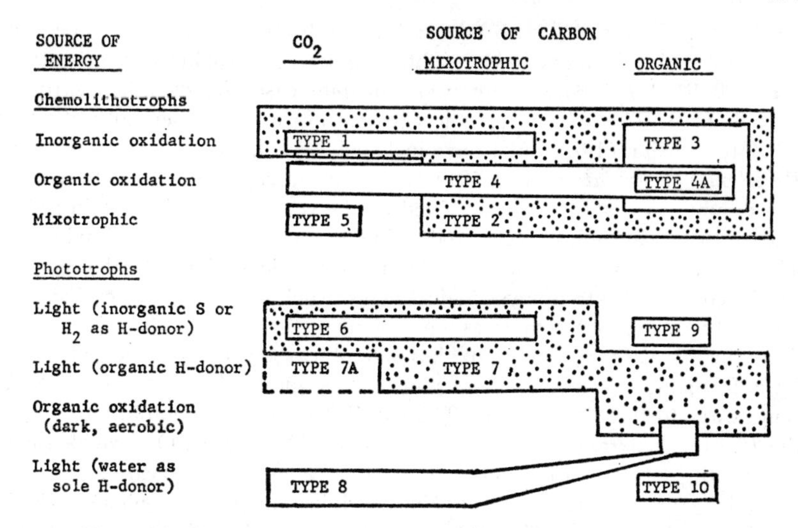

FIGURE 1. Sources of energy and carbon in lithotrophic bacteria.

Type 1: Obligate chemolithotrophs (some thiobacilli, possibly *Nitrosomonas, Spirillum* sp. (128), *Gallionella* (17, 125, 209)).

Type 2: "Facultative autotrophs" and mixotrophs (*Hydrogenomonas, Thiobacillus novellus, T.* A2 (232), Beggiatoa (188), *Nitrobacter agilis* (221), *T. ferrooxidans* (212), *T. intermedius* (143)).

Type 3: Chemolithotrophic heterotrophs (197) (*T. perometabolis* (145), *Desulfovibrio*)

Type 4: *Pseudomonas oxalaticus* (191).

Type 4A: The obligate methylotrophs (see text)

Type 5: None clearly shown

Type 6: Obligate photolithotrophs (*Chlorobium* spp., *Prosthecochloris* (75); probably some Thiorhodaceae and *Pelodictyon* (184)).

Type 7: Many Thiorhodaceae. The Athiorhodaceae.

Type 7A: Possibly some Athiorhodaceae on some substrates (63).

Type 8: Blue-green bacteria (Cyanophyta)

Type 9: None clearly shown, but physiologically feasible.

Type 10: None clearly shown among Cyanophyta. An example in the algae is a RDPCase-less *Chlamydomonas* (141).

ological adaptability exhibited by the individual species of each group. Thus, extreme metabolic conservatism is indicated for Types 1 and 6, the obligate chemo- and photolithotrophs. The "facultative autotrophs" of Types 2 and 7 represent organisms of much greater metabolic adaptability, including species whose physiological range spreads from the capacity to develop autotrophically to the ability to grow as a heterotroph in the case of some thiobacilli and *Hydrogenomonas,* or to behave as photoorganotrophs or even to obtain some energy and carbon from the aerobic or anaerobic oxidation of some organic compounds in the dark in the case of some Thiorhodaceae. Similarly, Type 8 (the blue-green bacteria) span the range from autotrophic development in the light, through mixotrophic carbon metabolism (90, 91, 175–178) to the capacity in some cases for dark heterotrophic growth (60). The scheme for the chemolithotrophs is essentially like that tabulated earlier (197), but I have included Type 4, which is represented by *Pseudomonas oxalaticus* (20, 191), an organism capable of growth on carbon dioxide as a carbon source, fixed by the reactions of the Calvin reductive pentose phosphate cycle, and using energy from formate oxidation. This organism can doubtless exhibit mixotrophic carbon metabolism and can be converted into a physiological heterotroph on alternative carbon substrates (20). Type 4A comprises the obligate methylotrophs, obtaining energy from methane or methanol oxidation and carbon from these substrates by a formaldehyde-fixing ribose phosphate cycle (119, 135), analogous in some ways to the Calvin cycle. The parallel among the photolithotrophs of the *P. oxalaticus* metabolism would be Type 7A, where CO_2 provided carbon at the expense of photoorganotrophic energy conservation. Such a physiology was inferred by van Niel's original analysis of photosynthetic bacteria and was exemplified by oxidations in which organic substrates seemed to serve solely as hydrogen donors (63, 245). Similarly, some alcohols, propionate and formate, support much better growth of *Rhodopseudomonas* in the presence of bicarbonate than in its absence (259), suggesting a prime role as hydrogen donors and perhaps that their dissimilation requires CO_2. In both the chemo- and photolithotrophs a continuous physiological spectrum is seen between obligate lithotrophy and complete heterotrophy. In some cases, one organism may be adaptable to life over the whole of the appropriate spectrum, and it is noteworthy that no rigid division can be made between strict photolithotrophy and photoorganotrophy as members of the Thiorhodaceae and Athiorhodaceae may exhibit varying capacity to effect both sorts of metabolism.

The clear establishment of autotrophs as components of the integrated system of "biochemical unity" among microbes, rather than as metabolic oddities possessed of unique physiologies, leaves a number of conceptual problems with regard to the obligate lithotrophs. Peck (181) and I (111) have already pointed out that the autotrophs may be regarded biosynthetically as the most "complete" among organisms, having a metabolic system basically like that of a common heterotroph, with which are integrated enzyme systems for the production of all cell components from CO_2 at the expense of the oxidation of unique inorganic substrates. Seeming inability to develop at the expense of organic oxidations thus remains a problem to be resolved in the study of organisms like *Nitrosomonas,* some *Thiobacillus* spp., *Chlorobium,* and *Anacystis.*

In subsequent sections of this review I have considered aspects of the metabolism and metabolic regulation of lithotrophs that concern their organic nutrition and may aid in an understanding of the cause of obligate lithotrophy.

ARE THERE OBLIGATE LITHOTROPHS?

Numerous workers have established that all the lithotrophs so far examined are able to incorporate exogenous organic compounds, even if they cannot develop as facultative heterotrophs (30, 35, 36, 48, 99, 104, 105, 108–112, 120, 177, 205, 225, 254). In some cases no development is achieved if attempts are made to grow the organisms in media lacking their usual energy source. These are the putatively obligate lithotrophs. Among the chemolithotrophs. *Nitrosocystis, Nitrosomonas, Nitrobacter winogradskyi, Thiobacillus neapolitanus, T. thioparus, T. denitrificans, T. concretivours,* and some strains of *T. ferrooxidans* and *T. thiooxidans* are generally regarded as obligate examples. Recently Pan (171, 172) has claimed to have grown the first three thiobacilli on glucose in continuously dialyzed culture media. This system was presumed to remove toxic materials normally detrimental to heterotrophic growth (22–24), a view that has recently been criticized (197). In any case, adaptation of these strains and of the facultative strain of *Nitrobacter agilis* (221) to heterotrophy has so far been achieved only with one medium or cultural condition, and in the case of *N. agilis,* the generation time increased from 20 hr to 90 hr on adaptation from nitrite to heterotrophic growth. Obligate lithotrophs are thus characterized by at least an extreme reluctance to develop on organic media.

Several attempts have been made to explain this reluctance (111, 197, 222, 242), the most popular being based on concepts of inadequate energy coupling mechanisms or of specific biosynthetic deficiencies. One cause of failure to grow in heterotrophic media can be the absence of a specific growth factor requirement. *Chlorobium* (142), *Chromatium,* and *Thiopedia,* for example, cannot reduce sulfate for assimilation and so cannot grow unless sulfide (the lithotrophic energy source) is available as a sulfur source. This may also apply to *T. neapolitanus,* whole cells of which will not assimilate sulfate. Similarly, *Nitrosomonas* could require ammonium ion as a nitrogen source as well as for energy metabolism. Such factors do not provide a unitary explanation of obligate lithotrophy.

OCCURRENCE AND REGULATION OF CARBON DIOXIDE-FIXING MECHANISMS AMONG LITHOTROPHS AND ORGANOTROPHS

The presence of a functional Calvin-Benson reductive pentose phosphate cycle has come to be regarded as a characteristic feature of all lithotrophs after growth under autotrophic conditions (2, 32, 67, 238, 253). Ribulose diphosphate carboxylase (RDPCase) and phosphoribulokinase (PRK) can occur in all lithotrophs that are capable of autotrophic development (49, 53, 149), but do not occur in nonphotosynthetic heterotrophs, or apparently in the methylotrophs (119). RDPCase is also absent from the sulfate-reducing lithotrophs (Fig. 1, Type 3), but does occur in the photoorganotrophic Athiorhodaceae after photosynthetic growth (13, 134, 226) and is inducible in

Micrococcus denitrificans which seems capable of development as a hydrogen bacterium (124).

Although the enzymes of the Calvin cycle are present in *Chlorobium* (220) it seems likely that a major mechanism for its CO_2 fixation is by means of the reactions of a tricarboxylic acid cycle (TCA) operating in opposite direction to that in aerobic respiration (16, 59, 247). New CO_2-fixing reactions in *Chlorobium* are the ferredoxin-dependent carboxylations of acetyl CoA to pyruvate and of succinyl CoA to α-oxoglutarate (16, 27, 58). These reactions also occur in other photosynthetic bacteria (28, 57). CO_2 is fixed into the cycle also by the carboxylation of pyruvate (or PEP), succinate, and α-oxoglutarate. The product of the cycle is effectively acetate, that is partly recycled and partly removed for biosynthesis. This cycle supplies all the necessary intermediates for the synthesis of lipid, amino acids, porphyrins, sugars, and polysaccharides (59, 218). A further CO_2-fixing reaction has recently been shown in *Chromatium,* in which a ferredoxin-dependent carboxylation of propionyl CoA by α-ketobutyrate synthase occurs in cells grown autotrophically or with malate or propionate (26).

Little is yet known of the regulation of the TCA cycle of CO_2-fixation, although feedback effects by products such as amino acids are likely (218, 219). Considerable information is, however, becoming available on the regulation of RDPCase synthesis and thus of the Calvin cycle. In the facultative chemolithotrophs (Type 2), RDPCase synthesis is often suppressed by organic nutrients or heterotrophic growth. This suppression may be virtually complete in heterotrophically-grown *Thiobacillus novellus* (6, 250) and probably in some facultative strains of *T. ferrooxidans* in which glucose decreases the capacity to fix CO_2 (216). In *Nitrobacter agilis,* acetate represses RDPCase but 51 percent of the autotrophic level remains after mixotrophic growth with 10 m*M* acetate (221). In *Hydrogenomonas* a complex relation to organic suppression has been found (197). The general finding is that organic compounds repress RDPCase to various extents, but that mixotrophic energy and carbon metabolism can occur when hydrogen, CO_2, and organic compounds are available. I shall summarize these views briefly. Quite early experiments (133) showed CO_2 fixation to be supported by lactate oxidation. Subsequently, Hirsch et al (88) found that suspensions of autotrophically grown *Hydrogenomonas* H-16 fixed CO_2 principally by the Calvin cycle when either hydrogen or succinate was supplied, energy for CO_2 fixation by preformed Calvin cycle enzymes was thus available from the organic oxidation. In contrast, fixation by the Calvin cycle was negligible in succinate-grown H-16 (88). Mixotrophy in *Hydrogenomonas* is well documented (197). Energy may be obtained simultaneously from the oxidation of H_2 + fructose and H_2 + lactate and probably H_2 + various amino acids, to give growth yields exceeding those of either substrate alone, and sometimes giving additive yields (39, 46, 72, 197, 198, 263). Carbon is obtained from the organic substrate during mixotrophic growth but there is good evidence that the Calvin cycle also functions in some strains during growth on some organic compounds, such as in *H. facilis* on ribose (151, 152). The syntheses of the key enzymes of the cycle are variously affected by organic nutrients. Numerous other factors affect the synthesis and degradation of RDPCase. For example, in *H. eutropha,* RDPCase was synthesized during growth on fructose only if low aeration was used and iron was supplied. Repression occurred with high aeration (127). This may explain the low levels reported by Lascelles and Rittenberg (197). Moreover, there was a heat-labile factor, possibly a protease, in heterotrophically grown cells that appeared to catalyze the rapid degradation of RDPCase both in vivo and in vitro (127). Current data available on the levels of RDPCase and PRK in *Hydrogenomonas* spp. are listed in Table 1. Often growth of *H. eutropha* under mixotrophic conditions produced either biphasic growth resembling diauxie or continued lithotrophic growth following the exhaustion of the organic substrate (227). Biphasic growth with acetate

TABLE 1. Activity of RDPCase and PRK in *Hydrogenomonas* After Autotrophic, Mixotrophic and Heterotrophic Growth

Growth conditions	Reported Enzyme levels (Autotrophic = 100)		
	H. eutropha RDPCase	*H. facilis* RDPCase	*H. facilis* PRK
A (log phase)	100	100	100
A (stationary phase)	132	—	—
H Fructose	12, 64, 94	76	97
M Fructose	77		
H Succinate	—	3	54
H Ribose	—	36	61
H Glucose	—	76	94
H Lactate (log or stationary)	4	15	52
H Lactate	56	—	—
M Lactate	58	—	—
H or M Pyruvate	1–4	—	—
H or M Acetate	0	2	11
H Glutamate	36	3	52
M Glutamate	74	—	—
H Alanine	27		
M Alanine	84		

Unless indicated, data is for organisms harvested during logarithmic growth. A—autotrophic culture (flask atmosphere contained $H_2+O_2+CO_2=7/2/1$ or $6/2/1$); M—mixotrophic culture (autotrophic atmosphere, organic supplement indicated); H—heterotrophic culture on indicated substrate; ——not determined. (Autotrophic levels reported for RDPCase by all authors fell within the range 25–67 nmoles/min/mg protein).

Individual data calculated from references 126, 152, 197, 198, 227.

or pyruvate was indicated to be due to the necessary adaptation period for RDPCase synthesis following pyruvate exhaustion before autotrophic growth began (227). With the exception of glutamate, substances allowing RDPCase synthesis did not produce biphasic growth. Glutamate partly repressed RDPCase but was believed to repress inorganic nitrogen assimilation (227). While the specific activity of RDPCase (and hydrogenase) in *H. eutropha* declined relative to autotrophically grown inocula, during growth either heterotrophically or mixotrophically with lactate, the total activity of the enzymes increased in both cases, indicating that RDPCase synthesis might proceed at reduced rates even under heterotrophic conditions (198). This requires critical examination as it was suggested that the observed extra synthesis might have occurred at the end of the growth cycle or during laboratory manipulations. Increase in RDPCase activity was found by Lascelles and Rittenberg (197) in stationary phase culture of H_2- or fructose-grown *H. eutropha,* but not in lactate or acetate cultures. The observed increase of RDPCase activity in lactate or lactate + H_2 cultures was 30–50 percent over the initial levels, which would seem a very large amount of enzyme to have arisen during laboratory manipulation at 0–4°C (198). Moreover, the data presented by Rittenberg (197) suggest that the level of RDPCase in *H. eutropha* grown for at least two transfers in lactate contained less than 4 percent of the autotrophic level. RDPCase synthesis undoubtedly occurs during growth on fructose, for its specific activity did not decrease during eight transfers on heterotrophic fructose media (127). Mixotrophic growth of *T. intermedius* resulted in repression of RDPCase and CO_2 utilization (144). Glucose caused approximately 60 percent repression of RDPCase but almost completely suppressed CO_2 assimilation. Yeast extract almost completely suppressed both activities while glucose caused 20–30 percent repression of RDPCase, and 80–90 percent depression of CO_2 used. Organic substrates thus regulate CO_2 fixation in a manner more complex than solely by repression of RDPCase.

In *Chromatium* and *Thiopedia* RDPCase was depressed maximally by about 75 percent after growth on various organic acids, fructose, or glycerol (97). RDPCase in Athiorhodaceae is depressed by high light intensity or oxygen, and during dark aerobic growth (13, 134). Levels reported in photoorganotrophically grown *Rhodopseudomonas spheroides* and *R. palustris* (134) were similar to the levels in pyruvate-grown *Chromatium* (97) but Anderson (13) reported equivalent levels in autotrophic *R. spheroides.* Anderson & Fuller (13) showed 97 percent repression of RDPCase in malate-grown *Rhodospirillum rubrum,* although only 75 percent was repressed during phototrophic growth on acetate. The cycle did not function during growth on malate, but did operate when acetate was the substrate (13).

Little is known of the effect of organic substrates on RDPCase synthesis in obligate lithotrophs, although growth inhibition in *T. neapolitanus* by acetate was suggested to be related to RDPCase repression (206). Depression of CO_2 fixation in this organism by organic substrates could simply indicate competition for available lithotrophic energy (110). It will be interesting to learn if RDPCase in *Chlorobium* is under any regulation by exogenous organic molecules: the endogenous controls of the two possible fixation cycles are yet to be established.

Control of the existing Calvin cycle seems to be governed in part at least by the cellular energy metabolism. AMP is a potent inhibitor of CO_2 fixation in thiobacilli and photolithotrophs (103, 149, 159). In *T. ferrooxidans* phosphoribulokinase is competitively inhibited by AMP, which is antagonized by ATP (66). A 1000-fold purified PRK from *T. thioparus* (149) seemed to be an allosteric regulatory protein with separate binding sites for AMP and ATP. PRK is also subject to AMP inhibition in *Hydrogenomonas* (152). Inhibition by AMP, ADP, and ATP of fructose-1,6-diphosphatase in crude extracts of *T. thioparus* and *T. neapolitanus* seems to be a second means of controlling the cycle (150). Regulation of the Calvin cycle by feedback inhibition by endogenous organic compounds is also a possible means of fine control, and is currently under investigation in this laboratory. Stimulation or inhibition of pyruvate kinase in *T. neapolitanus* by various metabolic intermediates presumably indicates a regulatory role for this enzyme (41).

Role of the Tricarboxylic Acid Cycle and Glyoxylate Cycle

In many aerobic heterotrophs and some facultative autotrophs (*Nitrobacter, Hydrogenomonas, Micrococcus denitrificans*) these cycles can provide energy, reducing power and carbon for biosynthesis. In the obligate lithotrophs there is growing evidence that the TCA cycle reactions serve solely a biosynthetic function, and that the obligate chemolithotrophs, blue-green bacteria, and *Chromatium* lack α-oxoglutarate (α-OG) dehydrogenase. The incomplete cycle thus operates in thiobacilli and blue-green bacteria to produce α-OG from acetate, which gives rise only to glutamate, proline and arginine (90, 110–112, 178, 222, 225, 233). No evidence yet exists to show that amino acid synthesis from acetate, pyruvate, or aspartate in thiobacilli or blue-green bacteria occurs by any mechanisms other than those known in heterotrophs (162). The importance of the TCA cycle reactions necessary to effect the acetate to α-OG sequence during autotrophic growth was indicated by the extreme toxicity of fluoroacetate to cultures of *T. neapolitanus* (112). The enzymes of the sequence are present in numerous photo- and chemolithotrophs. In general, as well as lacking α-OG dehydrogenase, the obligate chemolithotrophs and blue-green bacteria contain low levels of succinic and malic dehydrogenases (Table 2), while facultative species like *Hydrogenomonas* and *T. intermedius* have 10 to 1000 times as much of these enzymes. *T. neapolitanus* strains incorporate only trace amounts of succinate from the medium (110) although succinate and malate have been shown to have stimulatory effects on some strains (249) but not on others (206). The very large quantities of cytochromes synthesized by some thiobacilli and nitrifiers (14, 55, 163, 235) indicate a need for large

TABLE 2. Levels of Some TCA Cycle Dehydrogenase Enzymes in Some Lithotrophs and Heterotrophs

Organism	Range of activities reported (units[a])							References
	Dehydrogenases for					Oxidases for		
	Isocitrate (NAD)	Isocitrate (NADP)	α-OG	Malate	Succinate	NADH	NADPH	
Thiobacillus thioparus (A)	0–76	18–204	0	28–224	8	0–40	2	40, 105, 155, 222
Thiobacillus neapolitanus (A)	147–203	12–21	0	0–15		0–59	3–39	4, 8, 86, 105, 110, 155, 237
Thiobacillus thiooxidans (A)	117–380	0–80	0–300	11–75	8	0–72	++[b]	31, 81, 155, 222, 229
Thiobacillus denitrificans			0	low	low			207, 233
Thiobacillus novellus	0–20	480–1330						34, 155
Thiobacillus ferrooxidans (A)			0			+		12, 212
Thiobacillus intermedius (A)	5	15–103	<1	1600	<1–70	<1–12	1	153, 222
(H)	82–233	12–18	<1–2		1–3	3–7	1–3	
Thiobacillus perometabolis (H)	30	390				27	6	153
Hydrogenomonas (A)	<1	38–101	1–8	1690–8300	86–90	22		222, 239
(H)	<1	73–187	2–37	1370–13000	45–91	15		
Nitrobacter (A)	0	78–130	0–13	1095	0 to +	22–31		5, 221, 253, 254
(M)		72	13	1045	+	28		
Nitrosomonas	0	56–213	0	99	0–2	11 (−61)		94, 253, 254
Nitrosocystis	0	34	11	540	<1	3		262
Chromatium okenii	0	7	0	24	+			238
Anacystis	0	17	0	8	7	0-ca 10		19, 91, 95, 176
Gloeocapsa		36	0	17	3	0		222
Coccochloris		36	0	7	2	0		222
Phormidium		36	0	low	low	0	2	19
Anabaena		5	0	2	<1	0-ca 9	4	95, 136, 137, 175, 178, 222
Escherichia coli		530–1420	17–176	2120–3740	170	198		31, 77
Bacillus subtilis		351	13–644	5200	40			62, 203

A—autotrophically cultured organisms; M—mixotrophic culture: H—heterotrophic culture.
a—nmoles substrate converted/min/mg protein. b, +, or ++—activity or strong activity shown by some workers.

amounts of succinate as a porphyrin precursor. Succinate is probably produced by the reactions of the other limb of the TCA cycle, using the sequence from oxaloacetate to succinate. Oxaloacetate is produced in all lithotrophs by the carboxylation of phosphoenolpyruvate and is the precursor of the aspartate family of amino acids at least in *T. neapolitanus* (115). Evidence for succinyl CoA formation by this sequence and a CoA-transferase was obtained with *Anabaena,* which lacks succinyl CoA synthetase as well as α-OG dehydrogenase (178). The TCA cycle reactions thus operate using

PEP from CO_2 fixation to supply the two limbs of a biosynthetic carboxylic acid "horseshoe pathway," one limb leading to C_4 acids for aspartate family and tetrapyrrole synthesis, the other to the glutamate family and leucine (from pyruvate-acetate). This biosynthetic pathway occurs also in *Methylococcus* (173), *Clostridium kluyveri* (224), and in facultative anaerobic heterotrophs during anaerobic growth, when α-OG dehydrogenase is repressed (11, 156). Recently, most interesting parallels have been drawn between the control mechanisms operating on citrate synthase in heterotrophs and thiobacilli (231). NADH and α-OG are inhibitors of the enzyme from strictly aerobic Gram-negative bacteria and from facultative anaerobes, respectively (256–258). NADH regulates the cycle in organisms in which it acts as an energy-yielding process, while α-OG regulates it as a biosynthetic sequence. In facultative anaerobes like *Escherichia coli* the enzyme is regulated by both NADH and α-OG (256–258), demonstrating the dual role of the cycle. *Thiobacillus* A2 (232), a facultative autotroph using the complete TCA cycle for energy generation, has a NADH-sensitive, α-OG-insensitive citrate synthase, while those from *T. denitrificans* and *T. neapolitanus,* obligate lithotrophs, are inhibited only by α-OG (231). The enzyme from *T. novellus* is inhibited by neither compound (231).

Isocitrate dehydrogenases from several thiobacilli have been investigated and reports on their coenzyme specificity are confusing. That from *T. neapolitanus* seems to be NAD-specific (105, 155), while those from *T. thiooxidans, T. thioparus,* and *T. novellus* have been separately reported as principally specific for both NAD and NADP (31, 34, 81, 105, 155) (see Table 2). The enzymes from *T. novellus* (34) and *T. thiooxidans* (81) were both subject to concerted inhibition by glyoxylate and oxaloacetate. Purines and pyrimidines inhibited the *T. novellus* enzyme, as did ADP and ATP for *T. thiooxidans.* Biosynthesis in *T. thiooxidans* may thus be partly regulated by control over isocitrate metabolism through endogenous oxaloacetate levels, although there is probably no carbon flow from isocitrate to oxaloacetate (222). In *T. novellus* isocitrate lyase is also a constitutive enzyme (34) so control by glyoxylate and oxaloacetate could govern the relative flow of carbon into the glyoxylate cycle or the normal TCA cycle.

Isocitrate lyase also occurs in nitrifying bacteria (221, 254), *Hydrogenomonas* (239) and *Chromatium* (238), the two latter organisms also having malate synthase, that was reported absent from nitrifiers (254, 262). Recent work suggests a key role for isocitrate lyase in some, at least, of the lithotrophs. *Nitrobacter agilis* grows slowly in media containing acetate, which increases the isocitrate lyase activity from the autotrophic level of 2 units (i.e., 2 nmoles/min/mg protein) to 206 (221). Carbon from acetate supplied to *N. agilis* is not restricted to the four amino acids labeled in the obligate lithotrophs, but is distributed to all the protein amino acids (221, 254). Moreover, Smith & Hoare (221) showed α-OG dehydrogenase (13 units) in this organism and proposed that heterotrophic growth was made

TABLE 3. Amino Acids Inhibiting Thiobacilli

Amino acid tested (1–10 mM)	*T. thioparus*	*T. neapolitanus* *C*	*T. neapolitanus* *X*	*T. ferro-oxidans*	*T. thio-oxidans*	*T. concre-tivorus*
Phe	++	++	++	+	+	++
Glu	+	−	+	−	±	nt
Cys	+	+	+	nt	++	nt
Tyr	+	−	−	±	±	−
Thr	±	−	±	++	−	nt
His	−	−	++	+	++	nt
Met	−	±	+	++	++	nt
Val	−	−	−	nt	++	(±)
$AspNH_2$	−	nt	++	nt	++	nt
Ser	−	−	−	nt	++	−
Reference	147	110, 111, 114,	102, 147	106, 197	30, 147, 197	114.

nt—not tested; ++—strong inhibition; +—inhibition; ±—slight inhibition; −—no effect on development.

possible partly by its presence. Aleem (5) and Wallace and Nicholas (253, 254) failed to find any such activity, although labeled acetate gave rise to large amounts of labeled aspartate, as would occur if a complete TCA or glyoxylate cycle was functioning (254). It seems likely, therefore, that isocitrate lyase is present at considerably higher levels than α-OG dehydrogenase. This might indicate that the step from α-OG to succinate is bypassed even in the facultatively heterotrophic growth of *Nitrobacter*. Isocitrate dehydrogenase would then be primarily a biosynthetic enzyme in this as in the obligate chemolithotrophs. Similarly, *Chromatium* lacks α-OG dehydrogenase but has a functional glyoxylate cycle (238), so an exactly parallel situation may exist. The suggestion that aspartate might arise from acetate in *Nitrobacter* by a ferredoxin-dependent pyruvate synthase (254) seems highly improbable, as ferredoxin is not known in the aerobic lithotrophs, and such carboxylation is probably restricted to the anaerobic photosynthetic bacteria and clostridia (59). Although low levels of isocitrate lyase and malate synthase have been found in some obligate lithotrophs, there is no evidence from ^{14}C-tracer studies that they function significantly in the metabolism of exogenous acetate by thiobacilli or blue-green bacteria.

Lack of α-OG dehydrogenase has been proposed as one of the factors causing obligate lithotrophy (94, 111, 222). It would seem, however, that isocitrate lyase could compensate for this deficiency. It is noteworthy that although *Nitrobacter* may contain both these enzymes, only isocitrate lyase has been shown in *Nitrosomonas* (254) and only α-OG dehydrogenase occurs in *Nitrosocystis* (94). As yet, neither of these ammonia oxidizers has been grown heterotrophically, but further efforts to obtain such growth might be worthwhile, although the extremely low level of succinate dehydrogenase in the strain used by Williams & Watson (262) could be a barrier to heterotrophy in *Nitrosocystis*. α-OG dehydrogenase and isocitrate lyase could not be detected in *T. neapolitanus* (116).

NADH Oxidation and Energy Coupling in Aerobic Obligate Lithotrophs

It is indisputable that lithotrophic bacteria have essentially a heterotrophic metabolism with additional capacities for obtaining energy from inorganic oxidations and for using CO_2. If this concept is accepted, one expects to find an endogenous metabolism being used to obtain energy for cell maintenance (87) when exogenous substrates are absent, and dependent on the use of stored metabolic products. Endogenous respiration has been observed in obligate chemolithotrophs (54, 248, 249) and photolithotrophs (19). Storage materials demonstrated include poly-β-hydroxybutyrate (PHB) in *Nitrobacter* (236) and *T. ferrooxidans* (255), probably polysaccharides in *T. thiooxidans* (139, 140) and *Chlorobium* (218, 219), and polyphosphate in *Chlorobium* (183) and probably thiobacilli (249). It seems reasonable to assume that cell maintenance and survival in starved cells depends on the continuous use of these compounds. Indeed, in both *Thiobacillus* (118) and blue-green bacteria (92), ATP constitutes 0.05–0.2 percent of the dry weight of starving cells. Although turnover studies have not been reported, this ATP is presumably being continuously produced and consumed. There is also no reason to suppose that prolonged starvation does not result in the use of other cell components such as the known lipids and proteins (21, 42, 107, 187, 208, 213, 215), as sources of energy for survival. The occurrence of PHB among the thiobacilli is interesting. It is apparently not formed by *T. neapolitanus* (110, 206) or in autotrophically grown *T. novellus* (179), although it is present in heterotrophic *T. novellus* and in *T. ferrooxidans* under both growth conditions (255). *Chromatium* synthesizes a storage polysaccharide (68) which is converted to PHB and CO_2 in darkness, with the production of ATP that presumably provides maintenance energy (68).

Numerous NAD or NADP-linked dehydrogenases occur in lithotrophs, indicating that reduced NAD for a respiratory electron transport chain should be available. Cytochrome reduction by succinate and ether-soluble yeast extract fractions has been shown in *T. neapolitanus* (206) and by formate in several lithotrophs (1, 74, 99, 148, 217, 265).

Smith, London & Stanier (222) proposed that failure to oxidize NADH was a prime cause of obligate lithotrophy. If this were the case, the endogenous respiration would presumably either not be energy-yielding, which seems very unlikely, or would depend on substrate-level phosphorylations, or would result from the coupling of the oxidation of endogenous nutrients to flavin, quinone, or cytochrome reduction. If the latter were the case, growth on exogenous compounds should be achieved using the same electron transport mechanisms. As yet any such unique oxidation-reduction reactions have not been shown. It is also unlikely that all ATP generation would be by substrate phosphorylation since uncouplers of oxidative phosphorylation depress the endogenous ATP levels of *Anacystis nidulans* and *Phormidium luridum* (19). This suggests that electron transport phosphorylation is involved in dark aerobic respiration. Smith, London & Stanier's (222) claim is in any case no longer tenable as an unqualified explanation of obligate lithotrophy as several workers have also shown that NADH oxidation is effected by several obligate lithotrophs (Table 2) (31, 95, 136, 137, 204, 229, 237, 260). Data demonstrating "NADH oxidase" must, however, be considered with caution for disappearance of NADH does not necessarily indicate its oxidation by the respiratory chain, but could result from its consumption by enzymes such as malate dehydrogenase if oxaloacetate were present (40, 253). The rather doubtful value of data about the rate of disappearance of NADH from a cell-free reaction mixture may best be covered by the use of the name "NADH disappearase" (173) when the ultimate hydrogen acceptor is in doubt! Moreover, if its oxidation were by oxygen, the oxidation might not be coupled to an energy-trapping mechanism. For example, in *Anabaena* and *Anacystis* a cyanide-sensitive NADH:O_2 oxidoreductase was shown but NADH:cytochrome *c* and cytochrome *c*:O_2 oxidoreductases seemed to be absent (95). *Chlorobium thiosulfatophilum* contains a soluble NADH:cytochrome *c* oxidoreductase (260). There is evidence that in some lithotrophs NADH oxidation may supply a respiratory chain basically like that of the mitochondrion. The possession of the enzymes necessary for such a complete respiratory chain by *Nitrosomonas, Nitrobacter, Thiobacillus,* and *Ferrobacillus* was indicated by the demonstration of Aleem et al (3–5, 7, 182, 235) of a system of energy-dependent, NAD reduction by electron transport from reduced cytochromes. While this "reverse flow" system operates in lithotrophic growth, it is not reasonable to assume that NADH oxidation and the normal mode of electron flow never operate in obligate lithotrophs (253, 254). The demonstration of cytochrome *c* reduction by NADH in *T. neapolitanus* (237), by NADPH in *T. thiooxidans* (229), and of cytochromes *c* and *a,* by NADH in extracts of *N. winogradskyi* (73, 74) established that such a normal electron flow can occur in obligate chemolithotrophs. It was further indicated that NAD accelerated cytochrome *c* reduction by endogenous substrates in *T. neapolitanus* extracts (237). Detailed studies on electron transport in thiobacilli during sulfur compound oxidation have shown cytochrome *c* to be a principal component and that oxidative phosphorylation accompanies this electron transport (45, 86, 117, 118, 200, 202). Flavins and quinones may also be involved in lithotrophic respiration (5, 165, 182). Thus, cytochrome *c* reduced by NADH should subsequently be oxidized with concomitant oxidative phosphorylation. Moreover, Aleem's studies (4, 5) indicate that three coupling sites should exist for phosphorylation between NADH and oxygen. Hempfling & Vishniac (86) were unable to find phosphorylation during NADH oxidation by cell-free extracts of *T. neapolitanus,* and even with *T. novellus* the best P/O ratio obtained was about 0.25 (38) compared with 0.96 for cytochrome *c* oxidation and 1.9 for succinate oxidation. Possibly coupling of NADH oxidation to energy-yielding electron transport is inefficient in the living cell, but it has yet to be demonstrated that low phosphorylation efficiencies are not due to procedures used in the preparation of cell-free extracts. The phosphorylation demonstrated in *T. novellus* was catalyzed by the 144,000 *g* supernatant fraction of the cells (38). A membrane component might be required in the intact cell for effective energy-coupling during NADH oxidation. Biggins (19) failed to obtain NADH oxidation by photolithotrophically grown *Anacystis* and *Phormidium,* but did show NADH oxidation and infer the presence of an NADPH:O_2 oxidoreductase. ATP-P_i exchange reactions also indicated the possibility of oxidative phosphorylation in these extracts. As anaerobiosis of the whole organism resulted in decreases in the intracellular levels of both ATP and NADPH, but not of NADH, oxidative phosphorylation seemed more likely to accompany NADP-linked substrate oxidations. Leach & Carr (137) similarly found that twice as much phosphorylation accompanied NADPH oxidation as NADH oxidation in extracts of *Anabaena.* Only about 1 nmole of ATP was formed for each 10 NADPH oxidized, indicating even poorer efficiency than obtained with *T. novellus.* These low efficiencies may be irrelevant as equally low values have been obtained with extracts of heterotrophic bacteria (137), while yield studies on heterotrophs (79) suggest an efficiency more like that of mitochondrial phosphorylation. What is clear is that NADH and NADPH can be oxidized by obligate chemolithotrophs at rates quite comparable to those found with heterotrophs and facultative autotrophs (31, 201) (Table 2) and that these oxidations may be coupled to ATP synthesis. If inability to grow aerobically on organic substrates is related in obligate lithotrophs to respiratory or energetic shortcomings, these are more complex than can be explained by assuming total lack of an enzyme or coupling factors. Failure of exogenous organic substrates to stimulate the endogenous respiration of many lithotrophs (19, 111, 206, 222) might be related to poor permeation into resting cells, rather than to failure of the respiratory apparatus. Much more needs to be learned about the regulation of these respiratory systems. Perhaps the most significant observation yet made is that the level of "NADH oxidase" in *T. ferrooxidans* is some 140 times greater in cells grown on glucose than in iron-grown autotrophic cells (212).

Heterotrophic Growth Potential of Putatively Obligate Lithotrophs

Recent reports have described conditions under which growth can be achieved using some organisms previously regarded as obligate lithotrophs. It has become well established that *Nitrobacter agilis* (221) and some strains of *Thiobacillus ferrooxidans* (193, 212) will grow heterotrophically. *N. agilis* grows only very slowly on amino acids and acetate, while *T. ferrooxi-*

dans, in contrast to *T. intermedius,* (143) grows faster on glucose than on iron (212). *T. ferrooxidans* requires a period of adaptation for growth to occur on glucose, and Shafia & Wilkinson (212) found that of eight isolates, only two could adapt to grow heterotrophically. *Para*-aminobenzoic acid was also required for growth on glucose. In the presence of yeast extract (0.01 percent), growth of one strain was supported almost equally well by glucose, mannitol, or fructose, and progressively less well by sucrose, glutamine, ribose, salicin, or arginine (212). In contrast to *Nitrobacter,* the ability of *T. ferrooxidans* to grow autotrophically was lost after 7–10 transfers on glucose. Its growth on glucose was not stimulated by also supplying ferrous iron, again in contrast to *T. intermedius* and *Nitrobacter,* which grew faster mixotrophically than heterotrophically. The evidence that the organisms growing under these heterotrophic conditions were indeed *N. agilis* and *T. ferrooxidans* seems acceptable (212, 221), and these strains must be included in Group 2 of Figure 1.

Borichewski & Umbreit (23, 24) demonstrated growth of *T. thiooxidans* on glucose as a sole energy source, but showed that growth occurred only if high concentrations of sucrose were also present, or if the cultures were continuously dialyzed against fresh medium. They attributed failure to grow in more "normal" conditions to toxicity of pyruvate accumulated from glucose. Rittenberg has indicated discrepancies in the published data, suggesting that the dialysis conditions employed did not lower the pyruvate concentration to noninhibitory levels. It is not disputed that pyruvate is accumulated during growth by *T. thiooxidans* on sulfur or glucose (22–24) and by *T. ferrooxidans* (211), but acceptable evidence for the accumulation of a toxic material that prevents heterotrophic growth of *T. thiooxidans* on glucose is, in fact, lacking.

Recently, Pan and Umbreit (171, 172) demonstrated growth in continuous flow dialysis cultures on glucose media, using not only *T. thiooxidans* but also *T. thioparus, T. neapolitanus, T. denitrificans,* and *Nitrobacter winogradskyi.* Again, the view that dialysis removed inhibitory materials was proposed as the explanation for growth only under the special conditions employed. Numerous attempts by other workers to grow these strains in complex and defined organic media have failed, and further study of this dialysis technique is urgently required. It must be rigorously established that the growth observed was due to the lithotrophs and not to any contaminant. It has been inferred that *T. thiooxidans* might not have been the sole component in Umbreit's dialyzed cultures (197), but sulfur-oxidizing thiobacilli could still be grown from glucose cultures after prolonged heterotrophic growth (24). Serological tests also indicated the presence of *T. thiooxidans* in such cultures. Matin and Rittenberg (197) failed, however, to grow the same strain on glucose. The strain of *T. thiooxidans* used by Umbreit seems to differ from other presumed obligate strains of *Thiobacillus* in having a high level of α-OG dehydrogenase (Table 2), and deserves further comparative study.

In addition to the toxic products theory, three principal causes of the absence or elusiveness of heterotrophic growth in some lithotrophs can be suggested. The first concerns uptake of, and energy generation from, organic substrates. Such substrates may be unable to enter the cell because their uptake requires energy from a lithotrophic oxidation. *T. neapolitanus* fails to incorporate acetate unless thiosulfate is also supplied (110). *Methylococcus* similarly requires methane (174). *Nitrosomonas,* however, incorporates alanine in the absence of ammonia, although uptake is nearly doubled if ammonium ion is also supplied (36). Incorporation of low concentrations of aspartate could be increased as much as fivefold by ammonia, but less so at high concentrations of aspartate, perhaps suggesting that aspartate entry into the cell was accelerated by ammonia oxidation. The total amount of acetate metabolized by washed cells of the facultative *N. agilis* was unaffected by the presence of nitrite, but it decreased the proportion of acetate subsequently oxidized (221). In contrast, glucose assimilation by the facultative *T. intermedius* occurs only when thiosulfate is present (144, 153), perhaps indicating an energy requirement for uptake or metabolism that is not supplied by glucose itself. The problem of energy generation from organic compounds has been discussed, and it has been suggested that oxidative phosphorylation may be of low efficiency in obligate chemolithotrophs during organic oxidations (102, 222). It is indeed observed that for *T. ferrooxidans* (212) and for thiobacilli grown in dialysis culture (171) the growth yields are little or no better than for chemolithotrophic culture: that is, of the order of 150 μg dry weight per ml of culture in 0.5 percent glucose medium (212). If such yields are maximal, they could indicate the production of as little as 1 mole of ATP per mole of glucose metabolized, possibly indicating a fermentative pathway for glucose metabolism. Johnson & Abraham (105) proposed that the only mechanism available for energy generation in *T. thioparus* and *T. neapolitanus,* if oxidative phosphorylation were negligible, would be substrate level phosphorylation of the type accompanying the triose phosphate dehydrogenase reaction normally found in glycolysis. The glycolytic pathway was proposed to be incomplete in these organisms and in *Nitrosocystis* (262) because of the absence of phosphofructokinase (105, 262). An alternative route for the formation of glyceraldehyde-3-phosphate from glucose would be the Entner-Doudoroff pathway which is common in pseudomonads (47). The enzymes of this particular pathway are induced by glucose in *T. perometabolis* (155), *T. intermedius* (153), and *T. ferrooxidans* (228) which also possesses phosphofructokinase (12) and probably a functional glycolytic sequence. Fructose induces these enzymes in *Hydrogenomonas* (126). The Entner-Doudoroff pathway is absent from *T. neapolitanus, T. thioparus, T. thiooxidans,* and *T. novellus* (155). The absence of this pathway from nitrifiers has not been established. However, if oxidative phosphorylation is insufficient, glycolysis not operative, and the Entner-Doudoroff pathway absent, growth on glucose might be impossible or very slow. The presence of the hexose monophosphate pathway has been indicated in *T. ferrooxidans* (67), *T. thioparus,* and *T. neapolitanus* (195). This might effect complete glucose oxidation in the absence of a complete TCA cycle. More information is needed about ATP formation by oxidative or substrate level phosphorylation during organic metabolism before the problem of energy coupling in relation to facultative heterotrophy can be resolved.

The second possible basis for obligate lithotrophy could be in the low levels of some enzymes of the TCA cycle and of carbohydrate metabolism or some other enzymes essential for cell biosynthesis during growth on organic substrates instead of CO_2. If all such possible deficiencies could be established a medium suitable for growth of an otherwise obligate lithotroph might be devised.

A third cause could be the simple one noted earlier, of the organic medium being made nutrient element-deficient through omission of the normal inorganic reductant, which would also be used as a source of sulfur by thiobacilli (116, 170) or of nitrogen by *Nitrosomonas.*

Obligate chemolithotrophy could depend on a combination of factors rather than a single common one.

Among the photolithotrophs, the blue-green bacteria seem generally to be unable to grow without light as aerobic heterotrophs. Such growth as has been shown has been much slower than autotrophic growth, and is reliably established only for some *Nostoc* spp. (9, 84) and *Chlorogloea fritschi* (60, 92), which developed best and fixed nitrogen heterotrophically on a sucrose medium. Blue-green bacteria can assimilate large amounts of organic nutrients during photosynthetic growth. *Anabaena variabilis* oxidizes glucose by the pentose phosphate cycle and obtains up to 46 percent of its dry weight from it (177), but the presence of glucose has little or no effect on the growth rate, respiration, or levels of glycolytic and pentose cycle enzymes (177). Similarly, acetate contributed up to 8–15 percent of the carbon of several blue-green bacteria, but did not affect factors such as growth rate, levels of enzymes for acetate activation, the TCA cycle, or the glyoxylate cycle in *Anabaena, Anacystis, Nostoc, Coccochloris,* or *Gloeocapsa* (10, 91, 176, 178, 222).

Other nutrients were also assimilated in the light by blue-green bacteria, including substantial amounts of leucine and pyruvate by *Anacystis* and *Coccochloris* and of leucine and aspartate by *Gloeocapsa* (222). These organisms seem to remain dependent on substantial amounts of CO_2 for growth to occur and for acetate to be assimilated (91) as occurs in *Thiobacillus neapolitanus* (110). Two aspects of these data are significant. First, organic assimilations by blue-green bacteria require both CO_2 and light energy. Second, no induction of enzymes for organic metabolism has been demonstrated. Organic compounds are thus incorporated by introduction into a metabolic system that is functioning to convert CO_2 to cell material. They are not recognized by obligate chemolithotrophs or blue-green bacteria as full growth substrates when either CO_2 or lithotrophic energy supply is removed. It has been suggested that metabolic control by enzyme repression and depression that is so highly developed in some heterotrophic organisms, is lacking in *Anabaena variabilis* (93, 176–178). The only means of control is believed to be by feedback inhibition of existing enzymes (178). Apart from the observed lack of effect of environmental conditions on enzyme levels, the excretion of 18 amino acids by *Anabaena* during growth (93) is indicative of a rather coarse control over the metabolism.

Failure of exogenous nutrients to influence enzyme levels in obligate lithotrophs could be a general feature of obligate photo- and chemolithotrophs. The metabolic pathways of obligate lithotrophs would then have to be regarded as sequences of enzymes strictly regulated at levels enabling growth with carbon dioxide to proceed at an efficient rate. Failure to increase levels of enzymes in response to a nutrient in the environment could prevent use of that nutrient at a rate sufficient to support growth. As yet little eivdence exists for invariable levels of enzymes in obligate lithotrophs in general or on the nature of any regulatory processes that would maintain such levels. Certainly, the differential rates of incorporation of organic compounds by cultures of obligate lithotrophs do not increase with time as would be expected if enzymes were induced that accelerated the use of compounds such as acetate or fructose (110, 114, 222). Solutions to some of these problems may be obtained by seeking enzyme derepression in analogue-resistant mutants or studying auxotrophic mutants if the seemingly low frequency of the latter enables their isolation in any number.

Facultative heterotrophy among the photolithotrophs requires consideration of organic compounds in two ways. We need to establish firstly whether organisms like *Chlorobium* or *Chromatium* can be grown as heterotrophs in the dark; and secondly, whether organic materials can be used in place of sulfur compounds and hydrogen as electron donors for photosynthesis, and CO_2 as carbon sources, even if dark heterotrophy is impossible.

The Thiorhodaceae and Chlorobacteriaceae are strictly anaerobic phototrophs. *Chlorobium* seems incapable of aerobic metabolism or of phototrophic growth without CO_2 and sulfide or thiosulfate. It has been suggested to have a purely anabolic carbon metabolism, depending on stored polyphosphate as its major reserve product (183). Organic compounds such as acetate (89, 167, 205), propionate (131, 132), glucose and amino acids (116) are assimilated by *C. thiosulfatophilum* growing otherwise photolithotrophically, and may stimulate development (18, 89). Numerous Thiorhodaceae can use various organic compounds in place of inorganic sulfur as hydrogen donors (52, 234), but some strains of *Chromatium, Thiopedia, Thiocystis,* and *Rhodothece* (97, 183, 234) required sulfide as a sulfur source, being incapable of assimilatory sulfate reduction. Significant amounts of organic compounds such as acetate and sugars can be incorporated by Thiorhodaceae in addition to CO_2. Some *Chromatium* strains can use acetate as sole source of carbon (97, 183). Under such conditions their metabolism is thus comparable with that of the Athiorhodaceae, indicating that a sharp division between these two groups does not exist, rather, there is a physiological overlapping. This view is supported by the adaptive ability of some Athiorhodaceae to use thiosulfate as an electron donor and to grow as photolithotrophs (199, 246) with CO_2 as sole carbon source.

It is noteworthy that although *Chlorobium* and many Thiorhodaceae are obligate photolithotrophs they frequently exhibit a requirement for vitamin B_{12} (183, 185, 234). Moreover, although the chemolithotrophs, *Chlorobium,* and at least some Thiorhodaceae must use CO_2 as a major source of cell carbon, in many cases they can obtain all necessary nitrogen from organic nitrogen compounds in the absence of ammonium salts. Urea is used by *Thiobacillus* (25), *Chlorobium* (18), and some Thiorhodaceae (123, 234). Some amino acids serve as sole nitrogen sources for *Hydrogenomonas* (64), proline and arginine serve for *Chlorobium* (18), alanine, aspartate, valine and others for some methylothrophs (174), and peptone and casamino acids for some Thiorhodaceae (234).

The physiological range exhibited by Thiorhodaceae extends from complete photolithotrophy in most to complete photoorganotrophy in some, but with little or no capacity for dark aerobic heterotrophy. Among the Athiorhodaceae, photoorganotrophy is universal, dark heterotrophy common, and complete photolithotrophy probably rare and allowing slower growth than under photoorganotrophic conditions (199). *Chromatium* can consume oxygen and carry out some aerobic metabolism of organic compounds (70, 168). It contains *c*-type cytochromes that are oxidized by oxygen (65) and oxygen is apparently not toxic (96), but the low rates of aerobic metabolism and phosphorylation do not seem to allow measurable growth (65, 70).

Toxicity of Organic Nutrients in Relation to the Organic Metabolism of Obligate Lithotrophs

The fallacious concept of a general toxicity of organic compounds to obligate chemolithotrophs, implicit in Winogradsky's original definition of the inorgoxidant (197, 264), was demolished by Rittenberg (197). A substantial list remains, however, of organic nutrients that depress the development of some lithotrophs. Rittenberg (197) has tabulated some of the more recent reliable data on chemolithotrophs. I believe that study of the mechanism of these inhibitions will enable us to learn more about the regulation of endogenous metabolism of the lithotrophs and may lead to a solution of the problem of obligate lithotrophy. It is apparent from the literature that the physiological condition of a given organism determines its response to any material added to its medium. Thus, while Winogradsky (264) found very variable responses to many substances (including glucose, urea, glycerol, and acetate which retarded nitrification at relatively low concentrations), later workers (100, 160) found *Nitrosomonas* to be relatively insensitive to the same compounds. Glucose is tolerated at 0.56 *M* (100), although mannose (formed during autoclaving of glucose) was toxic. Some amino acids (glycine, alanine, asparagine, methionine, and cysteine) and peptone depressed the nitrifiers (100, 101, 264), although glutamate and aspartate were relatively nontoxic.

Thiobacillus ferrooxidans seems particularly sensitive to inhibition by numerous exogenous nutrients (130, 216). The lag in its growth on iron was extended by glucose, sucrose, maltose, galactose, starch, or urea, and development was completely suppressed by fructose, lactose, or complex materials like peptone or yeast extract (all supplied at 0.5 percent) (240). Primary alcohols also depress the development of *T. ferrooxidans,* inhibition increasing with increasing carbon chain lenth, so that 0.01 percent octanol is rather more inhibitory than 0.1 percent pentanol or 1 percent ethanol (240). Inhibition by organic acids similarly increased with chain length in a homologous series, and this was attributed to possible absorption by a lipid fraction of the cell envelope (241). Inhibition of microbial growth by organic nutrients is not an uncommon phenomenon and amino acids have frequently been found to exert complex effects on growth. *Bacillus anthracis* was inhibited by isoleucine or valine or leucine (71), but the three together were stimulatory. Similarly, inhibition of *E. coli* by valine was relieved by isoleucine (230). *Pseudomonas denitrificans* is also inhibited by amino acids (43). These inhibitions are often brought about by feedback inhibition by the added compound of an early step in a branched biosynthetic pathway, thus "starving" the organism of the other products of the pathway and hence depressing growth unless the other product compounds are also added to the growth medium (114, 147). Such feedback inhibition could be the cause of the growth depression caused by organic nutrients added to several lithotrophs.

Several studies have been made of inhibition by amino acids of several lithotrophs. Marked inhibition (37–90 percent) of *Nitrosomonas* was produced by 0.04 m*M* thr, his, lys, val, met, and arg (36), while *Nitrobacter* was inhibited similarly by 1 m*M* valine or threonine (48). The *Thiobacillus* spp. are inhibited by various amino acids (Table 3) and recent studies have indicated probable mechanisms of their action (102, 114, 147). Relatively few amino acids inhibited when supplied at concentrations of 1 m*M* or less (102, 114, 147, 197), although *T. thiooxidans* appears to be about the most susceptible to inhibition (147). Rittenberg's (197) report of complete inhibition of *T. thioparus* by 10 m*M* valine was erroneous (147). The apparently greater sensitivity of *T. neapolitanus* strain X (which can be inhibited by 0.1 m*M* thr or his (102)) than of strain C probably reflects the fact that the acids were added to strain C after exponential growth had started (114), while in Johnson's (102) and Lu's (147) experiments, media already containing amino acids were freshly inoculated with bacteria. The sensitivity of transferred or lag phase organisms might be greater than those already established in log growth in a medium that was subsequently altered only by adding an amino acid. Indeed, it is important to distinguish between effects on the lag phase and on the subsequent logarithmic growth rate. In the nitrifying bacteria the lag phase is shortened or increased by some additives (197), and in *T. ferrooxidans* some organic compounds probably increase the lag period but do not depress the logarithmic growth rate (106).

Phenylalanine toxicity has been extensively studied using *T. neapolitanus* and *T. thioparus* (102, 102a, 114, 147). Growth inhibition could be completely relieved by tyrosine or tryptophan (102, 102a, 114, 147) supplied separately at concentrations higher than that of the phenylalanine. Concentrations of tyr or trp that alone only partially relieved the inhibition of *T. neapolitanus* produced a more or less additive relief when supplied together (114, 147). These results can be interpreted in relation to the control mechanisms demonstrated for aromatic amino acid biosynthesis in heterotrophs (69). Phe could produce amino acid imbalance in *Thiobacillus* through preventing synthesis of try and trp by inhibiting an early step in the common pathway for the synthesis of these compounds. Relief of phe-induced growth inhibition by tyr and trp could then be expected. I have demonstrated (114) that phe is indeed a potent inhibitor of the DAHP synthetase enzyme of this pathway and provided evidence that a mutation affecting this enzyme might confer phe resistance on *T. neapolitanus*. This study was the first to establish that growth inhibition of an obligate chemolithotroph could be related directly to inhibition of a specific enzyme. It was also demonstrated that such depth of analysis is essential and that deductions from amino acid antagonism experiments (102, 114, 147) could be misleading. This deduction follows from the observation that inhibition of *T. neapolitanus* by phe was reversed by a number of amino acids which were chemically and metabolically unrelated to the aromatic amino acid pathway (114). These acids were competitive inhibitors of phe uptake and thus relieved growth inhibition by preventing permeation of phe to the sensitive sites within the cell. These experiments indicated that two groups of amino acids exist with respect to their uptake by growing cells, and that members of each group competed strongly only with other members of the same group (114). The aromatic amino acids competed with each other, but the effect of phe on the incorporation of tyr and trp indicated that phe inhibited both their uptake and biosynthesis, in contrast to its affecting only the uptake of other amino acids such as leucine or serine (114).

It is noteworthy that Johnson (102) found that inhibition of *T. neapolitanus* by 0.1 m*M* histidine was relieved by ala, gly, leu, ser, trp, tyr, val, or ileu (0.3 m*M*) and that these compounds (except ileu) or met also relieved inhibition by threonine (102). In *T. neapolitanus* strain C all these acids compete with each other for entry into the cell and the relief of inhibition of strain X doubtless resulted in some cases from the failure of an inhibitory level of his or thr to accumulate within it as a consequence of competitive inhibition of uptake.

Lu further investigated the inhibition of *T. thiooxidans* by serine and valine. Inhibition by ser of *Bacillus anthracis* (71) or some lactic acid bacteria (161) was reversed by thr or by val plus leu in *B. anthracis*. Serine inhibits ser phosphatase and phosphoglycerate dehydrogenase in several organisms (78) and its effect in *T. thiooxidans* could thus be to depress synthesis of compounds derived from phosphoglycerate. As predicted, inhibition by 10mM ser was partly reversed by 10 mM thr and completely by val plus ileu (10 mM each). An analysis of serine toxicity at the enzyme level would now be of interest.

Lu further proposed that inhibition by valine could result from inhibition of synthesis of ileu and leu and demonstrated that, separately, these partially relieved inhibition and together even stimulated growth in the presence of val to a level exceeding the control (147). The val and leu pathways have a common precursor, α-acetolactate derived from two pyruvate molecules, and the equivalent precursor for ileu is α-aceto-α-hydroxybutyrate, similarly formed by condensation of α-ketobutyrate and the "active acetaldehyde" of pyruvate (162). There is evidence that in *E. coli* the α-keto-acid-"active acetaldehyde" condensing enzyme (162) forming acetolactate also forms acetohydroxybutyrate (162) and that the synthesis of both compounds by *E. coli* extracts was inhibited by valine (162). Inhibition of this enzyme in *Thiobacillus* by valine could thus result in growth inhibition by depression of leu or ileu synthesis. Acetolactate synthesis by crude extracts was, in fact, shown to be depressed by valine (147). Leucine did not affect acetolactate synthesis or valine inhibition, although ileu did inhibit and added to the inhibition caused by val (147).

Inhibition of *T. thiooxidans* by valine and of *T. neapolitanus* by phenylalanine can thus both be attributed to end product inhibition of specific enzymes at the start of separate branched pathways.

Certain amino acids are also toxic to *Chlorobium* (116) and *Methylococcus* (50, 56). In liquid cultures exponential growth of cultures of *M. capsulatus* is markedly depressed only by threonine or phenylalanine (50). Amino acid uptake by *M. capsulatus* exhibits characteristics similar to *T. neapolitanus* in that threonine incorporation is antagonized by numerous amino acids (gly, ser, ala, val, leu, ileu, met, phe, tyr, homoserine), some of which also tend to relieve the inhibition of growth caused by threonine (50). Threonine may act as an inhibitor of an early enzyme, such as aspartokinase, in the branched pathway leading to lys, met, thr, and ileu in numerous bacteria (37, 76, 164, 223) and probably in *Methylococcus*.

Photolithotrophic growth of *Chlorobium* on thiosulfate is prevented or retarded by thr, ser, met, trp, gly, try (0.1–1mM) added to lag phase cultures (116). Labeled leu, phe, and tyr are incorporated into *Chlorobium* protein and contributed no carbon to other amino acids. Serine gives rise also to glycine and some labeling in alanine and cysteine. Incorporation of leucine was depressed by ileu, val, ala, phe, trp, and met, indicating broad specificity permeation mechanisms for amino acids in this lithotroph also. Met and thr (0.1 mM) prevented growth whether provided separately or together. Lys + ileu (1 mM) allowed normal development with both thr and met. Lysine reversed inhibition by thr. Preliminary experiments suggest that aspartokinase in *Chlorobium* extracts is inhibited by thr and the non-toxic lys, and that the two exert a combined effect on enzyme activity (116). The aspartokinase of *Rhodospirillum rubrum* (29) was inhibited by the concerted action of threonine and lysine, which also inhibited growth. This growth inhibition was reversed by methionine (29). For comparison, growth and aspartokinase of the heterotroph *Arthrobacter globiformis* (129) were also inhibited by thr. Lys or met relieved growth inhibition but did not affect aspartokinase or its inhibition by thr. Inhibition of *Chlorobium* by tyrosine or tyr plus trp was antagonized by phe. Valine prevented growth at 10 m*M* but 1 m*M* ileu reversed this effect (116).

Possibly all the observed inhibitory effects of amino acids on lithotrophs may be traced to end product inhibition of essential enzymes. The suggestion that phe might cause the synthesis of protein containing abnormally high proportions of phe (110) has not been substantiated and clearly did not occur in *T. neapolitans* (114). This view was construed by Rittenberg (197) as implying an alteration in protein composition induced at the transcription level rather than by general imbalance of protein synthesis. Alteration of the genetic code by a toxic amino acid is undoubtedly highly unlikely. It is, however, notable that coding ambiguities have been observed during in vitro protein synthesis in which quite high levels of leu are introduced into the polyphenylalanine produced in poly-U-directed synthesis (121). Evidence for in vivo coding ambiguity, where the same codon leads to the incorporation of two different amino acids was presented by Kindler & Ben-Gurion (121) who used a mutant of *E. coli* auxotrophic for phenylalanine. Leucine was lethal in the absence of phe, possibly through erroneous incorporation of leu into protein in place of phe, although such abnormal proteins have not yet been proved in vivo (121). It is certainly not yet to be discounted that some cases of amino acid toxicity, perhaps in some lithotrophs also, could result from such a phenomenon, brought about perhaps by transfer-RNA molecules of low specificity being loaded with the wrong amino acid when massive amounts of that acid are available from the environment.

The action of keto acids (e.g., pyruvate) as inhibitory and growth-limiting factors in autotrophic and glucose-grown *T. thiooxidans* has caused considerable debate (22, 24, 197). Pyruvate added to lithotrophic cultures of *T. thiooxidans* (24) or *T. ferrooxidans* (197) undoubtedly inhibits growth, whereas growth of *T. thioparus* (197) and *T. neapolitanus* is unaffected (115). Rao & Berger (192) have recently shown that *T. thiooxidans* rapidly accumulates pyruvic acid at pH 2.3, and that sulfur oxidation, CO_2 fixation and O_2 uptake are rapidly suppressed. At pH 7 pyruvate is not accumulated and does not affect O_2 uptake or inhibit cell-free CO_2 fixation at pH 7.4. They concluded that *T. thiooxidans* is passively permeable to undissociated molecules at low pH and accumulated it to a demonstrated internal concentration of 58.5 m*M* at pH 2.3. Ionization of this within the cell resulted in a progressive drop in intracellular pH as net flow of pyruvate into cells continued, and eventually prevented cellular functions. This may be a correct interpretation but does not allow the assumption that pyruvate is a growth-limiting factor in autotrophic thiobacillus cultures (22). Limitation by accumulated toxic products does not occur in *T. neapolitanus* for which thiosulfate is generally growth-limiting in batch culture (87, 117) or continuous culture (87), or with *Hydrogenomonas,* with which logarithmic growth to yields of 1.5×10^{11} viable cells/ml (21.6 g dry wt/liter) have been reported (195).

To conclude this section I wish to remark on some observations on the interaction of lithotrophic and organotrophic processes in facultative heterotrophic chemolithotrophs. Repression by organic nutrients of lithotrophic energy metabolism and CO_2 fixation have been considered previously (see above and 138, 250). Until recently, less information was available on the effect on the organic metabolic pathways of growth on lithotrophic substrates. In *Hydrogenomonas,* growth on hydrogen results in repression of TCA cycle and glyoxylate cycle enzymes (239) and of enzymes for the Entner-Doudoroff system even when fructose is present (210). Similarly, in *Thiobacillus novellus* (154), thiosulfate represses glucokinase, glucose-6-phosphate, phosphogluconate, and isocitrate dehydrogenases and NADH oxidase. Thiosulfate also results in repression of Entner-Doudoroff pathway enzymes, isocitrate dehydrogenase and α-OG dehydrogenase in *T. intermedius* (153). Growth of *Thiobacillus* A2 on thiosulfate resulted in decreased capacity to oxidize formate, acetate, *p*-hydroxybenzoate, and protocatechuate and repressed formate dehydrogenase (232). Acetate oxidation by *Nitrobacter* is prevented by nitrate (221). All such repressive effects of lithotrophic substrates result in reduced use of organic compounds in energy generation and tend to restrict their use to anabolic sequences only. If, as was discussed in an earlier section, levels of the "heterotrophic" enzymes mentioned here are present at invariable levels in obligate lithotrophs, those levels are likely to be equivalent to repressed levels found in lithotrophically grown facultative species and inadequate for heterotrophic growth to be possible. Repression of the TCA cycle enzymes is known also in heterotrophs (61, 82 ,83), and, in *E. coli* at least, glucose-grown cells require an adaptation period and to synthesize higher levels of TCA cycle enzymes before growth on succinate becomes possible (80). Recently, α-OG dehydrogenase and succinyl CoA synthetase were shown to be repressed in *Thiobacillus* A2 grown autotrophically, but derepressed during heterotrophic growth (182a).

Metabolic and Evolutionary Aspects of One-Carbon Compound Metabolism in Lithotrophs and Methylotrophs

The metabolic features shared by obligate methylotrophs and lithotrophs may have their origins in the metabolism of some organisms ancestral to the whole range of modern types. Concepts of the chronology of life evolution have discarded the earlier view of chemolithotrophs as primitive or ancestral types (133, 169, 252) and the earliest organisms were probably anaerobes fermenting organic compounds in the presence of a reducing atmosphere containing methane (146, 169, 186). The problems of subsequent evolution have been widely discussed by van Niel, Peck, and others (133, 180, 247) and it is possible that anaerobic chemolithotrophy was developed early, and predated photolithotrophy. Photoorganotrophy and photolithotrophy with inorganic sulfur doubtless preceded plant-type, oxygen-producing photosynthesis. Only after oxygen was available could aerobic chemolithotrophic metabolism or aerobic methylotrophy appear. Certain reasonable postulates can be made concerning the origin of specific metabolic reactions. Methane oxidation could be an ancestral process developed in anaerobic organisms as a fermentation; a precursor of the ribose phosphate cycle of formaldehyde fixation (135) could thus have preceded the Calvin cycle. In an atmosphere enriched in CO_2 by anaerobic fermentation, organisms could have arisen that fixed CO_2 to supplement their carbon metabolism. The TCA cycle and the Calvin cycle might then have both arisen as anabolic processes, perhaps both for the fixation of CO_2, in anaerobes. The TCA cycle might have developed into an aerobic energy-yielding process much later in evolution. The existence of sulfate-reducing bacteria, perhaps also clostridial types, and sulfide-oxidizing photolithotrophs probably in very ancient times (15, 180) does not allow speculation as to whether forward or reverse functioning of the TCA cycle arose first or whether they appeared separately. One implication is that the absence of α-OG dehydrogenase from some lithotrophs and methylotrophs may be an ancestral feature rather than a more recent loss of function.

Peck (181) has speculated "that growth on formate represents an intermediate stage in the evolution of autotrophy." It is notable that although the ability to use methane oxidation to support growth is retained today only by a few methylotrophs, the use of formate by lithotrophs is a strikingly common feature. Ribbons et al (196) refer to the methylotrophs as obligate heterotrophs but the metabolic links connecting them with lithotrophs include the use of one-carbon substrates by some lithotrophs. *T. thiooxidans* will grow on carbon disulfide as a sole source of carbon and energy (201). The stoichiometry of this process is not available and one cannot say if this is technically an autotrophic or heterotrophic process! Growth on carbon monoxide may pose a similar problem (133). Methane is assimilated by photosynthetic bacteria (251) and by *Rhodopseudomonas* after oxidation to CO_2 (259). Unequivocal demonstration of growth with methane as carbon and energy donor is lacking, but *R. palustris* (190, 226) could grow with formate as both carbon source and hydrogen donor. Formate carbon was fixed by the Calvin cycle after liberation as CO_2. *R. palustris* thus behaves with formate in a manner paralleling *Pseudomonas oxalacticus* (191). Recent work with *P. oxalaticus* showed the autotrophic pathway to occur exclusively with formate, and that, in the presence of substrates allowing a faster growth rate, formate was only oxidized as an ancillary energy source (20). In mixtures of formate and oxalate (which allowed growth at a rate 30 percent slower than growth on formate), autotrophic use of formate predominated (20). Control of the Calvin cycle thus occurs in this as in *Hydrogenomonas* by a mixture of induction and catabolite repression. The recently described facultative *Thiobacillus* A2 (232) can grow on formate in a mineral medium, perhaps with a metabolism then resembling *P. oxalaticus.* Slow growth on formate has been obtained also with *Nitrobacter agilis* (99) and *N. winogradskyi* (73), which showed a generation time of ca 18 hr in media containing nitrite and formate, but it continued to grow with a generation time of 144 hr when nitrite was exhausted. Moreover, formate was oxidized by cells of *T. thiocyanoxidans* (265), by *Nitrobacter* (168, 217), in which cytochromes *c* and *a* became reduced (74), and by autotrophically cultured *T. novellus* (1). When *T. novellus* was cultured heterotrophically, it lost not only the ability to oxidize thiosulfate but formate oxidase also (1) as did *Thiobacillus* A2 (232). Heterotrophic growth of *N. agilis* resulted in partial repression of both formate and nitrite-oxidizing enzymes (221).

Formate can contribute essentially all the cell carbon to *Thiobacillus* A2, *Pseudomonas oxalacticus,* and *Rhodopseudomonas palustris,* although yields with *R. gelatinosa* with formate alone were similar to those with methane and growth was somewhat stimulated by carbon dioxide (259). In contrast, ^{14}C-formate added to otherwise autotrophic cultures of *T. neapolitanus* (115) or *Chlorobium thiosulfatophilum* (116) was incorporated principally into adenine and guanine, with some labeling of lipid and protein. Formate increased the lag before growth of *Chlorobium,* although even 1 m*M* formate probably contributed only a small fraction of the total cell carbon.

Possible phylogenetic relationships between methylotrophs and lithotrophs are unresolved, but in the current reshaping of the terminology of these organisms (197, 261) it may be argued that the term autotroph might be extended to include all those organisms dependent on one-carbon compounds (chemo- and photolithotrophs, methylotrophs, and formate-users) and to classify heterotrophs as requiring compounds containing 2- or more linked carbon atoms.

Concluding Remarks

I hope to have illustrated how widespread is the capacity for autotrophic metabolism among microbes and to have shown the metabolic complexity and variability exhibited by lithotrophs. The facultative heterotrophic lithotrophs are proving remarkably interesting organisms for the study of adaptive and regulatory processes in bacteria. The problem of obligate lithotrophy seems to have a number of possible basic causes and study of such organisms in future years should be most valuable.

ACKNOWLEDGMENT

I wish to thank the following authors for allowing me to read the results of recent research prior to publication: Prof. S. C. Rittenberg, Prof. D. S. Hoare, Dr. Corinne Johnson, and Mr. O. H. Tuovinen.

LITERATURE CITED

1. Aleem, M. I. H. 1965. *J. Bacteriol.* 90:95–101
2. Aleem, M. I. H. 1965. *Biochim. Biophys. Acta* 107:14–28
3. Aleem, M. I. H. 1966. *Biochim. Biophys. Acta* 113:216–224; 128:1–12
4. Aleem, M. I. H. 1969. *Antonie van Leeuwenhoek J. Microbiol. Serol.* 35:379–91
5. Aleem, M. I. H. 1970. *Ann. Rev. Plant Physiol.* 21:67–90
6. Aleem, M. I. H., Huang, E. 1965. *Biochem. Biophys. Res. Commun.* 20:515–20
7. Aleem, M. I. H., Lees, H., Nicholas, D. J. D. 1963. *Nature (London)* 200:759–61
8. Aleem, M. I. H., Ross, A. J., Schoenhoff, R. L. 1968. *Bacteriol. Proc.* 140
9. Allison, F. E., Hoover, S. R., Morris, J. H. 1937. *Bot. Gaz.* 98:433
10. Allison, R. K. et al 1953. *J. Biol. Chem.* 204:197–205
11. Amarasingham, C. R., Davis, B. D. 1965. *J. Biol. Chem.* 240:3664–68
12. Anderson, K. J., Lundgren, D. G. 1969. *Can. J. Microbiol.* 15:73–79
13. Anderson, L., Fuller, R. C. 1967. *Plant Physiol.* 42:487–502
14. Aubert, J.-P., Milhaud, G., Moncel,, C., Millet, J. 1958. *C. R. Acad. Sci.* 246:1616–19
15. Ault, W. V., Kulp, J. L. 1959. *Geochim. Cosmochim. Acta* 16:201–35
16. Bachofen, R., Buchanan, B. B., Arnon, D. I. 1964. *Proc. Nat. Acad. Sci. USA* 51:690–94
17. Balashova, V. V. 1968. *Mikrobiologiya* 37:580–88 (Engl. trans.)
18. Belousova, A. A. 1968. *Mikrobiologiya* 37:1010–16 (Engl. trans. 855–60)
19. Biggins, J. 1969. *J. Bacteriol.* 99: 570–75
20. Blackmore, M. A., Quayle, J. R., Walker, I. O. 1968. *Biochem. J.* 107:699–704, 705–13
21. Blumer, M., Chase, T., Watson, S. W. 1969. *J. Bacteriol.* 99:366–70
22. Borichewski, R. M. 1967. *J. Bacteriol.* 93:597–90
23. Borichewski, R. M., Umbreit, W. W. 1964. *Bacteriol. Proc.* 92
24. Borichewski, R. M., Umbreit, W. W. 1966. *Arch. Biochem. Biophys.* 116:97–102
25. Brierley, J. A., Brierley, C. L. 1968. *J. Bacteriol.* 96:573
26. Buchanan, B. B. 1969. *J. Biol. Chem.* 244:4218–23
27. Buchanan, B. B., Evans, M. C. W. 1965. *Proc. Nat. Acad. Sci. USA* 54:1212–18
28. Buchanan, B. B., Evans, M. C. W., Arnon, D. I. 1967. *Arch. Mikrobiol.* 59:32–40
29. Burlant, L., Datta, P., Gest, H. 1965. *Science* 148:1351–53
30. Butler, R. G., Umbreit, W. W. 1966. *J. Bacteriol.* 91:661–66
31. Butler, R. G., Umbreit, W. W. 1969. *J. Bacteriol.* 97:966–67
32. Campbell, A. E., Helleburt, J. A., Watson, S. W. 1966. *J. Bacteriol.* 91:1178–85
33 Carr, N. G., Craig, I. W. 1970. *Phytochemical Phylogeny,* ed. J. B. Harborne. New York: Academic Press. Chap. 7
34. Charles, A. M. 1970. *Can. J. Biochem.* 48:95–103
35. Clark, C., Schmidt, E. L. 1966. *J. Bacteriol.* 91:367–73
36. Clark, C., Schmidt, E. L. 1967. *J. Bacteriol.* 93:1302–8, 1309–15
37. Cohen, G. N., Stanier, R. Y., Le Bras, G. 1969. *J. Bacteriol.* 99: 791–801
38. Cole, J. S., Aleem, M. I. H. 1970. *Biochem. Biophys. Res. Commun.* 38:736–43
39. Cook, D. W., Tischer, R. G., Brown, L. R. 1967. *Can. J. Microbiol.* 13:701–9
40. Cooper, R. C. 1964. *J. Bacteriol.* 88: 624–29
41. Cornish, A. S., Johnson, E. J. 1970. *Bacteriol. Proc.* 133
42. Crum, E. H., Siehr, D. J. 1967. *J. Bacteriol.* 94:2069–70
43. Daniels, H. J. 1966. *Can. J. Microbiol.* 12:1095–98
44. Davies, S. L., Whittenbury, R. 1970. *J. Gen. Microbiol.* 61:227–32
45. Davis; E. A., Johnson, E. J. 1967. *Can. J. Microbiol.* 13:873–84
46. DeCicco, B. T., Stukas, P. E. 1968. *J. Bacteriol.* 95:1469–75
47. DeLey, J. 1962. *Symp. Soc. Gen. Microbiol.* 12:164–95
48. Delwiche, C. C., Finstein, M. J. 1965. *J. Bacteriol.* 90:102–7

49. Din, G. A., Suzuki, I., Lees, H. 1967. *Can. J. Microbiol.* 13:1413–19
50. Eccleston, M., Kelly, D. P. In preparation
51. Echlin, P., Morris, I. 1965. *Biol. Rev., Cambridge Phil. Soc.* 40: 143–87
52. Eimhjellen, K. E., Steensland, H., Traetteberg, J. 1967. *Arch. Mikrobiol.* 59:82–92
53. Elsden, S. R. 1962. *The Bacteria,* ed. I. C. Gunsalus, R. Y. Stanier. III, Chap. 1. New York: Academic Press
54. Engel, H., Schön, G. 1964. *Z. Naturforsch.* 19b:52
55. Ermachenko, V. A., Lisenkova, L. L., Lozinov, A. B. 1966. *Mikrobiologiya* 35:240–9 (Engl. Trans.)
56. Eroshin, V. K., Harwood, J. H., Pirt, S. J. 1968. *J. Appl. Bacteriol.* 31:560–67
57. Evans, M. C. W. 1968. *Biochem. Biophys. Res. Commun.* 33:146
58. Evans, M. C. W., Buchanan, B. B. 1965. *Proc. Nat. Acad. Sci. USA* 53:1420–25
59. Evans, M. C. W., Buchanan, B. B., Arnon, D. I. 1966. *Proc. Nat. Acad. Sci. USA* 55:928–34
60. Fay, P. 1965. *J. Gen. Microbiol.* 39: 11–20
61. Flechtner, V. R., Hanson, R. S. 1969. *Biochim. Biophys. Acta* 184:252–62
62. Fortnagel, P., Freese, E. 1968. *J. Bacteriol.* 95:1431–38
63. Foster, J. W. 1940. *J. Gen. Physiol.* 24:123–34
64. Fraser-Smith, E. C. B., Austin, M. A., Reed, L. L. 1969. *J. Bacteriol.* 97:457–59
65. Frenkel, A. W. 1970. *Biol. Rev. Cambridge Phil. Soc.* 45:569–616
66. Gale, N. L., Beck, J. V. 1966. *Biochem. Biophys. Res. Commun.* 24: 792–9
67. Gale, N. L., Beck, J. V. 1967. *J. Bacteriol.* 94:1052–59
68. Gemerden, H. van, 1968. *Arch. Mikrobiol.* 64:111–17, 118–24
69. Gibson, F., Pittard, J. 1968. *Bacteriol. Rev.* 32:465–92
70. Gibson, J. 1967. *Arch. Mikrobiol.* 59: 104–12
71. Gladstone, G. P. 1939. *Brit. J. Expl. Pathol.* 20:189–200
72. Goodman, N., Rittenberg, S. C. 1964. *Bacteriol. Proc.* 112
73. Gool, A. van, Laudelout, H. 1966. *Biochim. Biophys. Acta* 127:295–301
74. Gool, A. van, Laudelout, H. 1967. *J. Bacteriol.* 93:215–20
75. Gorlenko, V. M. 1970. *Z. Allg. Mikrobiol.* 10:147–49
76. Gray, B. H., Bernlohr, R. W. 1969. *Biochim. Biophys. Acta* 178:248–61
77. Gray, C. T., Wimpenny, J. W. T., Mossman, M. R. 1966. *Biochim. Biophys. Acta* 117:33–41
78. Greenberg, D. M. 1961. *Metabolic Pathways,* Vol. 2. New York: Academic Press
79. Hadjipetrou, L. P., Gerrits, J. P., Teulings, F. A. G., Stouthamer, A. H. 1964. *J. Gen. Microbiol.* 36:139–50
80. Halpern, Y. S., Even-Shoshan, A., Artman, M. 1964. *Biochim. Biophys. Acta* 93:228–36
81. Hampton, M. L., Hanson, R. S. 1969. *Biochem. Biophys. Res. Commun.* 36:296–305
82. Hanson, R. S., Blicharska, J., Arnaud, M., Szulmajster, J. 1964. *Biochem. Biophys. Res. Commun.* 17:690–95
83. Hanson, R. S., Cox, D. P. 1967. *J. Bacteriol.* 93:1777–87
84. Harder, R. 1917. *Z. Botan.* 9:145
85. Harwood, J. H. 1970. PhD thesis. *Studies on the physiology of* Methylococcus capsulatus *growing on methane.* Univ. of London
86. Hempfling, W. P., Vishniac, W. 1965. *Biochem. Z.* 342:272–87
87. Hempfling, W. P., Vishniac, W. 1967. *J. Bacteriol.* 93:874–78
88. Hirsch, P., Georgiev, G., Schlegel, H. G. 1963. *Arch. Mikrobiol.* 48: 79–95
89. Hoare, D. S., Gibson, J. 1964. *Biochem. J.* 91:546–49
90. Hoare, D. S., Moore, R. B. 1965. *Biochim. Biophys. Acta* 109:622–25
91. Hoare, D. S., Hoare, S. L., Moore, R. B. 1967. *J. Gen. Microbiol.* 49:351–70
92. Holm-Hansen, O. 1968. *Ann. Rev. Microbiol.* 22:47–70
93. Hood, W., Leaver, A. G., Carr, N. G. 1969. *Biochem. J.* 114:12–13P
94. Hooper, A. B. 1969. *J. Bacteriol.* 97:776–79
95. Horton, A. A. 1968. *Biochem. Biophys. Res. Commun.* 32:839–45
96. Hurlbert, R. E. 1967. *J. Bacteriol.*

93:1346–52
97. Hurlbert, R. E., Lascelles, J. 1963. *J. Gen. Microbiol.* 33:445–58
98. Hutchinson, M., Johnstone, K. I., White, D. 1969. *J. Gen. Microbiol.* 57:397–410
99. Ida, S., Alexander, M. 1965. *J. Bacteriol.* 90:151–56
100. Jensen, H. L. 1950. *Nature (London)* 165:974
101. Jensen, H. L., Sörensen, H. 1952. *Acta Agr. Scand.* 2:295–304
102. Johnson, C. L. 1969. *PhD thesis,* University of Rochester.
102a. Johnson, C. L., Vishniac, W. 1970. *J. Bacteriol.* 104:1145–50
103. Johnson, E. J. 1966. *Arch. Biochem. Biophys.* 114:178–83
104. Johnson, E. J., Abraham, S. 1969. *J. Bacteriol.* 97:1198–208
105. Johnson, E. J., Abraham, S. 1969. *J. Bacteriol.* 100:962–68
106. Jones, C. A., Kelly, D. P. Unpublished data
107. Jones, G. E., Benson, A. A. 1965. *J. Bacteriol.* 89:260–61
108. Kelly, D. P. 1965. *J. Gen. Microbiol.* 34:ix
109. Kelly, D. P. 1966. *Biochem. J.* 100: 9P
110. Kelly, D. P. 1967. *Arch. Mikrobiol.* 56:91–105, 58:99–116
111. Kelly, D. P. 1967. *Sci. Progr. (London)* 55:35–51
112. Kelly, D. P. 1968. *Arch. Mikrobiol.* 61:59–76
113. Kelly, D. P. 1968. *Aust. J. Sci.* 31: 165–73
114. Kelly, D. P. 1969. *Arch. Mikrobiol.* 69:330–42, 343–59, 360–69
115. Kelly, D. P. 1970. *Arch. Mikrobiol.* 73:177–92
116. Kelly, D. P. Unpublished data
117. Kelly, D. P., Syrett, P. J. 1964. *J. Gen. Microbiol.* 34:307–17
118. Kelly, D. P., Syrett, P. J. 1966. *J. Gen. Microbiol.* 43:109–18
119. Kemp, M. B., Quayle, J. R. 1966. *Biochem. J.* 99:41–8
120. Kiyohara, T., Fujita, Y., Hattori, A., Watanabe, A. 1960. *J. Gen. Appl. Microbiol.* 6:176–82; 8:165–68
121. Kindler, S. H., Ben-Gurion, R. 1965. *Biochem. Biophys. Res. Commun.* 18:337–40
122 Kocur, M., Martinec, T., Mazanec, K. 1968. *J. Gen. Microbiol.* 52: 343–45
123. Kondrat'eva, E. N. 1965. *Photosynthetic Bacteria.* Jerusalem: Israel Program for Sci. Translations. (Engl. trans.)
124. Kornberg, H. L., Collins, J. F., Bigley, D. 1960. *Biochim. Biophys. Acta* 39:9–24
125. Kucera, S., Wolfe, R. S. 1957. *J. Bacteriol.* 74:344–49
126. Kuehn, G. D., McFadden, B. A. 1968. *Can. J. Microbiol.* 14:1259–61
127. Kuehn, G. D., McFadden, N. A. 1968. *J. Bacteriol.,* 95:937–46
128. Kuenen, J. G., Veldkamp, H. 1970. *Antonie van Leeuenhoek, J. Microbiol. Serol.* 36:186
129. Lam, J. W., Walker, R. W. 1970. *Bacteriol. Proc.* 28
130. Landesman, J., Duncan, D. W., Walden, C. C. 1966. *Can. J. Microbiol.* 12:957–63
131. Larsen, H. 1951. *J. Biol. Chem.* 193: 167–73
132. Larsen, H., 1953. *Kgl. Nor. Vidensk. Selsk. Skr.* Nr. 1
133. Larsen, H. 1960. *Encylopaedia Plant Physiology,* ed. W. Ruhland, V/ 2:613–48. Berlin, Heidelberg: Springer Verlag
134. Lascelles, J. 1960. *J. Gen. Microbiol.* 23:499–510
135. Lawrence, A. J., Kemp, M. B., Quayle, J. R. 1970. *Biochem. J.* 116:631–39
136. Leach, C. K., Carr, N. G. 1968. *Biochem. J.* 109:4P
137. Leach, C. K., Carr, N. G. 1969. *Biochem. J.* 112:125–26
138. LeJohn, H. B., Caeseele, L. van, Lees, H. 1967. *J. Bacteriol.* 94: 1484–91
139. LePage, G. A. 1942. *Arch. Biochem.* 1:255–62
140. LePage, G. A., Umbreit, W. W. 1943. *J. Biol. Chem.* 147:263–71
141. Levine, R. P., Togasaki, R. K. 1965. *Proc. Nat. Acad. Sci., USA* 53: 987–90
142. Lippert, K. D., Pfennig, N. 1969. *Arch. Mikrobiol.* 65:29–47
143. London, J. 1963. *Arch. Mikrobiol.* 46:329–37
144. London, J., Rittenberg, S. C. 1966. *J. Bacteriol.* 91:1062–69
145. London, J., Rittenberg, S. C. 1967. *Arch. Mikrobiol.* 59:218–25
146. Lowe, C. U., Rees, M. W., Markham, R. 1963. *Nature* 199:219–22
147. Lu, M. C. 1969. *The mechanism of amino acid inhibition of obligate chemolithotrophs.* PhD thesis, Univ. of California, Los Angeles

148. Macdonald, R., Michael, T. 1959. *Bacteriol. Proc.* 44
149. McElroy, R. D. 1969. *Bacteriol. Proc.* 64
150. McElroy, R. D. 1970. *Bacteriol. Proc.* 142
151. McFadden, B. A., Kuehn, G. D., Homann, H. R. 1967. *J. Bacteriol.* 93:879–85
152. McFadden, B. A., Tu, C.-C. L. 1967. *J. Bacteriol.* 93:886–93
153. Matin, A., Rittenberg, S. C. 1970. *J. Bacteriol.* 104:234–38, 239–46
154. Matin, A., Rittenberg, S. C. 1970. *Bacteriol. Proc.* 137
155. Matin, A., Rittenberg, S. C. 1971. In press
156. Mahler, H. R., Cordes, E. H. 1966. *Biological Chemistry.* New York: Harper & Row
157. Mahoney, R. P., Edwards, M. R. 1966. *J. Bacteriol.* 92:487–95
158. Malavolta, E., Delwiche, C. C., Burge, W. D. 1961. *Biochim. Biophys. Acta* 57:347–51
159. Mayeux, J. V., Johnson, E. J. 1967. *J. Bacteriol.* 94:409–14
160. Meiklejohn, J. 1954. *Symp. Soc. Gen. Microbiol.* 4:68–83
161. Meinke, W. W., Greenberg, B. R. 1968. *J. Biol. Chem.* 173:535–45
162. Meister, A. 1965. *Biochemistry of the amino acids.* New York: Academic Press
163. Milhaud, G., Aubert, J.-P., Millet, J. 1958. *C. R. Acad. Sci.* 246: 1766–69
164. Miyajima, R., Otsuka, S.-I., Shiio, I. 1968. *J. Biochem. (Tokyo)* 63: 139–48
165. Moriarty, D. J. W. 1969. *Inorganic sulphur compound oxidation in* Thiobacillus concretivorus. PhD thesis, Univ. of Adelaide, So. Australia
166. Murray, R. G. E., Watson, S. W. 1965. *J. Bacteriol.* 89:1594–1609
167. Nesterov, A. I., Gogotov, I. N., Kondrat'eva, E. N. 1966. *Mikrobiologiya* 35:163–68 (Engl. trans.)
168. Olson, J. M., Chance, B. 1960. *Arch. Biochem.* 88:26–39
169. Oparin, A. I. 1957. *The Origin of life on the earth.* Chap. 6 Edinburgh, London: Oliver & Boyd
170. Ostrowski, W., Skarzyński, B., Szczepkowski, T. W., 1954. *Proc. U.N. Int. Conf. Peaceful Uses At. Energy, 1st* 24:42
171. Pan, P. C. 1970. *Bacteriol. Proc.* 125
172. Pan, P. C., Umbreit, W. W. 1969. *Bacteriol. Proc.* 64
173. Patel, R., Hoare, D. S., Taylor, B. F. 1969. *Bacteriol. Proc.* 128 and personal communications, 1970
174. Patel, R., Hoare, D. S. 1971. In press
175. Pearce, J., Carr. N, G. 1967. *Biochem. J.* 105:45P
176. Pearce, J., Carr, N. G. 1967. *J. Gen. Microbiol.* 49:301–13
177. Pearce, J., Carr, N. G. 1969. *J. Gen. Microbiol.* 54:451–62
178. Pearce, J., Leach, C. K., Carr, N. G. 1969. *J. Gen. Microbiol.* 55:371–78
179. Pearce, P. M., Levin, R. A. 1969. *Bacteriol. Proc.* 63
180. Peck, H. D. 1967. *Lectures Theoret. Appl. Aspects Mod. Microbiol.* (Univ. Maryland) 1–22
181. Peck, H. D. 1968. *Ann. Rev. Microbiol.* 22:489–518
182. Peeters, T., Aleem, M. I. H. 1970. *Arch. Mikrobiol.* 71:319–30
182a. Peeters, T. L., Liu, M. S., Aleem, M. I. H. 1970. *J. Gen. Microbiol.* 64:29–35
183. Pfennig, N. 1967. *Ann. Rev. Microbiol.* 21:285–324
184. Pfennig, N., Cohen-Bazire, G. 1967. *Arch. Mikrobiol.* 59:226–36
185. Pfennig, N., Lippert, K. D. 1966. *Arch. Mikrobiol.* 55:245–56
186. Ponnamperuma, C., Sagan, C., Mariner, R. 1963. *Nature (London)* 199:222–26
187. Pope, L. M., Hoare, D. S., Smith, A. J. 1969. *J. Bacteriol.* 97:936–39
188. Pringsheim, E. G. 1967. *Arch. Mikrobiol.* 59:247–54
189. Proctor, H. M., Norris, J. R., Ribbons, D. W. 1969. *J. Appl. Bacteriol.* 32:118
190. Qadri, S. M. H., Hoare, D. S. 1968. *J. Bacteriol.* 95:2344–57
191. Quayle, J. R., Keech, D. B. 1959. *Biochem. J.* 72:623–30, 631–37
192. Rao, G. S., Berger, L. R. 1970. *J. Bacteriol.* 102:462–66
193. Remsen, C. C., Lundgren, D. G. 1963. *Bacteriol. Proc.* 33
194. Remsen, C. C., Valois, F. W., Watson, S. W. 1967. *J. Bacteriol.* 94: 422–33
195. Repaske, R., Ambrose, C. A. 1970. *Bacteriol. Proc.* 66
196. Ribbons, D. W., Harrison, J. E., Wadzinski, A. M. 1970. *Ann. Rev. Microbiol.* 24:135–58
197. Rittenberg, S. C. 1969. *Advan. Microbiol. Physiol.* 3:159–96
198. Rittenberg, S. C., Goodman, N. S. 1969. *J. Bacteriol.* 98:617–22
199. Rolls, J. P., Lindstrom, E. S. 1967.

J. Bacteriol. 94:784–85, 860–66
200. Ross, A. J., Schoenhoff, R. L., Aleem, M. I. H. 1968. *Biochem. Biophys. Res. Commun.* 32:301–6
201. Rothschild, B., Butler, R., Keller, J. R. 1969. *Bacteriol. Proc.* 64
202. Roy, A. B., Trudinger, P. A. 1970 *The Biochemistry of Inorganic Compounds of Sulphur.* Cambridge Univ. Press
203. Rutberg, B., Hoch. J. A., 1970. *J. Bacteriol.* 104:826–33
204. Sadler, M. H., Johnson, E. J. 1970. *Bacteriol. Proc.* 135
205. Sadler, W. R., Stanier, R. Y. 1960. *Proc. Nat. Acad. Sci. USA* 46: 1328–34
206. Saxena, J., Vishniac, W. 1970. *Antonie van Leeuwenhoek J. Microbiol. Serol.* 36:109–18
207. Schaefer, A. J., Aleem, M. I. H. 1969. *Bacteriol. Proc.* 15
208. Schaeffer, W. I., Umbreit, W. W. 1963. *J. Bacteriol.* 85:492–93
209. Schlegel, H. G. 1960. *Encyclopaedia Plant Physiology,* ed. W. Ruhland, V/2:649–63. Berlin, Heidelberg: Springer-Verlag
210. Schlegel, H. G., Trüper, H. G. 1966. *Antonie van Leeuwenhoek J. Microbiol. Serol.* 32:277–92
211. Schnaitman, C., Lundgren, D. G. 1965. *Can. J. Microbiol.* 11:23–27
212. Shafia, F., Wilkinson, R. F. 1969. *J. Bacteriol.* 97:256–60
213. Shively, J. M., Benson, A. A. 1967. *J. Bacteriol.* 94:1679–83
214. Shively, J. M., Decker, G. L., Greenawalt, J. W. 1970. *J. Bacteriol.* 101:618–27
215. Shively, J. M., Knocke, H. W. 1969. *J. Bacteriol.* 98:829–30
216. Silver, M., Margalith, P., Lundgren, D. G. 1967. *J. Bacteriol.* 93:1765–69
217. Silver, W. S. 1960. *Nature (London)* 185:555–56
218. Sirevåg, R., Ormerod, J. G. 1970. *Biochem.* J. 120:399–408
219. Sirevåg, R., Ormerod, J. G. 1970. *Science* 169:186–88
220. Smillie, R. M., Rigopoulos, N., Kelly, H. 1962. *Biochim. Biophys. Acta* 56:612–14
221. Smith, A. J., Hoare, D. S., 1968. *J. Bacteriol.* 95:844–55
222. Smith, A. J., London, J., Stanier, R. Y. 1967. *J. Bacteriol.* 94:972–83
223. Stadtman, E. R., Cohen, G. N., LeBras, G., Robichon-Szulmajster, H. de 1961. *J. Biol. Chem.* 236: 2033–38
224. Stern, J. R., Bambers, G. 1966. *Biochemistry* 5:1113–18
225. Still, G. G. 1965. *The role of some of the Krebs cycle reactions in the biosynthetic function of* Thiobacillus thioparus. PhD thesis, Oregon State University, Corvallis, Ore.
226. Stokes, J. E., Hoare, D. S. 1969. *J. Bacteriol.* 100:890–94
227. Stukas, P. E., DeCicco, B. T. 1970. *J. Bacteriol.* 101:339–45
228. Tabita, F. R., Lundgren, D. G. 1970. *Bacteriol. Proc.* 125
229. Tano, T., Imai, K. 1968. *Agr. Biol. Chem.* 32:401–4
230. Tatum, E. L. 1946. *Cold Spring Harbor Symp. Quant. Biol.* 11:278–84
231. Taylor, B. F. 1970. *Biochem. Biophys. Res. Comm.* 40:957–63
232. Taylor, B. F., Hoare, D. S. 1969. *J. Bacteriol* 100:487–97
233. Taylor, B. F., Hoare, D. S., Hoare, S. L. 1969. *Bacteriol. Proc.* 63
234, Thiele, H. H. 1968. *Arch. Mikrobiol.* 60:124–38
235. Tikhonova, G. V., Lisenkova, L. L., Doman, N. G., Skulachev, V. P. 1967. *Biochemistry (Moscow)* 32: 599–605
236. Tobback, P., Laudelout, H. 1965. *Biochim. Biophys. Acta* 97:589–90
237. Trudinger, P. A., Kelly, D. P. 1968. *J. Bacteriol.* 95:1962–63
238. Trüper, H. G. 1964. *Arch. Mikrobiol.* 49:23–50
239. Trüper, H. G. 1965. *Biochem. Biophys. Acta* 111:565–68
240. Tuovinen, O. H. Niemelä, S., Gyllenberg, H. G. In press
241. Tuttle, J. H., Dugan, P. R. 1969. *Bacteriol. Proc.* 64
242. Umbreit, W. W. 1962. *Bacteriol. Rev.* 26:145–50
243. van Niel, C. B. 1932. *Arch. Mikrobiol.* 3:1–112
244. van Niel, C. B. 1936. *Arch. Mikrobiol.* 7:323–58
245. van Niel, C. B. 1941. *Advan. Enzymol.* 1:263–328
246. van Niel, C. B. 1944. *Bacteriol. Rev.* 8:1–118
247. van Niel, C. B. 1949. *Photosynthesis in plants,* ed. J. Franck, W. E. Loomis, p. 437. Ames: Iowa State College Press
248. Vishniac, W. 1952. *J. Bacteriol.* 64: 363–73
249. Vishniac, W., Santer, M. 1957. *Bacteriol. Rev.* 31:195–213
250. Vishniac, W., Trudinger, P. A. 1962. *Bacteriol. Rev.* 26:168–75
251. Vishniac, W. et al. 1966. *Biology and*

the exploration of Mars, p. 235, ed. Pittendrigh, Vishniac, Pearman. Washington, D.C.: Nat. Res. Counc.

252. Wald, G. 1964. *Proc. Nat. Acad. Sci. USA* 52:595–611
253. Wallace, W., Nicholas, D. J. D. 1969. *Biol. Rev. Cambridge Phil. Soc.* 44:359–92
254. Wallace, W., Knowles, S. E., Nicholas, D. J. D. 1970. *Arch. Mikrobiol.* 70:26–42
255. Wang, W. S., Lundgren, D. G. 1969. *J. Bacteriol.* 97:947–50
256. Weitzman, P. D. J. 1966. *Biochim. Biophys. Acta* 128:213–15
257. Weitzman, P. D. J., Dunmore, P. 1969. *FEBS Lett.* 3:265–67; *Biochim. Biophys. Acta* 171:198–200
258. Weitzman, P. D. J., Jones, D. 1968. *Nature (London)* 219:270–72
259. Wertlieb, D., Vishniac, W. 1967. *J. Bacteriol.* 93:1722–24
260. Whale, F., Jones, O. T. G. 1970. *J. Gen. Microbiol.* 60:x
261. Whittenbury, R. 1969. *J. Gen. Microbiol.* 55:xxiv
262. Williams, P. J. LeB., Watson, S. W. 1968. *J. Bacteriol.* 96:1640–48
263. Wilson, E., Stout, H. A., Powelson, D., Koffler, H. 1953. *J. Bacteriol.* 65:283–87
264. Winogradsky, S. 1949. *Microbiologie du Sol.* Paris: Masson.
265. Youatt, J. B. 1954. *J. Gen. Microbiol.* 11:139–49

DISEASES OF OYSTERS[1]

Victor Sprague

University of Maryland Natural Resources Institute,
Chesapeake Biological Laboratory, Solomons, Maryland

Contents

INTRODUCTION

Definition and Scope

Oysters are among the many animal species which have shown a dramatic population decline in recent times. "Along with that of the American bison, the decline of the oyster population is one of the most striking cases of the depopulation of a once flourishing species . . ." (32). "Decline in

[1] Contribution No. 451, Natural Resources Institute of the University of Maryland.

abundance of oysters actually started in the 19th century, probably caused in large part by indiscriminate harvesting and destruction of beds. Extensive mortalities from unknown causes also contributed to decreased oyster production" (83). Decline continued at an accelerated rate and the present century has witnessed "mass mortalities of epic proportions" (82).

Depopulation of oyster species could, perhaps, be attributed almost entirely to disease in a broad sense. Disease is "deviation from a state of health" (22). It "is a process, not a thing and represents the response of the body to injury or insult. . . . When the range of easy tolerance within which an organism can function without too great a strain is passed the organism may be said to be diseased or in a pathic state" (quoted from Birkeland by Steinhaus & Martignoni, 94). The pathic state, if accompanied by sufficient strain, results in death of the organism.

With reference to a population, death is a mortality. This "can be either the basic annual mortality, below which losses never drop, or the peak waves of mortality which periodically sweep the oyster beds and which account for losses of high percentages of the total population. Or it may mean an increasing annual loss resulting in complete extinction of oysters over considerable parts of the original range" (51).

This review of oyster diseases might appropriately include almost any factor in the decline of oyster populations, but it is possible to consider only a few factors. I choose to deal almost exclusively with diseases known or thought to be infectious, possibly the most important factors in mortality (51), and to concentrate on developments of the past decade during which the greatest progress in study of oyster diseases has been made. Earlier studies, and some later ones, need be considered only very briefly, for most have been reviewed by many authors.

Brief Review of Early Developments

It is interesting to consider oyster diseases in the 20th century decade by decade, for in each period there has been a special mortality or a new emphasis in disease studies. Occasionally, there have been new insights into diseases long known.

In the first decade, "shell disease" in European oysters, known as early as 1878 (30), attracted much attention (3). This disease has been frequently studied and its etiological agent was specifically mentioned by Dollfus in 1922 in the first general review of oyster diseases (21), a well known and frequently cited paper. Strangely, the significance of his comments on shell disease seems to have been overlooked until quite recently (3) (see below).

One of the best known mortalities, affecting *Crassostrea virginica,* started in Malpeque Bay, Prince Edward Island, about 1915, but was first reported years later (62). It was caused by "Malpeque disease," etiology unknown, and has been the subject of a series of original studies and reviews. One of the most complete studies was made by Logie in 1958 (48) and an excellent review was given by Mackin in 1960 (50).

"The heaviest blow dealt to the oyster culture and oyster fishery in Europe was the widespread and devastating mortality of *Ostrea edulis* in the years 1920 and 1921" (44). This mortality was described in detail in 1924 by Orton (63, 64), but the cause was not determined. Authors (44, 51, 85) agree that the main factor was infectious disease. In 1926, Roughley (75) published a long report on "winter disease" in *Crassostrea commercialis* in New South Wales. He said that the cause of the mortality was not determined. However, he described a syndrome similar to that seen in oysters during several other mortalities. Macroscopic signs included yellowish spots on the body surface (due to accumulation of leucocytes), abcesses and ulcers, accompanied by severe damage to gills, palps, and other organs. Mackin (51) thought at one time that *Hexamita* was involved in some important way but later indicated in a very brief note (54) that Australian winter disease (apparently still endemic) and certain others are caused by *Labyrinthomyxa* spp.

The decade of the 1930's was a time of severe trial for the Dutch oyster (*Ostrea edulis*) industry because of shell disease which followed the planting of large quantities of *Cardium* shells for culch during the previous decade (43, 44). The same period saw the birth of oyster pathology in North America with reports by Needler (62) on "Malpeque Bay disease" in Canada and by Prytherch (70, 71) on mortalities of *Crassostrea virginica* in Mobjack Bay, Virginia, and other southern coastal areas of the United States. Needler found an organism which "may be a mycetozoan" (62) in oysters sick with Malpeque Bay disease, but the cause of this disease has not yet been established. Prytherch thought the mortalities he reported were caused by *Nematopsis* but Sprague & Orr (91) found this parasite to be relatively harmless to oysters. The principal factor in the Mobjack Bay mortality was probably *Dermocystidium* (53).

Before the 1940's, oyster diseases were studied sporadically by a few isolated (and probably poorly financed) individuals. In that decade extensive mortalities of *Crassostrea virginica* along the coast of the Gulf of Mexico ushered in an era of large-scale massive attack on oyster diseases. The studies were heavily financed for several years by different oil companies whose operations were alleged to be the cause of the mortality. The principal result of the studies was the definite incrimination, for the first time, of a microorganism as the main agent of mass mortalities in oysters. The agent was named *Dermocystidium marinum* by Mackin, Owen & Collier (55) and later transferred by Mackin & Ray (56) to genus *Labyrinthomyxa* Duboscq. The correct generic classification of this parasite is still in doubt (83). The studies in progress at the end of the first half of the 20th century contributed much to the growth of the new discipline of oyster pathology.

The decade of the 1950's was characterized mainly by a concentration of studies in *Dermocystidium marinum*. A voluminous literature on this parasite and its role in diseases has developed, mostly through the efforts of Mackin (53) and Ray (73) and their associates, working on the Gulf Coast of the United States, and Andrews & Hewatt (5) working on the Chesapeake Bay. A very important event in the late 1950's was the beginning and spread of extensive mortalities of *Crassostrea virginica,* caused by *Minchinia* spp. "Mortalities first appeared in Delaware Bay in 1957 and the pathogen was first observed by Stauber in 1958. . . ." (4, 11). The status of research on oyster diseases of North America at the end of this period was summarized by Mackin (50).

RECENT DEVELOPMENTS

The General Character of Oyster Disease Studies in the 1960's

Probably the past ten years have encompassed more activity and achievement in the area of oyster disease than all previous years. There was a great upsurge of research, largely due to the mass mortalities of *Crassostrea virginica* which occurred on the Atlantic Coast of the United States in association with *Minchinia nelsoni.* A large body of literature appeared, most of which has recently been reviewed by several authors. The research revealed a number of new diseases and disease agents, and produced unprecedented amounts of new data (which must be largely ignored in this short paper) concerning such special aspects of oyster disease as pathology, host defense reactions, development of immune strains, interaction of disease agents with other factors in the environment of the oyster, and epizooetiology.

It was during this period that oyster pathology developed into a mature science, flourished for a while as an isolated discipline, and then became integrated with other branches of invertebrate pathology. It was also during the latter part of this period that the study of infectious diseases of marine animals in North America entered an era of decline. This was due partly to financial difficulties brought about by the cost of the Indo-Chinese War and partly to a change of political climate in which the fashion is to promote the study of physical and chemical factors in disease rather than the biological factors.

Reviews and Bibliographies

Only a few publications dealt with oyster diseases in general before 1960, the main ones being those of Dollfus (21) and Korringa (44). Since then, more than a dozen publications either contained reviews or pertained to oyster diseases in some important general way. The nature of each of these recent publications will be briefly indicated now and some specific new developments included in them will be mentioned in other sections of this paper.

The first, mentioned earlier, was an excellent review by Mackin (50) in 1960. This paper dealt with known infectious diseases in North American oysters and their etiological agents, and diseases of unknown etiology believed to be infectious. In addition to diseases already well known, the new haplosporidan diseases were reported briefly and some other new ones were mentioned: "Mycelial disease" and "amber disease" of *C. virginica* along the Gulf Coast, "a species of *Dermocystidium* in Australian oysters, and a plasmodial parasite in *Crassostrea gigas* from Japan." Some problems of disease studies were discussed.

About a year later, the same author (51) gave an excellent discussion on mortalities of oysters. He classified mortalities into seven types, according to the kinds of agents primarily responsible for them. Type II dealt with mortalities due to disease (meaning infectious disease). This type was illustrated with the mortalities of 1919–1925 affecting *Ostrea edulis* in Europe, Australian winter disease affecting *Crassostrea commercialis,* and fungus disease caused by *Dermocystidium marinum* in *C. virginica* in the United States.

The next year, Mackin (53) gave a long and detailed review of disease caused by *Dermocystidium* in Louisiana. This paper contained also a review of studies of *Nematopsis,* plus the only published descriptions of "mycelial disease," "amber disease," "watery cysts" of oysters, "yellow pustule disease," and bacterial parasites in the gastric shield of *Crassostrea virginica.* Some of these diseases, not previously included in review papers, will be discussed below.

Laird (45) considered in detail the microecological factors in oyster epizootics with special reference to "Malpeque disease." He suggested that *Hexamita* plays an important role in this disease after the oysters are weakened by environmental factors.

Sindermann (80) wrote an important paper on Disease in marine populations. It is quite general in scope and concerns oysters only briefly. Its significance lies in the fact that it is one of the few papers wich have dealt primarily with the role of infectious disease in the fluctuation of marine populations.

Galtsoff, in his book, *The American Oyster* (29), gave a brief review of the principal diseases of *C. virginica* up to 1964. This included "Malpeque Bay disease," *Dermocystidium marinum,* the Haplosporida, "shell disease," "foot disease," *Hexamita, Nematopsis,* and metozoan associates.

Cheng (15) published a book, *Marine Molluscs as Hosts for Symbioses* which gave an excellent recent (1967) review of animal parasites of oysters.

In 1967, Sindermann & Rosenfield (85) published a good, brief resumé of infectious diseases of oysters, mostly those of known etiology. The summary included: bacteria, especially the one causing "focal necrosis" in *Crassostrea gigas;* under the fungi, *Dermocystidium,* shell disease, and fungus in oyster larvae; protozoans of the genera *Minchinia, Nematopsis* and *Hexamita,* and the ciliates; helminths; crustaceans. A paragraph was devoted to tumors in oysters.

Pauley (66) provided Acritical review of neoplasia and tumor-like lesions in molluscs.

In 1968, Andrews (4) reviewed about ten years of work done at Virginia Institute of Marine Science on the diseases of *Crassostrea virginica* caused by *Minchinia nelsoni* and *M. costalis.* The paper deals mainly with epizooetiology but includes also ecology, innate and acquired host resistances, various aspects of host-parasite relations, and the enigma of the suddenly great destructive capacity of an agent, *M. nelsoni,* which evidently has been endemic for several years.

As late as 1970, there have been at least two reviews of oyster diseases. Sprague (89) summarized Some protozoan parasites and hyperparasites in marine bivalve molluscs, and Sindermann (83) published a book on *The Principal Diseases of Marine Fish and Shellfish.* The latter contains an updated version of the review given previously by Sindermann & Rosenfield (85). Sindermann's book (83) is an excellent reference work. It is the most important single source of information on the haplosporidan diseases of oysters, which were discovered in the late 1950's and dominated the disease studies for the next ten years.

Two valuable bibliographies appeared in 1968: Sindermann's *Bibliography of Oyster Parasites and Diseases* (81) and Johnson's *An Annotated Bibliography of Pathology in Invertebrates Other Than Insects* (41). A supplement to the latter appeared in 1969 (42). In 1970, Sindermann circulated in the form of an informal report, a comprehensive Bibliography of diseases and parasites of marine fish and shellfish (84) which he is revising for publication.

The Newer Knowledge of Particular Infectious Diseases and Disease Agents

It is unnecessary to attempt in this short paper to present a comprehensive review of our knowledge of infectious diseases of oysters, since this has recently been well done by Sindermann (83). Therefore, I shall emphasize developments since publication of his book. In addition, I shall consider briefly some diseases which Sindermann did not discuss and shall mention other matters which seem to be worthy of emphasis. I have nothing to add to Sindermann's review of diseases associated with bacteria, *Dermocystidium,* fungi in oyster larvae, *Nematopsis, Hexamita,* the ciliate protozoans and the metazoans. Regarding species of *Minchinia,* I shall only add that Perkins (68) and Rosenfield et al (74) quite recently made some excellent electron microscope studies on *M. costalis* and *M. nelsoni.*

Shell disease.—A very significant recent development in the study of oyster diseases was the cultivation (2) and identification (3) of the etiological agent of a disease most commonly known as "shell disease" (maladie de la coquelle), but also known as "foot disease" (maladie du pied) and "hinge disease" (maladie de la charnier) (3). The different names merely refer to the parts of the shell (and adjacent tissues) most severely affected, although these different pathological manifestations have sometimes been attributed to separate diseases. Giard (30), Korringa (43), and Alderman & Jones (3) have described the disease syndromes in detail.

According to Alderman & Jones (3), shell disease is known for certain only on the Atlantic Coast of Europe and only in its two native oysters,

Ostrea edulis and *Crassostrea angulata*. However, Galtsoff (29) says, "Foot disease is found in *C. virginica,* particularly in oysters inhabiting muddy waters of the southern states, but in my experience it never reaches epidemic proportions." Quayle (72) also reported the typical foot disease syndrome in *C. gigas* in British Columbia. Shell disease is probably best known for its presumed role in the severe mortalities which hit the Dutch oyster industry in the 1930's. Many oysters died and others were so disfigured that they could never be marketed (43). "The Dutch oyster industry saw itself threatened with extinction . . ." (44).

The etiology of shell disease remained obscure until very recently when Alderman & Jones (3) identified the agent as the fungus *Ostracoblabe implexa* Bornet & Flahault, 1889 (9). They (2) had previously illustrated the shell disease mycelium, which they grew in culture, without mentioning its name. At that time they found sporangia which they considered to be stages of the same fungus. Later, however, Alderman (personal communication) said, "the sporangial stage described in 1967 is not now to be regarded as part of the *O. implexa* life cycle or of shell disease and will be described as a separate fungus. . . ." Korringa (43, 44) studied the disease in the Netherlands for a number of years before he concluded that it is caused by a fungus growing in the shell and stimulating the mantle to produce abnormal shell secretion (resulting in green or brown rubber-like warts on the shell). Much of his early work was based on the assumption that the seat of the trouble is in the secretory cells of the mantle itself, that these are somehow abnormal and therefore function abnormally to produce the unusual shell growth. It seems remarkable that the cause of the oldest scientifically recognized (3) infectious disease of oysters has remained so long obscure to even the most distinguished oyster biologists. As long ago as 1902, a Dutch biologist, F. A. C. Went, suggested that the disease might be due to a fungus similar to *Ostracoblabe* (3). He referred to *O. implexa* Bornet & Flahault, 1889 (9). These authors studied the flora of numerous molluscan shells, mostly dead ones. They found *O. implexa* growing in shells (only dead ones?) of *Ostrea edulis.* Dollfus (21), in his "Résumé de nos principales connaissances pratiques sur les maladies et les ennemis de l'huître," reviewed at length the relevant part of the paper by Bornet & Flahault. It is not completely clear that Dollfus identified "mycose de la coquelle" or "maladie de la coquille," which he attributed to *Ostracoblabe implexa,* with the serious disease syndrome known as shell disease in living oysters (not only in dead shells). Nevertheless, his paper certainly provided a clue which might, if pursued, have led to rapid discovery and identification of the agent responsible for the great epizootic of shell disease in the Netherlands a decade later. Instead, it was nearly a half-century before any one seems to have been fully aware that Dollfus clearly and correctly pointed to *O. implexa* as the agent of maladie de la coquelle. Probably the explanation for this slow progress in disease study is that no disease-oriented biologists, or pathologists, studied shell disease during that half-century.

Amoebae.—The best known amoeba associated with oysters is *Flabellula calkinsi* (Hogue, 1913) Schaeffer, 1926 [syn.: *F. patuxent* (Hogue, 1921) Schaeffer, 1926] in *Crassostrea virginica*. Hogue, obviously basing the name on that of Gary N. Calkins, consistently used the spelling *calkensi*. Most authors have followed Hogue, although Schaeffer, 1926, and Page (65), without comment on the spelling, used *calkinsi*. There is no evidence that Schaeffer intended to change the spelling. Therefore, he did not make an emendation. Page's (65) change, however, was clearly intended, for he used the original spelling when citing Hogue's work. For this reason, we can say that Page did make an emendation. Justification for this emendation depends on whether there is clear evidence that Hogue made an inadvertent error. Both Page (personal communication) and C. W. Sabrosky (personal communication) see such evidence. Therefore, the correct spelling is *calkinsi*, and *calkensi* has no status in nomenclature. Until recently (65, 76, 77) this amoeba was known mainly from the original work of Hogue (39, 40). That work was reviewed in detail by Cheng (15) and briefly by Sprague (89). Later work on this and a few other amoeboid organisms in oysters deserves special mention.

Sawyer (76), in an abstract, said, "Studies on *C. virginica* from Chesapeake Bay, Md. have revealed 4 different amoeboid organisms of which two show affinities to the Amoebidae and two to the Labyrinthulidae and Vampyrellidae. Microscopic studies of trophozoites, cysts, flagellate swarm cells, and zoospores suggest that at present all amoeboid organisms reported previously from *C. virginica* are of uncertain taxonomic status."

In a later short note, Sawyer (77) reported that "Two species of protozoa with morphologic characters similar to amoebae, proteomyxans, and slime molds have been isolated from tissues and mantle fluids of oysters from Maryland (*Crassostrea virginica*) and British Columbia (*Crassostrea gigas*)." The organism from *C. gigas* "shows affinities to the genus *Flabellula* (amoeba), *Biomyxa* (proteomyxan), and *Arachnula* (proteomyxan). The Maryland organisms resemble *Vahlkamfia calkensi* Hogue and *V. patuxent* Hogue."

It is noteworthy that Page (65) recently isolated a free-living amoebae from a marine habitat and identified it as *F. calkinsi*. After making a critical study of this amoeba, he concluded, "It is quite likely that Hogue, like some other workers of that period, was not working with the 'pure' (i.e., unispecific) cultures which she believed she had. The cysts which she figured for *V. calkensi* were quite clearly those of an *Acanthamoeba*. *V. patuxent* differed from *V. calkensi* only in the cysts present. Considering the present state of our knowledge, it behooves us to accept the opinion of so eminent an authority as Page and regard these two names as synonyms. It is also noteworthy that Page found "*F. calkinsi* is the most voraciously cannibalistic amoebae which I have observed." This is a matter of interest to oyster pathologists, since they frequently see unexplained instances of cells within cells as they study diseased oyster tissues.

Cheng (16) recently found amoebae in various tissues of moribund specimens of *Crassostrea commercialis* in Tahiti. He considered this organism, which he named *Hartmannella tahitiensis,* to be a secondary invader of oysters which have been rendered moribund by pollution. Sawyer (personal communication) thinks this amoeba must be studied in culture to reveal the type of mitosis its nucleus exhibits before its correct generic determination can be made. Furthermore, he thinks it is only a scavenger on moribund oysters, not an invader.

It will be noted elsewhere (6, 17, 27, 28) in this review that amoeboid organisms are associated with "gill disease" of Portuguese oysters. Also, *Dermocystidium marinum* or *Labyrinthomyxa marina,* one of the most studied oyster pathogens is said to have amoeboid stages (56), although it is not generally thought of as an amoeba-like organism because it is commonly recognized only in the hypnospore or prehypnospore stages. Likewise, amoeboid organisms are presumably associated with various other diseases said to be caused by *Labyrinthomyxa* or *Labyrinthomyxa*-like parasites. In fact, there is reason to suspect (54) that most of the great oyster mortalities (the notable exceptions being those caused by the fungus, *Ostracoblabe implexa,* and by the haplosporidans of genus *Minchinia*) are associated with organisms which have, or are said to have, amoeboid stages. When one considers that many and various protistan groups with amoeboid stages are abundant in the marine habitat, it is evident that an almost endless variety of "amoebae" may be found in association with oysters.

Associates of oysters which are most obviously amoeba-like in their more familiar stages have received remarkably little attention from oyster pathologists. This is illustrated by the fact that only those studied by Sawyer (77) and Page (65) seem to have been isolated as clonal cultures, while very few have been cultured at all. Probably the main reason is that Hogue's amoeba is generally thought to be harmless to the host and that available time might more profitably be spent on studying matters of more practical significance. A less obvious, but doubtless quite weighty, reason is that studying the amoeboid organisms associated with oysters presents numerous and very difficult problems which few students of oyster disease are able or willing to attack.

Gill disease.—In the summer of 1967, an abnormally high mortality struck the Portuguese oyster population, *Crassostrea angulata,* in the vicinity of Marennes and Archachon and, to a lesser extent, in other oyster-producing centers of the Altantic Coast of France. In previous months, examinations of oysters had revealed more or less conspicuous gill lesions. Different laboratories of the Institut des Pêches undertook to determine the nature, geographic distribution, frequency, and progression of the disease (58). Preliminary findings were published in a symposium on "La maladie des branchies des huîtres portugaises des côtes françaises de l'Atlantique." Contributors were Caty (13), Comps (17), Deltreil (20), Gras (31), His

(38), Lasserre (46), and Marteil (58). These papers constitute roughly half the literature on the subject. Franc & Arvy (28) quite recently wrote an important paper summarizing knowledge of the disease in continental Europe, while Alderman & Gras (1) briefly reported their observations on Portuguese oysters imported into Britain.

Gill disease was first noted in November, 1966, by Trochon (according to Martell, 58) who observed perforations in the gills of Portuguese oysters growing in the region of Marennes-Oléron. The gross pathology has been described by several authors (6, 17, 31, 46, 58). The contours of the gills became progressively altered, three stages being recognized and illustrated by Marteil (58): first, a notch in the margin of the gill; second, a succession of such lesions affecting one or more of the gill lamellae; the third stage, more rare, is an almost complete destruction of the gills, these organs being reduced to highly irregular shreds. Less obvious are perforations of gills and palps, surrounded by yellow patches. The condition of the oyster is not noticeably affected in the first two stages but in the third there is emaciation. In *C. gigas* from Japan or North America, as well as *Ostrea edulis* of the coast of Europe, the lesion may become a large notch but the gill is not completely destroyed. It is noteworthy that Arvy & Franc (6) said that the first sign of the disease is yellow spots on the gills, for these are also seen in several other oyster diseases.

Using gill damage as a criterion, the disease has been diagnosed in: *Crassostrea angulata* of France, Portugal and Spain; *C. gigas* of Japan, Korea, North America (British Columbia and State of Washington, USA); *Ostrea edulis* of France, the Adriatic, Norway and Denmark; *Mytilus edulis* of France (58). Roughley (75) may also have seen the gill disease syndrome in *Crassostrea commercialis* when he studied Australian winter disease. During one mortality he found surviving oysters with yellow abscesses and ulcers on the surface tissues. "In many cases pieces of tissue become eaten away, particularly in the gills and palps."

Franc & Arvy (27) found that oysters sometimes repair their lesions. Oysters which have resisted attack may contain cells with large granulations, which these authors consider to be stages of the etiological agent of gill disease (27). Later (28) they described the process of healing and repair in detail, in contrast to the developmental phase of the disease. Recovery from disease that is accompanied by gill damage, and repair of the lesions, is a phenomenon noted also by Roughley (75) in *Crassostrea commercialis* during his investigation of Australian winter disease.

Gill disease which seriously affected the Portuguese oysters along the Atlantic Coast of France, appeared strongly toward the end of 1967 and a good part of 1968 but seemed to be on the decline in 1969 (27). Franc & Arvy remarked that this affliction appears to have been effectively defeated by the oysters, since they exhibit a distinct ability to repair the lesions.

In April and May of 1969, there was a heavy mortality of Portuguese oysters imported into Britain from Portugal. Losses were more than 90 percent in some cases (1). Dying oysters showed grossly the typical gill disease syndrome as reported by the French workers. Damage to the adductor muscle was also noted. About 2 percent of one group of oysters showed active syndrome upon arrival, but in three weeks about 60 percent had active disease.

Gill disease seems not to have been regarded as a specific disease before 1967 although a similar syndrome was seen by Roughley (75) in Australian winter disease. Gill lesions were previously known also to be associated with *Pinnotheres* (92). This crab cannot be implicated in the present outbreak of gill disease because it is not harbored by the most seriously affected oyster species, *Crassostrea angulata* (58).

Besse, in 1968, noted the presence of numerous peritrich ciliates of the genus *Trichodina* in sick oysters and thought they might be causally related to the disease, but Marteil (58) pointed out that these ciliates were present in *C. angulata* of the Marennes and Arcachon areas many years before gill disease appeared.

Caty (13) cultured fragments of diseased oyster gill in sterile sea water and noted proliferations in the form of lenticular elements, filaments, coccoid elements in chains, and spirals within some of the lenticular elements. He could reach no conclusion regarding the nature of the forms observed.

Lasserre (46) studied histological sections of diseased oysters and found an augmentation of secretory cells in the filaments, but discovered no disease agent.

Gras (31) cultured pieces of oyster tissue (diseased and apparently healthy) in different modification of Sabraud's medium and in thioglycolate medium. Cells grew in the thioglycolate medium which, Gras thought, looked much like the more familiar stages of *Dermocystidium marinum* Mackin, Owen & Collier, 1950 (55). The author thinks the organism responsible for gill disease presents an analogy with *D. marinum* but is not the same, for the gill damage has not been mentioned as an accompaniment of infection with the latter parasite. Alderman & Gras (1) also observed in many oysters with gill disease certain cells resembling those of *D. marinum*. They isolated these cells in culture to determine their nature and possible pathogenicity to oysters. Alderman (personal communication) now says these are monocentric fungi of genera *Schizochytrium* and *Thaustochytrium* and that they seem to be nonpathogenic. Perhaps these are the same fungi cultured by Gras (31), although the authors do not say so.

Arvy & Franc (6) often found a ciliate, not the *Trichodina* observed by Besse, associated with the lesions in the gills and palps and possibly partly responsible for the lesions. However, they found another organism which they considered to be the true destructive agent. This they described without naming it at that time. This organism, a rhizopod which Franc & Arvy (27) later named *Thanatostrea polymorpha,* is seen in different forms. Along the walls of the pit in the lesions it appears in the form of large, irregular cells 30–60 μ long, connected to one another by long extension to form pseudoplasmodia. Each cell has a nucleus about 16 μ in its greatest dimension, with a large endosome, and frequently accompanied by a chromatoid body. Deeper in the tissue these cells are more rounded or pyriform. They constitute centers from which chromophobic extensions branch out. They may also develop into cysts covered with a yellowish envelope composed of scaly elements. The branches extend to the subepithelium and terminate in fusiform cells 10–20 μ long, which together form "filoplasmodes" somewhat similar to those described by Duboscq in *Labyrinthomyxa sauvageaui,* a parasite of *Laminaria lejolisi* Sauvageau. In recent perforations, there are small fusiform cells 9–10 μ, with small nuclei, and small amoebae 4–5 μ, both arising from the peripheral elements of the protist. These authors (6, 27, 28) described other peripheral elements in the form of cells with large granules. They supposed these were cells which Lasserre (46) had considered to be secretory cells. These cells were determined to be the terminal swellings of the filamentous extensions of the pseudoplasmodia. They occur any place in the oyster body, even in apparently healthy (recovered or resistant) oysters.

Comps (17), making a histological study of the lesions of diseased gills and palps, found large amoeboid cells and small fusiform cells like those described by Franc & Arvy (6, 27, 28).

It has been my privilege, through the courtesy of D. J. Alderman, to examine sections of four specimens of *C. angulata* said to have gill disease. One of these (No. 283) had a large gill lesion bordered by necrotic tissue but I could find nothing else unusual except a few trichodinid ciliates situated on the healthier part of the gill. The three other oysters contained many granular cells, which Alderman had called to my attention. Two of these oysters were sectioned through many organs, showing granular cells in all tissues. The third, 162(4), was represented only by a section of mantle with a large ulcer. The ulcer was packed with these granular cells, while similar cells were scattered throughout the mantle tissues. These cells seemed to be hemocytes full of spherical granules. There were no normal blood cells. Some of the granules appeared structureless but others, especially the larger ones, had a large vacuole somewhat eccentrically situated so that the sphere (cell ?) in optical section looked like a ring. In the thicker part of the ring a tiny spherule (nucleus ?) was sometimes distinguishable. Alderman (personal communication) found these ring stages in many oysters in 1969 but none in 1970. These do not seem to be the granular cells reported by Arvy & Franc (6, 27, 28) for the latter cells were much fewer in number, were apparently not blood cells, and seemed to be associated with or derived from the pseudoplasmodia in the lesions. These vacuoled cells (?), or rings, in Alderman's material reminded me of small cells of *Dermocystidium marinum.* These peculiar granules seem not to have been seen by any of the French workers, while Alderman (personal communication) says, "no evidence as yet found in material examined in England [is indicative] of the presence of any organism such as '*Thanatostrea polymorpha*'."

A review of the literature of gill disease does not reveal a clear picture of a specific disease attributable to an identified etiological agent. There is a certain consistency in the gross picture presented, consisting of yellowish pustules (abscesses and ulcers) on the external organs followed by marked necrosis of these organs. The pustules are not distinctive, for they are grossly characteristic of focal infections in a variety of diseases. The extensive disintegration of gill tissue seems to be the most distinctive of the gross features, for it has rarely been reported as a pathological accompaniment of an infectious disease. The microscopic picture of syndromes diagnosed as gill disease is full of ambiguities. The most striking contrast is in the results of Alderman and associates in England as compared with those of the French workers, and it is far from clear that the observations of Comps (17), Gras (31), Lasserre (46), Caty (13), Arvy & Franc (6), and Franc & Arvy (27) in France, are consistent with one another. Interpretation of all the microscopic observations is hampered by a number of major difficulties which are commonly encountered by oyster pathologists: distinguishing oyster cells from extraneous cells, deciding to what extent the latter are causally related to the disease and, sometimes, deciding whether objects observed are cellular or noncellular. Concerning the second of these difficulties, it is common knowledge that a great variety of protistans are associated with oysters and that many of them thrive as secondary invaders or opportunists taking advantage of an unhealthy condition, from whatever cause, in the oyster. The work of Arvy & Franc involves an additional serious difficulty, the very common problem for protozoologists of identifying observed objects as stages in a life cycle and putting them together to make a pattern. On the basis of present information, therefore, it seems necessary to regard the cause(s) of gill disease as not yet established.

Denman Island disease.—This disease, associated with mass mortalities of *Crassostrea gigas* at Denman Island in British Columbia, has not been included in previous reviews. It was reported by Quayle (72) and is "characterized by deep pustules on the surface of the body and mantle or by pus-filled sinuses . . . . One feature of this mortality that stands out is that only the older ages were affected and this in spite of the fact that they were in very excellent condition."

Concerning the etiology, the only published data are in a short note by Mackin (54). That note, relating not only to Denman Island disease but to several other diseases (not reviewed in this paper) of unknown etiology, is quoted here in its entirety. "Certain diseases of bivalve molluscs such as the Australian Winter Disease, Denman Island disease, Malpeque Bay disease [Gill disease, because of its pathological syndrome and alleged etiology, should now be added, although it may or may not be distinct from the others.] and several others caused by *Labyrinthomyxa* spp, have a common pathological syndrome, involving primary attack on gill, mantle, and gut epithelia, bacterial complication and bacteremia, and invasion and lysis of Leidig cell tissue. A new method of diagnosis of these diseases is presented.

Small populations of oysters are rid of bacteria by means of antibiotic treatment, and subjected to stress. Mantle from oysters in advanced disease is used to make impression mounts on slides and the parasites thus transferred are stained variously and diagnosis made under the microscope."

It is to be hoped that Mackin will soon follow up his note on this important group of diseases with more complete data. Meantime, it is pertinent to note an inherent difficulty in the diagnostic technique which Mackin used; direct smears of "mantle from oysters in advanced disease" possibly contain a great variety of organisms, including commensals, secondary invaders, saprobic organisms and primary pathogens of different kinds. An organism demonstrated by this method can not always confidently be regarded as the agent of a specific disease.

Amber disease.—Mackin (53) found oysters, *Crassostrea virginica,* with a light amber color in a population in Louisiana which suffered a small mortality. Sections showed the presence of an organism with amoeboid stages, plasmodia, and cysts containing spores. (Mackin illustrated the spores with rough sketches but it is impossible to interpret them.) He suggested a possible relation of the organism to the Haplosporida. Unfortunately, we do not know enough about the morphology of its spores to speculate intelligently about its systematic position.

Mycelial disease.—Mackin (53) reported this disease as the cause of mortalities of *Crassostrea virginica* in certain local areas in Texas and Louisiana and in aquaria. The disease appeared in the early spring when the oysters were in good condition and disappeared in the late summer when they were the poorest. Sections of dead or living oysters showed the presence of mycelium. In live oysters "the first infections were usually on the external epithelium of mantle, palps, or gills." Colonies of the mycelium were found loosely attached to the external epithelial surface. There was frequent penetration which involved the internal organs. The organism, which Mackin described in considerable detail, has been found in *C. virginica* in most of the middle and south Atlantic states and in *Ostrea lurida* in the state of Washington (53). "The mycelial organism may be related to the Actinomycetes," although it differs in some definite ways from the known Actinomycetes. It was studied only in the host organism, not in culture.

Chytridiopsis ovicola *Léger & Hollande, 1917.*—This egg parasite, found only in a single sample of *Ostrea edulis* from Marennes, France (47), is not believed to have much significance as an agent of disease. It is mentioned here in an attempt to clarify its systematic position. Its authors had no clear notion about its position but suggested a possible relation to the coccidians. Dollfus (21) treated it as an agent of coccidiosis in oysters, giving no indication that its systematic position might be in doubt. In an equally positive manner, Sinderman (83) called it a haplosporidan, while Sprague (89) argued that it belongs to the Microsporida. The latter view is based on the great similarity of *C. ovicola* to *C. mytilovum* (Field) in eggs of *Mytilus edulis* and to *C. socius* Schneider, type species, in the gut of a beetle. Sprague (87) found a PAS-positive polar cap, a distinctive feature of Microsporida, in the spore of *C. mytilovum,* while Manier & Ormières (57) demonstrated a polar filament in *C. socius* by means of electron microscopy. The latter work demonstrates conclusively that *Chytridiopsis* is a microsporidan genus. It follows that *C. ovicola* in Ostrea edulis belongs to the Microsporida if its generic assignment is correct, which there is no good reason to doubt (87).

Becker & Pauley (8) described in ova of *Crassostrea gigas* a parasite of uncertain systematic position that has cysts very much like those of *Chytridiopsis* but is different in other ways.

Miscellaneous parasites and diseases.—Probably everyone who has examined many oysters has seen evidence of diseases or parasites which he has never reported or has reported only very briefly in the literature or at meetings. It is not feasible to make a comprehensive list of these but a few are noteworthy. Mackin (50) said he "has observed a species of *Dermocystidium* in Australian oysters, and a plasmodial parasite in *Crassostrea gigas* from Japan." Later (53), he briefly reported yellow pustule disease in oysters all over the world. The conspicuous manifestations are yellow pustules or abscesses anywhere on the body and on the adductor muscle. As Mackin pointed out, this is a complex of diseases associated with certain organisms that penetrate the shell, such as *Polydora, Martesia,* and *Cliona.* At least one organism which penetrates the shell, *Ostracoblabe* (3), does not seem to cause yellow pustules. It should be recalled that yellow pustules also accompany a group of important diseases not associated with organisms which penetrate the shell, some conspicuous examples being Malpeque disease, gill disease and Australian winter disease.

Oyster Pathology Becomes a Mature Science

The decade just passed was the period during which oyster pathology grew to maturity, merged with the already flourishing discipline of insect pathology and became an integral part of the now firmly established field of invertebrate pathology.

Previously, most workers gave little more than superficial attention to the pathological manifestations of oyster disease. A noteworthy exception was Mackin (49), who gave details of the histopathology of *Democystidium* infection in *Crassostrea virginica.* "When oysters in Delaware Bay and Chesapeake Bay were subject to devastating mortalities [caused primarily by *Minchinia nelsoni*] beginning in the late 1950's, oyster biologists were unprepared to undertake the histopathological investigations required for the positive identification of the disease organism or to immediately establish its histopathological effect on the oyster . . ." (86).

The severe crisis in the oyster industry of the middle Atlantic states precipitated by the epizootic of haplosporidan disease provided a stimulus for intensified research on oyster diseases throughout the area. Because of the urgent need to obtain and disseminate information which might help to save the industry, annual oyster mortality conferences were held and all available information on the epizootic was freely exchanged. Nine such conferences were held between 1959–1967, inclusive. At first, the conferences consisted mostly of informal progress reports of studies on the etiology and epizootiology of the principal oyster diseases associated with the major current mortalities. Gradually, programs became more sophisticated, with presentation of many formal papers on all aspects of the pathology, not only of oysters but of other molluscs and of Crustacea. The changing character of the conferences was reflected in a change of name; the Oyster Mortality Conference became the Shellfish Pathology Conference.

Increasingly greater numbers of papers began to appear in print, concerning many subjects related to shellfish pathology. Limitations of space preclude mentioning more than a fraction of the noteworthy papers of that period. Stauber (93), Mackin (52), Cheng (14), Feng (26), Brooks (10), Tripp (95) and others published important papers on cellular reactions and defense mechanisms of oysters and other molluscs. Wood & Andrews (96) and Hasken et al (36) described, respectively, *Minchinia costalis* and *M. nelsoni*, new oyster pathogens of *Crassostrea virginica*. Couch et al (19), Farley (24), Couch (18), Perkins (67, 68), Rosenfield et al (74), and Barrow & Taylor (7) studied morphology or life cycle of species of *Minchinia*. Myhre (59, 60) and Eble & Rosenfield (23) have done histochemical studies on oysters infected with *Minchinia nelsoni*. Farley (25) summarized the disease syndrome of *Minchinia nelsoni* infection. Andrews and associates (4, and many other papers) and Haskin and associates (34, and many unpublished reports) carried out extensive studies on epizootiology of *M. nelsoni* infection. Haskin and associates (33, 35, 61) and Andrews (4) found innate resistance in progeny of oysters surviving epizootics of this disease, and Andrews (4) found acquired resistance in susceptible stocks exposed to infection at an early age. Many published papers mentioned observations on the interactions of oyster disease with physical factors in the environment, while Sprague et al (90) presented evidence from laboratory experiments that oysters held at lowered salinity recover from infection with *M. nelsoni*. Canzonier (12) summarized attempts to transmit *M. nelsoni* under controlled conditions.

The last Shellfish Pathology Conference was held in January, 1967. In May of that year a committee of seven with E. A. Steinhaus as Chairman, and including three insect pathologists and three oyster pathologists, organized the international Society for Invertebrate Pathology (37). Thus, shellfish pathology became an integral part of the relatively new and flourishing science of invertebrate pathology, second only to insect pathology as an area of activity within that field.

SOME PROBLEMS

At the outset it can truly be said that the oyster community is a part of man's environment, which he has altered so drastically. Many authors have commented on the dramatic decline of oyster populations "following in the wake of man's activities" (32). Oyster diseases must be conceived broadly to include not only the infectious diseases but all the interacting factors affecting oyster populations adversely. Thus, oyster diseases constitute one inseparable facet of the great problem arising in large part from the varied human activities which have drastically altered the global environment. Effective attack on oyster diseases (infectious or otherwise) may ultimately depend on solutions of the overall problem of the ravaged environment and must be attacked on many fronts. Some problems and suggested approaches to solutions have been listed by many authors, notably Mackin (50) and Sindermann & Rosenfield (85). Some of these are reiterated here, and some additional ones emphasized. It is not possible to mention all the problems or to be sure which are most important. Probably many of them are so interrelated that they are inseparable and of equal importance.

Biological factors in disease of marine animals must not be neglected while we concentrate on physical and chemical factors.—Sprague (88) has argued that biological factors in diseases of marine organisms, which have always received relatively little attention, are in danger of being further neglected as we become increasingly preoccupied with problems of pollution ecology. Therefore, it seems necessary to emphasize repeatedly the fact that agents of infectious disease, as well as physical and chemical pollutants, are important factors in the ecosystem. Furthermore, there is an interaction of all these factors in producing an effect on a marine population, and all must be taken into account if the system in which they interact is to be fully understood and brought under control.

Research on oyster diseases should be sustained, not sporadic.—A sustained effort in oyster disease studies by numerous individuals in many parts of the world is needed. "Only in recent years has more than sporadic attention been given to diseases of marine animals, and even now this attention is often restricted to periods of disease outbreak in a particular species" (80). It is only fair to say, eight years after Sinderman made this statement in 1963, that there are now centers of activity where oyster disease studies have been carried on intensively and without interruption for 10–20 years. Some other institutions have all but abandoned research on oyster diseases pending another crisis.

Basic research should be emphasized.—Other authors (50, 85) have emphasized the importance of basic research relative to oyster diseases, but this is a matter so important and so little appreciated by those who make policy and control finances that it can hardly be overemphasized. Furthermore, we biologists must recognize our responsibility to do basic research in spite of pressures which so often have led some of us to "exploit the urgency of disease problems in such a way as to capitalize on superficial matters" (Sawyer, personal communication). "We are much too prone to believe that basic taxonomic studies can be relegated to the background while studies of a 'practical' nature are pursued. This belief has delayed the solution of some problems of Malpeque disease for more than thirty years" (50). The basic problems are likely to be solved only by a sustained and concentrated effort on the part of many specially trained people over a long period of time.

Intensified study of amoeboid organisms may help to explain many mortalities due to unknown causes.—As already pointed out, amoebae such as *Flabellula calkinsi* have generally presumed to be nonpathogenic and consequently have received very little attention. Although some amoebae associated with oysters are probably harmless, amoeboid stages are said to occur in the life cycles of *Labyrinthomyxa* (56), and similar organisms are thought to be causally related to Gill disease (6, 17, 27, 28), "Australian winter disease, Denman Island disease, Malpeque disease, and several others . . ." (54). Thus, several of the great mortalities may largely be caused by protistans with amoeboid stages. Although bits of significant information have recently been accumulated on *Labyrinthomyxa* (51, 56, 69) and certain marine amoebae (65, 76–78), we have made little progress in such basic matters as sorting out and identifying the organisms in oysters which have amoeboid stages but which probably include a variety of completely unrelated species. One of the essential tools in studying these protists, the clonal culture, has been unused with one or two exceptions. We can confidently anticipate that an enormous amount of money, time, and effort will go into basic research on the amoeboid organisms which occur in the marine environment (78) before we acquire a good understanding of their biology and taxonomy and their relation to oyster diseases.

Increasing emphasis must be placed on searching for effective biological (rather than chemical) methods for control of infectious diseases.—Both Mackin (50) and Sindermann & Rosenfield (85) have emphasized this matter and have pointed ot a number of useful methods. These include: searching for weak links in the life cycles of the parasites, breeding disease-resistant stocks, producing oysters in an artificial environment where disease can be controlled, growing oysters in areas where natural conditions favor the oyster but not the parasite.

A comprehensive source of information on oyster disease is needed.—The literature on this subject is now voluminous and scattered. Some one could do a great service to oyster pathologists by bringing together this literature in a monographic work, somewhat as Schäperclaus (79) has done for fish pathologists in his "Fischkrankheiten." Sindermann (81, 84) has already taken an important step in that direction by preparing useful bibliographies. Such a monograph would greatly facilitate the work of oyster pathologists in solving some of the problems.

LITERATURE CITED

1. Alderman, D. J., Gras, P. 1969. *Nature (London)* 224:616–17
2. Alderman, D. J., Jones, E. B. G. 1967. *Nature (London)* 216:797–98
3. Alderman, D. J., Jones, E. B. G. In press
4. Andrews, J. D. 1968. *Proc. Nat. Shellfish. Assoc.* 58:23–36
5. Andrews, J. D., Hewatt, W. G. 1957. *Ecol. Monogr.* 27:1–26
6. Arvy, L., Franc, A. 1968. *C. R. Acad. Sci.* 267:103–5
7. Barrow, J. H., Jr., Taylor, B. C. 1966. *Science* 153:1531–33
8. Becker, C. D., Pauley, G. B. 1968. *J. Invert. Pathol.* 12:425–37
9. Bornet, E., Flahault, C. 1889. *Bull. Soc. Bot. Fr.* 1889:147–76
10. Brooks, W. M. 1969. *Immunity to Parasitic Animals,* ed. G. J. Jackson, R. Herman, I. Singer, 1:149–71. New York: Appleton. 2 vols. 292 pp.
11. Canzonier, W. J. 1960. *Nat. Shellfish. Assoc. Abstr. Tech. Pap.* 1960: Abstr 21
12. Canzonier, W. J. 1968. *Proc. Nat. Shellfish. Assoc.* 58:1
13. Caty, X. 1969. *Rev. Trav. Inst. Pêches Marit.* 33:167–70
14. Cheng, T. C. 1966. *J. Invert. Pathol.* 8:52–58
15. Cheng, T. C. 1967. Marine molluscs as hosts for symbioses. *Advan. Mar. Biol.* 5:424 pp.
16. Cheng, T. C. 1970. *J. Invert. Pathol.* 15:405–19
17. Comps, M. 1969. *Rev. Trav. Inst. Pêches Marit.* 33:151–60
18. Couch, J. A. 1967. *J. Parasitol.* 53: 248–53
19. Couch, J. A., Farley, C. A., Rosenfield, A. 1966. *Science* 153:1529–31
20. Deltreil, J.-P. 1969. *Rev. Trav. Inst. Pêches Marit.* 33:176–80
21. Dollfus, R. Ph. 1922. *Notes et Mem. Off. Sci. Tech. Pêches Marit.* 7: 1–58
22. Dorland, W. A. N. 1930. *American Pocket Medical Dictionary.* Philadelphia: Saunders. 837 pp.
23. Eble, A., Rosenfield, A. 1968. *Proc. Nat. Shellfish. Assoc.* 58:3
24. Farley, C. A. 1967. *J. Protozool.* 14: 616–25
25. Farley, C. A. 1968. *J. Protozool.* 15: 585–99
26. Feng, S. Y. 1967. *Fed. Proc.* 26: 1685–92
27. Franc, A., Arvy, L. 1969. *C. R. Acad. Sci.* 268:3189–90
28. Franc, A., Arvy, L. 1970. *Bull. Biol. Fr. Belg.* 104:3–19
29. Galtsoff, P. S. 1964. *The American Oyster* Crassostrea virginica *Gmelin. U. S. Fish Wildl. Serv. Fish. Bull.* 64:1–480
30. Giard, A. 1894. *C. R. Soc. Biol.* 46: 401–3
31. Gras, P. 1969. *Rev. Trav. Inst. Pêches Marit.* 33:161–64
32. Gross, F., Smyth, J. C. 1946. *Nature (London)* 157 :540–42
33. Haskin, H. H., Canzonier, W. J. 1969. *Proc. Nat. Shellfish. Assoc.* 59:4
34. Haskin, H. H., Canzonier, W. J., Myhre, H. L. 1965. *Am. Malacolog. Union Rep.* 32:20–21
35. Haskin, H. H., Krueger, F. E. 1970. *Proc. Nat. Shellfish. Assoc.* 60:4
36. Haskin, H. H., Stauber, L. A., Mackin, J. A. 1966. *Science* 153: 1414–16
37. Heimpel, A. M. 1967. *J. Invert. Pathol.* 9:iv [Editorial]
38. His, E. 1969. *Rev. Trav. Inst. Pêches Marit.* 33:171–75
39. Hogue, M. J. 1914. *Arch. Protist.* 35: 154–63, Pls. 16–18
40. Hogue, M. J. 1921. *Am. J. Hyg.* 1: 321–45
41. Johnson, P. T. 1968 *An Annotated Bibliography of Pathology in Invertebrates Other Than Insects.* Minneapolis: Burgess. 322 pp.
42. Johnson, P. T., Chapmann, F. A. 1969. *An Annotated Bibliography of Pathology in Invertebrates Other Than Insects Supplement.* Misc. Publ. 1. Center for Pathobiol., Univ. Calif., Irvine. 76 pp.
43. Korringa, P. 1948. *Nat. Shellfish. Assoc. Convention Addresses,* June: 86–94
44. Korringa, P. 1952. *Quart. Rev. Biol.* 27:266–308, 339–65
45. Laird, M. 1961. *Can. J. Zool.* 39: 449–85
46. Lasserre, C. 1969. *Rev. Trav. Inst. Pêches Marit.* 33:165–66
47. Léger, L., Hollande, A. C. 1917. *C. R. Soc. Biol.* 80:61–64
48. Logie, R. R. 1958. *Fish Res. Bd. Can. Ms. Rtp. Ser. (Biol)* 661. 91 pp.

49. Mackin, J. G. 1951. *Bull. Mar. Sci. Gulf Carib.* 1:72–87
50. Mackin, J. G. 1960. *Proc. Gulf Carib. Fish. Inst. 13th,* 98–109
51. Mackin, J. G. 1961. *Proc. Nat. Shellfish. Assoc.* 50:21–40
52. Mackin, J. G. 1961. *Am. Zool.* 1:371
53. Mackin, J. G. 1962. *Publ. Inst. Mar. Sci. Univ. Tex.* 7:132–229
54. Mackin, J. G. 1969. *Soc. Invert. Pathol. Newslett.* 2:16–17
55. Mackin, J. G., Owen, H. M., Collier, A. 1950. *Science* 111:328–29
56. Mackin, J. G., Ray, S. M. 1966. *J. Invert. Pathol.* 8:544–45
57. Manier, J., Ormières, R. 1968. *Protistologica* 4:181–85
58. Marteil, L. 1969. *Rev. Trav. Inst. Pêches Marit.* 33:145–50
59. Myhre, J. L. 1966. *Proc. Nat. Shellfish. Assoc.* 56:5–6
60. Myhre, J. L. 1968. *Proc. Nat. Shellfish. Assoc.* 58:6–7
61. Myhre, J. L., Haskin, H. H. 1970. *Proc. Nat. Shellfish. Assoc.* 60:9
62. Needler, A. W. H. 1931. *Biol. Bd. Can. Bull.* 22:1–30
63. Orton, J. H. 1924. *Min. Agr. Fish., Fish. Invest., London, Ser. 2* 6(3):1–199
64. Ibid. 6(4)1–69
65. Page, F. C. 1971. *J. Protozool.* 18:37–44
66. Pauley, G. B. 1969. *Nat. Cancer Inst. Monogr.* 31:509–39
67. Perkins, F. O. 1968. *J. Invert. Pathol.* 10:287–305
68. Perkins, F. O. 1969 *J. Parasitol.* 55: 897–920
69. Perkins, F. O., Menzel, R. W. 1966. *Proc. Nat. Shellfish. Assoc.* 56: 23–30
70. Prytherch, H. F. 1938. *Science* 88: 451–52
71. Prytherch, H. F. 1940. *J. Morphol.* 66:39–65
72. Quayle, D. B. 1961. *Fish. Res. Bd. Can. Ms. Rpt. Ser. (Biol.)* 713. 9 pp.
73. Ray, S. M. 1954 *Rice Inst. Pam. Spec. Issue,* Nov.:1–114
74. Rosenfield, A., Buchanan, L., Chapman, G. B. 1969. *J. Parasitol.* 55: 921–41
75. Roughley, T. C. 1926. *Proc. Linn. Soc. N. S. Wales* 51:446–91, Pls. 29–34
76. Sawyer, T. K. 1966. *J. Protozool.* (Suppl.) 13:23
77. Sawyer, T. K. 1968. *Trans. Am. Microsc. Soc.* 87:127
78. Sawyer, T. K. 1971. *Trans. Am. Microsc. Soc.* 90:43–51
79. Schäperclaus, W. 1954. *Fischkrankheiten.* Berlin: Akademie. 708 pp.
80. Sindermann, C. J. 1963. *Trans. N. Am. Wildl. Natur. Resources Conf., 28th,* 336–56
81. Sindermann, C. J. 1968. *U. S. Fish Wildl. Ser. Spec. Sci. Rpt., Fish.* 563. 13 pp.
82. Sindermann, C. J. 1968. *U. S. Fish Wildl. Serv. Spec. Sci. Rpt., Fish.* 569, 10 pp.
83. Sindermann, C. J. 1970. *Principal Diseases of Marine Fish and Shellfish.* New York: Academic. 369 pp.
84. Sindermann, C. J. 1970. *Trop. Atl. Biol. Lab. Inf. Rpt.* 11, 440 pp.
85. Sindermann, C. J., Rosenfield, A. 1967. *U. S. Fish Wildl. Ser. Fish. Bull.* 66:335–85
86. Sparks, A. K. 1970. *J. Invert. Pathol.* 15:i–iv. [Editorial]
87. Sprague, V. 1965. *J. Protozool.* 12: 385–89
88. Sprague, V. 1969. *J. Invert. Pathol.* 13:1–3
89. Sprague, V. 1970. *A Symposium on Diseases of Fishes and Shellfishes,* ed. S. F. Snieszko, 511–26. Washington, D. C.: Am. Fish. Soc. (Spec. Publ. 5). 526 pp.
90. Sprague, V., Dunnington, E. A., Jr., Drobeck, E. 1969. *Proc. Nat. Shellfish. Assoc.* 59:23–26
91. Sprague, V., Orr, P. E., Jr. 1955. *J. Parasitol.* 41:89–104
92. Stauber, L. A. 1945. *Biol. Bull.* 88: 269–91
93. Stauber, L. A. 1961. *Proc. Nat. Shellfish. Assoc.* 50:7–20
94. Steinhaus, E. A., Martignoni, M. E. 1967. *U. S. Dept. Agr. Pac. N. W. Forest Range Exp. Sta. Forest Serv.* 22 pp.
95. Tripp, M. R. 1970. *J. Reticuloendothel. Soc.* 7:173–82
96. Wood, J. L., Andrews, J. D. 1962. *Science* 136:710–11

THE PATHOGENICITY OF SOIL AMEBAS

CLYDE G. CULBERTSON, M.D.
Lilly Research Laboratories, Eli Lilly and Company, Indianapolis, Indiana

CONTENTS

The subclass Rhizopoda of the phylum Protozoa has been a difficult subject for study and understanding (34). Even the name "amoeba" (a–moiba—without form) denotes the difficult problem of description, and the spelling—amoeba or ameba—as well as the arguments about taxonomy, are items requiring arbitrary decisions in this discussion.

The first of these is the use of "ameba" except where "amoeba" is required as a proper name. Second, for reasons explained below, we will use the designation "H-A" to indicate the *Hartmannella-Acanthamoeba* group and "N" for the *Naegleria*. The laboratory code number will indicate the culture referred to.

In order to supply background for those who have not experienced any of the evolution of this subject, briefly outlined are the circumstances which explain why this subject emerged at such a late date in the development of biological science. Thus, in retrospect we can observe the preoccupations and misinterpretations of our predecessors. Those who follow us may have the same opportunity to judge our work.

Even limited knowledge of the voluminous literature leading to the establishment of an ameba as a cause of intestinal disease indicates the great difficulty encountered in this work. Direct microscopic examinations were limited in effectiveness as they are even today, and the very simple culture methods used initially appear to have resulted in the growth of the free-living amebas, not the amebas which were parasitic for the intestine (57, 81).

In 1903, Schaudinn finally convinced most workers that *Entamoeba his-*

tolytica was the only real pathogenic ameba for the human intestine (69). In 1911, Wells (85) clearly demonstrated that the free-living amebas which grew in the cultures were not the cause of dysentery, but they emerged because the cysts of these amebas were present in the dust of the air and in the food. He also showed that the cysts could pass through the intestine intact and become trophic forms when the feces were cultured.

Preoccupation with amebic intestinal disease effectively inhibited consideration of the role of free-living amebas as disease agents until the recent accidental observations which led to the demonstration that some of these amebas were pathogenic when introduced into the animal body by routes other than the gastrointestinal tract. It is of interest to note that when free-living amebas were encountered in bacterial and fungal cultures, tests were made for virulence only by inoculation of the animal via the intestine.

Suggesting that free-living amebas might cause extraintestinal disease was indeed a hazardous and difficult venture, since ill-fated claims, based on insufficient evidence, that an ameba was the cause of peridontal and joint disease dominated the minds of many workers in protozoology and medicine (51, 77).

Errors of the past wherein mononuclear cells were mistaken for amebas should remind us that caution is still required of all those who encounter a circumstance which suggests that an ameba may be the cause of a given pathologic condition. This caveat must never be forgotten—the dice of the amebas are loaded!

The purpose of this chapter is to discuss evidence which, according to Page, has led to the erosion of the taxonomic line between the free-living and parasitic amebas (61). The observations leading to these developments began when Jahnes et al reported that a small ameba classified as "*Acanthamoeba*" was found in a culture of monkey kidney tissue culture cells (48). Thereafter, Culbertson et al (26) observed a similar ameba in a tissue culture fluid. This tissue culture fluid, suspected of containing an unknown simian virus, was injected into the brains of both mice and monkeys to determine the virulence of the suspected virus. All the injected animals died and the amebas were found in the lesions of severe meningoencephalitis.

On direct microscopic examination of the tissue culture fluid used for the injections, motile amebas were found to have been present in the tissue culture itself. The effects of the amebas upon the tissue culture cells had been mistaken for viral cytopathogenic effects during study over a period of several weeks. After the amebas were removed from the fluid, no cytopathogenic agent was present in the culture. This ameba was designated as H-A, Strain A-1. After study of this culture, Dr. B. N. Singh classified it as *Hartmannella*. He found several other hartmannellid amebas which were identical and has recently proposed designating the new species *Hartmannella culbertsoni* (74). Other isolates discussed below, such as HN-3, which cause chronic granulomatous disease, were found by Dr. Singh to be *Hartmannella rhysodes*.

During the next ten years a number of instances were recorded in which amebas were found to have been responsible for changes in tissue culture that were at first believed to have been due to a virus (5, 19, 20, 39, 41, 64, 82, 84). In most cases pathogenicity studies were not reported for these amebas. The source of the amebas in the tissue culture is in some cases from air contamination; Kingston & Warhurst isolated 39 strains of soil amebas from the air that produced cytopathogenicity in tissue culture. These isolates were generally *Hartmannella,* but *Naegleria* sp. were also found (50).

Since I have been concerned primarily with the experimental infection and because I began following an observation involving a strain of H-A amebas and thus, with my associates Ensminger and Overton, have contributed additional information about this species, my discussion will draw more heavily upon our own experience and publications than would generally be appropriate. The N group of amebas has been proven by several workers to have caused human meningoencephalitis by numerous isolations of *Naegleria* sp., and consequently this species has been more widely studied.

Obviously, we have not included all the literature on the small soil amebas and there are many other important publications in addition to those listed. Dr. Warren Dolphin has assembled an extensive bibliography which has been of great value to us.

There are now only two species of small amebas of the soil about which there is sufficient information to indicate that a disease-producing ability is present in some member-isolated cultures. The first has been mentioned, namely the H-A group. Of this group, the isolate designated A-1 is the most virulent and the most studied. Other related isolates from the soil and from the nasal passages of humans have been studied also.

Since 1962, several reports of human infection have been added to the reports of Derrick (33) and Kernohan et al (49) to substantiate the fact that human disease due to soil amebas does exist and that the *Naegleria* sp. was implicated. An editorial in the *Medical Journal of Australia* of May 17, 1969, chronicles the reports of these cases in a concise fashion (40).

In January, 1966, Carter isolated an ameba, which he has designated *Naegleria fowleri,* from a fatal case of amebic meningoencephalitis (15). Independently, Butt et al (10) and our group (31) isolated a similar ameba from a fatal case in July, 1966. Dr. B. N. Singh has designated the HB-1 *Naegleria* as one of a new species, *Naegleria aerobia* (74).

I have tested isolates of several other species of small free-living amebas and have found no additional pathogenic cultures. The report of Sprague et al (78) concerning the isolation of a culture of *Paramoeba* from crabs suggests that other species might have pathogenic members. According to the literature, the H-A group and the N group of amebas are both common in the soil. By the methods regularly employed including our own (29), most of the isolates have been of the H-A type. Thus, more pathogenic cultures of H-A are known than of the N group.

Taxonomy and Biology of the Pathogenic Isolates

There are two schemes of classification in the literature which overlap to a considerable degree and consequently produce some difficulty in the selection of a name for the amebas we are describing. There are many publications on the subject, but those of Singh and of Page are most pertinent to this discussion. Singh's proposal (73) in 1952, based primarily upon the type of mitosis, was modified by Page (59, 60), who suggested an increased reliance upon other morphologic features. According to Singh's criteria, the A-1 Strain is classified as *Hartmannella* while by Page's method it is *Acanthamoeba.* Similar problems arise in the *Naegleria* and related species, but up to this time they have not concerned the N amebas encountered in the pathogenic group. Singh has re-emphasized his views and broadened the structure of his classification to include the new pathogenic members of both the H-A and N groups (74).

Discussion of the merits of the different classifications is beyond our purpose here and the merits are also beyond our competence to judge; those interested should read these excellent publications for the definitive considerations they contain. In our hands, the mitotic characteristics have been the most helpful and many of the other morphological features have been rather inconstant, depending upon differing conditions. Thus, we have found Singh's method very helpful, but we have also found Page's meticulous descriptions most interesting.

Singh has proposed the division of these amebas into two large families, Hartmannellidae and Schizopyrenidae, on the basis of the morphology of the mitosis. The main differentiating feature is the dissolution of the nucleolus in the Hartmannellidae and the persistence and division of it in the Schizopyrenidae. The two examples of pathogenic soil amebas found thus far represent both families; therefore, pathogenicity is not an attribute confined to the members of either family. In contrast to *Hartmannella, Naegleria* has a temporary flagellate form; but so far as is known, this form has no part in the disease-producing properties of N amebas.

Isolations from natural habitats, such as soil and water, have yielded H-A amebas, of which about 50 percent possessed some degree of intranasal virulence for mice. Similar isolations of *Naegleria* from nature have not been made extensively, although a number of unsuccessful attempts have been made to discover pathogenic *Naegleria* in the water of lakes from which it appeared that a patient had acquired amebic infection. Many of these attempts have resulted in the culturing of H-A amebas rather than *Naegleria,* and some of the H-A isolates have had low-grade mouse virulence.

Biological Characteristics

H-A and N amebas are common inhabitants of the soil and thus often are found in the water of lakes and rivers. Trophic forms ingest other microorganisms and survive drying and starvation by becoming cysts which are found in dry soil or dust. Excystation occurs when the cysts are moistened, and in response to the proper chemical stimulus which is best provided by living bacteria, such as *Escherichia coli* and related species (25).

On plates of 1.5 percent agar containing from 0–0.4 percent NaCl the amebas require no other nutrient than that provided by live bacilli which have been grown separately and smeared upon the plate, provided the bacteria used are not of a strain which inhibits the amebas (16, 23, 72).

H-A amebas are not affected by NaCl content (0.85 percent), but N amebas do not grow well in concentrations above 0.4 percent NaCl. Under the conditions described, the bacteria grow sufficiently to nourish the amebas, but do not overgrow to inhibit them (74).

In order to obtain a bacteria-free culture, the usual plate culture is prepared with the amebic inoculum placed upon a small circular area seeded with bacteria in the center of the plate. A paper disk containing neomycin is placed on the bacteria-free area at the periphery of the plate. The amebas migrate beyond the bacterial spot into the area around the disk where the bacteria will be absent in the clear zone around the disk. Single amebas are then picked by a suitable method from the bacteria-free area (29).

H-A amebas grow bacteria-free in many types of nutrient media, such as trypticase soy broth, but *Naegleria* require living cells, such as bacteria or tissue culture, for satisfactory growth from small inocula. After isolation, N amebas grow well upon several media, but these require larger inocula (6, 7). We have used sustained tissue culture cells (Lilly MK_2 and RK_1) for *Naegleria* and trypticase soy broth for H-A amebas. Many isolates grow very slowly in bacteria-free cultures and some resist all efforts to produce suitable bacteria-free growth.

The question as to whether a given culture is in fact bacteria-free is an important consideration. For bacterial testing of the culture, suitable aerobic and anaerobic methods are used. Some amebas are parasitized by other microbiological agents. Generally, these organisms fail to grow outside the amebas when ordinary methods are used, even though the parasitic form can be seen in the amebic cytoplasm (36). The HN-3 strain of H-A amebas, which we isolated from the nasopharynx of a child, contains a small bacillus that will grow for us only on prereduced anaerobic medium using a more sensitive method (56).

Some nonpathogenic strains of H-A amebas and *Naegleria,* but not all, fail to grow in tissue culture. In general, the amebas which are less pathogenic for mice grow more slowly, while those which are highly pathogenic for mice grow rapidly. Both types destroy the culture, but generally the rate of destruction parallels the animal virulence. A systematic study of the differences in the growth of these amebas in various tissue-cell-type cultures has not been reported, but in our experience the type of cell does not appear to be important.

The effect of pathogenic amebas upon tissue cells may be to mimic that of the cytopathogenic viruses. In the early stages there are minor nuclear changes; these are followed by destruction of all the tissue cells. The amebas usually multiply until their number is so great as to give the appearance of a sheet of "altered" tissue cells (5, 20).

Physical and chemical effects of the H-A and N amebas upon tissue culture cells have not been well studied, but the violent penetrating attack and the phagocytic activity upon the cells is readily demonstrable. The cells show cytoplasmic and nuclear changes which resemble those associated with virus cytopathogenicity, and these may be associated with enzymatic effects. There have been extensive studies of the metabolism of the H-A group, as well as of the enzymes elaborated (2, 20, 47, 58).

The trophic forms of isolates which do not form cysts readily survive for several days at 23°C or at 4°C. The HB-1 strain of *Naegleria* from a fatal case of meningoencephalitis survived in a cerebrospinal fluid for four days during mail transit. After three days of storage at 4°C, it caused fatal mouse encephalitis after intracerebral injection. Ordinary freezing at —20° C of amebas in tissue specimens in which no cysts are present will destroy the amebas (42). Preliminary information indicates that the H-A amebas can be preserved in liquid nitrogen by the use of dimethyl sulfoxide (DMSO) and, similarly, HB-1 *Naegleria* can be preserved by the use of N,N-dimethylacetamide (43).

MORPHOLOGY

The excellent descriptions of Page, Singh, and Rafalko are essential for an appreciation of the complex features upon which differentiation of these amebas is based (59, 60, 65, 73, 74).

This brief summary is based upon the literature and the study of only the pathogenic isolates. As stated above, the morphology may vary greatly, depending upon conditions of growth, the age of the culture, and the method of preparation. The live amebas can be seen well by placing a small loopful on the surface of a thin agar plate (1.5 percent agar, 0.85 percent NaCl), applying a cover slip, and viewing the amebas on a phase microscope using a long-distance condenser. (See Plates I, II, III.)

Bacteria can be added to stimulate division so that mitosis can be observed and photographed. Erythrocyte and bacterial phagocytosis can be studied. The conventional preparations, such as the hanging drop, will allow more unimpeded motility but less structural detail. Similar preparations on agar plates which contain no NaCl will often allow the demonstration of flagellate forms of N amebas which, after several hours incubation, will appear around the air bubbles beneath the coverslip.

Stained preparations must be made on slides upon which the amebas are well flattened. The technique of Sayk is very satisfactory for cerebrospinal fluids (52). When thin, well-fixed preparations are available, hematoxylin or trichrome stains are satisfactory (3). The best preparations we have were made from Bouin fixed tissue culture preparations of amebas grown in Leighton tubes and stained with hematoxylin and eosin.

The H-A amebas are sluggishly motile and typically show fanlike forward extension of clear ectoplasm from which the micro- and filiform pseudopodia extend. In preparations containing bacteria or other particles, there is an active contractile vacuole, but in clear fluid medium the vacuole may be inactive. The nucleus is spherical and there is a large nucleolus which varies in density, depending upon the stage of the reproductive cycle.

Naegleria are more actively motile and resemble the H-A amebas only when they are flattened, as in agar surface preparations. The *Naegleria* show eruptive lobate pseudopodia and no filiform pseudopodia. The nucleus is quite similar to that of the H-A amebas except that the nucleus divides during mitosis and never disappears (21, 65, 73, 74).

In cerebrospinal fluid the N amebas may develop long, slender, branching forms whose internal structural details do not demonstrate active, writhing motion. According to Dobell, these forms are associated with high concentrations of NaCl and may measure 30–40 microns in length and 4–5 microns in diameter (74). The higher concentration of NaCl present in the spinal fluid may, therefore, have a relationship to this unusual form of *Naegleria* that resembles a larval form of a nematode.

There is always a bulbous ectoplasmic zone on the tip of the pseudopod. After suitable stimuli, such as a sudden decrease in osmotic pressure, the *Naegleria* generate a varying number of temporary flagellate forms with two beating flagella. The ameba becomes rigid and smaller, it assumes a pear or oval shape, and becomes rapidly motile. After a few hours the flagella are lost and the amebic form is assumed. Unlike flagellate forms of other amebo-flagellates, the *Naegleria* flagellate never divides (22, 65).

Cysts of H-A amebas are quite different from those of N amebas. After prolonged culture of the A-1 strain of H-A amebas on trypticase soy broth cysts are no longer formed, but they will form if the culture is mixed with bacteria. When HB-1 *Naegleria* are grown on tissue culture, they also will not form cysts, but they will do so when grown on a plate seeded with bacteria. Details of cyst morphology are fully given by others and will not be attempted here in extenso (59, 60, 73, 74). The H-A cysts are large and have a double wall which shows various degrees of wrinkling that often produces a starlike effect. Some species, including the A-1, form cysts which are almost smooth on the surface. Many precystic forms show gradations that are difficult to identify. The nucleus can often be seen in the cyst. The ameba excysts via pores in the wall at the points where layers of the cyst wall overlap.

Cysts of *Naegleria* are much smaller than those of the H-A amebas and are generally smooth and round. In some cultures the wall shows small hyalin segments described as pores. The HB-1 *Naegleria* shows these structures only rarely and never as clearly as does *N. gruberi* from bacterial cultures

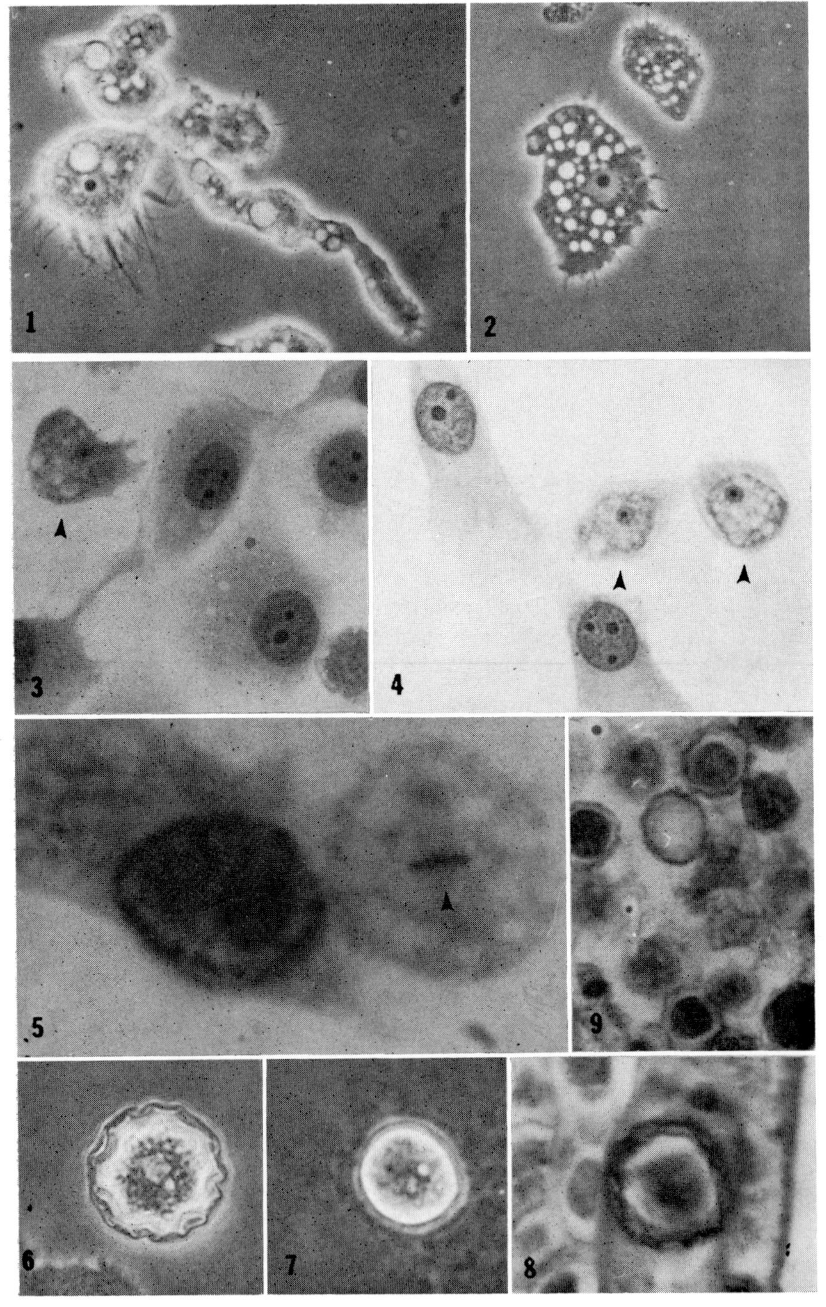

PLATE I

in nature. The excystment of the HB-1 has not been described (74).

Cysts of the H-A amebas are not adversely affected by short periods of treatment with 2 percent sodium desoxycholate solution or gall bladder bile (75). Such treatment destroys the trophic forms; therefore, this treatment can be used to obtain cysts without trophic forms. Cysts of A-1 prepared in this manner have not excysted when introduced by the usual routes in mice, and thus it appears that the cysts are not pathogenic. The possibilities that cysts might enter the nasal cavity in dust could explain the presence of these amebas in cultures from the nasopharynx. The question of whether excystment occurs in the nasal passage is under study.

Some strains of H-A amebas, such as HN-1 and HN-2, isolated from the nasal cavities of children, have been cultured on trypticase soy broth for five years without bacteria and continue to form typical cysts. These strains have lost some virulence but still produce chronic brain disease in 10–20 percent of the mice given 2–4 thousand amebas intranasally. In these infections cysts are formed in the lesions. Whether virulence depends upon the large numbers inoculated into the nose, or upon there being a very small number of virulent amebas in the culture, is not known.

The morphology of the organisms in tissue section is that of the rounded amebas. Here the trophozoites of both N and H-A amebas are similar except for the smaller size and often poor nuclear staining of the N amebas due to the dense endoplasm, a feature which leads to the amebas being mistaken for tissue cells. Conversely, the H-A amebas are larger and generally have a less dense endoplasm to obscure the nucleus in fixed tissue sections. This gives more prominence to the H-A nucleus with its large nucleolus, so that it is more easily observed by the histologist. Cysts of H-A amebas present in infections caused by less virulent strains stain well and are easily recognized, but in long-standing granulomata in mice they often require special stains, such as methanamine silver, which stain the polysac-

---

←

1. *Hartmannella culbertsoni* (A–1)—T.S.B. culture, slide and coverslip preparation. Phase X–800. Note various forms.
2. Same culture and optics—agar surface with coverslip.
3 and 4. *H. culbertsoni.* Two areas from same stained tissue culture coverslip. H. and E.—X–800. 3. Shows pseudopodia. 4. Shows endoplasm, ectoplasm, and characteristic nucleus.
5. *H. culbertsoni* in mitosis—X–1500. Chromosomes of dividing amebas (arrow). Note also changes in chromatin of tissue culture cell nucleus.
6. Cyst—*H. rhysodes* (HN–3). Phase—X–800.
7. Cyst—*H. culbertsoni* from culture. Phase—X–800.
8. *H. rhysodes* cyst. Nasal mucosa of mouse. PAS stain—X–1000.
9. Bovine lung. Amebic pneumonia. Note structures resembling cysts of *H. rhysodes* H. and E.—X–400.

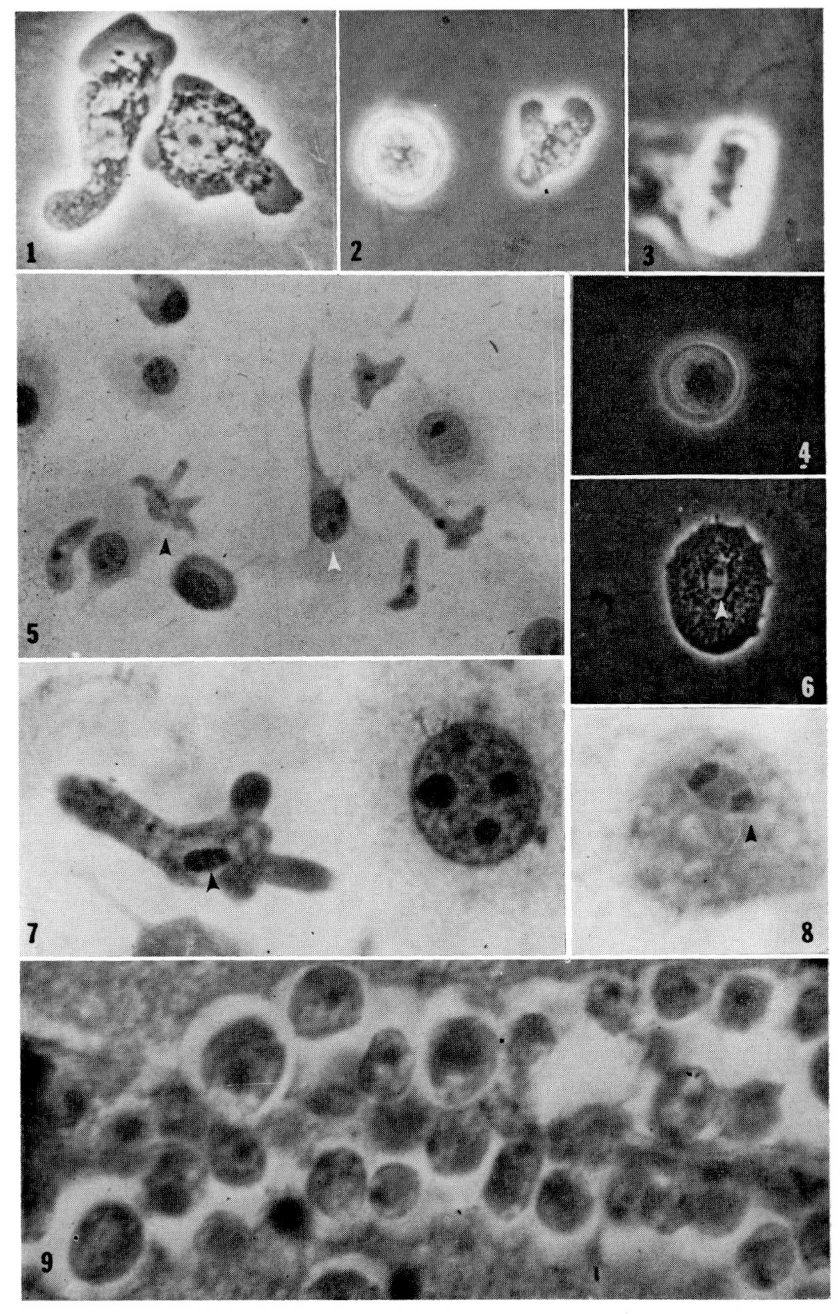

PLATE II

charide of the cyst wall. In such preparations the cysts may superficially resemble the cells of yeastlike fungi or even prototheca.

Experimental Animal Pathogenicity

After the observations that amebic encephalomyelitis and death in mice and monkeys followed intraspinal and intracerebral injections, tissue culture fluids containing the A-1 strain of H-A amebas were injected intravenously into the tail veins of mice. Intranasal, intraperitoneal, and subcutaneous injections were also made. With few exceptions, these experiments were carried out with amebas determined to be bacteria-free. The intravenous injections produced amebic pneumonia which was followed occasionally by fatal encephalitis. Intranasal instillation was followed by amebic rhinitis which progressed to invasion of the olfactory bulbs and the whole central nervous system. The pulmonary lesions which resulted from inocula entering the tracheobronchial tree also progressed so that a generalized bloodborne invasion occurred when the amebas gained entrance into the pulmonary veins. Intranasal instillations thus offer the amebas two portals of entry to the central nervous system, one by direct extension, and one by hematogenous spread via the pulmonary and systemic circulation.

After intraperitoneal or subcutaneous injection, the H-A amebas produce a local abscess which heals and *usually* does not spread and produce meningoencephalitis. It now appears that lesions caused by this ameba, other than those which involve the brain, do not cause death.

The production of local lesions from injection of bacterial flora together with H-A amebas has been reported by Wilson et al (87). This combination may be of importance in wounds contaminated by soil that contains both amebas and bacteria. Thus, there is need for investigation of dirt-contami-

---

1. *Naegleria aerobia* (HB–1) from tissue culture—agar plate with coverslip. Phase —X–800.
2. *N. aerobia*—flagellate form. Agar plate with coverslip. Also one small ameba. Phase—X–800.
3. *N. aerobia*—flagellate form. Agar plate with coverslip. Phase—X–800.
4. *N. aerobia*—cyst. Phase—X–800.
5. *N. aerobia* in tissue culture. H. and E.—X–300. Dividing ameba (black arrow). Damaged tissue cell (white arrow).
6. *N. aerobia*—mitosis. Agar plate with coverslip. Phase—X–800. Note divided nucleolus (polar mass) at pole (white arrow).
7. *N. aerobia*—tissue culture. Motile ameba in mitosis. H. and T.—X–1000. Note interzonal body (arrow).
8. *N. aerobia*—mitosis. Balamuth medium. Alger stain—X–1200. Note polar caps at tips of divided nucleolus (arrow).
9. *N. aerobia*—mouse brain encephalitis. Masses of amebas in brain tissue—X–800.

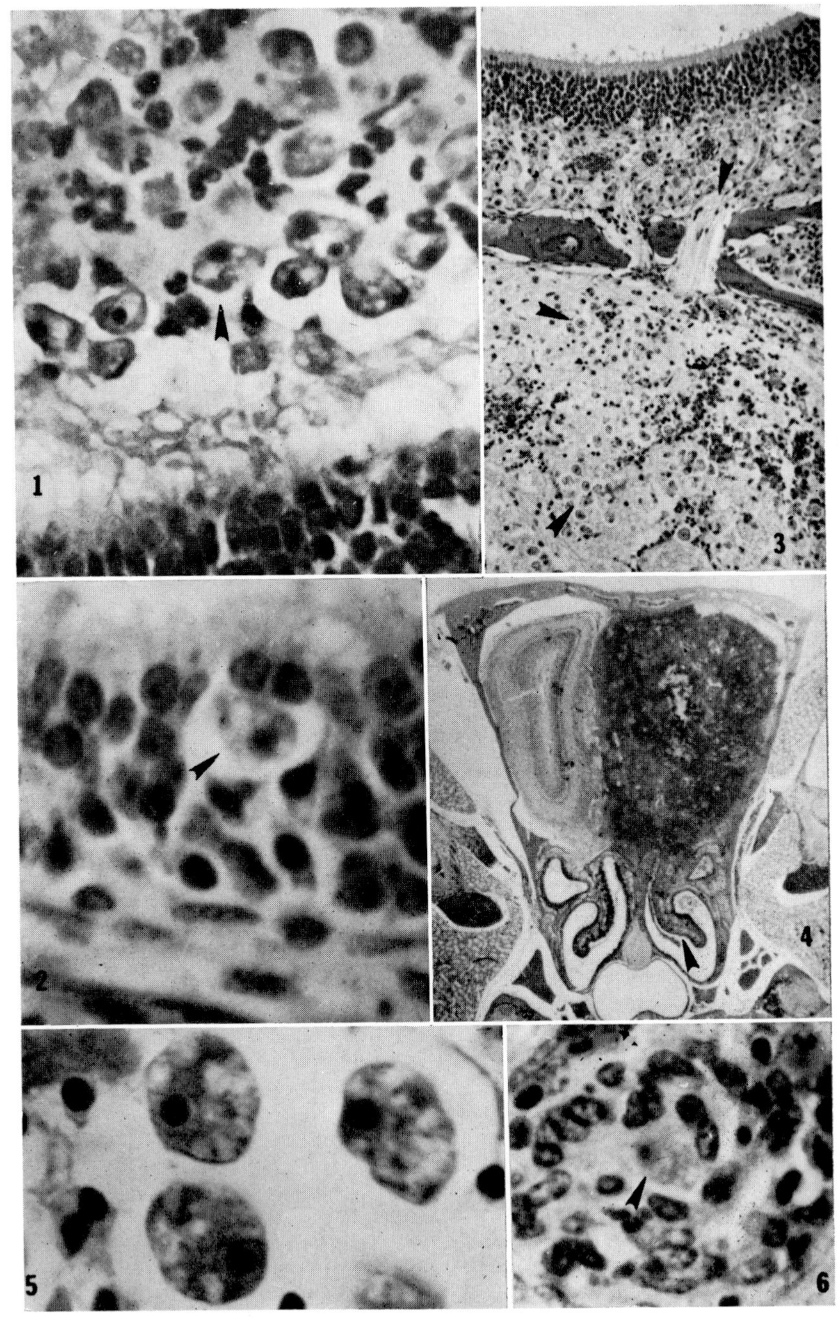

PLATE III

nated wound tissue for possible amebic infection in addition to the bacterial infection.

The A-1 strain of H-A amebas is so far unique in its degree of pathogenicity which has persisted for about thirteen years since it was first grown in tissue culture and transferred to trypticase soy broth. This ameba has an LD_{50} of between 20 and 50 amebas intranasally, and an LD_{100} of 1 to 5 amebas injected intracerebrally in specific pathogen-free (SPF) mice.

Other virulent strains have been isolated from soil from widely separated areas of the United States and some from the British Isles. These represent about 50 percent of the number of strains we have isolated. They have been less virulent than the A-1 strain, but some have caused acute encephalitis in mice that resulted in death in 20–30 percent of the animals inoculated intranasally. The most common lesions in this group, however, have been cerebral granulomata which have persisted for as long as eleven months and still contained viable amebas but also many cysts. The mortality rate in animals with cerebral granulomata was low during the period of observation, and in many instances the lesions became quiescent and healed (30).

After the isolation of *Naegleria* sp., Strain HB-1, from a patient in Florida, most of the animal experiments were repeated with that ameba in mice and monkeys in a manner similar to that used for the A-1 Strain of the H-A amebas. Results from intranasal instillation generally parallel those seen after similar instillation of A-1 (H-A) amebas. However, further studies showed a number of differences, most of which were probably related to the smaller size and greater motility of the *Naegleria* sp. Intravenous injection of *Naegleria,* Strain HB-1, caused few pulmonary lesions but did produce fatal meningoencephalitis. In all probability these amebas traverse the pulmonary capillaries more readily than do the H-A amebas. In recent as yet unpublished experiments when Strain HB-1 was injected into 150-g

---

←

1. Nasal mucosa—mouse inoculated intranasally with *H. rhysodes* (A–30). Colony of trophic forms in nasal mucus—ameba at arrow. H. and E.—X–1000.
2. Nasal mucosa—mouse. *H. culbertsoni* (A–1) penetrating mucosa (arrow). H. and E.—X–1000.
3. Cross section of mouse head illustrating route of amebas from nasal mucosa to brain (olfactory bulbs). Note olfactory nerve filament traversing bone (arrow). Amebas have entered brain substance (arrows). H. and E.—X–300.
4. Coronal section of mouse head showing unilateral hemorrhagic destruction of olfactory bulb. Note swelling of turbinate and exudate in nasal cavity (arrow).
5. *H. culbertsoni* (A–1) from amebic encephalitis in mouse brain. H. and E.—X–1200.
6. *H. rhysodes* (A–57). Granulomatous nodule in chronic encephalitis in a mouse. Note ameba in center (arrow). H. and E.—X–800.

guinea pigs intramuscularly, intraperitoneally, or subcutaneously, the animals exhibited a progressive invasion of the skin and muscle from which the regional lymph nodes were involved. Lung lesions developed presumably after hematogenous spread from the sites of primary infection. Cerebral disease did not occur in these animals. The infection following parenteral introduction of N amebas is less acute than that following intranasal instillation, being manifested by loss of weight of the animals over a period of several weeks, but some animals developed generalized infection. Under these same circumstances the H-A amebas, Strain A-1, caused local abscesses and inflammation, but generally did not show a tendency to spread (32).

When they invade the brain, both HB-1 and A-1 amebas injure capillaries and cause hemorrhage, and HB-1 *Naegleria* causes the more severe lesions. Both strains also penetrate larger blood vessels readily, and the injury causes thrombosis. Amebas are sometimes seen in the thrombi. Both strains phagocytize erythrocytes in a nibbling, piecemeal fashion which finally results in hemolysis of the red cell. Chi et al (24) found that the NIH isolate rejected the nuclei of chicken erythrocytes while ingesting the remainder of the cells. We have also observed this phenomenon using A-1 (H-A) amebas.

Cellular reaction to either strain of the amebas is similar, consisting of neutrophils in the early, acute state, followed by the appearance of many monocytes. Later, fibroblasts are evident. Eosinophils have been seen in guinea pig and monkey lesions, but usually they are not present in lesions in mice. In chronic lesions there are some lymphocytes, but many plasma cells and the typical epitheloid cells which form the granulomata. Cysts and trophozoites exist together. Because some cysts are empty, the possibility of in vivo encystment and re-excystment in the lesion is suggested.

Human Case Reports

There are now over 50 human deaths reported in which amebas have been observed in the brain tissue, and the number in which N amebas have been isolated and identified is increasing. Most instances are of retrospective diagnoses from tissue sections. The oldest known case, which was rediagnosed in 1969 by Symmers (79), died in 1909. The first probable case reported in the literature was by Derrick (33), who believed that it might represent an infection by *Iodamoeba williamsi* on the basis of the large nucleolus seen in the ameba. In a foreword to Derrick's report, Wenyon remarked that it might represent an infection by an ameba not previously known to have caused disease. The morphologic and other features of the amebas appearing in the sections of Derrick's case are, in our opinion, compatible with those of *Naegleria*.

A similar case reported by Kernohan et al (49) as being due to *Iodamoeba* appears, on further study, to have been caused by an ameba of the free-living group. However, some features of this case, including the granulomatous nature and the chronic course of the disease, plus the morphologic resemblance of the organisms to the H-A amebas, make it doubtful that this case is identical to Derrick's. The latter resembled the later-discovered *Naegleria* infections, except that Derrick's case was a unique example in the human series of widespread invasion in a starved, debilitated individual.

Subsequent cases are clustered in Florida (9, 10), Australia (13, 15, 44), Virginia (11, 12, 35, 37, 38, 80, 86), and Czechoslovakia (17, 18), with a smaller number observed in Texas (62), Georgia (55), and in England (4, 79). Almost all of these reported fatalities were associated with swimming, the two exceptions being in a man from Texas and in one of the Australian patients. The Czechoslovakian cases were almost all associated with a single swimming pool.

As stated above, the fatalities appear to have been caused by small amebas resembling *Naegleria,* with the possible exception of the case reported by Kernohan et al (49). The cases reported by Mandal et al in New Zealand (53) were believed by the authors to represent infections due to slime mold amebas. From the evidence presented, it is difficult to accept the slime mold ameba etiology, since the illustrations are compatible with *Naegleria* and the evidence presented for the slime mold derivation is not convincing.

The H-A amebas have been suspected of causing meningitis in several instances. Callicott et al (12) obtained a positive culture from the cerebrospinal fluid of a patient who recovered. The authors stated that the other features of the fluid were normal and that the etiologic relation was, therefore, open to question.

Hartmannellid ameba infections (44, 62) reported before the isolation of *Naegleria* are generally agreed now to have been caused by *Naegleria.* In my opinion these cases should not be counted as certain evidence for the human pathogenicity of H-A amebas. The question of the role in human disease of the H-A amebas is still unanswered. Animal experiments indicate that these amebas are potential pathogens. Since these experiments also indicate that they may cause nonfatal or chronic disease, further experience may confirm these amebas as human pathogens. The occurrence of the amebas in the nasopharynx is of interest in this regard, but sufficient knowledge as to whether the amebas exist in the nasopharynx as cysts or trophozoites or whether they are related to nasopharyngitis is not available (83). The presence of both H-A and N groups of amebas, presumably in cystic form, in the air (50) can easily account for the positive cultures from the nasopharynx. The important question is whether such organisms ever produce disease.

Skocil, Cerva & Serbus (76) have begun a long-term study on 1000 soldiers to determine the presence of limax amebas in the nasopharynx. Their first report on this study, published in 1970, indicates a considerable number of positive cultures, varying roughly from 3 percent up to 14 percent in some groups. The isolates were mostly of the H-A group and there was a statistically significant correlation between the positive cultures of limax amebas from the nasopharynx and a history of "cephalalgia, frequent rhinitis and epistaxis." The further results of this study should be of great interest.

There is also need for a study in which nasal secretions are searched for trophic forms of amebas. Work in progress on antibodies may aid in evaluating the importance of H-A amebas as human pathogens (43).

Natural Animal Infection

The experimental infections of laboratory animals strongly suggest that amebic disease in farm animals should be expected to occur. At present there is only one report of this in a bovine from the Azores, in which the histologic appearance of the amebas and their cysts in the lung lesions is such as to indicate that the causative organisms were of the H-A type. The brain in this case was not examined (54).

Another instance has come to our attention in a zoo-confined ovine (Nelson bighorn) in which typical amebic rhinitis and encephalitis was found (45). In this animal there was not sufficient evidence to suggest which species of amebas was present, and lesions were only in the nasal mucosa and brain.

There is an increased interest in the role of the H-A amebas in nonvertebrates. Richards (66) described two new species of *Hartmannella* as infectious for fresh-water mollusks. Pauley (63) observed amebas of the H-A type in sections of diseases mussels. *Paramoeba* have been found in diseases of crabs (67, 78). Sawyer (68) has reported on the influence of seawater on the biology and morphology of *Acanthamoeba,* which observations may prove to be of importance in the study of possible infections in marine forms.

The amebas of honey bees and grasshoppers do not appear to be either H-A or N amebas, and generally they are associated with another parasitic protozoal disease called Nosema. Consequently these infections in insects are probably parasitic and not caused by any free-living ameba species.

The question of the differentiation of soil amebas from *Entamoeba histolytica* in tissue sections is of importance in areas in which intestinal amebiasis is common. There is little or no difficulty in distinguishing these species when well-fixed stained sections are available; however, in poorly prepared tissue which has had delayed fixation, the differentiation may be difficult, if not impossible. Differences in the nuclei of these two species are the most striking feature. *E. histolytica* has a delicate nuclear membrane which tends to be slightly granular and indistinct, and a very small nucleolus surrounded by pale-staining nuclear substances. In contrast, in well-stained preparations the two species of soil amebas have a thin, but very distinct, nuclear membrane and a large nucleolus which stains deeply with most of the usual stains, such as hematoxylin and eosin. Modifications of Giemsa stain, Masson's trichrome, and iron hematoxylin stains are useful for the study of doubtful cases. Identification from tissue sections has great limitations; whenever possible, cultures should be made for identification from living amebas. For isolations from tissue or body fluid both intracerebral injection in mice and tissue culture are used. Care must be taken when making direct microscopic examination that the movement and form of neutrophilic leukocytes and active lymphocytes are not mistaken for amebas. Since the agar plate, *E. coli* cultural method is more easily available, it may be used; but it results in a mixed culture isolate which cannot be as easily studied or passed in animals, so that bacteria-free methods are preferable. In our hands, the axenic media have required rather large inocula, and it seems likely that intracerebral mouse injection, or tissue culture, or a combination of both, would be preferable (31).

HUMAN PATHOLOGY AND COMPARISON WITH EXPERIMENTAL DISEASE

The reports of fatal human cases in which necropsy studies have been possible now number about fifty. The gross pathological changes are generally similar in most of these cases, but the findings are not specific enough to permit differentiation of amebic meningitis from bacterial meningitis on gross examination. Thus far, there have been no recognized abscesses due to H-A or N amebas. Abscesses produced in monkeys either by intranasal infection or by direct intracerebral injection of H-A amebas have shown pathological changes identical to those described for *E. histolytica*. The fact that some of the published reports concerning the pathology of *E. histolytica* show microphotographs of amebas which in our opinion resemble more the soil amebas than *E. histolytica* causes a question to be raised as to whether some cerebral abscesses presumed due to *E. histolytica* are, in fact, due to soil amebas. The question deserves further study (8).

While brain involvement in fatal *Naegleria* meningoencephalitis has attracted the most attention, it seems probable that some other organs might also be involved, namely, lungs, liver, and kidney, since this is often the case in experimental animal disease (31). As outlined above, we have recently found that HB-1 *Naegleria* injected subcutaneously, intraperitoneally, or intramuscularly spreads rapidly to many organs and causes the death of the animal without causing meningitis. However, it appears that lack of involvement of organs other than the brain is, in fact, the case in the human with very few exceptions, because in most cases in which specific inquiry has been made, the pathologist after careful search has been unable to demonstrate lesions outside of the central nervous system.

The following is a brief summary of gross and microscopic descriptions from published necropsy findings. Some of the later reports vary from the earlier ones, because the course of the illness has probably been lengthened by chemotherapy and respirator treatment (4).

The first probable case, reported by Derrick (33), occurred in a starved Japanese soldier who was suffering from malaria, dysentery, and malnutrition with beriberi. Autopsy showed the brain to have a depression in the right parieto-occipital region overlying a soft hemorrhagic lesion that extended 2.5 cm into the brain. The marginal zone was brown in color, and beyond this petechiae were visible through the meninges. There were ulcers in the stomach, small intestine, and colon; and there was acute mesenteric lymphadenitis. The lungs showed hemorrhagic areas of consolidation. The liver contained no abscesses, but the kidneys showed purulent inflammation.

Microscopically, most of the lesions seen grossly contained many amebas, but there were none in the liver or kidney. The many excellent microphotographs showed an amebic morphology quite similar to that subsequently reported in the culturally confirmed cases of meningitis due to N amebas. Photographs of *E. histolytica* also shown in the report exhibited the morphologic differences mentioned above. This is the only instance, if it was actually an infection due to *Naegleria* as now seems probable, in which there was a generalized disease process. Such disease has been observed in the experimental infection of guinea pigs (32).

It is of interest that almost all the fatal human disease has occurred in otherwise healthy, usually young individuals, in contrast to Derrick's case and that reported by Patras & Andujar (62), in which the patients were debilitated. It should also be noted that Derrick's case had lingering illness for at least seven weeks. Since the patient had other serious diseases, it is difficult to know when the amebic disease began.

The next report, that of Kernohan et al (49), was of a 6-year-old girl who died following a head injury. In this case, watery fluid escaped from a scalp wound and the child was treated by craniotomy three weeks later. The return of neurological symptoms after five months necessitated a second operation, at which time a mass attached to the dura was found and removed. The child died three months later after developing left arm and speech paralysis. Study of the tissue from the second operation showed a granulomatous process. At necropsy only the brain showed a lesion. It was beneath the operative site, adherent to the dura, some 5 cm in diameter, and had a mottled appearance with a yellowish border zone. The center showed small gray foci throughout. The lesion was characterized by a granulomatous process in which there were moderate numbers of relatively large, well-staining amebas that sometimes showed pseudopodia.

In our own subsequent examination of tissue from this patient, there were not the small amebas in masses which have been seen in the brains of most of the human cases, but rather larger amebas with clearly staining nuclei and large nucleoli. There were some suggestions of precystic forms, but no mature cysts were found. Since glycogen vacuoles in the cyst are characteristic of *Iodamoeba,* there is no definitive method to confirm the identity in this case, nor in that of Derrick, because no cysts were present. Viewed in the light of present knowledge, this case is unique in that it is possibly representative of the H-A group. The pathogenesis of this case is far from clear.

All remaining reported cases are much alike in their clinical history, course, and necropsy findings. The cases diagnosed retrospectively from histologic studies are identical to those in which N amebas have been identified by culture.

Pathological findings in the usual acute *Naegleria* meningitis are not markedly different from those of acute bacterial meningitis with which many of the reported cases were at first confused, both on gross and microscopic examination. The brain shows swelling and redness, with a purulent exudate more extensive on the basilar surface of the cerebrum or cerebellum and over the brain stem. On sectioning, the cerebral and cerebellar gray substances show variable-sized lesions which tend to be hemorrhagic and quite soft when they are large. The existence of redness and destruction of the olfactory nerve may be a distinguishing feature between the bacterial and amebic disease, since the amebic disease appears to originate in the nasal cavity (12, 13, 37).

Histologically, ordinary tissue stains such as hematoxylin and eosin show the amebas in the neutrophilic exudate, but more and better preserved amebas are found beneath the pia and around the vessels of the Virchow-Robin spaces, where they tend to grow in great numbers without being surrounded by the neutrophils as they are in the subarachnoid space. The use of iron hematoxylin, Masson's trichrome, and tissue modifications of Giemsa stain are all of value in the demonstration of these amebas in tissue (17, 31).

If tissue fixation is delayed unduly, the amebas may not stain well even though the tissue cells may stain satisfactorily. Consequently, the amebas may not be evident and may be overlooked, as has been the case in many of the published reports.

Host Resistance: Immunology

Trophic forms of both the H-A and N groups of amebas are disrupted by gall bladder bile or by 2 percent sodium desoxycholate. Normal gastric juice has a similar effect. Fresh normal human blood serum will also cause lysis of the amebas. Heated serum does not. Bloods from other species have not been adequately tested.

Cysts have not been adequately tested, but it is known that they resist the effects of gastric juice and bile and may pass intact through the gastrointestinal tract. It is not known whether they excyst in the intestine and are perhaps lysed by bile salts and thus are not present in trophic form in the stools. When the amebas are mixed with normal stool, they are lysed. The effect of chemical and physical agents has been discussed by Carter (15).

The only evidence for specific resistance is survival of mice challenged intranasally with large inocula of A-1 amebas after previous intraperitoneal "vaccination" with the same amebas, as shown in Table 1. Table 2 shows a similar experiment using an avirulent ameba (H-A Strain A-5) as a vaccine. The A-5 strain is practically indistinguishable from the A-1 strain ex-

IMMUNIZATION OF 15-20 GM. SPF MICE WITH 9-DAY LIVE A-1 CULTURE

	No. of Mice	Intraperitoneal Immunization			0.03 ml. A-1 I.N. Challenge 2-5-65	Deaths	Survivors*
		1-15-65	1-22-65	1-29-65			
Group A	10	0.25 ml.	0.25 ml.	0.25 ml.	2500	7,9,S,S,S,S,S,S,S,S	8/10
Group B	10	0.25 ml.	0.25 ml.	—	2500	7,7,13,S,S,S,S,S,S,S	7/10
Group C	10	0.25 ml.	—	—	2500	5,11,S,S,S,S,S,S,S,S	8/10
Group D	10	Infected controls—no intraperitoneal dose			2500	6,6,6,6,7,8,9,9,11,12	0/10
Group A_1	10	0.25 ml.	0.25 ml.	0.25 ml.	5000	9,10,11,S,S,S,S,S,S,S	7/10
Group B_1	10	0.25 ml.	0.25 ml.	—	5000	6,9,S,S,S,S,S,S,S,S	8/10
Group C_1	10	0.25 ml.	—	—	5000	10,S,S,S,S,S,S,S,S,S	9/10
Group D_1	10	Infected controls—no intraperitoneal dose			5000	6,6,6,6,6,6,7,7,9,10	0/10

* Survivors were reinfected intranasally on March 6, 1965, with the same dose as in the first infection. One of the 47 survivors died on the 9th day after reinfection. The remainder survived three weeks.

cept that it shows no degree of virulence and does not grow in tissue culture. Thus, there is evidence that the virulent Strain A-1 has protective antigenic components, whereas the avirulent strain lacks such antigens. Similar "vaccination" using HB-1 *Naegleria* resulted in no resistance to subsequent intranasal challenge with that ameba. Potent antiserum treatment or prophylaxis also failed to protect mice against intranasal challenge with *Naegleria*.

Most studies on host resistance and immunology have pertained to the H-A amebas, Strain A-1. The mortality from the nasal infection can be decreased by previous treatment of the mice with various endotoxic preparations and pertussis vaccine. Mice given cortisone are generally more susceptible. The A-1 strain is attacked, surrounded, and phagocytized by normal neutrophilic leukocytes in vitro. The *Naegleria* HB-1 Strain is not phagocytized under identical conditions.

Serologic Reactions

Monkeys and rabbits develop complement-fixing (CF) antibodies after

IMMUNIZATION OF 20-25 GM. SPF MICE WITH UNDILUTED LIVE A-5 CULTURE

	No. of Mice	Intraperitoneal Immunization			0.03 ml. A-1 I.N. Challenge 2-16-65	Deaths	Survivors
		1-26-65	2-2-65	2-9-65			
Group A	10	0.25 ml.	0.25 ml.	0.25 ml.	2500	6,6,6,6,7,7,7,8,S,S	2/10
Group B	10	0.25 ml.	0.25 ml.	—	2500	6,6,6,6,6,7,8,8,S,S	2/10
Group C	10	0.25 ml.	—	—	2500	6,6,6,6,6,7,7,8,10,S	1/10
Group D	10	Infected controls—no intraperitoneal doses			2500	5,5,6,6,6,6,7,7,S,S	2/10
Group A_1	10	0.25 ml.	0.25 ml.	0.25 ml.	5000	6,6,6,6,7,7,7,10,12,S	1/10
Group B_1	10	0.25 ml.	0.25 ml.	—	5000	6,6,6,6,6,7,7,7,7,S	1/10
Group C_1	10	0.25 ml.	—	—	5000	5,6,6,6,6,S,S,S,S,S	5/10
Group D_1	10	Infected controls—no intraperitoneal doses			5000	5,5,5,5,5,6,6,6,7,13	0/10

intravenous injection of H-A amebas (Strain A-1). The titer after a single injection rises sharply and quickly falls as the amebas disappear from the lungs (27).

Intracerebral injection of A-1 amebas into one monkey caused a localized lesion which was accompanied by the development of CF antibody in both the blood and cerebrospinal fluid that persisted for several years before it finally declined.

Multiple large doses of intravenous H-A and N amebas produce high titer of CF as well as agglutinating and lytic antibodies in rabbits which survived the injections. Both direct and indirect immunofluorescence (IF) antibody tests have been devised, and hyperimmune antiamebic rabbit sera also show high titers by these methods.

Slides may be prepared with small deposits of four or more different amebas for simultaneous testing with small samples of serum for indirect immunofluorescence. Testing of human serum by CF and indirect IF tests is now being done, but as yet no conclusions as to the significance of these tests has been justified by the results (70).

The use of antigen from avirulent A-5 amebas to absorb the "nonspecific" amebic factors from serum in a fashion similar to that used for syphilis (46), and the use of Nairn's liquid nitrogen isopentane method for preparing the antigen slides, appear to provide for more reproducible serologic results. While such techniques give expected results with known specific hyperimmune sera, the value of testing unknown sera is yet to be determined. Both CF and IF tests indicate that the antiserum, produced as described, against H-A amebas does not react significantly with *Naegleria*. However, the different isolates of H-A amebas do cross-react to a considerable degree (1, 74).

Siddiqui & Balamuth (71) have performed agar diffusion antibody tests which agree in general with the results of CF and IF tests.

In limited exploration of human sera no reaction was noted from antiamebic antibodies with syphilitic reagin, from sera containing dye test antitoxoplasma antibody, or from some sera from patients with antibody to *E. histolytica*. Further testing is needed before valid conclusions can be reached in respect to this and other aspects of human serum testing.

Prevention and Treatment

The prevention of amebic encephalitis at this time can only be by abstinence from swimming in lakes, rivers, or swimming pools when the possibility of the presence of pathogenic amebas exists. The factors required to assure the safety of swimming pools are largely unknown, but general epidemiologic experience would suggest that careful management, using water of drinking quality and proper care to apply approved standard methods, would suffice to constitute safe conditions. Obviously, there is insufficient information to suggest other measures.

Experimental chemotherapy has indicated that sulfa drugs, particularly sulfadiazine, are very active in prevention and treatment of disease in mice infected intranasally with H-A amebas, Strains A-1 and HN-3 (28). Sulfa drugs have little or no effect on similar N ameba infections in mice. Amphotericin B has been found active in experimental *Naegleria* infection (14, 31). Several patients have now been treated with amphotericin B with and without sulfadiazine. As yet there are no certain beneficial effects known in the human cases (37). The one patient who survived for a longer period than the others was also treated with sulfadiazine and by respirator (4).

The typical case of *Naegleria* meningoencephalitis runs a very rapid course. After symptoms appear, it seems difficult to contemplate an effective treatment, since brain destruction occurs very quickly. Another discouraging feature is that when mice are treated with amphotericin, the acute destructive process is altered to a subacute or chronic disease in which abscesses and granulomata appear. These observations indicate an unpromising future for treatment of patients after symptoms develop. The alternative is vigorous treatment of patients if they develop nasal symptoms after known exposure to water which could possibly contain the amebas. Examination of nasal discharge might be of value in recognition of early phases of the disease which could then be treated not only with systemic drugs, but also with local agents.

Epidemiology and Relationship to Ecologic Conditions

Investigations into the natural habitat and ecology of amebas such as *N. aerobia* are just beginning, and as yet no solid information is at hand. Lakes and pools from which infections appear to have come should be carefully studied. Recent experiments have indicated that intracerebral injections of sterile muddy water containing very few *N. aerobia* were followed by fatal disease due to N amebas. That approach would provide a simple test procedure which would not involve the isolation of nonpathogenic amebas.

The question of the effect of pollution by sewage, nitrogen fertilizer, pesticides, either directly or by causing decay of vegetation, has been raised mostly in popular magazines. These elements could influence the growth of amebas, but as yet there is no information to evaluate their effects. Since there are now many valuable methods and many knowledgeable scientists and this problem has at least been defined, there is reason to expect rapid accretion of information regarding this subject.

ACKNOWLEDGMENT

In the preparation of this chapter I have had the assistance of many friends and associates. I wish particularly to express appreciation to Miss Phyllis Wente for collecting and cataloging material and for editorial assistance; to Messrs. Paul Ensminger and Willis Overton, who have been associated with this work from the beginning. I also owe much to the late Mrs. Philip Trexler who contributed greatly to the early exploration of this subject. I am indebted to Drs. B. N. Singh, C. T. Dolan, R. S. Chang, S. L.

Chang, and J. S. Rafalko for supplying me with cultures and material; to Captain Bruce Smith and Dr. Joe Griffin and their associates at the Armed Forces Institute of Pathology and to Dr. Seymour Hutner and Dr. William Balamuth for their support and advice in this work.

LITERATURE CITED

1. Adam, K. M. G. 1964. *J. Protozool.* 11:423–30
2. Adam, K. M. G., Blewett, D. A. 1967. *J. Protozool.* 14:277–82
3. Alger, N. 1966. *Am. J. Clin. Pathol.* 45:361–62
4. Apley, J. et al. 1970. *Brit. Med. J.* 1:596–99
5. Armstrong, J. A., Pereira, M. S. 1967. *Brit. Med. J.* 1:212–14
6. Balamuth, W. 1964. *J. Protozool.* 11(Suppl.):19–20
7. Band, R. N. 1959. *J. Gen. Microbiol.* 21:80–95
8. Brandt, H., Pérez Tamayo, R. 1970. *Human Pathol.* 1:351–85
9. Butt, C. G. 1966. *N. Engl. J. Med.* 274:1473–76
10. Butt, C. G., Baro, C., Knorr, R. W. 1968. *Am. J. Clin. Pathol.* 50:568–74
11. Callicott, J. H., Jr. 1968. *Am. J. Clin. Pathol.* 49:84–91
12. Callicott, J. H., Jr. et al. 1968. *J. Am. Med. Assoc.* 206:579–82
13. Carter, R. F. 1968. *J. Pathol. Bacteriol.* 96:1–25
14. Carter, R. F. 1969. *J. Clin. Pathol.* 22:470–74
15. Carter, R. F. 1970. *J. Pathol.* 100: 217–44
16. Castellani, A. 1932. *Trans. Roy. Soc. Trop. Med. Hyg.* 25:219
17. Cerva, L., Novak, K., Culbertson, C. G. 1968. *Am. J. Epidemiol.* 88:436–44
18. Cerva, L., Zimak, V., Novak, K. 1969. *Science* 163:575–76
19. Chang, R. S., Goldhaber, P., Dunnebacke, T. H. 1964. *Proc. Nat. Acad. Sci. USA* 52:709–15
20. Chang, R. S., Pan, I-Hung, Rosenau, B. J. 1966. *J. Exp. Med.* 124: 1153–66
21. Chang, S. L. 1958. *J. Gen. Microbiol.* 18:565–78
22. Chang, S. L. 1958. *J. Gen. Microbiol.* 18:579–85
23. Chang, S. L. 1960. *Can. J. Microbiol.* 6:397–405
24. Chi, L., Vogel, J. E., Shelokov, A. 1959. *Science* 130:1763
25. Crump, L. M. 1950. *J. Gen. Microbiol.* 4:16
26. Culbertson, C. G., Smith, J. W., Minner, J. R. 1958. *Science* 127: 1506
27. Culbertson, C. G., Smith, J. W., Cohen, H. K., Minner, J. R. 1959. *Am. J. Pathol.* 35:185–97
28. Culbertson, C. G., Holmes, D. H., Overton, W. M. 1965. *Am. J. Clin. Pathol.* 43:361–64
29. Culbertson, C. G., Ensminger, P. W., Overton, W. M. 1965. *Am. J. Clin. Pathol.* 43:383–87
30. Culbertson, C. G., Ensminger, P. W., Overton, W. M. 1966. *Am. J. Clin. Pathol.* 46:305–14
31. Culbertson, C. G., Ensminger, P. W., Overton, W. M. 1968. *J. Protozool.* 15:353–63
32. Culbertson, C. G., Ensminger, P. W., Overton, W. M. 1971. To be published
33. Derrick, E. H. 1948. *Trans. Roy. Soc. Trop. Med. Hyg.* 42:191–98
34. Dobell, C., O'Connor, F. W. 1921. *The Intestinal Protozoa of Man.* London:John Bale, Sons & Danielsson, Ltd.
35. dos Santos, J. G. 1970. *Am. J. Clin. Pathol.* 54:737–42
36. Drozanski, W. 1963. *Acta Microbiol. Polon.* 12:9–24
37. Duma, R. J. 1969. *Va. Med. Mon.* 96:546–48
38. Duma, R. J., Ferrell, H. W., Nelson, E. C., Jones, M. M. 1969. *N. Engl. J. Med.* 281:1315–23
39. Dunnebacke, T. H., Williams, R. C. 1967. *Proc. Nat. Acad. Sci. USA* 57:1363–70
40. Editorial. 1969. *Med. J. Aust.* 1:1036–38
41. Eldridge, A. E., Tobin, J. O'H. 1967. *Brit. Med. J.* 1:299
42. Ensminger, P. W., Culbertson, C. G. 1966. *Am. J. Clin. Pathol.* 46: 496–99
43. Ensminger, P. W., Culbertson, C. G. Unpublished data
44. Fowler, M., Carter, R. F. 1965. *Brit. Med. J.* 2:740–42

45. Grimer, L. A. Personal communication
46. Hunter, E. F., Deacon, W. E., Meyer, P. E. 1964. *Pub. Health Rep.* 79:410
47. Husain, M. M., Rao, V. K. 1969. *J. Gen. Microbiol.* 56:379–86
48. Jahnes, W. G., Fullmer, H. M., Li, C. P. 1957. *Proc. Soc. Exp. Biol. Med.* 96:484–88
49. Kernohan, J. W., Magath, T. B., Schloss, G. T. 1960. *Arch. Pathol.* 70:576–80
50. Kingston, D., Warhurst, D. C. 1969. *J. Med. Microbiol.* 2:27–36
51. Kofoid, C. A., Boyers, L. M., Swezy, O. 1922. *Proc. N.Y. Pathol. Soc.* 22:120–26
52. Kolar, O., Zeman, W. 1968. *Arch. Neurol.* 18:44–51
53. Mandal, B. N. et al. 1970. *N. Z. Med. J.* 71:16–23
54. McConnell, E. E., Garner, F. M., Kirk, J. H. 1968. *Pathol. Vet.* 5:1–6
55. McCroan, J. E., Patterson, J. 1970. *CDC Morbidity and Mortality Weekly Report.* 19:413–14 (Oct. 24) Atlanta, Ga: Center for Communicable Disease Control
56. McMinn, M. T., Crawford, J. J. 1970. *Appl. Microbiol.* 19:207–13
57. Musgrave, W. E., Clegg, M. T. 1906. *Philippine J. Sci.* 1:909–50
58. Neff, R. J., Neff, R. H., Taylor, R. E. 1958. *Physiol. Zool.* 31:73–91
59. Page, F. C. 1967. *J. Protozool.* 14: 499–521
60. Page, F. C. 1967. *J. Protozool.* 14: 709–24
61. Page, F. C. 1970. *J. Parasitol.* 56: 257–58, Sec. II, Part 1, Resumés 469 & 470
62. Patras, D., Andujar, J. J. 1966. *Am. J. Clin. Pathol.* 46:226–33
63. Pauley, G. B. 1968. *J. Invertebr. Pathol.* 12:321–28
64. Pereira, M. S., Marsden, H. B., Corbitt, G., Tobin, J. O'H. 1966. *Brit. Med. J.* 1:130–32
65. Rafalko, J. S. 1947. *J. Morphol.* 81: 1–44
66. Richards, C. S. 1968. *J. Protozool.* 15:651–56
67. Sawyer, T. K. 1969. *Proc. Nat. Shellfisheries Assoc.* 59:60–64
68. Sawyer, T. K. 1970. *Proc. Helminthol. Soc. Wash.* 37:182–88
69. Schaudinn, F. 1903. *Arb. Kaiserl. GesundhAmt.* 19:547–76
70. Schlaegel, T. F., Jr., Culbertson, C. G. 1970. *Invest. Ophthalmol.* 9:555 (Abstr.)
71. Siddiqui, W. A., Balamuth, W. 1965. *J. Protozool.* 13:175–82
72. Singh, B. N. 1942. *Ann. Appl. Biol., Cambridge,* 29:18
73. Singh, B. N. 1952. *Phil. Trans. Roy. Soc. London Ser. B* 236:405–61
74. Singh, B. N., Das, S. R. 1970. *Phil. Trans. Roy. Soc. London Ser. B* 259:435–76
75. Singh, B. N., Saxena, U., Iyer, S. S. 1965. *Indian J. Exp. Biol.* 3:110–12
76. Skocil, V., Cerva, L., Serbus, C. 1970. *J. Hyg. Epidemiol. Microbiol. Immunol.* 14:61–66
77. Smith, A. J., Middleton, W. S., Barrett, M. T. 1914. *J. Am. Med. Assoc.* 63:1746–49
78. Sprague, V., Beckett, R. L., Sawyer, T. K. 1969. *J. Invertebr. Pathol.* 14:167–74
79. Symmers, W. St. C. 1969. *Brit. Med. J.* 4:449–54
80. Wagner, W. P., Duma, R. J., McGehee, R. F., Suter, C. G. 1969. *CDC Morbidity and Mortality Weekly Report.* 18:241–42 (July 12) Atlanta, Ga: Center for Communicable Disease Control
81. Walker, E. L. 1908. *J. Med. Res.* 17:379–459
82. Wang, S. S., Feldman, H. A. 1961. *Antimicrob. Ag. Chemother.* 50–53
83. Wang, S. S., Feldman, H. A. 1967. *N. Engl. J. Med.* 277:1174–79
84. Warhurst, D. C., Armstrong, J. A. 1968. *J. Gen. Microbiol.* 50:207–15
85. Wells, R. T. 1911. *Parasitology.* 4: 204–19
86. White, P. C., Jr., Wagner, W. P., Duma, R. J. 1968. *CDC Morbidity and Mortality Weekly Report.* 17: 330 (Sept. 7) Atlanta, Ga: Center for Communicable Disease Control
87. Wilson, D. E., Bovee, E. C., Bovee, G. J., Telford, S. R., Jr. 1967 *Exp. Parasitol.* 21:277–86

THE ROLE OF VIRUSES AS CAUSES OF CONGENITAL DEFECTS

1570

LOUIS W. CATALANO, JR., M.D. AND JOHN L. SEVER, M.D., PH.D.

Department of Neurology, College of Physicians & Surgeons, Columbia University, New York, and National Institute of Child Health and Human Development and National Institute of Neurological Diseases and Stroke, Bethesda, Maryland

CONTENTS

Find out the cause of this effect,
Or rather say, the cause of this defect,
For this effect defective comes by cause
Hamlet

As the entire science of virology has developed, it has exerted a profound influence on virtually all medical and biological knowledge. The concern of virus researchers with infections of the developing fetus and viral teratogenesis has now resulted in a vast body of literature. Although still far from complete, this information has demonstrated the importance of virus infections of the human fetus and the adverse influence of such infections on postnatal physical and mental development.

This review has been written to bring together the presently available information on the role of viruses in the production of congenital defects. In so doing, we have chosen also to explore several hypothetical unifying mechanisms by which viruses infect and damage the fetus.

In the broadest sense, congenital infections can potentially produce three deleterious effects: malformations, abnormal function with or without histologic damage, and latent infection with subsequent induction of disease.

Clearly, certain viruses appear to have a propensity for producing malformations, and for the most part, research effort has explored viral teratogenesis from a gross physical deformity point of view. Recent knowledge has increasingly broadened the role of viruses in producing more subtle fetal effects. Thus, for example, after detailed prospective studies on human populations, one can begin to ascertain which viruses are instrumental in causing mental subnormality without producing clinical illness in either mother or infant. Indeed, investigations of congenital infection with rubella have demonstrated how critical these observations are, and how little data exists. Mental retardation or loss of hearing may be the only manifestation of congenital rubella. If other features of the rubella syndrome were absent, the etiology of the disease would not be recognized unless the proper studies had been performed. In addition, congenital virus infection has been further implicated in diseases manifest from months to years after birth, particularly in relation to various "immunological" diseases of animals, and, perhaps, even neoplasia.

All of these situations, then, may be considered *congenital viral diseases,* whether or not congenital viral infection is apparent by anatomic or histologic study of the fetus or newborn. One can be certain that future investigations will continue to reveal an ever-expanding role of viruses in this field.

MAGNITUDE OF THE PROBLEM

Although infant mortality has declined with improved ante and postnatal medical care, no such decline has been observed in the incidence of congenital malformations. Tragically, congenital malformations have become responsible for an ever-increasing proportion of all infant deaths. Approximately 14 percent (United States) to 20 percent (England and Wales, New Zealand) of all deaths within the first year of life are due to congenital malformations (14, 61, 108).

The number of prospective studies to ascertain the incidence of congenital malformations in large populations of the United States has been limited. In one prospective study of 6000 pregnancies, over 7 percent of newborns were found to be malformed (61). Another study of similar size found 4.7 percent of all liveborn children to have one or more "significant" congenital anomalies (92). A detailed study of 45,000 pregnancies from the Collaborative Project sponsored by the National Institute of Neurological Diseases and Stroke (National Institutes of Health) found a rate of 414 malformations per 10,000 births (4.1 percent). Nearly 0.6 percent of all births had multiple malformations (9).

Table 1 shows the rates of "significant" congenital anomalies found among children after two years of observation in New York (92). These rates are calculated from data presented in the original publication and children with multiple anomalies were counted for each abnormality. The rates probably approximate the total incidence of malformations of a particular

TABLE 1. Rates of Congenital Malformation or Maldeveopment for Various Parts of the Body. (Calculated from data of Shapiro et al (92); with permission of the authors and the *American Journal of Public Health.*)

"SIGNIFICANT" CONGENITAL ANOMALIES DIAGNOSED AMONG CHILDREN AFTER TWO YEARS OF OBSERVATION (NEW YORK)

Site	Rate/1000 Live Births	Examples of Major Anomalies
Skeletal	12.9	Poly or Syndactyly of Fingers and/or Toes, Jaw Deformities, Funnel Chest, Hip Dysplasia or Dislocation, "Clubfoot"
Genitourinary	8.0	Undescended or Ectopic Testicle, Hypospadias, Ureteral Duplication
Cardiovascular	6.2	ASD, PDA, PAS, Tetralogy, Coarctation, VSD, Other
Eye	4.1	Cataract, Micro or Anophtalmos, Ptosis, Nystagmus
Ear	4.1	Hearing Deficiency, Abnormal Position, Size, Shape or Location of Ears.
Face	2.9	Cleft Palate and/or Hare Lip, Absence of Nose, Tooth Abnormalities
Digestive	2.1	T-E Fistula, Pyloric Stenosis, Imperforate Anus
Nervous	2.1	Hydrocephalus, Mental Retardation, Anencephaly, Facial Palsy
Skin	<1.0	Ichthyosis, Dermal Sinus, Webbed Neck
Hernias	18.0	Inguinal, Umbilical, Other
Miscellaneous	4.1	Mongolism, Specific Hematological Abnormalities, Celiac Disease, Dysgammaglobulinemia
TOTAL	65.6	

part of the body for live births since the observation period permitted multiple examinations of the patients studied.

Three difficulties frequently arise in an attempt to determine the "true" incidence of congenital malformations, viz: severity of the defect, clinical recognition of the defect, and the considerable racial and geographic variations in the rate of occurrence of the defect (9, 101) (see Table 2). Overall, however, the world-wide incidence of clearly recognizable (and proba-

TABLE 2. Geographic Variation of Several Easily Recognized Congenital Malformations or Maldevelopments. [United States data from Berendes et al (9) used with permission; data from other countries and "world-wide" from Stevenson et al (101).]

VARIATION IN INCIDENCE OF SELECTED CONGENITAL ANOMALIES IN DIFFERENT COUNTRIES (RATE/1000 TOTAL BIRTHS)

Anomaly	No. Ireland	Egypt	Hong Kong	India	United States	"World-Wide"	Country With Highest Rate
Anencephaly	4.1	3.2	1.2	0.3	0.5	0.9	No. Ireland
Hydrocephalus	1.2	2.0	0.3	0.1	0.6	0.6	No. Ireland
All Neural Tube Defects	10.2	7.9	2.0	0.6	2.9	2.6	No. Ireland
Mongolism	1.1	0.0	0.2	0.0	0.8	0.8	Yugoslavia (3.9)
Polydactyly (Only)	0.4	0.3	0.6	0.4	0.6	1.0	So. Africa (6.2)
Harelip ± Cleft Palate	0.9	0.9	1.4	0.6	1.0	1.0	So. Africa (1.6)
Tracheo-Esophageal Fistula	0.14	0.10	0.10	0.10	0.25	0.17	Kuala Lumpur (0.56)

bly "significant") congenital malformations is at least 7-13/1000 births for a majority of countries (101).

At the present time, no data are available to estimate the relative importance of virus-induced malformations of various organ systems. Probably the most studied in this regard has been malformation or anomalous development of the central nervous system. In spite of the fact that certain viruses can and do produce central nervous system malformations in animals and humans, one recent study failed to demonstrate a correlation between serological evidence for any of the maternal viral infections studied during pregnancy and the occurrence of fetal central nervous system anomalies. Specifically, the number of pregnant women delivering malformed children did not have more antibody or seroconversion to the viruses studied than did the controls (24). In spite of these data, however, genetic studies clearly do not account for all cases of neural tube defects (97), and a major environmental contribution is probably present (101). This is supported by the large socioeconomic and geographic variations of nervous system defects within countries, as well as the low frequency in monozygotic twins of concordance for such defects, in particular for anencephalus (101).

More subtle, perhaps, but possibly more significant in the United States, is the problem of mental subnormality. Approximately 3 percent of the 3.5 million children born per year in the United States at the present time will be mentally retarded (83a). Although only 10 percent of cases with mental retardation can be clearly attributed to infectious diseases with our present state of knowledge (22), this obviously represents a large number of individuals.

INFECTION OF THE FETUS

The etiology of congenital defects involves two fundamental mechanisms, genetic and environmental. In this article we shall emphasize only the latter in regard to viruses, understanding that genetic susceptibility to infection or the manifestation and expression of disease may involve genetic-viral interaction. A recent publication has extensively reviewed the pathogenesis of viral infections of the fetus (69).

Basically, three principles are involved in the production of congenital viral diseases (*a*) the ability of the virus to infect the pregnant animal and be transmitted to the fetus; (*b*) the timing of infection in relationship to the stage of gestation; and (*c*) the nature of the virus and its capability of producing disease in the fetus or producing fetal damage indirectly, due to infection of the mother, e.g., fever. A schematic representation of the sequence of events which may occur prior to and following viral invasion of the fetus is shown in Figure 1. To establish a proper perspective, it is quite likely that an unaffected normal child is the most frequent outcome of maternal viral infection.

Infection of the Pregnant Animal and Transmission of Virus to the Fetus

The pregnant animal must be capable of being infected (or reinfected) in order for fetal infection to occur. Immunity gained by prior infection or immunization of the mother may protect the mother as well as the fetus. For example, pre-existing rubella antibody in a pregnant woman appears to prevent infection of the fetus completely upon subsequent exposure of the mother to rubella. Another potential situation exists, however, in that the fetus may become infected with a virus which has been inapparent or latent in either the tissues of the gravid animal or in the egg or sperm. Evidence has been accumulated demonstrating that vertical transmission of lymphocytic choriomeningitis (68), avian leukosis (21), and probably murine leukemia (33), may occur via the egg in the presence of persistent infection in the ovary. Infected ova have also been suspected of transmitting an agent responsible for spontaneous autoimmune hemolytic anemia in New Zealand

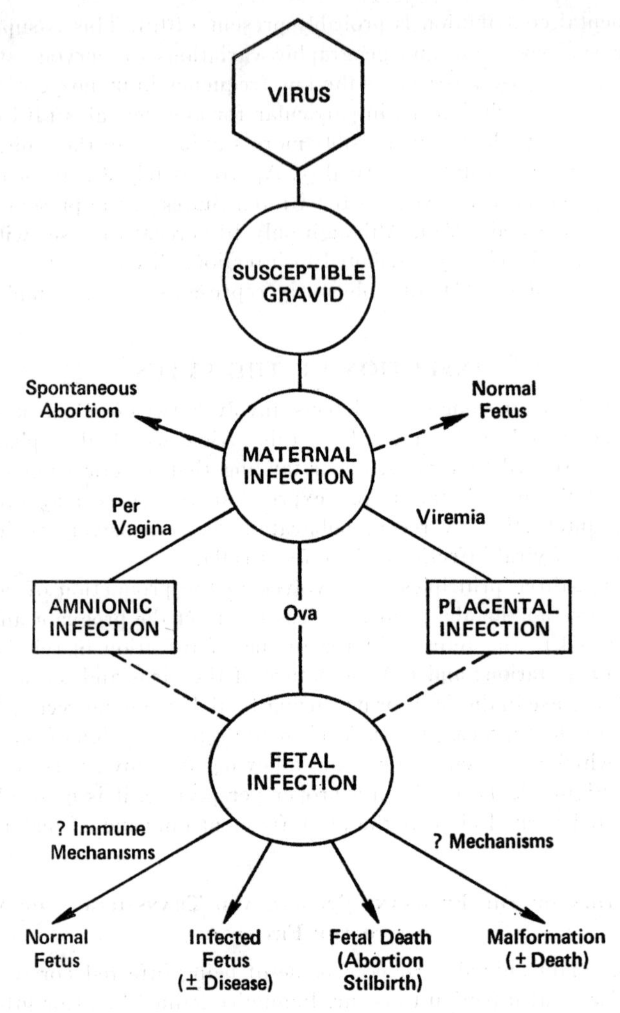

FIGURE 1. Pathogenesis of viral invasion of the fetus and fetal outcome.

Black (NZB) mice, although a transplacental factor has recently been incriminated (62).

As presently recognized, the important routes by which viruses invade the fetus are transplacental by hematogenous spread from the mother, transamniotic by invasion through the endocervical canal, as well as less common routes, such as transovarian (69).

Most of the well recognized in utero viral infections appear to occur transplacentally after hematogenous dissemination of the virus. The placenta presents a physical barrier to the transmission of virus to the fetus. In certain instances, virus probably replicates in placental cells, thereby establishing a focus for subsequent seeding of fetal tissues; alternatively, virus may cross directly to the fetus via infected leukocytes. With respect to rubella, virus may be isolated solely from the placenta, or from both the placental and most fetal organs (4, 72). Thus, placental infection appears to precede fetal infection but may not necessarily result in invasion of the fetus. Placental replication of the virus may not be necessary, however, if the agent is sufficiently small or is of a particular strain, e.g., rat virus (48).

Timing of Infection During Gestation

The gestational age of the pregnancy appears to alter the frequency of infection of fetal tissues with rubella (4), suggesting, at least with this virus, that a maturation of the placenta (or possibly the alteration of fetal tissue susceptibility) may account for the decreased incidence of fetal infection in the last two trimesters of pregnancy. It is well known that *Treponema pallidum* crosses the placenta only during the latter two-thirds of pregnancy. Similarly, equine rhinopneumonitis virus (probably a herpesvirus) appears incapable of crossing the placenta early in gestation. However, if infection occurs in the latter third of gestation, the fetus may develop generalized disease and die in utero or in the neonatal period (12).

Although *Herpesvirus hominis* (HVH, herpes simplex) has occasionally been isolated from the placenta of newborns with disseminated disease (28, 113), infection of the fetus probably occurs during passage through the birth canal at the time of delivery when maternal genital herpetic infection is present (76). However, ascending vaginitis with transamniotic fetal infection might also occur in certain cases (113). Since pregnant women have been known to excrete cytomegalovirus in their urine (27), it is entirely possible that infection of the newborn with this virus might also occur in a manner similar to that postulated for HVH.

The specific malformations which are clinically evident following fetal infection appear to be related to the stage of fetal development and the tissues affected. This is best documented for rubella and is considered in greater detail later. There is some evidence with rat virus (104) and H-1 virus (84) that cells which are rapidly dividing may be more susceptible to virus infection. This may, in part, account for the organ or tissue tropism of certain agents (see below) and the lesions produced by these viruses. In other instances, viral tropism may be more related to a particular stage of organogenesis or the development of immunological capabilities of the fetus.

The Nature of the Virus and its Ability to Invade and Damage the Fetus

There are at least five possible mechanisms by which viruses may exert adverse effects on the developing fetus. These include nonspecific placental interruption, generalized placental and fetal infection, localized fetal infection, chromosomal injury, and delayed (postnatal) disease (Table 3).

TABLE 3. Hypothetical Mechanisms of Congenital Viral Diseases.

THEORETICAL ADVERSE EFFECTS OF IN UTERO VIRAL INFECTION

Mechanism	Results	Possible Examples
Placental Interruption ("Nonspecific-Toxic")	Increased Frequency of Abortions and Stilbirths; Prematurity	Any Severe Viral Infection, Rubeola, Varicella
Generalized Placental and Fetal Infection		
–With NECROSIS	Severe Neonatal Infection and Death; Fetal Reabsorption. Possible Malformation	Herpesvirus, Vaccinia Variola
–Without NECROSIS	Alteration of Most Organ Functions and Embryogenesis. Malformation, Increased Fetal Mortality	Rubella, Cytomegalovirus
Vasculitis	Placental Insufficiency. Organ and Tissue Hypoxia with Hypoplasia and ?Malformation	Rubella, ?Cytomegalovirus and Herpesvirus
Localized Infection (Virus-Tissue "Specificity")	Isolated Organ System Disease, ± Malformation	Hepatitis, Poliovirus, Western Equine Encephalitis, ?Coxsackie B
Chromosomal Injury		
–Breaks, Pulverization	Decreased Mitoses, Organ Hypoplasia. Increased Abortion and Stillbirth	?Rubella , Rubeola, Herpesvirus
–Abnormal Mitoses	Aneuploidy, Deletion, Mosaicism	?Mongolism,
Delayed (Postnatal) Disease (Normal Intrauterine and Newborn Development)	Abnormal Postnatal Cell Migration or Differentiation	Rat Virus, H_1 Viruses
	Latent Infection with Subsequent Disease (± Immunological)	Lymphocytic Choriomeningitis, ? NZB Autoimmune Hemolytic Anemia
	Carcinogenesis (Viral Genome Incorporation Into Fetal Cells with Subsequent Tumor Induction)	Avian Leukosis, Murine Leukemia, ? Human Leukemia

Placental interruption.—Numerous mechanisms may result in abortion, stillbirth, or prematurity in the presence of maternal viral illness without infection of the fetus per se. Fever, circulating toxins, or changes in placental circulation have been proposed as nonspecific mediators of placental interruption in severe maternal virus infection. However, very little accurate data exist on this point.

Localized versus generalized fetal infection.—There are two basic patterns of fetal disease induced by viruses. The first pattern is one in which the virus produces an illness resembling the disease seen in the adult. The second is a distinctively fetal one in which the type of disease is unique and characteristically different from that of the adult.

The adult pattern diseases occur in occasional instances of in utero infection and generally appear to be due to viruses which have a propensity for infecting a particular organ. For example, poliovirus (25) and Western equine encephalitis (93) have been reported as producing central nervous system disease of the newborn following earlier maternal illness. Although studies have failed to demonstrate transmission of maternal infection with hepatitis to the newborn (1), a number of retrospective case reports exist in which neonatal hepatitis was found in infants born to mothers who had serum hepatitis during pregnancy (55, 102).

The fetal pattern viral diseases are caused by viruses which usually have limited pathogenicity in the adult, most notable rubella, cytomegalovirus and herpesvirus. Presumably this pattern of disease occurs because embryonic and fetal tissues are particularly susceptible to certain virus infections. Thus, if fetal infection occurs with one of these viruses, a characteristically generalized infection takes place which usually involves many organs, either by direct virus invasion or by secondary effects. If a virus is quite lytic, considerable tissue necrosis and damage will frequently occur without necessarily causing malformation. Herpesvirus, vaccinia, and variola are well known examples of highly lytic viruses which occasionally can produce severe intrauterine or neonatal infection and death. Viruses such as rubella and cytomegalovirus (CMV), which produce lesser degrees of cell lysis and tissue necrosis, on the other hand, may increase the frequency of fetal death, but more characteristically produce widespread malformation with variable degrees of tissue damage. These effects are presumably secondary to alteration of organ function and interference with embryogenesis.

The extent of fetal infection and the degree of damage produced is obviously spectral in nature and depends on many factors, most of which are not known. Although placental or fetal infection, or both, with either lytic or nonlytic viruses may be present, disease may be inapparent or subclinical in both the mother and the newborn. For example, herpesvirus (HVH) infection of the uterine cervix is frequently inapparent; newborn infection with this virus may occasionally result in only minimal skin lesions.

Because of a virtual absence of large, prospective virologic studies of specific infection of fetal membranes and placental tissue in man, the natural history and frequency of viral infections, and the extent of disease which is produced, is still largely unknown. To investigate this problem we recently analyzed data obtained from our laboratory study of approximately 2500 placentas collected prospectively following the 1964–65 rubella epidemic (see Table 4). The placentas were collected consecutively beginning five to seven months after the peak of the epidemics of each study location. Specimens were then obtained over a seven- to nine-month interval. In Los Angeles, where no clear epidemic occurred, specimens were collected for only a three-month interval. Of the 2488 placental homogenates cultured on African green monkey kidney cell cultures, 78 showed the presence of an interfering agent by inhibiting the cytopathogenic effect of Coxsackie A9. Among the 78 blocking agents isolated, 59 (75.6 percent) were tested with

TABLE 4. Isolation of Rubella Virus From Placentas Collected Prospectively Following the 1964–65 Epidemic.

City	Dates of placental collection	No. placentas cultured	No. blocking agents found	% "Agent" isolation	No. agents neutralized	No. rubella +	% Rubella + placentas
Memphis	12/64–6/65	317	22	6.9	20	2	0.63
Baltimore	9/64–5/65	417	10	2.4	5	3	0.72
Boston	11/64–5/65	759	22	2.9	12	3	0.40
Portland	11/64–7/65	318	10	3.1	10	1	(0.31)
	8/65–3/66	255	12	4.7	11	3[a]	1.18
Los Angeles	2/65–4/65	422	2	0.5	1	1	(0.24)
	TOTAL	2488	78	3.14	59	13	0.52

[a] One positive specimen obtained from one twin placenta (other twin placenta negative).

rubella antiserum to identify the positive isolation of this virus. A total of 13 specimens (22.0 percent of the blocking agents tested) were neutralized and confirmed as rubella virus. This established the minimum isolation rate of rubella from placentas during an epidemic year in this population as 0.52 percent. If one assumes that of all blocking agents isolated, the same percentage of proven rubella isolates were to be found in the other specimens not tested with rubella antiserum, approximately 0.69 percent of placentas would contain rubella virus during the epidemic year. Of the 14 infants (1 set of twins) which were identified as having rubella-positive placentas, one child died at the age of five months of congenital rubella. Another child has suspect neurological dysfunctioning at four years of age, but is otherwise normal. The overall rate of abnormality attributable to rubella in children having placental rubella isolates was therefore approximately 15 percent.

This study demonstrates four interesting points. Only two (15.4 percent) of the 13 mothers studied had a history of infection or exposure to rubella, yet all had virus isolated from their placentas. Secondly, there was an exceedingly high isolation rate of rubella from placentas (at least 1 out of 200 placentas cultured) following an epidemic. Third, there was a persistence of virus in the placenta to term, in spite of the presence of maternal antibody in all of the women; and fourth, the incidence of detectable abnormalities was relatively low (15 percent) even after four years of observation of these children.

Vasculitis and teratogenesis.—Many possible mechanisms have been proposed as explanations for the teratogenic capabilities of viruses. At the present time, there are only limited data to support any unifying hypothesis. However, generalized vasculitis of the placenta and fetus is a particularly appealing hypothetical mechanism which may prove to be an important cause of some types of congenital malformations.

Disseminated endothelial lesions with varying degrees of sclerosis or stenosis of large elastic blood vessels (including the aorta), as well as vasculitis of cerebral arteries, have been described in cases of congenital rubella (96). Vascular lesions of the cerebral circulation produce multiple microinfarcts of the cortex and lead to small foci of calcification (96). By deduction, this may be the mechanism by which intracerebral calcification so characteristically takes place in congenital toxoplasmosis and cytomegalic disease, as well as in herpesvirus type 2 malformations (99). Defective clotting tests indicative of disseminated intravascular coagulation have been clearly demonstrated in severe neonatal viral disease (7, 15a, 41). Intravascular coagulation may be manifested by clinical signs of a hemorrhagic diathesis such as thrombocytopenia and purpura. This mechanism could account for similarities in the clinical manifestations of several congenital infections (rubella, cytomegalovirus, toxoplasmosis, herpesvirus). Thrombosis and vasculitis of small blood vessels can produce organ hypoxia secondary to alterations in the microcirculation. Secondary hypoxia may then result in organ hypoplasia due to decreased protein synthesis and reduced rate of cellular division. In the case of the brain, cerebral hypoplasia can be manifest by microcephaly. Infants with congenital rubella are retarded in size due to retarded intrauterine growth, probably due to a reduced number of cells in many organs (73). This generalized hypoplasia has been attributed to a direct effect of the persisting virus-cell relationship; only limited data exist to support this concept in preference to hypoplasia induced by hypoxia.

Widespread vasculitis of placental vessels also exists in conjunction with disseminated virus infection of the fetus. Thus, adverse fetal effects could simultaneously be produced by placental insufficiency resulting in organ and tissue hypoxia, as well as from direct effects of virus on fetal cells. Abortion in mares due to equine viral arteritis virus, for example, may be due to placental infection without fetal involvement (69). A similar mechanism has been proposed for some malformations found in congenital rubella (73).

Chromosomal injury.—The possibility that viruses produce chromosomal injury in utero and thereby induce abortion or malformation remains speculative, primarily because of the limited data at the present time. Although chromosomal defects are found with high frequency in tissues cultured from abortions, few prospective studies have reported a relationship between virus infection of fetal tissues and abortion per se. Our laboratory has cultured 100 products of spontaneous abortion with isolation of cytomegalovirus in only one instance. Therefore, viral infection of the fetus appears to have little, if any, relationship to the frequent occurrence of spontaneous abortion, and presumably bears no relationship to the frequent chromosomal abnormalities which have been demonstrated in abortions.

However, since approximately 0.5 percent of newborns have gross chromosomal defects (86, 107) it is worthwhile considering the limited evidence that viruses may produce chromosomal injury and thereby induce congenital defects. Many viruses induce various kinds of chromosomal damage in vivo and in vitro (78, 80), although most of the damage is relatively nonspecific and has not been conclusively associated with any disease or anomaly. Rubella virus has been shown to inhibit mitosis and increase the number of chromosomal breaks in human embryonic cell cultures. It has been hypothesized that these effects may be factors in the production of some of the malformations observed in congenital rubella infection (82). Abnormal mitoses in vitro due to herpesvirus (13) and rubeola (78), and in vivo (rubeola, 78) have also been described. A recent article by Nichols has reviewed the literature on virus-induced chromosomal abnormalities (78a).

Significant differences in the distribution of patients with chromosomal aberrations (including Down's syndrome and sex chromosome abnormalities) in time and geographic area have been observed, suggesting a possible infectious agent (19). Another study showed that the geographic distribution of patients with mongolism was similar to that observed for hydrocephaly (16), although hydrocephaly and other central nervous system malformations have not been consistently associated with any chromosomal defect, excepting the holoprosencephalies (29, 81). Several studies have also shown a clustering of epidemics of infectious hepatitis and subsequent cases of mongolism. However, more recently other workers have not found such an association (6, 52). It is interesting to note that Australian antigen (probably a virus) has now been associated with serum hepatitis and appears with increased frequency in institutionalized patients with mongolism. However, since mongoloid patients not institutionalized fail to have an increased frequency of this antigen, it is unlikely that serum hepatitis is an etiologic factor in mongolism (103).

Inapparent and latent virus infection.—Three groups of delayed viral effects appear to be emerging at the present time in relationship to vertically transmitted virus infection: (*a*) abnormal postnatal morphogenesis (rat virus and H-viruses); (*b*) latent infection with potential induction of subsequent diseases, frequently immunological (lymphocytic choriomeningitis virus, spontaneous hemolytic anemia of NZB mice); and (*c*) carcinogenesis (avian leukosis, Gross leukemia). The concept that a virus may infect a fetus without producing recognizable disease in the newborn period is not new. However, recent evidence that a virus may remain inapparent or latent in various organs for variable periods of time, subsequently producing a clinical illness, has generated considerable speculation as to the importance of such an event in human disease.

A number of postnatal events occur in the development and maturation of the brain. Congenital or early postnatal infection with rat virus (Kilham) and feline panleukopenia virus interferes with cellular division and migration in the external granular layer of the cerebellum, producing cerebellar hypoplasia (48, 50). Although there is a type of cerebellar disease in man analogous to this, no associations with these or other viruses have been demonstrated. Recently, similar findings have been reported on neonatal rat infection with lymphocytic choriomeningitis virus and possibly other arenoviruses (72a).

Persistent infection with or without the production of infectious virus can clearly occur in utero. This has been demonstrated for a number of viruses, most notably lymphocytic choriomeningitis (LCM) virus, mouse leukemia, and avian leukosis. As previously mentioned, evidence indicates that all three of these agents may infect the ovum or developing fetus. In the case of LCM, congenitally infected animals display a persistent tolerant infection throughout life with somewhat retarded growth and variable degrees of reproductive failure (69). If mice are infected neonatally with LCM, the virus also persists but the animals appear to manifest an accelerated aging process (43). Recent evidence has shown that gnotobiotic, congenitally LCM-infected mice, although growing normally, display an immunopathologic dyscrasia resembling lupus erythematosis in man (83). The autoimmune hemolytic anemia of NZB mice and the lupus-like syndrome of NZB/W F_1 mice also appear to be related to transplacental infection with a murine leukemia-like virus (62).

Recent progress in viral oncology has increased the anticipation that some human malignancies will be found to be virally induced. At the present time only fragmentary evidence suggests that a common infectious mechanism with variable expression of disease may be responsible for human leukemia. Although genetic factors have long been known to greatly influence the incidence of leukemia in mice, the expression of disease involves genetic susceptibility interacting with one of several (known) leukemia viruses. In man, radiation exposure and chromosomal abnormalities (e.g., mongolism) are associated with increased rates of leukemia (67). In the study of large sibships or close relatives of leukemic patients, increased familial incidence of leukemia is observed (35, 66). Furthermore, major congenital defects (other than mongolism) appear more frequently among leukemic patients as well as in their siblings (66). Time-space clustering of childhood leukemia and significant seasonal variation in onset of leukemia in adults (34) suggest an infectious agent. Since a viral etiology is firmly established for all well-studied animal leukemias, and in utero transmission of leukemia viruses has been demonstrated, it seems highly probable that similar relationships with agents and mechanisms will be found for man. Recently a hypothesis has been proposed that genetic material from all animal species may carry information for producing tumor viruses (45).

One possible difficulty in future investigations of the role of latent and inapparent infections in man may be the problem of tolerance. Congenital or neonatal infection of mice with Gross leukemia virus induces specific immunological tolerance to the virus, making animals unable to respond to subsequent attempts at immunization with this virus (as contrasted to non-tolerant adults) (51). This phenomenon has also been demonstrated in animals congenitally infected with LCM virus (69). Thus, one may speculate that congenital or newborn infection in humans may, in some instances, also result in specific immunological tolerance to the agent. Therefore, later attempts to identify infection in such instances, using known specific serological techniques, could prove to be misleading, due to reduced or absent circulating antibody levels (as presently measured). Altered immunoglobin synthesis does occur in congenital rubella, although most infants with this disease have normal immunoglobulin levels by one year of age (98). Persistent dysglobulinemia characterized by low levels of IgG with or without elevation of IgM has been reported in rare instances (46, 98). Future studies may produce better evidence of immune tolerance to congenitally acquired viruses in man.

Summary

In summary, whether the embryo or fetus continues to mature and be delivered viable depends on the extent of tissue destruction or alteration of function, and the mechanisms by which the fetus alters viral replication. Thus, if viral infection of the mother takes place, the pregnancy may be altered in the following ways: (*a*) abortion with or without fetal infection; (*b*) fetal death and later abortion; (*c*) teratogenesis; (*d*) nonfatal infection with disease possibly being recognized after birth if any stigmata of in utero infection remain; or lastly, (*e*) the fetus remains unharmed (23). (Refer to Figure 1.)

FREQUENCY OF CLINICAL VIRAL INFECTIONS DURING HUMAN GESTATION

Data concerning the epidemiology of infections during pregnancy in humans has recently been reported from the Collaborative Perinatal Research Study. In general, attack rates have depended upon the population under observation, influenced by such factors as geographic location, season of the year, and occurrence of epidemics. The age and racial distribution of the population sample have also been found to influence susceptibility to infection and reflect multiple variables such as housing density, hygiene and sanitation, diet, travel, etc. From clinical studies conducted on 30,000 pregnant women during the period from 1959 to 1962, 4.8 percent were complicated by a single infection, 0.4 percent by two, and 0.03 percent by more than two infections (91). Upper respiratory infections accounted for 58 percent of the illnesses. Herpesvirus hominis (herpes simplex) infections occurred at the rate of 12:1000; laryngopharyngitis, 3.2:1000; gastroenteritis, 1.7:1000; mumps, 1.5:1000; rubella, 1.1:1000; varicella-zoster, 0.7:1000; rubeola, 0.4:1000; and conjunctivitis, 0.4:1000. Viral pneumonitis, infectious mononucleosis, pleurodynia, and hepatitis were reported less frequently. It is of interest that during nonepidemic periods, rubella occurs at the rate of one to three cases per 1000, whereas during the 1964 epidemic, there were 22 to 30 cases per 1000 (91).

Subsequent serological studies of clinically reported cases confirmed approximately 66 percent of the rubella cases, less than 20 percent of the rubeola cases (30 percent of these actually were rubella), and about 70 percent of mumps and varicella-zoster cases (111). Further analysis of these data in relation to the epidemiology of pregnancy outcome is now in progress.

Detection of Infection

It is beyond the scope of this review to consider all the immunological responses to perinatal infection. However, current studies have indicated the value of understanding these responses, since they may provide a means by which in utero viral infection may be detected. A review of this problem has recently been compiled at a conference held at the National Institutes of Health (87). We have elected to review briefly only the current status of IgM macroglobulin in perinatal viral infections.

IgM macroglobulin levels have been found to be elevated in a number of congenital infections including rubella, cytomegalic inclusion disease, herpes neonatorum, neonatal varicella, toxoplasmosis, syphilis, and various other bacterial diseases (87). Infections in the newborn may also result in significantly elevated levels of IgM in subsequent serum specimens (5). However, normal IgM levels have usually increased to a point by 6 to 12 months of age where significant elevations cannot be recognized. A large survey of IgM values in a low-income population (3) revealed elevated levels in 10 percent of the cord sera. Over 60 percent of these were falsely elevated because of contamination with maternal blood. Of the remaining 4 percent of the newborns with elevated IgM levels, infections (mostly subclinical) were detected in 34 percent, as contrasted to an infection rate of 0.8 percent in controls with normal IgM values. Nearly one-half of the diagnosed infections in the presence of elevated IgM levels were due to viruses, most frequently cytomegalovirus. Urinary tract infections accounted for the most frequent bacterial disease with elevated IgM values. Importantly, silent central nervous system involvement may be a frequent factor in congenitally infected, asymptomatic newborns with increased ($>$ 19.5 mg percent) IgM levels (3). Elevated protein values and pleocytosis of the cerebrospinal fluid were initially found in such children (1 per 200 deliveries), some of which had developed signs of central nervous system damage during the first year of life. Other studies of IgM in the neonatal period have not shown such dramatic effects in clinically well infants (54, 65). For example, Korones et al (54), studied 335 infants and found that 7.8 percent had elevated IgM levels, but only 1.2 percent (4 patients) were not falsely elevated. These four infants appeared to be normal. When specific clinical syndromes were studied (conjunctivitis, pneumonia, septicemia, diarrhea, and oral moniliasis), IgM elevations generally occurred several days after onset of physical signs which could be recognized by careful clinical surveillance.

Another recent observation has suggested the importance of abnormally low levels of IgM. Approximately half of the perinatal deaths at Johns Hopkins in the Collaborative Project population occurred among infants with very low IgM values; furthermore, 8 of 17 children with severe congenital malformations had cord IgM levels below 1.6 mg percent, including three of four anencephalic infants in the study (40). This would suggest the possibility that abnormally low IgM levels may be intimately related to, or caused by, an unexplained etiological agent, particularly in infants with anencephaly.

VIRUSES WITH ESTABLISHED OR PROBABLE ROLE AS TERATOGENS

At the present time, only five viruses can conclusively be said to induce congenital defects under natural (i.e., not strictly experimental) conditions. Two of these are in humans, rubella and cytomegalovirus. The other three are in animals, hog cholera, blue tongue, and the H-1, RV group of agents. Recently, evidence has indicated that *Herpesvirus hominis* (herpes simplex) and bovine viral diarrhea-mucosal disease virus have probable teratogenic capabilities and are also included here. Each of these agents is briefly considered below, particularly in light of recent information.

Teratogenic Viruses of Humans

Rubella.—Although rubella was first recognized as a clinical entity by Maton in 1815 (59), congenital malformations were not reported until Gregg published his classic paper in 1941 (30). Initially noting the association of maternal rubella with cataracts, a subsequent paper in 1945 reported that of 130 children with a history of maternal first trimester rubella, 111 were deaf, 38 had heart disease, and 23 had eye defects (31). Since the successful isolation of rubella virus in 1962 (79, 90, 110), a tremendous amount of knowledge has accumulated in regard to rubella as a teratogenic agent in animals and man. As a result, congenital rubella has become the prototype of viral-induced malformation.

It is well recognized that the frequency and type of defect depend on the time of maternal infection and gestational age of the fetus. The majority of presently recognized abnormalities occur with maternal infection in the first trimester (Fig. 2). Approximately 50 percent of such mothers produce abnormal children if infected in the first month of pregnancy, 22 percent in the second month, and 6–8 percent in the third month through the fifth month (20, 58, 63). In addition to deformities, spontaneous abortion may occur in 10 to 15 percent of rubella-affected pregnancies (95). Prior to the 1964 epidemic in the United States, congenital rubella syndrome was considered a constellation of structural anomalies, principally those of deafness, blindness and cataracts, congenital heart disease, microcephaly, and mental

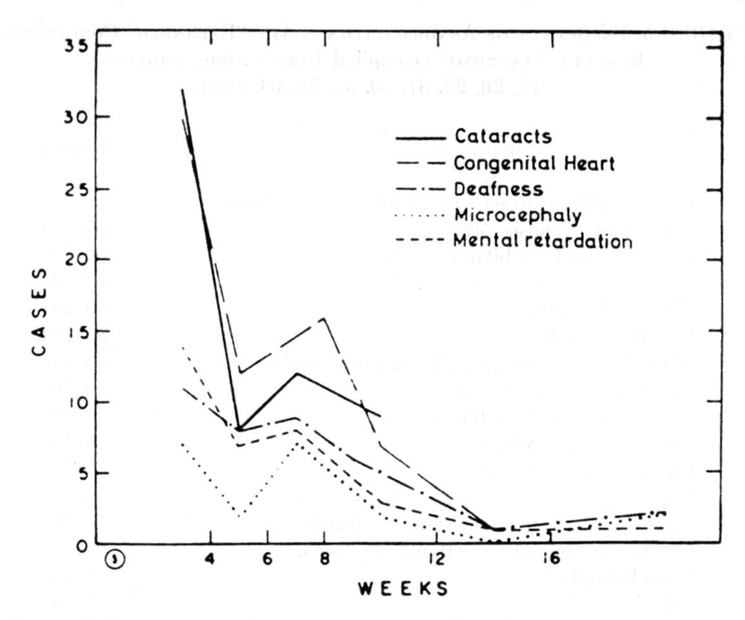

FIGURE 2. Frequency of abnormalities in 108 offspring related to time of maternal rubella [Dekaban et al (20); used with permission of the authors and the editors of *Neurology*].

retardation. With the 1964 epidemic, however, a number of new features were described constituting the "expanded" rubella syndrome (53). These included thrombocytopenia and petechiae, pneumonitis, hepatitis, hepatosplenomegaly, myocarditis with focal necrosis, encephalitis, anemia (occasionally hemolytic), and radiolucent lesions of the long bones. It has recently been suggested that this list also include serum immunoglobulin abnormalities (98). Table 5 summarizes the relative frequency of abnormalities in the presently recognized syndrome. Studies have demonstrated that these diverse clinical features were indeed evident in infants delivered prior to 1964, and therefore, the 1964 epidemic cannot be attributed to a "supervirulent" virus (112).

It is now clear that fetal infection following rubella in the second trimester produces more subtle damage than infection earlier in gestation. In a recent study, 24 women with clinical and laboratory evidence of rubella in the 14th and 31st weeks of pregnancy delivered 22 liveborn infants, 15 of which (69 percent) were suspected to be abnormal. Two-thirds of the abnormalities were problems of communication based on hearing loss, motor-mental retardation, and eye defects. Nearly 50 percent of the abnormal children had elevated IgM levels (39). Although these findings were present in mothers with clinical evidence of rubella, similar findings have been noted

TABLE 5. **Frequency of Abnormalities in the "Expanded" Congenital Rubella Syndrome (compiled from various sources— 17, 20, 23, 31, 39, 53, 58, 63, 112).**

Abnormality	Percent
Congenital heart disease (patent ductus arteriosus pulmonic stenosis, other)	60–100
Cataract (nuclear; bilateral in 60%)	60–70
Deafness (severe sensory-neural)	50–70
Hepatosplenomegaly	50–70
Low birthweight	50–70
Thrombocytopenia (petechiae & ecchymosis)	50–70
Psychomotor retardation	30–50
Radiolucency of long bones	30–50
Myocardial necrosis, focal	30–40
Interstitial pneumonitis	30–40
Jaundice	10–20
Excess fetal wastage and neonatal death	10–15
Chorioretinitis, micropthalmia, glaucoma, iris hypoplasia	5–10
Anemia	5–10
Abnormal cerebrospinal fluid, encephalitis	5–10
Delayed dentition & hypoplastic enamel	5–10
Total abnormal	70–90

in children of women with inapparent rubella after the first trimester of pregnancy (40, 70).

After birth, children with congenital rubella frequently remain infected for variable periods of time. By one year of age, fewer than 10 percent of children still have detectable virus in the nasopharynx (18); however, the virus may persist much longer in the spinal fluid and lens of the eye. At least 90 percent of children with congenital rubella have neutralizing (NEUT), hemagglutination-inhibition (HI), and fluorescent (FA) antibodies in their sera (Fig. 3) which apparently persist for life in most instances (88, 89). Although initially elevated, complement fixation (CF) antibody gradually declines over the first few months of life. The acquired infection (Fig. 4) differs somewhat in antibody response. Shortly after the onset of rash, NEUT, HI, and FA antibodies appear with CF antibody rising slightly later. All of these antibodies persist indefinitely except CF which tends to decrease and disappear in about 50 percent of patients within 10 to 20 years.

Although the IgM macroglobulin response in congenital infection has been previously discussed, it should be noted that IgM concentrations in excess of 20 mg percent has been found in only 18 percent of rubella-infected infants in our study population (60).

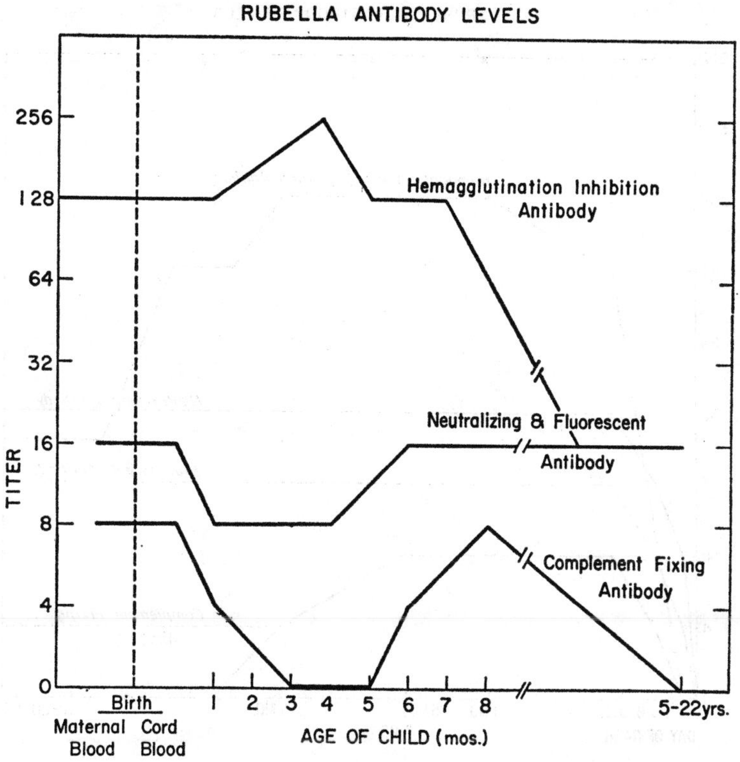

FIGURE 3. Antibody response of children with congenital rubella.

The recent development and licensing of live attenuated rubella vaccine in this country has afforded the opportunity of potentially eradicating congenital rubella by achieving and maintaining sufficient immunity throughout the population.

Cytomegalovirus.—The significance of severe cytomegalovirus (CMV) infection in the fetus and neonate is well appreciated by many physicians, particularly those in pediatrics. The gross manifestations of cytomegalic inclusion disease are characterized by the presence of microcephaly or hydrocephaly, chorioretinitis with micropthalmia and blindness, encephalitis with intracerebral calcifications, mental retardation and seizures, and hepatosplenomegaly. Occasionally pneumonia, anemia, thrombocytopenia with petechiae, jaundice, congenital heart disease, and kernicterus are present. Most cases were originally diagnosed at autopsy. However, it is now known that the prevalence of undiagnosed and even inapparent CMV infection is much greater than previously realized. Because of the large number of re-

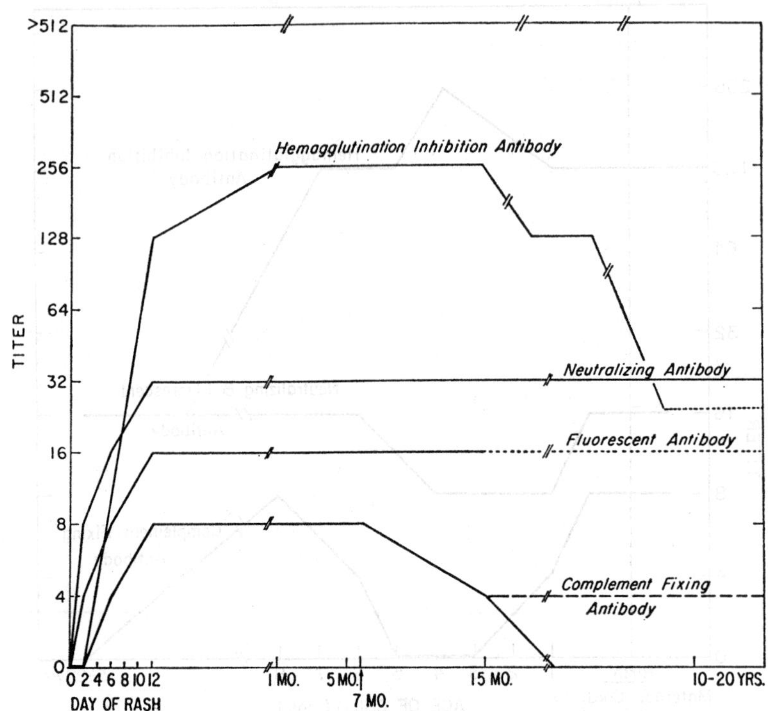

FIGURE 4. Antibody response of adults to acquired rubella infection.

views on this subject (36, 37, 100), the present discussion will consider only the most recent information.

A large serological survey has shown that the initial development of significant levels of antibody to CMV occurred in 3.5 percent of pregnant patients (91). Further studies have demonstrated that about 3 percent of pregnant women excrete virus in their urine (27) and that about 1 percent of newborns excrete cytomegalovirus in their urine at birth (10, 38). At present, an unknown percentage of such infants have minimal abnormalities such as transient petechiae, hepatosplenomegaly (10, 27) and later development of spasticity or mental retardation (10, 99a, 109). A recent report also suggests an association of this virus with inguinal hernias in males (56). Those newborns which develop frank cytomegalic inclusion disease (CID) are more likely to have low birthweight (38). About 15 to 20 percent of healthy, full-term infants excrete CMV in the first year of life, beginning at three months of age (57). Although excretors have lower birthweights as a group, growth and development may be completely normal (57).

The most sensitive available method for detecting congenital CMV infection is virus isolation (38). Detection of CMV complement fixation and neutralizing antibodies is more difficult and less reliable. Examination of the stained urine sediment shows a high rate of false negatives in the presence of frank CID (109). Specific CMV IgM antibody assay by indirect immunofluorescent microscopy detects most symptomatic children but is less reliable in detecting infected, but asymptomatic, infants (38).

As with rubella, CMV infection may persist from months to years, even in the presence of specific neutralizing antibody. At present, no specific therapy is available.

Herpesvirus.—Herpesvirus hominis (HVH) infection of older children and adults is usually subclinical or associated with recurring, localized infections, generally of the lip. Recent studies have clearly indicated that two strains of HVH exist, type 1 ("oral") and type 2 ("genital"). These strains differ considerably in their antigenic and biologic properties (74). A series of 28 cases of neonatal herpetic infection found that 25 were associated with type 2 herpesvirus (75). The finding of this strain of herpesvirus as the etiologic agent in herpes neonatorum is best explained by the fact that about 95 percent of strains isolated from the genital tract are type 2 (74).

The spectrum of clinical disease in the newborn may range from only a few cutaneous vesicles to fatal, generalized disease with cyanosis, jaundice, fever, hepatosplenomegaly, encephalitis, circulatory collapse, and death. Infection with the virus probably occurs in the great majority of cases at the time of delivery during passage through the birth canal; herpetic genital lesions may be present in the mother. Alternatively, transplacental or transamniotic infection prior to birth or intimate contact in the early newborn period with family members infected with oral-labial lesions are other possibilities which have been reported.

Evidence has now accumulated which indicates that genital infection of the mother with HVH in early or possibly mid-pregnancy may result in a congenitally malformed infant. Three such cases have been published (2, 85, 99) and two others exist (11, 15). All cases have had microcephaly. Micropthalmia, retinal dysplasia, intracerebral calcification, cutaneous vesicles, seizures, and mental retardation have been frequent but variable features. Infection was due to type 2 HVH in three patients. Infants which have survived the more typical neonatal disease have been observed to have microcephaly, chorioretinitis, and mental retardation (44, 75). It is also interesting to note that central nervous system damage and gross eye and auditory vesicle defects have been reported in early chick embryos infected with HVH (42).

These reports suggest that HVH has a teratogenic potential for man, which raises the possibility that a fetus may be at considerable risk for central nervous system damage when maternal HVH (type 2) infection occurs during any stage of gestation.

Present evidence indicates that transplacentally acquired type 1 or type 2 antibody does not protect infants from becoming infected with type 2 virus (75).

Treatment of HVH infection of the newborn is still experimental. Delivery of an infected mother by cesarean section has been proposed, but not proven, to avoid newborn infection. Systemic 5-iodo-2′-deoxyuridine (IUDR) has been tried without impressive results (106); furthermore, type 2 HVH is much more resistant to the action of IUDR in vitro than the type 1 strain. Although a synthetic inducer of interferon has been used to treat an older infant with HVH encephalitis (8), no newborn has received such treatment.

Teratogenic Viruses of Animals

Virtually no information is available concerning the incidence of most virus infections during gestation in animals. Only hog cholera virus and blue tongue virus, the parvoviruses (rat virus and H-viruses), and possibly bovine viral diarrhea-mucosal disease virus, have been shown to be teratogenic in animals.

Hog cholera virus.—It is now well established that abnormal swine fetuses may result from immunizing dams with rabbit-modified hog cholera virus (HCV) vaccine in early pregnancy. More limited evidence also indicates that natural outbreaks of swine fever may also result in disease of the fetus.

Early studies with HCV demonstrated a marked increase in fetal deformities of the nose, ears and legs, malformed kidneys, ascites, edema, and fetal or newborn death following injection of pregnant sows with modified HCV 10 to 16 days after breeding. Other abnormalities such as undescended testes and multiple septi of the gall bladder were also noted (114). Previous vaccination or immunity of the pregnant sow did not protect the fetus from the deleterious effects of HCV reinoculation of the dam (114).

Subsequent experiments confirmed these observations but also demonstrated the marked effect of HCV on the developing nervous system. Immunization of dams from 20 to 90 days in gestation produced cerebellar hypoplasia, hypomyelinogenesis, and congenital tremors (trembles) in newborn pigs (26). Other nervous system defects observed included hydrocephalus, small cerebral gyri, and cerebellar agenesis. These effects have been attributed to marked vasculitis and secondary hypoxia (26).

Newborn pigs, infected in utero with HCV, appear capable of harboring persistent virus in their viscera for extended periods of time, thus presenting the problem of secondarily transmitting infection to other members of a herd.

Bluetongue virus.—While bluetongue is a virus disease primarily affecting sheep, all ruminants are susceptible. In nature, the disease is transmitted by biting midges (*Culicoides* sp.) and for this reason the virus was originally grouped with the arboviruses. At the present time, bluetongue virus (BTV) is provisionally classified as a reovirus.

The disease produces a typical picture in adult animals, i.e., fever and hyperemia with cyanosis and edema of the tongue, mouth, eyes and nose, and respiratory tract. Mortality is low (1 to 5 percent) and is usually due to pneumonia; morbidity, on the other hand, is quite high (40 percent) and convalescence is prolonged.

Occasionally, fetal damage and malformations are noted with wild viruses in natural epidemics (32). Generally, however, the most serious episodes of fetal abnormalities have been associated with immunization of pregnant ewes at five to seven weeks gestation with attenuated chicken embryo BTV vaccine (94). Vaccination of pregnant ewes during this critical period produces abnormalities in 20 to 50 percent of the fetuses. Increased stillbirths and neonatal deaths occur, as well as extensive defects of the newborn, such as deafness, blindness, shortened limbs and mandible, and spine and neck deformities. Most severe are diffuse anomalies of the central nervous system, including anencephaly, hydranencephaly, porencephaly, hydrocephaly, and supranumerary gyral convolutions (26, 32). Vasculitis, encephalitis, and edema may be evident histologically. Lambs with such nervous system lesions walk with their heads down, ignore their dams, and walk into fences or into corners; appropriately, perhaps, they are called "dummies."

The virus appears to invade the fetus transplacentally, even in the presence of maternal immunity, raising the possibility that virus may be transported in the ewe inside circulating lymphocytes (or erythrocytes). Thus, the virus would not be neutralized in the presence of maternal antibody. Furthermore, IgG antibody does not cross the placenta in sheep (64), preventing limitation of fetal infection by this mechanism.

The parvoviruses.—This group of small DNA viruses includes H-viruses and rat virus (RV) which have in common their pathogenicity for newborn hamsters. Other viruses of the group include adeno-associated viruses (AAV) and feline panleukopenia virus (PLV). Toolan (105) has recently summarized much of the research with this group of agents.

We have previously considered the ability of certain strains of RV to produce congenital infection in pregnant rats resulting in fetal death and reabsorption, or fetal survival with hepatitis and cerebellar hypoplasia (48). Neonatal intracerebral inoculation of rats and hamsters with RV (48), of cats and ferrets with PLV, and hamsters with H-1 virus (50), has produced similar cerebellar alterations.

The most notable feature of the H-viruses and RV has been the "funny face" mongoloid-like deformity produced by injecting newborn hamsters with these agents (105). The tooth and facial abnormalities observed are due to a more generalized severe osteolytic action of these viruses. Although these hamsters superficially resemble patients with Down's syndrome, there are no chromosomal abnormalities present. Inoculation of H-1 virus subcutaneously in pregnant hamsters also results in mongoloid-like deformities in the newborn, whereas RV does not produce such an effect (105).

Recently, H-1 virus given orally to hamsters has been shown to penetrate the gut wall and thence the placenta, infecting the fetus within 24 hours after oral inoculation. This virus induces transplacental infection in rats only if given by parenteral inoculation. The situation is reversed for the different species and RV, i.e., rats are transplacentally infected with RV given orally, but hamsters must be inoculated parenterally to infect the fetus (49). These experiments indicate that the maternal oral route may lead to infection of the fetus with certain viruses and may be an important area of future study in man.

The relationship of H-1 virus to human abortions and other human disease is unknown. Previous studies have reported isolation of this virus from up to 20 percent of human tumors, placentas, and embryos (105). Neutralizing and hemagglutination-inhibiting antibody to H-1 and HB have been found in a few sera from 500 patients with cancer (105). Recent study of 50 spontaneous human abortions failed to recover H-1 virus from any of the specimens. None of the sera available (43 specimens) of the 50 women having abortions had detectable hemagglutination-inhibiting antibody to H-1 or RV viruses (77). Another study in man failed to detect any relationship between maternal antibody to H-1 or RV and pregnancies which terminated in spontaneous abortion, stillbirth, mongolism, or congenital skull defects (71).

Bovine diarrhea-mucosal disease virus.—A recent study has reported the association of bovine diarrhea-mucosal disease virus (BD-MD) infection of pregnant dams with congenital cerebellar hypoplasia and ocular defects in calves (47). This virus is endemic in cattle in all parts of the world. Although infection is usually subclinical, severe disease with mucous membrane ulceration of the mouth and gastrointestinal tract may occur.

The limited evidence available suggests that infection of pregnant cattle with BD-MD during the second trimester of gestation can result in ataxic calves and may increase the frequency of abortion. The ataxic calves are blind due to lenticular opacities and retinal degeneration. At necropsy, the cerebellum may be rudimentary or hypoplastic and degenerated. This type of cerebellar malformation appears similar to the lesion induced by rat virus, hog cholera virus, and feline panelukopenia virus (47).

SUMMARY

Presently available information has conclusively implicated at least five viruses in the production of congenital malformations. These include rubella and cytomegalovirus in humans, and blue tongue, hog cholera, and the H-1, RV group in animals. *Herpesvirus hominis* and bovine diarrhea-mucosal disease virus may be provisionally included as additional agents with this potential. Other human viruses have recently received a great deal of study in this regard, most notably Coxsackie B viruses and mumps. At the present time, only limited evidence is available to determine what role these two viruses may play in human teratogenesis. For example, only one study has reported the association of congenital heart disease and maternal Coxsackievirus B3 and B4 infection (13a); other studies have not reported such an association, but further investigation is clearly necessary to resolve this possible relationship. Large prospective population studies before and during natural epidemics are beginning to provide data on the frequency of various virus infections during pregnancy and will permit estimation of the teratogenic potential of many agents. Continued development of animal model systems, particularly in primates, may provide additional data and allow experimental therapeutic programs to be tested.

There has been an ever broadening interest in the role of viruses as causes of congenital viral diseases. In terms of the potential number of individuals affected, inapparent and latent viral infections may prove to be the most important group of deleterious agents which are congenitally acquired. The subtle manifestations of disease (such as subnormal mentality), and the possible relationship of these viruses in initiating human malignancy and immunological diseases, are difficult to study with present methodology. However, it is hoped that such technical problems will soon be solved and establish the role of viruses in many of these diseases.

LITERATURE CITED

1. Adams, R. H., Combes, B. 1965. *J. Am. Med. Assoc.* 192:195–98
2. Alford, C. A., Snyder, M. E., Stubbs, G. 1967. *Pediat. Res.* 1:209
3. Alford, C. A., Jr., Foft, J. W., Blankenship, W. J., Cassady, G., Benton, J. W., Jr. 1969. *J. Pediat.* 75:1167–78
4. Alford, C. A., Jr., Neva, F. A., Weller, T. H. 1964. *N. Engl. J. Med.* 271:1275–81
5. Alford, C. A., Schaefer, J., Blankenship, W. J., Straumfjord, J. V., Cassady, G. 1967. *N. Engl. J. Med.* 277:437–39
6. Baird, P. A., Miller, J. R. 1968. *Brit. J. Prev. Soc. Med.* 22:81–85
7. Bayer, W. L., Sherman, F. E., Michaels, R. H., Szeto, I. L. F., Lewis, J. H. 1965. *N. Engl. J. Med.* 273:1362–66
8. Bellanti, J. A., Catalano, L. W., Jr., Chambers, R. W. 1971. *J. Pediat.* 78:136–45
9. Berendes, H. W., Weiss, W., Miller, R. W. 1969. Presented at *Int. Conf. Congenital Malformations, 3rd, The Hague, Netherlands*
10. Birnbaum, G., Lynch, J. I., Margileth, A. M., Lonergan, W. M., Sever, J. L. 1969. *J. Pediat.* 75:789–95
11. Blanc, W. A. 1969. Unpublished observations
12. Blood, D. C., Henderson, J. A. 1963. *Veterinary Medicine,* 2nd ed. Chap. 20, 651–54. Baltimore: Williams & Wilkins
13. Boiron, M., Tanzer, J., Thomas, M., Hampe, A. 1966. *Nature* 209:737–38

13a. Brown, G. C. 1970. *Arch. Environ. Health* 21:362–65

14. Carter, C. O. 1967. *World Health Organ. Chron.* 21:287–92
15. Catalano, L. W., Jr. 1969. Unpublished data

15a. Catalano, L. W., Jr., Safley, G. H., Museles, M., Jarzynski, D. J. 1971. *J. Pediat.* In press

16. Collmann, R. D., Stoller, A. 1962. *J. Ment. Def. Res.* 6:22–37
17. Cooper, L. Z. 1968. *Orig. Art. Series,* 4, No. 7:23–35, Intrauterine Infections, ed. D. Bergsma, New York: Nat. Found.
18. Cooper, L. Z., Krugman, S. 1969. *Disease-A-Month* 1–38, February
19. Day, R. W. 1966. *Am. J. Hum. Genet.* 18:70–80
20. Dekaban, A., O'Rourke, J., Corman, T. 1958. *Neurology* 8:387–92
21. Di Stefano, H. S., Dougherty, R. M. 1966. *J. Nat. Cancer Inst.* 37:869–75
22. Edsall, G. 1966. In The Prevention of Mental Retardation Through Control of Infectious Disease, 325–34. *Pub. Health Serv. Publ. No. 1692,* Washington, D. C.
23. Eichenwald, H. F. 1965. *Orig. Art. Series* 1:74–76, Birth Defects
24. Elizan, T. S., Ajero-Froehlich, L., Fabiyi, A., Ley, A., Sever, J. L. 1969. *Arch. Neurol.* 20:115–19
25. Elliott, G. B., McAllister, J. E. 1956. *Am. J. Obstet. Gynecol.* 72:896–902
26. Emerson, J. L., Delez, A. L. 1965. *J. Am. Vet. Med. Assoc.* 147:47–54
27. Feldman, R. A. 1969. *Am. J. Dis. Child.* 117:517–21
28. Gagnon, R. A. 1968. *Obstet. Gynecol.* 31:682–84
29. Gorlin, R. J., Yunis, J., Anderson, V. E. 1968. *Am. J. Dis. Child.* 115:473–76
30. Gregg, N. M. 1941. *Trans Ophthalmol. Soc. Aust.* 3:35–46
31. Gregg, N. M. 1945. *Med. J. Aust.* 2:122–26
32. Griner, L. A., McCrory, B. R., Foster, N. M., Meyer, H. 1964. *J. Am. Vet. Med. Assoc.* 145:1013–19
33. Gross, L. 1961. *Proc. Soc. Exp. Biol. Med.* 107:90–93
34. Gunz, F. W., Spears, G. F. S. 1968. *Brit. Med. J.* 4:604–8
35. Gunz, F. W., Veale, A. M. O. 1969. *J. Nat. Cancer Inst.* 42:517–24
36. Hanshaw, J. B. 1966. *Pediat. Clin. N. Am.* 13:279–93
37. Hanshaw, J. B. 1968. *Virology Monographs* (No. 3), 1–23. New York: Springer-Verlag
38. Hanshaw, J. B. 1969. *J. Pediat.* 75:1179–85
39. Hardy, J. B., McCracken, G. H., Jr., Gilkeson, M. R., Sever, J. L. 1969. *J. Am. Med. Assoc.* 207:2414–20
40. Hardy, J. B., Monif, G. R. G., Sever, J. L. 1966. *Bull. Johns Hopkins Hosp.* 118:97–108
41. Hathaway, W. E., Mull, M. M., Pechet, G. S. 1969. *Pediatrics* 43:233–40
42. Heath, H. D., Shear, H. H., Imagawa, D. T., Jones, M. H., Adams,

J. M. 1956. *Proc. Soc. Exp. Biol. Med.* 92:675–82
43. Hotchin, J. 1962. *Cold Spring Harbor Symp. Quant. Biol.* 27:479–99
44. Hovig, D. E., Hodgman, J. E., Mathias, A. W., Jr., Levan, N., Portnoy, B. 1968. *Am. J. Dis. Child.* 115:438–44
45. Hubner, R. J., Todaro, G. J. 1969. *Proc. Nat. Acad. Sci. USA* 64: 1087–94
46. Huntley, C. C., Handcock, W. P., Sever, J. L. 1966. *Soc. Pediat. Res.* (Abstr.):142
47. Kahrs, R. F., Scott, F. W., deLahunta, A. 1970. *Teratology* 3: 181–84
48. Kilham, L., Margolis, G. 1966. *Am. J. Pathol.* 49:457–75
49. Kilham, L., Margolis, G. 1969. *Teratology* 2:111–24
50. Kilham, L., Margolis, G., Colby, E. D. 1967. *Lab. Invest.* 17:465–80
51. Klein, E., Klein, G. 1966. *Nature (London)* 209:163–65
52. Kogon, A., Kronmal, R., Peterson, D. R. 1968. *Am. J. Pub. Health* 58:305–11
53. Korones, S. B. et al. 1965. *J. Pediat.* 67:166–81
54. Korones, S. B., Roane, J. A., Gilkeson, M. R., Lafferty, W., Sever, J. L. 1969. *J. Pediat.* 75:1261–70
55. Krainin, P., Lapan, B. 1956. *J. Am. Med. Assoc.* 160:937–40
56. Lang, D. J. 1966. *Pediatrics* 38:913–16
57. Levinsohn, E. M., Foy, H. M., Kenny, G. E., Wentworth, B. B., Grayston, J. T. 1969. *Proc. Soc. Exp. Biol. Med.* 132:957–62
58. Lundstrom, R. 1962. *Acta Paediat. Scand.* 51 *Suppl. 133:*1–110
59. Maton, W. G. 1815. *Med. Trans., Roy. Coll. Phys. (London)* 5:149
60. McCracken, G. H., Jr. et al. 1969. *J. Pediat.* 74:383–92
61. McIntosh, R., Merritt, K. K., Richards, M. R., Samuels, M. H., Bellows, M. T. 1954. *Pediatrics* 14:505–22
62. Mellors, R. C. 1969. *J. Inf. Dis.* 120: 480–87
63. Michaels, R. H., Mellin, G. W. 1960. *Pediatrics* 26:200–9
64. Miller, J. F. A. P. 1966. *Brit. Med. Bull.* 22:21–26
65. Miller, M. J., Sunshine, P. J., Remington, J. R. 1969. *J. Pediat.* 75: 1287–91
66. Miller, R. W. 1963. *N. Engl. J. Med.* 268:393–401
67. Miller, R. W. 1964. *N. Engl. J. Med.* 271:30–36
68. Mims, C. A. 1966. *J. Pathol. Bacteriol.* 91:395–402
69. Mims, C. A. 1968. *Progr. Med. Virol.* 10:194–237
70. Monif, G. R. G., Hardy, J. B., Sever, J. L. 1966. *Bull. Johns Hopkins Hosp.* 118:85–96
71. Monif, G. R. G., Sever, J. L., Cochran, W. D. 1965. *J. Pediat.* 67:253–56
72. Monif, G. R. G., Sever, J. L., Schiff, G. M., Traub, R. G. 1965. *Am. J. Obstet. Gynecol.* 91:1143–46
72a. Monjan, A. A., Gilden, D. H., Cole, G. A., Nathanson, N. 1971. *Science* 171:194–96
73. Naeye, R. L., Blanc, W. 1965. *J. Am. Med. Assoc.* 194:1277–83
74. Nahmias, A. J., Dowdle, W. R. 1968. *Progr. Med. Virol.* 10:110–59
75. Nahmias, A. J. et al. 1969. *J. Pediat.* 75:1194–1203
76. Nahmias, A. J., Josey, W. E., Naib, Z. M. 1967. *J. Am. Med. Assoc.* 199:164–68
77. Newman, S. J., McCallin, P. F., Sever, J. L. 1970. *Teratology* 3: 279–81
78. Nichols, W. W. 1966. *Am. J. Hum. Genet.* 18:81–92
78a. Nichols, W. W. 1970. *Ann. Rev. Microbiol.* 24:479–500
79. Parkman, P. D., Buescher, E. L., Artenstein, M. S. 1962. *Proc. Soc. Exp. Biol. Med.* 111:225–30
80. Pavan, C. 1967. *Triangle* 8:42–48
81. Pfitzer, P., Müntefering, H. 1968. *Nature* 217:1071–72
82. Plotkin, S. A., Boué, A., Boué, J. G. 1965. *Am. J. Epidermiol.* 81:71–85
83. Pollard, M., Sharon, N., Teah, B. A. 1968. *Proc. Soc. Exp. Biol. Med.* 127:755–61
83a. Reports of the National Center for Health Statistics and the Bureau of the Census. 1967. *Statist. Bull.* 48:3
84. Ruffolo, P. R., Margolis, G., Kilham, L. 1966. *Am. J. Pathol.* 49:795–824
85. Schaffer, A. J. 1965. *Diseases of the Newborn,* 2nd ed., 733. Philadelphia: W. B. Saunders
86. Sergovich, F., Valentine, G. H., Chen, A. T. L., Kinch, R. A. H., Smout, M. S. 1969. *N. Engl. J. Med.* 280:851–55

87. Sever, J. L., Ed. 1969. *J. Pediat.* 75 (Part 2):1111–1294
88. Sever, J. L. 1970. *Int. J. Gynecol. Obstet.* 8 (part 2):763–69
89. Sever, J. L. et al. 1967. *Pediatrics* 40:789–97
90. Sever, J. L., Schiff, G. M., Traub, R. G. 1962. *J. Am. Med. Assoc.* 182:663–71
91. Sever, J., White, L. R. 1968. *Ann. Rev. Med.* 19:471–86
92. Shapiro, S., Ross, L. J., Levine, H. S. 1965. *Am. J. Pub. Health* 55:268–82
93. Shinefield, H. R., Townsend, T. E. 1953. *J. Pediat.* 43:21–25
94. Shultz, G., DeLay, P. D. 1955. *J. Am. Vet. Med. Assoc.* 127:224–26
95. Siegel, M., Greenberg, M. 1960. *N. Engl. J. Med.* 262:389–93
96. Singer, D. B., Rudolph, A. J., Rosenberg, H. S., Rawls, W. E., Boniuk, M. 1967. *J. Pediat.* 71: 665–75
97. Smithells, R. W., D'Arcy, E. E., McAllister, E. F. 1968. *Develop. Med. Child. Neurol.* (Suppl. 15): 6–10
98. Soothill, J. F., Hayes, K., Dudgeon, J. A. 1966. *Lancet* 1:1385–88
99. South, M. A., Tompkins, W. A. F., Morris, C. R., Rawls, W. E. 1969. *J. Pediat.* 75:13–18
99a. Starr, J. G., Bart, R. D., Jr., Gold, E. 1970. *N. Engl. J. Med.* 282:1075–78
100. Stern, H. 1968. *Brit. Med. J.* 1:665–69
101. Stevenson, A. C., Johnston, H. A., Stewart, M. I. P., Golding, D. R. 1966. *Bull. W. H. O.* 34, Suppl.
102. Stokes, J., Jr., Wolman, J. J., Blanchard, M. C., Farquhar, J. D. 1951. *Am. J. Dis. Child.* 82:213
103. Sutnick, A. I., London, W. T., Gerstley, B. S., Cronlund, M. M., Blumberg, B. S. 1968. *J. Am. Med. Assoc.* 205:80–84
104. Tennant, R. W., Layman, K. R., Hand, R. E., Jr. 1969. *J. Virol.* 4:872–78
105. Toolan, H. W. 1968. *Int. Rev. Exp. Pathol.* 6:135–80
106. Tuffli, G., Nahmias, A. 1970. *Am. J. Dis. Child.* In press
107. Walzer, S., Breau, G., Gerald, P. S. 1969. *J. Pediat.* 74:438–48
108. Warkany, J. 1957. *Pediatrics* 19: Part II, Suppl.:725–33
109. Weller, T. H., Hanshaw, J. B. 1962. *N. Engl. J. Med.* 266:1233–44
110. Weller, T. H., Neva, F. A. 1962. *Proc. Soc. Exp. Biol. Med.* 111: 215–25
111. White, L. R., Sever, J. L. 1969. Unpublished data
112. White, L. R., Sever, J. L., Alepa, F. P. 1969. *J. Pediat.* 74:198–207
113. Witzleben C. L., Driscoll, S. G. 1965. *Pediatrics* 36:192–99
114. Young, G. A., Kitchell, R. L., Luedke, A. J., Sautter, J. H. 1955. *J. Am. Vet. Med. Assoc.* 126:165–71

DESTRUCTION OF VIRUS-INFECTED CELLS BY IMMUNOLOGICAL MECHANISMS

1571

DAVID D. PORTER

Department of Pathology, University of California, Los Angeles School of Medicine, Los Angeles, California

CONTENTS

INTRODUCTION

During the course of replication, animal viruses may destroy or transform the host cell or may cause no obvious abnormality. Virus infected cells which are not directly destroyed by the effect of the virus may, however, be destroyed by the host immunologic defenses reacting with antigens induced by the virus. Such host-mediated cell destruction may aid in recovery from infection, or may lead to the production of lesions which would not occur without the immune response. I will review selected recent literature on the in vitro mechanisms of destruction of virus-infected cells by immunologic processes, and discuss some in vivo effects of such cell destruction.

ANTIBODY-MEDIATED CYTOLYSIS

Antiviral antibody alone added to a virus-infected cell culture may prevent the spread of virus through the medium, but such antibody does not cause cytolysis of infected cells. Anticellular antibody added to such a culture may inhibit the cytopathic effect of the virus, suppress viral multiplication, or show both effects. These effects have been interpreted as a blockage of viropexis or of viral receptor sites (4), an effect on the cells (24), or as neutralization of the virus (19) in different virus and cell systems, but such antibody does not result in cytolysis of the infected cells.

If serum complement (C) is added to cells sensitized by reaction of IgG or IgM antibody with an antigen on the surface of the cell, cytolysis may occur. Lysis of erythrocytes by C can be initiated by a single IgM molecule bound to the surface, or two IgG molecules bound in close proximity (9, 10); thus, cellular lysis requires perhaps 1000 times as many bound IgG molecules as IgM. The biology of C, a 9 component system (designated C1–C9) of 11 proteins has recently been reviewed by Müller-Eberhard (48). The production of a single C-mediated functional lesion in the cell membrane is sufficient to cause cell lysis. Apparently, the production of functional lesions requires all C components, although some increased cell fragility occurs after the C8 step (22). Discrete ultrastructural lesions have been observed on the surface of cells subsequent to the action of C (41). These lesions are produced at the C5 step and represent a true alteration of the cell membrane. The ultrastructural lesions, however, are not accompanied by a permeability defect until C6 to C9 have reacted with the cell membrane (89).

The C-mediated cytotoxic tests on cells other than erythrocytes have usually been performed by the method of Gorer & O'Gorman (5, 21) in which the proportion of cells killed during a reaction are evaluated by their inability to exclude a dye such as trypan blue. Better quantitation may be obtained by labeling the target cells with $Na_2{}^{51}Cr\ O_4$ and measuring the isotope released during cell lysis (25). A fluorochromatic test (12) may be applicable to the screening of serum samples for cytotoxic activity.

Complement-mediated immune cytolysis of virus-infected cells is well illustrated in a study of rabies virus by Wiktor et al (84). This RNA-containing, ether-sensitive rhabdovirus matures by budding from the cell plasma membrane 72 hours after infection (40). Immunofluorescence showed viral antigen in the cytoplasm of infected hamster Nil-2 or human WI-38 cells as early as 12 hours after infection, but the cytolytic effect of anti-rabies antibody plus C was first seen 16-18 hours after infection and became complete 24 hours after infection. Immunofluorescence showed the presence of rabies antigen on the surface of infected cells at the time they were sensitive to cytolysis, and after the cytolytic reaction was performed, fluorescein-labeled rabies antibody was able to penetrate the cells and stain intracytoplasmic viral antigen. Isolated human IgG rabies antibody produced cytolysis; an IgM antibody was not tested. The use of trypsin to suspend infected cells in preparation for the test abolished the cytotoxic reaction, presumably by destroying the rabies antigen on the cell surface.

A representative list of viruses in which C-mediated cytolysis of the infected cells or of virions by anti-viral antibody has been shown is given in Table 1. Cytolytic reactions have been widely used for the classification of murine leukoviruses (reviewed by Pasternak, 57). It is evident that at least one member of each group of enveloped, ether-sensitive viruses can render cells sensitive to C-mediated cytolysis. Maturation of the virion at the cell plasma membrane apparently is not required for sensitivity to cytolysis, since herpes simplex virus, which matures at the outer nuclear membrane and is released via channels through the cytoplasm (49), can nevertheless render the cell surface susceptible to the action of antibody plus C (62), or to other membrane effects such as mixed hemadsorption (65). Other agents of the herpesvirus group such as EB virus (43) and Marek's disease virus (13) may induce cell surface antigens; however, C-mediated cytolysis has not been tested.

Typical antibody-C-induced ultrastructural lesions have been shown in the plasma membranes of cells infected with bovine viral diarrhea virus (61), and of considerable interest, such lesions have been produced in the

TABLE 1. Representative Viruses With Which Antibody Plus Complement-Mediated Cytolysis of the Infected Cell or Virion has Been Shown in Vitro

Virus group[a]	Virus	Cytotoxic target	Reference
Myxovirus	Newcastle disease	Infected cell	17
Paramyxovirus	Sendai	Infected cell	17, 18
	SV5	Infected cell	39
Rhabdovirus	Rabies	Infected cell	84
Leukovirus	Gross	Infected cell	52, 69, 80
	Friend-Moloney-Rauscher	Infected cell	12, 25, 50, 51
Togavirus	Rubella	Virion	3
	Bovine viral diarrhea	Infected cell	61
Coronavirus	Avian infectious bronchitis	Virion	7
Arenovirus[b]	Lymphocytic choriomeningitis	Infected cell	55
Herpesvirus	Herpes simplex	Infected cell	62

[a] Classification according to Melnick (46)
[b] Proposed new group (64)

virions of rubella virus (3) and avian infectious bronchitis virus (7). Ultrastructural lesions have been noted in the plasma membranes of Gross lymphoma cells treated with mouse antiserum in the absence of added C; however, it was not stated whether the mouse C was inactivated (42).

The neutralization of herpes simplex virus (87) is markedly enhanced by the addition of C to the neutralizing antiviral antibody. This enhancement may be due to C-mediated lesions produced in the envelope of the virion, a structure which is necessary for infectivity of this virus (70). However, maximal enhancement of neutralization of antibody-sensitized herpes simplex virus has been found at the (early) C3 step in the complement activation sequence, (15), suggesting that simply covering the virion surface with C components results in enhancement of virus neutralization.

The effect of host cell type on cytotoxic sensitivity may be marked. When SV5 virus was grown in BHK-21 cells (virus yield 7–10 infectious particles/cell) the cells were very sensitive to the effect of antibody plus C; while primary monkey kidney cells, which yield 1500 infectious particles/cell, could be lysed only by high concentrations of antibody and C (39). Interestingly, the cytotoxic sensitivity of these cells parallels the sensitivity to cell fusion by SV5.

Passively administered cytotoxic antibody to Friend virus antigens was shown to be absorbed in vivo in mice bearing Friend leukemias (79). In vivo evidence for cytolysis of tumor cells infected with Newcastle disease or Sendai virus in the presence of anti-viral antibody has been presented (17). Antibody-mediated cytotoxicity has been shown for lymphocytic choriomeningitis virus-infected cells (55), and when such antibody is passively administered to persistently infected mice, cell necrosis followed by polymorphonuclear and then mononuclear cell infiltrates are seen (53). Passive administration of Aleutian disease virus antibody to acutely infected mink causes similar cell necrosis and inflammation, and antibody and host C can be found in cytoplasm of cells containing viral antigen (Porter, unpublished data). Murine cytomegalovirus infection is associated with similar cell necrosis and inflammation, which is largely blocked by irradiation of the host (32). In addition to in vivo cytotoxicity, it is likely that viral antigen-antibody complexes activate complement components leading to generation of C3, C5, or C$\overline{567}$ neutrophil chemotactic factors (83), or activate factors leading to mononuclear cell chemotaxis (82). Herpes simplex virus infection can release a neutrophil chemotactic-generating factor from rabbit kidney cells which activates C5 (11).

Cell-Mediated Cytolysis

Cellular immunity represents a class of immune responses implicated in various types of tissue destruction, for example, transplantation rejection, resistence to neoplasia, autoimmune disease, and delayed hypersensitivity. It has been postulated that the cellular immune systems represent the mechanism for surveillance or rejection of antigenically altered cells, especially cells altered by neoplastic events (73). Since viruses may antigenically alter cells with or without neoplastic transformation, the cellular immune systems may be equally important in ridding the body of virus-infected cells.

The first step in cellular immunity involves recognition of antigen by a hypothetical receptor on the immune cell; destruction of the target cell occurs subsequently. The abundant literature may be found in recent reviews of the recognition process and destructive phase of cellular immunity (58, 86). Various soluble mediators of cellular immunity have been identified, including transfer factor (44) for the recognition phase and lymphotoxin (23), macrophage migration inhibitory factor (8), and lymphocyte transforming factor (77) for the effector or destructive phase. Although the stimulus to produce a cellular immune response is highly specific, the target cell cytolysis produced by lymphotoxin is nonspecific, and appears to result from degradation of the target cell plasma membrane (85). Mouse lymphotoxin is equally effective on many types of cells (85), but human lymphotoxin is extremely active upon fibroblasts and epithelial cells, less active upon parenchymal cells, and least active upon lymphocytes (23).

Relatively few in vitro studies of cellular immunity in viral infections have been reported as yet. Tompkins et al (74) demonstrated that macrophages from rabbits infected with fibroma virus were inhibited from migrating upon monolayers of fibroma virus-infected rabbit kidney cells. Rabbits infected with fibroma virus developed cutaneous delayed sensitivity five days after inoculation, and peritoneal macrophages showed a decrease in migration upon infected monolayers six days after infection of the rabbit. The inhibition of macrophage migration upon fibroma virus-infected cell monolayers did not correlate with the amount of infectious virus present in the monolayer. These authors noted similar results with vaccinia virus infection in rabbits; inhibition of macrophage migration was specific for the virus to which delayed sensitivity had developed. Subsequently, Tompkins et al (75) have shown that fibroma virus-infected cells develop a nonvirion surface antigen demonstrable by immunofluorescence. The time of appearance of the surface antigen paralleled that of the one inhibiting macrophage migration, suggesting that the antigen may be the same. Although both cellular and humoral immune responses to cell surface antigens have been demonstrated in vitro with the fibroma system, the significance of either response in the process of in vivo tumor regression has not been established.

There is abundant evidence that lymphocytic choriomeningitis (LCM) virus is relatively noncytopathic and that cellular injury is largely due to the immune responses of the host (53, 59). Oldstone & Dixon (54) have demonstrated that lymph node or spleen cells from mice immunized with LCM virus five days previously could liberate a lymphotoxin upon exposure to LCM virus in vitro. The release of lymphotoxin required living lymphoid cells and was immunologically specific, since living or killed LCM virus could cause release of lymphotoxin, but unrelated viruses, or other antigens could not. The lymphotoxin was able to kill LCM-infected or normal cells from mice or monkeys, and could reduce the cloning efficiency of mouse fibroblasts or induce ^{51}Cr release from labeled target cells. This type of lymphotoxin assay appears to hold great promise for the study of cellular immunity to viral antigens. Lundstedt (45) has demonstrated statistically significant reduction of viability of LCM-infected L-cells when "syngeneic" LCM-immune lymphoid cells were added to the cultures.

In vivo studies with LCM virus and other agents of the arenovirus group have shown that treatment of mice with antithymocyte or antilymphocyte serum will prevent fatal disease in adult animals (37). Neonatal thymectomy followed by virus challenge of the adult results in low mortality as compared with events after challenge of normal mice (56, 63). Antithymocyte serum may permit the establishment of persistent LCM infection in mice more than one day old (36), and cyclophosphamide blocks lesion production in persistently infected mice (66). Adult mice which have recovered from a sublethal infection with LCM virus may develop viremia upon treatment with antilymphocyte serum, despite the presence of humoral antibody (78). The relative importance of cellular and humoral immune responses in the production of disease by LCM and related viruses is not clear at present.

Speel et al (71) have demonstrated that specifically sensitized mouse spleen cells could either inhibit or lyse cells persistently infected with mumps virus. The inhibiting effect of the spleen cells could be abolished by mumps antibody.

Tumors induced by both RNA and DNA oncogenic viruses have been shown to possess viral-specific transplantation antigens and to be rejected by cellular immune mechanism (30, 68). Hellström et al and Heppner (27–29, 31, 33) in a series of papers demonstrated immunologically specific cytotoxicity of lymphocytes upon cells infected with Shope papilloma, Moloney sarcoma, and murine mammary tumor viruses. The lymphocyto-toxicity could be abolished by addition of specific antibody to the systems. Transient cell-mediated cytotoxicity has been shown with Rous sarcoma virus (6). Of special interest, lymphoid cells from preleukemic AKR mice were found to be cytotoxic for syngeneic embryo cells in culture (81), and the cytotoxicity could be blocked by incubation of the target cells with antiserum against Gross antigen. This suggests the absence of "tolerance" to viral antigens in the leukovirus systems. Viral tumor induction may be enhanced by treatment of the host with antithymocyte or antilymphocyte serum (1, 2); however, induction of viral leukemia in mice may be blocked by neonatal thymectomy (2). Tumor rejection in a host with established specific cellular immunity may be blocked by antibody, presumably by covering of antigenic sites on the target cell (31). As a word of caution, cells transformed by SV40 virus may carry a derepressed host embryonic antigen on the cell surface (16, 26), which may participate in SV40 tumor cell rejection (14). Therefore, viral specificity should be carefully controlled in studies involving immune responses to virus infection.

The cellular and perhaps the humoral immune responses have been altered in a number of experimental viral infections by antithymocyte or antilymphocyte serum (tabulated by Hirsch, 34), thymectomy, irradiation, cytotoxic drugs, or steroids. Hirsch et al (38) demonstrated that antithymocyte serum given to vaccinia virus-infected mice markedly enhanced mortality. Widespread necrotic lesions of the viscera and brain were observed in treated mice, probably due to direct cytotoxicity of the virus. The treatment did not alter the humoral immune or interferon responses of the mice. Rabbits rendered unresponsive at birth to vaccinia virus show suppression of the local lesion at the site of cutaneous inoculation, but may develop disseminated vaccinia (20). Viral encephalitis is typically accompanied by lymphocytic infiltrates around vessels and in the brain substance. Treatment of animals infected with yellow fever (35), louping ill, and western equine encephalitis viruses (88) with antithymocyte serum or cyclophosphamide abolished the lymphocytic inflammatory response. However, the mortality of yellow fever was only slightly delayed, and the other two infections were converted from nonfatal to fatal diseases with marked cell destruction. In some cases in which mortality is increased by reducing or abolishing the cellular immune response, the tissue viral titers are enhanced (72).

Other Mechanisms of Cytolysis

Macrophages can play a role in the induction of the immune response (76) and are important in host susceptibility to and defense against viral infections (67). Infection of macrophages may be important in slow virus infections (60). At present, immunologically mediated destruction of virus-infected cells by macrophages has not been shown. An unusual technique involving conjugation of diphtheria toxin to mumps antibody was shown to result in destruction of mumps virus-infected cells (47).

Comment

Antibody plus complement-mediated cytotoxicity has the potential for destroying cells infected with enveloped viruses and, in some cases, the virions themselves. In the reported studies, inadequate attention has been paid to the markedly greater cytolytic efficiency of IgM antibody as compared with IgG antibody. Study of the production of active fragments of complement components produced during viral antigen-antibody-complement interactions may shed some light upon the pathogenesis of lesions of viral disease.

Cellular immune responses appear to affect or determine the outcome of a number of viral infections, and such reactions may destroy the infected cells. New in vitro methods of studying cellular immunity such as lymphotoxin production and inhibition of macrophage migration should lead to qualitative and quantitative descriptions of cellular immunity in a number of viral infections. The relative importance of humoral and cellular immune mechanisms in destroying virus-infected cells in vivo is not presently clear, and should be investigated.

LITERATURE CITED

1. Allison, A. C., Berman, L. D., Levey, R. H. 1967. *Nature (London)* 215:185–87
2. Allison, A. C., Law, L. W. 1968. *Proc. Soc. Exp. Biol. Med.* 127:207–12
3. Almeida, J. D., Laurence, G. D. 1969. *Am. J. Dis. Child.* 118:101–6
4. Axler, D. A., Crowell, R. L. 1968. *J. Virol.* 2:813–21
5. Batchelor, J. R. 1967. *Handbook of Experimental Immunology,* ed. D. M. Weir, 988–1008. Philadelphia: F. A. Davis, 1245 pp.
6. Bellone, C. J., Pollard, M. 1970. *Proc. Soc. Exp. Biol. Med.* 134:640–43
7. Berry, D. M., Almeida, J. D. 1968. *J. Gen. Virol.* 3:97–102
8. Bloom, B. R., Jimenez, L. 1970. *Am. J. Pathol.* 60:453–67
9. Borsos, T., Rapp, H. J. 1965. *J. Immunol.* 95:559–66
10. Borsos, T., Rapp, H. J. 1965. *Science* 150:505–6
11. Brier, A. M., Snyderman, R., Mergenhagen, S. E., Notkins, A. L. 1970. *Science* 170:1104–6
12. Celada, F., Rotman, B. 1967. *Proc. Nat. Acad. Sci. USA* 57:630–36
13. Chen, J. H., Purchase, H. G. 1970. *Virology* 40:410–12
14. Coggin, J. H., Ambrose, K. R., Anderson, N. G. 1970. *J. Immunol.* 105:524–26
15. Daniels, C. A., Borsos, T., Rapp, H. J., Snyderman, R., Notkins, A. L. 1970. *Proc. Nat. Acad. Sci. USA* 65:528–35
16. Duff, R., Rapp, F. 1970. *J. Immunol.* 105:521–23
17. Eaton, M. D., Scala, A. R. 1969. *Proc. Soc. Exp. Biol. Med.* 132:20–29
18. Eaton, M. D., Scala, A. R. 1970. *Proc. Soc. Exp. Biol. Med.* 133:615–19
19. Fernelius, A. L., Packer, R. A. 1969. *J. Gen. Virol.* 5:243–49
20. Flick, J. A., Pincus, W. B. 1963. *J. Exp. Med.* 117:633–46
21. Gorer, P. A., O'Gorman, P. 1956. *Transplant. Bull.* 3:142–43
22. Götze, O., Haupt, I., Fischer, H. 1968. *Nature (London)* 217:1165–67
23. Granger, G. A. 1970. *Am. J. Pathol.* 60:469–81
24. Habel, K., Hornibrook, J. W., Gregg, N. C., Silverberg, R. J., Takemoto, K. K. 1958. *Virology* 5:7–29
25. Haughton, G. 1965. *Science* 147:506–7
26. Hayry, P., Defendi, V. 1970. *Virology* 41:22–29
27. Hellström, I., Evans, C. A., Hellström, K. E. 1969. *Int. J. Cancer* 4:601–7
28. Hellström, I., Hellström, K. E. 1969. *Int. J. Cancer* 4:587–600
29. Hellström, I. et al. 1969. *Proc. Nat. Acad. Sci. USA* 62:362–68
30. Hellström, K. E., Hellström, I. 1969. *Advan. Cancer Res.* 12:167–223
31. Hellström, K. E., Hellström, I. 1970. *Ann. Rev. Microbiol.* 24:373–98
32. Henson, D., Smith, R. D., Gehrke, J. 1966. *Am. J. Pathol.* 49:871–88
33. Heppner, G. H. 1969. *Int. J. Cancer* 4:608–15
34. Hirsch, M. S. 1970. *Fed. Proc.* 29:169–70
35. Hirsch, M. S., Murphy, F. A. 1967. *Nature (London)* 216:179–80
36. Hirsch, M. S., Murphy, F. A., Hicklin, M. D. 1968. *J. Exp. Med.* 127:757–66
37. Hirsch, M. S., Murphy, F. A., Russe,

H. P., Hicklin, M. D. 1967. *Proc. Soc. Exp. Biol. Med.* 125:980–83

38. Hirsch, M. S., Nahmias, A. J., Murphy, F. A., Kramer, J. H. 1968. *J. Exp. Med.* 128:121–32
39. Holmes, K. V., Klenk, H.-D., Choppin, P. W. 1969. *Proc. Soc. Exp. Biol. Med.* 131:651–57
40. Hummeler, K., Koprowski, H., Wiktor, T. J. 1967. *J. Virol.* 1:152–70
41. Humphrey, J. H., Dourmashkin, R. R. 1969. *Advan. Immunol.* 11:75–115
42. Jakobsson, S. V., Wahren, B. 1965. *Exp. Cell Res.* 37:509–15
43. Klein, G. et al. 1968. *J. Exp. Med.* 128:1011–20
44. Lawrence, H. S. 1969. *Advan. Immunol.* 11:195–266
45. Lundstedt, C. 1969. *Acta Pathol. Microbiol. Scand.* 75:139–52
46. Melnick, J. L. 1970. *Progr. Med. Virol.* 12:337–41
47. Moolten, F. L., Cooperband, S. R. 1970. *Science* 169:68–70
48. Müller-Eberhard, H. J. 1969. *Ann. Rev. Biochem.* 38:389–414
49. Nii, S., Morgan, C., Rose, H. M. 1968. *J. Virol.* 2:517–36
50. Old, L. J., Boyse, E. A., Lilly, F. 1963. *Cancer Res.* 23:1063–68
51. Old, L. J., Boyse, E. A., Stockert, E. 1964. *Nature (London)* 201:777–79
52. Old, L. J., Boyse, E. A., Stockert, E. 1965. *Cancer Res.* 25:813–19
53. Oldstone, M. B. A., Dixon, F. J. 1970. *J. Exp. Med.* 131:1–19
54. Oldstone, M. B. A., Dixon, F. J. 1970. *Virology* 42:805–13
55. Oldstone, M. B. A., Habel, K., Dixon, F. J. 1969. *Fed. Proc.* 28:429
56. Parodi, A. S., Schmunis, G. A., Weissenbacher, M. C. 1970. *Experientia* 26:665–66
57. Pasternak, G. 1969. *Advan. Cancer Res.* 12:1–99
58. Perlmann, P., Holm, G. 1969. *Advan. Immunol.* 11:117–93
59. Petersen, I. R., Volkert, M. 1966. *Acta Pathol. Microbiol. Scand.* 67:523–36
60. Porter, D. D., Larsen, A. E., Porter, H. G. 1969. *J. Exp. Med.* 130:575–93
61. Ritchie, A. E., Fernelius, A. L. 1969. *Ges. Virusforsch.* 28:369–89
62. Roane, P. R., Roizman, B. 1964. *Virology* 22:1–8
63. Rowe, W. P., Black, P. H., Levey, R. H. 1963. *Proc. Soc. Exp. Biol. Med.* 114:248–51
64. Rowe, W. P. et al. 1970. *J. Virol.* 5:651–52
65. Salmi, A. A., Grönroos, J. A., Halonen, P. E. 1968. *Ann. Med. Exp. Biol. Fenn.* 46:103–8
66. Sharon, N., Pollard, M. 1969. *Nature (London)* 224:707–9
67. Silverstein, S. 1970. *Seminars Hematol.* 7:185–214
68. Sjögren, H. O. 1964. *J. Nat. Cancer Inst.* 32:361–74, 375–93
69. Slettenmark, B., Klein, E. 1962. *Cancer Res.* 22:947–54
70. Smith, K. O. 1964. *Proc. Soc. Exp. Biol. Med.* 115:814–16
71. Speel, L. F., Osborn, J. E., Walker, D. L. 1968. *J. Immunol.* 101:409–17
72. St. Geme, J. W., Jr., Brumbaugh, J. L., Pajari, K. L., Singer, A. D., Toyama, P. S. 1970. *J. Lab. Clin. Med.* 76:213–20
73. Thomas, L. 1959. (Discussion in) *Cellular and Humoral Aspects of the Hypersensitive States,* ed. H. S. Lawrence, 529–32. New York: Paul B. Hoeber, 667 pp.
74. Tompkins, W. A. F., Adams, C., Rawls, W. E. 1970. *J. Immunol.* 104:502–10
75. Tompkins, W. A. F., Crouch, N. A., Tevethia, S. S., Rawls, W. E. 1970. *J. Immunol.* 105:1181–89
76. Unanue, E. R., Cerottini, J.-C. 1970. *Seminars Hematol.* 7:225–48
77. Valentine, F. T., Lawrence, H. S. 1969. *Science* 165:1014–16
78. Volkert, M., Lundstedt, C. 1968. *J. Exp. Med.* 127:327–39
79. Wahren, B. 1963. *J. Nat. Cancer Inst.* 31:411–23
80. Wahren, B. 1966. *Exp. Cell Res.* 42:230–42
81. Wahren, B., Metcalf, D. 1970. *Clin. Exp. Immunol.* 7:373–86
82. Ward, P. A. 1968. *J. Exp. Med.* 128:1201–21
83. Ward, P. A. 1970. *Arthritis Rheum.* 13:181–86
84. Wiktor, T. J., Kuwert, E., Koprowski, H. 1968. *J. Immunol.* 101:1271–82
85. Williams, T. W., Granger, G. A. 1969. *J. Immunol.* 102:911–18
86. Wilson, D. B., Billingham, R. E. 1967. *Advan. Immunol.* 7:189–273
87. Yoshino, K., Taniguchi, S. 1966. *J. Immunol.* 96:196–203
88. Zlotnik, I., Smith, C. E. G., Grant, D. P., Peacock, S. 1970. *Brit. J. Exp. Pathol.* 51:434–39
89. Polly, M. J., Müller-Eberhard, H. J., Feldman, J. D. 1971. *J. Exp. Med.* 133:53–62

ISOLATION OF LYMPHOID CELLS WITH ACTIVE SURFACE RECEPTOR SITES[1]

1572

HANS WIGZELL, M.D. AND BIRGER ANDERSSON, M.D.
Department of Tumor Biology, Karolinska Institute, 104 01 Stockholm, Sweden

CONTENTS

INTRODUCTION

Lymphoid cells represent a most interesting group of cells, morphologically very similar, especially while in the resting stage (= the small lymphocyte), but yet displaying extreme variation in potential immune capacity when analyzed at the single cell level. It is likely that other organ groups of cells will show similar, preformed heterogeneity when isolated cells are studied, i.e., liver or central neuron cells are plausible candidates, but at present no suitable tests are available for these. In tumor immunology it is well established that chemical carcinogen-induced sarcomas, even if induced in the same individual, have unique and distinct surface changes for each

[1] This work was made possible through the support from the Swedish Cancer Society, the Karolinska Institute, the Cancer Association of Stockholm and Anders Otto Swärds Stiftelse.

tumor cell line (*27*). It is not know, however, whether clones of (normal) fibroblasts or other normal cells would manifest a similar variation in surface configuration(s) or whether the tumor-associated antigens are the outward expressions of necessary changes for oncogenesis to take place. In the lymphoid system we already know that variation exists in the normal cell population, and tests are available for the analysis of immunocompetence of the single lymphoid cell (3, 9, *26*). Thus, presently existing or yet fictional ways of segregating and isolating subgroups of lymphoid cells according to their surface properties have been chosen as the main topic of this article. The general reasoning and principles behind the fractionation procedures should, however, be applicable to any approach to isolate subgroups of cells from a population in which variation is reflected on the outer membrane. We have not attempted to make any complete coverage of existing literature on cell fractionation but, rather, have chosen isolated examples to illustrate various principal ways of approaching the problem(s).

Table 1 summarizes methods and underlying principles which will be discussed. They constitute a heterogeneous group by themselves, the order of presentation being arbitrary and not representing any ranking list as to

TABLE 1. Some Existing and Hypothetical Methods for the Isolation and Purification of Lymphoid Cells

Method	Principle	Reference No.
Selective lysis	Cytotoxic antibodies	31, 38
	Drugs with elective action	8
Isolation of clones	Limiting dilution and isolation of single according to the Poisson distribution	4, 10, 28, 29
Cell electrophoresis	Electrostatic charge	6, 36
Charged columns	—11—	41
Fluorescence selection	Physical separation of cells according to intensity of fluorescent stain	25
Rosette cell sedimentation	Different density of rosettes and non-rosette forming cells	5, 16, 35
Affinity columns	Specific retention of cells with receptor for the compound used for labeling the column (antigen, antireceptor antibody, lectin, complement factor 3)	43, 48
Micromanipulation	Mechanical isolation by picking out cells showing a particular property (e.g., hemolytic plaque formation)	11, 15, 18
Two-phase systems	Differences in distribution between two phases of polymer solutions. Surface properties of cell determine distribution patterns	1, 2, 45

suitability. It will be obvious from the presentation which methods we would advocate for further development.

Selective Killing Causing Passive Enrichment of Subgroups of Lymphoid Cells

This approach is applicable when cell populations can be subdivided into two or very few subgroups. It is thus possible to obtain populations of lymphoid cells devoid of the thymus-derived type by treatment in vitro with specific antisera against this kind of lymphocyte (31, 38). This leaves lymphocytes of bursa-type responsible for humoral antibody production undisturbed and thereby passively enriched. This has been possible because of the relative ease of production of specific antithymus lymphocyte antibodies, whereas, so far, it has been more difficult to do the reverse, namely, to select out thymus lymphocytes by the use of antibursa lymphocyte sera.

Alternatively, varying groups of lymphoid cells have marked differences in sensitivity to physical or chemical agents. Treatment of thymus cell populations with X rays or cortiocosteroid will selectively eliminate cells in the cortical areas of the thymus, leaving cells in the medullary regions intact. Cortical thymocytes have no detectable immunological activity, whereas medullary cells have distinct immunological activities (8). Thus, in this particular system, 95 percent of the cells of the thymus, which are cortical, are selectively destroyed, leaving a cell population largely consisting of purified medullary thymus cells for further analysis. It is likely that chemical agents capable of distinguishing one group of lymphoid cells from another and acting as selective killers will be developed. We consider the passive enrichment principle a primitive one suitable as a first step in obtaining "purer" cell populations for further fractionation steps, but providing little positive information as to active sites present on the enriched cells.

Mechanical Isolation of Cells Through Micromanipulation ("Catch-as-Catch-Can")

A most direct method for the purification of a cell minority within a population would be to pick out the relevant cells under visual inspection. This would imply making the lymphoid cells in question stand out from surrounding cells. Cells releasing antibody detected by hemolytic plaques surrounding the cell can be picked out and studied by micromanipulation (18). By this method it could be shown that plaque-forming cells play no part in immunological memory of the lymphoid cell population (18). Also, it has been possible to transfer serially high-rate antibody-forming cells, allowing the cell to produce plaques under varying experimental conditions (11). This would provide a direct approach to the study of such problems as whether a single cell is capable of secreting immunoglobulins of different classes but directed against the same antigenic specificity (11). The major obvious obstacle is the limiting number of cells that can be isolated by this method, but it constitutes the most direct approach in the study of certain problems.

Isolation of Clones of Immunologically Reactive Cells by Limit-Dilution Assays

A lymphoid cell suspension, containing precursor cells for either cell-bound immunity or humoral antibody formation, can be so diluted as to contain statistically only one single immunoreactive unit per portion. This is done in practice by inoculating gradually decreasing doses of lymphoid cells into lethally irradiated animals together with antigen, followed by a later in vitro assay of the development of immunity. At cell doses leading to positive results in 63 percent of the recipients, Poisson statistics indicate that 37 percent of the individuals will have been inoculated with a single potential immune unit. This approach has been used successfully for the isolation of antibody-forming cell units from lymphoid organs of animals. Evidence for the clonal nature of the antibodies formed by such units has been obtained from their physiochemical characteristics, i.e., the antibodies produced by such a clone are homogeneous in electrophoretic mobility (29) and allotype expression (10), and all have the same association constant for the antigen (4, 28). Such "biological" isolations of clones of immunocompetent cells are valuable as they might provide information on the possible selection for certain characteristics within the clone, i.e., they may be informative in the field of somatic cell genetics. Because of the complexity of the immune process, which frequently requires co-operation between different cell types, e.g., for humoral antibody formation to take place, purity of the isolated clones has been a functional one only. In fact, in most experiments, the relative proportion of specific cells in the "clone" experiments has been lower than 1 percent. Attempts to direct clones of antibody-forming cells to become malignant (= myeloma cells) have so far been unsuccessful. It would thus seem that the limiting dilution technique is suitable for obtaining functionally pure clones of immunocompetent cells, but is not good if the object is to obtain 100 percent pure cell clones, or for studying the characteristics of the active surface sites on these immunocompetent cells in a more direct manner.

Fractionation of Cells by the use of Two-Phase Systems

Macromolecules and particles with different physicochemical properties can be separated according to the counter-current distribution principle (1). For the separation of cells it is advisable to use a phase system consisting of two layers of water-soluble polymers (e.g., dextran, polyethyleneglycol, polyvinyl pyrrolidon, 1). There will be an uneven distribution of particles or cells with different surface properties between the phases. Using such techniques it has been possible to separate different species of bacteria (2) and to fractionate erythrocytes according to species of origin or age (45). More interesting, it was possible to obtain several fractions of nucleated cells (lymphoid cells?) from chicken spleen, and a change in distribution profile was observed after immunization (7).

This procedure may constitute a useful tool for the analysis of lymphoid cells with regard to surface characteristics. Cell populations which differ according to surface charge may be readily separated. This would mean that any induced change in surface charge caused by whatever means would be reflected in a changed distribution pattern of the cell in the counter-current test. It is possible to make a highly precise separation using automatic counter-current distribution machines consisting of any series of two-phase systems. We consider this approach to be technically a more difficult one than conventional gradient separation techniques now very much in vogue, but would recommend it for its potentially high resolving power.

Distribution Analysis of Reactive Lymphoid Cells According to Size or Density

Increase in cell size or density, or both, will allow a separation of the changed cell(s) from the rest of the population by centrifugation in gradients of varying density or by velocity sedimentation. The latter method achieves separation mainly on the basis of size. The following discussion will deal only with techniques that take advantage of a reaction between particles, usually antigenic, and lymphocytes with specifically reactive sites for these particles.

The following is an example of separation by difference in size: lymphoid cells with receptors for the surface antigens of a given bacterium can be detected by incubating the cells with the bacteria in vitro. The mixture is then spun at appropriate speed, after which the lymphoid cells are found in the pellet while the bacteria remain in the supernatant. After several washings the cells are plated onto agar; bacterial colonies then appear around those cells which brought bacteria down with them during the centrifugation. Thus, cells with specific receptors for the bacterial antigens can be studied as an isolated group and enumerated by the corresponding number of discrete bacterial colonies (20). However, this method will not permit separation within the cell population, as the additional mass and volume provided by the attached bacteria are too small.

If the particle size is increased to that of an erythrocyte, however, fractionation of the lymphoid cells binding red cells to their surfaces can be accomplished. Rosette formation by such lymphocytes can be shown to reflect the potential immune capacity of the cell in certain systems. Thus, if spleen cells are incubated with sheep erythrocytes, and the rosette-forming cells are then removed by gradient centrifugation, the remaining population can be shown to have lost its ability to produce humoral antibodies against sheep red blood cells while retaining the capacity to react against other erythrocytes, such as those of the chicken (5, 35). In principle, gradient centrifugation or velocity sedimentation could be applicable for any antigen that can be presented to immunocompetent cells in the form of particles of suitable size. This can then be used either to deprive or enrich a population with specific, particle-binding cells. An illustration of the potential as to enrichment can be given. Peripheral blood lymphocytes from chicken immunized against sheep erythrocytes were subjected to purification in an effort to obtain a "pure" population of rosette-forming cells. It was possible to achieve a purity of approximately 90 percent, a hitherto unsurpassed degree of homogeneity of immunoreactive cells in fractionation procedures (16). Using this cell population, it was possible to perform experiments directly demonstrating the restricted immune capacity of the isolated lymphoid cells, as these rosette cells for sheep erythrocytes were completely unable to react against a strong histocompatibility antigen.

Surface receptors of nonantibody nature are known to exist on certain lymphoid cells, and they can be demonstrated by the formation of rosettes. Primate erythrocytes and the lymphocytes of most mammals have receptors for antigen-antibody-complement complexes which have been called immune adherence receptors (33). These receptors have recently been suggested to constitute a marker for different subclasses of lymphocytes (24). Consequently, sedimentation of lymphocytes, together with erythrocytes sensitized within antibody-complement complexes, should allow the specific separation of this subgroup of immuno-competent cells. In a more specific way, the sedimentation of primate red cells together with lymphocytes, after incubating the latter with soluble antigen plus complement, should theoretically permit a depletion or enrichment of lymphocytes with specificity for the antigen used.

Lymphoid cells displaying immunoglobulin markers on their surfaces can be enumerated (and most likely separated with the above techniques) by the use of immunoglobulin-coated red cells incubated with the lymphoid cells in the presence of an anti-immunoglobulin antiserum (14). This should allow separation of lymphoid cells into subgroups of immunoglobulin producers.

The virtues of the gradient centrifugation and velocity techniques have been emphasized above, but the vices have not. One major disadvantage of the techniques is that we are left with cell populations contaminated with particles, mostly of antigenic nature, which might mean serious trouble for further analysis. The phenomenon of particles adhering to the reactive cells involves a large factor of all-or-none reactions at a discontinuous rate, making any sophisticated analysis of characteristics of active sites very difficult.

Cellular Electrophoresis

Cells placed in an electric field will move with characteristic speed under appropriate conditions, and the rate of migration is primarily dependent upon the charge of the cell (6). Using such technical devices, it has been possible to select mutants within malignant cell lines having a higher net charge than the original cell line, and with greater capacity to spread and metastasize (36). Circulating peripheral lymphoid cells have a very high negative surface charge under physiological conditions and this is seemingly a necessary prerequisite for cells to exist in the blood for any period of time (6). Cells from lymphoid organs have been shown to consist of some with lower net charge and others with surface charges similar to those of the peripheral blood, and there are reasons to believe that these differences correlate with potential immune capacities (6). Antibodies directed against immunoglobulins have demonstrated the existence of immunoglobulin-like material on the surfaces of lymphoid cells, as reflected by a reduction in the migratory speed of the cells due to diminished charge (6). The use of a charged antigen would conceivably alter the charge of lymphoid cells specifically reacting with it, with consequent changes in electrostatic potential.

Preparative cellular electrophoresis is a developing field and only matters of technicality are delaying the production of machines capable of separating huge numbers of cells into groups according to their electrostatic charges. The selective modification of cell charge in conjunction with an appropriate preparative cellular electrophoresis apparatus would constitute a powerful tool for analysis of the heterogeneity of lymphoid cells, making it possible to relate charge of sites to potential function. No doubt this technique will demonstrate a still larger complexity of lymphoid cell populations than has already been found with gradient or velocity sedimentation procedures. Isolation of clones of immunocompetent cells, however, might require a setup using isoelectric focusing. No such apparatus exists as yet; this would require the synthesis of ampholytes not harmful for living cells. Even so, such a method would only be able to distinguish cells of the same clone at the same stage of the cell cycle, since there is a wide variation in surface properties among cells of the same cell line in various stages of the replicative process (12, 50).

Fluorescence Selection

Labeling surface sites with a fluorescent substance makes it possible to sort out the labeled cells from the rest of the population (25). The apparatus for doing this is an electric photomultiplier tube which detects the fluorescence of single cells passing through a small glass nozzle in microdroplets. The droplets are then deflected by an electric field, the intensity of which is proportional to the fluorescence of the cell within the droplet. The small drops are thus collected in tubes according to the brilliance of the fluorescent cells. Problems with the highly complex electronics have so far hampered the development of this apparatus into a functioning unit.

Such an instrument would have obvious possible applications, e.g., for the selection of certain lymphoma cells with several kinds of virus-associated antigens expressing themselves in the nucleus or cytoplasm or on the outer cell membrane (27). These antigens can be labeled with fluorescein-tagged antibodies and it might thus be possible, for example, to select for antigen-negative cells as units that have reverted back to "normal" behavior. Any cells that can be selectively labeled can be analyzed by this approach in a highly enriched form. A questionable factor is the necessity of introducing a soluble fluorescent substance onto the surface sites which might interfere with certain functions of the cell as well as with further analysis of the characteristics of the membrane structures.

Specific Adsorption of Lymphoid Cells Onto Antigen-Labeled Slides

Films of activated gel (agarose), using the cyanogen bromide activation principle (42), have recently been used for making antigen-coated surfaces to study specific adsorption of cells through membrane-bound antibodies. It was found that lymphoid cells derived from immunized individuals showed selective affinity for the relevant antigen surface (42). This could be shown to be specific by dipping the film with the attached cells into melted agar containing antigen-coated red cells, followed by analysis of the number of hemolytic plaque-forming cells present on the film. Under appropriate conditions, including the use of heparin in the incubation medium, it is possible to reduce nonspecific attachment to close to nil (42). The number of cells attaching can easily be enumerated by ordinary light microscopy. This type of isolation of antigen-binding cells might prove useful in rapid diagnostic tests for certain allergic systems, a field in which in vitro methods are at present sorely needed.

Affinity Fractionation of Cells by Column Filtration

Separation of lymphoid cells according to their varying affinities for column material (usually beads coated with various substances) has been attempted with success in several experimental systems. The results obtained have so far provided more detailed information about the characteristics of certain surface structures on lymphoid cells than has hitherto been possible with the use of other fractionation techniques. Thus, we consider it appropriate to allot more space to this approach than to the others already mentioned.

A typical experiment consists of labeling beads with a given substance, making a bead column, washing this, and then applying a cell suspension to the column. Control, passed, and retained cells are then examined for their properties and activities.

A first illustration concerns the separation of lymphoid cells according to charge by column filtration and the potentialities for immune responses of the passing cells.

Immunoglobulin-like material can be shown to cover one-third or more of the surface of a certain percentage of small lymphocytes (37). A characteristic of serum immunoglobulins is their wide range of electrostatic charge easily visualized by their varying migration rates in electric fields even if they belong to the same immunoglobulin class and react against the same haptenic group. Since the immunoglobulins produced by a cell clone can be shown to migrate as a very narrow band (29), the wide variation in ordinary serum immunoglobins probably reflects a corresponding heterogeneity at the cellular level. If predominantly one type of immunoglobulin molecules were represented on the surface of the single immunocompetent cell this might affect the general overall charge of the cell, and this could be used as a distinguishing marker for separation. It is well established that positively charged antigens will mainly induce the formation of antibodies with a negative charge, and vice versa (32). In this system it was shown that in the immunoglobulin molecule the light chain, and particularly its variable part, was the major factor responsible for this variation in charge of the antibody molecule as a consequence of the corresponding basic or acidic nature of the antigen (32, 41). However, the charged sites on the antibody molecule in this system could be shown *not* to be situated within the actual antigen-combining site (39). It was considered likely that this would imply that the surface receptor for antigen present on immunocompetent cells might carry the same relative charge as the eventual product of that cell, the humoral antibodies.

In verification of this hypothesis are recent results obtained when a population of immunocompetent cells was allowed to filter through a column containing beads with a high negative charge (41). The cells which passed through the column displayed a selective decrease in their capacity to produce antibodies with a strong positive charge. This would be expected if the cell-bound receptor for antigen had the same relative charge as its humoral counterpart, and it would also mean that the charges of the surface-attached antibodies would be large enough to affect the net electrostatic behavior of the whole cell. It would seem likely that electrostatic separation according to charge of the immunoglobulin-like receptors on the membrane of immunocompetent cells is a possible way of approaching fractionation of subgroups of cells according to expressed V-genes, particularly as far as light chains are concerned (32). Separation of cells within the pH limits of cellular viability with exchange chromatographic techniques would be necessary for a more sophisticated approach to this problem, but this seems to be a very interesting possibility.

Instead of using general physicochemical characteristics of a cell such as the general net charge it should be possible to use column separation using more specific probes to pick up certain cells according to well-defined chemical characteristics. Two approaches will be discussed. The first is entirely fictional; as far as is known to the authors, nothing has been published on this subject. The second, however, relies on solid experimental evidence.

The first approach would have as a basic requirement access to a relatively homogeneous substance with affinity for lymphoid cells of a certain kind. One such substance would be anti-immunoglobulin antibodies which, if purified to a certain minimal degree (48), could be used for coating a bead column to provide a specific immunosorbant for cells carrying such immunoglobulin antigens on their outer surface. Normally, mere attachment to beads of antiserum or even of purified gamma globulins containing the relevant antibodies would be of little help; this would fail to function as a specific retaining factor for cellular separation in the column system. This owes to quantitative reasons, since a minimum of 10^3 specific binding sites for the cell per square micron area of the beads seem to be required if selective retention of cells is to be achieved when using conventional antigen-coated columns (47; as will be discussed later). Thus, highly purified anticellular antibodies would be required for coating the beads, and we suggest that the simplest approach would be to use a relevant myeloma protein-coated column through which the anti-immunoglobulin antiserum would be allowed to flow in excessive quantities, thereby creating a high specific density of antibody molecules with at least one (or, if IgM, more than one) site free for interaction with passing cells. Such an approach, if functional, would be of great importance if one would like selectively to eliminate a cell population that produces an antibody belonging to a particular immunoglobulin class. Thus, in the case of "blocking" antibodies in tumor immunology (44), IgM-producing cells might be removed to see whether this would influence the switch to antibodies of IgG nature. All this is based on the assumption that lymphoid cells differ in their surface expression of immunoglobulin classes, which might not be true at least for "normal" small lymphocytes.

There is another group of most interesting substances that can be used for coating bead columns and which can be shown to have binding affinities for lymphoid cells, namely, the large group of lectins. A number of these have distinct and specific binding requirements mostly for sugar groups, and many of these lectins can be purified and obtained in highly homogeneous form and in quantity. An example is the phytoagglutinin concanavalin A which shows a preferential binding for IgM of the various immunoglobulin classes (34), and which might be a useful probe for such questions as the expression of carbohydrate groups on the cell membrane. Various phytoagglutinins and mitogens can be shown to activate lymphoid cells in different manners as assessed by biological tests of cell function (23), and it is quite possible that this might represent both selective activation at the cellular level as well as at the level of the population. It is well established that transformation of cells into malignancy is accompanied by drastic changes in the expression of sites reactive with lectins, particularly concanavalin A (40). Thus, there is good reason to believe that heterogeneity at the cellular level can be recognized by suitable phytoagglutinins, and bead columns coated with lectins might constitute a highly fruitful system for the analysis of cell surface structure in relation to possible function of a particular site. There is little problem in the construction of such columns. Concanavalin A, for instance, has two identical binding sites reactive with glucose; commercially available dextran beads such as Sephadex could be coated with the lectin in excess, thereby allowing one of the sites to bind to dextran and the other to remain free for association with surface sites on passing cells. A more general approach for attaching macromolecules to beads made up of dextran or agarose is by using cyanogen bromide as an activator of the beads. It would thus be possible to construct bead columns coated with one or several lectins according to choice, with corresponding changes in binding qualities.

It has long been postulated that immunocompetent cells probably express their potential immune capacity by a display of preformed antigen-binding receptors on the outer cell surface (21). This hypothesis has been amply verified in the column separation systems in which the antigen-coated bead column can be shown to retain selectively immunocompetent cells of several types (47). By such techniques it has been possible to demonstrate the existence of cell-bound antibodies on cells potentially able to produce humoral antibodies. So far, however, the lymphoid cells participating in cell-bound immune reactions, i.e., thymus-dependent immune systems, show no evidence of specific retention in similar circumstances (48). When analyzing the surface-receptors on the lymphoid cells of the humoral antibody line, or B line (48), it was found that cells in all stages of differentiation, once immunocompetent, would display this capacity by the existence of membrane-bound antibodies. Thus, high-rate antibody-forming cells, immunological memory cells, and lymphoid cells from normal unimmunized individuals, all of B type, could be separated according to potential immune reactivity (48). This retention under certain circumstances approached 100 percent, indicating that in those particular experiments all immuncompetent cells were present in stages of the cell cycle where antigen-binding receptors were expressed (48). When analyzed further as to characteristics of antigen-binding specificity, the surface sites binding antigen were shown to have binding behavior very similar to that of the potential product of the cell, the humoral antibody. Thus, a cell destined to produce humoral antibody with high affinity for the antigen would also display high-affinity surface sites for that immunogen, and vice versa. Furthermore, by the use of hapten-carrier immunogens using the same haptenic group on different carrier macromolecules (49), it was possible to demonstrate that the surface antibody had antigen-combining sites of sizes very similar to those of the humoral antibody directed against the same haptenic group (49). Here, immunization with hapten A coupled to carrier X would induce immune cells directed against A *or* X sites shown by the fact that columns coated with A coupled to carrier Y or columns coated with carrier X alone would remove only anti-A or anti-X cells, respectively. Also, if immune anti-A-X cells were applied to a column coated with A-X molecules but in the presence of free A hapten, only the anti-A cells were selectively allowed to pass through, whereas anti-X cells were retained in the normal fashion.

An analysis of cell receptor sites for antigen has also been possible by the use of anti-lymphocyte and anti-immunoglobulin sera. Lymphoid cells which had been preincubated with anti-lymphocyte serum produced in another species, were subsequently retained on filtration through the specific antigen-coated column (46). This showed that mere combination between antibodies with any surface site would not hinder the binding between surface receptors for antigen and bead-attached antigen. However, if cells were preincubated with anti-immunoglobulin antibodies prior to column filtration, a significant degree of blocking of attachment could be demonstrated. Further, more sophisticated analysis using anti-immunoglobulin antibodies specific for a given immunoglobulin class showed that it was possible to block the relevant antigen-binding receptors carrying the appropriate class-specific antigenic determinants. This was done as follows: incubation with an anti-$\gamma 1$ or $\gamma 2a$ antiserum would allow the respective potential $\gamma 1$ or $\gamma 2a$-producing cells, with specific binding ability for the column antigen, to pass through the column undisturbed, whereas cells belonging to the other immunoglobulin classes, also having specificity for the column antigen, were retained. This directly indicates that the surface receptor for antigen, and the potential product of the cell with the same antigen-binding specificity, i.e., the humoral antibody, would have the same heavy-chain class determinants. Additional experiments using antilight-chain specific sera showed that the surface antibody also had the same light-chain type as the humoral antibody eventually to be synthesized by that cell (46). Thus, it seems likely that the surface receptor for antigen and the humoral antibody produced by the same cell represent different distribution forms of the same molecule, with only minor chemical differences to account for the difference in "solubility." In summary, it has been possible to characterize and describe active antigen-combining sites on lymphocytes, linking structural analysis with function by the use of affinity separation through antigen-coated columns.

As we consider column separation of heterogeneous cell populations a most promising way of eventually obtaining homogeneous subpopulations of lymphoid cells according to surface sites, we will dwell in detail on the virtues and drawbacks of this approach. The main virtues of the technique are two: (*a*) It allows a separation of cells according to surface structure using cheap material in a matter of minutes (48). A very large number of cells (10^9 or more) can be analyzed with one column, and fractionation of cells is easily measured using standard biochemical fraction collectors to allow isolated analysis of each fraction (see below). (*b*) It makes possible the separation of immunocompetent cells into cell populations of distinct immunoreactivity without adding any detectable contaminating antigen to the cells (47). This is a fact of importance for several aspects of further studies on the function and characteristics of cell surface sites. This latter fact that immunocompetent cells are not triggered into reactivity by the combination between surface-bound antigen and cell-bound antibodies in the column systems (48) is not necessarily true for other kinds of cell systems, however. Fat cells exposed for a short time to insulin-coated Sepharose beads in vitro, could be shown to be activated into glucose incorporation and metabolism in an insulin-specific manner (17). In these experiments there was no evidence of carry-over of insulin from the beads to the cells, which would indicate that mere combination between a receptor for insulin and the insulin molecule would induce the fat cell. The findings would also indicate that fat cells have their insulin-specific receptor available on their outer cell membrane.

What, then, are the disadvantages of using column separation for the isolation of subgroups of lymphoid cells? The major one so far has been the relatively low factor of enrichment of relevant, specific cells in the retained cell populations due to the high background of lymphoid cells retained for nonspecific reasons, such as general electric charge (41). This has so far meant that the major analysis of surface sites on lymphoid cells by this technique has been carried out by studying the selectively deprived cell populations—the passed cells. This kind of study has been quite fruitful (48), but for further detailed biochemical analysis it would be necessary to use highly enriched cell populations. Certain new bead materials suitable as immunosorbants do not have this high degree of cellular retention, and perhaps will remove only the relevant, immunocompetent cells (43). Yet it remains to be shown that cells retained for immunological reasons can be specifically eluted and enriched by the addition of free hapten or antigen to the column fluid and, is so, that they will not be interfered with by this treatment (43). Mechanical elution can be shown to function but will inflict damage and lysis to a certain percentage of the cells (47). We would, however, believe that both immunological and mechanical forces might be necessary for the elution of cells because of secondary surface-to-surface forces coming into play when cells adhere to most bead materials. In fact, there are reasons to believe that such nonserological forces are the major factors involved in actual retention of immunocompetent cells when using columns of antigen-coated beads of glass or polymetaacrylic plastic (47). Here it can be shown that mere changes in the flow rates of filtered cells through a column not coated with antigen will significantly alter the percentage of retained cells, and with a low enough flow rate very few cells will pass through. Preliminary experiments on the actual profile distribution of immunocompetent cells coming out at the bottom of the columns strongly suggest that if, for example, anti-A and anti-B cells are mixed and run together through a column coated with A, the actual anti-A cells that might sneak through the column will be delayed as compared to the average anti-B cell in the different fractions of the passed cells. Figure 1 depicts the outcome of such an experiment where it was actually possible to attain a selective enrichment of anti-A versus anti-B cells in some later fractions as contrasted to the average overall reduction. This would strongly suggest that the antigen-coated column, using at least this particular bead material, will function in a two-step manner in retaining immunocompetent cells. The first step would involve the actual combination between cell-attached antibodies and the column antigen, thereby retarding the rate of passage of the cell as compared to other cells not having this specific antibody type. This serologically specific reduction would automatically increase the likelihood of the cell being retained, actually binding the cell to the bead surface, whereas now after surface-to-surface contact the major part of the forces involved

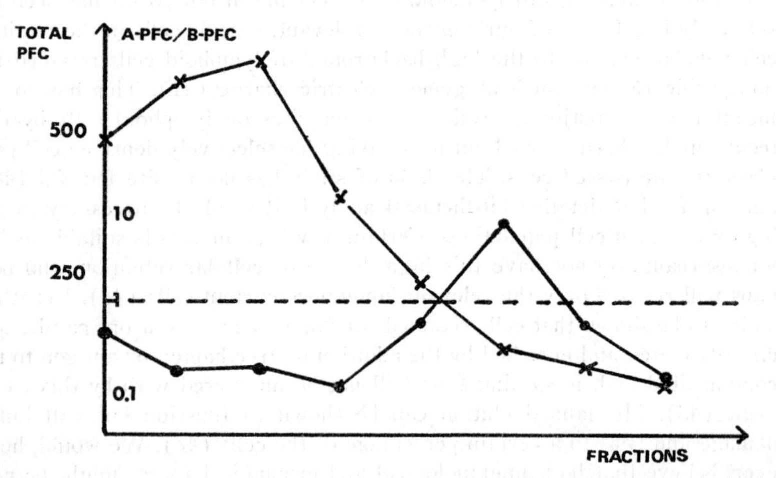

FIGURE 1. A mixture of anti-A and anti-B cells was filtered through an A-coated column and collected in fractions. The first fraction obtained is furthest to the left. The ratio 1 in A-PFC/B-PFC refers to the ratio in the control cell population. Total PFC denotes total number of plaque-forming cells (=A + B PFC) in each separate fraction.

would be of nonserological nature. As already stated, the use of other bead material might circumvent this problem of nonspecific retention (43).[1]

Criteria for an Ideal Technique for the Isolation of Lymphoid Cells With Distinguishing Active Surface Sites

This will summarize what we consider the critical needs to be fulfilled in developing an optimal technique for the separation and isolation of subgroups of lymphoid cells with different surface characteristics. To study the characteristics of a group of cells within a larger population, one of two ways may be chosen: (*a*) if the other cells in the population do not interfere, the functional characteristics of the cells with relevant sites can be analyzed by selectively removing them and studying the deprived population. These specific deprivation techniques are the simplest and will yield important information about such details as the characteristics of surface receptors for antigen, binding characteristics, biochemical buildup, etc. (48). The second way (*b*) is selectively to enrich and purify the cells with relevant surface sites. This has so far been more difficult for technical reasons, but it has obvious advantages compared to the depriving techniques for such purposes as detailed studies of cellular restriction in immunocompetence, detailed analysis of the biosynthesis and turn-over of the surface sites, etc. The optimal technique should thus be one of enrichment.

It is vital that the method of isolation itself should not change the site or the cell in any detectable way, thereby possibly clouding future analysis. Thus, the use of soluble substances to coat sites in order to allow cellular separation, or the admixture of particulate material to make physical separation between particle-binding and nonbinding cells possible, are undesirable factors on this account. For this reason, insoluble material such as beads or any other surface configuration with specific affinity for either nonrelevant or for relevant cells should be the material of choice. Obviously, material with affinity only for the relevant cells would be the simplest to prepare when trying to purify specific cells, and we would advocate the bead column as perhaps the most promising method at present for obtaining large numbers of highly purified, subpopulation of cells. The procedure for this purification, however, need not necessarily rely upon the elution of retained cells with specific free antigen. We believe that a cell once stuck to most surfaces will adhere to them by a variety of physicochemical forces that prevent elution by free antigen (which, in any case, would have to come from within the bead to come close enough to compete with the surfact-attached antigen for the membrane-bound antibodies. Hence, fractionation of cells in passage might be a better way of obtaining enriched popula-

[1] In a recent work by L. Wofsy, J. Kimura, and P. Truffa-Bachi, pure populations of antihapten-specific lymphocytes could be eluted from polyacrylamide bead columns (manuscript submitted to *Nature*).

tions. The best bead material would be of a kind to provide little nonspecific cellular retention (43) and the cells would have to travel through tall columns at low speed to permit enough fractionation according to affinity for the bead antigen *without* allowing the cells to adhere to the beads. The same approach could be used regardless of whether antigen or some other substance with affinity for only a minority of cells were used for coating the column.

Concluding Remarks

Much work is going on in the world in an effort to fractionate and isolate lymphoid as well as other cells as has been indicated briefly in this article. There are obvious reasons for doing this, and some practical implications from the medical point of view have already been reported (19). It seems likely that separation of normal from malignant cells will become feasible before long.

For any scientist trying to obtain a completely pure and homogeneous population of cells through fraction and isolation of a heterogeneous cell suspension as starting material, the obstacles are impressive. The surface of the mammalian cell is an exceedingly active material with rapid turnover and drastic changes in surface sites, depending upon the stage in the cell cycle (12). Most, if not all cells seem to express most of their "social" phenotype during the G_1 and early S phases of the cycle and, from the point of view of isolating cells with immunocompetence, it would be reassuring to know that most small lymphocytes under study are in G_1 (22). At any rate, this variation in cell surface expression in relation to the mitotic phase, and the fact that any group of active surface sites will represent only one of many groups, will largely restrict the possibility of obtaining complete purity without accepting very heavy losses of cells during the multistep fractionation procedures required.

Obviously, the most direct method for purification of a cell type within a heterogeneous population would be to pick out that cell by visual inspection, but this procedure will yield only a very limited number of cells (15, 18). Even so, the population obtained would most likely be heterogeneous, since cells synthesizing antibody against a given antigen at a high rate can be shown to constitute a mixture of entities with "similar" specificity for the antigen, but belonging to different immunoglobulin-producing cell lines, or producing antibodies of the same immunoglobulin class with different amino acid sequence, etc. Biological cloning in vivo of the relevant antigen-combining cells would thus have to be the precluding step prior to any attempt to isolate a population of cells with identical antigen-binding surface receptors. However, the need for cellular interaction to occur between lymphoid cells of thymus-derived and "bursa-derived" types in order for humoral antibody synthesis to take place against several immunogens (13, 30) would work against this purification process. We believe that it will be much simpler to select pure subgroups of lymphoid cells where purity refers to a given surface site not related to antigen-binding specificity. Such a minority cell type would be, for example, the small lymphocytes having receptors for activated complement components. There are good reasons to believe in this case that a high level of homogeneity of the isolated cell population could be obtained on the basis of this receptor site.

Finally, may it be stated that analysis of a superficially homogeneous cell population, the lymphocytes, by the use of various methods aiming to reveal heterogeneity have demonstrated a cellular society so complex that it is quite obvious that most experimental systems of cellular immunology are very similar to experiments in human sociology. It is to be hoped that purification of cells into categories will proceed faster than the human counterpart.

LITERATURE CITED

1. Albertsson, P. Å. 1962. *Methods Biochem. Anal.* 10:229
2. Albertsson, P. Å., Baird, G. D. 1962. *Exp. Cell Res.* 28:296
3. Armstrong, W. D., Diener, E. 1969. *J. Exp. Med.* 129:371
4. Asconas, B. A., Williamson, A. R., Wright, B. F. G. 1970. *Proc. Nat. Acad. Sci. USA* 67:1398
5. Bach, J. F., Muller, J. Y., Dardenne, M. 1970. *Nature (London)* 227: 1251
6. Bert, G., Massaro, A. L., Di Cossano, D. L., Maja, M. 1969. *Immunology* 17:1
7. Bäck, O. Unpublished experiments
8. Blomgren, H., Andersson, B. 1969. *Exp. Cell. Res.,* 57, 185
9. Bloom, B. R., Jimenez, L., Marcus, P. I. 1970. *J. Exp. Med.* 132:16
10. Bosma, M., Weiler, E. 1970. *J. Immunol.* 104:203
11. Bussard, A., Nossal, G. in *Cell Interaction in the Imune Response,* ed. A. Cross. London : Academic. In press
12. Cikes, M., Friberg, S., Jr. 1970. *Proc. Nat. Acad. Sci.* USA In press
13. Claman, H. N., Chaperon, E. A., Triplett, R. F. 1966. *Proc. Soc. Exp. Biol. Med.* 122:1167
14. Coombs, R. R. A., Feinstein, A., Wilson, A. B. 1969. *Lancet* 2:1157
15. Coffino, P., Laskov, R., Schaff, M. D. 1970. *Science* 167:186
16. Crone, M., Hála, K., Simonsen, M. 1969. In *Developmental Aspects of Antibody Formation and Structure.* ed. J. Sterzl, Prague.
17. Cuatrecasas, J. 1969. *Proc. Nat. Acad. Sci. USA* 63:450
18. Cunningham, A. J. 1969. *Immunology* 16:621
19. Dickie, K. A., van Hooft, J. I. M., van Beckum, D. W. 1968. *Transplantation* 6:562
20. Diener, E. 1968. *J. Immunol.* 100:1063
21. Ehrlich, P. 1900. *Proc. Roy. Soc. (London)* 66:424
22. Gelfant, S. 1966. *Methods Cell Physiol.* 2:359
23. Goldstein, I. J., Iyer, R. N. 1966. *Biochim. Biophys. Acta* 121:197
24. Huber, H., et al. 1968. *Science* 162: 1281
25. Hulett, H. R., et al. 1969. *Science* 166: 747
26. Jerne, N. K., Nordin, A. A., Henry, C. 1963. In *Cell Bound Antibodies,* ed. B. Amos, H. Koprowski. Philadelphia, Pa.: Wistar Institute Press
27. Klein, G. 1966. *Ann. Rev. Microbiol.* 20:223
28. Klinman, N. R. 1970. *Fed. Proc.* 29: 571
29. Luzzati, A. L., Tosi, R. M., Carbonara, A. O. 1970. *J. Exp. Med.* 132: 199
30. Miller, J. F. A. P., et al. 1970. *Transplant. Rev. Vol. I*
31. Martin, W. J., Miller, J. F. A. P. 1968. *J. Exp. Med.* 128:855
32. Mozes, E., Robbins, J. B., Sela, M. 1967. *Immunochemistry* 4:239
33. Nelson, R. A. 1952. *Science* 118:733
34. Nordin, A. A., Cosenza, H., Hopkins, W. 1970. *J. Immunol.* 103:859
35. Osoba, D. 1970. *J. Exp. Med.* 132: 368
36. Purdom, L., Ambrose, E. J., Klein, G. 1958. *Nature (London)* 181:1586
37. Raff, M. C., Sternberg, M., Taylor, R. B. 1970. *Nature (London)* 225: 553
38. Reif, A. E., Allen, J. M. 1964. *J. Exp. Med.* 102:413
39. Rüde, E., Mozes, E., Sela, M. 1968. *Biochemistry* 7:2971
40. Sachs, L., Inbar, M. 1969. *Nature (London)* 223:710
41. Sela, M., Mozes, E., Shearer, G. M., Karniely, Y. 1970. Manuscript to be published
42. Tallberg, T., Andersson, B. Data to be published.
43. Truffa-Bachi, P., Wofsy, L. 1970. *Proc. Nat. Acad. Sci. USA* 66:685
44. Voisin, G. A., Kinsky, R. G., Jansen, F. K. 1966. *Nature (London)* 210: 138
45. Walter, H., Selby, F. W., Brahe, J. M. 1964. *Biochem. Biophys. Res. Commun.* 15:497
46. Walters, C. S., Wigzell, H. 1970. *J. Exp. Med.* 132:1233
47. Wigzell, H., Andersson, B. 1969. *J. Exp. Med.* 129:23
48. Wigzell, H. 1970. *Transplant. Rev.* 5: 76
49. Wigzell, H., Mäkelä, O. 1970. *J. Exp. Med.* 132:110
50. Yata, J., Klein, G. 1969. *Int. J. Cancer* 4:767

PHYLOGENY AND FUNCTION OF THE COMPLEMENT SYSTEM[1]

1573

IRMA GIGLI, M.D.[2] AND K. FRANK AUSTEN, M.D.

Departments of Dermatology and Medicine, Harvard Medical School; and Department of Medicine, Robert B. Brigham Hospital, Boston, Massachusetts

CONTENTS

INTRODUCTION

The phylogeny of the immune response, particularly with regard to antibody production and the definition of the classes of immunoglobulins present in various animal species, has received increasing attention (27, 28, 45, 80, 86). The relation of the structure of these immunoglobulins to their function is also being analyzed and in this regard it is pertinent to trace the phylogeny of a nonspecific effector system of the immune reaction, namely, the complement system.

Before the turn of the century Buchner (14), von Fodor (31), and Nuttal (79) observed that fresh serum exerted a destructive influence upon bacteria, and that this effect was lost upon heating serum at 56°C. Bordet subsequently demonstrated (8) that small amounts of fresh serum, inactive alone,

[1] Supported by NIH Grant AI-07722, a grant from the John A. Hartford Foundation, Inc., and a grant from the American Cancer Society, Inc.
[2] Recipient of a Faculty Research Award from the American Cancer Society, Inc.

restored the bactericidal activity of heat-inactivated serum. The heat-labile substance present in fresh normal serum was named "complement," and the heat-stable factor, "sensitizing substance" or "bactericidin" (antibody). It was also established that lysis of erythrocytes from a variety of animal species required the presence of both factors, complement and specific antibody to the erythrocyte. These observations were followed by the recognition of four serum fractions which were inactive by themselves but produced lysis of a sensitized erythroycte when combined. These complement (C) constituents were referred to as C1, C4, C2, and C3. Treatment of serum by acidification and dialysis (26), heat inactivation, cobra venom (91), or ammonia (43) provided reagents, R reagents, lacking one or more of the individual complement components for use in serological studies of the complement system (5). Studies on guinea pig and human serum have established that the classical component referred to as C3 actually included six proteins which function sequentially, C3, C5, C6, C7, C8, C9 (1, 70, 75).

The Complement Reaction Sequence

Detailed studies of the nature of the complement proteins and their sequential activation have been concentrated to a large extent on the analysis of immune hemolysis. The immune hemolytic sequence shown in Figure 1 uses sheep erythrocytes (E) as the target cell, although similar reactions have been shown with other cell systems (3, 104), bacteria (42), and for soluble antigens (33). The union of IgG or IgM immunoglobulins with their corresponding antigen activates the complement system; in immune hemolysis the fixation and activation of C1 to $C\overline{1}$ usually requires either a doublet of IgG

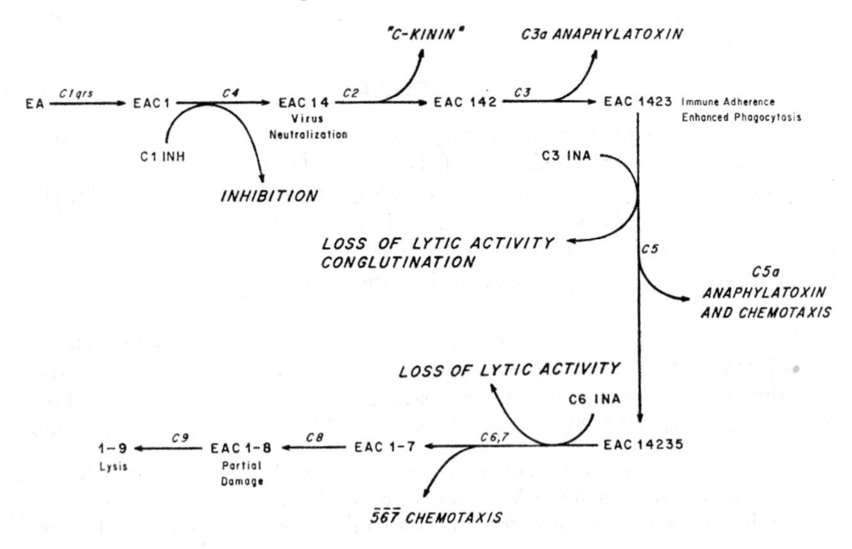

molecules or a single IgM molecule (11). The first component, C1, is an inactive complex of three subunits, C1q, C1r, and C1s (58, 71). The subunit C1q carries the binding site which physically combines with the immunoglobulin, thereby activating C1r (72), which in turn activates C1s to C$\overline{1}$s. The C$\overline{1}$s subunit alone or in the C$\overline{1}$s portion of the C$\overline{1}$ molecule demonstrates esterolytic activity on the synthetic substrates, *p*-toluene sulfonyl L-arginine methyl ester (TAMe), N-acetyl L-arginine methyl ester (AAMe), N-acetyl L-tyrosine methyl ester (ATMe) or ethyl ester (ATEe), and acetyl glycyllysyl methyl ester (AGLMe) (52, 72), and splits C4 into at least two portions (66). The major product, C4b, either is bound to the cell membrane forming the cellular intermediate EAC14 (107), or remains in the fluid phase as a hemolytically inactive product, C4i.

It has long been appreciated that C4 precedes C2 in the hemolytic sequence even though both are natural substrates of the C$\overline{1}$s subunit (57). The basis of this order has only recently become apparent (37, 38) with the finding that the activity of C$\overline{1}$ upon C2 is markedly enhanced by the prior interaction of C$\overline{1}$, free in the fluid phase or cell-bound (EAC1), with C4. It has been postulated the C$\overline{1}$-C4 interaction alters the configuration of the C$\overline{1}$ molecule, exposing the enzymatic site responsible for C2 utilization. The action of the C$\overline{14}$ enzyme on C2 leads to the physical uptake of the major fragment, C2a, by the EAC14 intermediate to generate the EAC142 complex (61, 63), which decays to release C2a^{d} into the fluid phase, thereby reverting to the original EAC14 (9) state. This intermediate is able to interact again with native C2 to regenerate the EAC142 intermediate. Major fragments not immediately taken up by the EAC142 complex become hemolytically inactive and are designated C2i. The assembly of the major fragments of C4 and C2 yields a new enzymatic activity, C$\overline{42}$, termed C3 convertase, which cleaves C3 into C3a and C3b. The major fragment, C3b, remains on the intermediate, EAC1423, or appears in the fluid phase as the hemolytically inactive byproduct, C3i. The lesser fragment, C3a, appears only in the fluid phase (67, 69).

The interaction of the intermediate EAC1423 with C5 releases a lesser fragment of C5, C5a, into the fluid phase and binds the major fragment, C5b, to the intermediate (99). The interaction of C5 with the EAC1423 intermediate is influenced by the supply of C6 and C7 even though these components act later in the sequence (78). Because the bound C5b fragment is unstable, the EAC14235 intermediate is a rate-limiting step in the sequence (100) similar to the earlier intermediate EAC142, which carries C2a.

Once the cell has reached the EAC1423567 state, neither the decay of C2 nor the decay of C5 prevents the effective interaction of the intermediate with C8 and C9. The interaction with C8, as demonstrated by C8 depletion from the fluid phase and formation of EAC14235678 sites on the cell membrane, initiates cell damage (102). Although such EAC14235678 cells tend to lyse upon incubation at 37°C, the introduction of C9 markedly increases their rate of lysis. The C9 disappears from the fluid phase and is found specifically bound to the cells (46). Electron microscopic studies have revealed the appearance of the characteristic ultrastructural lesions on the cell membrane following C9 action (10, 48); similar lesions have recently been reported to occur in cells in the EAC14235 state (87).

ALTERNATIVE PATHWAYS OF ACTIVATION OF THE COMPLEMENT SYSTEM

The complement system can be activated by pathways which apparently differ from the classical immunological mechanism of activation described above. These include: direct enzymatic activation of C1 (84); interaction of endotoxin with the complement sequence in a manner which achieves relative sparing of C1, C4, and C2 (33); by formation of an immune complex involving the electrophoretically "fast" 7S guinea pig immunoglobulin designated $\gamma 1$ (95); and by a bypass of a requirement for these early components by interaction of cobra venom with a protein cofactor (68).

Pillemer et al (84) reported that the addition of streptokinase to human serum inactivated complement by depletion of C4 and C2 as measured with the classical R reagents. Since streptokinase converts plasminogen to plasmin (88) and plasmin had no effect on C4 or C2 in the absence of C1, it was suggested that plasmin activated C1, which in turn depleted the serum of its natural substrates. More recently, Ratnoff & Naff (88) have shown that the C1s moiety of the C1 molecule develops esterase activity in the absence of C1q and C1r when treated with plasmin. The esterase activity uncovered by streptokinase-activated plasminogen inactivated C4 and C2 and was inhibited by the inhibitor of the first component of complement ($C\bar{1}INH$). Another proteolytic enzyme, trypsin, had a similar capacity to activate C1 directly.

Gram-negative bacterial polysaccharides from the outer membrane of the cell wall (endotoxin) have been shown to inactivate complement in vitro, with a preferential utilization of the classical C3 component complex of complement (85). Although this reaction was thought to be independent of the participation of antibody (85), and to depend on the participation of a substance termed "properidin," later studies indicated that specific antibody was involved (73). Endotoxin added to whole serum depleted whole hemoytic complement in guinea pig, rabbit, mouse, and human serum (33). The utilization profile of the complement components by endotoxin treatment was characterized by a pronounced consumption of C3, C5, C6, C7, C8, and C9 (33) despite only minimal apparent utilization of C1, C4, and C2. This contrasted to the consumption profile by classical antigen-antibody complexes at equivalence or slight antigen excess where there was substantial utilization of the early as well as late-acting components (33). Even though endotoxin can efficiently activate the late-acting components, the need for antibody and the early-acting components cannot be excluded. Purified preparations of any of the six terminal components failed to be consumed upon incubation with endotoxin while prior incubation of endotoxin in undiluted normal human or guinea pig serum formed a serum endotoxin complex which destroyed purified C3, C5, C8, or C9 (98). The serum-endotoxin complex contained at least six serum proteins including those sharing immunodeterminants with γ2-globulin and C3. In support of a requirement for small quantities of the early-acting components of complement in the activation of C3, C5, C6, C7, C8, and C9 was the evidence that sera from individuals genetically deficient in C2 (93) or Clr (83) failed to inactivate the late-acting components in the presence of endotoxin (34). In addition, destruction of C4 in guinea pig serum prior to incubation with endotoxin rendered the late-acting components stable. Thus, it is reasonable to suspect that antibody and C1, C4, and C2 can be involved in complement consumption by endotoxin, although a true bypass of the early components may also be a factor.

An additional mechanism of late-acting component activation has been demonstrated with guinea pig γ1-immunoglobulin. This immunoglobulin failed to activate the complement system when added to antigen in the presence of diluted whole guinea pig serum as a complement source (95). However, preformed immune aggregates composed of γ1-immunoglobulin and antigen do consume C3 and the other later acting components, while utilizing little or no C1, C4, and C2. This reaction was blocked by ethylenediaminetetraacetate (EDTA), suggesting a divalent cation requirement. Recent findings from the same laboratory indicate that a site on the Fab portion of the γ1-immunoglobulin activates the guinea pig complement system through the C3 proactivator to be described below.

Flexner & Noguchi (30) showed that cobra venom inactivated serum complement and concomitantly lysed red blood cells. A component was subsequently separated from the venom of *Naja naja* cobra which predominantly depleted C3 in vitro or in vivo without the attendant toxicity of the parent venom (74). The action of the highly purified venom factor required the participation of a β-globulin cofactor (68) to form a complex which in turn was responsible for the inactivation of C3. Formation of such a complex was apparently prevented by EDTA. The complex of cobra venom and serum cofactor not only depletes C3 but also will indirectly lead to lysis of unsensitized guinea pig red cells (4). The lytic effect requires the participation of the late-acting components, C3 through C9, but not C1, C4, or C2 (82, 97) and is therefore a true bypass. It is suggested that the cobra venom-cofactor complex can substitute for convertase in activating C3, thereby recruiting the biologic potential of the terminal complement components (17). This cofactor has been recently designated C3 proactivator.

PHYLOGENETIC STUDIES OF THE COMPLEMENT SYSTEM

The nine serum fractions constituting the hemolytic complement system have been recognized as functions (70, 76) and to some extent as particular proteins (76) in the guinea pig, and in terms of both physicochemical and functional characteristics in man (70). In other species, knowledge of the complement sequence is incomplete, but those about which there exists sufficient information to permit some conclusions are noted in Table 1.

The assay methods employed to detect the complement system or its

TABLE 1. Classification of Representative Species for Which Some Complement Data Exist

		Species
Invertebrates	Phylum Echinodermata	starfish (*Asterious forbesi*)
	Phylum Sipunculoidea	sipunculid worm (*Goldfingea* sp.)
	Phylum Arthropoda	horseshoe crab (*Limulus polyphemus*)
Vertebrates	Phylum Chordata, subphylum Vertebrata	
	Class	
	Agnatha	Sea lamprey
	Chondrichthyes	Nurse shark, lemon shark
	Pisces	Paddlefish, carp, bowfin, longnose gar, channel catfish, big mouth buffalo, and goldfish
	Reptilia	Snake, turtle
	Aves	Chicken
	Mammalian	Human, guinea pig, dog, rabbit, mouse, and pig

constituents are numerous. Whole complement activity has been assessed by the hemolysis of sensitized red cells by serial dilutions of serum as a complement source (62), or by hemolysis of unsensitized red cells by serial dilutions of serum as a source of both complement and natural antibody. The terminal components, C3 through C9, have been detected by cobra venom activation of the serum and indirect lysis of unsensitized red cells (19). Individual components or component groups have been measured by lysis of a suitable intermediate, EAC_n, where n defines the components bound to the intermediate and all but the component or component group being measured are supplied in great excess (75).

Different complement sources vary greatly in hemolytic activity when tested with erythrocytes derived from different animals (65, 101) or with erythrocytes sensitized with antibody produced in other species (Table 2). With antibodies to human erythrocytes produced in guinea pig, cat, and chicken, the strongest hemolytic activity was obtained with their homologous complement while antibody from the rabbit and the dog reacted more efficiently with certain heterologous complement sources. The highest hemolytic activity with rabbit antibody was obtained using guinea pig, goat, cat, and dog complement. Dog antibody is very effective with goat complement. The complement systems of the cow and of the sheep are poorly reactive with the antibodies depicted in Table 2, but are quite reactive with goat antibody (65). Sera from all the mammalians used in the above studies, except the pig, are devoid of any complement activity for the chicken antibodies, while

TABLE 2. Reactivity of Antibodies From Different Species With Different Sources of Complement

Complement Source	Antibody Source				
	Rabbit	G. Pig	Dog	Cat	Chicken
Rabbit	200	665	400	250	0
Guinea Pig	400	1000	40	65	0
Pig	25	40	65	65	14
Sheep	20	20			
Ox	25	25	25	20	0
Rat	100	100		0	0
Goat	400	400	1000	200	0
Chicken	0	100	20	14	33
Cat	400	400	40	1000	0
Dog	200	200	40	27	0

Inverse of the dilution of complement required to produce 100% lysis of human erythrocytes optimally sensitized with antibody made in different species. Heated sera, 56°C, 30 min, was used for sensitization.

chicken serum contains complement of some suitability for the antibodies derived from those mammals. Thus, it is clear that the titer of a given serum complement may vary greatly according to the antibody used for sensitization.

It should also be noted that guinea pig complement is not always the most efficient complement, the titer being influenced by the antibody source as shown in Table 2 or by the nature of the target cell. Guinea pig erythrocytes sensitized (65) with rabbit antibody, and rat mast cells prepared with homologous or heterologous antibody (104) fail to experience appreciable lysis upon the addition of guinea pig serum, although they undergo lysis in the presence of rabbit or human complement. In addition, a given serum complement titer is markedly influenced by the number of erythrocytes used and the volume of the reaction mixture prepared. Thus, attempts to identify lytic complement in phylogenetic studies employing sensitized target cells are greatly complicated by variations in results due to the complex relationship in the test system chosen between target cells and their concentration, the antibody source and the complement.

COMPLEMENT IN INVERTEBRATES

While agglutinins and lysins have, from time to time, been reported in invertebrates, their significance or specificity is unclear. The complement studies available in invertebrate species include a representative of the Arthropoda, horseshoe crab (*Limulus polyphemus*), a Sipunculoidea, sipunculid worm (*Golfingea* sp.), and the Echinodermata, starfish (*Asterius forbesi*).

Although no lytic activity has been demonstrated in the hemolymph of these species by classical methods (19, 32), studies by Day et al (19) have revealed that invertebrates possess a lytic "complement system" which can be activated by a cobra venom factor. Lysis-inducing activity of purified cobra venom factor was observed in hemolymph of two invertebrates, the horseshoe crab and the sipunculid worm, but not in the third, the starfish. To elucidate further the effect of the cobra venom in these species the reaction was divided into two steps. In the first step a complex of the cobra venom factor with the cofactor in the hemolymph was apparently formed; and in the second step erythrocytes were introduced along with a source of terminal complement components, C3 through C9. Cobra venom–horseshoe crab complex and cobra venom–starfish complex induced lysis of sheep erythrocytes in the presence of frog serum in EDTA. The magnitude of lysis observed was similar to that obtained when the complex used in the first step consisted of cobra venom–frog serum. However, when cobra venom–frog serum complex was added to various hemolymphs in EDTA, lysis of sheep erythrocytes was observed only with horseshoe crab hemolymph. EDTA and heat inactivation inhibited the reaction by preventing the formation of the complex, while salicylaldoxime (64) prevented participation of the late acting complement components (19).

Two interesting aspects of this study should be emphasized: A serum cofactor and late-acting components are present in horseshoe crab and sipunculid worm, while only the cofactor has been demonstrated in starfish. It should also be noted that the horseshoe crab terminal components are suitable for the cobra venom–frog complex, while those of the sipunculid worm are not.

COMPLEMENT IN VERTEBRATES

Complement in Agnatha

The sea lamprey (*Petromyzon marinus*, a cyclostomata) is related to the primitive ostraderms and immunologically represents a transitional species. It is the lowest form to develop specific antibody although this response is poor (32, 45). Sea lamprey serum has a natural heat-stable hemolysin for rabbit erythrocytes (32) which has a temperature dependence similar to that of other poikilotherms and is potentiated by the introduction of carbowax. No potentiation of lysis occurred by prior sensitization of rabbit cells with sheep antibody or pretreatment with lamprey serum. The hemolysin found in lamprey serum cannot be absorbed and its action is only partially inhibited by EDTA. Further examination of lamprey serum revealed agglutinating activity for red cells which had cathodal mobility on electrophoresis and appeared in the first peak on Sephadex G-200 gel filtration. This agglutinin increased in titer with immunization but could not be examined for its capacity to mediate lysis with lamprey serum because of the natural lytic activity of such serum. The lytic activity in lamprey serum appeared throughout the gel filtration peaks and was not further characterized.

TABLE 3. Some Properties of the Lytic Activity in Sera From Different Species

	Agnatha	Chondrichthyes[a]	Pisces	Amphibia	Reptilia	Mammals
Natural lysins	rabbit, sheep E	goldfish, turtle, chicken, rabbit, sheep E	goldfish, turtle, rabbit, sheep, dog E	goldfish, turtle, duck, rabbit, sheep, dog, human E	sheep and human	variety of mammalian
Temperature dependency	4°C	25°–30°C	4°–28°C	4°–28°C	15°–37°C	30°–37°C
Potentiation: antibody	no	shark and turtle Ab	variable	rabbit and turtle Ab	rabbit and snake Ab	see Table 2
Potentiation: carbowax 4000	yes	yes	moderately	yes	—	—
Heat stability	rel. stable 2.5 hr 53°C	labile 20 min 48°C	labile 15 min 53°C	labile 20 min 48°C	labile 6 min 56°C	labile 56°C
Blocked by EDTA	rel.	yes	yes	yes	yes	yes
Restored by Ca^{++}, Mg^{++}		no	yes	yes	yes	yes
Components of complement						
Classical (EA lysis)	no	C1, 4, 2, 3[b]; 9	C1, 4, 2, 3[b]	C1, 4, 2, 3[b]	—	C1–9
Cobra venom activation						
E lysis	yes[c]	no	no[d]	yes	no	yes
Complement consumed	—	yes	no[d]	yes	yes	yes

[a] Data for nurse and lemon shark.

[b] Components recognized in reconstitution experiments of heat-treated serum, deficient in C1 and C2, and hydrazine treated serum deficient in C4 and C3.

[c] Data with hagfish, none with sea lamprey.

[d] Noted only by table in Reference 19.

Complement in Chondrichthyes

Some elasmobranchii, lemon shark (*Negaperion brevirostis*) and nurse shark (*Ginglymostoma cirrutum*), considered to be quite primitive on the basis of their cartilaginous skeleton, possess serum with extremely high titers of lytic activity against unsensitized erythrocytes of several other species (Table 3). Since the degree of lysis depended upon the erythrocytes used, it was suggested that the results represented a characteristic complement-dependent antibody system (55). This possibility is supported by absorption experiments in which antibody was removed by incubation with sheep erythrocyte stomata. Such absorbed serum had no hemolytic activity alone but served as a complement source when added to sheep erythrocytes sensitized with either heat-inactivated shark serum as a source of natural antibodies, partially purified nurse shark hemolysin,[3] or turtle antisheep erythrocyte (55). The hemolytic activity of absorbed shark serum was greater with cells sensitized with partially purified shark hemolysin obtained by Sephadex G-200 gel filtration[3] than with heat-inactivated shark serum, suggesting the presence of some interfering material under the latter experimental circumstances.

The complement activity of elasmobranchii is extremely labile to dilution, freezing, and storage at −20°C and −70°C. Further, the complement activity is inactivated by treatment with carrageenin, hydrazine, or heat and is blocked by EDTA, maneuvers known to inhibit the well-defined complement activity of guinea pig serum. However, in contrast to the experience with guinea pig serum, the hemolytic activity of elasmobranchii serum treated with EDTA was not restored by the addition of excess calcium or magnesium ions. A combination of heat-treated shark serum, presumably deficient in C1 and C2, and hydrazine-treated serum, presumably lacking functional C4 and C3, manifested hemolytic activity against sensitized target cells (55). In addition, Nelson & Gigli[3] have observed the presence of C9 in nurse shark serum, namely, a component reacting with the cellular intermediate EAC14235678 prepared with guinea pig components. Thus, it is reasonable to assume that these elasmobranchii possess a complement system consistent with that described for guinea pig serum (Table 3).

Shark serum treated with cobra venom failed to induce lysis of unsensitized erythrocytes, although the hemolytic activity of the serum was reduced by 95 percent, indicating that an alternative pathway of complement activation was activated (19).

Jensen (49) has observed a low molecular weight material in shark serum capable of destroying the hemolytic activity of human or guinea pig C4 in whole serum or in purified fractions. The characteristics of the C4 inactivator in sharks, such as a sedimentation constant of 5.3S and enzymatic activity on TAMe, are similar to those described for human $C\bar{1}s$. Further, Gigli & Austen (38) have demonstrated a marked species dependency in the enzyma tic activity of $C\bar{1}$ on C2 and a dissociation between C4 and C2 inactivating

[3] Nelson, R. A., and Gigli, I. Unpublished observations.

activity after heating $C\overline{1}$. Thus, it is possible that the C4 inactivator of the shark is a subunit of $C\overline{1}$.

In contrast to the data presented for the lemon and nurse sharks, the sting ray (*Dasyatis americana*), another member of the chondrichthyes, presented a negligible degree of lytic activity against sheep erythroyctes which was not potentiated either by carbowax, turtle, or shark hemolysins (55). The discrepancies between shark complement and sting ray serum indicate that there may be no orderly increase in complement levels occurring as the phylogenetic scale is ascended; however, additional studies of sting ray serum treated with cobra venom or versus some other sensitized target cell are essential before any conclusions are drawn.

Complement in Bony Fish (Osteichthyes)

The species of bony fish which have been studied represent certain key branches of the vertebrate phylogenetic tree. The bony fish evolved in three stages, with the early chondrostean fish represented by the paddlefish and carp; followed by holostean fish, typified by bowfin and longnose gar; and the more advanced teleost fish represented by channel catfish, big-mouth buffalo, and goldfish. The latter are known to possess a high degree of immune capability (24, 25).

In 1911, Liefmann (60) reported that sera from certain fish hemolyzed erythrocytes from sheep, rabbit, dog, turtle, and goldfish. The hemolytic activity varied according to the species of erythrocytes used. This natural, temperature-dependent lytic activity, optimal at 28°C (32, 56), could not be potentiated by pretreatment of the erythrocytes with rabbit, turtle, or channel catfish serum (56). Carbowax slightly increased lysis against sheep and rabbit erythrocytes. More recently, however, Gewurz et al (32) have obtained potentiation of the lytic action of paddlefish and carp serum by presensitization of the erythrocytes with mammalian or heat-inactivated paddlefish serum.

The natural lytic activity of the fish sera was blocked by EDTA and restored by the addition of calcium and magnesium ions. All fish sera possessed activities which were labile to freezing, dilution, or heating at 53°C for 15 minutes. None of the sera was decomplemented by an ovalbumin–rabbit antiovalbumin precipitate, perhaps consistent with the lack of potentiation of the lytic activity by mammalian antibody (56). Carrageenin inactivated channel catfish, paddlefish, bowfin, gar, and big-mouth buffalo sera but did not affect the activity of goldfish sera. All sera were rendered hemolytically inactive by hydrazine treatment and, in combination with heat-treated serum, manifested hemolytic activity; heat-inactivated channel catfish serum was an exception in this regard. Attempts to combine heated fish sera with guinea pig fractions were unsuccessful (32), except for fish $C\overline{1}$ (32) which could be demonstrated with intermediates prepared with guinea pig reagents (Table 3). Fractionation of longnose gar serum in Sephadex G-200 showed three major protein peaks, with sedimentation constants of approximately 19S, 8S and 5S. No hemolytic activity was recovered from individual peaks, but combination of the 19S and 8S mixtures restored the hemolytic capacity for sensitized erythrocytes with no requirement for the 5S protein (56). The lack of requirement for material in the third peak, the location of C9 in other species studied including shark, is puzzling but could be explained by contamination of the other fractions with this component.

Complement in Amphibia

The complement system has been studied in four species of amphibia: marine toad (*Bufo marinus*), leopard frog (*Rana pipiens*), bullfrog (*Rana catesbieana*), and mud puppy (*Necturus macolusus*). Serum from each amphibian showed a natural hemolytic effect on erythrocytes from a broad number of species (sheep, rabbit, dog, human, duck, turtle, and goldfish) but the pattern was not uniform. Marine toad and leopard frog serum had no hemolytic effect on dog or duck erythrocytes but lysed rabbit and goldfish red cells. The natural hemolytic activity of amphibian serum, except for mud puppy serum could be markedly potentiated by sensitizing the sheep erythrocyte with rabbit antibody, while turtle antibodies or carbowax 4000 mediated a lesser potentiation. The total complement values reported in the literature vary from 44 to 500 CH_{50} units (54) according to the species and number of erythrocytes used in the test system. Heat inactivation showed the presence of a heat-labile component with a critical temperature of 48°C, lower than that required to inactivate mammalian complement. Inhibition by EDTA (54) was reversible by the addition of calcium and magnesium ions. Carrageenin inhibited the hemolytic activity in all sera except that of the leopard frog. Recombination experiments with depleted sera reconstituted the lytic activity. Thus, inactivation procedures indicate that amphibian complement is a multiple component system which behaves similarly to human or guinea pig complement.

Complement activity was removed from amphibian serum, except in the mud puppy, by exposure to antigen-antibody precipitates of ovalbumin-rabbit antiovalbumin. This finding is consistent with the observation that rabbit antiserum to sheep erythrocytes potentiated the hemolytic activity of the marine toad and leopard frog but not that of the mud puppy.

A higher lytic activity was obtained when frog serum was activated by cobra venom in the presence of unsensitized erythrocytes (1000–1500 CH_{50} units) than when cells sensitized with rabbit antibody were exposed to untreated serum (300–500 CH_{50} units) (19). Frog serum can provide the cofactor for cobra venom and the lytic terminal components in a homologous system. Further, EDTA–frog serum can provide the lytic terminal components for the complex of cobra venom and hemolymph of invertebrates. On the other hand, it fails to provide terminal components for a cobra venom–guinea pig serum cofactor complex or to form a complex with cobra venom that can employ guinea pig terminal components in a lytic system (19). Thus, although frog serum possesses an active source of cofactor and late-acting components, these two are not necessarily interchangeable with different species (Table 5).

Complement in Reptilia

Flexner & Noguchi (29) reported that fresh snake serum is hemolytic for erythrocytes of warm-blooded animals and that this activity was lost upon heat treatment at 56°C for 6 minutes. Bond & Sherwood (7) studied the natural hemolytic activity for human and sheep erythrocytes in 38 specimens of snakes, representative of 13 species. Hemolytic activity was found in all species with titers varying from 12 to 32 CH_{50} units. The hemolytic activity of snake serum was markedly increased when erythrocytes sensitized with heated snake sera or rabbit antibody were used, indicating that the hemolytic activity observed with unsensitized cells was limited by the amount of natural antibodies present in the sera. Indeed, the complement titers of snake sera assayed with sensitized erythrocytes were comparable to the complement content of guinea pig sera, 80 to 200 CH_{50} units.

Cobra and turtle sera treated with cobra venom failed to produce lysis of unsensitized erythrocytes, although the hemolytic activity of both sera on sensitized cells was abolished, implying complement consumption. On the other hand, turtle and snake sera can provide an effective cofactor for a cobra venom complex which employs turtle terminal components in a lytic system[4], suggesting the presence of a control step in cobra and turtle sera which can be circumvented by carrying out the reaction in two steps.

Complement in Aves

It has been mentioned earlier in this review that avian complement fails to interact with erythrocytes sensitized with a variety of mammalian antibodies (65). When used with guinea pig or chicken antibody, complement in chicken serum is hemolytically active. Further, avian serum can be divided by acidification and dilution into pseudoglobulins and euglobulins which are not lytic alone when tested against erythrocytes sensitized with chicken antibody but are lytic upon recombination. The fact that guinea pig complement does not restore the hemolytic activity of heat-inactivated chicken serum is in agreement with the general observation that chicken antibody fails to fix guinea pig Cl (89). Since most chicken sera show a hemolytic property, it would seem logical to conclude that they contain all the components present in other species; the failure to detect them with reagents prepared from mammalian complement may represent species incompatibilities in the reagents used.

Complement in Mammals

Nine distinct components of the hemolytic reaction sequence have been recognized functionally in the guinea pig (75) and man (70), and characterized as specific proteins in the case of man. In other mammalian species only the early components of the complement system have been isolated in a

[4] Day, N. K. B. Personal communication.

state of functional purity (Table 4). As is noted in the next section, isolation of the nine components of the complement system has been difficult because heterologous components available from guinea pig or man are not necessarily suitable for the detection of certain components in other species.

In addition to the nine serum proteins required for the complete action of the complement system, control proteins of this system have also been recognized. Four naturally occurring inhibitors or inactivators acting at different steps of the complement sequence have been noted in various mammalian species, and all of them are demonstrable in human serum (94).

The inhibitor of $C\overline{1}$ ($C\overline{1}$INH) was first described by Levy & Lepow as an $\alpha 2$-globulin in human serum (59) which inhibits the esterolytic and hemolytic (36, 59) activities of $C\overline{1}$s and $C\overline{1}$ either in the fluid phase or bound to a cell. $C\overline{1}$INH has also been reported to inhibit the activation of Cls by Clr and the esterolytic activity of the Clr subunit (72). Antigenic determinants cross-reacting with human $C\overline{1}$INH have been noted only in other primate sera, although inhibition of the esterolytic activity of human $C\overline{1}$ was obtained with sera from a variety of species including cow, horse, rabbit, guinea pig, and other primates (21). Inhibition of human $C\overline{1}$ esterase activity was not obtained with mouse or rat serum. The fact that purified human $C\overline{1}$INH failed to inhibit the hydrolysis of acetyl–L–tyrosine–ethyl ester in fresh serum from rat and mouse suggests a species incompatibility between human and rat or mouse in regard to $C\overline{1}$ and $C\overline{1}$INH rather than an absence of the inhibitor. In addition to its action on $C\overline{1}$, the $C\overline{1}$INH of human origin can effectively inhibit the action of kallikrein (39) either on its natural substrate, kininogen, or on synthetic substrates; this interaction, like that with $C\overline{1}$, is stoichiometric. Plasmin also has been shown to be inhibited by $C\overline{1}$INH; however, as a result of this interaction the inhibitor is enzymatically degraded (47).

An inactivator of the third component of complement (C3bINA) described in rabbit, guinea pig (103), and man (92) destroys the hemolytic, immune adherence, and phagocytosis-enhancing activity (35) associated with cell-bound C3 of the cellular intermediate EAC1423. The action of C3bINA is time- and temperature-dependent and is accompanied by the release of a protein fragment into the fluid phase. The data available suggest that C3bINA is an enzyme identical with the conglutininogen-activating factor described by Lachmann & Müller-Eberhard (53).

A small molecular weight fragment C3a, termed anaphylatoxin, split off from C3 during its enzymatic conversion to C3b, was observed to be readily produced from isolated human components but not in whole human serum, suggesting that human serum contained an inactivator of anaphylatoxin. This inactivator, C3aINA, has been characterized as an αl-globulin with a sedimentation rate of 9.5S and a molecular weight of approximately 310,000. C3aINA removes the C-terminal residue from C3a and may be identical to the carboxypeptidase which splits a C-terminal arginine residue from bradykinin (6).

TABLE 4. Components of Complement Identified in Different Mammalian Species

Complement	C1 q	C1 r	C1 s	C4	C2	C3	C5	C6	C7	C8	C9	$C\bar{1}INH$	C3bINA	C6INA	Reference
Human[+]															
Elec. mobility	$\gamma2$	β	$\gamma2$	$\beta1$	$\beta2$	$\beta1$	$\beta1$	$\beta2$	$\beta2$	$\alpha1$	α	$\alpha2$	$\beta1$	$\beta2$	70, 92
Sw 20	11S	7S	4S	10S	5.5S	9S	8.7S	6.6S	5.6S	8S	4.5S	4S	5S	6.6S	70
Serum conc.	100–			200–	8–	900–	51–				1–	12–			
ug/ml	200	—	—	600	12	1500	99	—	—	—	2	30	—	—	70
Guinea pig[+]		+		+	+	+	+	+	+	+	+	+	+	+	75
Rabbit[++]		+		+	+	+*	+*	+*	+*	+*	+*	+	+	+	50, 77
						late-acting components C3–9 tested as C-EDTA									
Dog[++]		+		+	+	+°						+			96
Rat[++]		+		+	+	+°						+			12(*a*)
Mouse*		+		+	+	+°						+			12
Horse, sheep*, cow		+		±	±										12, 89, 90

[+] Components isolated in state of chemical or functional purity.
[++] Components isolated in state of relative functional purity.
* Components identified during chromatographic procedures.
° Group of terminal components recognized by function.
(*a*) Dr. Jacques Caldwell, personal communication.

An inactivator of the sixth component of complement is present in rabbit, guinea pig, and human serum (103). As with C3bINA, C6INA acts on cell-bound C6, destroying the hemolytic activity of the cellular intermediate EAC142356.

SPECIES INCOMPATIBILITIES AMONG COMPONENTS OF COMPLEMENT

Absolute incompatabilities among complement components of heterologous sera have been long recognized (90) while partial incompatibilities were apparently masked by the use of R reagents. For example, Ecker et al (23) observed no incompatibility among R reagents of human and guinea pig origin, whereas recent studies with purified components reveal a marked incompatibility among certain components of these two different species. Detailed studies of early component incompatibilities have been conducted with guinea pig, rabbit (50), dog, and human (41) serum (Table 5). The principle that the binding of a component is not necessarily followed by conversion to a hemolytically active form has become clear. For example, a cell-bound complement enzyme binds an incompatible heterologous complement protein but may fail to produce those alterations essential to its efficient participation in the progression of the hymolytic sequence (40).

The preferred specificity of an enzyme for the homologous complement protein can be demonstrated either in the fluid phase or on a cellular intermediate. $C\overline{1}$ of human origin inactivates fluid phase homologous C4 and C2 more rapidly than does $C\overline{1}$ of guinea pig origin (37). Similarly, cell-bound $C\overline{1}^{dog}$,[5] $EAC1^{dog}$, is much more efficient than cell-bound $C\overline{1}^{gp}$, $EAC1^{gp}$, in effectively titrating $C4^{dog}$ or producing an efficient $EAC14^{dog}$ intermediate capable of reacting with $C2^{dog}$ (96). The distinction between binding of a component and the conversion to a hemolytically active site is clearly seen in studies of heterologous C4 and C2. The interaction of $C2^{hu}$ with $SAC14^{gp}$, for example, yields a hemolytically inactive intermediate designated $EAC14^{gp}2x^{hu}$ (41, 51). The postulated hemolytically inactive intermediate fixes $C3^{hu}$ generating the $EAC14^{gp}2x^{hu}3x^{hu}$ state which will lyse upon introducing $C2^{gp}$ and the addition of all terminal components other than $C3^{hu}$. Since only the introduction of C2 homologous to the C4 was necessary for lysis of $EAC14^{gp}2x^{hu}3x^{hu}$ by the terminal components, it seemed clear that $C3^{hu}$ was present on the complex, and the binding enzyme $4^{gp}2x^{hu}$ has been termed "counterfeit convertase." All the C4, C2 incompatibilities noted in Table 5 are presumably explained on the same basis, namely, the inability of the counterfeit convertase to activate effectively the source of terminal components available.

[5] The species of origin of a given component is indicated by a superscript (hu, human; gp, guinea pig; dog, dog). SAC14 refers to the proportion of hemolytically active sites formed per erythrocyte during the interaction of EAC1 with C4.

TABLE 5. Known Partial and Total Species Incompatibilities Among Complement Components

Species of Complement	C$\bar{1}$	C4	C2	C-EDTA (C3, 5–9)	C8	Ref.
Human	$C4^{gp}$ [a]	$C2^{ra}$	$C4^{gp}$			37
	$C2^{gp}$ [a]	$C2^{dog}$	$C4^{ra}$		$EAC1\text{-}7^{gp}$	41
		$C2^{rat}$	$C4^{dog}$			50
			$C4^{rat}$			75, 96
Guinea pig	$C4^{hu}$ [a]	$C2^{hu}$	$C4^{ra}$	$EAC142^{dog}$		37
	$C2^{hu}$ [a]	$C2^{ra}$	$C4^{dog}$			41
		$C2^{dog}$	$C4^{rat}$			50
		$C2^{rat}$				96(*a*)
Rabbit		$C2^{hu}$	$C4^{gp}$		$EAC1\text{-}7^{gp}$	50
		$C2^{gp}$	$C4^{hu}$			77
Dog	$C4^{gp}$	$C1^{gp}$	$C4^{hu}$			
		$C2^{hu}$	$C4^{gp}$			96
		$C2^{gp}$				
Rat		$C2^{hu}$	$C4^{hu}$			
		$C2^{gp}$	$C4^{gp}$			(*a*)
	direct lysis by CV	lysis by CV by II steps[b]	step I[c] cofactor CV	step II[d] C-EDTA		
Frog	+	+	gp, starfish & sp worm	gp–CV, sp worm–CV		19
Horseshoe crab	+	+				19
Starfish	−	+		frog–CV		19
Sipunculoid worm	+	−	frog	frog–CV		19

[a] Incompatibilities demonstrated in fluid phase reactions.

[b] Complexes formed with the various species of serum or hemolymph and cobra venom, reacted with the same serum or hemolymph in 0.05 *M* EDTA and E.

[c] Step I. Complexes formed with the various species of serum or hemolymph and CV, reacted with frog serum in 0.05 *M* EDTA and E.

[d] Step II. Complexes formed with frog serum and CV, reacted with hemolymph from various species in EDTA in the presence of E. Frog serum was also tested with gp-EDTA or gp-CV complex.

(*a*) Dr. Jacques Caldwell, personal communication.

In addition to this incompatibility at the C4, C2 step, the failure of the late-acting components, C3-9 (C-EDTA), prepared from guinea pig serum to lyse $EAC14^{dog}2^{dog}$ cells introduces an additional species incompatibility (96). This incompatibility may reside in the inability of $\overline{42}^{dog}$ convertase to activate $C3^{gp}$ for its role in the hemolytic sequence (41), or alternatively between $C2^{dog}$ and the ill-defined role of C2 in the fixation of C5. In contrast, rat C-EDTA has been reported to provide a highly effective source of late-acting components (13), capable of reacting with a variety of EACl42 intermediates. The effectiveness of rat C-EDTA manifests itself mainly in C2 titrations where a limited amount of C2 is provided.

The ability of a heterologous component to block the completion of the reaction sequence without providing a hemolytically active site is also evident when $C8^{hu}$ interacts with $EAC1\text{-}7^{gp}$. The resulting intermediate is rendered hemolytically inactive to lysis by $C8^{gp}$ and $C9^{gp}$, or $C9^{hu}$. Presumably, the $EAC1\text{-}7^{gp}$ has bound $C8^{hu}$, producing a site that is blocked for the further action of the homologous system.

FUNCTION OF THE COMPLEMENT SYSTEM

The products and byproducts of the complement reaction sequence depicted in Figure 1 have all been demonstrated by in vitro techniques, and conclusions as to their in vivo significance are only speculative. Nonetheless, it is reasonable to point out that the potential biologic meaning of the system is no longer only in terms of completion of the whole sequence with consequent death of some antigen-carrying cells, but also includes the elaboration of various ingredients of the inflammatory response.

Under certain conditions, antisera against herpes simplex virus produces little or no neutralization unless complement is present. The mechanism by which complement adds to the action of γ-globulin suggests that the presence of $C\overline{1}$ and C4 on the sensitized virus results in neutralization by more extensive coverage of the surface of the virus. Although $C\overline{1}$ and C4 are sufficient to enhance virus neutralization, under certain circumstances C2 and C3 can further augment the reaction (18). No effect of later components has yet been demonstrable.

A vasoactive peptide distinguishable from bradykinin has been isolated from the plasma of patients with angioedema of the hereditary type (22). This material may represent a previously unrecognized split product of C2, and has been temporarily termed C-kinin; it differs from the known kinins in being susceptible to inactivation by trypsin as well as chymotrypsin and in producing a pressor effect in the rat.

The major fragments resulting from the sequential action of $C\overline{1}$ on C4 and C2 generate C3 convertase (69); through the action of this enzyme a small fragment of the C3 molecule, C3a, is released into the fluid phase; this fragment anaphylatoxin (20) causes a local wheal when injected intracutaneously into man, releases histamine in vitro from rat peritoneal mast cells, and causes the isolated guinea pig ileum to contract (20). C3b, the major fragment of C3, either is bound to the cellular intermediate EAC1423 or remains free in the fluid phase as the hemolytically inactive C3i. The presence of C3b on the cell confers upon it the ability to participate in the immune adherence phenomenon, a reaction in which the C3b-coated cell binds to a specific receptor site on erythrocytes, polymorphonuclear leukocytes, or platelets (76). This phenomenon enhances phagocytosis by polymorphonuclear leukocytes (35).

The interaction of the EAC1423 intermediate with C5 liberates a small molecular weight fragment C5a with anaphylatoxin properties (16) and chemotactic activity for polymorphonuclear leukocytes (106).

In the course of the formation of EAC1-7 an additional chemotactic factor for polymorphonuclear leukocytes (105) is generated. This activity resides in a high molecular weight complex formed by the interaction of C5, C6 and C7. The activation of the complement sequence is associated with the appearance of characteristic membrane discontinuity which has been observed by electron microscopic studies of negatively stained membranes from lysed erythrocytes and other target cells (10, 48).

Whether the immunologic activation of the complement sequence is considered beneficial or detrimental to the host depends at least in part on the location of the antigens against which the mediating antibody is directed. Antibodies directed against a soluble foreign protein result in the deposition of aggregates in vessel walls, activation of the complement system with a local increase in vascular permeability and the elaboration of principles chemotactic for polymorphonuclear leukocytes, and enhanced phagocytosis of the aggregates with release of lysosomal enzymes and secondary focal necrosis of the vessel wall; this detrimental hypersensitivity reaction is termed the Arthus reaction (15). In contrast, the same noncytotoxic reaction sequence directed against an invading pyogenic bacterium with resultant enhanced phagocytosis would be considered a beneficial aspect of the immune response. Similarly, when the complete sequence is activated to yield a cytotoxic reaction directed against the host's own red cells or those received by transfusion, the event is considered hypersensitivity; whereas if the target cell were some Gram-negative bacterium, the bactericidal reaction would be considered immunity (2).

The possible immunobiological significance of the complement system has recently been extended by several additional observations. The recent report that target cells carrying the first four reacting components undergo enhanced ingestion by macrophages (70), while those carrying the first seven components experience accelerated damage upon exposure to a suspension of purified lymphocytes (81) introduces a role for complement in reactions in which mononuclear cells predominate. Further, the appreciation that activation of the complement system can be cytotoxic for bystander cells not sensitized to be the primary target cell broadens the possible effects of a specific activation (44). Finally, bypass and pseudo-bypass activation mechanisms may have important implications for the recruitment of the complement component sequence to a wide range of inflammatory reactions of diverse etiologies.

CONCLUSION

In conclusion, numerous difficulties have been encountered by workers in carrying out phylogenetic studies of the complement system. Negative results do not exclude the existence of a complement component sequence since they may reflect an unsuitable choice of the detection system in terms of target cells, antibody source, or developing reagents, or all these factors. Whereas a specific antibody response is poorly defined in invertebrates, certain invertebrate species have been shown to possess functional terminal components of the complement system and the C3 proactivator (cofactor) as demonstrated by cobra venom activation of their hemolymph.

Primitive vertebrates, such as the lamprey, have been shown to produce antibody but have not as yet had a complement system identified. However, the methods used do not exclude the possibility that such a system exists in the lamprey. The natural antibodies of other primitive vertebrates have been used in the sensitization of the erythrocytes; such antibodies may have low affinity and a limited capacity to activate the complement system. The immunoglobulin class of these natural antibodies has not as yet been defined, and studies have not yet been conducted with antibodies elicited by specific immunization with the target cell. Nonetheless, the data available (Table 3), using classic inactivation procedures, are consistent with the presence of a complete complement system beginning with the elasmobranchii.

In mammalians, the nine components of the complement sequence have been clearly defined in the guinea pig and man, and partially in other species. Through studies with functionally pure components of guinea pig and man as well as certain other species (Table 4) the problems of species incompatibilities have become apparent. Incompatibilities can be recognized uniformly between C4 and C2 of heterologous species, the two proteins forming a counterfeit convertase which will bind but not activate C3 for the completion of the hemolytic sequence. Additional incompatibilities appear to exist between guinea pig C7 and human C8 and possibly between dog C2 and guinea pig C5; these incompatibilities have not been sought in other combinations. Incompatibilities between late-acting components have also been observed in the activation of the complement system following the interaction of cobra venom with the C3 proactivator; for example, the cobra venom–frog C3 proactivator complex cannot express its lytic potential when the terminal components are supplied by the sipunculid worm hemolymph or guinea pig serum.

The studies reviewed demonstrate the early development of the complement system, certainly as early as the specific humoral immune response. It will be interesting to observe whether further studies of complement with a full range of detection systems will reveal the counterpart of all nine mammalian components in these vertebrate species and whether a complement-like system, rudimentary or complete, will be recognized in still lower invertebrate forms.

LITERATURE CITED

1. Austen, K. F., Becker, E. L., Biro, C. E. et al 1968. *Bull. W. H. O.* 39:935–38
2. Austen, K. F. 1970. *Harrison's Principles of Internal Medicine*, ed. R. D. Adams, I. L. Bennett, E. Braunwald, K. J. Isselbacher, G. W. Thorn, M. M. Wintrobe, 342. New York: McGraw-Hill.
3. Bach, J. F., Gigli, I., Dardenne, M., Dormont, J. Manuscript in preparation.
4. Ballow, M., Cochrane, C. G. 1969. *J. Immunol.* 103:944–52
5. Bier, O. G., Leyton, G., Mayer, M. M. 1945. *J. Exp. Med.* 81:449–68
6. Bokisch, V. A., Müller-Eberhard, H. J. VIth Immunopathology Symp. In press
7. Bond, G. C., Sherwood, N. P. 1939. *J. Immunol.* 36:11–16
8. Bordet, J., Gargou, O. 1901. *Ann. Inst. Pasteur* 15:209–13
9. Borsos, T., Rapp, H. J., Mayer, M. M. 1961. *J. Immunol.* 87:326–29
10. Borsos, T., Dourmashkin, R. R., Humphrey, J. H. 1964. *Nature (London)* 202: 251–52
11. Borsos, T., Rapp, H. J. 1965. *Science* 150:505–6
12. Borsos, T., Rapp, H. J. 1965. *J. Immunol.* 94:510–13
13. Borsos, T., Rapp, H. J., Colten, H. R. 1970. *J. Immunol.* 105:1439–46
14. Buchner, H. 1889. *Centralbl. Bakteriol. Parasitol.* 5:817–23
15. Cochrane, C. G. 1965. *The Inflammatory Process*, ed. B. W. Zweifach, L. Grant, R. T. McCluskey, 613. New York: Academic Press
16. Cochrane, C. G., Müller-Eberhard, H. J. 1968. *J. Exp. Med.* 127:371–86
17. Cochrane, C. G., Müller-Eberhard, H. J., Aikin, B. S. 1970. *J. Immunol.* 105:55–69
18. Daniels, C. A., Borsos, T., Rapp, H. J., Snyderman, R., Notkins, A. L. 1969. *Proc. Nat. Acad. Sci. USA* 65:528–35
19. Day, N. K. B., Gewurz, H., Johannsen, R., Finstad, J., Good, R. A. 1970. *J. Exp. Med.* 32:941–50
20. Dias da Silva, W., Lepow, I. H. 1967. *J. Exp. Med.* 125:921–46
21. Donaldson, V., Pensky, J. 1970. *J. Immunol.* 104:1388–95
22. Donaldson, V. H., Ratnoff, O. D., Dias da Silva, W., Rosen, F. S. 1969. *J. Clin. Invest.* 48:642–53
23. Ecker, E. E., Pillemer, L., Seifter, S. 1943. *J. Immunol.* 47:181–93
24. Evans, E. E. 1963. *Fed. Proc.* 22:1132–37
25. Evans, E. E., Kent, S. P., Attleberger, M. H., Seibert, C., Bryant, R. E., Booth, B. 1965. *Ann. N. Y. Acad. Sci.* 126:629–46
26. Ferrata, A. 1907. *Berlin. Klin. Wochenschr.* 44:366–68
27. Finstad, J., Good, R. A. 1964. *J. Exp. Med.* 120:1151–67
28. Finstad, J., Papermaster, B. W., Good, R. A. 1964. *Lab. Invest.* 13: 490–512
29. Flexner, S., Noguchi, H. 1902–1903. *J. Pathol. Bacteriol.* 8:379–410
30. Flexner, S., Noguchi, H. 1903. *J. Exp. Med.* 6:277–301
31. Fodor, J. von. 1886. *Deut. Med. Wochenschr.* 13:745–47
32. Gewurz, H., Finstad, J. Muschel, I., Good, R. A. 1967. *Phylogeny of Immunity*, ed. R. T. Smith, P. A. Miescher, R. A. Good, 105. Gainesville, Fla: Univ. of Florida Press
33. Gewurz, H., Shin, H. S., Mergenhagen, S. E. 1968. *J. Exp. Med.* 128: 1049–1137
34. Gewurz, H. 1970. *Proc. Symp. Biological Activity of Complement*, ed. R. Ingram. In press
35. Gigli, I., Nelson, R. A. 1968. *Exp. Cell Res.* 51:45–67
36. Gigli, I., Ruddy, S., Austen, K. F. 1968. *J. Immunol.* 100:1154–64
37. Gigli, I., Austen, K. F. 1969. *J. Exp. Med.* 129:679–96
38. Gigli, I., Austen, K. F. 1969. *J. Exp. Med.* 130:833–46
39. Gigli, I., Mason, J. W., Colman, R. W., Austen, K. F. 1970. *J. Immunol.* 104:574–81
40. Gigli, I., Austen, K. F. 1970. *Fed. Proc.* 29:304
41. Gigli, I., Austen, K. F. *Proc. Int. Congr. Microbiol., 10th.* In press
42. Goldman, J., Ruddy, S., Feingold, D., Austen, K. F. 1969. *J. Immunol.* 102:1379–87
43. Gordon, J., Whitehead, H. R., Wormall, A. 1926. *Biochem. J.* 20:1028–35
44. Götze, O., Müller-Eberhard, H. J. 1970. *J. Exp. Med.* 132:898–915
45. Grey, H. M. 1969. *Advan. Immunol.* 10:51–104
46. Hadding, U., Müller-Eberhard, H. J. 1969. *Immunology* 16 :719–35
47. Harpel, P. C. 1970. *J. Clin. Invest.* 49: 568–75

48. Humphrey, J. H., Dourmashkin, R. R. 1965. *Int. Symp. Immunopathology*, 5th, 209. Ed. P. Graber, P. A. Miescher. Basel: Schwabe and Co.
49. Jensen, J. A. 1969. *J. Exp. Med.* 130: 217–41
50. Kempf, R. A., Gigli, I., Austen, K. F. 1969. *J. Immunol.* 102:795–803
51. Koethe, S., Gigli, I., Austen, K. F. Manuscript in preparation
52. Kondo, M., Gigli, I., Austen, K. F. Manuscript in preparation
53. Lachmann, P. J., Müller-Eberhard, H. J. 1968. *J. Immunol.* 100:691–98
54. Legler, D. W., Evans, E. E. 1966. *Proc. Soc. Exp. Biol. Med.* 121: 1058–1062
55. Legler, D. W., Evans, E. E. 1967. *Proc. Soc. Exp. Biol. Med.* 124:30–34
56. Legler, D. W., Evans, E. E., Dupree, H. K. 1967. *Trans. Am. Fisheries Soc.* 96:237–42
57. Lepow, I. H., Wurz, L., Ratnoff, O., Pillemer, L. 1954. *J. Immunol.* 73: 146–58
58. Lepow, I. H., Naff, G. B., Todd, E. W., Pensky, J., Hinz, C. F. *J. Exp. Med.* 117:983–1008
59. Levy, L. R., Lepow, I. H. 1959. *Proc. Soc. Exp. Biol. Med.* 101:608–11
60. Liefmann, H. 1911. *Berlin. Klin. Wochenschr.* 48:1682–83
61. Mayer, M. M., Levine, L., Rapp, H. J., Marucci, A. A. 1954. *J. Immunol.* 73:443–54
62. Mayer, M. M. 1961. *Experimental Immunochemistry*, ed. E. A. Kabat, M. M. Mayer, 133. Springfield, Ill: C. C Thomas
63. Mayer, M. M., Miller, J. A. 1965. *Immunochemistry* 2:71–93
64. Mills, S. E., Levine, L. 1959. *Immunology* 2:368–83
65. Muir, R. 1911. *J. Pathol. Bacteriol.* 16:523–34
66. Müller-Eberhard, H. J., Lepow, I. H. 1965. *J. Exp. Med.* 121:819–27
67. Müller-Eberhard, H. J., Dalmasso, A. P., Calcott, M. A. 1966. *J. Exp. Med.* 123:33–54
68. Müller-Eberhard, H. J. 1967. *Fed. Proc.* 26:744
69. Müller-Eberhard, H. J., Polley, M. J., Calcott, M. A. 1967. *J. Exp. Med.* 125:359–80
70. Müller-Eberhard, H. J. 1968. *Advan. Immunol.* 8:1–80
71. Naff, G. B., Pensky, J., Lepow, I. H. 1964. *J. Exp. Med.* 119:593–613
72. Naff, G. B., Ratnoff, O. D. 1968. *J. Exp. Med.* 128:571–94
73. Nelson, R. A., Jr. 1958. *J. Exp. Med.* 108:515–35
74. Nelson, R. A., Jr. 1966. *Surv. Ophthalmol.* 11:498–505
75. Nelson, R. A., Jr., Jensen, J., Gigli, I., Tamura, N. 1966. *Immunochemistry* 3:111–35
76. Nelson, R. A., Jr. 1965. *The Inflammatory Process*, ed. B. W. Zwiefach, L. H. Grant, R. T. McCluskey, 818. New York and London: Academic Press
77. Nelson, R. A., Jr., Biro, C. E. 1968. *Immunology* 14:527–40
78. Nilsson, U. R., Müller-Eberhard, H. J. 1967. *Immunology* 13:101–17
79. Nuttal, G. 1888. *Z. Hyg.* 4:353–94
80. Papermaster, B. W., Condie, R. M., Finstad, J., Good, R. A. 1964. *J. Exp. Med.* 119:105–30
81. Perlmann, P., Perlmann, H., Müller-Eberhard, H. J., Manni, J. A. 1969. *Science* 163:937–39
82. Pickering, R. J., Wolfson, M. R., Good, R. A., Gewurz, H. 1969. *Proc. Nat. Acad. Sci. USA* 62:521–27
83. Pickering, R. J., Naff, G. B., Stroud, R. M., Good, R. A., Gewurz, H. 1970. *J. Exp. Med.* 131:803–15
84. Pillemer, L., Ratnoff, O., Blum, L., Lepow, I. H. 1953. *J. Exp. Med.* 97. 573–89
85. Pillemer, L., Schoenberg, H., Blum, L., Wurz, L. 1955. *Science* 122:545–49
86. Pollard, B., Finstad, J., Good, R. A. 1966. *Phylogeny of Immunity*, ed. R. T. Smith, P. A. Miescher, R. A. Good, 88. Gainesville: Univ. of Florida Press
87. Polley, M. J. Müller-Eberhard, H. J., Feldman, J. D. 1971. *J. Exp. Med.* 133:53–62
88. Ratnoff, O. D., Naff, G. B. 1967. *J. Exp. Med.* 125:337–58
89. Rice, C. E. 1950. *Can. J. Comp. Med.* 14:369–76
90. Rice, C. E., Crowson, C. N. 1950. *J. Immunol.* 65:201–10
91. Ritz, H. 1913. *Z. Immunitätsforsch.* 13:62–83
92. Ruddy, S., Austen, K. F. 1969. *J. Immunol.* 102:533–43
93. Ruddy, S., Klemperer, M. R., Rosen, F. S., Austen, K. F., Kumate, J. 1970. *Immunology* 18:943–54
94. Ruddy, S., Austen, K. F. 1970. *Proc. Symp. Biological Activity of Complement*, ed. R. Ingram. In press
95. Sandberg, A. L., Osler, A., Shin, H. S., Oliveria, B. 1970. *J. Immunol.* 104:329–34
96. Sargent, A., Austen, K. F. 1970. *Proc.*

Soc. Exp. Biol. Med. 133:1117–22
97. Shin, H. S., Gewurz, H., Snyderman, R. 1969. *Proc. Soc. Exp. Biol. Med.* 131:203–7
98. Shin, H. S., Snyderman, R., Friedman, E., Mergenhagen, S. E. 1969. *Fed. Proc.* 28:485
99. Shin, H. S., Pickering, R. J., Mayer, M. M. 1971. *J. Immunol.* 106:473–79
100. Shin, H. S., Pickering, R. J., Mayer, M. M. 1971. *J. Immunol.* 106:480–93
101. Shrigley, E. W., Irwin, M. R. 1937. *J. Immunol.* 32:281–90
102. Stolfi, R. L. 1968. *J. Immunol.* 100: 46–54
103. Tamura, N., Nelson, R. A. 1967. *J. Immunol.* 94:582–89
104. Valentine, M. D., Bloch, K. J., Austen, K. F. 1967. *J. Immunol.* 99: 98–110
105. Ward, P. A., Cochrane, C. G., Müller-Eberhard, H. J. 1966. *Immunology* 11:141–53
106. Ward, P. A., Newman, L. J. 1969. *J. Immunol.* 102:93–99
107. Willoughby, W. F., Mayer, M. M. 1965. *Science* 150:907–8

INTERFERON INDUCTION AND ACTION

1574

CLARENCE COLBY AND MICHAEL J. MORGAN[1]

Microbiology Section, The University of Connecticut, Storrs, Connecticut 06268

CONTENTS

INTRODUCTION

Interferon is a protein produced by cells in response to a viral infection. The biological activity of interferon is to convert cells of the same species into a viral refractory state. Since the discovery of interferon 14 years ago (86), much experimental effort has been spent probing the mechanism of induction and mechanism of action of this protein. While neither mechanism is understood on a molecular basis, several cellular proteins have been postulated to be involved in the interferon system. This system may be thought of as composed of the following components: (*i*) the interferon inducer, (*ii*) the cell in which the interferon is to be induced, (*iii*) interferon itself, (*iv*) the cell on which the interferon is to be assayed, and (*v*) the challenge virus used in the interferon assay. In this review we shall consider the experiments which led investigators to postulate the existence of four cellular proteins; a receptor site protein responsible for recognizing an interferon-inducing molecule, interferon itself, a repressor of interferon synthesis, and an antiviral protein postulated to be the mediator of interferon action. Thus, the emphasis of this review is on the mechanism of induction of interferon. There are several recent reviews that cover a variety of aspects of the interferon system (25, 35, 55, 73, 78, 165, 173).

[1] Harkness Fellow of the Commonwealth Fund of New York. Present address: Department of Biochemistry, University of Leicester, Leicester, England.

TABLE 1. In Vivo Inducers of Interferon

A. Animal viruses
   1. DNA-viruses: papova-, herpes-, adeno-, and pox virus
   2. RNA-viruses: picorna-, myxo-, arbo-, and reo virus.
B. T4 coliphage (102)
C. Bacteria
   B. abortus, E. coli, S. marcescens, S. typhimurium, H. influenzae, B. pertussis, L. monocytogenes, F. tularensis.
D. Protozoa
   Plasmodium berghei, Toxoplasma gondii, Trypanosoma cruzi
E. Rickettsiae
   Rickettsia tsutsugamushi, Rickettsia prowazeki, Coxiella burneti.
F. Chlamydia
G. Bacterial and Rickettsial extracts
   endotoxins, exotoxins, etc.
H. Plant extracts
   phytohemagglutin
I. bis-DEAE-fluorenone (105, 124)
J. Fungal extracts
   double-stranded RNA of various mycophages
K. Naturally occurring double-stranded RNAs
   reovirus, MS 2 and MU 9 coliphage replicative form, rice dwarf virus, cytoplasmic polyhedrosis virus, mengo virus (50, 51)
L. Synthetic polymers
   polycarboxylates, polysulfonates, helical polyribonucleotides (25).

Unless indicated otherwise, the papers describing these inducers are referenced in the recent review by De Clercq & Merigan (35).

MECHANISMS OF SYNTHESIS AND RELEASE OF INTERFERON

Inducers of Interferon

Heat-inactivated influenza virus acted as the interferon inducer in the initial experiments by Issacs & Lindemann (86). For the next six years experimenters successfully employed a variety of infectious and inactivated animal viruses as interferon inducers. The first report of a nonviral inducer of interferon was made in 1963 when certain ribonucleic acid preparations were found to be active (85, 149). Currently, the list of interferon inducers includes all classes of animal viruses, microorganisms, extracts of microorganisms and plants, polyanions, and a few small molecules (Tables 1 and 2).

We would call the reader's attention to the striking difference between the number of interferon inducers in animals as compared with those in growing cells in culture. While there are many interferon inducers that are

TABLE 2. In Vitro Inducers of Interferon

A. Animal viruses
   DNA and RNA viruses
B. Fungal extracts (containing double-stranded RNA)
C. Naturally occurring double-stranded RNAs
D. Helical polyribonucleotides (25)

Unless indicated otherwise, the papers describing these inducers are referenced in the recent review by De Clercq & Merigan (35).

active only in intact animals but not in growing cells in culture, the reverse statement does not hold. Of the former, many require intravenous or intraperitoneal injections of relatively large doses. Stinebring & Youngner (162) first suggested that the interferon which appeared in the serum of virus-infected animals was synthesized de novo, whereas the nonviral interferon inducers were stimulating the release of interferon from a preformed pool. Support for this hypothesis came from studies on the kinetics of interferon production, the sensitivity of interferon production to metabolic antagonists and the sensitivity to body temperature and presence or absence of corticosteroids (77, 78, 95, 141, 142, 162, 174, 177, 178). We shall consider these two classes of inducers separately.

Inducers active only in vivo.—The interferon inducers that are active only in intact animals are microorganisms, extracts of microorganisms, extracts of plants, polycarboxylates, T4 coliphage, and the various small molecules. The microorganisms that induce interferon in animals vary widely including bacteria, rickettsiae, protozoa, and chlamydia. The only obvious common characteristic they have is that they are all intracellular parasites. Yet, it seems unlikely that there is one or more intracellular events which they all share and which is essential for the interferon response since intracellular growth is mandatory for the rickettsiae but not for the bacteria (74, 80).

Endotoxin is the most studied of the bacterial extracts that stimulate interferon production in animals. It has been used successfully to treat corneal lesions in rabbits infected with influenza, Newcastle disease (NDV), and herpes simplex viruses (136, 139, 140). Since the endotoxins are incapable of inducing cells growing in culture to synthesize interferon, it is interesting to note the recent reports that mouse peritoneal leukocytes and mouse spleen lymphocytes, mononuclear cells, and polymorphonuclear leukocytes in cell suspension may be stimulated to release interferon when treated with endotoxin (103, 171). This is similar to the earlier report by Wheelock (172) that phytohemagglutinin stimulates interferon release by human leukocytes in suspension but not by human lung fibroblasts in culture.

Following Regelson's (143) initial experiments with synthetic copolymers of maleic acid anhydride, a variety of polycarboxylated (41, 42, 127–129) and polysulfonated polymers (13) have been shown to stimulate interferon release in treated animals. The most recent concerns polyacetal carboxylic acid (24) which is similar to the above polyanions in that it exhibits antiviral activity both in vivo and in vitro and that the interferon system plays a role only in vivo.

In an interesting series of experiments, Kleinschmidt et al (102) found that T4 coliphage stimulated serum-level interferon in mice. The animals were injected intravenously with 10^9 to 3×10^{11} phage particles. The possibility that extraneous endotoxin was responsible for interferon production was eliminated by showing that following heat treatment at 98°C for 15 min and sonication for 15 min endotoxin was still an active inducer while heat and sonication obliterated the inducing activity of the phage. The activity does not appear to reside with the protein coat of the phage since T4 ghosts were inactive. Purified T4 DNA in amount equivalent to 10^9 phage particles was also inactive. However, at a DNA concentration equivalent to 3×10^{11} phage particles, circulating interferon was detected at 2 hr after injection. Unfortunately, no quantitative data were presented and the authors do not mention whether the amount of circulating interferon increased with time (the peak interferon titer with intact phage particles was 18 hr post-injection). When T4 phages were added to monolayers of mouse L-cells there was no protection of the monolayers against infection with vesicular stomatitis virus (VSV). Primary rabbit kidney cell cultures showed 50 percent inhibition of VSV when treated with 7.5×10^{11} phage particles per ml of culture medium. In neither case could exogenous interferon be detected in the supernatant fluids.

Three substances of low molecular weight have been reported to stimulate interferon production in vivo. The first is cycloheximide (177). Unfortunately, the amount of this antibiotic required for interferon production results in the irreversible inhibition of protein synthesis leading to the death of the animal. The second is the antibiotic, kanamycin (118), which also requires relatively high dosage levels. The third, and most recent, is bis-DEAE-fluorenone (105, 124). The particularly interesting property of this interferon inducer is that subtoxic doses administered orally stimulate impressive amounts of circulating interferon in mice and rats. The compound is active in mice against infection with at least nine viruses of both RNA and DNA groups (104). The peak interferon titer occurs 16 hr after administration of the drug (38). The therapeutic index, expressed as the ratio of toxic to effective dose, is greater if the route of administration is oral as compared to intraperitoneal (38). Kaufman et al (94) studied the interferon-inducing properties and the toxic properties of bis-DEAE-fluorenone in humans. Subjects were given 1000 mg orally per day and their sera were tested for interferon activity. None was found. Other patients received one drop of a 200 mg/ml solution of the drug per eye per day. Tears were collected from the patients and again no interferon was found. Patients who had received oral bis-DEAE-fluorenone developed gastrointestinal problems including diarrhea and vomiting. These problems ceased when the drug was withdrawn. Patients subjected to topical ocular treatment were troubled by blurred vision apparently caused by the deposition of the drug in the epithelium of the cornea. Approximately two months were required for the drug to disappear from the epithelium (94). Thus, it appears that bis-DEAE-fluorenone, even at high doses, is ineffective in inducing interferon in man. Furthermore, it appears that even if it were effective in man, the toxic effects would preclude its therapeutic or prophylactic clinical use. Despite its high effectiveness in inducing circulating interferon in mice, the drug is quite ineffective as an interferon inducer in mouse L-cells, mouse embryo fibroblasts, and human skin fibroblasts (38).

All of the interferon inducers which are active only in intact animals have one property in common. They appear to have a toxic effect, particularly on the membranes of cells of the reticuloendothelial system. Thus, the search for substances that can stimulate the release of interferon in animals, and therefore serve as antiviral agents, must continue until agents are found that have very high ratios of efffective dose to toxic dose.

Inducers active in vivo and in vitro.—This class of inducers includes infectious and inactivated viruses, certain fungal extracts, and natural and synthetic polynucleotides (25). Isaacs (85) hypothesized that the nucleic acid component of the virus is responsible for triggering the interferon response and that the mechanism involved the cell's capacity to recognize a "foreign" nucleic acid. This hypothesis arose from his finding that chick liver RNA was more effective in inhibiting vaccinia virus growth in mouse cells in culture than in chick cells and that mouse liver RNA was more effective in chick cells (85).

The most important series of experiments in the area of interferon induction were carried out in 1967 at the Merck Institute. Field et al (52) discovered that very low concentrations of double-stranded synthetic polynucleotides induce interferon in rabbits and primary rabbit kidney cells in culture. Poly(I)·poly(C), poly(A)·poly(U) and poly(I)·n(CpC) were active in microgram quantities while the single-stranded polymers, poly(I), poly(C), poly(A) and poly(U) were not active at significantly higher concentrations. The group also showed that naturally occurring double-stranded RNA's induce interferon. There are several fungal extracts that are active inducers. Lampson et al (107) found that the active component in helenine, a mycelial extract of *Penicillium funiculosum,* is double-stranded RNA and they suggested that the preparation of helenine contained a fungal virus that has double-stranded RNA either as its genome or its replicative form.

Statalon, the interferon-inducing extract of *Penicillium stoloniferum,* was shown to contain hexagonal particles (100, 101). These were named mycophage PS1 and were shown to contain complementary RNA (101). The presence of virus particles in statalon and helenine was confirmed by Banks et al (1). Double-stranded RNA has been shown to be the interferon-inducing principle in extracts of *Penicillium cyaneo-fulvum* (2) and in the mushroom *Cortinellus shitake* (164).

The Specificity of Interferon Induction in Cell Culture

As noted above, most of the interferon inducers are active only in animals and only at relatively high doses. The exception appears to be double-stranded RNA. Accordingly, Field et al (53) postulated that the double-stranded replicative form of RNA present in cells infected with RNA-containing viruses is responsible for triggering the interferon response. The induction of interferon by natural and synthetic polynucleotides has been confirmed for a great variety of virus-cell systems by many laboratories and has recently been reviewed in detail (25).

In many of these studies it was found that the concentrations of poly(I)·poly(C) required for the detection of extracellular interferon in the culture medium far exceeded those concentrations necessary for the detection of the state of viral resistance in the treated cultures. That is, 10–20μg/ml of poly(I)·poly(C) will stimulate both the synthesis and release of interferon, while less than a factor of 10^{-4} is required for the development of cellular resistance (26, 34, 52). Thus, the question arose as to whether or not the state of viral resistance in cells treated with very small amounts of double-stranded polynucleotide was the result of the stimulation of an intracellular interferon response. The question was experimentally attacked by Schaefer & Lockart (151) using two lines of African green monkey cells. One line, LLC-MK_2, can synthesize and respond to monkey interferon. The other line, VERO, cannot be induced to synthesize interferon but it will respond to exogenous interferon (40). Cultures of both cells lines were incubated with monkey interferon and other cultures of both lines with poly(I)·poly(C). All cultures were then infected with Sindbis virus. Both lines treated with interferon showed a 99.9 percent reduction in virus yield. There was a 99 percent reduction of viral replication in LLC-MK_2 cells treated with $(I)_n \cdot (C)_n$ while the $(I)_n \cdot (C)_n$-treated VERO cultures exhibited full virus production. Controls were done to show that both cell lines were capable of taking up the helical polynucleotides. They therefore concluded that the viral resistant state of poly(I)poly(C)·treated cells is mediated by interferon (151).

Many species of animal cells in culture respond to double-standed polynucleotides. These include rabbit, mouse, chicken, dog, calf, hamster, monkey, and human cells (25). However, the concentrations of polynucleotide required for a given level of response differ considerably. Dianzani et al (43) found that the presence of DEAE-dextran in the induction medium caused a 100-fold stimulation in $(I)_n \cdot (C)_n$-induced interferon in mouse L-cells. Colby & Chamberlin (26) reported a twenty-fold increase in the rate of uptake of P^{32}-labeled $(I)_n \cdot (C)_n$ in chick embryo cells treated simultaneously with DEAE-dextran, and Dianzani et al (44) concluded that the polycation increases the permeability of L-cells to polynucleotides rather than protecting the polynucleotides from ribonuclease degradation.

A very complete study of the absorption spectra, thermal denaturation, susceptibility to enzymatic degradation, and ability to induce interferon of complexes of poly(I)·poly(C) and DEAE-dextran has recently been carried out by Pitha & Carter (138). Aggregation occurs between these two polyions and is maximal at electrostatic equivalency. The presence of DEAE-dextran decreased the initial rate of hydrolysis of $(I)_n \cdot (C)_n$ by pancreatic RNase, while the initial rate of phosphorolysis, catalyzed by polynucleotide phosphorylase, was increased. A greater enhancement of interferon production was observed when preformed polynucleotide-DEAE-dextran complexes were used than when the cells were pretreated with DEAE-dextran followed by polynucleotide. The maximum interferon production was observed under conditions which led to maximum aggregation, that is when the phosphorus-to-nitrogen ratio was close to electrostatic equivalence. This complex was adsorbed by cells twice as effectively as when the P/N ratio was 0.25, and 40 times more effectively than $(I)_n \cdot (C)_n$ alone (138).

The first step in the sequence of events leading to the production of interferon by cells treated with poly(I)·poly (C) appears to be the binding of the polynucleotide to the cell surface (5). This can be demonstrated at either 4°C or 37°C and is greatly enhanced by DEAE-dextran. A second, temperature-dependent step occurs rapidly thereafter and is followed by cellular resistance to virus infection and interferon production (5).

What are the specified physical and chemical properties which the inducer molecule must satisfy in order to be effective? Some possibilities are (*a*) primary structure; that is, the particular sequence of bases in the nucleic acid; (*b*) secondary structure; for example, random coil, associated hairpin, A helix, B helix, etc.; (*c*) chemical structure of the nitrogenous bases; (*d*) chemical structure of the sugar moieties.

Isaacs' theory that virus-infected cells can respond to a "foreign nucleic acid" (149) intimated that the induction mechanism involves the recognition of a specific sequence of bases. It now seems quite clear that this notion is incorrect. Naturally occurring double-stranded RNA's from a variety of sources (see Tables 1 and 2) are very efficient inducers of interferon. Secondly, synthetic double-stranded polyribonucleotides containing homosequences of I, C, A, U, G, and X are also active inducers (26, 34, 52). The associated homopolymers poly (I)·poly(C) are approximately as efficient at inducing the interferon response as the alternating polymer, poly(I-C) (26). Thus, the specificity of induction does not involve either a run of purines or a run of pyrimidines. Finally, recent experiments done with double-stranded RNA isolated from uninfected cells (27, 131) have settled the question. Harel & Montagnier (71) isolated and characterized double-stranded RNA from rat liver cells. De Maeyer et al (39) then showed that this double-stranded RNA induced interferon in rat, mouse, and chick embryo fibroblasts. Kimball & Duesberg (99) purified double-stranded RNA from rabbit kidney cells and chick embryo cells and showed that the purified RNA's induced interferon in both homologous and heterologous systems. Since normal cells appear to contain small quantities of double-stranded RNA, and since this RNA can clearly act as an interferon inducer, we must conclude that the normal uninduced state of cells reflects either the compartmentalization of this RNA away from the interferon-triggering mechanism or that the intracellular concentration of this RNA is below the threshold level required for induction.

Now we turn to the secondary structure of the interferon-inducing nucleic acid. In their initial report, Field et al (52) emphasized that the single-stranded polyribonucleotides, poly(I), poly(A), poly(C), and poly(U) were inactive at concentrations more than 10,000 times greater than that which allowed detectable viral interference with the helical homopolymer pair, poly (I)·poly(C). This result has been confirmed in a number of laboratories (25). An apparent conflict arose when Baron et al (3) found that particular batches of commercially available synthetic "single-stranded" polynucleotides were active inducers both in vivo and in vitro. Only an occasional batch of poly(I) or poly(C) was found to behave in this maverick fashion and then only if extremely large doses were used. The authors ruled out the possibility of contamination of their samples with double-stranded RNA leaving us to suppose that the polynucleotide might be in a conformation intermediate between that of a helix and a random coil. De Clercq & Merigan (34) studied the effects of hydrogen and magnesium ion concentrations on the physicochemical and interferon-inducing properties of several synthetic polynucleotides. Poly(I)·poly(C) and poly(A)·poly(U) are much more active than the homopolymer pairs poly(U)·poly(X), poly(A)·poly(I), poly(I)·poly(X) and poly(A)·poly(X). The activity of the latter duplexes was found to increase with decreasing pH and with increasing magnesium ion concentration. However, their increased specific activity remained significantly below poly(I)·poly(C). They also found a slight Mg^{++}-induced activity with poly(A) and poly(C). Again, these were used at very large concentrations.

There are several series of experiments using viruses as the interferon inducer that appeared to challenge the importance of the secondary structure of the inducer molecule. In all cases the investigators attempted to eliminate the production of any double-stranded replicative form or replicative intermediate RNA. Lockart et al (116) investigated the interferon-inducing capacity of wild-type Sindbis virus (W^+) and two classes of temperature-sensitive mutants, one of which can produce viral RNA but is blocked in some later function (*ts-RNA*$^+$) and another which cannot synthesize significant amounts of viral RNA (*ts-RNA*$^-$). Sindbis W^+ induced interferon at both 29°C and 40°C. Sindbis (*ts-RNA*$^-$) would induce interferon only at the permissive temperature (29°C). If cycloheximide was present when the infected culture was held at 29°C and then was washed out when the culture was shifted to 40°C, there was no interferon production. This suggested that early viral protein synthesis, presumably of a viral RNA polymerase, was required for interferon induction. Thus far their results are consistent with viral RNA replication being a necessary prerequisite for interferon induction. However, *ts-RNA*$^+$ mutants also failed to elicit the interferon response at the nonpermissive temperature even though significant quantities of viral RNA were being made. Lockart et al (116) therefore concluded that the accumulation of viral RNA alone is not enough to trigger the interferon response, in agreement with Skehel & Burke (155).

Marcus (120) has repeated these experiments using the plaque reduction assay for interferon which is 100–400 times more sensitive than measuring the inhibition of cytopathic effects of a challenge virus as did Lockart et al (116). Marcus found that all three classes of Sindbis virus (W^+, *ts-RNA*$^+$, and *ts-RNA*$^-$) induced interferon at both the permissive and the nonpermissive temperatures. This confusing result was explained by Marcus' finding that crude stocks of all of these viruses contain an interferon inducer exclusive of the virion. When the experiments were again repeated with purified virus he found that Sindbis W^+ and *ts-RNA*$^+$ induce interferon at both temperatures while *ts-RNA*$^-$ mutants are inducers only at the permissive temperature.

The second series of experiments referred to above are those of Dianzani et al (45). One of the most efficient systems for producing large quantities of extracellular interferon is the mouse L-cell-Newcastle disease virus system. Again, the object of the experimental design was to attempt to eliminate the production of double-stranded RNA. Mouse L-cells were treated with cycloheximide and infected with very high multiplicities of NDV. After 4 hr actinomycin D was added. The cycloheximide was washed out at 5 hr postinfection and the cultures were assayed for viral resistance and interferon production at 8 hr postinfection. The authors reasoned that the initial cycloheximide block, by inhibiting the production of a viral RNA polymerase, would not allow viral RNA synthesis. During this time host DNA could be transcribed but the resulting mRNA could not be translated. Host transcription was blocked at 4 hr and then all species of mRNA could be translated between the 5th and 8th hours. No viral resistance developed in these cells but significant amounts of interferon were produced. This result demanded that the mRNA for interferon had been synthesized prior to the addition of actinomycin D. Since cycloheximide had been present during that time, Dianzani et al (45) concluded that the stimulus for interferon induction was provided by a component of the input virus, and that this component was the molecule of single-stranded RNA which is the genome of this virus. They therefore suggested that, in some virus-infected cells, interferon may be induced by single-stranded RNA.

Several alternative explanations have been suggested (25). These included high concentrations of RNA with a secondary structure intermediate between a random coil and the helical configuration as well as helical RNA formed by intracellular annealing of the viral RNA in accord with the findings of Robinson (147).

The explanation is probably even more straightforward. Huang et al (82) have recently discovered an NDV virion-associated RNA transcriptase. Thus, the original conclusion of Dianzani et al (45) need be modified only to change the phrase input component to input components, these being the viral genome and the virion-associated transcriptase. The biochemical interaction between these input components would, of course, result in the synthesis of double-stranded RNA.

A third enigma concerns the well-established observation that ultraviolet-inactivated virus preparations can induce interferon. Many workers have assumed that cells infected with such preparations would not contain any new viral RNA. Huppert et al (83) showed that while a preparation of ultraviolet-inactivated NDV was unable to produce any infectious progeny in chick embryo cells, viral RNA, including RNase-resistant RNA was synthesized. Bratt (9) has recently found that the virion transcriptase is 50-fold more resistant to ultraviolet irradiation than viral infectivity.

In the case of cells infected with Semliki Forest virus (SFV), Burke et al (12) found that heat-inactivated SFV induced interferon in the absence of viral RNA synthesis. This result was confirmed by Goorha & Gifford (63), and extended to show that the inactivated virus preparations were better inducers than active ones since the latter caused an inhibition of host RNA synthesis (64). Skehel & Burke (154) found that hydroxylamine treatment of SFV caused a decrease in infectious virus, virus haemagglutin, virus-reduced RNA synthesis, virus-induced RNA polymerase, and interferon production all with first-order kinetics. They concluded that functional nucleic acid synthesis is essential for interferon production in this system. Lomniczi & Burke (117) characterized the abilities of wild type, *ts-RNA*$^{+}$, and *ts-RNA*$^{-}$ SFV to induce interferon. At high multiplicites, both classes of mutants and the wild type induced interferon production without requiring virus RNA synthesis, while at lower multiplicities interferon production depended on the synthesis of virus RNA. The authors concluded that at high multiplicities the inducer was input virus RNA. An alternative explanation involves the possibility that crude stock preparations of SFV contain an inducer of interferon exclusive of the virion (as in the case of the stocks of Sindbis virus mentioned above) which would be active at high multiplicities but too dilute at low multiplicities. With the exception of the hydroxylamine experiments, all of the above mentioned studies were done with unpurified virus stocks.

As of the time of writing of this review, there are no reports which unequivocally demonstrate that interferon induction in cells infected with live or killed RNA-containing viruses is not mediated via double-stranded RNA. Furthermore, in the case of Mengo virus infection, once the replicative form or the replicative intermediate is formed, no further virus functions are required for interferon production (51).

Double-stranded RNA is more resistant to a variety of ribonucleases than single-stranded RNA. If the essential factor that determines the relative specific activity of a nucleic acid to stimulate interferon production is its ability to survive nuclease degradation, then there are several predictions that one may make. First, we should expect a direct correlation between the efficiency of induction and the relative nuclease sensitivities of a variety of interferon-inducing polynucleotides. Secondly, we should expect any structurally stable nucleic acid to be an efficient inducer.

There is some experimental evidence that is consistent with the first point. Double-stranded polynucleotides containing I·C or G·C base pairs are more active than those containing A·U base pairs (26, 34, 52). Colby & Chamberlin (26) studied the rates of uptake and breakdown of P^{32}-labeled poly(I-C) and poly(A-U) in chick embryo cells. Both polymers were taken up at the same rate, but poly(A-U) was broken down more rapidly inside the cell than poly (I-C) (29). Not only was poly(I-C) active at lower concentrations, but a much shorter time of exposure to the cells was required (28).

Chemical modification of polynucleotides often leads to changes in thermal stability and nuclease sensitivity. De Clercq et al (33) compared the interferon-inducing capacity and ribonuclease sensitivity of two helical copolymers, poly(A-U) and poly ($A^{S}U$). Both polymers have alternating riboadenosyl and ribouridyl nucleoside units. They differ from each other in that poly($A^{S}U$) has one of the phosphate oxygens replaced by sulfur. This polymer was 100 times more resistant to pancreatic ribonuclease and 100–1000 times more active as an interferon inducer.

Colby & Chamberlin (26) determined the ribonuclease sensitivities of the seven helical polynucleotides which they had found to be active interferon inducers. Poly(G)·poly(C) is extremely resistant to ribonuclease, yet it is no better than poly(I)·poly(C) as an inducer. Poly(I-BrC) is 10 times as resistant to ribonuclease as poly(I-C); the latter is the more potent inducer. Poly(A-U) and poly(A-BrU) are equivalent inducers; the latter is twofold more resistant to ribonuclease. Finally, poly(A)·poly(U) is degraded by RNase 10-fold more slowly than its alternating analog, poly(A-U); poly(A-U) is by far a more efficient inducer than poly(A)·poly (U).

Carter & Pitha (17) demonstrated the importance of secondary structure most clearly using a helical polymer composed of poly(C) complexed with hexainosate, poly (I_6)·poly(C), which has a melting temperature of 5° C. This polynucleotide induced interferon in cultures of human fibroblasts only when the cells were exposed to the complex below, but not above, its melting temperature. Thus, a stable secondary structure is quite necessary for a polynucleotide to act as an interferon inducer. However, the results of the experiments mentioned above do not support the idea that the *sole* requirement for interferon-inducing activity is that the polynucleotide survive RNase degradation.

The second prediction (i.e., that any structurally stable nucleic acid should act as an inducer) has also been tested experimentally. This prediction suggests that helical DNA should be an inducer. Lampson et al (107) found calf thymus DNA to be inactive as an interferon inducer in rabbits, Colby & Chamberlin (26) found no antiviral activity in chick cells treated with DNA from coliphage λ, and Kleinschmidt et al (102) found no activity with T4 phage DNA. Three DNA-like polynucleotides, poly (dA-dT), poly(dI)·poly (dC), and poly(dG)·poly(dC) were reported to be inactive at 10 μg/ml in chick cells (26). However, De Clercq et al (32) found that high concentrations of poly(dG)·poly(dC), poly(dA-dT), and poly (dA)·poly(dT) were active in human fibroblasts and that poly(dA-dT), poly(dG-dC), and poly (dI-dC) could be activated by heating. Colby et al (29) attempted thermally to activate the following helical polynucleotides without success: $(rI_n)\cdot(rC)_n$, $(rI\text{-}rC)_n$, $(rI\text{-}rBrc)_n$, $(rG)_n\cdot(rC)_n$, $(rA\text{-}rU)_n$, $(rA\text{-}rBrU)_n$, $(rA)_n\cdot(rU)_n$, $(rI)_n\cdot(dC)_n$, $(dI)_n\cdot(rC)_n$, $(dI)_n\cdot(dC)_n$ and $(dA\text{-}dT)_n$. Their failure could be related to the fact that they used the chick embryo cell system that requires DEAE-dextran (26) which tends to obscure the thermal activation effect (37). However, Pitha & Carter (138) also failed to confirm the thermal activation of poly(I)·poly(C) using human cells.

While Field et al (53) suggested that double-stranded RNA could be the inducer of interferon in cells infected with RNA-containing viruses, they also postulated that a viral DNA-RNA complex might induce interferon in DNA-virus infected cells. A DNA-RNA hybrid would also be resistant to ribonuclease degradation. Vilcěk et al (167) found that very high concentrations of poly (rI)·poly(dC) induced some resistance to vesicular stomatitis virus in rabbit kidney cells. Poly (rI)·poly(dC), poly(dI)·poly (rC) and poly(rA-dU) are inactive at 10μg/ml in chick embryo cells (26). Using P^{32}-labeled poly(I-C) and poly(rI)·poly(dC), Colby & Chamberlin (26) excluded the possibility that permeability and intracellular stability differences accounted for the 10,000-fold difference in the interferon-inducing capacity of these two polynucleotides. Thus, it appears that at low concentrations of synthetic polynucleotides something other than a stable helical structure is required for interferon induction.

Nemes et al (134) reported that a preparation of DNA-RNA hybrid prepared by transcribing f1 phage single-stranded DNA with *Escherichia coli* RNA polymerase was an active interferon inducer in rabbits and rabbit kidney cells in culture. Robertson & Zinder (144), and Robertson (145) found that the products of such a transcription reaction mixture contain three species of nucleic acids; DNA-RNA hybrid, single-stranded RNA, and double-stranded RNA. Colby et al (31) separated the products in Cs_2SO_4 gradients, identified the products by physical, biochemical, and immunological methods and showed that the antiviral activity resides solely with the contaminating double-stranded RNA. Purified DNA-RNA hybrid was inactive at 10 μg/ml, whereas the unfractionated products were active at 0.1 μg/ml. They therefore concluded that a purified DNA-RNA hybrid containing all eight of the mononucleotide subunits is as impotent an interferon inducer as the synthetic hybrid-like polynucleotides previously tested (26).

Thus, the presence of deoxyribose in one or both chains of a helical nucleic acid lowers the specific activity of that nucleic acid as an interferon inducer by a factor of 10,000. Colby & Chamberlin (26) therefore postulated that DNA viruses which induce interferon production do so by some agent other than a DNA-RNA hybrid. Subsequently, Colby & Duesberg (27) looked for and found RNase-resistant RNA in chick embryo cells infected with vaccinia virus. The RNA was characterized as double-stranded RNA by a variety of physicochemical methods and by its ability to induce resistance to Sindbis virus (27, 47). The double-stranded RNA is vaccinia virus-specific (27) and appears to be made via a DNA-dependent mechanism (27, 30). Vaccinia virus double-stranded RNA may also be made in an in vitro reaction using the virion-bound transcriptase (30). Double-stranded RNA has also been found in *E. coli* cells infected with coliphage T4 (89). There are no reports of virus-specific, double-stranded RNA in animal cells infected with DNA-containing viruses that replicate in the nucleus. Nothing is known concerning the possible role(s) of these double-stranded RNA's in the replicative cycle of the DNA viruses.

All of the evidence presented above is consistent with the notion that the mechanism of interferon induction is a highly specific one demanding definite chemical and physical requirements that a potential interferon-inducing molecule must meet. For low concentrations of nucleic acids, the specificity appears to involve a stable helical configuration and the presence of 2′-hydroxyl groups on the sugar moieties. The latter point is reinforced by the recent finding by Levy (110) that polynucleotides containing 2′-O-methy-ribonucleotide units are inactive interferon inducers.

Receptor Site Protein

Colby & Chamberlin (26) postulated that the specificity of interferon induction is mediated by the combination between a specific intracellular receptor site and the inducer molecule, and that the extent of induction is proportional to the amount of complex formed. The molecular nature of the receptor site was postulated to be a protein rather than a nucleic acid. Nucleic acids interact with each other in a specific way through base pairing, a mechanism demanding a particular sequence of bases in the inducer molecule while completely ignoring the nature of the sugar residues. Both of these restrictions are contrary to the experimental evidence. However, there is a variety of proteins that have the capacity to recognize the secondary structure of a nucleic acid. Secondly, proteins are known to be able to distinguish RNA from DNA from DNA–RNA hybrids. The receptor site was therefore postulated to be a protein, the primary structure of which is presumably dictated by a cellular gene. There is no direct evidence for the receptor site protein, i.e., no one has identified and purified such a molecule. Its existence was postulated to provide a mediator for the highly specific interferon induction mechanism that is evident at low inducer concentrations.

The receptor site protein hypothesis also allows us to interpret many of the observations concerning interferon production stimulated by high concentrations of other polyanionic macromolecules. If helical polyribonucleotides interact with the receptor site most efficiently (and are therefore the best interferon inducers), and if single-stranded polyribonucleotides with a secondary structure of a totally random coil do not interact at all with the receptor, then there should be a series of intermediate secondary structures that could interact with the receptor to an intermediate extent. One would expect that much higher concentrations of such molecules would be required for activity. The results of De Clercq & Merigan (34) and of Baron et al (3) are consistent with these ideas.

Secondly, one would expect the specificity of interferon induction, with respect to the presence of 2′-hydroxyl groups, to be overcome by high concentrations of polynucleotides containing deoxyribose. Such results have been reported by Vilček et al (167) and De Clercq et al (34). Finally, the discovery that the thiophosphate analog of poly(A-U) is a more potent interferon inducer than poly(A-U) itself was causally related to the observation that poly($A^{s}U$) is much more resistant to ribonuclease (33). Since the T_m of the two polymers is the same, indicating that the stability of the helix is unchanged, we may offer an alternate explanation within the framework of this discussion. That is, we may suggest that the increased resistance to RNase of poly($A^{s}U$) may be due to an inhibitory action of the sulfur atom at that portion of the active site of the enzyme responsible for the recognition of the phosphodiester linkage. Similarly, poly($A^{s}U$) may have a higher affinity than poly(A-U) for the receptor site protein responsible for triggering interferon induction.

PROPERTIES OF INTERFERON

Interferon may be defined as a cellular protein that confers antiviral activity. Since there are many ways to inhibit virus growth, Lockart (114) has suggested the following criteria for the acceptance of a viral inhibitor as an interferon: (i) The inhibitor must be a protein and be formed as a result of the addition of an inducing substance to cells or animals. (ii) The antiviral effect must not result from nonspecific toxic effects on the cells. (iii) The proposed interferon must inhibit the growth of viruses in cells through some intracellular action involving both RNA and protein synthesis on the part of the cells. (iv) The inhibitor must be active against a range of unrelated viruses. If an inhibitor shows marked specificity for homologous cells, this is highly suggestive that it is, in fact, an interferon. We will now consider the experimental evidence on which each of these criteria of acceptance is based.

Interferon is a protein. Of all the putative cellular gene products involved in the interferon system, only interferon itself has been purified, characterized, and studied in some detail (see reviews 25, 55, 165). Interferon contains most of the common amino acids and some carbohydrate including glucosamine. Amino and disulfide groups and the methyl groups of methionine may not be derivatized without loss of antiviral activity. Substitution on sulfhydryl and hydroxyl groups is inconsequential to its biological activity. One of the more remarkable properties of interferon is its stability in the pH range 2.0 to 10.0. This property is the basis for the most commonly used first step in purification. The most highly purified interferons are those from chick, mouse, and human.

Interferons of the same species origin have been reported to have extensive variabilities in molecular weights depending on the inducer used and the type of cell or animal used. Many workers therefore thought of interferons as a family of proteins (some investigators have even suggested that the interferons are analogous in complexity to the products of the immune system). Recent experiments done by Carter (14) on purified interferon preparation of human and mouse origin directly test this suggestion. Mouse interferon was purified 500-fold and subjected to isoelectric focusing. Two equally active molecular forms were found; molecular weights were 38,000 and 19,000. The former could be dissociated to form the latter, but the reverse reaction occurred to a much lesser extent. Human interferon, purified 1500-fold, was also found to exhibit two molecular species of molecular weight 24,000 and 12,000. Thus, the native molecules appear to be dimers of similar or identical subunits. These experiments were done using a viral inducer, NDV. When poly(I)·poly(C) was used as the inducer of human interferon, a 96,000 molecular weight species of interferon was isolated

(17). This molecule could be dissociated into 24,000 and 12,000 molecular weight species. Thus, it seems likely that wide variabilities in molecular weights of interferon merely reflects the state of aggregation of the molecule rather than a complex variety of different molecules.

Interferon is a cellular protein. The only alternative, i.e., that it is a viral protein, is ruled out by the following experiments. First, nonviral agents induce interferon. Secondly, Lampson et al (106) showed that chick interferon induced by herpes simplex virus has the same physicochemical properties as that induced by influenza virus. Thirdly, actinomycin D, which blocks DNA-dependent RNA synthesis inhibits interferon synthesis in cells infected by RNA-containing viruses whose replication is insensitive to the drug (72, 170).

Finally, the interferon synthesized by one cell type possesses antiviral activity only when assayed on cells of the same species. To demonstrate this point, the use of purified interferon is particularly important. For instance, Buckler & Baron (11) obtained reduced yields of vaccinia virus from chick cells treated with chick interferon, normal allantoic fluid, mouse serum interferon, and normal mouse interferon. If the cultures were washed carefully after interferon treatment, only the chick interferon exhibited antiviral activity. Merigan (126) demonstrated the species specificity of chick and mouse interferon purified 6000-fold.

The third criterion is discussed below in the next section. The fourth criterion considering the lack of virus specificity of a particular interferon is also well documented (25, 35, 55, 165). The most definitive study done recently is by Stewart et al (161). They determined the relative sensitivities of Sindbis, vesicular stomatitis, Semliki Forest, and vaccinia viruses to interferon-treated mouse embryo, hamster kidney, rabbit kidney, human embryonic lung, and bat embryo cells. An example of the cell-dependent variabilities is their finding that vaccinia was the most sensitive virus to interferon-treated mouse and hamster cells and the least sensitive in human, rabbit and bat cells.

ANTIVIRAL ACTIVITY OF INTERFERON

It is generally accepted that the establishment of the antiviral state conferred by interferon treatment depends on the synthesis of another protein, since inhibitors of both protein synthesis and mRNA synthesis can prevent the antiviral activity of interferon (62, 72, 109, 113, 163, 169) and since interferon itself has no effect on viruses (66, 75). It is conceivable that this merely promotes the action of interferon or aids in its transport into the cell or to its site of action (152). We know of no experiments that establish beyond a shadow of doubt that an antiviral protein (AVP), distinct from interferon, actually exists. Such proof will await the physical isolation and characterization of AVP. For the purposes of this review, however, it is necessary to discuss the synthesis, nature, and mode of action of the hypothetical AVP.

Dianzani et al (45) have shown that by a judicious choice of inhibitors one can induce the synthesis and release of interferon in mouse L-cells without establishing an antiviral state. This certainly supports the existence of AVP. Recent work (160) claims to refute the proposal that interferon per se has antiviral activity. Stewart & Lockart (160) showed that human cells incubated with human interferon or poly(I)·poly(C) became more resistant to vesicular stomatitis virus than to Semliki Forest virus whereas monkey cells incubated with monkey interferon or poly(I)·poly(C) became more resistant to Semliki Forest virus than to vesicular stomatitis virus. However, monkey cells incubated with the heterologous human interferon developed an antiviral state identical to that induced with the homologous monkey interferon and vice versa for human cells. They concluded from this that interferon treatment induces the synthesis of AVP and does not itself have antiviral properties.

We would agree that the data tend to show that the relative antiviral activity is a property of the host cell; we would not agree, however, that it disproves the hypothesis that interferon itself is antiviral. The specificity could reside with the inhibitor (AVP or interferon) or in the reaction that is inhibited. Until we have a better understanding of the molecular events involved in the antiviral state we should consider all possibilities and reserve judgment. Since the authors measured antiviral activity using multiple-cycle inhibition methods rather than a single-cycle inhibition method (70), it is also possible that the relative antiviral activity is artifactual.

What do we know of AVP, its synthesis and mode of action? As stated above, AVP has not yet been isolated and little is known about its characteristics, other than that it may, like interferon, be species-specific (135).

AVP is synthesized in response to interferon produced within the cell (endogenous) or administered to the cell (exogenous). At present, no one has reported the synthesis of AVP in response to an inducer other than interferon. Using the Vero line of Green Monkey kidney cells (40) which cannot synthesize interferon, Schaefer & Lockart (151) showed that interferon synthesis was a prerequisite to antiviral activity even under circumstances (low inducer concentration) where interferon was not released into the culture medium. When cells are treated with interferon, only minute quantities are removed from the medium by the cells (10); the first step in this removal is an energy-independent binding of interferon to the cell membrane (58, 115) followed by a presumably energy-dependent transport into the cell. Little is known of the transport of interferon through the membrane. However, it has been shown that the degree of effectiveness of the antiviral state is concentration-dependent (4) rather than dependent on the absolute amount of interferon, suggesting that the transport of interferon is also concentration-dependent, i.e., transport by facilitated or simple diffusion. In binding to the cell membrane, interferon recalls the action of polypeptide hormones (58), which affect the activity of adenyl cyclase and subsequent levels of cyclic 3′,5′-adenosine monophosphate (CAMP). Developing this idea further, the effect of CAMP on interferon action in chick embryo cells was tested (61) and found to potentiate the antiviral state. However, CAMP alone had no effect on AVP synthesis in contrast to results obtained in systems sensitive to polypeptide hormones (146). Evaluation of the effect of CAMP on the interferon system must await further experiments.

On reaching the nucleus (if, indeed, it does), interferon activates the cell genome leading to the synthesis of mRNA (actinomycin D-sensitive) for AVP synthesis which is then translated (puromycin-sensitive) to AVP.

The action of AVP is to block viral-dependent protein synthesis; the mechanism of its action is still open to question. In 1966, two papers were published (88, 122) that together tended to establish that AVP so modified the structure of the host cell ribosome that viral RNA was bound and translated far less efficiently. Marcus & Salb (122) examined the ribosomes from interferon-treated and -untreated chick embryo fibroblasts in a cell-free protein synthesizing system with purified Sindbis (RNA-virus) RNA as message. They found that Sindbus RNA interacted with normal chick ribosomes (74S) to form aggregates (240S) at 4°C. These aggregates broke down at 37°C with a concomitant incorporation of amino acids. The breakdown was energy-dependent and sensitive to cycloheximide. In marked contrast, Sindbis RNA bound to ribosomes isolated from interferon-treated chick fibroblasts with only two-thirds the efficiency of binding to normal chick ribosomes. In addition, such aggregates did not break down on incubation at 37°C nor were amino acids incorporated. Siliar differences in the rates of translation have been found with Mengo RNA and mouse ribosomes (16). The authors went on to show that ribosomes from normal and interferon-treated chick cells would bind RNA isolated from noninfected chick cells with identical efficiency and that ribosomes isolated from such cells pretreated with actinomycin D before interferon treatment behaved similarly to normal ribosomes. Joklik & Merigan (88) showed that in vaccinia (DNA-virus) infected mouse L-cells, viral RNA synthesis was more rapid in interferon-treated cells than in controls, but that the rate of protein synthesis was considerably lower in interferon-treated cells. They showed that infection of L-cells with mengo or vaccinia leads to a rapid disaggregation of polyribosomes. In normal cells, mengo or vaccinia RNA-ribosome complexes appear after such disaggregation but in interferon-treated cells no such viral-RNA ribosome complexes were found. The acute cytotoxic effects on L-cells of vaccinia virus particles described in this paper are such as to cast doubts on the significance of the data described. However, on the basis of these papers it was concluded that AVP [translation inhibitory protein (TIP) (122)] bound in some manner to the ribosome enabling it to discriminate between host and viral RNA. Numerous other reports (119, 125, 157, 158) were soon forthcoming, all tending to support the suggestion that AVP had its effect at the level of translation, and by 1968 Marcus & Salb (123) were able to present a more unified picture of TIP as a factor involved in the regulation of translation in the normal cell cycle. An earlier paper (150) had shown that ribosomes isolated from HeLa cells in metaphase were very much less active in a cell-free protein-synthesizing system than ribosomes isolated from interphase cells, and that "metaphase" ribosomes could be converted to "interphase" ribosomes on incubation with trypsin. Expanding this concept (123), they showed that partial restoration with viral RNA of activity of ribosomes isolated from interferon treated cells could be achieved by incubating with trypsin. Unfortunately, they were unable to demonstrate a new protein (or absence of a protein) in ribosomes from interferon-treated cells by the sensitive technique of dual-labeling with ^{14}C and ^{3}H-labeled amino acids.

Carter & Levy (15) found that ribosomes obtained from interferon-treated L-cells would bind in vitro to rapidly labeled RNA extracted from noninfected L-cells but not to rapidly labeled RNA from mengo-infected L-cells. These results are in agreement with their in vivo studies (111). They examined the binding of mengo RNA with the 40S ribosomal subunit in interferon-treated and -untreated L-cells infected with mengo and were able to isolate a 50S particle bearing mengo RNA in untreated cells but not in interferon-treated cells. Interferon treatment apparently blocked the entry of mengo 37S RNA into a 50S cytoplasmic form, possibly a preliminary step in polyribosome formation.

It is distressing and perplexing that no real evidence has been obtained to correlate an alteration in the function of ribosomes isolated from interferon-treated cells with an alteration in the proteins making up the structure of the ribosome (98, 123, 158, 159). This inability to demonstrate the physical existence of TIP (AVP) had led to a re-examination of the action of AVP and a new attempt has been made, with scant success, to repeat the experiments of Marcus & Salb (122) and Carter & Levy (15). Kerr et al (98) examined the effect of chick interferon on chick embryo fibroblasts by studying the interaction of RNA from encephalomyocarditis (EMC) virions and ribosomes from chick embryo fibroblasts. They were unable to obtain a discrete peak of virus-specific polysomes from control or interferon-treated cells and found that there was not necessarily a correlation between formation and breakdown of the viral ribosome complex and amino acid incorporation (a finding with which Marcus agrees). Neither could they find any alteration in the ribosomes from interferon treated cells.

Although Marcus & Salb were using RNA from a virus (Sindbis) particularly sensitive to AVP and obtained their ribosomes from cells so aged that they were also particularly sensitive to interferon, it seems unlikely that this would explain the inability of others to repeat the work. More likely, perhaps, it is the suggestion (98) that AVP is only transiently bound to the ribosomes or, indeed, is only associated with the ribosome after certain isolation procedures. There is evidence from other systems for such factors (46, 139) involved in, for example, the initiation of protein synthesis, but there is also an abundance of evidence from other systems that alterations in ribosomes occur in response to (or precede) alteration in host cell metabolism (79, 84, 108, 156).

Many of the papers purporting to show differences in the rate of synthesis of specific proteins in interferon-treated and -untreated cells can be explained on the basis of a more general inhibition of protein synthesis at the level of the ribsome (59, 60, 65, 130), but there are some data that cannot so easily be explained in this manner.

Kerr & Martin (97) have developed an in vitro protein synthesizing system from chick and mouse cells with RNA from EMC virus as message. Kerr (96) has used the system to compare the response to EMC RNA of cell sap and ribosomes from interferon-treated and control chick embryo fibroblasts in mixed chick-mouse and purely chick systems. Ribosomes from interferon-treated cells supported 30–70 percent and cell sap fractions 50–100 percent of the level of EMC RNA-stimulated incorporation observed with corresponding controls, whereas poly U-directed protein synthesis was equally active with ribosomes obtained from interferon-treated or -untreated chick embryo fibroblasts.

The data can be questioned on at least two counts. The protein synthesizing system itself depends on very high concentrations of ribosomal and cell sap protein, and the peptides synthesized by the system could not be demonstrated to be the same as those isolated from EMC-infected cells. However, the data do throw grave doubts on the data already referred to (15, 88, 122). Unlike these authors, Kerr was unable to show any difference in the binding of viral RNA to ribosomes from interferon-treated and control cells. Nor was he able to confirm the conclusion (122) that the breakdown of the viral RNA ribosome complex was necessarily correlated with protein synthesis as measured by amino acid incorporation. Since, in his system, both ribosomes and cell sap from interferon-treated cells were less able to respond to EMC RNA than preparations from control cells, Kerr was unable to conclude at what level of translation protein synthesis was inhibited.

Recently, Horak et al (81) repeated the experiments of Joklik & Merigan (88) but used an improved system. They studied the effect of purified interferons on the synthesis of pox virus RNA and its ability to form polysomes in both mouse L-cells (vaccinia virus) and chick embryo fibroblasts (cow pox virus). Synthesis of viral RNA and polysome formation was detected by short-pulse labeling with ^{3}H-uridine. They found that chick (but not mouse) interferon treatment markedly inhibited viral RNA synthesis and viral polysome formation in chick embryo fibroblasts at all times studied. In L-cells, mouse interferon did not reduce viral RNA synthesis or viral polysome formation until at least 1 hr past infection. Chick interferon had no effect on the mouse L-cells. Treatment with poly(I)·poly(C) inhibited both viral RNA synthesis and polysome formation in chick embryo fibroblasts but not in L-cells. The conclusion seems inescapable that viral RNA synthesis and viral polysome formation is always similarly influenced by interferon treatment and, importantly, that the binding of newly synthesized viral RNA to cellular ribosomes to form viral polysomes is also possible in interferon-treated cells. The experiments do not, however, allow one to discriminate between the possibilities that AVP exerts its action by inhibiting the core transcriptase (90, 91) or the newly synthesized DNA-dependent RNA polymerase. These results are in marked contrast to those observed previously (88) and the authors suggest that the different method of cell disruption that they use may be the reason for this difference.

Using chick embryo cells treated with both actinomycin D and cycloheximide, Marcus et al (121) have recently demonstrated a functional VSV virion-transcriptase activity in infected cells. They were therefore able to directly test the action of interferon on the first viral synthetic activity thought to occur within the infected cell under conditions that uncoupled transcription from viral mRNA translation. VSV transcriptase activity was inhibited, in a dose-dependent manner, in cells pretreated with chick interferon or with poly(I)·poly(C), while mouse interferon had no effect. Marcus et al (121) concluded that interferon appears to act by inducing a cellular function that inhibits the transcription of VSV virion RNA.

Other authors (69) also favor an inhibition at a level other than that of the ribosome. Establishment of the correct mechanism of action of AVP will await its isolation in a pure form and study in an acceptable in vitro system.

REGULATION OF INTERFERON SYNTHESIS AND ACTIVITY

The past year has witnessed an abundance of publications dealing with the regulation of interferon synthesis in cell culture systems. We shall confine our discussion, in the main, to these papers. For a more detailed review of *inverse interference* (112), "blocker" (87), "stimulon" (21), "enhancer" (93), and "tolerance" (76, 176), the reader is referred to the excellent monograph of Vilček (165).

In 1966, Friedman (57) reported that an increase in the rate of protein synthesis in chick cells could be correlated with a decrease in interferon synthesis induced with SFV. He postulated that the cells were synthesizing a repressor of interferon synthesis at an increased rate. Vilček et al (168) reported that the effect of actinomycin D on interferon synthesis in rabbit kidney cells depended on the time of treatment with the inhibitor. If given 30 min prior to induction with poly(I)·poly(C) no interferon was released into the medium; if given 3.5 hr postinduction, interferon synthesis was stimulated. They postulated that actinomycin D blocked the synthesis of an inhibitor or repressor of interferon synthesis. In a more recent paper (166), Vilček has shown that in rabbit kidney cells the synthesis of this repressor is more sensitive to the effects of cycloheximide than is the synthesis of interferon. Poly(I)·poly(C)-induced interferon synthesis continued for approximately 18 hr longer in the presence of cycloheximide than in its absence. The data have been confirmed for interferon production in human cells (5) and mice (36).

The refractory state (137, 175) in which cells are unable to respond to a second induction stimulus by synthesizing interferon might be explained by supposing that the first induction of interferon also resulted in the induction of repressor systems which might then block a second induction. However, Billiau (7) has shown that in rabbit kidney cells cycloheximide did not prevent the establishment of the refractory state after poly(I)·poly(C) induction. Release of interferon was necessary for the establishment of the refractory state and there was a correlation between the amount of interferon released and the degree of refractionness to a second dose. This suggests a feedback by interferon on its own synthesis, as suggested previously (56). Borden & Murphy (8) reported that the refractory state in L-cells was due to a repressor which was released by the cells late in interferon synthesis. Pretreatment of L-cells with interferon preparations collected 36 hr after induction repressed interferon synthesis to a greater extent than interferon preparations collected 8 hr after induction, suggesting that the repressor activity can be separated from the interferon activity.

Further (but also indirect) evidence of the existence of a repressor of interferon synthesis comes from investigation of the effects of the in vivo or in vitro, or both, aging of cells on the synthesis of interferon. Carver & Marcus (18) reported that Sindbis-induced interferon synthesis was much greater in chick cells aged for 7 days in vitro than in cells only 1–2 days old. They showed that protein synthesis was depressed in the aged cells and suggested that the synthesis of repressor would likewise be depressed. In agreement with this the cells lost their aged effect if they were trypsinized and replated, a procedure which also stimulated protein synthesis. In contrast, it has been reported (92) that aged chick cells release a factor into the medium that enhances interferon synthesis, i.e., aged cells may synthesize an enhancer rather than not synthesize a repressor of interferon synthesis. More difficult to evaluate are the studies on the influence of cell-aging with cells that are still growing (133, 148) but they also show that "aged" cells are more sensitive to the action of interferon.

Not only interferon synthesis may be regulated; Grossberg & Morahan

(67, 68) have presented data suggesting that the antiviral activity of interferon may be under cellular control. Older (13 day) chick embryos appear to have a greater ability than young (6 day) embryos to express interferon activity as well as to respond to interferon inducers (132). Cells obtained from such embryos reflect these properties in vitro (68). Cell cultures derived from 13-day-old embryos were 20 to 30-fold more sensitive to the action of interferon than cells from 6-day-old embryos. In addition, VSV plaqued 2 to 3-fold more efficiently in the young cells than in the old. More impressive and potentially more enlightening was the observation that mixed cultures of young and old cells containing as little as 12 percent of 6-day-old cells behaved as if the cultures were composed entirely of young cells. The implication, of course, is that the young cells release a product that is able to inhibit the synthesis or action, or both, of interferon. Disrupted 6-day embryonic cells, medium from 6-day cells or extracts of whole 6-day embryos reduced the titer of interferon on 13-day cells. Preliminary experiments suggest that the substance does not have proteolytic activity, but efforts to isolate and characterize it have so far been unsuccessful. Similar conclusions can be made for human fetal tissue (153).

Further evidence for the existence of repressor(s) of interferon synthesis and activity come from the work of Chany and his co-workers (22, 23). They isolated a strain of mouse cells that no longer exhibited a refractory state. The strain was obtained by culturing a transformed line of mouse embryo cells in the presence of interferon. After 200 passages, a line exhibiting contact inhibition was isolated and has been characterized (23). The line produced 10-fold more interferon than its "wild-type" parent when induced with NDV, and released interferon into the medium on repeated stimulation with NDV, i.e., it failed to exhibit a refractory state. In contrast to this high rate of interferon synthesis (derepressed?) the line was insensitive to the action of exogenous interferon: treatment with mouse interferon failed to protect the cells from infection. The authors postulated that the mutant line is unable to produce or to use AVP and wondered if there was a correlation between the high rate of interferon production, its reinducibility and the absence of the antiviral state. They suggested that the likeliest explanation of this postulated correlation "is the presence of a repressor of interferon synthesis which appears after the first induction with NDV and which could be located on the cell genome close to the site responsible for the synthesis of antiviral protein." This repressor is inactive or is not produced in the mutant line. No conclusions can be drawn at present and there is at least one other explanation for the data, viz: the mutant line is unable to transport exogenous interferon across the cell membrane, and is therefore unable to establish an antiviral state. If interferon itself, rather than another product, feeds back to inhibit its own synthesis (56) then an inability to transport interferon back into the cell would result in an absence of the refractory state. Whether or not the mutant establishes an antiviral state could be easily determined by treating the cell with a nonviral inducer such as poly I·poly C.

Recently, Chany et al (22) have reviewed the status of the various inhibitors and activators of the interferon system and have devised a model system for its regulation. Interested readers are referred to it for the details.

The regulation of interferon synthesis and activity is of obvious importance and will no doubt be the subject of much research in the next few years. We would predict that work, already begun (19, 20), with animal cell hybrids will be of increasing importance in studying this phenomenon.

LITERATURE CITED

1. Banks, G. T. et al. 1968. *Nature (London)* 218:542–45
2. Banks, G. T., Buck, K. W., Chain, E. B., Darbyshire, J. E., Himmelweit, F. 1969. *Nature (London)* 223:155–58
3. Baron, S. et al. 1969. *Proc. Nat. Acad. Sci. USA* 64:67–74
4. Baron, S., Buckler, C. E., Levy, H. B., Friedman, R. M. 1967. *Proc. Soc. Exp. Biol. Med.* 125:1320–26
5. Bausek, G. H., Merigan, T. C. 1969. *Virology*. 37:491–98
7. Billiau, A. 1970. *J. Gen. Virol.* 7: 225–32
8. Borden, E. C., Murphy, F. A. In press
9. Bratt, M. A. Personal communication
10. Buckler, C. E., Baron, S. 1966. *J. Bacteriol.* 91:231–37
11. Buckler, C. E., Baron, S., Levy, H. 1966. *Science* 152:80–82
12. Burke, D. C., Skehel, J. J., Low, M. 1967. *J. Gen. Virol.* 1:235–37
13. Came, P. E., Lieberman, M., Pascale, A., Shimonaski, G. 1969. *Proc. Soc. Exp. Biol. Med.* 131: 443–46
14. Carter, W. A. 1970. *Proc. Nat. Acad. Sci. USA* 67:620–28
15. Carter, W. A., Levy, H. B. 1967. *Science*. 155:1254–57
16. Carter, W. A., Levy, H. B. 1968. *Biochim. Biophys. Acta* 155:437–43
17. Carter, W. A., Pitha, P. M., *Structural Requirements of Ribopolymers for Induction of Human Interferon: Evidence for Interferon Subunits. Biological Effects of Polynucleotides,* ed. Beers and Braun. New York: Springer-Verlag. In press
18. Carver, D. H., Marcus, P. I. 1967. *Virology* 32:247–57
19. Carver, D. H., Seto, D. S. Y., Migeon, B. R. 1968. *Science* 160: 558–59
20. Cassingena, R., Chany, C., Vignal, M., Estrade, S., Suarez, H.-G. 1970. *C. R. Acad. Sci. Paris* 270: 1189–91
21. Chany, C., Brailovsky, C. 1967. *Proc. Nat. Acad. Sci. USA* 57: 87–94
22. Chany, C., Fournier, F., Rousset, S. 1970. *Ann. NY Acad. Sci.* 173: 505–15
23. Chany, C., Vignal, M. 1970. *J. Gen. Virol.* 7:203–10
24. Claes, P. et al. 1970. *J. Virol.* 5: 313–20
25. Colby, C. 1971. *Progr. Nucleic Acid Res. Mol. Biol.* 11:1–32
26. Colby, C., Chamberlin, M. J. 1969. *Proc. Nat. Acad. Sci. USA* 63: 160–67
27. Colby, C., Duesberg, P. H. 1969. *Nature (London)* 222:940–44
28. Colby, C., Chamberlin, M. J., Duesberg, P. H. *The Induction of Interferon. Viruses Affecting Man and Animals,* ed. Sanders and Schaeffer. St. Louis, Mo: W. H. Green. Inc. In press
29. Colby, C., Chamberlin, M. J., Duesberg, P. H., Simon, M. I. *The Specificity of Interferon Induction. Biological Effects of Polynucleotides,* ed. Beers and Braun. New York: Springer-Verlag. In press
30. Colby, C., Jurale, C., Kates, J. R. 1971 *J. Virol.* 7:71–76
31. Colby, C., Stollar, B. D., Simon, M. I. 1971. *Nature (London)* 229: 172–74
32. De Clercq, E., Eckstein, F., Merigan, T. C. 1969. *Ann. NY Acad. Sci.* 173:444–61
33. De Clercq, E., Eckstein, F., Merigan, T. C. 1969. *Science* 165: 1137–39
34. De Clercq, E., Merigan, T. C. 1969. *Nature (London)* 222:1148–52
35. De Clercq, E., Merigan, T. C. 1970. *Ann Rev. Med.* 21:17–46
36. De Clercq, E., Merigan, T. C. 1970. *Virology* 42:799–802
37. De Clercq, E., Merigan, T. C. 1971. *J. Gen. Virol.* 10:125–30
38. De Clercq, E., Merigan, T. C. *J. Infect. Dis.* In press
39. De Maeyer, E., De Maeyer-Guignard, J., Montagnier, L. Personal communication
40. Desmyter, J., Melnick, J. L., Rawls, W. E. 1968. *J. Virol.* 2:955–61
41. De Somer, P., De Clercq, E., Billiau, A., Schonne, E., Claesen, M. 1968. *J. Virol.* 2:878–85
42. De Somer, P., De Clercq, E., Billiau, A., Schonne, E., Claesen, M. 1968. *J. Virol.* 2:886–93

43. Dianzani, F., Cantagalli, P., Gagnoni, S., Rita, G. 1968. *Proc. Soc. Exp. Biol. Med.* 128:708–10
44. Dianzani, F., Gagnoni, S., Cantagalli, P. 1969. *Ann. NY Acad. Sci.* 173:727–35
45. Dianzani, F., Gagnoni, S., Buckler, C. E., Baron, S. 1970. *Proc Soc. Exp. Biol. Med.* 133:324–28
46. Dube, S., Rudland, P. S. 1970. *Nature (London)* 226:820–23
47. Duesberg, P. H., Colby, C. 1969. *Proc. Nat. Acad. Sci. USA* 64: 396–403
48. Ellis, L. F., Kleinschmidt, W. J. 1967. *Nature (London)* 215:649–50
49. Engelhardt, D. Personal communication
50. Falcoff, R., Falcoff, E. T. 1969. *Biochim. Biophys. Acta* 182:501–10
51. Falcoff, R., Falcoff, E., Catinot, L. 1970. *Biochim. Biophys. Acta* 217: 195–98
52. Field, A. K., Tytell, A. A., Lampson, G. P., Hilleman, M. R. 1967. *Proc, Nat. Acad. Sci. USA* 58: 1004–10
53. Field, A. K., Lampson, G. P., Tytell, A. A., Nemes, M. M., Hilleman, M. R. 1967. *Proc. Nat. Acad. Sci. USA* 58:2102–8
54. Field, A. K., Tytell, A. A., Lampson, G. P., Hilleman, M. R. 1968. *Proc. Nat. Acad. Sci. USA* 61: 340–46
55. Finter, N. B., Ed. 1966. *Interferons*. Philadelphia: W. B. Saunders Co.
56. Friedman, R. M. 1966. *J. Immunol.* 96:872–77
57 Friedman, R. M. 1966. *J. Bacteriol.* 91:1224–29
58. Friedman, R. M. 1967. *Science* 156: 1760–61
59. Friedman, R. M. 1968. *J. Virol.* 2: 1081–85
60. Friedman, R. M., Fantes, K. H.. Levy, H. B. Carter, W. B. 1967. *J. Virol.* 1:1168–73
61. Friedman, R. M., Pastan, I. 1969. *Biochem. Biophys. Res. Commun.* 36:735–40
62. Friedman, R. M., Sonnabend, J. A. 1964. *Nature (London)* 203:366–67
63. Goorha, R. M., Gifford, G. E. 1970. *Proc. Soc. Exp. Biol. Med.* 134: 1142–47
64. Goorha, R. M., Gifford, G. E. *J. Gen. Virol.* In press
65. Gordon, I., Chenault, S. S., Stevenson, D., Acton, J. D. 1966. *J. Bacteriol.* 91:1230–38
66. Grossberg, S. E., Holland, J. J. 1962. *J. Immunol.* 88:708–14
67. Grossberg, S. E., Morahan, P. S. 1971. Inducible Antiviral Resistance Against Myxoviruses During Embryogenesis. *Proc. Intern. Symp. Interferon, 2nd*
68. Grossberg, S. E., Morahan, P. S. 1971. *Science* 171:77–79
69. Guggenheim, M. A., Friedman, R. M., Rabson, A. S. 1969. *Proc. Soc. Exp. Biol. Med.* 130:1242–45
70. Hallum, J. V., Thacore, H. R., Youngner, J. S. 1970. *J. Virol.* 6: 156–62
71. Harel, L., Montagnier, L. Personal communication
72. Heller, E. 1963. *Virology* 21:652.56
73. Hilleman, M. R. 1968. *J. Cell. Physiol.* 71:43–60
74. Ho, M. 1964. *Science* 146:1472–74
75. Ho, M., Enders, J. F. 1959. *Virology* 9:446–77
76. Ho, M., Kono, Y. 1965. *J. Clin. Invest.* 44:1059–60
77. Ho, M., Kono, Y. 1965. *Proc. Nat. Acad. Sci. USA* 53:220–24
78. Ho, M., Postic, B., Ke, Y. 1968. The *systemic induction of interferon. Ciba Found. Symp. Interferon, 1967* 19–35
79. Hoagland, M. B., Scornik, O. A., Pfefferkorn, L. C. 1964. *Proc. Nat. Acad. Sci. USA* 51:1184–91
80. Hopps, H. E., Kohno, S., Kohno, M., Smadel, J. E. 1964. *Bacteriol. Proc.* 115
81. Horak, I., Hilfenhaus, J., Siegert, W., Jungwirth, C. Personal communication
82. Huang, A., Baltimore D., Bratt, M. 1971. *J. Virol.* In press
83. Huppert, J., Hillova, J., Gresland, L. 1969. *Nature (London)* 223: 1015–17
84. Hsu, W. T., Weiss, S. B. 1969. *Proc. Nat. Acad. Sci. USA* 64:345–51
85. Isaacs, A., Cox, R. A., Rotem, Z. 1963. *Lancet* 2:113–16
86. Isaacs, A., Lindenmann, J. 1957. *Proc. Roy. Soc. London* 147: 258–67
87. Isaacs, A., Rotem, Z., Fantes, K. H. 1966. *Virology* 29:248–54
88. Joklik, W. K., Merigan, T. C. 1966. *Proc. Nat. Acad. Sci. USA* 56: 558–65

89. Jurale, C., Kates, J. R., Colby, C. 1970. *Nature (London)* 226: 1027–29
90. Kates, J. R., Beeson, J. 1970. *J. Mol. Biol.* 50:1–18
91. Kates, J. R., McAuslan, B. R. 1967. *Proc. Nat. Acad. Sci. USA* 57: 314–20
92. Kato, N., Eggers, H. J. 1969. *Virology* 37:545–53
93. Kato, N., Okada, A., Ota, F. 1965. *Virology* 26:630–37
94. Kaufman, H., Centifanto, Y., Ellison, E., Brown, D. *Proc. Soc. Exp. Biol. Med.* In press
95. Ke, Y., Singer, S. H., Postic, B., Ho, M. 1966. *Proc. Soc. Exp. Biol. Med.* 121:181–83
96. Kerr, I. M. In press
97. Kerr, I. M., Martin, E. M. In press
98. Kerr, I. M., Sonnabend, J. A., Martin, E. M. 1970. *J. Virol.* 5:132–44
99. Kimball, P., Duesberg, P. H. Personal communication
100. Kleinschmidt, W. J., Ellis, L. F. 1968. *Ciba Found. Symp. Interferon*
101. Kleinschmidt, W. J., Ellis, L. F., Van Frank, R. M., Murphy, E. B. 1968. *Nature (London)* 220:167–68
102. Kleinschmidt, W. J., Douthart, R. J., Murphy, E. B. 1970. *Nature (London)* 228:27–30
103. Kobayashi, S., Yasui, O., Masuzumi, M. 1969. *Proc. Soc. Exp. Biol. Med.* 131:487–94
104. Krueger, R. F., Mayer, G. D. 1970. *Science* 169:1213–14
105. Krueger, R. F., Yoshimura, S. 1970. *Fed. Proc.* 29:635
106. Lampson, G. P., Tytell, A. A., Nemes, M. M., Hilleman, M. R. 1965. *Proc. Soc. Exp. Biol. Med.* 118:441–47
107. Lampson, G. P., Tytell, A. A., Field, A. K., Nemes, M. M., Hilleman, M. R. 1967. *Proc. Nat. Acad. Sci. USA* 58:782–89
108. Levine, E. M., Becker, Y., Boone, C. W., Eagle, H. 1965. *Proc. Nat. Acad. Sci. USA* 53:350–56
109. Levine, S. 1964. *Virology* 24:586–88
110. Levy, H. B. Personal communication
111. Levy, H. B., Carter, W. A. 1968. *J. Mol. Biol.* 31:561–77
112. Lindenmann, J. 1960. *Z. Hyg. Infektionskr.* 146:287–309
113. Lockart, R. Z., Jr. 1964. *Biochem. Biophys. Res. Commun.* 15:513–18
114. Lockart, R. Z., Jr. 1967. *Interferons,* ed. N. B. Finter, 14. Amsterdam: North-Holland Publ.
115. Lockart, R. Z., Jr. 1967. *J. Virol.* 1: 1118–63
116. Lockart, R. Z., Jr., Bayliss, N. L., Toy, S. T., Yin F. H. 1968. *J. Virol.* 2:962–65
117. Lomniczi, B., Burke, D. C. 1970. *J. Gen. Virol.* 8:55–68
118. Lukas, B., Hruskova, J. 1968. *Acta Virol.* 12:263–67
119. Magee, W. E., Levine, S., Miller, O. V., Hamilton, R. D. 1968. *Virology* 35:505–11
120. Marcus, P. I. Personal communication
121. Marcus, P. I., Engelhardt, D. L., Hunt, J. M., Sekellick, M. J. 1971. *Bacteriol. Proc.* 275
122. Marcus, P. I., Salb, J. M. 1966. *Virology* 30:502–16
123. Marcus, P. I., Salb, J. M. 1968. *The Interferons,* 111–27. New York: Academic Press
124. Mayer, G. D., Fink, B. A. 1970. *Fed. Proc.* 29:635
125. Mécs, E., Sonnabend, J. A., Martin, E. M., Fantes, K. H. 1967. *J. Gen. Virol.* 1:25–40
126. Merigan, T. C. 1964. *Science* 145: 811–13
127. Merigan, T. C. 1967. *Nature (London)* 214:416–17
128. Merigan, T. C., Finkelstein, M. S. 1968. *Virology* 35:363–74
129. Merigan, T. C., Regelson, W. 1967. *N. Engl. J. Med.* 277:1283–87
130. Miner, N., Ray, W. J., Jr., Simon, E. H. 1966. *Biochem. Biophys. Res. Commun.* 24:264–68
131. Montagnier, L., 1968. *C. R. Acad. Sci.* 267:1417–20
132. Morahan, P. S., Grossberg, S. E. In press
133. McLaren, C. 1970. *Arch. ges Virusforsch.* In press
134. Nemes, M. M., Tytell, A. A., Lampson, G. P., Field, A. K., Hilleman, M. R. 1969. *Proc. Soc. Exp. Biol. Med.* 132:784–89
135. Officer, J. E., Stevenson, D., Gordon, I. In press
136. Oh, J. O. 1965. *Am. J. Pathol.* 46: 117–19
137. Paucker, K., Boxaca, M. 1967. *Bacteriol. Rev.* 31:145–56
138. Pitha, P. M., Carter, W. A. *Virology* In press
139. Pollack, Y., Groner, Y., Aviv (Greenshpa) H., Revel, M. 1970.

FEBS Lett. 9:218–21
140. Pollikoff, R., Jankauskas, P., DiPurro, A., Nazario, R. N. 1968. *Invest. Ophthalmol.* 7:397–404
141. Postic, B., DeAngelis, D., Breinig, M. K., Ho, M. 1966. *J. Bacteriol.* 91:1277–81
142. Postic, B., DeAngelis, D., Breinig, M. K., Ho, M. 1967. *Proc. Soc. Exp. Biol. Med.* 125:89–92
143. Regelson, W. 1967. *The Reticulo-endothelial System and Atherosclerosis,* ed. N. R. Di Luzio, R. Paoletti, 315–32. New York: Plenum Press
144. Robertson, H. D., Zinder, N. 1968. *Fed. Proc.* 27:296
145. Robertson, H. D. *Nature (London)* In press
146. Robinson, G. A., Butcher, R. W., Sutherland, E. W. 1968. *Ann. Rev. Biochem.* 37:149–74
147. Robinson, W. S. 1970. *Nature (London)* 225:944–45
148. Rossman, T. G., Vilček, J. 1969. *J. Virol.* 4:7–11
149. Rotem, Z., Cox, R. A., Isaacs, A. 1963. *Nature (London)* 197:564–66
150. Salb, J. M., Marcus, P. I. 1965. *Proc. Nat. Acad. Sci. USA* 54: 1353–58
151. Schaefer, T. W., Lockhart, R. Z., Jr. 1970. *Nature (London)* 226:449–50
152. Sheaff, E. T., Stewart, R. B. 1969. *Can. J. Microbiol.* 15:941–53
153. Siewers, C. M. F., John, C. E., Medearis, D. N., Jr. 1970. *Proc. Soc Exp. Biol. Med.* 133:1178–83
154. Skehel, J. J., Burke, D. C. 1968. *J. Gen. Virol.* 3:35–42
155. Skehel, J. J., Burke, D. C. 1968. *J. Gen. Virol.* 3:191–99
156. Soeiro, R., Amos, H. 1966. *Science* 154:666
157. Sonnabend, J. A., Martin, E. M., Kerr, I. M. 1968. *The Interferons,* 161–71. New York: Academic Press
158. Sonnabend, J. A., Martin, E. M., Mécs, E., *Ciba Found. Symp. Interferon, 1968* 143–56
159. Sonnabend, J. A., Martin, E. M., Mécs, E., Fantes, K. H. 1967. *J. Gen. Virol.* 1:41–48
160. Stewart, W. E., II, Lockart, R. Z., Jr. 1970. *J. Virol.* 6:795–99
161. Stewart, W. E., II, Scott, W. D., Sulkin, S. E. 1969. *J. Virol.* 4: 147–51
162. Stinebring, W. R., Youngner, J. S. 1964. *Nature (London)* 204:712
163. Taylor, J. 1964. *Biochem. Biophys. Res. Commun.* 14:447–51
164. Tsunoda, Suzuki, F., Sato, N., Miyazaki, K., Ishida, N. Personal communication
165. Vilček, J. 1969. *Interferon.* New York: Springer-Verlag
166. Vilček, J. 1970. *Ann. NY Acad. Sci.* 173:390–403
167. Vilček, J., Ng, M. H., Friedman-Kien, A. E., Karwciw, T. 1968. *J. Virol.* 2:648–50
168. Vilček, J., Rossman, T. G., Varacalli, F. 1969. *Nature (London)* 222:682–83
169. Wagner, R. R. 1961. *Virology* 13: 323–37
170. Wagner, R. R. 1963. *Trans. Assoc. Am. Physicians* 76:92–101
171. Waschke K., Borecký, L., Lackovič, V. 1969. *Acta Virol.* 13:393–400
172. Wheelock, E. F. 1965. *Science* 149: 310–11
173. Wheelock, E. F., Larke, R. P. B., Caroline, N. L. 1968. *Progr. Med. Virol.* 14:286–347
174. Youngner, J. S. 1968. *Interferon production in mice injected with viral and nonviral stimuli. Medical and Applied Virology,* 210–22. Ed. M. Sanders, E. H. Lennette. St. Louis, Missouri: W. H. Green, Inc.
175. Youngner, J. S., Hallum, J. V. 1969. *Virology* 37:473–75
176. Youngner, J. S., Stinebring, W. R. 1965. *Nature (London)* 208:456–58
177. Youngner, J. S., Stinebring, W. R. 1966. *Virology* 29:310–16
178. Youngner, J. S., Stinebring, W. R., Taube, S. E. 1965. *Virology* 27: 541–50

BIOCHEMICAL ECOLOGY OF MICROORGANISMS 1575

Martin Alexander

Laboratory of Soil Microbiology, Department of Agronomy, Cornell University, Ithaca, New York

Contents

Every organism has an ecological *raison d'être.* Natural selection acting on so many generations of microorganisms and under conditions in which the selective pressures are almost invariably quite keen and highly effective dictates that the survivors possess sets of characteristics that allow them to endure the struggle for existence among the region's potential inhabitants, many of which are physiologically similar and hence may be directly interacting. Each organism possesses a specific biochemical trait or group of traits that account for its existence in its typical habitats, and it is these which are of interest in the present essay.

A critical task for the biochemical ecologist is to define, characterize, or explain the biochemical basis for a population living where it does, multiplying solely at given times in certain regions and carrying out in nature only a few of the many physiological processes it inherently is capable of performing. The chemistry of microbial cells and the major metabolic pathways are remarkably similar, yet critical biochemical differences unquestionably do exist—physiological idiosyncrasies and peculiarities often overlooked in the search for the common biochemical constituents and pathways. These variations in the general biochemical themes undoubtedly account for the differences in microbial distribution, abundance and function in nature.

The point of focus for the biochemical ecologist is nature and not the laboratory, the characteristic environments where the species grows and affects its surroundings and not the artificial conditions imposed on the isolate, which is pampered or mishandled—depending on the investigator's propensities—to perform this or that reaction. The biochemical ecologist is an *in vivoologist* applying chemistry to explain the functions of an organism in nature or its restriction to given regions, not an *in vitroologist* attempting to describe the functioning of an organism in isolation. Granting that the laboratory is the chief professional home of the microbial ecologist and that he may, in fact, deal with axenic cultures, his efforts are nonetheless directed to problems in natural habitats and to accounting for phenomena in heterogeneous communities.

Any one or frequently a group of physiological traits or biochemical properties may be the basis for a population's *raison d'etre.* The trait may allow for survival in the face of a recurrent or occasional environmental stress. The property may be a unique enzyme endowing the possessor with the ability to grow at the expense of a substrate few neighboring species can utilize or with the capacity to colonize or penetrate a region few others are able to exploit. The characteristic may be a structure allowing for active dispersal through air or water or for passive dispersal through fluid environments or on the surfaces of suitable vectors, thereby permitting the bearer of the attribute to become introduced into an environment where propagules of the species once again can initiate replication. In some instances, the critical trait allows the organism to participate in some interaction with a second organism or population, as in those species owing their existence to their ability (*a*) to cope with phagocytes, phytoalexins, or other host defense mechanisms; (*b*) to gain a greater measure of some limiting resource than cohabitants of the area; (*c*) to couple together with a second individual in an association having a unique advantage at that place and time; or (*d*) to excrete a toxin which holds in check organisms that might otherwise displace the toxin producer. Each of these traits, and others too, has been the subject of cursory or intensive inquiry in some microbial species. Space does not permit an extensive chronicling of these investigations, but an elementary review of several facets of this aspect of ecology has been published recently (5).

Considering the diversity of microorganisms, the large number of reactions they catalyze in nature and the countless kinds of distinctly different communities, it would be naive to believe that only a small number of fitness traits exists or that merely a few patterns composed of several traits would be of consequence in nature. To some biochemists accustomed to thinking of one or, at best, a few catabolic or anabolic pathways for the metabolism of a given molecule, this abundance may be a source of frustration; to others, the very diversity underlying uniqueness in the world of microorganisms may be a most striking challenge. Needless to say, however, the multitude of characteristics does mean that it is necessary to seek out a large number of individual or sets of biochemical traits that are crucial for the existence of a species in natural ecosystems.

Because of the provincialism of some microbiologists and many plant and animal ecologists, it is unfortunately still necessary to reiterate what are habitats of legitimate concern to the microbial ecologist. In addition to the seas, fresh and polluted waters and soil, traditional subjects for ecological investigations, natural habitats for microorganisms—be the community multi- or monospecific—include the tissues or body fluids of animals, lesions or the vascular system of higher plants, the rumen of the bovine animal, man's gastrointestinal tract, the oral cavity, and the surfaces of rocks, to mention but a few sites which have provoked ecological exploration.

Also within the province of the biochemical ecologist is concern with the influence of environmental factors on resident populations, interactions among the populations of an ecosystem and the chemical changes microorganisms bring about in their surroundings (2). Recent years have seen a dramatic upsurge in interest in the substances microorganisms elaborate in nature and the ways by which particular chemical substances modify the composition or activities of a community. Limitations of space dictate that no more than a handful of these investigations can be cited.

Natural Selection and Fitness Traits

The continued presence of a species in an environment suggests that it possesses some ecological advantage at that place not possessed by aliens. This advantage may be related to nutrition, tolerance to abiotic or biologically generated toxins, or resistance against elimination by parasites or predators. The specific fitness trait, or set of characteristics, underlying the dominance or survival of the species serves as the basis for its success in coping with the forces of natural selection operating in that region. The environment favors the bearers of the fitness traits useful in that locality, and these species survive and, at appropriate times, multiply. Individuals lacking these properties are either unable to withstand the significant abiotic stresses operating at that locus or they succumb in the presence of organisms better attuned to the local situation; they disappear and leave little or no trace of their having been a transitory resident of the area.

The identities of only a few of the fitness traits have been well established, but a reasonable case can be made for the importance of others, such as cell retention to surfaces, buoyancy, organism size, or its tolerance to biologically produced toxins or to abiotic determinants (5). From the biochemical viewpoint, however, the possession of carotenoids, melanin or melanin-like pigments, heteropolysaccharide wall components, and capsules has been of greatest recent interest. In selected environments, species with these characteristics gain the ascendancy, and the climax community shows the importance of an organism having the particular property.

Carotenoids provide a striking illustration of an important fitness trait. Prolonged exposure of cells or filaments to visible or ultraviolet light results in death, yet bacteria, fungi, and algae survive in fresh water, the oceans, salt brines, the atmosphere, and on leaf surfaces, environments in which photoinactivation is likely to be prominent because of the daily and usually prolonged bathing of the area with sunlight. A high proportion of the individuals in these ecosystems were known to contain carotenoid pigments, but the reason for the abundance of such organisms was not understood. Several lines of evidence now point to the carotenoids serving a protective role, shielding the cells against photoinactivation and allowing them to persist as the carotenoidless cells are eliminated. For example, Sistrom, Griffiths & Stanier (176) noted that a mutant of *Rhodopseudomonas spheroides* devoid of colored carotenoids was rapidly killed in air in the presence of light while cells of the carotenoid-containing wild type grew readily.

A similar study with the extreme halophile, *Halobacterium salinarium,* provides a basis for understanding why most bacteria from highly saline waters receiving considerable sunlight, such as solar evaporation ponds, are richly endowed with colored carotenoids. In this instance, the colorless mutant was not killed when incubated in the light but rather it grew far more slowly than its red parent strain, so that in mixtures of mutant and parent strains the pigmented wild type became ever more prominent with time, in effect displacing its light-sensitive daughter population (58). Carotenoid shielding against photodynamic injury likewise occurs in a *Micrococcus* (135) and in hyphae of the fungus *Dacryopinax spathularia* (75). The carotenoids of the cell or filament shield the organisms from being killed or prevent an inhibition of growth by sunlight, making them fit for growth in sun-drenched ecosystems and allowing them to become dominant in inter- or intraspecific competition.

A comparable inquiry, unfortunately with far too few strains, supports the contention that there is a natural selection for dark-pigmented fungi in locales subject to high intensities of ionizing radiation, for the colored propagules are much less readily inactivated than the nonpigmented ones (142).

Melanins or melanin-like pigments have a unique ecological role in environments where either light or certain types of lytic heterotrophs are important selective forces. The significance of melanins as a fitness trait in regions with intense sunlight is suggested by the frequency in desert soils of fungi containing melanins in their spore walls or hyphae. Light-colored fungal spores seem to be particularly prone to inactivation by ultraviolet light (59). In vitro studies also suggest a protection against light inactivation afforded by melanin; thus, although tyrosinase, ATPase, acid phosphatase, and succinic dehydrogenase are quite sensitive to inactivation by ultraviolet light, destruction of the enzymatic activity is prevented in the presence of melanin (171).

Melanins act to enhance fitness in an entirely different manner in fungi inhabiting regions containing lytic bacteria and actinomycetes. Soil, for example, teems with such parasites, and lytic populations with the requisite enzymes soon bring about the destruction of hyphae of many genera as the filaments emerge from an appropriate resting stage. Lysis frequently results from the excretion by the parasite of chitinase and one or another glucanase, these enzymes depolymerizing structural polysaccharides of the hyphal walls of susceptible fungi. Nevertheless, the resting structures of these very same organisms are refractory and persist. With some subterranean fungi, the mycelium itself is not readily eliminated by components of the soil community. Moreover, not a few of the resistant structures do, in fact, contain chitin or a glucan wall component, or both, yet they are refractory. Clearly, the sclerotia, conidia, chlamydospores or other resting bodies as well as the refractory hyphae either contain a substance that protects the susceptible polysaccharide backbone or the composition of the polysaccharide is such that it is not readily hydrolyzed enzymatically. In the underground microflora, populations having structures of this sort have a particularly useful fitness trait—the shielding substance or the resistant polysaccharide—because the fungi can maintain themselves, while less favorably endowed organisms are eliminated by mycolytic parasites.

Several lines of evidence indicate that melanin is just such a fitness trait and is responsible for the resistance of some of the refractory structures. First, *Rhizoctonia solani,* a fungus existing in nature in the hyphal form, has a surface-localized melanin-like material, and not only is this fungus not readily lysed in soil but no microorganism actively digesting its hyphae has been isolated (156); because of the absence of a specialized resting structure, the mycelium must be resistant to enzymatic digestion if the fungus is not to be eliminated. Second, although melanin-free hyphae of *Sclerotium rolfsii* are digested by lytic organisms or by a mixture containing glucanase and chitinase, the fungus remains an important soil-borne plant pathogen by virtue of its melanin-covered sclerotia. The sclerotium is not attacked by microorganisms that decompose hyphal walls, yet, under the melanin-containing sclerotial rind are coiled the presumably susceptible filaments. Third, by contrast with the glucose- and N-acetylhexosamine-containing walls of *Aspergillus phoenicis,* which are extensively digested by lytic enzyme preparations, the surfaces of the spicule-covered conidia of the fungus are resistant to microbial digestion; yet when the melanin-rich spicules are removed from the spores, glucanase and chitinase can, in fact, act on the underlying structural components (22). Fourth, the susceptibility of heterolysis of hyphae of an *Aspergillus nidulans* wild-type strain is inversely related to their melanin content, but walls of a melanin-less mutant of *A. nidulans* show essentially no resistance to enzymatic destruction (115). The inhibition of polysaccharases by melanin has been investigated by Bull (32). It is not yet clear whether the polyaromatic acts by complexing with wall carbohydrates so they are not depolymerized enzymatically, by coating the polysaccharides with a largely impervious barrier, or by inhibiting the enzymes potentially able to destroy the backbone of the surface structure, but little doubt remains that melanins or related polyaromatics have pronounced ecological significance.

Nevertheless, many soil fungi which do not succumb readily to lysis have neither melanin-containing resting structures nor hyphae. These organisms must have structural or physiological features to account for their durability in ecosystems that support highly heterogeneous communities which perennially encounter insufficiencies of organic nutrients. Studies of one species of *Mortierella,* a genus containing representatives that often exist in soil in the vegetative form, revealed that its hyphal walls are resistant to digestion by the soil microflora and by a glucanase-chitinase mixture. The walls apparently contain a heteropolysaccharide, and the resistance has been attributed to this heteropolymer (154). A wall-localized heteropolysaccharide rich in fucose has been assigned an analogous role in protecting *Zygorhynchus vuilleminii* against rapid elimination from soil; this polysaccharide is itself quite slowly degraded in nature (J. P. G. Ballesta and M. Alexander, unpublished observations). A polymer, be it a polysaccharide or a polyaromatic like melanin, made up of different linkages or dissimilar monomers, or both, might be expected to be depolymerized quite slowly inasmuch as several enzymes are probably required for its destruction, and the enzymes likely would often be excreted by dissimilar organisms, each of which would have to be located in the immediate vicinity of and alight on the substrate.

Other fitness traits have been known for some time, but the number which has been well-documented is surprisingly small. The significance of the capsule to the bacterium residing in an animal or human host ecosystem is fully appreciated, the encapsulated organism having an advantage not present in related nonencapsulated bacteria when the environmental stress is the phagocyte, a common stress factor in body fluids. Bulmer & Sans (33) recently provided evidence that a nonencapsulated mutant of *Cryptococcus neoformans* is similarly far more subject to phagocytosis than the encapsulated organism. Reversion of the mutant to the encapsulated state is accompanied by a decreased susceptibility to phagocytosis. Internal osmotic pressure may be deemed as a fitness trait in an aquatic organism introduced into the marine environment; species of blue-green algae with low internal osmotic pressures are apparently selected against in inter- or intraspecific selection in salt-rich waters, while blue-greens with higher osmotic pressures in their cells are favored and assume dominance (69).

Effect of Environment on Microorganisms

An exhaustive or even a representative survey of the ways in which factors in an environment may affect the biochemistry of its inhabitants is impossible within the confines of this brief essay. The literature is so vast, the types of organisms so numerous, and the kinds of responses so varied that a voluminous compendium would be necessary to do justice to the field. The difficulty in selecting pertinent references is eased somewhat because much of the literature is derived from investigations in vitro, and extrapolation to conditions in vivo would be risky at best, foolhardy at worst. The few illustrations cited herein are designed not to be typical but rather to exemplify the in vivo, as contrasted with the in vitro, approach to microbial ecology and to show a few of the areas of current concern. The examples are chosen from among those investigations which are designed to explain or account for community structure and metabolism in natural ecosystems, not those defined by the confining walls of the flask, test tube, and petri dish.

Considerable effort is currently being devoted to exploring the relationship between nutrient sources and community composition or the appearance of individual kinds of organisms in an ecosystem. Some of this interest represents basic science, but much is prompted by the ever-increasing problems of man-made eutrophication. It is a laboratory truism that an organism will not grow if an essential nutrient or a particular organic substrate is lacking, but it is surprising how few microbiologists realize the in vivo importance of this truism. An organism having a nonuniversal nutrient requirement—a sterol, coenzyme, or vitamin, for example—is restricted to habitats where this compound is either constantly or intermittently generated or introduced. A simple illustration is the localization of certain fatty acid-requiring species of *Bacteroides* and *Ruminococcus* in the rumen, a region in which carbohydrate fermentation is a perpetual source of the substances demanded by these bacteria (51). This dependency is also well illustrated—with important environmental, economic, and esthetic consequences—by the fresh-water algae appearing in greater quantities in impounded and flowing bodies of water as society's activities cause an enrichment of these waters. Identification of the nutrients triggering or responsible for the aquatic blooms is particularly important in attempts to control these nuisance algae and to regulate the source of the nutrient input. In many, possibly most, but surely not all lakes of this country, the limiting nutrient controlling phytoplankton density is either phosphorus or nitrogen, and a great deal of study therefore is being devoted to these two nutrient elements (61, 67). However, it is manifestly clear that in occasional lakes, rivers, or streams a third element or a factor other than a nutrient governs the extent of algal colonization. Thus, during seasons of the year when diatoms and other silicon-requiring photoautotrophs are abundant, silicate may stimulate the rate of photosynthesis (82). Evidence for the importance of temperature and manganese in regulating the diatom and blue-green algal flora recently has also been obtained (152), and reports suggesting a regulatory function related to the heterotrophic degradation of organic compounds and the consequent evolution of CO_2 have prompted active discussion (114, 116, 117). To assume that a single nutrient element controls algal development in all rivers, streams, and lakes is naive and, taking into account the need for managing or minimizing phytoplankton blooms, action based on this assumption might lead to valueless and expensive control measures. Additional meaningful information and carefully balanced ecological assessments are needed to prevent further deterioration of fresh waters caused by proliferation of the bloom algae.

In situations where primary productivity, the size or biomass of the community, or the presence of a selected population is dependent on the presence of trace metals, and evidence exists that inorganic nutrients may, in fact, act in one or more of such ways (5), the concentration of the available forms of the element takes on particular relevancy. A large number of natural products or microbial excretions may act as chelating agents and, if indeed formed and released in nature, serve to keep the critical substance in solution in an available state. Hence, the relation of algal growth and phytoplankton production to natural organic chelators and complexing agents continues to attract interest (14, 18), and it seems quite probable that microbial excretions do have a major impact on neighboring populations and their activities. To date, however, few of the molecules involved in nature have been characterized.

Despite the wealth of laboratory data showing the ability of various algae to grow heterotrophically in the dark, little is known about the actual heterotrophic development of these organisms in their typical habitats. The absence of information on the organisms' behavior in situ has not dampened speculation, conjectures usually based solely on studies of axenic cultures in artificial media. In this light, it is exciting to read of experiments designed to show directly whether such heterotrophic patterns do actually occur in natural ecosystems, work suggesting that algal heterotrophy is unlikely in the many fresh waters and areas of the open ocean in which bacterial metabolism keeps the level of organic materials too low for algal heterotrophy to be of consequence (177, 210).

Morphological changes in microorganisms induced by discrete organic compounds generated by plants and animals are quite widespread, and one microbial population may also provoke its neighbors to undergo morphogenetic alterations. These changes may be related to the synchrony between the excystation of ciliates and the molting period of their crustacean hosts, a synchronization without which species of *Gymnodinoides* could not continue to exist (193), or they may be dependent on the exudation by plant roots of substances promoting germination of fungal cysts, conidia, or sclerotia. In many instances, the chemical responsible for triggering the morphogenetic sequence is not exotic, inasmuch as many simple organic compounds suffice, but the prevalence of the substance should not cloud the uniqueness of the response or the issue of the biochemical basis for the restriction of the protozoa or fungi involved to that particular milieu. In other instances, if in vitro investigations are used as a guide, the chemical agent of morphogenesis is likely to be rare in nature and novel in structure, but the ecological role of these exotic molecules has yet to be explored.

Growth factors probably are synthesized and appear in solution to a lesser or greater extent in nearly every microbial ecosystem. Yet, the presence or absence of individual growth factors may be a major regulator of the composition of diverse communities. As stated above, growth of a species needing a specific vitamin, amino acid, purine base, or sterol is restricted to localities where that compound is present in the ambient fluid in amounts sufficient for the organism's use, and the ubiquity of auxotrophs is a priori evidence that the release of these compounds to the liquid is far more than a laboratory curiosity. It has been widely accepted that a large percentage of soil bacteria cannot grow in the absence of one or several B vitamins of amino acids, and the work of Dias & Bhat (54) revealed the growth factor demands of sewage inhabitants. Similarly, many marine bacteria have a need for biotin and thiamine, smaller numbers demand vitamin B_{12} and nicotinic acid for replication, while pantothenate and riboflavin requirements are reasonably infrequent (36).

Numerous investigators have demonstrated that isolated microorganisms obtained from many ecosystems excrete the growth factors in question, in vitro at least, and these biosynthetic reactions presumably account for the frequency of auxotrophy in nature. The liberation of vitamins, amino acids, purines, and pyrimidines thus apparently has an appreciable impact, the metabolites regulating the microbial community and its functions much as secretions of the animal regulate that organism's functioning. Furthermore, direct assays of samples of soil and sea water, as cases in point, have revealed the presence of thiamine, biotin, vitamin B_{12}, and an assemblage of amino acids known to be nutritionally significant, and many investigators, Curl (49) as an example with *Skeletonema costatum* and vitamin B_{12}, have either clearly correlated or attempted to link the occurrence of individual organisms or a succession of populations with growth factor fluctuations. Not only the absolute concentration but also the turnover of growth factors is of interest, and Hobbie, Crawford & Webb (89) examined the flux of amino acids in estuarine water, demonstrating a range of turnover rates for these nutrients.

Organic molecules functioning as growth factors are also synthesized by animals or plants that support microbial life. The effect of this synthesis is strikingly evident among the auxotrophic parasites which rely on their hosts for nutrients. One might expect that host tissues would be sufficiently rich in growth factors that the parasite would obtain enough of all of its requisites, but that this is not the case was disclosed most dramatically in studies using auxotrophic mutants of plant and animal pathogens. For example, an adenine-less mutant of *Klebsiella pneumoniae* is avirulent for mice, but reversion of the mutant to adenine independence restores the virulence of the bacterium (71). Similarly, *Agrobacterium tumefaciens* mutants requiring adenine, asparagine, or methionine are less harmful than the wild type, and addition to the host plant of the compound needed or reversion to prototrophy leads to increased infectivity (122). The nutritional inadequacy of hosts insofar as their potential parasites are concerned is also evident in the relationship between *Prunus domestica* and one of its parasites, *Rhodosticta quercina,* a fungus requiring myo-inositol; the ecological importance of the need for this compound is suggested by the observation that the bark, on which the fungus typically forms cankers, of susceptible plants is rich in inositol, whereas the inositol content of the bark of resistant plant varieties is reasonably low. Furthermore, the resistance of the latter plants to colonization by *R. quercina* is overcome if they are treated with exogenous myo-inositol (126). Hence, the capacity of a microorganism to invade, become established, and do damage sometimes is regulated by the availability of discrete metabolites in the tissue or body fluid ecosystem. The situation is often not quite this simple, as revealed in studies of adenine-requiring mutants of *Bacillus anthracis* which failed to kill mice; though reversion of the mutants to prototrophy fully restored their virulence, additions of adenine to the animal did not lead to killing of the host by the original mutants (96).

Substances responsible for chemotaxis in nature represent still another category of compounds having a unique ecological role. The movement toward or away from a chemical may allow an organism to find a congenial environment or usable food or to avoid injurious circumstances. Therefore, such chemotactic responses may represent a fitness trait of enormous value in attaining a favorable outcome in natural selection and even be an essential attribute for certain organisms in given localities. Protozoa, bacteria, and the amebal stages of slime molds such as *Dictyostelium* exhibit chemotaxis, but much of the current effort in seeking to establish the ecological importance of chemotaxis has centered on fungal zoospores, especially those of *Pythium, Phytophthora,* and *Aphanomyces.* Positive chemotaxis of their zoospores is induced by products exuded by plant roots, and this movement assumes prominence because many of these fungi parasitize plants. By moving along the gradient of the exuded product, the pathogen may locate host tissues suitable for its proliferation, although nonhosts may liberate attractants too. Among the agents responsible for this chemotaxis are gluconic acid and a number of sugars (88, 161).

The establishment of innumerable heterotrophs and autotrophs reaching a site that contains all nutrients they need is prevented because the arrival cannot tolerate the level of some abiotic factor. The localization of *Oscillatoria terebriformis* in hot springs provides a clear illustration of the control by temperature. The site in the springs where the algal population is to be found varies with the seasonal temperature trends, but it is absent from regions where the water temperature is above 54°C, the maximum for its growth (39). The control of algal community composition by salinity has also been examined in solar salt works (50), and the limits for microbial existence have been assessed in terms of oxidation-reduction potential, temperature, and pH (11, 29). Numerous additional reports on these and other abiotic factors have been published in the last few years.

More directly pertinent to a discussion of biochemical ecology, however, is the presence of toxic chemicals that affect community structure or function. In many environments, the nutrient supply is adequate and the intensity of abiotic factors is not excessive, but a biologically produced compound prevents a species of interest from multiplying or surviving. Inhibitors excreted by plants, present in or on bodies of animals, or generated by microorganisms may determine the identities of species able to colonize the area, and they may regulate the course of natural selection in the microbial community. Microbial inhibitors released by plants and toxic to soil bacteria include gallotannic, gallic, and chlorogenic acids (205). Man-made chemicals entering water or soil, and drugs introduced into the body likewise modify or occasionally devastate the resident community. Such changes have been amply documented; for example, the intraspecific alterations when drug-resistant populations replace drug-sensitive strains, and the interspecific modifications when a chemotherapeutic agent entirely eliminates sensitive species and allows for the appearance of a community of resistant individuals (66, 188). These population shifts are as much within the realm of ecology as are the microfloral changes mediated by root exudates. In this regard, Anderson's review (8) on the ecology of transferable drug resistance in the enterobacteria is particularly illuminating.

The number and variety of selective stresses imposed on microorganisms in one region or another are immense, yet few biochemical investigations have been designed to understand how these stresses influence natural communities. Braun & Firshein (26) reviewed the evidence for intraspecific changes in *Brucella abortus* and other bacteria resulting from the presence of oligodeoxyribonucleotides in artificial conditions, and it would be interesting to determine whether analogous modifications occur in vivo. Genetic effects resulting from environmental challenges are known in several habitats; for example, much effort, albeit little from a physiological standpoint, has been devoted to analyzing for the appearance of new races of fungal pathogens in response to the introduction of new varieties of crop plants resistant to the currently widespread pathogens. The possible occurrence of other genetic changes in vivo is suggested by the demonstration of transformation in mice infected with two genetically distinct strains of *Diplococcus pneumoniae* (48) and by the induction of mutations in two *Aspergillus* species by the pesticides Ferbam and 3′,4′-dichloropropionanilide (157, 158).

The effect of light intensity in controlling the vertical distribution of fresh-water and marine algae has been reported repeatedly, and considerable attention has been given to the light requirements and sensitivities of many species. Gilmartin (72), in a continuation of this line of investigation, proposed that the inhibition of *Tydemania expeditionis* by high light intensities directly regulates this alga's localization in water columns, the need being for a microenvironment with a low light intensity.

The enormous surface area and great reactivity of clays contribute immeasurably to their impact on soil organisms. The surface area of certain clay minerals ranges from 10 to more than 100 square meters per gram, and it should not be surprising, therefore, that such reactive colloidal materials appreciably influence the activities and possibly the composition of subterranean communities. Montmorillonite, for example, stimulates bacterial respiration, in part because of its effect in maintaining pH, whereas the clay inhibits mycelial respiration (183, 184). Clays likewise modify the apparent pH range and pH optimum for microbial activity, the result of a difference in hydrogen ion concentration at the clay surface and in the ambient solution. This pH differential is reflected in the behavior of *Nitrobacter agilis* in sterile soil and solution culture (138).

Biochemical barriers to colonization are omnipresent, and an organism, to be successful in a particular locality, must be able to breach the prevailing barriers in one way or another. Several of the barriers have now been identified, and a few of the ways or mechanisms by which organisms destroy or bypass these obstacles to their successful establishment have been characterized. The mechanism of destroying or bypassing the barrier is merely a fitness trait useful to the organism in its struggle for existence.

The presence of a chemical barrier in a site microorganisms are attempting to colonize is often revealed by the existence of a toxicity to the potential colonist, but rarely has the responsible inhibitor been isolated and identified. A few examples from the large literature will suffice to show the variety of natural products barring intrusion, rapid proliferation, or pathogenicity of microorganisms. Cow's milk contains thiocyanate and peroxidase, and lactoperoxidase in the presence of thiocyanate catalyzes the formation of a bactericidal substance. Some streptococci bypass this particular barrier because they destroy the toxicant (162). Peroxidase activity is also related to the resistance of potato plants to infection by *Phytophthora infestans* (65). Several of the inhibitors that are deleterious to microbial invaders or inhabitants of foods have recently been characterized also (46, 104). Not surprisingly, the antimicrobial defense mechanisms of man and animals have been carefully scrutinized; recent reviews have dealt with pulmonary antibacterial mechanisms and antibacterial factors in the urinary bladder (45, 77). Antimicrobial proteins have likewise been observed in the teat canal of the cow (87). How parasites react to and occasionally overcome these barriers or defense mechanisms has been thoroughly explored, as in studies of the significance of the capsule as a protective device against phagocytosis (33) or of the production of enzymes by *Toxoplasma gondii* so that it can penetrate mammalian cells and enter the intracellular habitat required for its reproduction (128).

Notable progress has been made in defining the chemical barriers of plants that preclude the penetration, replication, and pathogenesis by assorted fungi and occasional bacteria. Such substances apparently form part of the plant's armament against organisms which otherwise would find the tissues to be a most satisfactory environment. Turpentine components in the oleoresin of *Pinus ponderosa* xylem have been implicated in the tree's resistance to infection by species of *Fomes, Ceratocystis,* and other fungi (44), α-methylenebutyrolactone in the resistance of tulip tissues to penetration by *Fusarium oxysporum* (20), and quinones in the capacity of plants to hold *Pseudomonas* species in check (144). Antifungal agents in barley, potato tubers, and *Pinus monticola* have also been the subject of recent inquiry (7, 91, 182). The remarkable longevity of *Ginkgo biloba,* moreover, has been ascribed to its content of substances that are detrimental to viruses, bacteria, fungi, and other organisms (132).

By contrast with the static chemical barriers which retard or prevent the microbial occupation of an environment or site, "environmental feedback" refers to a change in the habitat resulting from the presence of one or several populations and affecting the abundance of invaders or their very existence (5). Defense mechanisms of animals and plants which respond to invading microorganisms illustrate the operation of such feedback. Interferon, peroxidase, antibodies, phytoalexins, and phagocytes contribute to or are responsible for the feedback. In terms of specific low molecular weight chemicals involved in host response to infection, plants serve as excellent examples. As one case in point, antifungal phenolics are synthesized in soybean seedlings in response to the presence of *Helminthosporium carbonum* or *Alternaria* sp., fungitoxins deleterious to these and other fungi, and it has been suggested that the phenols function either by inhibiting fungal polysaccharidases that are important in the destruction of plant tissue or by polymerizing to form a barrier which is difficult for the invader to bypass (21). Kosuge (111) reviewed the significance of phenolic compounds in the response of plants to microbial colonization.

Phytoalexins are either metabolites generated de novo in plants or compounds whose concentration rises markedly in response to microbial infection. These substances are apparently important in the plant's ability to retard or prevent colonization by potential parasites. Among the characterized antimicrobial agents thus synthesized by plants are 3-hydroxy-7-methoxy-4′, 5′-methylenedioxy chromanocoumarin (pisatin), 3-methyl-6-methoxy-8-hydroxy-3,4-dihydroisocoumarin, steroid glycoalkaloids, chlorogenic acid, caffeic acid, and ipomeamarone (113). An antifungal terpenoid was found to be formed in potato tubers exposed to *Phytophthora infestans,* the concentration rising to levels sufficiently high to inhibit vegetative growth of the fungus (190). The formation and importance of phytoalexins in plants of economic importance have been subjects of active inquiry (17, 120, 153).

Upon exposure of higher plants to pathogens, the activities of certain enzymes increase, and these increases are correlated with—and, possibly directly or indirectly, account for—the plant's capacity to prevent establishment of the pathogen; for example, peroxidase activity markedly rises in resistant wheat varieties exposed to *Puccinia graminis* f. sp. *tritici,* and this enzyme in media containing H_2O_2 likewise suppresses hyphal development; by contrast, the activity in susceptible varieties changes only to a modest extent (130). Environmental feedback is also evident in invertebrates, as in the bactericidin appearing in the lymph of the spiny lobster injected with a Gram-negative bacterium (64).

Effect of Microorganisms on Their Surroundings

Microorganisms residing within or upon living or inanimate objects may modify, drastically or modestly, their immediate surroundings. Countless of these changes have been observed in living animals and plants, lakes, rivers, the open sea, soil, manufactured products, and foods. The modifications are described in terms often readily understandable only by the physician, veterinarian, plant pathologist, aquatic or soil scientist, engineer, or food technologist, but rarely is the underlying biochemical mechanism defined. The direct agent of environmental change may be one or several enzymes or excreted products, but the specialist is generally more interested in the response itself, or ways of preventing it, than in understanding why and how the reaction occurs. Although certain of these areas may be deemed to be outside the realm of conventional ecology, they surely are not alien to microbial ecology since these sites are, in fact, microbial habitats, just as much as is the forest or a remote portion of the ocean.

Turning first to the living plant as an environment for microbial life, it is possible to distinguish distinct biochemical changes brought about by microorganisms living in immediate proximity to or directly on higher plants or by species residing within the tissues themselves, either as symbionts or parasites. A pronounced effect is exerted by heterotrophs on the absorption of inorganic nutrients through the roots, the variety of populations in the rhizosphere either enhancing or diminishing the uptake of a number of different ions. Certain nutrients enter the roots quickly while others penetrate more slowly because of the presence of the rhizosphere microflora. The uptake of phosphorus, sulfur, calcium, potassium, zinc, and manganese are thus enhanced or retarded. Typical recent studies have dealt with the effect of rhizosphere populations on molybdenum uptake (124) and on nutrients assimilated by soybeans (141). How microorganisms thus influence the physiology of the plants together with which they live is still largely a matter of speculation, for few of the necessary critical studies have been conducted.

The investigations of Libbert and colleagues (118, 119) provide some insight into how epiphytic bacteria may alter the physiology of the plant. By comparing sterile and nonsterile corn, cucumber, and pea plants, they demonstrated the higher content of auxin in the nonsterile tissue and reported that the introduction of indoleacetic acid-producing bacteria onto sterile plants led to a rise in their auxin content. Bacteria on the surfaces of the tissue presumably metabolize tryptophan or other organic exulates and in turn excrete indoleacetic acid or related substances. Auxins may also be involved in the stimulation of rice by the blue-green alga, *Cylindrospermum muscicola* (197), and of mulberry seedlings by rhizosphere bacteria (198).

Heterotrophs residing near roots may profoundly affect the plants by the excretion of simple toxic compounds. In waterlogged soils, the reduction of sulfate by *Desulfovibrio* or related anaerobes generates sulfide, and should the free sulfide not be detoxified, as by its precipitation as ferrous sulfide, harm may be done to roots developing nearby (186). Ammonia evolved during the decomposition of organic remains likewise may cause injury, as suggested by the work of Gilpatrick (73) with *Persea indica* seedlings. Organic phytotoxins are also produced frequently in waterlogged soils, but often all that is needed is a microhabitat where, though the surrounding soil is well aerated, the decomposition of carbonaceous matter results in a local O_2 depletion. Among the organic phytotoxins formed in soil are benzoic, phenylacetic, 3-phenylpropionic, 4-phenylbutyric, *p*-hydroxybenzoic, *p*-coumaric, vanillic, ferulic, and syringic acids as well as methyl, ethyl, *n*-propyl, and *n*-

butyl alcohols (121, 202, 203). The evidence that some of these metabolites are the agents of the phytotoxicity observed in nature is sound; in other instances, the evidence rests solely on studies of model systems. Nevertheless, it does seem that simple organic and inorganic inhibitors are important ways by which microorganisms do damage to roots and retard their development.

The biochemically fascinating symbiotic and parasitic associations between microorganisms and plants have yet to be adequately unravelled, but a few physiological aspects are now beginning to be clarified. One of the two interactants benefits, and the second either is aided or harmed in some way. The partnership in the symbiotic association is subject to natural selection and the partners succeed in the struggle with their neighbors because the bilateral relationship has some striking fitness trait. The plant-microorganism symbioses most thoroughly characterized are those wherein there exists a capacity for N_2 fixation and those involving a fungus residing within a mycorrhizal root. Among the many recent additions to our knowledge of the N_2-assimilating symbiosis are the reports of N_2 fixation in the *Nostoc*-containing leaf glands of *Gunnera arenaria* (175), and the demonstration of marked nitrogen gains in an alder ecosystem (200). Mycorrhizae, too, continue to be topics for considerable physiological research, and one of the more interesting findings in terms of microbial products of ecological importance is the demonstration that the antibiotic formed by *Cenococcum graniforme* in vitro is apparently generated in conifer mycorrhizae containing this fungus and is translocated through the roots and into the needles (112). This observation is of especial significance in light of observations that mycorrhizae contain substances which suppress nearby bacteria and actinomycetes and protect against *Phytophthora cinnamomi* infection (134, 149). Inasmuch as the physiology of mycorrhizae has been considered in detail (83), no attempt will be made here to deal with the topic.

Plant pathologists have been actively pursuing the biochemical bases of fungal pathogenesis. The viable habitat of these microorganisms undergoes a number of biochemical transformations mediated by the invading populations, but much remains to be learned about the precise details and the responsible agents, by they enzymes or nonprotein toxins, of tissue destruction and symptom expression. The relationship of certain characterized polysaccharide depolymerases to infection and disease is becoming more well-defined, and the significance of cellulases and pectic enzymes to pathological processes is attracting much attention. Bateman and his associates (127, 195) have provided information which suggests that phosphatidases also may be of consequence in the destruction of host tissue invaded by *Thielaviopsis basicola, Sclerotium rolfsii,* and *Botrytis cinerea,* presumably because the enzymes destroy the membrane permeability of the plant cells. Arabanase, by contrast, does not seem to be important, at least to tissue maceration by *S. rolfsii* (47). A number of injurious nonproteinaceous metabolites synthesized by microorganisms within the plant host ecosystem have been identified, two of which have attracted recent interest: the cyclic peptide of *Helminthosporium carbonum* which acts as a host-specific toxin (160) and the ammonia formed on leaves by species of *Pseudomonas* and *Erwinia* and causing tissue necrosis (125).

Beneficial and, quite frequently, obligate symbiotic associations between animals and either heterotrophic or autotrophic microorganisms are widespread in nature. Such relationships are found in the simplest to the most complex animals. The frequency and ubiquity of these associations attest to their usefulness to one or both of the interacting species in their struggle to maintain themselves. Many of the animals would not survive were it not for the microorganisms inhabiting them, and the essentiality of the symbiosis for the microscopic partners is seen in the fact that many exist in nature only in these relationships. Few of these animal-microorganism relationships have been adequately studied, and in only a scant number is the chemical or biochemical basis for the symbiosis clearly described. The tie between the two organisms probably has frequently a nutritional basis, one organism providing the second with substances without which the latter could not grow. However, a uni- or bilateral nutrient exchange surely cannot be the sole explanation for the symbiosis in many interactions, for the capacity to excrete the metabolites in question is widespread yet the partners are quite specific for one another.

Although the benefits accruing to the animal are often poorly delineated in biochemical terms, occasionally unequivocal evidence is at hand to account for the good in the sense of provision of given metabolites. For example, the reasons that protozoa or corals grow faster when harboring a microbial symbiont are unknown but, conversely, the dependency of many metazoa or the benefit they obtain is the result, at least in large part, of the vitamins, amino acids, or other growth factors excreted by their endosymbionts (5). Sometimes microorganisms contribute individual enzymes or enzyme complexes to their partners, as exemplified by the microbial community found in the gastric mucosa of dogs, cats, and rats, a community which makes the urease that the host tissues cannot synthesize (52). Comparisons of the kidneys of germ-free mice, germ-free animals injected with D-alanine, and conventional mice have provided evidence that the gastrointestinal microflora may be important in inducing enzymes in metazoan tissue; thus, D-amino acid oxidase appears to be induced by the microbiota, possibly as a consequence of the degradation of the cell walls of the enteric flora (129).

A wealth of data is available on animal-microorganism symbioses in which the microorganisms act on foods ingested by the vertebrate or invertebrate, which the metazoan is itself unable to digest, and convert them to molecules that are directly assimilable. The microscopic agents in the most notable of these associations are the populations in the bovine rumen or termite which excrete cellulolytic and other polysaccharide-hydrolyzing enzymes to depolymerize the carbohydrates the ruminant or termite cannot use. In the alliance between attine ants and poorly described fungi, the fungi contribute the enzyme complex for depolymerizing cellulose, and the ants supply fecal material that provides the microorganisms with proteolytic enzymes and simple nutrients including nitrogenous substrates; the fungi have much to gain from the partnership because they compete poorly away from the insects, whereas they are dominant in the microbial gardens cultivated by the ants (133).

Several other biochemical contributions made to animals by their microscopic partners have been described in the last few years. Kok, Norris & Chu (109) reported that the ergosterol needed for pupation by the ambrosia beetle, *Xyleborus ferrugineus,* is provided by its symbiotic fungus, *Fusarium solani.* The symbiosis involving wood-invading beetles has always been of great interest because wood is quite deficient in sterols, vitamins, and other growth factors essential for the insects, and the microsymbionts have often been proposed, with a minimum of supporting data, to be the provider of the nutrients lacking in the wood. A species of *Pseudomonas* may serve an unexpected protective function for the apple maggot with which it resides, since the bacterium is able to degrade pesticides employed for insect control (24). A protective function was also ascribed some time ago to terpenoid-synthesizing dinoflagellates, the antimicrobial terpenoids being suggested as preventing microbial invasion of the gorgonian in which the dinoflagellates dwell (42).

Considerable progress has been made in increasing our understanding of the physiological relationship between algae and the marine animals they inhabit. The algae assimilate CO_2 photosynthetically, and a portion of the carbon thus fixed is transferred and apparently has nutritive value for the protozoan or metazoan host. A significant portion of the products formed in photosynthesis and excreted by the endosymbiotic algae was found to be converted to polysaccharides by the animal tissues of *Chlorohydra viridissima* (168). The algal symbionts of *Paramecium bursaria,* a sponge, and *C. viridissima* liberate much of the carbon gained in photosynthesis in the form of maltose and glucose (146), while the zoochlorellae of reef corals and *Tridacna crocea* release considerable glycerol (145). In support of the view that the algae may be beneficial by providing O_2 to corals, Roffman (167) noted that endozoic algae of reef-forming stony corals evolved O_2 far in excess of the amount consumed by the host and microorganism jointly during the daylight hours.

Some of the toxic compounds synthesized in nature by microorganisms may affect the functioning or very existence of animals. A few, like the H_2S generated in soil, have a simple structure; toxins of selected *Clostridium* spp., by contrast, are complex proteins. Some act on a narrow range of animals, but many injure a broad cross-section of species. Research in the last few years has induced a flurry of excitement about mycotoxins and a more modest degree of interest in phycotoxins. These metabolites are not mere laboratory curiosities, for mycotoxins have been reported to be of consequence in domesticated animals, and phycotoxins in fish mortalities. The extensive literature on mycotoxins will not be considered here since it has been the subject of thorough reviews (13, 41). Penicillic acid, a metabolite of several *Penicillium* species, can also be formed on agricultural commodities stored for animal feeding (40). Estrogenic substances are likewise synthesized by fungi, and one produced on stored corn by *Fusarium gramineum* has been implicated in reduced size of swine litters and in infertility problems in dairy cattle (143). Differing from the well-studied mycotoxins, the injurious metabolite of *Prymnesium parvum,* a phytoflagellate responsible for fish mortalities in a number of regions, is a highly complex substance containing protein, fatty acids, and sugars (196).

Continued effort, understandably, is directed toward the biochemistry of pathogenic bacteria residing within the animals they harm. Many of these pathogens owe their continued existence solely to their having acquired fitness for life within the animal body yet, despite the practical importance of these properties, astonishingly little is known about the fitness traits—be they major biochemical alterations or minor but still profound idiosyncrasies—which underlie the adaptation of the bacteria to multiply within the tissue or body-fluid ecosystem. Limitations of space require that reported progress in understanding the biochemical bases for microbial pathogenicity and of fitness for life in these habitats be omitted, but the illuminating discussion by Smith (178) must nevertheless be cited in this regard.

Among the more traditional habitats for ecological exploration are soil and water, and it is the terrestrial, fresh-water, and marine environs that have been foci for much of the research on microbial activities. The literature is truly vast and adequate coverage in this brief essay is not possible. Several introductory texts are available, however (5, 55, 137, 208). The concern with what microbial communities do in water and soil has markedly increased, moreover, with the growing awareness of the role of microorganisms in pollution abatement and in the genesis of hazardous compounds. Considering the numbers of environmental problems, the abundance of approaches, and the variety of processes effected by microorganisms in water and on land, I have chosen to select a few illustrative studies, hoping that the reader will delve further into the related literature.

The role of natural communities and selected populations in the weathering of rocks, the destruction of minerals, and the formation of soils continues to attract the interest of scores of researchers. Individual minerals or rocks have been found to be acted on and altered by fungal isolates, and elements contained therein are rendered soluble as a consequence; in the case of *Penicillium simplicissimum,* which excretes citric acid, the release of silicon, iron, aluminum, and magnesium from basalt apparently results entirely from the attack on the rock by the acid generated by the fungus (25, 174, 204). On the other hand, in vitro evidence suggests that 2-ketogluconic acid serves as a chelating agent to solubilize silicon, calcium, magnesium, manganese, aluminum, potassium, and iron bound in minerals and related materials (57), so that doubt remains as to whether acid formation, chelation, or some other process is the major mechanism of microbiological weathering. The work of Visser, Theisen & Mehlich (199) shows, moreover, that such transformations may lead to permanent alterations in crystal structure and the formation of entirely new aluminosilicates. Though the precise biochemical details remain to be unravelled, investigations on possible mechanisms of humus formation continue (81).

The microbial transformations of sulfur stand out in biogeochemistry because its oxidation and reduction are accompanied by the release of certain elements or their precipitation as insoluble sulfides and because the sulfuric acid evolved in the oxidation solubilizes a number of ions. The responsible thiobacilli, the physiologically related ferrobacilli, as well as ore and soil samples have been scrutinized to establish the reactions, the products, and the environmental conditions necessary for a rapid conversion. The activity of *Thiobacillus ferrooxidans* in vitro is illustrated in studies of its oxidation of sulfide minerals and its solubilization of such ores as chalcopyrite and pyrite (16). The bacterial leaching of ores, which may mimic what occurs in natural circumstances, has been examined with regard to the release of elements such as copper, zinc, and uranium (105, 192, 212). The enormous quantities of sulfuric acid that are created as the autotrophs, together with possible abiotic processes, cause the oxidation of sulfides in mining regions provide impetus to studies of the mechanism, the relative significance of biological and nonbiological agencies (12), and possible means of controlling this devastating kind of water pollution.

The consumption of O_2, the synthesis of products with reducing properties, the reduction of inorganic electron acceptors or the elaboration of inorganic and organic acids in fresh and polluted waters, aquatic sediments, and poorly drained soils are involved in the reduction of several elements capable of existing in more than one oxidation state. Sulfate reduction has always been an exciting topic, largely by virtue of the unique bacteria involved and the geochemical impact of the product, and Bloomfield (23) has examined factors governing the release of free H_2S from waterlogged soil, his data showing the importance of iron and alkalinity in determining H_2S loss. Sulfate reduction in soils of the arid zone which occasionally become waterlogged may lead to production of equivalent quantities of carbonates, and this process has an effect in decreasing soil salinity as the soluble or exchangeable calcium is precipitated as $CaCO_3$ (148). Deuser (53) contrasted the relative contribution of the microbial mineralization of organic sulfur and the bacterial reduction of sulfate as sources of the enormous quantities of H_2S in Black Sea waters and concluded that only about 3 to 5 percent comes from organic sulfur. The bacterial reduction of iron and manganese has also been dealt with (194), although the function of these widespread reductions in the physiology of the responsible organisms remains unresolved. Ottow (150, 151), however, suggested that nitrate reductase may be implicated in the reduction of ferric iron and that the ferric iron may indeed serve as an alternate electron acceptor for the responsible bacteria.

Stimulated by the International Biological Program, the discovery of new organisms or symbioses effecting the process, and the tremendous need for additional protein in the developing nations, ecologists have intensified their scrutiny of nitrogen fixation. The acetylene-reduction, Kjeldahl, and ^{15}N techniques have been employed to assess the rate of N_2 assimilation by the indigenous populations of a variety of ecosystems. Brezonik & Harper (27), for example, reported that the rates of N_2 fixation at O_2-depleted sites in lakes ranged from 0 to 83 ng per liter per hour, and they suggested that the rates might be too low to be of ecological importance. MacRae & Castro (131) observed appreciable N_2 fixation in tropical rice soils exposed to the light or kept in the dark, and Voigt & Steucek (200) noted an average nitrogen gain of about 85 kg per hectare in an alder community. Other investigators have examined the biological assimilation of N_2 by populations of soil (173, 180), subtropical marine waters (35), and the phyllosphere of conifers (102). A most interesting nitrogen balance was calculated by Brezonik & Lee (28) for Lake Mendota; they estimated that 14 percent of the nitrogen entering the lake came from N_2 fixation while about 11 percent of that lost was the result of denitrification.

Microorganisms act on a number of other elements, some being micronutrients, others not being required for life. These conversions are frequently of geochemical importance, and often they are essential for plant growth. The responsible organisms and the biochemical mechanisms of many of these reactions have yet to be established. Indeed, biologists are still discovering reactions in mineral metabolism hitherto unknown either in nature or in the laboratory. The biological volatilization of selenium from soil was explored by Abu-Erreish, Whitehead & Olson (1), but neither the substrates nor the products were identified. Selenium volatilization has been known for some time as a laboratory curiosity, but the significance to geochemistry or environmental pollution of the volatilization of this and other heavy elements has scarcely been examined. The decomposition of litter and the accompanying release of sodium, potassium, calcium, magnesium, and phosphorus were evaluated in a forest ecosystem by Attiwell (10), and the turnover of cesium in decomposing litter was explored by Witkamp (206).

Metabolism of Polluting and Hazardous Chemicals in Nature

Natural communities of terrestrial and aquatic ecosystems are enormously versatile. A fantastic array of biologically produced or synthetic chemicals and a large assortment of natural products enter soil and water, and one or a group of indigenous populations frequently multiplies and makes use of the introduced substrates. A species with the requisite enzymes to catalyze a change in an introduced molecule may not be present initially, but either invaders with the appropriate physiological traits or individuals of some indigenous population which have undergone a genetic change begin to grow so that the compounds entering the ecosystem are modified to a greater or lesser extent. Nowhere is this impressive catabolic versatility so evident as in the steps in mineralization or nutrient regeneration. Provided that O_2 is available and that conditions are not too extreme, the appropriate populations proliferate, degrade the substances, and release the organically bound elements in inorganic forms. The literature on mineralization and nutrient regeneration in soil and, to a lesser degree, fresh-water and oceanic habitats is extensive, and no attempt will be made to delve into it here. Johannes (100) has reviewed the topic with reference to lakes and oceans.

This catabolic versatility and the limitations thereof are of particular relevance with reference to polluting chemicals. Many of these pollutants are organic, and a high percentage of them are destroyed largely or solely by microbiological means. The mere fact that certain pollutants are not readily destroyed, indeed, they are often excruciatingly persistent, demonstrates that components of natural microfloras are not as active as might have been expected. Furthermore, attention may be focused on these communities because not only do they not destroy chemicals at a rapid rate but in fact occasionally create new pollutants, converting one unwanted substance into a metabolite that likewise has an adverse affect. Many of the ways by which microorganisms contribute to environmental deterioration or help abate pollution are exemplified in the field of pesticide biochemistry. Many pesticides are detoxified rapidly, and the pathways of degradation and products generated have been identified in a few instances. Some, but these are the compounds particularly disturbing in terms of environmental quality, are recalcitrant, and they are not enzymatically modified at a significant rate, if at all (3); hence, these chemicals may accumulate in nature, exert effects on desirable species for prolonged periods, or more from place to place with eroding soil, moving water, or streams of air. A few are activated; that is, a compound having little or no toxicity is enzymatically modified to give rise to toxicants. Some may be defused, a term applied to a process wherein a substrate potentially metabolizable by one population to yield an inhibitor is, instead, converted to innocuous products by a second population. Finally, evidence exists that pesticides that selectively suppress one group of organisms can be converted to compounds inhibitory to a second group.

Because pesticide metabolism has recently been examined in detail (4, 85), only a few highlights will be cited to illustrate the types of investigations being conducted. Needless to say, the pathway of DDT degradation has been studied intensively in natural habitats, but the fate in soil and water of only one of the 14 carbon atoms in the insecticide molecule is known. Therefore, to provide a biochemical model for predicting what products might be made in nature, individual cultures have been used. As a result, it has been shown that *Mucor alternans* converts DDT to a series of metabolites not previously detected in soil or water (9) and that *Hydrogenomonas* sp., in the presence of O_2, cleaves one of the diphenylmethane rings of a known intermediate in DDT decomposition to form *p*-chlorophenylacetic acid (68). The same bacterium will effect initial steps in DDT degradation anaerobically, and the subsequent introduction of O_2 allows the organism to cleave the ring and make *p*-chlorophenylacetic acid; the latter, in turn, is further degraded by other microorganisms (F. K. Pfaender and M. Alexander, unpublished observations). The pathways of decomposition of other chlorinated hydrocarbon insecticides—problem chemicals because of their persistence in soil and aquatic sediments—are unknown, but modest progress has been made with isolates from soil. For example, a large number of bacteria and fungi epoxidize heptachlor, some organisms dechlorinate this pesticide to yield chlordene (139), and strains of *Trichoderma, Pseudomonas,* and *Bacillus* metabolize the insecticide dieldrin (136). A novel set of reactions, outlined by Bartha & Pramer (15), takes place as microorganisms metabolize N-(3,4-dichlorophenyl)propionamide; the herbicide is apparently converted to 3,4-dichloroaniline, part of the aniline formed is oxidized biologically to a phenylhydroxylamine and then the phenylhydroxylamine reacts nonbiologically with its aniline precursor to yield 3,3′,4,4′-tetrachloroazobenzene. The in vivo conversion of a halogenated benzonitrile to the corresponding benzamide and benzoic acid is exemplified in a study of the herbicide ioxynil (213) and the formation of a toxin from a pesticide in an investigation of Benlate (43). Microbial reduction of the nitro group of parathion (76), degradation of the herbicide chloral hydrate (170), and the destruction of the herbicide pyrazon by a Gram-negative coccus (62) have been described recently. Steps involved in the complete mineralization of one of the most widely used herbicides, 2,4-dichlorophenoxyacetic acid (2,4-D), have been defined for an *Arthrobacter* (60, 189), but lacking still is information on how this compound is transformed in nature.

The proposed widespread use of nitrilotriacetic acid (NTA) to replace part of the phosphates in detergent formulations raised fears concerning the possible, though unproven, formation by microorganisms of nitrosamines, a class of compounds with representatives that are carcinogenic, mutagenic, and teratogenic. NTA is biodegradable (63) and though the mechanism of its destruction has yet to be established, it is likely that a secondary amine, such as iminodiacetic acid, is an intermediate. Should an amine of this sort ever accumulate, it seems plausible to believe that nitrosamines may be synthesized where microorganisms are simultaneously creating nitrite, as in the gastrointestinal tract of man, the bovine rumen, or sewage; ample precedent exists, for dimethylnitrosamine has been found in nitrite-treated fish meal, presumably coming from a nitrosation of secondary amines derived from the fish (169). Moreover, nitroso compounds are known to be elaborated by cultures of *Streptomyces* (86) and *Pseudomonas* (187).

Mercury is now recognized as a significant pollutant, and though microbiologists have grown up with the knowledge that inorganic and organic mercurials are potent germicides, only in the last few years have they gained an awareness that both inorganic and organic mercury may be substrates too. Organic mercurials used for the control of soil-borne fungi pathogenic for plants are subject to microbial attack; in soil treated with phenylmercuric acetate, for example, the microscopic residents effect the release of metallic mercury (108). Tonomura and his associates (70, 191) examined the metabolism of such substrates by cells and extracts of a soil pseudomonad, and they reported that phenylmercuric acetate and ethylmercuric phosphate were cleaved to metallic mercury and either benzene or ethane in a reaction requiring NADH. Microbial alkylation of cationic inorganic mercury occurs too and is of enormous concern since it seems that bacteria in aquatic sediments or on the surfaces of fish are responsible for formation of the methylmercury found in fish. Methylation of mercuric ions can be brought about by unidentified aquatic bacteria (99) and by cell extracts of a methanogenic bacterium (209). Recognizing the appearance of disturbing quantities of cations of other heavy metals in rivers and streams, an urgent need exists to establish whether the laboratory demonstration of the microbial methylation of arsenic, selenium, and tellurium is indeed a model of what takes place in fresh or oceanic waters. Moreover, the possible methylation of cations of additional elements, especially those of importance as pollutants, and the genesis of other organo-metal complexes ought to be determined.

The examples cited above illustrate how microorganisms gain ecological notoriety because of their ability to metabolize pollutants. They have acquired notoriety, too, because of what they do not, or cannot, do. This author has already devoted many words to a consideration of the kinds of recalcitrant molecules, the mechanisms of recalcitrance, and practical problems resulting from such microbial failings (3). Recently, new synthetic refractory chemicals have come to the forefront, and additional data have been gathered on both synthetic and natural materials which persist intact for long periods in localities teeming with bacteria, fungi, and actinomycetes. For example, the miticide isopropyl 4,4′-dibromobenzilate was recovered in almost quantitative yield from soil 14 months after it was applied (38). Such durability must suffer by comparison with those noted for toxaphene, DDT, endrin, or chlordane, the recovery from soil treated 14 years earlier being 45, 39, 41, and 40 percent, respectively (147). A new class of long-lived and presumably slowly or nonbiodegradable pollutants is made up of the polychlorinated biphenyls (80). Dwarfing the persistence time of the synthetic chemicals, although likely durable in nature for entirely different reasons, are organic fractions of soil, which have been assigned ages in excess of a millenium (37), and the materials known as kerogen and sporopollenin, which Brooks & Shaw (30) believe to be polymers of carotenoids and carotenoid esters.

The mechanisms of recalcitrance have been further examined. Tannins or melanins may inactivate or complex enzymes, thereby possibly protecting not only the tannin or melanin itself but also substances with which they are associated; this may help explain the retarded decomposition of certain organic substances and the slow digestion of melanized fungal surfaces in soil (19, 22, 32). The formation of a complex between a resistant inorganic or organic material like melanin (115) and a readily biodegradable polysaccharide or protein may protect the normally susceptible substrate. Heteropolysaccharides and other heteropolymers may be inherently resistant because the degradation requires an array of different enzymes. The effect of chemical structure on the susceptibility of a variety of relatively simple compounds to microbial attack has also been examined (6, 68, 85, 106). On the basis of his finding that certain marine bacteria fail to grow at very low carbon concentrations and leave a small amount of the substrate unutilized, Jannasch (97) proposed that the essentially constant organic matter level of the seas results not from the intrinsic resistance of the molecules to breakdown but rather from their presence in too low a concentration for utilization.

Although they may be unable to degrade certain pollutants, heterotrophs may still be able to metabolize them. On the basis of numerous attempts, and as numerous failures, to isolate microorganisms which can grow on certain insecticides, fungicides, and halogenated aromatic compounds, and the scattered but real successes in obtaining cultures which modify the molecules, albeit to a slight extent, it seems that the phenomenon of *cometabolism* is widespread. The in vitro cometabolism of a class of halogenated aromatics (92, 93) and of DDT and related compounds (68) has been investigated, and these observations on cometabolism probably illustrate what occurs in the slight transformation of chlorinated hydrocarbon insecticides which, as far as is known, do not serve the metabolizing organism as a source of carbon or energy (9, 136, 139). Totally unclear at present, however, is how common is cometabolism, whether it results solely from varying degrees of substrate specificity among the enzymes in a metabolic sequence or from the toxicity of products, and whether cometabolism can be exploited to enhance the rate of degradation of slowly biodegradable pollutants.

Biochemistry of Interspecific Relationships Among Microorganisms

Progress in unravelling the biochemical or physiological bases of interspecific relationships involving two or more microorganisms or populations has been spotty. With the selected species which have become veritable pets of a goodly collection of biologists, progress has been notable and the relationships in which they participate have been blessed with a wealth of data. Interactions involving species that have not received comparable degrees of scientific attention remain ill-defined as expected although the scant information available suggests that some of these associations may be excellent models for developing an understanding of the way microorganisms interact.

The parasitic relationship involving two microorganisms which has been subject to the most thorough biochemical scrutiny is that between the bacteriophage and its bacterial host. So many reviews and monographs have considered the topic that no attempt will be made here to deal with it. Similarly, the literature on the lysis of bacteria is vast, but the excellent essay of Strominger & Ghuysen (185) highlights the key biochemical facets. Yet, the ecological relevancy of the fascinating biochemical information on the bacteriophage-bacterium association and bacteriolysis is not readily evident. Much the same can be said for the data on the biochemistry of lysis of yeasts and filamentous fungi, although the data on lysis in these instances are far less complete. Important, indeed, and ecologically relevant too, would be a biochemical explanation for the apparent resistance to lysis of bacterial endospores, protozoan cysts, and the many nonmelanized resting structures of fungi. The contribution of melanin to the resistance of sclerotia, chlamydospores, and conidia has been considered above. Worth citing in this connection are the reports that a *Streptomyces* lyses arthrospores while causing only minor change to sporangiospores of *Mucor ramannianus* (101), and that hyaline fungi are generally more quickly digested than melanized species (94).

The bases for prey selection and for the attraction of predator to its living food have prompted much speculation but little meaningful experimentation. A significant step in learning of the biochemicals implicated in allowing a predator to find its prey is the demonstration that the cyclic 3′,5′-adenosine monophosphate released by bacteria attracts the amebae of slime molds, aiding the amebae in finding cells to feed on (110). The predator's selectivity among potential prey cells is of ecological concern inasmuch as it may affect the relative abundance of individual prey species in the community. It may also explain why aliens, as a rule, do not persist for long in foreign habitats. Groscop & Brent (78) found that three of four inedible bacterial species were toxic to the potential predator. Heal & Felton (84), though providing no supporting evidence, suggested that stimulatory or inhibitory exudates, size, and the presence of capsules may be responsible for the susceptibility, or lack thereof, of bacteria to predation. In this connection, it is worth citing the report that an encapsulated *Cryptococcus neoformans* strain, which is more resistant to phagocytosis than nonencapsulated mutants derived from it, produces a polysaccharide inhibitory to phagocytosis by human leukocytes (34).

Competition has often been postulated as the cause of the displacement of one species by a second, and competition frequently seems to be indeed the likely explanation in many instances. Still, few well-defined illustrations of this kind of interaction exist, and fewer examples are available in which the factor for which the rivalry takes place or the reason for the displacement is known, although differential growth rates are usually the stated cause. Jannasch (98) argued that the low growth rates of Enterobacteriaceae are a key factor in their competitive displacement from sea water. A rivalry for nicotinamide was observed to be crucial in a two-membered culture containing *Staphylococcus aureus* and streptococci (95). An interesting consequence of competition for organic carbon is suggested by the work of Wright & Hobbie (210), who proposed that the rapid bacterial utilization of glucose and acetate keeps their concentrations so low as to preclude heterotrophy among planktonic algae. Nevertheless, the significance of competition in situ has hardly been explored as yet.

Considerable advancement has been made in showing amensalism to be a reality in nature. The microbial production of toxins affecting inhabitants of an ecosystem has been proposed countless times, but the evidence to support these contentions has usually been meager, flimsy, or nonexistent. However, by seeking an actual activity in nature rather than by isolating or enumerating organisms having the potential for an activity, several investigators have established unequivocally that toxins are actually a fitness trait for certain species and a means for the exclusion of other species. Background information and part of the controversy in regard to these kinds of interactions have recently been considered (5). A noteworthy demonstration of the ecological value of antibiotic production can be found in a study of *Cephalosporium gramineum,* a parasite which persists only in host tissue although it forms no resting structures and competes poorly there; apparently the fungus elaborates an antibiotic which allows it to survive in wheat tissue invaded by organisms that otherwise would eliminate it (31). Similarly, *Rhizopus nigricans* produces an antibacterial compound as it ferments certain oriental foods (201), and mycorrhizae elaborate antimicrobial agents harmful to organisms attempting to colonize adjacent sites (112, 134, 149).

Chemical characterization of the ecologically significant toxins has centered largely on inorganic and simple organic metabolites. Ammonia, for example, seems to be important in inhibiting *Phytophthora cinnamomi* in soil (73), possibly hydrogen peroxide in suppressing bacteria in foods colonized by *Lactobacillus* (159), sulfide in holding Athiorhodaceae in check in anaerobic aquatic habitats (155), and very likely CO_2 liberated by hordes of organisms in retarding growth of selected fungi. Only a few organic agents of amensalism have been characterized and a smaller number identified. Thus, the protection by avirulent staphylococci of chick embryos infected with virulent staphylococci seems to result from the formation of a substance interfering with nicotinamide metabolism (163, 164), and the inhibition of potato virus X may be attributable to a polysaccharide made by *Phytophthora infestans* in the potato plant (90). Volatile fatty acids, especially acetic, propionic, butyric, and formic, have a striking role in interspecific interactions as indicated by evidence which suggests that such metabolites are responsible for the decline of *Salmonella typhimurium* during the curing of cheese (74) and for the failure of *Escherichia coli* to become established in the rumen (207). Short- and long-chain organic acids have also been implicated in the disappearance of aliens entering other ecosystems.

Extracellular products of one population quite commonly have a beneficial rather than a detrimental influence on nearby microorganisms. The metabolite excreted may serve as a carbon source, growth factor, or inducer of a morphogenetic response in the adjacent cells, and the provision by one species of molecules needed by a second serves as the basis for commensalism, protocooperation, and symbiosis. These metabolites make the environment suitable for an organism which otherwise would be unable to grow, carry out a particular function, or undergo a specific morphological alteration, and the compounds probably are responsible for instances of succession and for defining the composition of innumerable communities.

Aquatic, soil, and fermentation microbiologists as well as microbial physiologists have gathered together an impressive list of compounds released by fungi, bacteria, actinomycetes, and photosynthesizing and N_2-fixing algae, and a select few of these compounds have so far been found to participate in interspecific interactions. Smith et al (179) identified a variety of compounds in culture filtrates of *Chlorella vulgaris* and found that bacteria utilized certain of these substances and, in turn, generated additional excretion products. Similarly, nitrogenous excretions of the N_2-fixer, *Calothrix scopularum,* were reported to be used by organisms living together with the alga (103). A high percentage of soil, water, and sewage bacteria, marine algae, and soil fungi have been noted to be dependent on one or several B vitamins, amino acids, or additional growth factors, and these were presumed to have their needs satisfied by dissimilar individuals at the same site. The rate of production and use of amino acids in estuarine waters has also been estimated (89). Laboratory models have been created to demonstrate commensalistic relationships involving one population making a biochemical stimulatory to or essential for another; e.g., *Saccharomyces cerevisiae* synthesizing niacin or a related compound for use by *Proteus vulgaris* (172) or *Colpidium campylum* making a ribonucleic acid that increased the growth rate of *Tetrahymena* (181).

Protocooperation represents a higher degree of mutual dependence than commensalism. In some kinds of protocooperation, the two species probably function as an integrated unit, and one partner cannot survive in nature without its associate or another organism able to provide the same compounds or conditions as the usual associate. The two are selected together and are essential components of each other's environs. Bilateral exchanges have now been amply documented in two-membered cultures. Protocooperation based on vitamin exchange is evident in the observation that a niacin-requiring strain of *Proteus vulgaris* and a biotin-requiring strain of *Bacillus polymyxa* grow well in a medium with neither vitamin, one population supplying the needs of its associate (211). Synergism, though possibly sometimes of little value to the two organisms effecting the synergistic response, is somewhat analogous to protocooperation because a reaction proceeds more rapidly or occurs only when the two populations reside in proximity. Synergism between species of *Arthrobacter* and *Streptomyces* in the degradation of the insecticide diazinon was demonstrated by Gunner & Zuckerman (79), and synergism in mixed infections has also been considered recently (123, 166).

Symbiosis represents the ultimate in bilateral dependency. The two dissimilar organisms are dependent exclusively on one another, and their coexistence is obligate for the carrying out of certain reactions or for the very existence in a particular ecosystem of one or both of the symbionts. When the dependency of one individual on the second is absolute, probably several biochemical factors serve as the basis for the cooperation. Little progress has been made, however, in detailing precisely what each organism contributes to the symbiosis. A number of plausible hypotheses have been advanced (5), but sound experimental support is largely lacking. Considerable advancement has been made, on the other hand, in characterizing the movement of carbohydrates from the algal to the fungal component of the lichen symbiosis (56, 165) and in showing the transfer of the N_2 fixed in lichens containing *Nostoc* (107, 140).

LITERATURE CITED

1. Abu-Erreish, G. M., Whitehead, E. I., Olson, O. E. 1968. *Soil Sci.* 106:415–20
2. Alexander, M. 1964. *Ann. Rev. Microbiol.* 18:217–52
3. Alexander, M. 1965. *Advan. Appl. Microbiol.* 7:35–80
4. Alexander, M. 1969. In *Soil Biology*, ed. Anonymous, 209–40. Paris: UNESCO
5. Alexander, M. 1971. *Microbial Ecology*. New York: Wiley
6. Alexander, M., Lustigman, B. K. 1966. *J. Agr. Food Chem.* 14:410–13
7. Allen, E. H., Kuc, J. 1968. *Phytopathology* 58:776–81
8. Anderson, E. S. 1968. *Ann. Rev. Microbiol.* 22:131–80
9. Anderson, J. P. E., Lichtenstein, E. P., Whittingham, W. F. 1970. *J. Econ. Entomol.* 63:1595–99
10. Attiwell, P. M. 1968. *Ecology* 49:142–45
11. Baas Becking, L. G. M., Kaplan, I. R., Moore, D. 1960. *J. Geol.* 68:243–84
12. Baker, R. A., Wilshire, A. G. 1970. *Environ. Sci. Technol.* 4:401–17
13. Bamburg, J. R., Strong, F. M., Smalley, E. B. 1969. *J. Agr. Food Chem.* 17:443–50
14. Barber, R. T., Ryther, J. H. 1969. *J. Exp. Mar. Biol. Ecol.* 3:191–99
15. Bartha, R., Pramer, D. 1970. *Advan. Appl. Microbiol.* 13:317–41
16. Beck, J. V., Brown, D. G. 1968. *J. Bacteriol.* 96:1433–34
17. Bell, A. A. 1969. *Phytopathology* 59:1119–27
18. Bender, M. E., Matson, W. R., Jordan, R. A. 1970. *Environ. Sci. Tech.* 4:520–21
19. Benoit, R. E., Starkey, R. L. 1968. *Soil Sci.* 105:203–8
20. Bergman, B. H. H., Beijersbergen, J. C. M., Overeem, J. C., Sijpesteijn, A. K. 1967. *Rec. Trav. Chim. Pays-Bas* 86:709–14
21. Biehn, W. L., Williams, E. B., Kuc, J. 1968. *Phytopathology* 58:1261–64
22. Bloomfield, B. J., Alexander, M. 1967. *J. Bacteriol.* 93:1276–80
23. Bloomfield, C. 1969. *J. Soil Sci.* 20:207–21
24. Boush, G. M., Matsumura, F. 1967. *J. Econ. Entomol.* 60:918–20
25. Boyle, J. R., Voigt, G. K., Sawhney, B. L. 1967. *Science* 155:193–95
26. Braun, W., Firshein, W. 1967. *Bacteriol. Rev.* 31:83–94
27. Brezonik, P. L., Harper, C. L. 1969. *Science* 164:1277–79
28. Brezonik, P. L., Lee, G. F. 1968. *Environ. Sci. Technol.* 2:120–25
29. Brock, T. D., Darland, G. K. 1970. *Science* 169:1316–18
30. Brooks, J., Shaw. G. 1968. *Nature (London)* 220:678–79
31. Bruehl, G. W., Millar, R. L., Cunfer, B. 1969. *Can. J. Plant Sci.* 49:235–46
32. Bull, A. T. 1970. *Arch. Biochem. Biophys.* 137:345–56
33. Bulmer, G. S., Sans, M. D. 1967. *J. Bacteriol.* 94:1480–83
34. Bulmer, G. S., Sans, M. D. 1968. *J. Bacteriol.* 95:5–8
35. Bunt, J. S., Cooksey, K. E., Heeb, M. A., Lee, C. C., Taylor, B. F.

1970. *Nature (London)* 227:1163–64
36. Burkholder, P. R., Lewis, S. 1968. *Can. J. Microbiol.* 14:537–43
37. Campbell, C. A., Paul, E. A., Rennie, D. A., McCallum, K. J. 1967. *Soil Sci.* 104:81–85
38. Cannizzaro, R. D., Cullen, T. E., Murphy, R. T. 1970. *J. Agr. Food Chem.* 18:728–30
39. Castenholz, R. W. 1968. *J. Phycol.* 4:132–39
40. Ciegler, A., Kurtzman, C. P. 1970. *Appl. Microbiol.* 20:761–64
41. Ciegler, A., Lillehoj, E. B. 1968. *Advan. Appl. Microbiol.* 10:155–219
42. Ciereszko, L. S. 1962. *Trans. N.Y. Acad. Sci. Ser.* 2, 24:502–3
43. Clemons, G. P., Sisler, H. D. 1969. *Phytopathology* 59:705–6
44. Cobb, F. W., Krstic, M., Zavarin, E., Barber, H. W. 1968. *Phytopathology* 58:1327–35
45. Cobbs, C. G., Kaye, D. 1967. *Yale J. Biol. Med.* 40:93–108
46. Cogan, T. M., Gilliland, S. E., Speck, M. L. 1968. *Appl. Microbiol.* 16: 1220–24
47. Cole, A. L. J., Bateman, D. F. 1969. *Phytopathology* 59:1750–53
48. Conant, J. E., Sawyer, W. D. 1967. *J. Bacteriol.* 93:1869–75
49. Curl, H. 1962. *Limnol. Oceanogr.* 7:422–24
50. Davis, J. S. 1968. *J. Phycol.* (Supplement) 4:6–7
51. Dehority, B. A., Scott, H. W., Kowaluk, P. 1967. *J. Bacteriol,* 94: 537–43
52. Delluva, A. M., Markley, K., Davies, R. E. 1968. *Biochim. Biophys. Acta* 151:646–50
53. Deuser, W. G. 1970. *Science* 168: 1575–77
54. Dias, F. F., Bhat, J. V. 1964. *Appl. Microbiol.* 12:412–17
55. Dommergues, Y., Mangenot, F. 1970. *Écologie Microbienne du Sol.* Paris: Masson
56. Drew, E. A., Smith, D. C. 1967. *New Phytol.* 66:389–400
57. Duff, R. B., Webley, D. M., Scott, R. O. 1963. *Soil Sci.* 95:105–14
58. Dundas, I. D., Larsen, H. 1962. *Arch. Mikrobiol.* 44:233–39
59. Durrell, L. W. 1964. *Mycopathol. Mycol. Appl.* 23:339–45
60. Duxbury, J. M., Tiedje, J. M., Alexander, M., Dawson, J. E. 1970. *J. Agr. Food Chem.* 18:199–201
61. Edmondson, W. T. 1970. *Science* 169:690–91
62. Engvild, K. C., Jensen, H. L. 1969. *Soil Biol. Biochem.* 1:295–300
63. Erickson, S. J., Maloney, T. E., Gentile, J. H. 1970. *J. Water Pollut. Contr. Fed.* 42:R329–35
64. Evans, E. E., Weinheimer, P. F., Painter, B., Acton, R. T., Evans, M. J. 1969. *J. Bacteriol.* 98:943–46
65. Fehrmann, H., Dimond, A. E. 1967. *Phytopathology* 57:69–72
66. Finland, M. 1970. *J. Infect. Dis.* 122: 419–31
67. Fitzgerald, G. P. 1970. *J. Phycol.* 6: 239–47
68. Focht, D. D., Alexander, M. 1970. *Science* 170:91–92
69. Fulco, L., Karfunkel, P., Aaronson, S. 1967. *J. Phycol.* 3:51–52
70. Furukawa, K., Suzuki, T., Tonomura, K. 1969. *Agr. Biol. Chem.* 33:128–30
71. Garber, E. D., Hackett, A. J., Franklin, R. 1952. *Proc. Nat. Acad. Sci. USA* 38:693–97
72. Gilmartin, M. 1966. *J. Phycol.* 2: 100–5
73. Gilpatrick, J. D. 1969. *Phytopathology* 59:973–78
74. Goepfert, J. M., Olson, N. F., Marth, E. H. 1968. *Appl. Microbiol.* 16: 862–66
75. Goldstrohm, D. D., Lilly, V. G. 1965. *Mycologia* 57:612–23
76. Graetz, D. A., Chesters, G., Daniel, T. C., Newland L. W., Lee, G. B. 1970. *J. Water Pollut. Contr. Fed.* 42:R76–94
77. Green, G. M. 1968. *Yale J. Biol. Med.* 40:414–29
78. Groscop, J. A., Brent, M. M. 1964. *Can. J. Microbiol.* 10:579–84
79. Gunner, H. B., Zuckerman, B. M. 1968. *Nature (London)* 217:1183–84
80. Gustafson, C. G. 1970. *Environ. Sci. Technol.* 4:814–19
81. Haider, K., Martin, J. P. 1970. *Soil Biol. Biochem.* 2:145–56
82. Hamilton, D. H. 1969. *Limnol. Oceanogr.* 14:579–90
83. Harley, J. L., Lewis, D. H. 1969. *Advan. Microbial Physiol.* 3:53–81
84. Heal, O. W., Felton, M. J. 1965. *Progress in Protozoology,* ed. Anonymous, 121. Amsterdam: Excerpta Medica
85. Helling, C. S., Kearney, P. C., Alexander, M. 1971. *Advan. Agron.* In press
86. Herr, R. R., Jahnke, H. K.,

Argoudelis, A. D. 1967. *J. Am. Chem. Soc.* 89:4808–9
87. Hibbitt, K. G., Cole, C. B., Reiter, B. 1969. *J. Gen. Microbiol.* 56:365–71
88. Hickman, C. J., Ho, H. H. 1966. *Ann. Rev. Phytopathol.* 4:195–220
89. Hobbie, J. E., Crawford, C. C., Webb, K. L. 1968. *Science* 159: 1463–64
90. Hodgson, W. A., Munro, J., Singh, R. P., Wood, F. A. 1969. *Phytopathology* 59:1334–35
91. Hoff, R. J. 1970. *Can. J. Bot.* 48: 371–76
92. Horvath, R. S., Alexander, M. 1970. *Appl. Microbiol.* 20:254–58
93. Horvath, R. S., Alexander, M. 1970. *Can. J. Microbiol.* 16:1131–32
94. Hurst, H. M., Wagner, G. H. 1969. *Soil Sci. Soc. Am. Proc.* 33:707–11
95. Iandolo, J. J., Clark, C. W., Bluhm, L., Ordal Z. J. 1965. *Appl. Microbiol.* 13:646–49
96. Ivanovics, G., Marjai, E., Dobozy, A. 1968. *J. Gen. Microbiol.* 53:147–62
97. Jannasch, H. W. 1967. *Limnol. Oceanogr.* 12:264–71
98. Jannasch, H. W. 1968. *Appl. Microbiol.* 16:1616–18
99. Jensen, S., Jernelöv, A. 1969. *Nature (London)* 223:753–54
100. Johannes, R. E. 1968. *Advan. Microbiol. Sea* 1:203–13
101. Jones, D., Bacon, J. S. D., Farmer, V. C., Webley, D. M. 1968. *Antonie van Leeuwenhoek J. Microbiol. Serol.* 34:173–82
102. Jones, K. 1970. *Ann. Bot.* 34:239–44
103. Jones, K., Stewart, W. D. P. 1969. *J. Mar. Biol. Assoc. UK* 49:701–16
104. Juven, B., Samish, Z., Henis, Y. 1968. *Isr. J. Agr. Res.* 18:137–38
105. Kamalov, M. R., Kreines, R. Z., Ilyaletdinov, A. N. 1969. *Mikrobiologiya* 38:505–10
106. Kearney, P. C. 1967. *J. Agr. Food Chem.* 15:568–71
107. Kershaw, K. A., Millbank, J. W. 1970. *New Phytol.* 69:75–79
108. Kimura, Y., Miller, V. L. 1964. *J. Agr. Food Chem.* 12:253–57
109. Kok, L. T., Norris, D. M., Chu, H. M. 1970. *Nature (London)* 225:661–62
110. Konijn, T. M., van de Meene, J. G. C., Bonner, J. T., Barkley, D. S. 1967. *Proc. Nat. Acad. Sci. USA* 58:1152–54
111. Kosuge, T. 1969. *Ann. Rev. Phytopathol.* 7:195–222
112. Krywolap, G. N., Grand, L. F., Casida, L. E., Jr. 1964 *Can. J. Microbiol.* 10:323–28
113. Kuc, J. 1966. *Ann. Rev. Microbiol.* 20:337–70
114. Kuentzel, L. E. 1969. *J. Water Pollut. Contr. Fed.* 41:1737–47
115. Kuo, M.-J., Alexander, M. 1967. *J. Bacteriol.* 94:624–29
116. Lange, W. 1968. *Nature (London)* 215:1277–78
117. Lange, W. 1970. *J. Phycol.* 6:230–34
118. Libbert, E., Kaiser, W., Kunert, R. 1969. *Physiol. Plant.* 22:432–39
119. Libbert, E., Manteuffel, R. 1970. *Physiol. Plant.* 23:93–98
120. Lim, S. M., Paxton, J. D., Hooker, A. L. 1968. *Phytopathology* 58: 720–21
121. Linderman, R. G., Toussoun, T. A. 1968. *Phytopathology* 58:1571–74
122. Lippincott, B. B., Lippincott, J. A. 1966. *J. Bacteriol.* 92:937–45
123. Loosli, C. G. 1968. *Yale J. Biol. Med.* 40:522–40
124. Loutit, M. W., Brooks, R. R. 1970. *Soil Biol. Biochem.* 2:131–35
125. Lovrekovich, L., Lovrekovich, H., Goodman, R. N. 1970. *Can. J. Bot.* 48:167–71
126. Lukezic, F. L., DeVay, J. E. 1964. *Phytopathology* 54:697–700
127. Lumsden, R. D., Bateman, D. F. 1968. *Phytopathology* 58:219–27
128. Lycke, E., Norrby, R., Remington, J. 1968. *J. Bacteriol.* 96:785–88
129. Lyle, L. R., Jutila, J. W. 1968. *J. Bacteriol.* 96:606–8
130. Macko, V., Woodbury, W., Stahmann, M. A. 1968. *Phytopathology* 58:1250–54
131. MacRae, I. C., Castro, T. F. 1967. *Soil Sci.* 103:277–80
132. Major, R. T. 1967. *Science* 157:1270–73
133. Martin, M. M. 1970. *Science* 169: 16–20
134. Marx, D. H., Davey, C. B. 1967. *Nature (London)* 213:1139
135. Mathews, M. M., Krinsky, N. I. 1965. *Photochem. Photobiol.* 4: 813–17
136. Matsumura, F., Boush, G. M. 1967. *Science* 156:959–61
137. McLaren, A. D., Peterson, G. H. 1967. *Soil Biochemistry*. New York: Dekker
138. McLaren, A. D., Skujins, J. J. 1963. *Can. J. Microbiol.* 9:729–31
139. Miles, J. R., Tu, C., Harris, C. R. 1969. *J. Econ. Entomol.* 62:1334–38

140. Millbank, J. W., Kershaw, K. A. 1969 *New Phytol.* 68:721–29
141. Miller, R. H., Chau, T. J. 1970. *Plant Soil* 32:146–60
142. Mirchink, T. G., Kashkina, G. B., Abaturov, Yu. D. 1968. *Mikrobiologiya* 37:865–69
143. Mirocha, C. J., Christensen, C. M., Nelson, G. H. 1968. *Biotechnol. Bioeng.* 10:469–82
144. Moustafa, F. A., Whittenbury, R. 1970. *Phytopathol. Z.* 67:214–24
145. Muscatine, L. 1967. *Science* 156: 516–19
146. Muscatine, L., Karakashian, S. J., Karakashian, M. W. 1967. *Comp. Biochem. Physiol.* 20:1–12
147. Nash, R. G., Woolson, E. A. 1967. *Science* 157:924–27
148. Ogata, G., Bower, S. C. A. 1965. *Soil Sci. Soc. Am. Proc.* 29:23–25
149. Ohara, H., Hamada, M. 1967. *Nature (London)* 213:528–29
150. Ottow, J. C. G. 1969. *Z. Pflanzenernaehr. Bodenk.* 124:238–53
151. Ottow, J. C. G. 1970. *Z. Allge. Mikrobiol.* 10:55–62
152. Patrick, R., Crum, B., Coles, J., 1969. *Proc. Nat. Acad. Sci. USA* 64: 472–78
153. Paxton, J. D., Chamberlain, D. W. 1969. *Phytopathology* 59:775–77
154. Pengra, R. M., Cole, M. A., Alexander, M. 1969. *J. Bacteriol.* 97: 1056–61
155. Pfennig, N. 1967. *Ann. Rev. Microbiol.* 21:285–324
156. Potgieter, H. J., Alexander, M. 1966. *J. Bacteriol.* 91:1526–32
157. Prasad, I. 1970. *Can. J. Microbiol.* 16:369–72
158. Prasad, I., Pramer, D. 1968. *Phytopathology* 58:1188–89
159. Price, R. J., Lee, J. S. 1970. *J. Milk Food Technol.* 33:13–18
160. Pringle, R. B. 1970. *Plant Physiol.* 46:45–49
161. Rai, P. V., Strobel, G. A. 1966. *Phytopathology* 56:1365–69
162. Reiter, B., Oram, J. D. 1967. *Nature (London)* 216:328–30
163. Ribble, J. C. 1967. *J. Immunol.* 98: 716–23
164. Ribble, J. C., Shinefeld, H. R. 1967. *J. Clin. Inves.* 46:446–52
165. Richardson, D. H. S., Smith, D. C. 1968. *New Phytol.* 67:69–77
166. Roberts, D. S. 1969. *J. Infect. Dis.* 120:720–24
167. Roffman, B. 1968. *Comp. Biochem. Physiol.* 27:405–18
168. Roffman, B., Lenhoff, H. M. 1969. *Nature (London)* 221:381–82
169. Sakshaug, J., Sögnen, E., Hansen, M. A., Koppang, N. 1965. *Nature (London)* 206:1261–62
170. Schuette, H. R., Stephan, U. 1969. *Z. Pfanzenernaehr. Bodenk.* 123: 212–19
171. Seiji, M., Iwashita, S. 1965. *J. Biochem.* 57:457–59
172. Shindala, A., Bungay, H. R., Krieg, N. R., Culbert, K. 1965. *J. Bacteriol.* 89:693–96
173. Silver, W. S., Mague, T. 1970. *Nature (London)* 227:378–79
174. Silverman, M. P., Munoz, E. F. 1970. *Science* 169:985–87
175. Silvester, W. B., Smith, D. R. 1969. *Nature (London)* 224:1231
176. Sistrom, W. R., Griffiths, M., Stanier, R. Y. 1956. *J. Cell. Comp. Physiol.* 48:473–515
177. Sloan, P. R., Strickland, J. D. H. 1966. *J. Phycol.* 2:29–32
178. Smith, H. 1968. *Bacteriol. Rev.* 32: 164–84
179. Smith, H. C., Brown, H. E., Moyer, J. E., Ward, C. H. 1968. *Develop. Ind. Microbiol.* 9:363–69
180. Steyn, P. L., Delwiche, C. C. 1970. *Environ. Sci. Technol.* 4:1122–28
181. Stillwell, R. H. 1967. *J. Protozool.* 14:19–22
182. Stoessl. A., Unwin, C. H. 1970. *Can. J. Bot.* 48:465–70
183. Stotzky, G., Rem, L. T. 1966. *Can. J. Microbiol.* 12:547–63
184. Stotzky, G., Rem, L. T. 1967. *Can. J. Microbiol.* 13:1535–50
185. Strominger, J. L., Ghuysen, J.-M. 1967. *Science* 156:213–21
186. Takai, Y., Kamura, T. 1966. *Folia Microbiol.* 11:304–13
187. Tamura, S., Murayama, A., Hata, K. 1967. *Agr. Biol. Chem.* 31:758–59
188. Thayer, J. D., Samuels, S. B., Martin, J. E., Jr., Lucas, J. B. 1965. *Antimicrob. Ag. Chemoth. 1964,* 433–36
189. Tiedje, J. M., Duxbury, J. M., Alexander, M., Dawson, J. E. 1969. *J. Agr. Food Chem.* 17:1021–26
190. Tomeyami, K. et al. 1968. *Phytopathology* 58:115–16
191. Tonomura, K., Kanzaki, F. 1969. *Biochim. Biophys. Acta* 184:227–29
192. Torma, A. E., Walden, C. C., Branion, R. M. R. 1970. *Biotechnol. Bioeng.* 12:501–17
193. Trager, W. 1957. *Biol. Bull.* 112: 132–36

194. Troshanov, E. P. 1969. *Mikrobiologiya* 38:634–43
195. Tseng, T.-C., Bateman, D. F. 1968. *Phytopathology* 58:1437–38
196. Ulitzur, S., Shilo, M. 1970. *Biochim. Biophys. Acta* 201:350–63
197. Vasantharajan, V. N., Bhat, J. V. 1967. *Plant Soil* 27:261–72
198. Venkataraman, G. S., Neelakantan, S. 1967. *J. Gen. Appl. Microbiol.* 13:53–61
199. Visser, S. A., Theisen, A. A., Mehlich, A. 1965. *Soil Sci.* 100: 232–37
200. Voigt, G. K., Steucek, G. L. 1969. *Soil Sci. Soc. Am. Proc.* 33:946–49
201. Wang, H. L., Ruttle, D. I., Hesseltine, C. W. 1969. *Proc. Soc. Exp. Biol. Med.* 131:579–83
202. Wang, T. S. C., Chuang, T.-T. 1967. *Soil Sci.* 104:40–45
203. Wang, T. S. C., Yang, T.-K., Chuang, T.-T. 1967. *Soil Sci.* 103: 239–46
204. Weed, S. B., Davey, C. B., Cook, M. G. 1969. *Soil Sci. Soc. Am. Proc.* 33:702–6
205. Whittaker, R. H. 1970. *Chemical Ecology*, ed. E. Sondheimer, J. B. Simeone, 43–70. New York: Academic
206. Witkamp, M. 1969. *Soil Biol. Biochem.* 1:177–84
207. Wolin, M. J. 1969. *Appl. Microbiol.* 17:83–87
208. Wood, E. J. F. 1965. *Marine Microbial Ecology*. London: Chapman and Hall
209. Wood, J. M., Kennedy, F. S., Rosen, C. G. 1968. *Nature (London)* 220: 173–74
210. Wright, R. T., Hobbie, J. E. 1966. *Ecology* 47:447–64
211. Yeoh, H. T., Bungay, H. R., Krieg, N. R. 1968. *Can. J. Microbiol.* 14:491–92
212. Zajic, J. E., Ng, K. S. 1970. *Develop. Ind. Microbiol.* 11:413–19
213. Zaki, M. A., Taylor, H. F., Wain, R. L. 1967. *Ann. Appl. Biol.* 59: 481–91

ION TRANSPORT BY ENERGY-CONSERVING BIOLOGICAL MEMBRANES

1576

Peter J. F. Henderson[1]

Institute for Enzyme Research, University of Wisconsin,
Madison, Wisconsin

Contents

[1] Present address: Department of Biochemistry, University of Leicester, Leicester LE 1 7RH, United Kingdom.

INTRODUCTION

In only one preliminary report has a "soluble" preparation from a biological source, *Thiobacillus novellus,* been shown to catalyze oxidative phosphorylation (45). In all other biological systems, the preservation of insoluble membranous vesicles has been necessary to demonstrate coupling of ATP synthesis to electron transport in vitro. Whether the vesicles be intact mitochondria, sub-mitochondrial particles, chromatophores derived from bacterial plasma membranes, or sub-chloroplast preparations from plants, gross "leakiness" or disruption of the vesicle membranes leads to loss of phosphorylating ability (75, pp. 6–7). Even when the integrity of these membranes appears to be undisturbed, ATP synthesis can be abolished by inducing active ion transport processes or by changing the passive permeability of the vesicles towards various ions; in particular, preservation of active proton transport appears to be correlated with the retention of phosphorylating ability, and some current research is guided by the perhaps desperate hope that these correlations may provide a clue to the mechanism by which the energy from electron transport is transferred to the phosphorylation of ADP.

It is the task of this review to explain the effects of cation-transporting antibiotics on such energy transfer systems, evidently a difficult one if the mechanism of energy transfer is itself not understood. Accordingly, it is proposed 1. to make an assumption that is felt to be sufficient to explain many or all of the antibiotic effects, and 2. to describe the compatibility of the assumption with the experimental observations and leave the reader to judge the veracity of the initial proposition.

The assumption is that a trans-membrane pH gradient or electrical potential, or both, is generated between electron transport and phosphorylation of ADP, as proposed by Mitchell (160, 162, 165). This "proton motive force," or PMF (162) may be the only high-energy intermediate required for ATP synthesis, or it may be on a sidepath of the energy transfer reactions, which is contrary to Mitchell's proposals because the PMF is then only incidental to the energy transfer process, not an obligatory part of it (75, 86, 235, 248). It may be on the direct path but result from chemical or conformational intermediates generated by electron transport, and it may generate chemical or conformational intermediates before phosphorylating

ADP; these proposals have been the basis of compromise positions (51, 74, 75, 272) between the "chemical," "conformational," and "chemiosmotic" hypotheses (247, 21, 160). Only the existence of the pH and electrical gradient is necessary to explain the antibiotic effects, and if other high-energy intermediates are found to exist, an important premise of the chemiosmotic hypothesis will cease to exist, but our assumption will not necessarily be invalidated. The evidence that electron transport-generated H^+ translocation is indeed associated with ATP synthesis will be presented for each biological system in turn.

I shall simplify the discussion by referring only to the antibiotic valinomycin (213) when it may be inferred that the actins (143) and enniatins (190, 244) exhibit essentially identical experimental effects, even though this has not necessarily been demonstrated for each of the phenomena associated with valinomycin. Also nigericin will be cited most frequently when the antibiotics dianmycin, X-537A, X-206, and monensin (86, 138, 143, 144) may be considered similar in effect; high concentrations of nigericin and X-537A do exhibit differences which will be referred to when relevant. Each of the nigericin types contains a single carboxyl group, and they will be referred to as the "monocarboxylic" antibiotics as a group, whereas valinomycin, enniatin, the actins, and gramicidin are the "neutral" antibiotics. Pressman (213, 218) has coined the now commonly used term "ionophore" to describe all antibiotics which induce cation permeability, and Skulachev (246) recently introduced the novel "protonophore" specifically for antibiotics that induce H^+ permeation. Normally, in the experiments considered, potassium will be the only alkali metal cation present in significant quantity, which is obviously a good practice for mimicking physiological conditions. However, in this treatment, any of the other alkali metal cations could substitute for K^+, provided that it is transported at rates comparable to K^+ by the antibiotic under discussion. A review of the cation selectivities has been published by Kinsky (138)—all the antibiotics of both the valinomycin and nigericin groups readily transport K^+ ions. Gramicidin will be used as a blanket term for gramicidins A, B, or C (D is a mixture of these) but does not include gramicidin S, which has distinctly different properties (244, 245) and is, in fact, a tyrocidine (214, 264).

The word "electrogenic" will be used to describe trans-membrane movement of an ion as a charged species so that an electrical potential across the membrane is generated (163, 165). The term "electro-impelled" will be employed to describe the converse process, namely that the ion is drawn across the membrane by an existing electrical potential so that the potential is diminished or even abolished. Only an ion transport process that is electrogenic in the absence of an imposed electrical potential can be electro-impelled by a potential. The term "model" membranes should be restricted to experiments on truly artificial membranes of the monolayer, bilayer, and liposome type (17, 138, 182), and also experiments measuring partition of ions into, or translocation across, organic bulk phases (59, 208, 213, 217,

218, 266). However, both erythrocytes and mitochondria have been utilized as model membranes by using suitable inhibitors to prevent their metabolic processes interfering with passive cation exchanges (see, e.g. 39), by the same token, passive cation exchanges can be measured in photosynthetic membrane systems in the absence of light. Accordingly, the term "artificial membranes" will be used to cover monolayers, bilayers, liposomes, and organic solvent systems; and "model membranes" to cover both these and also biological membranes that have been suitably treated to inhibit metabolic effects.

This fruitful field of inquiry was created by the theoretical and experimental concepts introduced by Mitchell (160, 162, 165) and by the discovery of antibiotic-mediated ion transport by Pressman (180, 211, 212).

The present review owes a great deal to the clear discussions of antibiotic effects on model membranes by Mueller & Rudin (182) and Kinsky (138), to the scrutinies of the chemiosmotic hypothesis published by Robertson (223) and Greville (75), and to reviews of bacterial (68, 86, 168), chloroplast (195, 265), and mitochondrial (142, 165) metabolism in relationship to antibiotic effects and the chemiosmotic theory. More specialized, but equally helpful, have been the reports of Shemyakin et al (245) and Ivanov & Shkrob (116) on antibiotic-induced ion transport across membranes.

THE CHEMIOSMOTIC VIEW OF ENERGIZED ION TRANSPORT

According to the chemiosmotic theory the primary function of oxidative or photosynthetic electron transport is the electrogenic translocation of protons across the membrane containing the electron transport chain (160, 162, 165; cf 75, 223). A similar electrogenic H^+ translocation is postulated to result from ATP hydrolysis, so that electron transfer reactions can reverse the ATPase, i.e., drive phosphorylation of ADP, via the "high energy intermediate" of the proton gradient. The third and fourth postulates require that the protons are prevented from re-equilibrating across the membrane except by stoichiometric exchange with a cation (antiport) or accompanied by a stoichiometric amount of anion (symport) (162, 163, 165, 168). If the rapidity of the latter processes approaches that of the initial electron transport driven H^+ translocation, the chemical potential of protons on each side will return to the initial value, i.e., the electron transport-imposed pH difference across the membrane approaches zero, but the membrane potential component retains a maximal value (160, 165). These processes are illustrated in Figure 1. If, at this point, a transportable cation is introduced on the positive side, or a transportable anion on the negative side, and if each crosses the membrane as the charged species (uniport), i.e., not accompanied by an oppositely charged "carrier," then the ion will be drawn down and will discharge the membrane potential; this will relieve the "back pressure" on the proton pump, and the primary H^+ extrusion will accelerate (162, 165, 166) see Figure 2. Provided that the H^+ re-equilibrating pathways

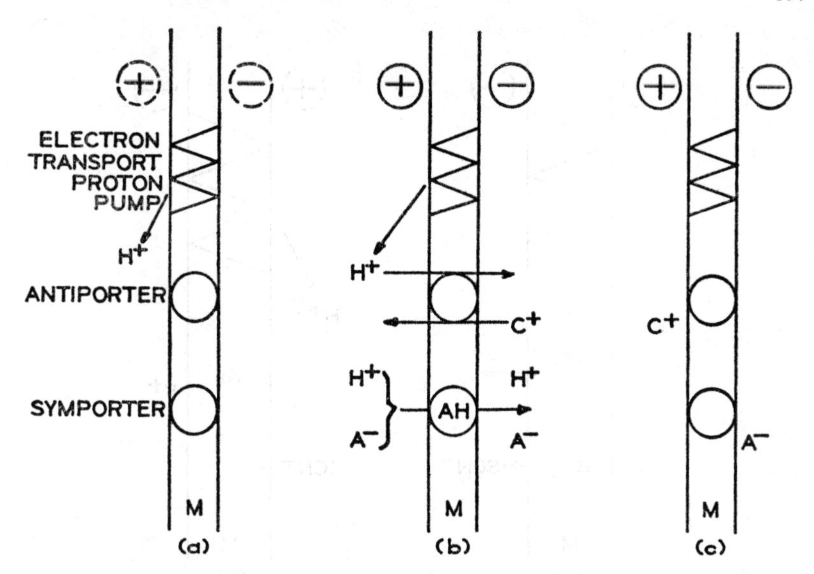

FIGURE 1. The generation of a trans-membrane electrical potential by proton transport. (a) The initiation of electron transfer by light or oxygen + oxidizable substrate causes a proton to be transported across the membrane, thereby setting up an electrical and pH gradient. (b). Specific translocation systems re-equilibrate the protons in a nonelectrogenic fashion. (c). The "back pressure" from the membrane potential and the enhanced electrochemical gradients of the transported ions prevents further proton ejection; the total electrical potential is conserved with only a minimal pH gradient. Symbols used: M = membrane phase; C^+ = cation; A^- = anion. A circle around the positive or negative sign indicates the presence of a maximal electrical potential; a broken circle indicates a submaximal potential. For further explanation, see references 162, 165 and the text.

do not increase in activity, a new equilibrium will be reached when the enhanced electrochemical gradient of the transported ion causes sufficient "back pressure" to prevent further increase of the pH gradient (165). Thus, three limiting situations may arise after the initial "pumping" of protons across the membrane by the electron transfer reactions:

1. The presence alone of systems for catalyzing proton transport back across the membrane, but in an electrically neutral manner by virtue of cation antiport or anion symport, maximizes the membrane potential and minimizes the pH gradient.

2. The electro-impelled penetration of ions as the charged species minimizes the membrane potential and maximizes the pH gradient.

3. Both effects together diminish the algebraic sum of pH + electrical gradient towards zero.

If the sum of these two components, i.e., the total "proton motive force" (162) is the "high-energy intermediate" for ATP synthesis, then only the

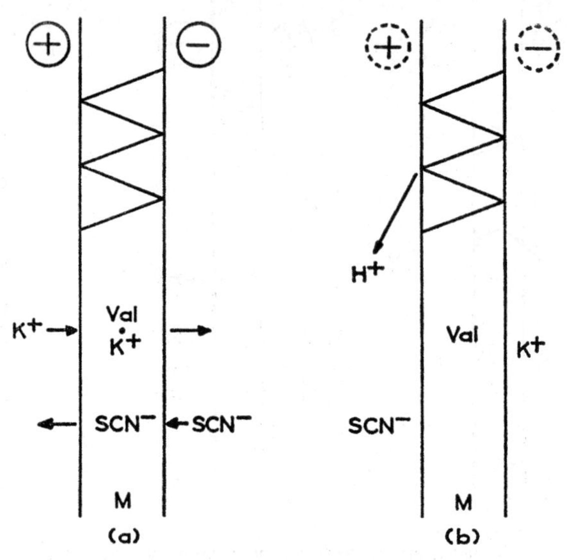

FIGURE 2. The replacement of an electrical potential difference by an enhanced pH gradient as a result of electro-impelled ion transport. (a). An electrical potential has been generated as in Figure 1. K^+ added to the positive side (in the presence of valinomycin) or SCN^- added to the negative side are drawn down the electrical gradient. (b). This results in a dissipation of the electrical potential —partial or complete—and the resulting relief of the "back pressure" allows the proton pump to enhance the pH gradient. N.B. a dynamic equilibrium is achieved in which the translocases of Figure 1 also participate.

third case will lead to uncoupling of phosphorylation from electron transport. Of course, if a single added agent accomplishes translocation of both electrical charge and H^+, then uncoupling will not require two synergistic agents (160).

Therefore, our approach to considering the effects of induced ion transport on ATP synthesis will be as follows. First, does a particular permeability-inducing agent catalyze electrical charge transferring or electrically neutral ion translocation in model membrane systems? Second, does the agent translocate protons, and if so, is the translocation electrical? Third, are the permeation characteristics induced by a particular agent the same on both model and energy-conserving membrane systems? In this regard it will be necessary to catalogue the inherent permeability properties of each energy-conserving system, so far as they have been examined. Fourth, assuming that the answer to the third query is "yes" (hopefully supported by experimental evidence!), do the observed effects of each agent on ion transport and phosphorylation reflect the presence of a pH or electrical gradient, or both, across the biological membrane; also, does the accompanying disappearance or preservation of ATP synthesis correlate with the disappearance or presence of the pH or electrical gradient? The experimental investigation of these points is facilitated by the ability to measure a pH gradient, and is hampered by an inability to measure the trans-membrane electrical gradient because of the small size of the energy-conserving vesicles (see 257, discussed in 246 and 253). Note that the assumption that ionophorous antibiotics have identical effects on both model and ATP-synthesizing membranes is not always explicitly expressed in chemiosmotic interpretations of experimental effects (but see 165).

THE CHARACTERISTICS OF ANTIBIOTIC-INDUCED CATION TRANSPORT ACROSS MODEL MEMBRANES

Charge Transfer

When valinomycin was added to an aqueous phase containing K^+ and separated from another aqueous phase without K^+ by an hydrophobic barrier, an electrical potential difference appeared (149, 181, 250, 252). It was proposed (218) that valinomycin solubilized a K^+ ion in the hydrophobic membrane interior as a 1:1 antibiotic-cation complex, a proposal that has received much additional support (3, 42, 58, 115, 189, 210, 220, 221, 252, 256, 269; reviews 86, 138). Solubilization allows the diffusion of K^+ down the concentration gradient, but movement of only a minute amount produces an opposing electrical charge (diffusion potential, 49, 93) sufficient to prevent the bulk transfer of K^+. However, net cation transfer will occur if one of three conditions are met to preserve electrical neutrality on each side of the membrane (see also 11): (*a*) an anion is transported in the same direction as the cation (49, 93); (*b*) a cation is transported in the opposite direction (163, 259); (*c*) an electrical potential opposite in polarity to that resulting from cation transport is applied (49, 93, 163).

Using a variety of model membranes, valinomycin-catalyzed net cation transport has been shown to occur with each of these conditions. Thus, Pressman and Eisenman and their co-workers (59, 208, 213, 216, 218, 219) demonstrated that valinomycin and the actins catalyzed net cation transfer with condition (*a*); McGivan (see 40, 41), Harris & Pressman (98), and Mitchell & Moyle (171) with condition (*b*); and Pressman (217) and Wipf & Simon (270) with condition (*c*). We now make the important additional assumption that valinomycin induces net cation transport in biological energy transfer systems only under the same rules. This contrasts with the proposal of Pressman and co-workers (213, 214) that valinomycin acts by increasing the supply of K^+ to an energy-linked K^+ pump, at least in mitochondria.

At concentrations which profoundly affect mitochondrial metabolism (143, 144, 218), nigericin did not reduce the resistance of artificial membranes (217, 218, 256). Added permeant anions were not necessary for nigericin to promote net translocation of K^+, but translocation of a cation in the opposite direction always occurred in both artificial systems (4, 104) and across biological membranes (98, 103, 104), and it was effective only above a pH of about 6 (218)—all the nigericin class of compounds are weak carboxylic acids (143). The conclusion drawn by Pressman (213, 218; cf 165) and now generally accepted is that the nigericin anion forms a 1:1 complex with K^+ (129, 207) and that the neutral complex, but not the free anion, traverses the hydrophobic barrier. Thus, net K^+ transport by nigericin does not require another anion, nor does it influence or initiate an electrical potential. In order for the negatively charged nigericin to continue "ferrying" K^+, it must be neutralized for the return trip, by either a second alkali cation or a proton from the aqeuous bulk phase (4, 98, 103, 104, 165).

Two incidental reports indicate that significant deviation from this ideal exchange diffusion (259) behavior occurs at high concentrations of nigericin (greater than about 1 $\mu g/ml$ or 0.15 μM) when it decreases the resistance of bilayer membranes (182, 256). This phenomenon was predicted on chemiosmotic grounds by Henderson et al (104), and will be discussed in more detail. Experiments on bilayers in this laboratory have shown that nigericin X-464 (identical to nigericin, 143, 251) and X-537A, but not dianmycin, monensin, or X-206 lower the resistance of bilayer membranes, and can generate bi-ionic potentials across the membrane, i.e. nigericin and X-537A appear to promote electrogenic cation transport but only at high concentrations. The cation selectivity for the effect is $K^+ > Rb^+ > Na^+ > Cs^+ \simeq Li^+$ for nigericin, while selectivity is much less marked for X-537A; this corresponds very well with their selectivity patterns derived from partition experiments (213, 219). On bilayers, 1–2 × 10^{-7} M nigericin or X-537A was necessary for a detectable decrease of membrane resistance and about 100 times this for maximal effect.

Proton Translocation

It is firmly established that uncoupling agents such as dinitrophenol, tetrachlorosalicylanilide, substituted carbonyl cyanide phenylhydrazones, and dicumarol electrically transport protons, but not other cations, across bilayer, liposome, erythrocyte, mitochondria, chloroplast and bacterial membranes (reviewed in 86, 138, 168, but see 29). Valinomycin, the actins, and enniatins transport some of the alkali metal cations, but not protons at physiological pH values (104). Gramicidin does transport protons but less efficiently than it transports K^+ (39, 104, 173). By analogy with gramicidin's transport of cations (181) and the similarity of its H^+ translocation to that by uncouplers (173), it is reasonable to assume that the gramicidin-induced H^+ translocation is electrogenic and can be electro-impelled. Experiments that measure the pH dependence of gramicidin's ability to lower bilayer membrane resistance (by analogy with those of Lehninger and co-workers, 18, 108) should provide a firmer experimental basis for this assumption.

The ability of gramicidin to transport protons, its linear molecular structure (264), and its failure to give satisfactory results in cation partition experiments (266) are all differences between gramicidin and the other neutral ionophorous antibiotics. Perhaps the explanation of its singular behavior lies in the recent report (258) that it can adopt a helical conformation capable of interacting with cations.

All members of the monocarboxylic class of antibiotics transport protons with at least equal facility to their induced cation transport (103, 104). Their failure to lower membrane resistance at concentrations sufficient for rapid cation transport clearly indicates that the H^+ translocation is nonelectrogenic. At higher concentrations of nigericin and X-537A, it may be that electrogenic H^+ transport occurs, although the alkali metal cation requirement for their effects on bilayer membranes indicates it does not (Henderson, unpublished observations).

To paraphrase the differences which will be most significant in the ensuing discussion: valinomycin transports electrical charge but not protons; nigericin transports protons, but not charge; and gramicidin and the classical uncoupling agents transport both charge and protons.

Special Case of Ammonia and Low Molecular Weight Amines

As long ago as 1911, Harvey (99) demonstrated that primary, secondary, and tertiary ammonium salts caused rapid alkalinization of the interior of plant and animal cells when strong bases—NaOH, KOH, or quaternary amines—were ineffective. Jacobs (121) and Jacobs & Parpart (123) extended these results and concluded that diffusion of neutral NH_3 across the membrane was responsible for the apparent alkali transport (122, 123). The mechanism is illustrated in Figure 3—the net effect is nonelectrogenic H^+ translocation across the barrier toward the side on which the NH_4^+ salt was added. It will be apparent from Figure 3 that NH_4^+ ions will only discharge a pre-existing pH difference when added to the alkaline side. Considering the general acceptance of this mechanism and the wide use of ammonium salts in studying uncoupling phenomena, especially in photosynthe-

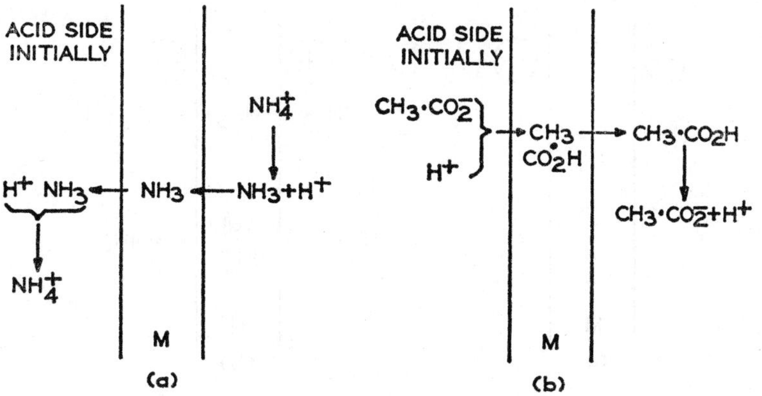

FIGURE 3. The dissipation of a previously established pH gradient by (a) NH_4^+ ions added to the alkaline side, and (b) acetate ions added to the acid side.

sis, it is unfortunate that this early evidence for the mechanism of NH_4^+ translocation has been largely ignored.

From studies on mitochondrial and erythrocyte membranes it appears that gramicidin, the actins, and valinomycin (descending order of effectiveness) catalyze NH_4^+ translocation (38, 39, 104). A single report indicates that the nigericin-type antibiotics do not promote ammonium transport (207). This means that valinomycin and the actins could catalyze H^+ transport (Fig. 4) especially in the presence of a membrane potential (positive on the acid side), provided that a catalytic amount of ammonium is present. In the absence of a potential, an accompanying anion or exchanging cation will be necessary. Substituted primary, secondary, and tertiary amines could substitute for NH_4^+, provided only that the neutral amine could diffuse through the membrane, and that the protonated species combined with the antibiotic. Thus, gramicidin, but not valinomycin, nonactin, or dinactin, is effective with methylamine (39, 104). It should be noted that valinomycin can discharge both a membrane potential and a pH gradient in the presence of NH_4^+ ions (Fig. 4) (cf 176, 179), and that there is then no restriction as to which side of the membrane the NH_4^+ ions must be added.

PENETRATION OF IONS ACROSS MEMBRANES IN THE ABSENCE OF ANTIBIOTICS

Nonspecific Permeation

Nonspecific permeation is translocation that can be expected to occur across any membrane system, for example, ammonium ion movement by the mechanism described in the last section. Skulachev, Liberman and their coworkers selected or designed a series of compounds possessing a single charge, either positive or negative, and sufficient hydrophobicity to solubilize the charge in a variety of model membranes, as evidenced by their abil-

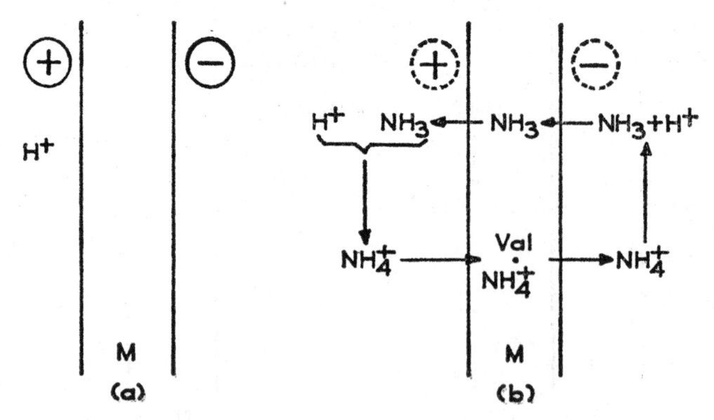

FIGURE 4. The simultaneous dissipation of an established pH gradient and an established electrical potential. (a) The initial situation; (b) after addition of valinomycin and NH_4^+ ions.

ity to induce trans-membrane potentials across artificial membranes, like valinomycin in the presence of K^+, or uncouplers with H^+ (150–152, 246). The cations are N,N-dibenzyl N,N-dimethyl ammonium, tetrabutylammonium, and triphenyl methyl phosphonium, all of which were accumulated by mitochondria in an energy-dependent fashion, causing H^+ extrusion and swelling, i.e., identical effects to those given by K^+ and valinomycin, or Ca^{++} (13). Also, the anions phenyldicarbaundecaborane (76), tetraphenylboron (76, 176, 178, 179), picrate (58, 59, 151, 176, 178), thiocyanate (175, 213, 218), and iodide (126, 127, 145, 151) were shown to undergo electrogenic and electro-impelled translocation, so that net anion movement required the presence of an electrical potential (Fig. 2), an accompanying cation, or an exchanging anion. The application of all these compounds to energy-transfer systems will be described below.

Special Case of Weak Acids

Neutral acetate salts are able to cause a rapid apparent uptake of acid into cells of higher plants and animals (49, 121, 123). The postulated mechanism is shown in Figure 3, and involves penetration of acetic acid (neutral acids are more soluble in a hydrophobic environment than their dissociated salts), with release of a proton as the acidifying process on the inside (see also 39, 40). In the presence of a pre-existing pH gradient across a membrane, it can be seen that addition to the the acid side of acetate salts (at the pH of the acid side) will cause rapid dissipation of the gradient (38–40, 175) (Fig. 3), whereas addition of acetate (at the pH of the alkaline side) to the alkaline side will serve only to enhance the gradient.

Some Specific Permeability Properties

Na^+ ions appear to penetrate mitochondria much more readily than other alkali metals, especially in the presence of acetate anions, the probable mechanism being a Na^+/H^+ exchange (19, 24, 25, 173, 175). In the presence of EDTA, presumably sufficient to remove Mg^{++} bound to the outside of the membrane, the other alkali cations also penetrate, in the order $Li^+ > Na^+ > K^+ > Rb^+ > Cs^+$ (2, 8, 10, 52, 237). Since the Mg^{++} ions are not added during the normal isolation procedure, they are probably present under physiological conditions. Incidentally, reports that K^+ transport is obligatory for mitochondrial oxidative phosphorylation should be regarded with caution when EDTA has been present throughout the entire preparation procedure (73, 209, 228). EDTA is normally added only in the initial stages of mitochondrial isolation procedure in order to remove Ca^{++} ions; its proclivity to deplete the mitochondria of beneficial Mg^{++} can be avoided by using instead EGTA (85). Changing the pH outside the physiological range also appears to induce K^+ permeability in mitochondria (30, 31).

Ca^{++}, Mn^{++}, and Sr^{++} penetrate mitochondria rapidly under energized conditions (reviewed by Chance, 34; Lehninger, 147). Azzone & Piemonte (12) and Selwyn et al (236) have produced evidence that Ca^{++} transport is electrogenic in nonenergized mitochondria and therefore may be electro-impelled by energized mitochondria (162, 165, 167). It is noteworthy that Ca^{++} can alter the permeability properties of artificial membranes (28, 188, 204, 227). Scarpa & Azzone (230) invoke a specific carrier to explain mitochondrial Ca^{++} transport phenomena, but its operation seems compatible with a net electrogenic transfer of Ca^{++} ions. In general, the effects of Ca^{++} on mitochondria are very similar to those induced by valinomycin plus K^+, or the synthetic cations mentioned above (p. 402).

Cl^- ions appear to penetrate mitochondrial membranes at elevated pH levels (8, 12, 22, 23). The mechanism is not entirely clear—it may be by Cl^- /OH^- exchange (24) or by electrogenic Cl^- movement (24, 175). In chloroplasts, however, Cl^- movement occurs without any special conditions and is probably electrogenic (50, 195, 265); this also seems to be the mechanism of a specific translocation of NO_3 ions into sub-mitochondrial particles (44, 176) and beef heart mitochondria (24).

There is good evidence that mitochondria possess exchange-diffusion systems that catalyze uptake of substrate anions in exchange for internal anions (39–41; reviewed in 37, 139). In the case of P_i uptake, the process almost certainly involves symport with H^+, or antiport against OH^- (39–41, 37, 139; also 168, 175, 201, 203). However, it appears that nearly all of the other anions can enter as the free acid (197) as well as by the exchange mechanism, and could therefore also influence a pH gradient. Klingenberg and co-workers have demonstrated that uptake of ADP occurs in 1:1 exchange for ATP, but the excess negative charge on the latter renders the process electrogenic (139, 140). Hence, in mitochondria, at least, distribution of the substrates and products of oxidative phosphorylation can affect a trans-membrane pH or electrical gradient and vice versa (38, 39, 139, 162, 165, 169, 235). Specific translocation of anionic substrates has not been so thoroughly investigated in the case of chloroplasts (but see 101, 102) or bacteria, and should be borne in mind as a possible complication of our further discussions.

The response of ATP synthesis of bacteria, chloroplasts, and mitochondria to these ion transport processes will now be considered in the light of the charge- and proton-transporting characteristics of each agent, or combination of agents. As discussed earlier, it will be assumed that these characteristics are equally operative during energized ion transport by the biological membrane system.

THE RELATIONSHIP OF ION TRANSPORT TO ATP SYNTHESIS BY SOME BACTERIA

Antibiotic-Catalyzed Ion Movements in the Absence of ATP Synthesis

There is little doubt that uncoupling agents such as dinitrophenol, substituted carbonyl cyanide phenylhydrazones and salicylanilides, or dicumarol catalyze the electrogenic transport of protons across the cytoplasmic membranes of *Micrococcus lysodeikticus* (161), *Streptococcus faecalis* (89, 91,

92, 206), *Rhodospirillum rubrum* and *Rhodopseudomonas spheroides* (233), *Bacillus megaterium* (82–84) and several other bacterial species (reviewed in 86, 168). In detailed studies on *S. faecalis,* Harold and co-workers demonstrated that gramicidin, valinomycin, monactin, and nigericin, but not uncouplers, promoted passive alkali cation exchange across the plasma membranes (87, 88) with a cation selectivity of each antibiotic similar to that described for other membrane systems (138). Gramicidin and nigericin, but not valinomycin, enhanced proton translocation (89, 206) although it was earlier reported that valinomycin, and also monactin, induced an H^+/K^+ exchange (87, 88); perhaps H^+ permeation in these experiments occurred by other proton-conducting pathways in the membrane (see later discussion). Gramicidin, but not valinomycin, induced H^+ permeation across *B. megaterium* membranes and was also selective for Na^+ and Li^+ in comparison with valinomycin (82, 83). Unfortunately, nigericin was not tried in this system, so a complete comparison of electrogenic and protonophorus abilities cannot be made.

Jackson et al (120) measured antibiotic-catalyzed passive K^+ extrusion from chromatophores of *R. rubrum* suspended in medium of very low K^+ content in the dark (to eliminate the light-dependent H^+ pump). Valinomycin increased K^+ efflux, presumably accompanied by an anion because little H^+ uptake occurred; in the presence of an uncoupler, FCCP,K^+/H^+ exchange was rapid. Both gramicidin and nigericin catalyzed K^+/H^+ exchange without added uncoupler, thereby confirming their ability to translocate H^+ in comparison with valinomycin.

Interestingly, nigericin, gramicidin, and uncoupling agents, but not valinomycin or monactin, inhibited metabolic accumulation of K^+, P_i or alanine by *S. faecalis* (86, 89). The inference drawn was that induction of H^+ permeation through the membrane inhibited the activity of the K^+ pump of the organism (87–89); this led to discovery of an ATP-requiring outward secretion of H^+ and Na^+ as the *modus operandi* of the K^+ accumulation (89–92, 206). It should be emphasized that this H^+ pump is not related to oxidative phosphorylation (absent from this strain of *S. faecalis*), and I will not further discuss these interesting phenomena; their significance and appearance in other microorganisms have been discussed by Harold (86) and by Mitchell (168).

Hamilton (84) utilized salt-induced lysis of *S. aureus, B. megaterum, M. lysodeikticus, E. coli,* and *P. aeruginosa* protoplasts as a criterion for passive cation or anion permeation. Similarly to mitochondria (39, 40), the protoplasts swell rapidly in NH_4-acetate, indicating that rapid penetration of CH_3CO_2H neutralizes the pH change due to NH_3 penetration and vice versa (40; cf 273). Stimulation of lysis in NH_4NO_3, i.e., the salt of a strong acid, by uncoupler or gramicidin is probably due to their facilitation of electro-impelled H^+ or NH_4^+ transport to discharge the potential resulting from NO_3^- permeation (driven by the concentration gradient, 0.4 *M* NO_3^- outside the cells). The lesser ability of valinomycin to transport H^+ or NH_4 in comparison with gramicidin or uncouplers (104) would correlate with its lower lytic ability in NH_4NO_3 (84). As expected of electro-impelled K^+ transporters, valinomycin and gramicidin both induced maximal lysis in KNO_3, again assuming NO_3^- to be the penetrating species. Several experiments demonstrated that both valinomycin and gramicidin catalyzed K^+ transport, but only gramicidin transported Na^+. The results in chloride salts were more equivocal, possibly because of an unrecognized requirement of Cl^- ions to exchange with OH^- ions, a situation which is known to occur in erythrocytes (93, 229) and liposomes (158).

Up to this date, then, the characteristics of passive cation and proton permeation induced in bacterial membranes by valinomycin, monactin, gramicidin, nigericin, uncoupling agents, and NH_4^+ (273) are similar to those induced in model membranes (reviewed in 138). More experiments confirming these similarities and establishing the electrogenic nature of the antibiotic-induced ion transport would lay a firmer foundation for the experimental house already being constructed on the sand of experiments dealing with energy-requiring, "active" H^+ and induced cation transport by bacteria.

Light-Induced H^+ Transport by Bacteria

The term "chromatophore" is commonly used to describe membranous vesicles derived from certain photosynthetic bacteria (54, 67), so named because they contain the pigments that catalyze light-induced electron transport (64). Chromatophores also synthesize ATP and carry out "active" ion transport in response to light (reviewed in 67, 86), even though there are good topological and experimental reasons to believe that chromatophores are "inside out" with respect to the intact cell cytoplasmic membrane from which they are derived (54, 164, 165, 233). Furthermore, the discovery that illumination of chromatophores (from *R. rubrum*) resulted in a proton uptake which slowly reversed in the dark (36, 164, 261–263), showed them to be an ideal system for examining the validity of the chemiosmotic hypothesis (75, 262). The internal acidification is prevented by the inhibitors of electron transport, antimycin and hydroxyquinoline N-oxide (14, 186, 261, 262), and uncoupling agents dissipate the H^+ gradient coincident with their inhibition of ATP formation (20, 120), so the pH gradient can be inferred to be closely related to photophosphorylation, if not an actual intermediate (discussed in 75). The conclusion that formation of the H^+ gradient is too slow to be a high-energy intermediate (36, 186) is not necessarily valid because development of an electrical potential will prevent release of H^+ ions to reach a measuring electrode (165), and because the bromthymol blue method of measuring H^+ secretion is subject to uncertainties (78).

The H^+ translocation can also be demonstrated in intact cells of *R. rubrum* suspended in media of high ionic strength (55, 56, 164, 262), but the polarity is reversed—H^+ is secreted outward—and again the susceptibility of the intact cell H^+ secretion to inhibitors indicates a role in photophosphorylation. However, Edwards & Bovell (55, 56) were not able to rule out Na^+

or Cl^- transport (utilizing energy supplied by a separate system) being the main physiological function of the primary H^+ secretion, similar to *S. faecalis* (see p. 405).

I shall now summarize the effects of the ionophorous antibiotics on H^+ and K^+ transport and ATP synthesis induced by light in chromatophores, bearing in mind the characteristics of each antibiotic on H^+/K^+ exchange in the nonenergized system (i.e., in the dark) that have already been described.

Effect of antibiotics on light-induced pH gradient.—Valinomycin enhanced the rate and extent of light-induced H^+ uptake (120, 187, 233, 255, 263), and its effectiveness was increased by raising the external (and probably internal) K^+ concentration (120, 262). Its failure to discharge the pH gradient reflects the inability of valinomycin to transport protons, while its facility in enhancing the gradient indicates that it discharges an electrical disequilibrium set up by the H^+ translocation; consistent with this is the observation that in the presence of valinomycin and at low external K^+ concentrations, light induces a K^+ extrusion opposite in direction to, and presumably electro-impelled by, the enhanced H^+ uptake (120). These observations, therefore, suggest that the light-induced H^+ translocation by chromatophores is electrogenic.

Supporting this is the observation that both gramicidin (120, 263) and uncoupling agents (120, 186) prevented H^+ gradient formation, or accelerated dissipation of an established gradient, and added K^+ had little influence on the gramicidin effect. In the absence of other ion movemements, the established ability of these agents to catalyze electro-impelled proton transport would lead them to dissipate readily the chromatophore pH gradient if the original light-induced H^+ translocation was itself electrogenic.

Despite its ability to promote H^+ transport, nigericin did not collapse the light-induced H^+ gradient unless the external KCl concentration exceeded that inside (120, 185, 243, 255). At each KCl concentration the extent of H^+ equilibration was balanced by an equal K^+ uptake (120). Clearly, because nigericin at the concentrations used does not transport ions as the charged species it is unable to translocate H^+ outward continuously unless an external cation is available to neutralize the negatively charged nigericin during its return to "pick up" more protons; also, net transport of this cation will be reduced if the opposing cation concentration gradient is greater than the "driving" H^+ gradient (120, 165). Thus, with sufficient added KCl, nigericin discharges the chemical gradient of protons, but is unable to affect any electrical imbalance.

Valinomycin abolished or reversed the K^+ uptake promoted by nigericin, presumably by electro-impelled outward K^+ transport, without affecting the ability of nigericin to dissipate the H^+ gradient, and nigericin prevented the enhancement of the pH gradient by valinomycin (120, 255). In both cases, a maximal synergistic effect required the amounts of KCl necessary for nigericin to dissipate the pH gradient (120). The relevance of this synergism to the chemiosmotic hypothesis should become apparent in the next section.

Effects of antibiotics on bacterial photophosphorylation.—Valinomycin inhibited somewhat variably the incorporation of P_i into ATP in response to light. When the chromatophore electron transfer reactions were inhibited by hydroxyquinoline N-oxide and reactivated by phenazine methosulfate, valinomycin did not inhibit phosphorylation at all (120, 262), but in the absence of these it inhibited about 20 percent (120, 255) and with succinate as electron donor about 50 percent (14, 262). Thus, valinomycin is not a potent uncoupler of photophosphorylation. Agents which induce electro-impelled H^+ transfer—the classical "uncoupling" agents and also gramicidin—readily prevented light-induced formation of ATP by chromatophores (15, 20, 137, 263).

On its own, nigericin did not affect ATP synthesis, even with high KCl concentrations (120, 135, 243, 255). This means that replacement of the pH gradient by a K^+ gradient does not lead to loss of phosphorylation efficiency. However, nigericin and valinomycin together prevented phosphorylation (120, 255) provided that (*a*) the same sufficiency of KCl was present as that required for nigericin to discharge the pH gradient completely; (*b*) valinomycin was sufficient to discharge the K^+ accumulated as a result of nigericin addition (120, 255). The failure of Nishimura & Pressman (187) to obtain complete synergistic uncoupling by valinomycin plus nigericin was probably due to their use of insufficient amounts of each antibiotic to fulfill conditions (*a*) and (*b*).

The chemiosmotic theory proposes that the high-energy intermediate between electron transport and phosphorylation is a trans-membrane potential accompanied by a pH difference (162, 165):

$$\Delta p = \Delta\psi - Z\,\Delta pH$$

Δp = proton motive force
$\Delta\psi$ = trans-membrane electrical differential
ΔpH = trans-membrane pH difference (negative because $pH = -\log[H^+]$ see ref. 75)
$Z = 2.303\ RT/F$

It is further suggested that discharge of $\Delta\psi$ will cause an enhancement of Δ pH so that Δp tends to remain constant as long as electron transport can accelerate to compensate for the rate of discharge of $\Delta\psi$ (Fig. 2) (75, 162, 165). Therefore, in the chromatophore experiments, the known ability of valinomycin to discharge a membrane potential and not translocate H^+ would result in its enhancing the pH gradient if Δp is not to decrease. This is observed experimentally. The enhanced pH gradient could still be utilized for ATP synthesis, and hence phosphorylation would be relatively unaffected, as observed experimentally. The appearance of some uncoupling by valinomycin could reflect the leakiness of a particular preparation toward protons, or in the succinate system (14, 77) could result from the ability of succinate anions to associate with H^+ ions and transport them by diffusion of the free acid across the membrane.

Even where sufficient K^+ was available for nigericin to dissipate any chemical component of the gradient of H^+ ions completely, it would not be able to affect the electrical component of the original H^+ translocation, and should not therefore affect Δp, as reflected by its failure to prevent phosphorylation. If valinomycin was added after the nigericin, the K^+ accumulated by nigericin-induced K^+/H^+ exchange could be transported outward down the membrane potential. Now, both membrane potential and pH gradient would be dissipated, and ATP synthesis should be abolished. Experimentally, it is observed that valinomycin does discharge the K^+ accumulated by nigericin in negating the pH gradient, and that both agents together, but not separately, uncouple ATP synthesis from electron transport (cf ref. 75, p. 61). Also, gramicidin, which can discharge both a membrane potential and an H^+ gradient, uncouples on its own. Thus, the physical characteristics of the antibiotic-induced ion transport are consistent with their uncoupling ability if energy is conserved in chromatophores mainly as a membrane potential rather than a pH gradient (120, 165).

Fleischman & Clayton (66) suggested that a light-dependent absorption change of the carotenoid pigments of *R. rubrum* had kinetic characteristics and responded to energy transfer inhibitors as if it were a high-energy intermediate (see ref. 75, p. 68). An absorption change of the carotenoid identical to the light-induced change was observed by Jackson & Crofts (118, 119), when they induced artificial membrane potentials, positive inside the chromatophore, with pulses of H^+ in the presence of uncoupler, or pulses of K^+ in the presence of valinomycin. This observation lends support to the thesis that light induces a membrane potential, positive inside, and that it is to the potential that the carotenoid responds (118, 119; cf 185). A similar effect has been reported in chloroplasts (100, 105, 131, 225, 271; reviewed in 75, 265). A direct generation of membrane potential without chemical or conformational high-energy intermediates and its direct interaction with the carotenoid is the simplest interpretation—á la chemiosmotic theory—but the additional presence of chemical or conformational intermediate states cannot be ruled out (75).

Effects of other ions on photophosphorylation.—5 m*M* NH_4Cl prevents light induction of a pH gradient across the chromatophore membrane (27) but does not uncouple photophosphorylation (27, 109, 179). However, the addition of valinomycin as well as NH_4^+ ions completely prevented phosphorylation (136, 179). The rationale would be that, in the absence of valinomycin, NH_4^+ dissociates to acidify the external environment—alkaline with respect to the chromatophore interior by virtue of the light-driven H^+ pump—and then the neutral NH_3 produced diffuses across the membrane to associate with internal protons, so decreasing ΔpH as observed experimentally, but presumably not affecting the membrane potential (Fig. 3). Provided that $\Delta\psi$ is the major component of the proton motive force, NH_4^+ would not uncouple. However, added valinomycin would transport the internal NH_4^+ ions outwards, diminishing $\Delta\psi$ and setting up a cycle of NH_3 and NH_4^+ diffusion capable of discharging the total proton motive force (Fig. 4), and so producing the observed uncoupling (46, 47, 179). This explanation for uncoupling by NH_4^+ in the presence of antibiotics was first advanced by Chappell & Crofts (38, 39) in their studies of energy conservation by mitochondria.

A reasonable induction from these experimental observations is that combinations of any H^+ transporting agent with any charge-transporting agent will bring about uncoupling. This is also a corollary of the chemiosmotic hypothesis (75, 160, 162, 165), as discussed in the first part of this paper, that has been extensively pursued by Montal et al (176–179) and Skulachev and co-workers (reviewed by Skulachev, 246). Thus, using anions capable of electro-impelled charge transport (p. 402) Isaev et al (114) demonstrated that phenyl dicarbaundecaborane (PCB^-) uptake by chromatophores was stoichiometric with light-induced H^+ uptake and concluded that movement of PCB^- and also tetraphenyl boron$^-$ and I^- occurred in response to a light-induced membrane potential, positive inside. In separate experiments, tetraphenyl boron, and to a lesser extent, picrate, both of which are also lipid-soluble anions capable of discharging a trans-membrane potential (p. 402), were potent uncouplers of photophosphorylation only in the presence of either NH_4^+ or nigericin, i.e., H^+ transporters which fail to affect an electrical imbalance and do not uncouple by themselves (179). Cl^-, NO_3^- and HCO_3^- were not effective in such synergistic uncoupling (27, 179) but 5 m*M* SO_4^- was about as efficient as .05 in m*M* picrate (179). By corollary, one may deduce that SO_4^-, but no Cl^-, NO_3^- or HCO_3^- undergoes electro-impelled transport across the chromatophore membrane, but the more important deduction is that the chemiosmotic theory provides an excellent rationale for synergistic uncoupling by these reagents.

Some preliminary experiments have established that the reversal of photophosphorylation by chromatophores, i.e., ATP hydrolysis in the dark, also generates an H^+ gradient as required by the chemiosmotic hypothesis. Consistent with the light-induced gradient, it is acid inside, enhanced by valinomycin, dissipated by uncoupling agents (233), and is capable of driving PCB^- accumulation, which indicates the presence of a membrane potential, positive inside (114). Continuous hydrolysis of ATP, i.e., uncoupling, was obtained with gramicidin and with nigericin or NH_4^+ plus valinomycin, but not with each of nigericin, NH_4^+, or valinomycin on its own (136, 262); this is consistent with the H^+ and charge-conducting properties of these agents, as already described.

Aerobic transport of H^+ by bacteria.—Addition of O_2 to anaerobic suspensions of intact cells of *M. lysodeikticus* (161), *Micrococcus denitrificans* (164, 232, 233), and *R. rubrum* and *Rhodopsedomonas spheroides* (233, grown aerobically in the dark to prevent formation of the photosynthetic apparatus) results in a small H^+ secretion. With the last two organisms, valinomycin enhanced the O_2-induced pH gradient, and DNP or FCCP accelerated its decay (164, 233), hence the H^+ extrusion is probably electrogenic.

In more detailed studies on *M. denitrificans,* valinomycin, gramicidin, or SCN^- all enhanced the H^+ extrusion, and gramicidin appeared to be most effective of these in accelerating decay, especially at higher concentrations (232). The enhancement by gramicidin may be contrasted with its dissipating effect in chromatophores (120, 263) and probably occurs because its K^+-transporting ability is much greater than its H^+-transporting ability at the concentration of antibiotic used (89, 104). FCCP was most effective in accelerating decay (231, 232). Thus, agents catalyzing electro-impelled ion transport enhance secretion of H^+, and those catalyzing electro-impelled proton transport (gramicidin and FCCP) most efficiently accelerate decay, supporting the conclusion that the O_2-induced H^+ movement is electrogenic (164, 232). However, these experiments are not as clear-cut as those on mitochondria (171–173) because the rate of H^+ decay was not always exponential, and also a variability with pH and time of prior incubation with the different agents was observed (232). These difficulties could possibly be overcome by use of lysozyme- and tris-EDTA-treated cells (148) to facilitate penetration of the agents to the plasma membrane; for example, tris EDTA lysozyme treatment was necessary to obtain a rapid effect of uncouplers and antibiotics on the cell membranes of *S. faecalis* (206), and valinomycin and tetrachlorosalicylanilide were much more effective on protoplasts than on intact cells of *B. megaterium* (82, 83).

Reports on the properties of subcellular membrane systems that are capable of oxidative phosphorylation (57, 113, 128, 130) lead one to expect in the near future a proliferation of publications on the relationships of bacterial ion transport to the mechanism of phosphorylation.

LIGHT-INDUCED ION TRANSPORT AND PHOSPHORYLATION BY CHLOROPLASTS

Proton Transport

This subject and its relationship to the chemiosmotic theory have been expertly described in several reviews (72, 75, 124, 194, 265), perhaps in most detail by Packer et al (195), and only those points bearing on the previous arguments will be emphasized here.

First, there is a light-induced inward translocation of protons by chloroplasts (124, 125, 184) leading to a pH differential of 2.5–3 units across the grana membrane (reviewed in 75, 265). Both artificial and light-induced pH gradients are intimately related to ATP synthesis (46, 117, 124, 159, 184, 234, 241; reviews, 75, 132, 195, 265), and this evidence is excellent support for Mitchell's suggestion that the pH gradient is the high-energy intermediate (162, 165). Nevertheless, it must still be regarded as equivocal, because undetected chemical or conformational high-energy intermediates related to all three of the electron transport, H^+ transport, and phosphorylation processes could still be operative (75), and because the pH gradient may not be generated at a sufficient rate to be the actual high-energy intermediate (reviewed by Packer et al, 195). The light-induced uptake of protons is accompanied by a stoichiometric amount of Cl^- ions (50, 75), and there is not a stoichiometric exchange with cations under the usual experimental conditions (265). Thus, anion symport seems to neutralize any major electrical component of the H^+ translocation, but it is possible that the anion movement lags behind sufficiently for there to be a significant electrical gradient, especially in the initial phase of H^+ movement (131, 225, 271; reviewed in 75, 195, 265). From the chemiosmotic viewpoint, the characteristics of photophosphorylation by isolated chloroplasts indicate that the high-energy intermediate of ATP synthesis is a trans-membrane pH gradient, acid inside the grana with only a small, possibly zero, contribution by an electrical potential gradient (124, 162, 165). We shall now examine other ion transport effects to analyze their consistency with this proposal.

Effects of Ionophorous Antibiotics

Valinomycin renders the grana membranes permeable to K^+ in the dark, but it does not increase or decrease the light-induced H^+ translocation (192, 193, 238, 240) or significantly uncouple electron transport from phosphorylation (5, 16, 72, 133, 134, 136, 265; but see 226). Thus, from our previous arguments, it can be inferred that the chloroplast pH gradient is not accompanied by a significant electrical potential.

Gramicidin, by contrast, readily collapses the chloroplast pH gradient and uncouples phosphorylation from light-induced electron transport (80, 81, 238–240). However, it has been stated that H^+ transport by gramicidin is very probably electro-impelled and apparently there is no membrane potential in chloroplasts to facilitate this H^+ transport. (In very short-term experiments, a sufficient potential may exist to facilitate the gramicidin effect (271).) Nonetheless, there is sufficient alkali metal cation under the experimental conditions (see, for example, 238) for gramicidin to catalyze exchange diffusion of H^+ with cation (134) as it has been demonstrated to do on erythrocytes (39) and liposomes (104)—assuming the absence of an electrical potential in such model membranes. This antiport explanation for the ability of gramicidin to transport H^+ in the apparent absence of an electrical potential will be referred to again below.

As was the case with chromatophores, nigericin dissipated the light-induced pH gradient of chloroplasts provided there was sufficient external alkali metal cation to be transported inward (192, 193, 238, 240) but, unlike its failure on chromatophores, nigericin was a very potent uncoupler of photophosphorylation by chloroplasts (80, 238–242). Clearly, if an electrical potential was not part of the high-energy intermediate, nigericin could be an uncoupler by virtue of its ability to discharge the pH component. Thus, under conditions where a nigericin-type antibiotic had only partial or no effect on the pH gradient—due to provision of a cation for which it had a low affinity—the phosphorylation of ADP was affected to an identical degree (238). But, because the ratio of net ΔK^+/net ΔH^+ varied between 6–10, Shavit et al (238) surmised that stoichiometric K^+/H^+ exchange was not responsible for discharging the pH gradient. However, in experiments on nigericin-induced passive K^+/H^+ exchange across mitochondrial, erythrocyte, and liposome membranes, it was apparent that the initial rates were nearly always identical, i.e., $K^+/H^+ = 1$, while the ratios of net exchanges were always greater (see, for example, Figure 7 of ref. 104; also, 98). This increase in net ΔK^+/net ΔH^+ versus initial rate ΔK^+/initial rate ΔH^+ can be due to subsequent anion equilibration (e.g., 158). Chloroplasts swell during nigericin-induced K^+ uptake in the light, which is indicative of simultaneous anion uptake (192, 193, 240); swelling could also lead to increased "leakiness" toward protons, allowing H^+ equilibration by pathways other than nigericin. The even lower H^+/Na^+ ratio reported by Shavit et al (240) may well reflect Na^+ uptake (driven by the concentration gradient) in exchange for internal K^+ instead of internal H^+:K^+/Na^+ exchange can decrease the H^+/K^+ ratio induced by nigericin in erythrocytes (104). In their discussion of these effects, Shavit et al (238) also point out that the conclusion that nigericin does not primarily act by inducing stoichiometric cation/H^+ exchange is not necessarily warranted from the observation that net cation/H^+ exchange is greater than 1. Thus far, the proton-translocating ability of nigericin and gramicidin is sufficient to explain why they, but not valinomycin, uncouple chloroplast photophosphorylation, provided that the light-induced pH gradient is indeed a vital high-energy intermediate (165).

Nigericin uncoupled photophosphorylation of ADP at concentrations below those capable of affecting trans-membrane potentials (see, for example, 239). However, in one case, that of pH-jump-induced ATP synthesis, higher concentrations were necessary (239), inferring that an electrical potential is a significant component of the proton motive force induced by the pH jump. Further experimental support for this would be provided if a high concentration of X-537A but not of dianmycin or monensin, also uncoupled. In connection with some different observations, Shavit et al (238) have also pointed out that nigericin at higher concentrations may be catalyzing charge transfer across chloroplast membranes.

Dinitrophenol and FCCP are rather poor uncouplers of chloroplast photophosphorylation (72), in contrast to their effectiveness on mitochondria and chromatophores (above), but Karlish and co-workers (133, 134) demonstrated that valinomycin greatly enhanced the potency of these uncouplers. To explain this, Mitchell (165, 231, 232) has suggested that the presence of the positive valinomycin-K^+ complex within the hydrophobic membrane augments the solubility of the anionic uncoupler (cf 151), so facilitating the probable rate-limiting step of the uncoupling process (165, 168). However, it also seems plausible that the absence of a membrane potential in chloroplasts retards the essentially electro-impelled proton translocation by these agents, and that valinomycin-induced K^+ translocation sets up an exchange diffusion of K^+ and H^+, circumventing macroscopic charge transfer (134, 265), as discussed earlier for gramicidin. Thus, uncoupler + valinomycin achieves the effectiveness of gramicidin or nigericin alone.

Effects of Weak Bases or Weak Acids

Ammonium ions and substituted amines prevent the formation of a chloroplast pH gradient in the light, and also uncouple phosphorylation from electron transport (32, 33, 46, 47, 72, 157, 194). The proposed mechanism (same references; also, 75) is the discharge of the pH gradient as a result of uptake of neutral NH_3 and its internal conversion to NH_4^+; in contrast to chromatophores, this results in uncoupling, because the pH gradient, and not an electrical potential, is the main component of the proton motive force in chloroplasts (124, 162, 165).

Ammonium ions can discharge a pH gradient only when they are added to the alkaline side, and weak acid ions only when added to the acid side as explained in the section on Nonspecific Permeation (Fig. 3). Thus, the accumulation of P_i, or succinate inside chloroplasts in the dark is reversed when light causes acidification of the internal phase, presumably because of increased efflux of the protonated anion (50, 194). Also, chloroacetate and acetate are expelled from chloroplasts by light, and expulsion is accompanied by a decrease, but not complete reversal, of the pH gradient (48, 79, 194); acetate and butyrate cause a transient uncoupling during the initial stages of illumination (71). All these observations support a mechanism of uncoupling whereby the anionic form of the weak acid must be supplied on the acid side of the light-induced pH gradient in order for it to transport protons outwards as the free acid; hence, uncoupling is not complete because of the very limited amount of anion available within the small volume of the chloroplast interior. As pointed out by Crofts (46, 47, 75, 265), the observed NH_4^+ and weak acid movements and uncoupling are consistent with a pH gradient being the high-energy intermediate, and NH_4^+ ions prove a more potent uncoupler by virtue of the necessity for them to be supplied in the external alkaline phase, which has the much larger volume.

Sub-Chloroplast Particles

Particles derived from chloroplasts also exhibit a light-induced H^+ uptake, but they are not uncoupled by nigericin, valinomycin, or NH_4^+ (154, 155, 157, 183), even though NH_4^+ ions or nigericin (+ K^+) completely suppressed the light-induced pH gradient (100, 154). However, valinomycin + NH_4^+, or valinomycin + nigericin, were combinations possessing potent uncoupling ability (27, 100, 155). These authors pointed out that their results can be explained by assuming that an electrical potential is a significant component of the particle proton motive force, and that the preservation of the proton motive force is required to maintain ATP synthesis (156, 183). Hence, the situation for sub-chloroplast particles is analogous to that for chromatophores (page 408, 409). Furthermore, the assumption that the chloroplast absorption change at 518 nm reflects the presence of a membrane potential (195, 265, 271) is supported by the susceptibility of this signal to valinomycin + NH_4^+ or amines in sub-chloroplast particles (100; cf discussion on p. 409). Possibly the sub-chloroplast particles differ from the intact chloroplasts in that Cl^- ions fail to penetrate the sub-chloroplast membrane and so do not discharge the membrane potential component of the proton motive force (155, 183).

AEROBIC ION TRANSPORT AND PHOSPHORYLATION BY SUB-MITOCHONDRIAL PARTICLES

Proton Transport

Vesicles prepared by sonication under standardized conditions are inside out with respect to the original intact mitochondria (146). In the presence of oxygen and a respiratory substrate, or with ATP, they catalyze uptake of H^+ ions into the vesicles (35, 162, 164, 170), i.e., in the same direction as chromatophores and chloroplasts but opposite to the intact mitochondria (164, 171–173). Subinhibitory amounts of oligomycin were previously reported to enhance respiratory control and ATP synthesis by these particles (146), and the discovery that oligomycin enhanced the H^+ translocation by preventing the "leakage" of translocated protons back through the membrane (work of C. P. Lee, reported in 164, 233; also 110, 198) lent support to the thesis that the H^+ gradient is a high-energy intermediate of mitochondrial oxidative phosphorylation (162, 165). With the experimental conditions established for obtaining a maximal H^+ gradient, subsequent studies have measured ion transport in relation to the pH gradient and phosphorylation, in order to further investigate the mechanism of ATP synthesis.

K^+ ions alone seemed to decrease the efficiency of ATP synthesis (44, 202, 209), possibly because they were taken up in exchange for internal protons (44), or because they lowered the affinity of the system for ADP (202). The nature of the anion of the K^+ salt had a profound influence on particle ATP synthesis (199), a phenomenon we shall return to later, but first the effects of the ionophorous antibiotics in the absence of a significant influence of anions will be considered.

Effects of Ionophorous Antibiotics

In the presence of K^+ ions, valinomycin enhanced the electron transport-dependent pH gradient with or without oligomycin present (198) and also caused K^+ expulsion (44), indicating that the original H^+ translocation into the particles is electrogenic. Valinomycin only slightly reduced the P/O ratio (202, 249) or respiratory control index (44, 178), and hence is a poor uncoupling agent. In a comprehensive investigation, Montal et al (176,

178) confirmed these results but also showed that valinomycin can cause up to 50 percent inhibition of other indices of coupling ability, namely the energy-requiring reductions of NAD by succinate, or NADP by NADH, and the energy-requiring changes in bromthymol blue absorption or anilinonaphthalene sulfonate fluorescence. It may be concluded that, at best, valinomycin causes only partial uncoupling of energy-linked reactions of sub-mitochondrial particles.

Gramicidin depressed the pH gradient established by electron transport in sub-mitochondrial particles (110, 177) and readily caused release of respiratory control and suppression of both the P/O ratio (independently of the alkali metal cation concentration) and bromthymol blue response (97, 177). As pointed out in (97, and 177), this uncoupling ability of gramicidin, by contrast with valinomycin, can be attributed to the former's facility for electro-impelled H^+ translocation; this would dissipate both ΔpH and $\Delta\psi$ components of a proton motive force, thus uncoupling (cf 179), whereas with valinomycin discharge of $\Delta\psi$ can be compensated by an enhancement of ΔpH (Fig. 2).

Although nigericin caused dissipation of the H^+ gradient by promoting uptake of added K^+ ions, it was not an uncoupling agent, and neither were NH_4^+ ions (44, 110, 176, 178, 198). It may be inferred that an electrical potential is probably the major component of the proton motive force that results from electron transport (see especially, 176, 179). Also, one can predict that a combination of charge-transferring and proton-transferring agents should uncouple synergistically, as demonstrated in the chromatophore experiments (p. 408). Indeed, nigericin + valinomycin $\pm$ K^+, and valinomycin + NH_4^+ are combinations that cause potent uncoupling of sub-mitochondrial particle oxidative phosphorylation (44, 176, 178, 179).

Other Ion Transport Phenomena

Montal et al (176, 179) and Skulachev and co-workers (76, 150, 246) reasoned that electro-impelled intrusion of lipid-soluble anions should be able to replace valinomycin-induced K^+ extrusion as a means of discharging the membrane potential. In confirmation they showed that, in response to energization by ATP or respiratory substrates, sub-mitochondrial particles accumulated the anions phenyldicarbaundecaborane, tetraphenyl boron, SCN^-, and picrate, but not lipid-soluble cations such as tetrabutyl ammonium (76, 176); also the anion accumulation enhanced the pH gradient (76, 176; Fig. 2). These results indicate that the primary H^+ transport is electrogenic, positive toward the inside. Electro-impelled anion accumulation caused some uncoupling on its own (76), but a combination of penetrant anion with nigericin (and K^+) or NH_4^+ ions induced maximal uncoupling (176, 177), consistent with the synergism already described for the chromatophore system (150, 179; p. 410). It should be mentioned that NO_3^- ions could replace SCN^- and the synthetic anions for these uncoupling effects, and it seems that sub-mitochondrial particles translocate NO_3^- ions (44, 176, 198, 199).

Azzi and co-workers (6, 9) have pointed out that their observations on 8-anilino-1-napthalene-sulfonic acid (ANS) fluorescence in energized sub-mitochondrial particles indicate that a conformational change is a high-energy intermediate of the system. However, they do conclude that generation of a trans-membrane potential could lead to identical observations if it changed the distribution of the charged ANS molecule into the hydrophobic membrane region; such changes have been observed experimentally (6), and are discussed by Skulachev (246). Most other authors who have studied induced and endogenous ion transport in relation to energy conservation by sub-mitochondrial particles have deduced that the existence of an electrogenic H^+ pump as defined by the chemiosmotic theory is a sufficient basis to predict and explain their experimental observations (see, e.g., 44, 176, 179, 246). It is apparent that the observations are remarkably analogous to those obtained with chromatophores, and for both systems it may be concluded that the vesicles generate a proton motive force consisting of both an electrical and pH gradient, positive inside (162, 165), maximum development of which is hindered by an inherent leakness to K^+ ions and protons, although the latter can be mitigated by addition of oligomycin (233).

OXIDATIVE PHOSPHORYLATION AND ION TRANSPORT BY INTACT MITOCHONDRIA

The characterization of the effects of valinomycin on mitochondria by Pressman (180, 211–219) using innovative techniques, is the basis of the still expanding use of the valinomycin- and nigericin-type antibiotics as tools for the study of mitochondrial metabolism. Not the least of these efforts has been the study of the relationships of the antibiotic effects to the chemiosmotic theory, as discussed in detail by Mitchell (165, 166, 174) and summarized by Greville (75) and by Harold (86). Accordingly, we will limit ourselves to discussing experiments similar to those we have already described for the bacterial and plant systems.

Ion Translocation by Mitochondria in the Absence of Ionophorous Antibiotics

In the presence of an oxidizable substrate and oxygen, there is only a minimal pH gradient across the mitochondrial membrane (1, 142), although the system readily phosphorylates added ADP. Mitchell (162, 166), therefore, suggests that the high-energy intermediate of mitochondria is almost entirely an electrical potential with only a minimal pH component (cf 75, 142). The chemiosmotic theory requires that such a potential must arise from an initial electrogenic translocation of protons by the respiratory chain, and in order to account for the dissipation of the pH component of an initial H^+ translocation, it should be recalled that mitochondria possess very active translocation systems in which anion transport, particularly that of P_i (37, 69, 196, 197, 200, 201, 203, 222), and possibly also an Na^+/H^+ exchange system (24, 173, 175) can rapidly neutralize a difference in the chemical potential of protons across the membrane without affecting an electrical gradient, (Fig. 1) (162, 165).

Mitchell & Moyle (171–174) have demonstrated the existence of very rapid, electron transport-dependent H^+ translocation by adding pulses of oxygen to anaerobic suspensions of mitochondria. Subsequently, it became apparent that Ca^{++} ions moved inward during the H^+ extrusion (35, 40, 254) and, in opposition to the chemiosmotic proposals, it was suggested that a primary Ca^{++} uptake may be responsible for a secondary H^+ extrusion. Mitchell (165–167, 174) has pointed out that the electrical component of the H^+ translocation would have to be discharged in order for a pH gradient to be detected by a pH electrode in the suspension, and that electro-impelled ion movement (in this case, Ca^{++} uptake) is therefore obligatory for measurement of the proton pump activity. This question of the primacy of a mitochondrial cation pump versus mitochondrial proton pump has been admirably discussed by Greville (75), and will possibly be resolved when unequivocal techniques for measuring very rapid cation and proton movements are devised.

Skulachev and co-workers investigated the response of the mitochondrial proton pump to synthetic lipid soluble ions (246). The anions were extruded by energized mitochondria while cations were taken up, and the latter process was accompanied by proton ejection (effects very similar to Ca^{++} uptake, 13, 150). These researchers therefore concluded that mitochondria conserve energy in the form of a trans-membrane potential that is positive on the outside, in contrast with chromatophores and sub-mitochondrial particles which are negative outside by this criterion. Also, as required by the chemiosmotic mechanism (166, 174; Fig. 2), the electrical gradient is seen to be replaced by a trans-membrane pH gradient under the condition of induced electro-impelled ion transport in these experiments.

There exist many experimental results consistent with the ability of NH_3 and CH_3CO_2H to penetrate the mitochondrial membrane and affect any pH gradients by the nonelectrogenic mechanisms illustrated in Figure 3 (24–26, 39, 40, 111, 175). Neither NH_4^+ nor $CH_3CO_2^-$ causes loss of respiratory control or uncoupling of mitochondria when present individually (26, 38, 97). This indicates that a trans-membrane pH gradient is not a form in which energy is conserved by isolated mitochondria, and may be contrasted with the situation for isolated chloroplasts (see p. 414). However, agents that catalyze electro-impelled H^+ translocation are excellent uncouplers of mitochondrial oxidative phosphorylation as predicted by Mitchell (75, 160, 161, 172, 173).

In summary, all these ion transport effects are consistent with the chemiosmotic proposal that mitochondria normally conserve energy as a trans-membrane electrical potential with a minimal pH gradient component.

Effects of Ionophorous Antibiotics

Valinomycin and gramicidin.—When mitochondria are "primed" with a high-energy intermediate, e.g., in the presence of oxidizable substrate under aerobic conditions and omitting ADP and P_i, then the addition of valinomycin causes accumulation of K^+ ions, even against the concentration gradient, and extrusion of protons (180, 211, 212). In experiments where pulses of O_2 are added to a valinomycin-treated anaerobic suspension, the H^+/K^+ ratio is unity (166, 174), but when oxygen is continuously available and at higher K^+ concentrations the ratio falls, (7, 180, 211) possibly because subsequent ion movements have time to re-equilibrate the protons and also because of increased leakiness due to swelling. In such experiments, valinomycin does not of itself (see p. 421) induce maximal rates of respiration, i.e., loss of respiratory control, nor is the synthesis of ATP seriously impaired (97, 107). These effects may be contrasted with the acceleration of respiration and uncoupling produced by agents which catalyze electro-impelled proton, but not alkali cation, movement (reviewed in 75, 162).

The difference can be rationalized by the prediction of the chemiosmotic theory that electro-impelled alkali cation movement would reduce the electrical potential, but that this could be replaced by a compensatory pH gradient (as observed experimentally) (Fig. 2). The pH gradient would maintain the total proton motive force, and hence coupling ability, but a compensatory pH gradient would not be possible in the presence of protonophores. Thus, although gramicidin produced very similar effects to valinomycin (38, 212) it had a much greater propensity to uncouple (97); once again, in chemiosmotic terms this is readily explicable by gramicidin's capacity for electro-impelled proton transport. Incidentally, inward K^+ movement and outward H^+ extrusion induced by valinomycin further emphasizes the opposite polarity of the intact mitochondria to sub-mitochondrial particles or chromatophores.

Pressman and co-workers have proposed an alternative explanation for the action of valinomycin, namely, that it facilitates the supply of K^+ ions to a cation pump situated within the mitochondrial membrane (213–219; see also 153). The normal function of this pump would be to initiate substrate anion uptake by a K^+ -symport mechanism (95, 96, 211, 215, 216), in contrast to the OH^--antiport or H^+-symport mechanisms preferred by Chappell (37) and Mitchell (165), and adopted in this paper. Proton extrusion would then be a secondary result of the primary K^+ uptake (94, 95). Again, this chicken/egg problem has not yet been resolved and the existence of a primary cation pump must be allowed as a possible explanation for ion transport phenomena of mitochondria. Presumably, the energy-requiring transport of Ca^{++}, K^+ (in the presence of valinomycin), and lipid-soluble cations reflects a rather broad specificity of such a pump, although the operation of several highly specific pumps can also be invoked. The chemiosmotic mechanism regards such diverse forms of cation transport all as manifestations of a single phenomenon—the presence of a membrane potential, negative inside.

Nigericin.—Proponents of both the chemiosmotic and cation pump mechanisms of ion transport suggest that the nigericin antibiotics induce an electrically neutral exchange-diffusion of cations across the mitochondrial membrane, in the same way as for model membrane systems (104, 165, 213, 217, 218; but see 143). Thus, when added to mitochondria respiring in a medium containing 5–10 m*M* K^+, nigericin induces K^+ extrusion (down the K^+ concentration gradient) in exchange for an uptake of protons, but despite the proton movement, nigericin on its own does not cause loss of respiratory control or coupling efficiency (143, 213, 218). This would be expected if $\Delta\psi$ is the major component of the high-energy intermediate, as described above (cf 104).

At this point it is relevant to point out that concentrations of nigericin above about 2 μg/ml induce loss of respiratory control and accelerate the ATPase activity of intact mitochondria i.e. exhibit uncoupling activity, provided that about 80 m*M* K^+ is present (62, 63, 141). X-537A also uncouples, but with no rigorous alkali cation requirement (141) and dianmycin, monensin and X-206 are ineffective (62, 63). According to the chemiosmotic theory an uncoupling agent must translocate electrical charge as well as protons, and as pointed out on page 400, high concentrations of nigericin and X-537A, but not the other monocarboxylic antibodies, lower the resistance of bilayer membranes in the presence of suitable cations, consistent with a facility of the elevated antibiotic concentrations for electrogenic cation transfer. This observation provides a plausible mechanism for the uncoupling activity of nigericin and X-537A, if it is assumed that the high-energy intermediate is an electrical potential (the nigericin or X-537A would prevent formation of a pH gradient at high as well as low concentrations). However, it seems feasible that X-537A can combine with Ca^{++} ions since it is isolated as the barium salt (129, 205, 267), and it could therefore uncouple by promoting an energy-dependent cyclic flux of Ca^{++} ions. These points are receiving further attention in this and other laboratories (Estrada-O. and Pressman, personal communications).

ATP synthesis by antibiotic-induced cation movements.—In an elegant set of experiments, Cockrell et al (43) demonstrated that K^+ efflux from mitochondria induced by valinomycin produced ATP synthesis, but equivalent K^+ movement by nigericin did not. They rationalized the synthesis by invoking reversal of the energy-linked K^+ pump, by analogy with experiments that demonstrated P_i incorporation into ATP and ADP as a result of reversal of the erythrocyte Na^+/K^+ pump (43, 70). However, Glynn (70) pointed out that K^+ efflux from mitochondria would be electrogenic if catalyzed by valinomycin, and that the resulting electrical potential could cause ATP synthesis by a chemiosmotic mechanism. More recently, Rossi & Azzone (224) devised experiments that increased the efficiency of the mitochondrial process to 4 K^+ extruded per ATP synthesized. They deduced that 16.6 kcal were required for synthesis of one ATP molecule, as opposed to the requirement of 8.3 kcal from the ratio of 2 K^+ per ATP predicted by the chemiosmotic hypothesis (165, 167, 174). However, assuming that the already swollen mitochondria of Rossi & Azzone are operating at a maximal efficiency, it is still necessary to consider that efflux of ATP in exchange for external ADP translocates a negative charge outward (page 404), which would presumably alter the required ratio to 3 K^+ per ATP made. Perhaps the moral to be drawn is that a definitive conclusion as to the mechanism of these effects is not justified by the evidence now available.

Synergistic Uncoupling of Mitochondria

As already pointed out, antibiotics capable of inducing electro-impelled cation transport cause only a slight decrease in respiratory control of mitochondria (38, 39, 97, 180), the chemiosmotic explanation being that an enhanced gradient of protons replaces the depressed membrane potential as the high-energy intermediate (38, 162, 165). The subsequent addition of any agent promoting proton transport across membranes by an electrically neutral mechanism would be predicted to induce uncoupling (38, 39). Thus, P_i or acetate ions (38, 39, 97, 180), NH_4^+ ions (38–40), and nigericin (143, 215, 218), none of which uncouple on their own, all cause loss of respiratory control provided that valinomycin, the actins, or gramicidin are already present (see also discussion in ref. 11).

Succinate, β-(OH)-butyrate or proline have to be the respiratory substrates in order to demonstrate this synergism with nigericin and NH_4^+ clearly, because an inhibition of oxidation of other substrates occurs, probably as a result of phosphate efflux from mitochondria (60, 104, 222). The significance of these interesting effects on the mechanism of substrate accumulation by mitochondria is the subject of continuing investigations, and will not be discussed here.

Synergistic uncoupling of mitochondria by nigericin, ammonium, P_i, or acetate can clearly be rationalized by the chemiosmotic theory in terms of their capacity for electrically neutral proton transfer. However, in the case of nigericin, an energy-dependent cyclic flux of K^+ ions could provide an alternative explanation (218), and in the case of P_i and acetate the observed swelling of the mitochondria could possibly dislocate the coupling enzymes. Hence, the chemiosmotic theory provides a catholic explanation, but not one unique for each individual case.

CONCLUDING REMARKS

In order to cover a wide range of biological systems, it has been necessary in this review to omit some less completely studied ionophorous antibiotics. They are monazomycin (61, 65, 143, 182), antamanide (219, 268), alamethicin (182, 213, 219), and boromycin (112, 191; A. Yoshimoto (observations to be published)), all of which exhibit interesting characteristics of their own, and which may lead to novel developments in the field. Also, new approaches to examining the plausibility of the chemiosmotic theory are appearing: for example, investigation of the correlation between energy state and midpoint potential of the cytochromes (53, 106), new methods for following rapid translocation reactions (234, 260), and more stringent examination of the stoichiometry and thermodynamic aspects (51, 165, 167, 224, 248, 265).

It should be apparent from this review that many very diverse ion transport phenomena can be rationalized by the chemiosmotic assumption that biological systems conserve energy as a trans-membrane pH or electrical gradient, or both. Furthermore, the disappearance of ATP synthesis correlates with the predicted reduction of the total proton motive force by the induced ion transport, so that the role of the proton motive force as a direct high-energy intermediate for ATP synthesis is very feasible. In assessing the validity of Dr. Mitchell's chemiosmotic theory, therefore, a comment of Albert Einstein seems remarkably apt: "A theory is more impressive the greater is the simplicity of its premises, the more different the kinds of things it relates, and the more extended is its range of applicability."

ACKNOWLEDGMENTS

The author is indebted to Professor Henry A. Lardy for much advice and helpful criticism that has been incorporated into this article. Discussions with Miss Shelagh Ferguson and Dr. Peter Reed also raised many cogent points that have been of use. I wish to thank Mr. Brian A. Warnecke and Miss Margaret Taylor for their expertise and patience during the preparation of the manuscript for publication.

LITERATURE CITED

1. Addanki, S., Sotos, J. F. 1969. *Ann. N. Y. Acad. Sci.* 19:756–804
2. Anagnosti, E., Tedeschi, H. 1970. *J. Cell Biol.* 47:520–25
3. Andreoli, T. E., Tieffenberg, M., Tosteson, D. C. 1967. *J. Gen. Physiol.* 50:2527–45
4. Ashton, R., Steinrauf, L. K. 1970. *J. Mol. Biol.* 49:547–55
5. Avron, M., Shavit, N. 1965. *Biochim. Biophys. Acta* 109:317–31
6. Azzi, A. 1969. *Biochem. Biophys. Res. Commun.* 37:254–60
7. Azzi, A., Azzone, G. F. 1966. *Biochim. Biophys. Acta* 113:445–56
8. Azzi, A., Azzone, G. F. 1967. *Atti del Seminario di Studi Biol. Vol. III,* ed. E. Quagliariello, 57–67. Bari:Adriatica Editrice. 280 pp.
9. Azzi, A., Chance, B., Radda, G. K., Lee, C. P. 1969. *Proc. Nat. Acad. Sci. USA* 62:612–19
10. Azzi, A., Rossi, E., Azzone, G. F. 1966. *Enzymol. Biol. Clin.* 7:25–37
11. Azzone, G. F., Rossi, E., Scarpa, A. 1968. *Regulatory Functions of Biological Membranes,* ed. J. Järnefelt, 236–46. Amsterdam: Elsevier. 311 pp.
12. Azzone, G. F., Piemonte, G. 1969. *The Energy Level and Metabolic Control in Mitochondria,* ed. S. Papa, J. M. Tager, E. Quagliariello, E. C. Slater, 115–24. Bari: Adriatica Editrice. 466 pp.
13. Bakeeva, L. E., Grinius, L. L., Jasaitis, A. A., Kuliene, V. V., Levitskii, D. O., et al. 1970. *Biochim. Biophys. Acta* 216:13–21
14. Baltscheffsky, H., Arwidsson, B. 1962. *Biochim. Biophys. Acta* 65: 425–28
15. Baltscheffsky, H., Baltscheffsky, M. 1960. *Acta Chem. Scand.* 14:257–63

16. Baltscheffsky, H., Baltscheffsky, M. 1964. *Abst. Int. Congr. Biochem., 6th, New York,* X:773B
17. Bangham, A. D. 1968. *Progr. Biophys.* 18:29–95
18. Bielawski, J., Thompson, T. E., Lehninger, A. L. 1966. *Biochem. Biophys. Res. Commun.* 24:948–54
19. Blondin, G., Green, D. E. 1970. *J. Bioenerget.* 1:193–213
20. Bose, S., Gest, H. 1963. *Energy-linked Functions of Mitochondria,* ed. B. Chance, 207–18. New York:Academic. 282 pp.
21. Boyer, P. D. 1965. *Oxidases and Related Redox Systems,* ed. T. E. King, H. S. Mason, M. Morrison, II:994–1008. New York:John Wiley & Sons. 1144 pp.
22. Brierley, G. P. 1969. *Biochem. Biophys. Res. Commun.* 35:396–402
23. Brierley, G. P. 1970. *Biochemistry.* 9:697–707
24. Brierley, G. P., Jurkowitz, M., Scott, K. M., Merola, A. J. 1970. *J. Biol. Chem.* 245:5404–11
25. Brierley, G. P., Settlemire, C. T., Knight, V. A. 1968. *Arch. Biochem. Biophys.* 126:276–88
26. Brierley, G. P., Stoner, C. D. 1970. *Biochemistry.* 9:708–13
27. Briller, S., Gromet-Elhanan, Z. 1970. *Biochim. Biophys. Acta* 205:263–72
28. Butler, K. W., Dugas, M., Smith, I. C. P., Schneider, H. 1970. *Biochem. Biophys. Res. Commun.* 40:770–76
29. Caswell, A. H. 1969. *J. Membr. Biol.* 1:53–78
30. Carafoli, E., Rossi, C. S. 1967. *Biochem. Biophys. Res. Commun.* 29:153–57
31. Carafoli, E., Rossi, C. S., Gazzotti, P. 1969. *Arch. Biochem. Biophys.* 131:527–37
32. Carmeli, C. 1969. *Biochim. Biophys. Acta* 189:256–66
33. Carmeli, C. 1970. *FEBS Lett.* 7:297–300
34. Chance, B. 1965. *J. Biol. Chem.* 240:2729–48
35. Chance, B., Mela, L. 1967. *J. Biol. Chem.* 242:830–44
36. Chance, B., Nishimura, M., Avron, M., Baltscheffsky, M. 1966. *Arch. Biochem. Biophys.* 117:158–66
37. Chappell, J. B. 1968. *Brit. Med. Bull.* 24:150–57
38. Chappell, J. B., Crofts, A. R. 1965. *Biochem. J.* 95:393–402
39. Chappell, J. B., Crofts, A. R. 1966. *Regulation of Metabolic Processes of Mitochondria,* eds. J. M. Tager, S. Papa, E. Quagliariello, E. C. Slater, 293–316. Amsterdam:Elsevier. 582 pp.
40. Chappell, J. B., Haarhoff, K. N. 1967. *Biochemistry of Mitochondria,* eds. E. C. Slater, Z. Kaniuga, L. Wojtczak, 75–91. New York:Academic. 122 pp.
41. Chappell, J. B., Henderson, P. J. F., McGivan, J. D., Robinson, B. H. 1968. *The Interaction of Drugs and Subcellular Organelles in Animal Cells,* ed. P. N. Campbell, 71–95. London:Churchill. 130 pp.
42. Ciani, S., Eisenman, G., Szabo, G. 1969. *J. Membr. Biol.* 1:1–36
43. Cockrell, R. S., Harris, E. J., Pressman, B. C. 1967. *Nature (London)* 215:1487–88
44. Cockrell, R. S., Racker, E. 1969. *Biochem. Biophys. Res. Commun.* 35:414–19
45. Cole, J. S., Aleem, M. I. H. 1970. *Biochem. Biophys. Res. Commun.* 38:736–43
46. Crofts, A. R. 1967. *J. Biol. Chem.* 242:3352–59
47. Crofts, A. R. 1968. *Regulatory Function of Biological Membranes,* ed. J. Järnefelt, 247–63. Amsterdam:Elsevier, 311 pp.
48. Crofts, A. R., Deamer, D. W., Packer, L. 1967. *Biochim. Biophys. Acta* 131:97–118
49. Davson, H., Danielli, J. F. 1943. *The Permeability of Natural Membranes,* 194–203. Cambridge:The University Press. 361 pp.
50. Deamer, D. W., Packer, L. 1969. *Biochim. Biophys. Acta* 172:539–45
51. Dilley, R. A. 1970. *Arch. Biochem. Biophys.* 137:270–83
52. Dow, D. S., Walton, K. G., Fleischer, S. 1970. *J. Bioenerget.* 1:247–71
53. Dutton, P. L., Wilson, D. F., Lee, C. P. 1970. *Biochemistry* 9:5077–82
54. Echlin, P. 1970. *Symp. Soc. Gen. Microbiol.* 20:221–48
55. Edwards, G. E., Bovell, C. R. 1969. *Biochim. Biophys. Acta* 172:126–33
56. Edwards, G. E., Bovell, C. R. 1970. *J. Membr. Biol.* 2:95–107
57. Eilermann, L. J. M., Slater, E. C. 1970. *Biochim. Biophys. Acta* 216:226–28
58. Eisenman, G. Ciani, S. M., Szabo, G. 1968. *Fed. Proc.* 27:1289–1304
59. Eisenman, G., Ciani, S., Szabo, G.

1969. *J. Membr. Biol.* 1:294–345
60. Estrada-O., S., Calderon, E. 1970. *Biochemistry* 9:2092–99
61. Estrada-O., S., Gomez-Lojero, C. 1971. *Biochemistry* (In Press)
62. Estrada-O., S., Graven, S. N., Lardy, H. A. 1967. *J. Biol. Chem.* 242:2925–32
63. Estrada-O., S., Graven, S. N., Lardy, H. A. 1967. *Fed. Proc.* 26:610
64. Evans, M. C. W., Whatley, F. R. 1970. *Symp. Soc. Gen. Microbiol.* 20:203–20
65. Ferguson, S. M. F., Lardy, H. A. 1968. *Fed. Proc.* 27:1751
66. Fleischman, D. E., Clayton, R. K. 1968. *Photochem. Photobiol.* 8: 287–98
67. Frenkel, A. W. 1970. *Biol. Rev.* 45: 569–616
68. Gelman, N. S., Lukoyanova, M. A., Ostrovskii, D. N. 1967. *Respiration and Phosphorylation of Bacteria.* New York:Plenum Press. 238 pp.
69. Ghosh, A. K., Chance, B. 1970. *Arch. Biochem. Biophys.* 138: 483–92
70. Glynn, I. M. 1967. *Nature (London)* 216:1318
71. Good, N. E. 1962. *Arch. Biochem. Biophys.* 96:653–61
72. Good, N. E., Izawa, S., Hind, G. 1966. *Curr. Top. Bioenerget.*, 1: 75–112
73. Gomez-Puyou, A., Sandoval, F., Chavez, E., Tuena, M. 1970. *J. Biol. Chem.* 245:5239–47
74. Green, D. E. 1970. *Proc. Nat. Acad. Sci. USA* 67:544–49
75. Greville, G. D. 1969. *Curr. Top. Bioenerget.*, 3:1–78
76. Grinius, L. L., Jasaitis, A. A., Kadziauskas, V. P., Liberman, E. A., Skulachev, V. P., et al. 1970. *Biochim. Biophys. Acta* 216:1–12
77. Gromet-Elhanan, Z. 1970. *Biochim. Biophys. Acta* 223:174–82
78. Gromet-Elhanan, Z., Briller, S. 1969. *Biochem. Biophys. Res. Commun.* 37:261–65
79. Gross, E. L., Packer, L. 1967. *Arch. Biochem. Biophys.* 122:237–45
80. Gross, E., San Pietro, A. 1969. *Arch. Biochem. Biophys.* 131:49–56
81. Gross, E., Shavit, N., San Pietro, A. 1968. *Arch. Biochem. Biophys.* 127:224–28
82. Hamilton, W. A. 1970. *Biochem. J.* 118:46P–47P
83. Hamilton, W. A. 1970. *Inhibition and Destruction of the Microbial Cell,* ed. W. B. Hugo. New York: Academic. In Press
84. Hamilton, W. A. 1970. *Membranes —Structure and Function,* ed. J. R. Villanueva, F. Ponz, 71–79. New York:Academic. 153 pp.
85. Hansford, R. G., Chappell, J. B. 1969. *Preparation and Fractionation of Subcellular Components,* ed. G. D. Birnie, S. M. Fox, 100–20. New York:Plenum Press. 173 pp.
86. Harold, F. M. 1970. *Advan. Microb. Physiol.* 4:45–104
87. Harold, F. M., Baarda, J. R. 1967. *J. Bacteriol.* 94:53–60
88. Harold, F. M., Baarda, J. R. 1968. *J. Bacteriol.* 95:816–23
89. Harold, F. M., Baarda, J. R. 1968. *J. Bacteriol.* 96:2025–34
90. Harold, F. M., Baarda, J. R., Baron, C., Abrams, A. 1969. *J. Biol. Chem.* 244:2261–68
91. Harold, F. M., Baarda, J. R., Pavlasova, E. 1970. *J. Bacteriol.* 101: 152–59
92. Harold, F. M., Pavlasova, E., Baarda, J. R. 1970. *Biochim. Biophys. Acta* 196:235–44
93. Harris, E. J. 1954. *Transport and Accumulation in Biological Systems.* London:Butterworth. 291 pp.
94. Harris, E. J. 1969. *FEBS Lett.* 5: 50–52
95. Harris, E. J. 1970. *FEBS Lett.* 11: 225–28
96. Harris, E. J., Berent, C. 1970. *FEBS Lett.* 10:6–12
97. Harris, E. J., Hofer, M. P., Pressman, B. C. 1967. *Biochemistry* 6: 1348–60
98. Harris, E. J., Pressman, B. C. 1967. *Nature (London)* 216:918–20
99. Harvey, E. H. 1911. *J. Exp. Zool.* 10:507–56
100. Hauska, G. A., McCarty, R. E., Olson, J. S. 1970. *FEBS Lett.* 7: 151–56
101. Heldt, H. W. 1969. *FEBS Lett.* 5: 11–14
102. Heldt, H. W., Rapely, L. 1970. *FEBS Lett.* 10:139–42
103. Henderson, P. J. F., Chappell, J. B. 1967. *Biochem. J.* 105:16P
104. Henderson, P. J. F., McGivan, J. D., Chappell, J. B. 1969. *Biochem. J.* 111:521–35
105. Hildreth, W. W. 1970. *Arch. Biochem. Biophys.* 139:1–8
106. Hinkle, P., Mitchell, P. 1970. *J. Bioenerget.* 1:45–60

107. Hofer, M. P., Pressman, B. C. 1966. *Biochemistry* 5:3919–25
108. Hopfer, U., Lehninger, A. L., Lennarz, W. J. 1970. *J. Membr. Biol.* 3:142–55
109. Horio, T., Yamashita, J. 1964. *Biochim. Biophys. Acta* 88:237–50
110. House, D. R., Packer, L. 1970. *J. Bioenerget.* 1:273–85
111. Hunter, G. R., Kamishima, Y., Brierley, G. P. 1969. *Biochim. Biophys. Acta* 180:81–97
112. Hutter, R., Keller-Schierlein, W., Knüsel, F., Prelog, V., Rodgers, G. C., et al. 1967. *Helv. Chim. Acta* 50:1533–39
113. Imai, K., Asano, A., Sato, R. 1967. *Biochim. Biophys. Acta* 143:462–76
114. Isaev, P. I., Liberman, E. A., Samuilov, V. D., Skulachev, V. P., Tsofina, L. M. 1970. *Biochim. Biophys. Acta* 216:22–29
115. Ivanov, V. T., Laine, I. A., Abdulaev, N. D. 1969. *Biochem. Biophys. Res. Commun.* 34:803–11
116. Ivanov, V. T., Shkrob, A. M. 1970. *FEBS Lett.* 10:285–91
117. Izawa, S. 1970. *Biochim. Biophys. Acta* 223:165–73
118. Jackson, J. B., Crofts, A. R. 1969. *FEBS Lett.* 4:185–89
119. Jackson, J. B., Crofts, A. R. 1969. *Abst. FEBS Meeting, 6th Madrid* 299
120. Jackson, J. B., Crofts, A. R., von Stedingk, L. V. 1968. *Eur. J. Biochem.* 6:41–54
121. Jacobs, M. H. 1922. *J. Gen. Physiol.* 5:181–88.
122. Jacobs, M. H. 1940. *Cold Spring Harbor Symp. Quant. Biol.* 8:30–39
123. Jacobs, M. H., Parpart, A. K. 1938. *J. Cell. Comp. Physiol.* 11:175–92
124. Jagendorf, A. T. 1967. *Fed. Proc.* 26:1361–69
125. Jagendorf, A., Hind, G. 1963. *Photosynthetic Mechanisms of Green Plants,* 599–610. Washington, D.C.:*Natl. Acad. Sci. Nat. Res. Counc. Publ. 1145*
126. Jain, M. K., Strickholm, A., White, F. P., Cordes, E. H. 1970. *Nature (London)* 227:705–7
127. Jendrasiak, G. L. 1970. *Chem. Physiol. Lipids* 4:85–95
128. John, P., Whatley, F. R. 1970. *Biochim. Biophys. Acta* 216:342–52
129. Johnson, S. M., Herrin, J., Liu, S. J., Paul, I. C. 1970. *J. Am. Chem. Soc.* 92:4428–35
130. Jones, C. W., Erickson, S. K., Ackrell, B. A. C. 1971. *FEBS Lett.* 13:33–35
131. Junge, W., Witt, H. T. 1968. *Z. Naturforsch.* 23B:244–54
132. Karlish, S. J. D., Avron, M. 1968. *Comparative Biochemistry and Biophysics of Photosynthesis,* ed. K. Shibata, A. Takamiya, A. T. Jagendorf, R. C. Fuller, 214–21. Pennsylvania: University Park Press. 445 pp.
133. Karlish, S. J. D., Avron, M. 1968. *FEBS Lett.* 1:21–24
134. Karlish, S. J. D., Shavit, N., Avron, M. 1969. *Eur. J. Biochem.* 9:291–98
135. Keister, D. 1970. *Electron Transport and Energy Conservation,* ed. J. Tager. Bari:Adriatica Editrice. In Press
136. Keister, D. L., Minton, N. J. 1970. *J. Bioenerget.* 1:367–77
137. Keister, D. L., Minton, N. J. 1969. *Biochemistry* 8:167–73
138. Kinsky, S. C. 1970. *Ann. Rev. Pharmacol.* 10:119–42
139. Klingenberg, M. 1970. *FEBS Lett.* 6:145–54
140. Klingenberg, M., Wulf, R., Heldt, H. W., Pfaff, E. 1969. *Mitochondria Structure and Function,* ed. E. Ernster, Z. Drahota, 59–77. New York: Academic. 393 pp
141. Lardy, H. A., Estrada-O., S., Graven, S. 1967. *Abstr. Int. Congr. Biochem., 7th, Tokyo.* 906
142. Lardy, H. A., Ferguson, S. M. F. 1969. *Ann. Rev. Biochem.* 38: 991–1034
143. Lardy, H. A., Graven, S. N., Estrada-O., S. 1967. *Fed. Proc.* 26: 1355–60
144. Lardy, H. A., Johnson, D., McMurray, W. C. 1958. *Arch. Biochem. Biophys.* 78:587–97
145. Lauger, P., Lesslauer, W., Marti, E., Richter, I. 1967. *Biochim. Biophys. Acta* 135:20–32
146. Lee, C. P., Ernster, L. 1966. *Regulation of Metabolic Processes in Mitochondria,* ed. J. M. Tager, S. Papa, E. Quagliariello, E. C. Slater. 218–34. New York:Elsevier. 582 pp.
147. Lehninger, A. L. 1970. *Biochem. J.* 119:129–38
148. Leive, L. 1967. *J. Biol. Chem.* 243: 2373–80
149. Lev, A. A., Buzhinskii, E. P. 1967. *Cytology* (USSR) 9:102–6
150. Liberman, E. A., Skulachev, V. P.

1970. *Biochim. Biophys. Acta* 216:30–47
151. Liberman, E. A., Topaly, V. P. 1965. *Biochim. Biophys. Acta* 163:125–36
152. Liberman, E. A., Topaly, V. P., Tsofina, L. M. 1969. *Nature (London)* 222:1076–78
153. Massari, S., Azzone, G. F. 1970. *Eur. J. Biochem.* 12:310–18
154. McCarty, R. E. 1968. *Biochem. Biophys. Res. Commun.* 32:37–43
155. McCarty, R. E. 1969. *J. Biol. Chem.* 244:4292–98
156. McCarty, R. E. 1970. *FEBS Lett.* 9: 313–16
157. McCarty, R. E., Coleman, C. H. 1970. *Arch. Biochem. Biophys.* 141:198–206
158. McGivan, J. D. 1969. *Abstr. FEBS Meeting, 6th, Madrid* 220
159. Miles, C. D., Jagendorf, A. T. 1970. *Biochemistry* 9:429–34
160. Mitchell, P. 1961. *Nature (London)* 191:144–48
161. Mitchell, P. 1963. *Cell Interface Reactions,* ed. H. D. Brown, 33–56 New York:Scholar's Library. 107 pp.
162. Mitchell, P. 1966. *Oxidative and Photosynthetic Phosphorylation* Bodmin:Glynn Research Ltd. 192 pp.
163. Mitchell, P. 1967. *Comprehensive Biochemistry,* ed. M. Florkin, E. H. Stotz, 167–97. New York:Elsevier. 214 pp.
164. Mitchell, P. 1967. *Fed. Proc.* 26: 1370–79
165. Mitchell, P. 1968. *Chemiosmotic Coupling and Energy Transduction.* Bodmin:Glynn Research Ltd. 111 pp.
166. Mitchell, P. 1969. *Mitochondria—Structure and Function,* ed. L. Ernster, Z. Drahota, 219–32. New York:Academic. 393 pp.
167. Mitchell, P. 1969. *The Molecular Basis of Membrane Function,* ed. D. C. Tosteson, 483–518. New Jersey:Prentice Hall. 598 pp.
168. Mitchell, P. 1970. *Symp. Soc. Gen. Microbiol.* 20:121–66
169. Mitchell, P. 1970. *Biochem. J.* 116: 5P–6P
170. Mitchell, P., Moyle, J. 1965. *Nature (London)* 208:1205–6
171. Mitchell, P., Moyle, J. 1967. *Biochemistry of Mitochondria,* ed. E. C. Slater, Z. Kaniuga, L. Wojtczak, 53–74. New York: Academic. 122 pp.
172. Mitchell, P., Moyle, J. 1967. *Biochem. J.* 104:588–600
173. Mitchell, P., Moyle, J. 1967. *Biochem. J.* 105:1147–62
174. Mitchell, P., Moyle, J. 1969. *Eur. J. Biochem.* 7:471–84
175. Mitchell, P., Moyle, J. 1969. *Eur. J. Biochem.* 9:149–55
176. Montal, M., Chance, B., Lee, C .P. 1970. *J. Membr. Biol.* 2:201–34
177. Montal, M., Chance, B., Lee, C. P. 1970. *FEBS Lett.* 6:209–12
178. Montal, M., Chance, B., Lee, C. P., Azzi, A. 1969. *Biochem. Biophys. Res. Commun.* 34:104–10
179. Montal, M., Nishimura, M., Chance, B. 1970. *Biochim. Biophys. Acta* 223:183–88
180. Moore, C., Pressman, B. C. 1964. *Biochem. Biophys. Res. Commun.* 15:562–67
181. Mueller, P., Rudin, D. O. 1967. *Biochem. Biophys. Res. Commun.* 26:398–404
182. Mueller, P., Rudin, D. O. 1969. *Curr. Top. Bioenerget.* 3:157–249
183. Nelson, N., Drechsler, Z., Neumann, J. 1970. *J. Biol. Chem.* 245:143–51
184. Neumann, J., Jagendorf, A. T. 1964. *Arch. Biochem. Biophys.* 107: 109–19
185. Nishimura, M. 1970. *Biochim. Biophys. Acta* 197:69–77
186. Nishimura, M., Kadota, K., Chance, B. 1968. *Arch. Biochem. Biophys.* 125:308–17
187. Nishimura, M., Pressman, B. C. 1969. *Biochemistry.* 8:1360–70
188. Ohki, S., Goldup, A. 1968. *Nature (London)* 217:458–59
189. Ohnishi, M., Urry, D. W. 1970. *Science* 168:1091–92
190. Ovchinnikov, Y. A., Ivanov, V. T., Evstratov, A. V. 1969. *Biochem. Biophys. Res. Commun.* 37:668–76
191. Pache, W., Zaehner, H. 1969. *Arch. Mikrobiol.* 67:156–60
192. Packer, L. 1967. *Biochem. Biophys. Res. Commun.* 28:1022–27
193. Packer, L., Allen, J. M., Starks, M. 1968. *Arch. Biochem. Biophys.* 128:142–52
194. Packer, L., Crofts, A. R. 1967. *Curr. Top. Bioenerget.* 2:23–64
195. Packer, L., Murakami, S., Mehard, C. W. 1970. *Ann. Rev. Plant Physiol.* 21:271–304
196. Palmieri, F., Quagliariello, E. 1967. *Eur. J. Biochem.* 8:473–81

197. Palmieri, F., Quagliariello, E., Klingenberg, M. 1970. *Eur. J. Biochem.* 17:230–38
198. Papa, S., Guerrieri, F., Rossi-Bernardi, L., Tager, J. M. 1970. *Biochim. Biophys. Acta* 197:100–3
199. Papa, S., Guerrieri, F., Lorusso, M., Quagliariello, E. 1970. *FEBS Lett.* 10:295–98
200. Papa, S., Lofrumento, N. E., Quagliariello, E., Meijer, A. J., Tager, J. M. 1970. *J. Bioenerget.* 1:287–307
201. Papa, S., Quagliariello, E., Chance, B. 1970. *Biochemistry.* 9:1706–15
202. Papa, S., Tager, J. M., Guerrieri, F., Quagliariello, E. 1969. *Biochim. Biophys. Acta* 172:184–86
203. Papa, S., Zanghi, M. A., Paradies, G., Quagliariello, E. 1970. *FEBS Lett.* 6:1–4
204. Papahadjopoulos, D. 1970. *Biochim. Biophys. Acta* 211:467–77
205. Paul, I. C., Johnson, S. M., Herrin, J., Liu, S. 1969. *Acta Crystallog. Sect. A* 25:S196
206. Pavlasova, E., Harold, F. M. 1969. *J. Bacteriol.* 98:198–204
207. Pinkerton, M., Steinrauf, L. K. 1970. *J. Mol. Biol.* 49:533–46
208. Pinkerton, M., Steinrauf, L. K., Dawkins, P. 1969. *Biochem. Biophys. Res. Commun.* 35:512–18
209. Pinto, E., Gomez-Puyou, A., Sandoval, F., Chavez, E., Tuena, M. 1970. *Biochim. Biophys. Acta* 223:436–38
210. Pioda, L. A. R., Wachter, H. A., Donner, R. E., Simon, W. 1967. *Helv. Chim. Acta* 50:1373–76
211. Pressman, B. C. 1963. *Energy-linked Functions of Mitochondria,* ed. B. Chance, 181–203. New York: Academic. 282 pp.
212. Pressman, B. C. 1965. *Proc. Nat. Acad. Sci. USA* 53:1076–83
213. Pressman, B. C. 1968. *Fed. Proc.* 27:1283–88
214. Pressman, B. C. 1970. *Membranes of Mitochondria and Chloroplasts,* ed. E. Racker, 213–50. New York:Reinhold Co. 322 pp.
215. Pressman, B. C. 1969. *Mitochondria Structure and Function,* ed. L. Ernster, Z. Drahota, 315–33. New York:Academic. 393 pp.
216. Pressman, B. C. 1969. *The Energy Level and Metabolic Control in Mitochondria,* ed. S. Papa. J. M. Tager, E. Quagliariello, E. C. Slater, 87–96. Bari:Adriatica Editrice. 465 pp.
217. Pressman, B. C. 1969. *Ann. N. Y. Acad. Sci.* 147:829–41
218. Pressman, B. C., Harris, E. J., Jagger, W. S., Johnson, J. H. 1967. *Proc. Nat. Acad. Sci. USA* 56: 1949–56
219. Pressman, B. C., Haynes, D. H. 1969. *The Molecular Basis of Membrane Function,* ed. D. C. Tosteson, 221–46. New Jersey: Prentice-Hall. 598 pp.
220. Prestegard, J. H., Chan, S. I. 1969. *Biochemistry* 8:3921–27
221. Prestegard, J. H., Chan, S. I. 1970. *J. Am. Chem. Soc.* 92:4440–46
222. Quagliariello, E., Palmieri, F. 1970. *FEBS Lett.* 8:105–8
223. Robertson, R. N . 1968. *Protons, Electrons, Phosphorylation, and Active Transport,* London: Cambridge Univ. Press. 96 pp.
224. Rossi, E., Azzone, G. F. 1970. *Eur. J. Biochem.* 12:319–27
225. Rumberg, B., Siggel, U. 1969. *Naturwissenschaften* 56:130–32
226. Saha, S., Izawa, S., Good, N. E. 1970. *Biochim. Biophys. Acta* 223:158–64
227. Saha, J., Papahadjopoulos, D., Wenner, C. E. 1970. *Biochim. Biophys. Acta* 196:10–19
228. Sandoval, F., Gomez-Puyou, A., Tuena, M., Chavez, E., Pena, A. 1970. *Biochemistry* 9:684–89
229. Scarpa, A., Cecchetto, A., Azzone, G. F. 1970. *Biochim. Biophys. Acta* 219:179–88
230. Scarpa, A., Azzone, G. F. 1970. *Eur. J. Biochem.* 12:328–35
231. Scholes, P., Mitchell, P. 1970. J. *Bioenerget.* 1:61–72
232. Scholes, P., Mitchell, P. 1970. *J. Bioenerget.* 1:309–23
233. Scholes, P., Mitchell, P., Moyle, J. 1969. *Eur. J. Biochem.* 8:450–54
234. Schwartz, M. 1968. *Nature (London)* 219:915–19
235. Selwyn, M. J., Dawson, A. P. 1969. *Biochem. J.* 114 :90P
236. Selwyn, M. J., Dawson, A. P., Dunnet, S. J. 1970. *FEBS Lett.* 10:1–5
237. Settlemire, C. T., Hunter, G. R., Brierley, G. P. 1968. *Biochim. Biophys. Acta* 162:487–99
238. Shavit, N., Degani, H., San Pietro, A. 1970. *Biochim. Biophys. Acta* 216:208–19
239. Shavit, N., Dilley, R. A., San Pietro, A. 1968. *Comparative Biochemistry and Biophysics of Photosynthesis,* ed. K. Shi-bata, A, Ta-

kamiya, A. T. Jagendorf, R. C. Fuller, 253–65. Pennsylvania:University Park Press. 445 pp.
240. Shavit, N., Dilley, R. A., San Pietro, A. 1968. *Biochemistry* 7: 2356–63
241. Shavit, N., Herscovici, A. 1970. *FEBS Lett.* 11:125–28
242. Shavit, N., San Pietro, A. 1967. *Biochem. Biophys. Res. Commun.* 28:277–83
243. Shavit, N., Thore, A., Keister, D. L. 1968. *Proc. Nat. Acad. Sci. USA* 59:917–22
244. Shemyakin, M. M., Antonov, V. K., Bergelson, L. D., Ivanov, V. T., Malenkov, G. G., et al. 1969. *The Molecular Basis of Membrane Function,* ed. D. C. Tosteson, 173–210. New Jersey:Prentice-Hall. 598 pp.
245. Shemyakin, M. M., Ovchinnikov, Yu. A., Ivanov, V. T., Antonov, V. K., Vinogradova, E. I., et al. 1969. *J. Membr. Biol.* 1:402–30
246. Skulachev, V. P. 1970. *FEBS Lett.* 11:301–8
247. Slater, E. C. 1953. *Nature (London)* 172:975–78
248. Slater, E. C. 1969. *Mitochondria—Structure and Function,* eds. L. Ernster, Z. Drahota, 205–17. New York:Academic. 393 pp.
249. Smith, E. H., Beyer, R. E. 1967. *Arch. Biochem. Biophys.* 122: 614–20
250. Stefanac, Z., Simon, W. 1967. *Microchem. J.* 12:125–32
251. Stempel, A., Westley, J. W., Benz, W. 1969. *J. Antibiot.* 22:384–85
252. Szabo, G., Eisenman, G., Ciani, S. 1969. *J. Membr. Biol.* 1:346–82
253. Tedeschi, H., Racker, E. 1969. *J. Gen. Physiol.* 54:72s
254. Thomas, R. C., Manger, J. R., Harris, E. J. 1969. *Eur. J. Biochem.* 11:413–18
255. Thore, A., Keister, D. L., Shavit, N., San Pietro, A. 1968. *Biochemistry* 7:3499–3507
256. Tosteson, D. C., Andreoli, T. E., Tieffenberg, M., Cook, P. 1968. *J. Gen. Physiol.* 51:3735–45
257. Tupper, J. T., Tedeschi, H. 1969. *Science* 166:1539–40
258. Urry, D. W. 1971. *Proc. Nat. Acad. Sci. USA* 68:672–767
259. Ussing, H. H. 1949. *Physiol. Rev.* 29:127–55
260. Vainio, H., Mela, L., Chance, B. 1970. *Eur. J. Biochem.* 12:387–91
261. von Stedingk, L. V. 1967. *Arch. Biochem. Biophys.* 120:537–41
262. von Stedingk, L. V. 1968. *Ion Movements Related to Electron Transport and Energy Conservation.* Dissertation, Univ. Stockholm, Sweden. 20 pp.
263. von Stedingk, L. V., Baltscheffsky, H. 1966. *Arch. Biochem. Biophys.* 117:400–4
264. Waley, S. G. 1966. *Advan. Protein. Chem.* 21:1–112
265. Walker, D. A., Crofts, A. R. 1970. *Ann. Rev. Biochem.* 39:389–428
266. Wenner, C. E., Hackney, J. H. 1969. *Biochemistry* 8:930–38
267. Westley, J. W., Evans, R. H., Jr., Williams, T. H., Stempel, A. 1970. *Chem. Commun.* 72:71–72
268. Wieland, T. H., Faulstich, H., Burgermeister, W., Otting, W., Moehle, W., et al. 1970. *FEBS Lett.* 9:89–92
269. Wipf, H. K., Pache, W., Jordan, P., Zaehner, H., Keller-Schierlein, W., et al. 1969. *Biochem. Biophys. Res. Commun.* 36:387–93
270. Wipf, H. K., Simon, W. 1969. *Biochem. Biophys. Res. Commun.* 34:707–11
271. Witt, H. T. 1968. Fast Reactions and Primary Processes in Chemical Kinetics. *Nobel Symp.* 5:81, 261
272. Young, J. H., Blondin, G. A., Vanderkooi, G., Green, D. E. 1970. *Proc. Nat. Acad. Sci. USA* 67: 550–59
273. Zarlengo, M. H., Abrams, A. 1963. *Biochim. Biophys. Acta* 71:65–77

STOMACH MICROBIOLOGY OF PRIMATES

1577

T. BAUCHOP[1]

Rowett Research Institute, Bucksburn, Aberdeen, Scotland

CONTENTS

INTRODUCTION

In the alimentary canal of warm-blooded animals, wherever gastrointestinal stasis occurs, microorganisms accumulate with a resultant fermentation of digesta. In ruminants and ruminant-like animals the physiological importance of this fermentation is well recognized (16). In animals with a simple stomach a microbial fermentation typically is found in the distal regions of the alimentary tract. In mammals, the acid secretions of the normal stomach have been regarded as creating an inhospitable environment for microbial colonization. This has long been the accepted position in man (25). However, in a wide range of mammalian species it has more recently been shown that a considerable bacterial flora can exist in the stomach (24).

The microbiology of the alimentary tract in nonhuman primates has not been widely studied but from the limited studies which have been made on the stomach of some species it is clear that a variety of microbe-host relationships exist. The present review will attempt to evaluate the status of the primate stomach as a microbial habitat.

SIMPLE STOMACHS

Nonhuman primates.—All primates, apart from the important subfamily Colobinae, appear to possess a relatively simple stomach. However, both quantitative and qualitative studies on the microbial flora of nonhuman primates with simple stomachs have been extremely limited; information is available only on the subfamily Cercopithecinae of the Old World monkeys,

[1] The author was previously on the staff of the National Center for Primate Biology, University of California, Davis, California.

and even this is limited to the genus *Macaca*. Thus, there is a lack of information on the prosimians, the New World monkeys and the apes.

In a study of the microbiology of the alimentary tract of the rhesus monkey (*Macaca mulatta*), Haenel, Grützner & Henneberg (14) examined eleven animals maintained on a diet of commercial dried monkey food, fruit, and rice. The animals were killed and samples of intestinal contents were removed for pH measurements and bacteriology. The average pH of the stomach contents in eight healthy monkeys was 4.6. The predominant organisms found were lactobacilli at a level of 5×10^6 per g of contents. Lower numbers of enterococci (10^4/g), "putrefactive" bacteria (10^4/g) and coliform bacteria (10^3/g) were also found. Small numbers of fungi were also isolated. The predominant lactobacilli isolated belonged to the acidophilus group. Bifidobacteria, identified on the basis of morphology, were found to comprise between 20 and 30 percent of the total numbers of bacteria present.

In a comparative study of the bacteria in the alimentary tract of a wide range of animals, Smith (24) examined three specimens of *Macaca irus* maintained on a diet of cereal and fruit. Stomach contents were removed following sacrifice. The pH in the cardiac region was found to be 4.8 but fell to a value of 2.8 in the pyloric region. Nevertheless, high numbers of viable bacteria were isolated from contents obtained from both of these areas, and similar types of organisms were isolated from the two areas. In the cardiac region, the bacterial counts obtained per gram of digesta were: lactobacilli, 5×10^8; streptococci, 3×10^6; yeasts, 2×10^5; *Escherichia coli*, 1×10^3; *Clostridium welchii*, 50. *Bacteroides* spp. were not isolated from samples of gastric contents in any of these animals.

The microbial ecology of the gastrointestinal tract of the rhesus monkey has also been studied by Bauchop & Martucci (3). Their subjects were maintained on a diet based on cereals. Samples of stomach contents were obtained directly from sacrificed animals and by stomach tube from live animals. Identical results were obtained using these two methods. The range of pH values found in stomach contents was 2.2 to 6.0 although most samples were within the pH range 4.0 to 5.5. High numbers of viable bacteria were always present in the stomach even after animals had fasted for up to fifteen hours. The highest bacterial counts were obtained using the anaerobic techniques of Hungate (16). Counts obtained range from 3×10^8 to 1×10^{10} per g of contents. The wide range of values obtained was found to be a function of the time elapsed since the last feeding. Highest numbers of bacteria were found to be present prior to feeding when the lowest pH values were also usually obtained. The predominant bacteria found by Bauchop & Martucci (3) to be present in the stomach of the rhesus monkey were lactobacilli, streptococci, *Bifidobacterium* and anaerobic *Corynebacterium* spp. *Lactobacillus leichmanii, L. plantarum* and *L. fermenti* were the most frequently isolated *Lactobacillus* spp. *Streptococcus* spp., on the basis of physiological tests, appeared to belong to Group D, with *S. bovis* predominating.

In keeping with the predominance of the lactic acid bacteria in the gastric microbial population lactic acid was found to be the major fermentation product formed in the stomach of the rhesus monkey. The concentration found in stomach contents ranged from 1.2 to 10.6 mg/g of contents and, as with the bacterial counts, was directly related to the time elapsed since the last feeding. In an experiment to investigate in greater detail the course of events subsequent to food intake, gastric contents were removed by stomach tube and incubated anaerobically. Under these conditions bacterial growth continued for six hours. During this time the pH fell from 4.9 to 3.9 and there was a concomitant increase in lactic acid from a value of 1.2 to 34.6 mg/g of contents. This upper value was considerably greater than the highest values found in vivo, indicating that the limiting factor in the gastric fermentation may be the gastric HCl secreted by the host animal and not the acid produced during fermentation. In addition to lactic acid, small amounts of carbon dioxide, ethanol, and acetic acid were found to be produced in vitro. Thus, in the rhesus macaque ingested food and salivary secretions diluted the bacteria present in the stomach contents and also elevated the pH to a level which permitted bacterial growth to recommence. Bacterial growth continued until the stomach contents became excessively acid; this presumably was due to gastric secretions rather than to an accumulation of microbial fermentation products. This pattern of events was found to recommence with the next intake of food. Although the bacterial fermentation is limited there is general agreement (3, 14, 24) that a stable and characteristic bacterial flora does exist in the macaque and, clearly, sufficient growth can occur even under the prevailing conditions to permit the maintenance of a relatively stable and steady population.

Man.—It has long been thought that the lumen of the normal human stomach is sterile or only sparsely populated with microorganisms (24). This view has been confirmed by a number of recent studies involving a variety of cultural conditions for isolation of bacteria.

During a study of the microflora of the small intestine, Gorbach et al (12) examined samples of gastric contents, obtained by aspiration, from seven normal subjects. One sample was found to be sterile, four samples yielded culture counts of 10 to 10^2 bacteria per g and the remaining two samples contained in the range 10^2 to 10^3 bacteria per g. Similar counts were obtained with fasting subjects by Moore, Catto & Holdeman (19) but when the stomach was full, bacterial counts of 10^4 per g were obtained. In a study of 42 samples obtained from normal fasting subjects, Drasar, Shiner & McLeod (7) found no bacteria in any of the samples although five samples were positive for yeasts. Gorbach (13) has pointed out that the lowest counts which Drasar et al (7) could obtain by their dilution methods were 10^2 per g. This may explain the high proportion of apparently sterile samples which they found. Nevertheless, it confirms the general finding that numbers of viable bacteria in normal human gastric contents are extremely low. Franklin & Skoryna (10) examined 154 samples obtained by nasogastric tube from 149 fasting subjects. Twenty-two of these samples contained less than 100 bacteria per ml and were considered as negative by the authors, who allowed for possible contamination of the nasogastric tube. The remaining samples yielded culture counts which ranged from 10^2 to 10^8 per ml. The pH of the samples obtained were as high as 8.5, and the higher bacterial counts were obtained from samples with high pH. However, from the data presented by these workers (10) it was not possible to determine the exact number of samples which produced high counts of bacteria. Bacterial counts of 10^6 per g have been obtained also by Drasar et al (7) in fasting achlorhydric subjects.

In general, the antibacterial activity of gastric juice has been attributed to its acidity. A consistent finding has been that low bacterial counts are associated with the lowest pH values (7, 10, 11, 12). In an experiment to study the survival of bacteria in the human stomach, Drasar et al (7) withdrew serial samples of gastric juice at fifteen minute intervals commencing immediately after a meal. The total number of bacteria fell sharply with time from an initial count of 10^5 bacteria per g until after one hour the contents were sterile. The decrease in bacterial numbers reflected the fall in pH of the contents; lactobacilli and yeasts were found to survive longest. Apart from achlorhydric subjects such a course of events would seem to be in keeping with the bacteriological results obtained by other workers. The normal human stomach cannot therefore be described as a habitat for microbial proliferation. Instead, the adverse environment would appear merely to select those organisms with the ability to survive longest. This concept would seem to be supported by the types of organisms isolated most frequently, which were streptococci of the viridans group, lactobacilli and fungi (7, 12). Moore et al (19) found lactobacilli to predominate. In the study of Franklin & Skoryna (10) the viridans group *Streptococcus mitis* and *Streptococcus salivarius* were found to outnumber all others both in frequency of occurrence and in total count.

Nelson & Mata (20) attempted to examine the possibility that bacteria might be associated with the human stomach mucosa as has been reported in the gastric mucosa of the mouse (9, 22). Unfortunately, only one sample of normal gastric tissue was available for study and this proved to be negative.

It is of interest to note that in the 19th century, *Sarcina ventriculi,* an organism well known to proliferate under extremely acid conditions (4, 6, 23), was frequently reported to be present in high numbers in human gastric contents (23). It was considered that this anaerobic sarcina multiplied in the stomach under conditions of pyloric stenosis (23). In 1960, samples of stomach contents were examined from six patients with clinical signs of pyloric stenosis and one sample was found to contain the typical large packets of *S. ventriculi* in high numbers (Bauchop, unpublished data). However, the sarcina could not be cultured from this source and thus Beijerinck (5) remains the only worker who has succeeded in isolating *S. ventriculi* from gastric contents.

Complex Stomachs

Old World monkeys of the subfamily Colobinae differ from all other primates in the large size and anatomical complexity of the stomach (15). These differences are related to a diet consisting mainly of leaves, hence the name "leaf-eaters" commonly used to describe these primates. Members of the Colobinae are found in large numbers widely distributed across Africa and southern Asia. This subfamily comprises the colobus monkeys of Africa, the langurs of Asia, and the proboscis monkeys of Borneo.

Considerable literature has accumulated on the peculiarities of the gastric anatomy of the Colobinae (15). A gross anatomical resemblance of the stomach to the rumen of herbivorous animals has been noted, but a number of authors (1, 17, 21) have stated that rumination does not occur. Although rumination is the obvious characteristic of ruminants, recent work has placed greater emphasis on the fermentative processes occurring in the rumen. The demonstration of an extensive microbial fermentation (2, 17), in addition to the known anatomical facts, have established that members of the Colobinae possess a highly evolved ruminant-like digestion (18).

The essential adaptation of the stomach in the Colobinae is that the anatomical structure permits excellent separation of the ingesta in the proximal parts from the distal acid pyloric region. The pH is thus normally maintained at a value between 5.0 and 6.7 in the langurs examined (2), and around pH 7.0 in the colobus monkeys (17). This range of pH values permits an active fermentation of ingesta by the large numbers of anaerobic bacteria present. As in ruminants, gastric contents constitute a large proportion of the body weight of these monkeys. In colobus monkeys, values obtained for gastric contents were 11.5 to 20.6 percent of the total body weight (17), and a value of 17 percent of body weight was obtained for gastric contents of a langur monkey (2). The large capacity of the stomach allows the accumulation of ingesta and a slow rate of passage essential for extensive fermentation of plant materials.

In direct smears of stomach contents obtained from colobus monkeys Kuhn (17) found bacteria to be present, although he did not consider that they compared either numerically or in diversity with ruminant flora. Characteristic protozoan fauna were absent. The availability of langur monkeys in a laboratory environment permitted Bauchop & Martucci (2) to investigate the bacterial flora in these animals in greater detail. Anaerobic culture counts were found to range between 2×10^{10} and 1×10^{11} per g of contents,[2] and the ratio of anaerobic to aerobic cultures counts was in the range of 100:1 to 1000:1. As langurs thrive on a leafy diet it was significant that large numbers of cellulose-digesting bacteria were demonstrated in gastric contents (2). The counts obtained using Hungate's techniques (16) ranged from 8×10^{6} to 1×10^{8} per g of contents. Two forms of cellulose-

[2] Bacterial counts were originally (2) expressed in terms of dry matter content of samples. In the present paper, the more normal method has been adopted and bacterial counts are expressed in terms of wet weight of sample.

digesting bacteria were isolated, a Gram-positive coccus and a Gram-negative *Bacteroides* spp. Although these organisms appeared similar to the major cellulose-digesting bacteria isolated from the bovine rumen (16), they were not adequately described for detailed comparisons to be made. These results indicate that cellulose digestion may be important in the Colobinae, but the extent of this digestion in these animals remains to be more fully investigated.

In the colobus monkey, fermentation products include a gas mixture of carbon dioxide and methane (17). Similar results were found in the langurs except that small amounts of hydrogen were also found to be produced (2). A search for methanogenic bacteria resulted in the isolation of *Methanobacterium ruminantium* at high dilutions of stomach contents (2). The counts obtained ranged from 3×10^8 to 1×10^9 per g of contents.

Drawert, Kuhn & Rapp (8) analyzed samples of gastric contents from colobus monkeys and found high concentrations of short-chain volatile fatty acids (VFAs) similar in concentration and character to these fermentation products formed in rumen contents. In two species of langurs, Bauchop & Martucci (2) also found VFAs in concentrations and molecular proportions essentially similar to those found in ruminants. Although the presence of isobutyric acid has not been reported in stomach contents of the Colobinae, in the langurs at least this acid was found to comprise up to 6.0 percent of the total VFA content of some samples (Bauchop & Martucci, unpublished observations).

In ruminants the VFAs are absorbed directly from the rumen, and in ruminant-like animals, from the stomach (18). Although this has not been demonstrated directly in the Colobinae, Drawert et al (8) found in *Procolobus verus* that the VFA concentration of the contents decreased from 230 mmoles/liter in the mid-stomach region to 24 mmoles/liter in the pyloric region. Although this indicates that absorption of VFAs does occur in the stomach, further work on this aspect of the digestive physiology would be of value. The gastric concentrations of these acids therefore represent an equilibrium between their rate of production and their absorption or passage out of the upper stomach. A constant VFA concentration was demonstrated by analysis of gastric samples taken from a specimen of *Presbytis cristatus* (2). After feeding fresh alfalfa, samples of gastric contents were withdrawn by stomach tube at regular intervals. Within two hours of feeding the VFA concentration attained a high level which was maintained relatively constant between 162 and 197 mmoles per liter throughout the 6.5 hours of the experiment. The constancy of the gastric pH at 5.6 during the course of this experiment confirmed that langurs possess an excellent mechanism for controlling, within narrow limits, the pH of their stomach contents.

A high in vitro rate of VFA production was found in fresh samples of gastric contents of both colobus and langur monkeys (2, 17). In addition, in the langurs (2) samples obtained before the morning feeding, 10.5 hours after the previous feeding, also produced VFAs at a high rate. Thus, the microbial fermentation in these monkeys would appear to be a continuous process. On the basis of the fermentation rates obtained and assuming that the VFAs were absorbed, Bauchop & Martucci (2) calculated that the gastric fermentation could make important contributions to the energy metabolism of these monkeys. In addition, these workers suggested that the bacterial biosynthetic capacities may also benefit the vitamin and nitrogen economy of these unusual primates. Moir (18) has concluded that the stomach form in this group of monkeys is a highly evolved ruminant-like one which could permit these animals to invade ecological niches difficult or even impossible for other primates.

LITERATURE CITED

1. Ayer, A. A. 1948. *The Anatomy of Semnopithecus entellus.* Madras: Indian Pub. House
2. Bauchop, T., Martucci, R. W. 1968. *Science* 168:698–700
3. Bauchop, T., Martucci, R. W. To be published
4. Bauchop, T., Dawes, E. A. 1968. *J. Gen. Microbiol.* 52:195–203
5. Beijerinck, M. W. 1911. *Proc. Sect. Sci. K. ned. Acad. Wet.* 13:1237–40
6. Canale-Parola, E. 1970. *Bacteriol. Rev.* 34:82–97
7. Drasar, B. S., Shiner, M., McLeod, G. M. 1969. *Gastroenterology* 56: 71–79
8. Drawert, F., Kuhn, H.-J., Rapp, A. 1962. *Z. Physiol. Chem.* 329:84–89
9. Dubos, R., Schaedler, R. W., Costello, R., Hoet, P. 1965. *J. Exp. Med.* 122:67–76
10. Franklin, M. A., Skoryna, S. C. 1966. *Can. Med. Assoc. J.* 95:1349–55
11. Gray, J. D. A., Shiner, M. 1967. *Gut* 8:574–81
12. Gorbach, S. L. et al. 1967. *Gastroenterology* 53:856–67
13. Gorbach, S. L. 1969. *Gastroenterology* 57:231–32
14. Haenel, H., Grützner, L., Henneberg, G. 1960. *Zentralbl. Bakteriol. Parasitenk. Infektionskr. Hyg. Abt. 1* 178:42–50
15. Hill, W. C. O. 1952. *Proc. Zool. Soc. London* 122:127–86
16. Hungate, R. E. 1966. *The Rumen and Its Microbes.* New York: Academic Press
17. Kuhn, H.-J. 1964. *Folia Primat.* 2: 193–221
18. Moir, R. J. 1968. *Handbook of Physiology.* Sect. 6, V:2673–94, ed. C. F. Code. Am. Physiol. Soc., Washington, D.C.
19. Moore, W. E. C., Catto, E. P., Holdeman, L. W. 1969. *J. Infect. Dis.* 119:641–49
20. Nelson, D. P., Mata, L. J. 1970. *Gastroenterology* 58:56–61
21. Owen, R. 1835. *Trans. Zool. Soc. London.* 1:65–70
22. Savage, D. C., Dubos, R., Schaedler, R. W. 1968. *J. Exp. Med.* 127: 67–76
23. Smit, J. 1933. *J. Pathol. Bacteriol.* 36:455–68
24. Smith, H. W. 1965. *J. Pathol. Bacteriol.* 89:95–122
25. Wilson, G. S., Miles, A. A. 1964. *Topley & Wilson's Principles of Bacteriology and Immunity.* 2464 Baltimore: Williams & Wilkins Co.

TOWARD A METABOLIC INTERPRETATION OF GENETIC RECOMBINATION OF *E. COLI* AND ITS PHAGES

1578

Alvin J. Clark

Department of Molecular Biology, University of California, Berkeley, California

Contents

INTRODUCTION

Five excellent reviews of the recent work on genetic recombination have appeared within the past year (24, 52, 84, 98, 117). All aspects of genetic recombination have thereby been summarized for those who want a thorough and complete bibliography of the field. There is one point which, although not neglected by these reviews, is not given the emphasis this author considers appropriate. Hence, the following will be a limited and highly speculative discussion of the field emphasizing the view that genetic recombination is a complex area of nucleic acid metabolism. If this view is accepted, the goal of the study of recombination must be to detail the biochemical steps of recombination and to show the relation of recombinational intermediates and enzymes to those involved in repair, transcription, and replication. Considerable progress has been made toward this goal through the use of molecular biological methods of labeling and centrifugation, genetic methods of mutant detection and characterization, enzymological methods of protein purification and reaction characterization, and intellectual methods of imagination and illustration. To illustrate the progress and the metabolic point of view the author will draw mainly upon studies with *Escherichia coli* and its viruses, not because other studies are less exemplary but because brevity demands that the author rely essentially upon his own expertise.

Studies on the recombination of other bacteria and their viruses are of great importance in ascertaining the range of differences in recombination pathways to be found within the biological world. Several such studies are worthy of note even though they cannot be reviewed in this brief article. *Diplococcus pneumoniae* and *Bacillus subtilis,* for example, are subject to transformational gene transfer (53). The structure of recombinant molecules following transformation has been extensively studied in these two species and correlations can therefore probably be drawn between mutational defects affecting recombinational enzymes and structures of intermediate products (52). The results may be compared with those obtained from conjugational studies of *E. coli* to ascertain if there are any similarities between the pathways of recombination following transformation and those following conjugation. *Salmonella typhimurium,* on the other hand, represents a close relative to *E. coli* and *Pseudomonas aeruginosa* represents an unrelated species. Both species engage in conjugation and transduction. It will be interesting to observe whether or not the pathways of recombination in these species are similar to those in *E. coli* (29, 49, 76, 122, 126).

Regrettably, space limitation also prevents our discussing the involvement of recombination in the repair of irradiation-induced lesions in DNA. Fortunately, two reviews (54, 124) have appeared recently and one more will appear this year (102) so that an up-to-date survey of this topic is available.

A Definition of Recombination

"Recombination" refers to a set of phenomena some of which have been known for 60 years (83). In addition to crossing-over, these include the integration of plasmids with one another or with larger chromosomes, high negative interference, and gene conversion. Still other phenomena have been attributed by hypothesis to recombination, e.g., repair of DNA damage caused directly or indirectly by radiation (57, 58, 102), ultraviolet-induced mutation (81, 125), and the production of chromosomal aberrations including inversions, duplications, and deletions. With all of these phenomena in mind it becomes increasingly difficult to define recombination so that the definition is appropriately sharp and yet broad enough to include the phenomena of uncertain explanation. The following seems an appropriate definition of recombination, although changes are to be expected as research continues.

> Recombination: any of a set of pathways in which elements of nucleic acid interact with a resultant change of linkage of genes or parts of genes.

This definition includes several features that are generally accepted. First of all there is no a priori reason to exclude phenomena involving RNA recombination although the recombinational phenomena under study currently involve DNA. Secondly, different "elements" of nucleic acid conceivably may be involved in recombination: i.e., single-stranded, wholly or partially double-stranded elements. The elements also may have different cellular or viral parents, or they may be sister replicas, or they may be more or less distantly linked regions of a single chromosome. Thirdly, the elements of nucleic acid interact by a process of controversial nature termed synapsis. Finally, the definition excludes phenomena that result from independent assortment of different linkage groups because no linkage changes are involved in these phenomena.

The definition, however, omits the commonly accepted notion that recombination occurs by a single mechanism and substitutes the idea that there is a set of recombinational pathways. This omission and substitution require justification.

Essentially three mechanisms of recombination have been conceived (25): 1. breakage and reunion (or breakage-and-joining); 2. copy-choice; and 3. fragmented copy-choice (or breakage-and-copying). To a first approximation the mechanisms can be distinguished by two criteria: the exchange of parental fragments of DNA characterizes mechanisms 1 and 3 while the necessity of nucleic acid synthesis characterizes mechansisms 2 and 3. Thus far, evidence has been obtained by radioactive labeling techniques for the exchange of parental fragments (63, 80, 89, 90), implicating mechanisms 1 and 3. Evidence for DNA synthesis has also been obtained, implicating mechanisms 2 and 3 (104). Conservative DNA synthesis has been used to formulate the copy-choice (25) and the breakage-and-copying (79) mechanisms. Conservative synthesis seems unlikely to be of widespread occurrence, however, so that formulations involving semiconservative synthesis are to be preferred. Such formulations seem to require fragmentation or breakage of at least one parental molecule (25); it therefore seems appropriate to refer to all of these by the term "breakage-and-copying."

Are breakage-and-joining and breakage-and-copying mutually exclusive? Must one assume that recombination in all its manifestations and in all organisms and viruses proceeds by a single mechanism? Is it still valid to couch discussions of recombination around the idea of "mechanism" as if there were one enzyme which performed one reaction known as recombination? Testing the nature of the final products of recombination cannot answer these questions. To answer them requires a study of mutants of bacteria and viruses deficient in recombination. Although the recombination intermediates or the activities of recombination enzymes have not yet been fully characterized, it is possible to formulate hypothetical pathways each of which contains several steps. A framework of this formulation will be presented in the next section before we comment on the questions raised here.

Enzymes and Pathways of Recombination

Introduction.—Speculations on the enzymes and pathways of recombination serve the purpose of illustrating and clarifying the complexities of the field and providing testable hypotheses. A general knowledge of DNA and the known or imagined activities and substrate specificities of DNA enzymes is all that is necessary to propose a seemingly endless series of pathways. For this reason it seems desirable to organize the proposals by arbitrarily dividing recombination into a series of stages. Since the concept of synapsis is central to the concept of recombination, it seems appropriate to consider the stages to be Presynapsis, Synapsis, and Postsynapsis. The DNA that passes through these stages becomes a recombinant product, which can be replicated directly or can be subject to postrecombinational changes prior to replication. The diagram below illustrates the sequence of these stages.

Nucleic Acid Elements ⇆ Presynapsis ⇆ Synapsis → Postsynapsis → Recombinant Product → Replication

Synapsis ↓ Lengthening the Synaptic Region ↑ Postsynapsis

Recombinant Product ↓ Postrecombinational Changes ↑ Replication

Between two of the main stages a possible intermediate stage is shown, involving lengthening of the synaptic region.

Presynapsis.—The major theories of synapsis require the formation of a single-stranded portion of one or both of the two interacting parental elements. There are numerous events that might lead to single-stranded regions in a double-stranded molecule. Thomas (109) has classified these events in the manner illustrated by Figure 1. Double-stranded DNA is

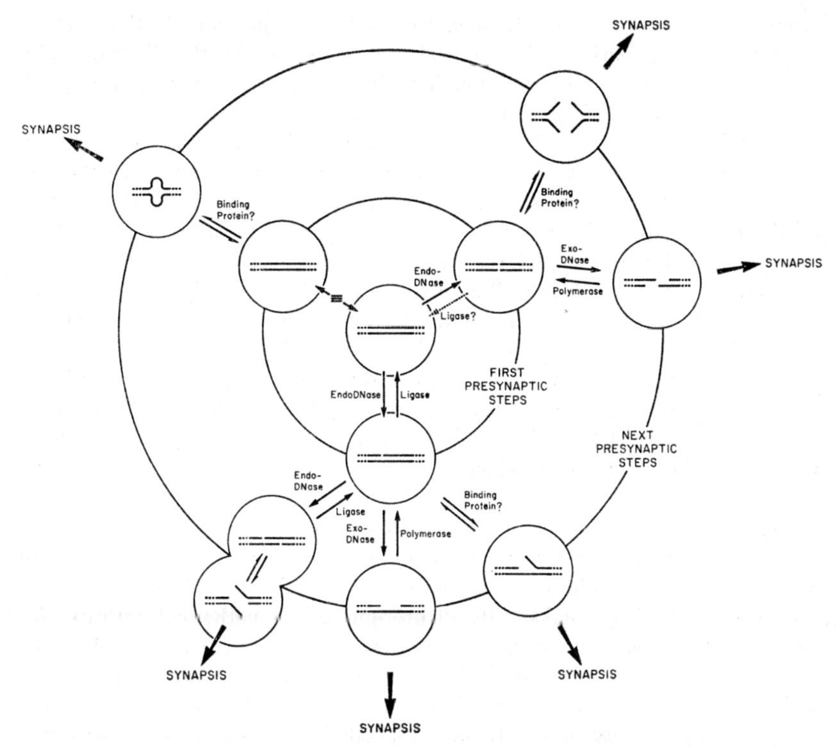

FIGURE 1. Six possible presynaptic pathways of recombination. Double-stranded native DNA is represented by closely spaced parallel lines. Dotted lines indicate that the helix is indefinite in extent. The final products of presynapsis are partially single-stranded. Reactions hypothesized to be reversible are represented by opposite half arrows; cyclic processes are represented by opposite whole arrows. Enzymes or proteins that might carry out the indicated reactions are named next to the appropriate arrows. In the repair of a double-strand the hypothesized ligase step is indicated by a dotted arrow because of the uncertainty that the reaction which occurs in vitro may occur in vivo (96a).

shown in the center circle of the diagram. Possible first steps are shown on the first concentric circle. Beginning at the upper left and proceeding clockwise the DNA may remain unchanged, may have both of its strands broken, or may be nicked, i.e., broken in only one strand. Possible subsequent steps are shown on the outermost circle. These involve either unwinding of the DNA, exonucleolytic action, or endonucleolytic action followed by unwinding.

The six molecules with single-stranded regions shown in Figure 1 are the products of six possible presynaptic pathways. If each type of molecule could synapse with itself or with any of the others to form a duplex there would be twenty-one possible synaptic products; consequently there are at least twenty-one pathways of recombination implied by the diagram. This number is increased (but not doubled) if certain of the presynaptic products may have either 5′ or 3′ terminated single-stranded ends. In addition, a completely single-stranded DNA element may be available for synapsis following transfer by transformation or conjugation, and this further increases the number of possible recombination pathways. Many, if not most, of these possibilities have been used singly by various authors to illustrate their proposed pathways of recombination. Two recently suggested pathways are indicated in Figure 2 as illustrations. Broker & Lehman (15) use a presynaptic series beginning with the introduction of a nick in double-stranded DNA and continuing with exonucleolytic degradation of one strand beginning at the nick. Cassuto & Radding (19), on the other hand, use a presynaptic series probably beginning with a double-stranded endonucleolytic scission and continuing with an exonucleolytic degradation from the ends of the fragments produced.

Synapsis.—Speculations on the nature of the presynaptic and postsynaptic stages depend upon the imagined nature of synapsis. We are aware of only two major hypotheses for synapsis in bacterial and phage recombination. One is used by Lacks (70) in presenting a hypothetical pathway for recombination following transformation in *D. pneumoniae*. It involves the insertion of a single strand of DNA into the major groove of a native bihelical element and the formation of trios of hydrogen-bonded bases at every station along the resultant triplex. Specificity of synapsis is permitted through the proposed specificity of the hydrogen bonds in each trio of bases. The second and more frequently encountered hypothesis involves the formation of a DNA duplex whose specificity is insured by the complementary base sequence of the participating single strands. The pathways in Figure 2 exemplify the duplex hypothesis. Since DNA occurs naturally as a stable duplex while stable triplex structures are formed only by some synthetic polymers (94), it is obvious why the duplex hypothesis is more appealing. Two thoughts should temper any dogmatic assertion of the duplex hypothesis, however: 1. a triplex might be an unstable intermediate between presynaptic and stable synaptic forms: 2. a triplex might be stabilized in some situations by particular proteins.

The triplex and duplex hypotheses are both useful in explaining the random occurrence of recombination and the site-specific recombinational phenomena which have also been reported (28, 115). To explain site-specific recombination, an additional step may be added to duplex or triplex formation. In this hypothetical step synapsis would occur indirectly between two native and unbroken duplexes of DNA through the medium of proteins with two DNA affinity sites and a high degree of base-sequence specificity in their attachment to the DNA. Analogies for the base-sequence specificity of

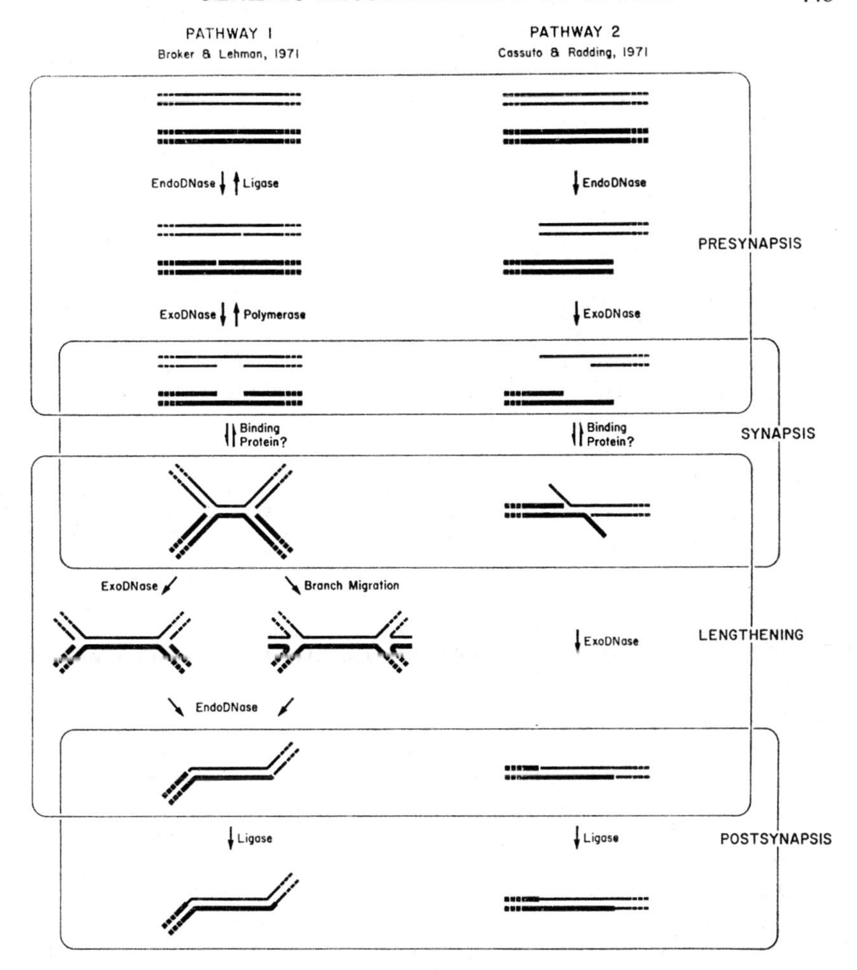

FIGURE 2. An analysis of two possible recombination pathways recently suggested to account for phage recombination as described in the text. Double-stranded DNA of indefinite length is represented as in Figure 1 except that molecules of different parental origin are indicated by different thickness of the lines. The steps in the pathways that correspond to the four stages described in the text are circumscribed and the stage involved is named on the right of the figure on the appropriate circumscribing line. Note that the starting materials and the recombinant products of the two pathways are identical although the intermediate steps are different.

the hypothetical synaptic proteins are found in the high degree of base-sequence affinity exhibited by repressor proteins, certain RNA polymerase subunits, and restriction-modification enzymes.

The fact that none of the three possible types of synapsis have been demonstrated complicates any speculative discussion of recombination pathways. To simplify, we consider here only those pathways involving the formation of a synaptic duplex.

Lengthening the synaptic region.—Regardless of its nature the synaptic complex is likely to be a relatively transient intermediate. Once formed it could conceivably meet any one of three possible fates: 1. decay to form the original presynaptic intermediates; 2. joining the parental DNA sequences to form the recombinant products; or 3. lengthening the synaptic region prior to either of the above. The stability of any duplex synaptic product has been the subject of a number of speculations and investigations. The evidence appears to indicate that duplexes of between 10 and 50 nucleotides are thermodynamically stable in vitro at 37°C (78, 108). The half-life of such short duplexes under physiological conditions would probably depend partly on the enzymes or proteins present and partly on the physical conditions of temperature and salt concentration. The stability of such duplexes in vivo might increase if they become longer. To lengthen synaptic duplexes two processes have been suggested: 1. branch migration (or strand displacement), and 2. exonucleolytic degradation. Branch migration was first noticed in vitro (71) and is thought to involve the sequential displacement of the bases of a DNA strand from their hydrogen-bonded association with a complementary strand by the bases of an adjacent redundant or counterpart strand. Broker & Lehman (15) have pointed out that branch migration following synapsis in vivo would serve to extend the hybrid region of interaction between the parental elements. Lengthening ceases when this branch reaches strategically placed nicks or double-strand breaks. Broker & Lehman (15) and Cassuto & Radding (19) have proposed that exonucleolytic degradation of redundant strands would also lengthen the hybrid region. These two proposals, indicated in Figure 2, differ mainly in the way the degradation-induced lengthening is terminated. In the first case endonuclease-introduced nicks in the molecule cause degradation to cease. In the second case, degradation ceases when the redundant portion of the duplex has been completely removed and replaced by its counterpart strand.

Postsynapsis.—During postsynapsis the intermediates are committed to form recombinant molecules. Lengthening the synaptic region should be considered a possible but not necessary route for synaptic DNA prior to postsynapsis. Postsynaptic events are those which are directly involved in the formation of the first covalent bond between base sequences of the two interacting elements; all subsequent events may also be called postsynaptic. Two processes would serve to link covalently base sequences from the two interacting elements: 1. DNA synthesis, and 2. ligase or ligase-like joining reactions. Figure 3 shows these two processes producing alternative products from a single hypothetical synaptic intermediate.

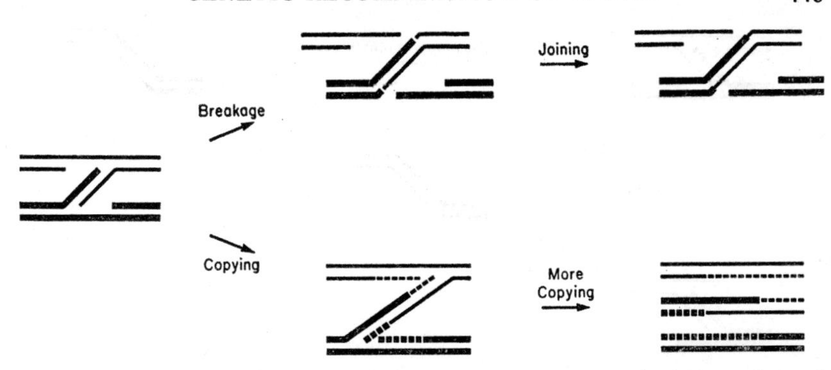

FIGURE 3. Two possible postsynaptic fates of one hypothetical product of synapsis. Native DNA is represented as in Figures 1 and 2 except that the broken lines in this case stand for newly synthesized DNA. The linkage of the two parental base sequences is accomplished by an endonuclease and a ligase step in the upper pathway and by DNA synthesis in the lower pathway. Genetically the two recombinant products are identical but the additional products are different.

The sequence involving synthesis is similar to that proposed by Boon & Zinder (13). DNA synthesis is initiated in the region of synapsis and serves to link covalently each of the parent strands with a base sequence copied from the other parent. If linear molecules interact as drawn in Figure 3 then synthesis can continue to the ends of the molecules and the postsynaptic steps are concluded when synthesis is completed. If one or both of the interacting molecules are circular then a series of steps involving endonuclease, ligase, and possibly polymerase are required to complete the recombination process (13).

The sequence involving a polynucleotide ligase reaction is shown in only one of its variations in Figure 3. In this variation an endonuclease has cut the parent molecule precisely so that a nicked intermediate is formed, i.e., the bases at the ends of the parental fragments are hydrogen-bonded to adjacent bases on a complementary strand. Polynucleotide ligase is then shown closing the nick to join the two parental DNA elements. This sequence is repeated in Figure 4 along with two other possibilities for this kind of joining process. One of these possibilities envisions an exonuclease step in addition to the endonuclease and ligase steps. The exonuclease step is required because endonuclase action has left redundant single strands on the postsynaptic intermediate. Another possibility suggested in essence by Lacks (70) combines endonuclease and joining reactions into a single transfer reaction. The hypothetical enzyme involved in such a reaction might be called a polynucleotide transferase because of its ability to break a phosphodiester bond and transfer the end of one of the strands directly to an adjacent end, thereby restoring the continuity of the double helical backbone. A polynucleotide transferase would perform recombinational joining events in the absence of a polynucleotide ligase.

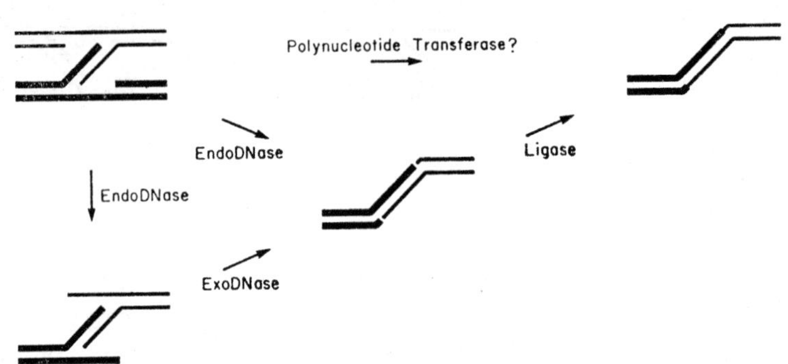

FIGURE 4. Three possible postsynaptic pathways involving no DNA synthesis. DNA is represented as indicated in the legends to Figures 1 and 2. Only the recombinant and prerecombinant products are shown because the secondary products of each pathway are different. The pathways differ by the substrate specificities and reaction mechanisms imagined for the enzymes indicated.

Recombinant products and their postrecombinational fates.—The recombinant products produced by any of the possible pathways discussed are similar to the extent that they carry a hybrid region formed by complementary single strands of the two parental elements. This heteroduplex region may contain mismatched base pairs because of genetic differences in the two parents. Hence, the recombinant is in a sense heterozygous. The progeny of this heterozygous recombinant molecule are expected to be heterogeneous since the parental base sequences from the heteroduplex region would segregate into different replicas. "Het" recombinant molecules may therefore be recognized by their production of heterogeneous progeny upon replication. Prior to replication, however, any recombinant product may undergo a postrecombinational repair process. If such a process occurs, it will excise one or the other mismatching base and replace the excised region with matching (i.e., complementary) bases. Such postrecombinational repair events may give rise to gene conversion (48) and to marker effects (84, 86) on recombination frequencies. Although some believe that postsynaptic recombinational events are involved in marker effects (32, 39), both gene conversion and marker effects may reflect postrecombinational rather than recombinational phenomena. However, since both gene conversion and marker effects modify the frequency of recombinant clones, it is clear that the detection of recombinant clones may be an imprecise method for the detection of recombinant molecules.

A comment on the notion of recombination mechanisms.—It seems that there should be some way to typify the myriad possible recombination pathways. The small number of categories termed recombination mechanisms offers the possibility that any conceivable pathway may be classified as exemplifying one or another mechanism. If one ignores (perhaps unjustifiably) any mechanism predicated on conservative DNA synthesis there appear to be only two possibilities: breakage-and-joining and breakage-and-copying. It is tempting to classify all possible pathways that involve no DNA synthesis as examples of the breakage-and-joining mechanism and all possible pathways that involve some DNA synthesis as examples of the breakage-and-copying mechanism. Such a classification is difficult to maintain, however, because some possible pathways, although involving DNA synthesis, lead to recombinant products that contain no newly synthesized DNA. These would be mistaken for breakage-and-joining products. Also there has been a tendency to consider the extent of DNA synthesis or the particular kind of synthetic process, rather than the presence or absence of DNA synthesis, as determinative in separating breakage-and-joining from breakage-and-copying. This means that even breakage-and-joining pathways have been considered to involve some copying. What is more, there is the additional confusion that breakage-and-copying pathways involve joining reactions. All of these points militate against the use of the terms breakage-and-joining and breakage-and-copying to typify the pathways of recombination. To make matters worse, each step in a pathway may be carried out by a different enzymic mechanism; therefore to categorize multistep pathways as examples of a particular mechanism seems equivocal. Consequently, in this article, the idea that all recombination can be described by one or two overall mechanisms has been abandoned in favor of the view that recombination can be described by a set of pathways. The number of recombinational pathways eventually demonstrated to operate may be small enough that categories in which they fall can be discerned and named accurately.

GENETIC ANALYSIS OF RECOMBINATION PROFICIENCY IN *E. COLI*

A genetic analysis involves the isolation of mutants and the characterization of the enzymatic reactions affected by the mutations. Such a genetic analysis of recombination proficiency of *E. coli* has been in progress for about seven years. In addition, a similar genetic analysis of recombination of the bacteriophages T4 and λ, which attack *E. coli,* has been in progress for nearly as long. The phages have one decided advantage in this study because it is easy to isolate and detect their recombination products (e.g., phage hets). Phages have one decided disadvantage, however, in that they perform recombination in host cells which have their own recombination enzymes. Because of the interplay between the host and phage enzymes the physiological complexity of phage recombination is greater than for bacterial recombination.

Isolation of recombination-deficient mutants.—Detection methods for recombination-deficient (i.e., Rec^-) mutants are generally nonrigorous and require screening of all survivors of mutagenic treatment for transmission deficiency (22), or some phenotypic property correlated with recombination deficiency such as sensitivity to ultraviolet irradiation (87, 91) or X irradiation (31, 59). To narrow the search somewhat another method has been developed which involves plating the mutagen-treated survivors on medium containing methyl methane sulfonate at a predetermined concentration, which causes some Rec⁻ mutants to form smaller-than-normal colonies (85). Another method, used successfully for enrichment (26, 64), depends on the pleiotropic phenotype produced by mutations in one of the *E. coli rec* genes (*recA*), namely the interference with some forms of induction of prophage λ (16, 45). In one of these cases the detection method is rigorous enough that spontaneous *Rec⁻* mutants can be uncovered (64); unfortunately, mutations in only the *recA* gene can be detected in this way.

In every case the detection methods employed are not specific for recombination-deficient mutants. Those mutants in which processes other than recombination have been affected must be eliminated by an additional test. Usually this test consists of comparing the abilities of the mutants and their immediate ancestors to produce progeny by recombination following a cross with an Hfr donor strain and to produce progeny by repliconation following a cross with an F′ donor strain. In essence, mutants are checked for their ability to produce zygotes following conjugation, the zygotes being detected by the effects of repliconation [i.e., a process of inheritance in which a DNA fragment is established as an autonomously replicating unit (20)]. Repliconation is involved both in inheritance of F′ elements following transfer and in lysis of zygotes following zygotic induction of prophage λ. At present there is no firm evidence that recombination is a necessary stage in repliconation, although a 50 percent to 90 percent reduction in the inheritance of F*lac* by putative recombination-deficient mutants is common (20), and with one F′ element a more severe effect was noted (91).

Mapping the mutations and phenotypically grouping the mutants.—Mutants suffering from decombination deficiency may be grouped phenotypically (20). For example λ lysogens of *recA* mutants show no spontaneous production of λ prophage nor are they inducible by ultraviolet (16, 45). The λ lysogens of other *rec* mutants (e.g., those carrying *recB⁻* and *recC⁻*), on the other hand, are normal in their spontaneous production or ultraviolet induction of λ phage (20). Another test of difference between *recA* mutants and *recB* and *recC* mutants is the degree to which endogenous labeled DNA is degraded following ultraviolet irradiation (30, 56, 120).

The phenotypic properties of a mutant may be misleading, however, if it carries additional mutations. Additional mutations are not uncommon and may produce effects attributable by reasonable hypothesis to recombination deficiency. Strains AB2463 and AB2462, both isolated following mutagenic treatment of AB1157, carry, in addition to *recA13* and *rec-12,* respectively, independently produced secondary mutations which prevent acridine-curing of F′ elements (7). AB2462 may also carry a third mutation leading to a high reversion rate of three auxotrophic characteristics to prototrophy

(23). In addition, four mutants were studied following their appearance as strains unable to inherit F*lac* and to perform recombination following transduction and conjugation. All four are at least double mutants, carrying, in addition to *recA* or *recC* mutations, one or more mutations preventing conjugational transfer and zygote formation when the strains are used as recipients (23).

The best way to confirm the nature of the phenotype determined by a *rec* mutation is to map the mutation. A recombinant inheriting the *rec* mutation but not inheriting distantly linked adventitious mutations may then be characterized by its phenotype. Reversion may also be used to determine the effects of a *rec* mutation but the occurrence of suppressor mutations in the revertants may lead to erroneous conclusions.

Determination of enzymatic and physiological defects in the mutants.—When an effort is made to separate *rec* mutations from adventitious mutations, it is clear that most *rec* mutants fall into two phenotypic groups. The first group carries mutations in *recA* and the second group carries mutations in *recB* and *recC*. Previously, three phenotypic groups were proposed (20) but two of those groups have been amalgamated because they were based on differences in apparent dominance of mutations all of which are probably in *recA* (23, 121). Meanwhile recombination-deficient mutants have been reported that do not carry *recA*, *recB*, or *recC* mutations. Phenotypic recognition of groups of these mutants must await more extensive characterization. The discovery of indirect suppression of *recB* and *recC* mutations has led to the recent discovery of some other groups of recombination-deficient mutants but characterization of these is also incomplete (50).

recA *mutants.*—Many recombination-deficient mutants show a much more extensive breakdown of endogenous DNA following ultraviolet irradiation than does the wild-type strain (23a, 56, 96). The same mutants, when lysogenic for wild-type λ prophage (i.e., λ^{++}), show greatly reduced spontaneous production of phage and greatly reduced capacity for induction of λ multiplication following ultraviolet irradiation, thymine starvation, or mitomycin treatment (20, 96). Mutants with these two properties carry mutations weakly cotransducible with *cysC* and *pheA*, at 53 and 50, respectively, on the standard map of *E. coli* (121). Presumably, these mutations lie in one gene but complementation studies are difficult for two reasons. First, no F′ element is available to construct permanent merodiploids. Our attempts to isolate such an F′ element have been unsuccessful (23) and our attempts to use the KLF8 element (75) have been equally unsuccessful, presumably because of the tendency of this F′ element to lose the *recA* region by deletion (119). Secondly, the mutations cotransducible with *cysC* appear to grade continuously from almost recessive to almost dominant when tested for their effect on recombinant formation in transient rec^+/rec^- conjugational merozygotes (20, 23, 121). Unfortunately, the measurement of dominance and recessiveness in transient zygotes suffers from the possible complication that differences in the rate or extent of degradation of the exogenote may also lead to differences in the number of recombinants and consequently may be mistaken for differences in dominance and recessiveness. Whichever explanation for the differences in recombinant formation is correct, the fact remains that, in the absence of an appropriate stable F′ element, we cannot reliably determine the number of complementation groups into which the *rec* mutations cotransducible with *cysC* fall. As a working hypothesis based on the phenotypic similarity of the mutants, we propose that all these mutations lie in one gene, i.e., *recA*.

Two *recA* mutations have been checked for their quantitative effects on recombination. Although progeny are produced in crosses with Hfr strains at 10^{-3} to 10^{-4} the frequency observed with a Rec$^+$ recipient, all the progeny have apparently been formed by repliconation of F′ elements (75). The ability of *recA* mutants to perform recombination is therefore undetectable at this level and presumably is virtually nonexistent. Recombination following transduction has also not been detected (23, 45). The product of the *recA* gene is consequently of great importance to recombination. Its absence leads to pleiotropic effects on prophage induction, survival following ultraviolet irradiation, mitomycin, and X-ray treatment, and ultraviolet-mutagenesis (81, 125). The *recA* product may be an enzyme (such as a polynucleotide kinase or an endonuclease), a synapsis protein [such as a DNA-binding protein (1)], a membrane protein determining the structure of the bacterial chromosome (112), or some other. The effect of the *recA* products in limiting or inhibiting the action of the *recB-recC* enzyme on endogenous DNA (120) has led to the finding that less *recB-recC* enzyme activity can be measured in lysates of *recA*$^+$ than in lysates of *recA*$^-$ cells (53a, 67). It is too early to tell whether this difference represents an action of the *recA*$^+$ product in vitro after cell lysis or an action in vivo prior to cell lysis.

recB *and* recC *mutants.*—The primary detection of *recB* and *recC* mutants is more difficult than the primary detection of *recA* mutants, since considerable residual recombination occurs in *recB* and *recC* mutants and their sensitivity to ultraviolet irradiation is considerably less than that of *recA* mutants. Consequently, the great majority of *recB* and *recC* mutations have been detected by their X-ray sensitivity, which is similar to that produced by *recA* mutations (30, 120a).

All the *recB* and *recC* mutations studied are highly cotransducible with *thyA* on one side and *argA* on the other (30, 120a). Since F15 covers these two reference genes the recessiveness of the mutations determined by transient merozygote analysis has been confirmed by permanent merozygote analysis (120a). As a result the mutations assigned to *recB* have all been shown to give no complementation with *recB21* and those assigned to *recC* to give no complementation with *recC22*. In addition, the *recC* mutations have all been found to be more highly cotransducible with *thyA* than with *argA* while the reverse is true for *recB* mutations. These results are sufficient to classify *recB* and *recC* as two different cistrons and to make it likely that they represent different genes (120a).

From a study of DNA degradation initiated when a temperature-sensitive DNA synthesis mutant was incubated at the nonpermissive temperature, Buttin & Wright (17) were led to conclude that the product of the *recB* and *recC* cistrons was a DNase. The same conclusion was reached by Willetts & Clark (120) as a result of their study of the ultraviolet-induced DNA degradation occurring in multiple *rec* mutants. Both studies showed that *recB* or *recC* mutants, or both, degraded endogenous DNA far less than wild-type strains did. The conclusions of these studies were verified by the detection of an ATP-dependent DNase activity in crude lysates of $recB^{+}$ $recC^{+}$ strains that was absent from lysates of *recB* and *recC* mutants (6, 88, 125a). This activity has now been purified 10,000-fold by Goldmark & Linn (36, 37). In this preparation four activities are found. Three of these activities are ATP-dependent and one is not. The three that are ATP-dependent are a double-stranded DNA exonuclease, a single-stranded DNA exonuclease, and a single-stranded DNA endonuclease. The one ATP-independent activity is a single-stranded DNA endonuclease. All four of these activities are absent from partially purified extract of *recB* and *recC* mutants (36). There is a distinct possibility that the activities are due to an enzyme made up of different subunits, or made up of one subunit with more than one active site, such as the DNA polymerase of *E. coli* may be (65).

Unpublished genetic results of Hoekstra, Storm, and their co-workers are consistent with the complex nature of the ATP-dependent DNase. They have characterized several mutations cotransducible with *thyA* that reduce the levels of the ATP-dependent DNase (105). Some of these mutations complement both *recB21* and *rec22* and hence fall into at least one additional cistron (46). Complex complementation behavior among all of the mutations indicates no clear picture and may reflect intragenic complementation or a complex interaction of subunits (46).

Other rec *mutants.*—Mutations affecting recombination that lie in genes other than *recA, recB,* and *recC* have been detected in *E. coli*. As many as six other *rec* genes may be recognizable: 1. One report showed four *rec* mutations only one of which was mapped in the *recA-recB-recC* region (91). Subsequent work has shown that two of the mutations are in *recA* (*recA36, recA35*) (23, 95) and one is in *recC* (*recC38*) (120a). One of the mutations (*rec-34*) does not, however, map as if it were a *recA, recB,* or *recC* mutation (44). It is essential to map this mutation before its characteristics are determined, but we are reserving *recD* as its designation when such tests are completed. 2. One mutation causing temperature-sensitive DNA synthesis was for several years thought to affect conjugational transfer (11, 12). Subsequent work has shown, however, that recombination and not transfer may be prevented (77, 82). 3. Mutations in *cetC* have been detected that cause refractivity to colicin E2 and reduce conjugational recombination but do not affect transductional recombination (47, 110). 4. Among the mutations detected by Storm and his co-workers is at least one that is not in *recA, recB,* or *recC*. This mutation apparently lies between *ilv* and *pyrE* and has been called *recG162* (105). 5. Mount (85) has detected two dominant mutations which reduce recombination about three- to five-fold, cause X-ray sensitivity and are located near *uvrA*. These mutations therefore resemble the *lex* and *exr* mutations which, in addition, have been found to prevent the production of ultraviolet-induced mutations (124, 125). Mount's mutants and *lex* and *exr* mutants have properties qualitatively but not quantitatively similar to those of *recA* mutants including lowered ability to produce lambda phage following ultraviolet induction (26a, 85), and an abnormally large extent of *recB-recC-* enzyme-determined ultraviolet-induced DNA breakdown (85). The hypothesis should be tested that lex^{+} determines a repressor of the *recA* gene and that the *lex* mutations detected determine a more effective repressor which results in lower *recA* product levels.

Presumably these mutations either directly or indirectly affect the major pathway of recombination in *E. coli.* Alternative pathways have been detected and the genes that participate in them are under investigation (see section on Some Physiological Complexities of Recombination). It thus seems reasonable to expect that a complete picture of the genetics of recombination proficiency in *E. coli* will involve at least a dozen genes.

GENETIC ANALYSIS OF RECOMBINATION PROFICIENCY IN SOME PHAGES OF *E. COLI*

Extensive genetic work has been done on the recombination proficiency of several phages of *E. coli.* Because of the ease with which phage DNA may be isolated and characterized, this genetic work can be correlated with the structure of recombination intermediates and end products.

Lambda phage.—Recombination-deficient mutants of λ phage have been detected by their reduced ability to recombine with a mutant prophage which has lost by deletion the repressor gene responsible for immunity from superinfection by λ and the regulatory, integration, and excision genes required for inducibility of the prophage (27, 100). Detection of the remaining genes occurs when they are rescued by recombination with appropriate superinfecting phage mutants producing wild-type progeny. Fortunately, the genes responsible for rescue recombination have been deleted from the mutant prophage; consequently, mutations in the superinfecting phage causing recombination deficiency are detectable because they reduce the production of wild-type phage. Many recombination-deficient λ mutants have been detected by their rescue deficiency (97).

Recombination-deficient mutants carry either point mutations or deletion mutations in which the bacterial *bio* genes among others have been substituted for the deleted λ genes. The point mutations affect either or both of two proteins: an exonuclease and an immunologically detected protein known as "β protein" (92). A complex complementation relationship among the point mutations and the multiple enzymatic and protein defects caused by some nonpolar point mutations has prevented a conclusive description of the genetic situation. A simplified picture may, however, be presented. The structural gene for the exonuclease is called *exo* or *redX*. *X* fulfills a double function in the gene symbol since it stands for exonuclease on one hand and indicates on the other that there are two cistrons *redA* and *redC* subsumed by the structural gene. The structural gene for the β protein is called *redB*. Thus far, no other λ genes involved in general recombination have been detected.

The exonuclease determined by *redX* degrades DNA from the 5′ end provided that a 5′ phosphoryl residue is present; 5′ mononucleotides are released (74). The exonuclease is much more active on double-stranded DNA than on single-stranded DNA (74) but it will not degrade double-stranded DNA from a nick, i.e., a break in the phosphodiester backbone of one strand consisting of adjacent 3′ hydroxyl and 5′ phosphoryl residues (18a). These properties have led Cassuto & Radding (19) to propose that exonuclease acts both presynaptically and in lengthening the synaptic region, as shown Figure 2. The β protein has no known enzymatic activity but it does bind to the exonuclease and appears to enhance the affinity of the exonuclease for DNA (93).

Lambda phage is capable of site-specific recombination as well as the general recombination that has just been described (28, 115). The difference between the two kinds of recombination is detected by determining recombination frequencies between markers at various locations on the λ chromosome. General recombination can be detected between any two markers but site-specific recombination can be detected only if the two markers are separated by the attachment site at which λ phage DNA recombines with the host chromosome to become integrated prophage DNA. The *int* gene determines site-specific recombination and the integration of λ and *E. coli* chromosomal DNAs. The *xis* gene, which is necessary in addition to *int* for the detachment of λ prophage DNA from the chromosome, shows involvement in only that site-specific recombination which mimics detachment (41, 42). Site-specific and general recombination are apparently separate pathways since mutations in *int* do not affect general recombination and *red* mutations do not affect site-specific recombination (28, 115).

The nature of recombinant molecules produced by general and site-specific recombination has been under investigation for about ten years (63, 80). From the first reports of the use of density and radioactivity labeling techniques it has been clear that a large fraction and possibly all λ recombination involves the exchange of original parental DNA. Recent work by

Stahl & Stahl (104) indicates that DNA synthesis is often associated with general recombination of λ. Recent work by Kellenberger-Gujer (62, 116) indicates that there may be two different kinds of general recombination, i.e., double-strand and single-strand insertions. These ideas form a context within which to explain the discrepancy noticed by Jordan & Meselson (61) between recombination frequency and physical distance in one locale of the λ chromosome and to explain, in addition, the finding by Stahl & Stahl (104) that DNA synthesis is associated with recombination to different degrees depending on the region of the chromosome involved. It would therefore not be surprising to find at least two pathways of general recombination occurring in λ crosses.

T4 phage.—Genetic analysis of T4 recombination has occurred as an offshoot of the saturation mapping of temperature sensitivity and nonsense codon mutations (33). Several of the genes detected in this mapping effort were concerned with DNA metabolism and several of these have been shown to be involved with recombination. Mutations in genes 30 and 43, affecting the thermal stability of the DNA ligase and DNA polymerase, respectively, raise the frequency of recombination at temperatures that approach those at which the proteins are inactive (9). This finding was confirmed by using amber mutations in genes 30 and 43 (8). Thus, polymerase and ligase activity seem to inhibit recombination. This can be rationalized on the basis of hypothetical pathways of recombination such as that proposed by Broker & Lehman (15). In figure 2 it can be seen that both polymerase and ligase would remove presynaptic intermediates from the pathway, reducing the concentration of molecules able to synapse and consequently reducing the amount of recombination possible.

The enhancing effect of polymerase and ligase mutations on recombination is correlated with their enhancing effect on the formation of branched-joint molecules formed by hydrogen-bonding fragments of infecting-phage DNA (15). Joint molecules have been considered recombinational intermediates since their discovery by Tomizawa & Anraku (111). Recently, the nature of the joint molecules has been studied by enzymatic techniques (2, 3) and by electron microscopy (15). The latter study is of particular interest because the joint molecules were seen to consist in large part of branched DNA molecules. To explain the occurrence and nature of the branched molecules Broker & Lehman (15) proposed the recombination pathway illustrated in Figure 2.

The influence of mutations in genes other than 30 and 43 has been measured both on recombination frequencies and on the frequency of occurrence of branched-joint molecules. Mutations in genes 32, 46, and 47 reduce both recombination frequencies (8, 9) and branched-joint molecule frequencies (15). Since the product of gene 32 greatly accelerates annealing of complementary single-stranded DNA at physiological temperatures (1), it may mediate the synaptic and branch migration steps indicated in Figure 2.

The products of genes 46 and 47 are said to be associated with exonuclease activity (66, 69) and may act in both synaptic events and the exonucleolytic lengthening of the synaptic region illustrated in Figure 2 (15).

The pathway hypothesized by Broker & Lehman (15) can be used to explain various recombinational phenomena, depending only on the position and frequency of the endonucleolytic nicks that terminate the synaptic region lengthening step. It is too early to ascertain whether this kind of one-pathway hypothesis is correct or whether the multiple-pathway hypothesis proposed by other workers is correct (84).

S13 phage.—Recombination deficiency of S13 phage is caused primarily by *recA* mutations in the host; a *recB* mutation and various phage mutations do not reduce normal levels of recombination by the primary recombination pathway (4, 107). The residual recombination, which presumably occurs in *recA* host mutants by a secondary pathway, seems to require the S13 gene A product because gene A mutations reduce the level of residual recombination (4). The gene A product is involved in the multiplication of the double-stranded replicative form that is formed following infection with the closed circular single-stranded virion DNA (106). The precise role of the gene A product in the secondary pathway is not clear.

f1 Phage.—Recombination of f1 phage has been reported to result in the production of one parental and one recombinant molecule (13). This has been interpreted to occur by a pathway utilizing DNA synthesis (13). It remains to be seen if this is the only recombination pathway operative for f1 phage. No genetic analysis of f1 recombination has yet been published.

SOME PHYSIOLOGICAL COMPLEXITIES OF RECOMBINATION

The isolation of recombination-deficient mutants of bacteria and of viruses provides us with a method for examining the physiological complexity of recombination. Three methods are now employed for this examination: 1. isolation of multiple mutants, 2. isolation of indirectly suppressed revertants, and 3. determination of the effects on cell and virus of single and mixed infection. Two parameters for detecting complexity are used in connection with each of these methods: 1. lethality (i.e., failure to multiply normally), and 2. change in recombination frequency between two markers.

Isolation of multiple mutants.—Mutants carrying single *recA, recB,* or *recC* mutations have populations containing a higher fraction of cells unable to multiply than rec^+ strains (18, 120a). This is consistent with the higher rate at which the mutants divide to produce inviable cells (43). Both of these phenomena indicate that the *recA, recB,* and *recC* products contribute to the viability of *E. coli* cells but are not absolutely essential. Curiously, the *recA* mutants, in which DNA degradation following UV ultraviolet irradiation is much greater than in *recB* or *recC* mutants, have a smaller fraction of inviable cells than *recB* and *recC* mutants (18). The metabolic abilities of the inviable *recB* and *recC* mutant cells are now under study (5).

One way to discover the reason for the loss in viability upon mutating the *rec* genes is to block other pathways of DNA metabolism to see if the viability increases or decreases. Thus far, mutants carrying *uvrA*⁻, *uvrB*⁻, and *uvrC*⁻, together with *recA*⁻, have been made (35, 51, 60). These strains grow more slowly than the single mutants presumably because a larger proportion of their cells are inviable (34, 44a, 55). Hence, it appears that the *rec* and *uvr* products may contribute to viability in a supplementary way.

A similar conclusion has been reached as the result of unsuccessful attempts to produce *polA*⁻ (i.e., deficient in DNA polymerase I) *recA*⁻, and *polA*⁻ *recB*⁻ strains (40, 118, 123). Neither strain can be constructed unless the *polA* mutation is conditional and the double mutant is grown under permissive conditions (40). Thus, it appears that the *recA, recB,* and *recC* products supplement DNA polymerase in their contribution to viability.

Indirect suppression.—Two cases of indirect suppression of *rec* mutational effects reveal a great deal of unsuspected physiological complexity. The first involves the occurrence of Mit^R (mitomycin-resistant) revertants of Mit^s *recB*⁻*recC*⁻ double mutants. These strains regain the recombination proficiency and resistance to both ultraviolet irradiation and mitomycin treatment exhibited by the *recB*⁺*recC*⁺ wild-type strains; yet these Rec^+ revertants are still genetically *recB*⁻*recC*⁻ and lack the ATP-dependent DNase determined by *recB*⁺*recC*⁺ (6a, 21, 68). The latter fact indicates that the suppression of the *recB* and *recC* mutations is probably indirect [i.e., it does not restore the mutant enzyme activity (38)]. Two groups of indirectly suppressed (i.e., genetically *sbc*⁻) mutants have been isolated. One group carries *sbc* mutations that are about 50 percent cotransducible with the *his* operon. These mutations all inactivate Exonuclease I (68). Although complementation tests have not yet been performed, we are operating under the working hypothesis that all these *sbc* mutations are in the *sbcB* gene. The *sbcA* mutants are characterized by normal levels of Exonuclease I, absence of an *sbc* mutation cotransducible with *his,* and the presence of an abnormally high level of an ATP-independent DNase active on double-stranded DNA (6a). This ATP-independent DNase is different from the ATP-dependent enzyme in substrate specificity, pH optimum, and cation dependence as well as in its independence of ATP (6a).

The *sbcA* mutations may derepress the synthesis of the ATP-independent DNase, which in turn may replace the *recB-recC* enzyme in its role in recombination. This is a shaky but attractive working hypothesis to explain indirect suppression by *sbcA*⁻. We denote the structural gene for the ATP-independent DNase as *recE* and denote the pathway of recombination operative in *sbcA*⁻ strains as the RecE pathway (21). The *sbcB*⁻ strains carry out recombination by another pathway which is permitted to operate or is stimulated by the absence of Exonuclease I. Genetic analysis shows that this pathway depends on the *recA*+ gene and on at least one and perhaps three additional genes (50, 96). We have nicknamed this alternative the "RecF pathway" to indicate the participation of a *rec* gene we call *recF* (21). Thus, in *E. coli* at least two and possibly three pathways of recombination may occur: the RecBC pathway and the alternative RecE and RecF pathways.

Our understanding of the RecE pathway is incomplete because we are uncertain how similar or different this pathway is from the RecBC pathway. The RecF pathway seems to be clearly different. What we know now leads us to suspect that single-stranded ends of DNA are important in the RecF pathway since active Exonuclease I reduces the amount of recombination by about 99 percent in *recB* and *recC* mutants. Exonuclease I has been reported to act exclusively on 3′OH single-stranded termini (72). If its activity in vivo were that specific, then such 3′OH terminated single strands would be important in the RecF pathway. They would be degraded by Exonuclease I and consequently the RecF pathway would in essence be inhibited. The RecBC pathway might involve 5′OH or 5′P or 3′P terminated single strands or else it might not involve single-stranded termini at all. Much more work is required before drawing any conclusions about the differences between the RecBC and the RecF pathways of recombination.

A second case of indirect suppression has been reported by Low (75a) and is implicit in the recombination proficiency of *recB* mutants crossed with Hfr B7 (75). The *rac* gene, which has been mapped four minutes clockwise of *trp* on the standard map of *E. coli* (105a), when transferred from a donor into certain *recB* mutants activates recombination in the zygotes. It is unclear whether this type of indirect suppression by zygotic induction bears any relationship to the *sbcA* or *sbcB* types of indirect suppression.

Phage-infected cells.—The *old*+ gene of P2 phage elaborates an unknown product which has two detectable effects (73). First, it prevents the normal multiplication of wild-type λ phage, a phenomenon known as "interference" (10); wild-type λ phage are therefore sensitive to P2 interference (i.e., Spi +). Second, the *old*+ gene may prevent the growth of P2 lysogens of *recB*− and *recC*− strains, thus explaining why such lysogens cannot be isolated (73). These two effects of the *old*+ product are related because of an unsuspected effect of several wild-type λ genes on the level of *recB*-*recC* enzyme in their hosts. In an as yet unexplained way, several λ genes (including *redX, redB*, γ and δ) seem to be involved in reducing the levels of the ATP-dependent DNase determined by *recB*+ and *recC*+ (114). The reduction causes *recB*+ *recC*+ cells infected by wild-type λ to become phenotypically RecB− RecC− (i.e., recB− recC− phenocopies). Such phenocopies are subject to the inhibitory action of the *old*+ product of P2, which results in the cessation of wild-type λ multiplication (101). If the infecting λ phage is mutant in λ, δ, and either *redX* or *redB* (i.e., is λ Spi− rather than wild type) then there is no interference by P2 *old*$^{+}$, presumably because the host cells do not become *recB*$^{-}$ *recC*$^{-}$ phenocopies (1).

The effect of the Spi^{+} genes of λ on the levels of *recB-recC*-determined ATP-dependent DNase can be inferred from direct enzyme measurements and from recombination frequency studies (114). The latter consist of determining the frequency of recombinants formed between the two λ genes *susA* and *susJ* in a variety of hosts. Normally the amount of recombination detected between these two genes seems to be largely dependent on the *redX* and *redB genes.* This is indicated in two ways (27, 100): 1. in *recA*$^{-}$, *recB*$^{-}$, or *recC*$^{-}$ hosts the frequency of recombinants measured is the same as in wild-type *rec*$^{+}$ hosts; 2. *redX* or *redB* mutations cause up to a tenfold decrease in the amount of recombination in a *rec*$^{+}$ host. When Spi^{-} strains (e.g., *redX*$^{-}$, *redB*$^{-}$, λ^{-}, δ^{-} deletion mutants) are crossed in a *rec*$^{+}$ host the amount of recombination between *susA* and *susJ* is greater than that observed when single *redX* or *redB* mutants are crossed. Presumably, this increase is due to the failure of the λ Spi^{-} phage to reduce the *recB*$^{+}$-*recC*$^{+}$ enzyme levels. Hence, the low frequency of recombination of *red*$^{-}$ λ phage in a *rec*$^{+}$ host is due both to the absence of functional *red* products and to a reduction in *recB-recC* enzyme levels.

There is additional interaction of λ phage with its host's recombination enzymes. Mutants of λ phage are found that will not multiply normally in a *recA*$^{-}$ *E. coli* host (128). These mutants (known as Fec^{-}) are double mutants carrying *redX* or *redB* mutations and a mutation in γ. If the host is *recA*$^{-}$*recB*$^{-}$, the Fec^{-} λ strains multiply normally. One implication is that the *recA*$^{+}$ product may be necessary to protect λ *red*$^{-}$ γ^{-} from the activity of the *recB-recC* nuclease, possibly because the λ mutant is unable to reduce this enzyme to tolerable levels (98).

A similar phenomenon is observed in infections of *Salmonella typhimurium* by the temperate phage P22. Rec^{-} mutants of *S. typhimurium* fail to support the normal mutiplication of certain mutants of P22 known as *erf* mutants (14, 127). A rationale developed to explain this phenomenon is that certain essential recombination functions permit the circularization of infecting P22 phage DNA prior to its replication and these functions may be filled either by bacterial or by phage gene products (14, 127).

METABOLIC INTERPRETATION OF RECOMBINATIONAL PHENOMENA

If recombination refers to a set of metabolic pathways, it should be possible to interpret recombinational phenomena with terms and concepts currently used in understanding intermediary metabolism. In that light the following are brief discussions of recombination frequency, high negative interference, and site-specific and general recombination.

1. The frequency of recombination of two genetic markers has always been interpreted as a measure of the distance between the markers. This interpretation is adequate as long as other things are equal. A metabolic

interpretation would hold that enzyme levels and the particular pathways of recombination followed are important in determining the frequency of recombination. Furthermore, postrecombinational pathways of repair may act on recombinant DNA to modify the recombinant frequency ultimately detected. Thus, to compare recombination frequencies to obtain relative distances requires control of these metabolic variables.

2. High negative interference refers to the excessive occurrence of recombination in an interval when recombination has also occurred in an adjacent interval. One proposal is stated in terms of the structures of recombinant products (84). In metabolic terms this hypothesis might be rephrased to indicate a proportionally greater contribution of one pathway to recombination between very closely linked markers than between distant markers. An alternative explanation is that high negative interference reflects the action of the postrecombination process of repair on the heteroduplex region of recombinant DNA.

3. General and site-specific types of recombination have been recognized because of the recent work with λ phage. General recombination is more or less random in its occurrence on the λ chromosome. This feature is shared by most recombination and its explanation can be expected to reflect some general property of most recombinational pathways. It is possible that presynaptic events such as those diagrammed in Figure 1 are random in their occurrence along the chromosome, and hence synaptic and postsynaptic events should also be random. The randomness of presynaptic events would not, however, be expected to be detrimental to overall DNA metabolism because the steps may be reversible or the intermediates may be engaged in a cyclic set of reactions. Site-specific recombination may reflect site-specific synapsis through protein mediation, site-specific presynaptic events, like so or both.

The occurrence of lethal combinations of mutations such as *polA⁻ recA⁻* and *polA⁻ recB⁻* may also be explained metabolically. What is necessary for this explanation is a view of the interactions of recombinational metabolism and the other metabolic processes that act on DNA; namely, replication, transcription, and repair. It is clear that there must be interactions because the four metabolic processes all use the bacterial or phage chromosome as substrate. At present the details of the interactions are just as much under investigation as the details of each of the processes separately. Since the processes cannot be studied in vivo in isolation from each other, it may be necessary to have in mind some overall hypothesis concerning the interactions to perform in vivo studies most effectively. The following is the author's current view on this subject.

The entire bacterial or phage chromosome is subject at least once each generation to replication. This process begins with one chromosome and it ends with two chromosomes presumably in the same condition as the original starting material. The state of each segment of the chromosome that enables it to be replicated is not known but there must be some requirements; for example, a segment may have to be unbroken, unbranched, and inactive in protein synthesis to be replicated. Repair, recombination, or transcription of a chromosomal segment may inhibit the segment's replication by changing its structure. If replication is to be complete and periodic then the inhibitory changes in segment structure must be transitory. Repair, recombination, and transcription must, therefore, be cyclic processes; i.e., they must restore an affected segment to a state in which it can be replicated. If the cycle is interrupted, the chromosomal segment affected may be unreplicatable or may lead to an abnormal product if replicated. In either case the chromosome may be irreversibly prevented from multiplication. There are many ways in which the cycle may be interrupted: e.g., loss of an enzyme activity through mutation, repression, or inhibition, or gain of an enzyme activity through mutation, deprepression, or loss of inhibition. Since there are so many ways of interruption or perturbation of the physiological balance of cyclic processes, one might expect a proliferation of enzyme activities or control of enzyme concentrations to meet every exigency. Thus, recombination pathways may be complexly branched, with certain intermediates subject to removal from recombination pathways through recycling reactions or reactions leading to repair or transcription.

This overall hypothesis leads to a convenient explanation of the lethality of *polA* mutations in combination with *recA* or *recB* mutations. DNA polymerase I, on one hand, and the *recA* and *recB* products on the other, may act in different pathways of restoring a replicatable structure to one kind of unreplicatable chromosome segment which has been produced in the course of recombination, repair, or transcription. If either pathway is blocked by mutation, the other is available for restoring the replicatable structure. When both pathways are blocked, however, no restoration occurs and lethality results from the production of unreplicatable chromosome segments by normal metabolism.

Unreplicatability of changed chromosomal segments is, however, only one way of explaining lethality stemming from recombination deficiency and other DNA involved mutations. It is also possible that changed chromosomal segments may be replicated but may form abnormal replication products that are lethal. Whatever the explanation, it appears that a complex set of interacting metabolic pathways is emerging from work on DNA metabolism, in general, and on recombination, in particular. The necessity for such complexity may stem from the metabolic versatility of chromosomal segments and the precision and inevitability of their replication.

ACKNOWLEDGMENTS

The author is particularly indebted to the following colleagues for the discussions and criticisms that have helped to clarify the ideas and their presentation in this article: R. Calendar, M.-C. Clark, H. Echols, M. Guyer, J. Hegyi, S. Kushner, L. Margossian, J. Roth, R. Unger, and C.

Wilde. In addition, the author is grateful to many other colleagues who provided information of their work prior to its publication.

This investigation was supported by US Public Health Service research grant AI-05371 from the National Institute of Allergy and Infectious Diseases.

LITERATURE CITED

1. Alberts, B. M., Frey, L. 1970. *Nature* 227:1313–18
2. Anraku, N., Lehman, I. R. 1969. *J. Mol. Biol.* 46:467–79
3. Anraku, N., Anraku, Y., Lehman, I. R. 1969. *J. Mol. Biol.* 46:481–92
4. Baker, R., Doniger, J., Tessman, I. 1971. *Nature.* 230:23–25
5. Barbour, S. D. Personal communication
6. Barbour, S. D., Clark, A. J. 1970. *Proc. Nat. Acad. Sci. USA* 65: 955–61

6a. Barbour, S. D., Nagaishi, H., Templin, A., Clark, A. J. 1970. *Proc. Nat. Acad. Sci. USA* 67: 128–35

7. Bastarrachea, F., Willetts, N. S. 1968. *Genetics* 59:153–66
8. Berger, H., Warren, A. J., Fry, K. E. 1969. *J. Virol.* 3:171–75
9. Bernstein, H. 1968. *Cold Spring Harbor Symp. Quant. Biol.* 33: 325–31
10. Bertani, G. 1958. *Advan. Virus Res.* 5:151–93
11. Bonhoeffer, F. 1966. *Z. Vererbungslehre* 98:141–49
12. Bonhoeffer, F., Hösselbarth, R., Lehmann, K. 1967. *J. Mol. Biol.* 29: 539–41
13. Boon, T., Zinder, N. D. 1969. *Proc. Nat. Acad. Sci. USA* 64:573–7
14. Botstein, D., Matz, M. J. 1970. *J. Mol. Biol.* 54:417–40
15. Broker, T. R., Lehman, I. R. 1971. *J. Mol. Biol.* Submitted for publication
16. Brooks, K., Clark, A. J. 1967. *J. Virol.* 1:283–93
17. Buttin, G., Wright, M. R. 1968. *Cold Spring Harbor Symp. Quant. Biol.* 33:259–69
18. Capaldo-Kimball, F., Barbour, S. D. 1971. *J. Bacteriol.* 106:204–12

18a. Carter, D. M., Radding, C. M. 1971. *J. Biol. Chem.* 246:2502–10

19. Cassuto, E., Radding, C. M. 1971. *Nature New Biol.* 229:13–16
20. Clark, A. J. 1967. *J. Cell. Physiol.,* 70 Suppl. 1:165–80
21. Clark, A. J. *Proc. Int. Congr. Microbiol., X* In press
22. Clark, A. J., Margulies, A. D. 1965. *Proc. Nat. Acad. Sci. USA* 53: 451–59
23. Clark, A. J., Templin, A. Unpublished results

23a. Clark, A. J., Chamberlin, M., Boyce, R. P., Howard-Flanders, P. 1966. *J. Mol. Biol.* 19:442–54

24. Davern, C. I. 1971. *Progr. Nucl. Acid Res. Mol. Biol.* 11:229–58
25. Delbruck, M., Stent, G. S. 1957. In *The Chemical Basis of Heredity,* ed. W. D. McElroy and B. Glass, 699–736. Baltimore: Johns Hopkins
26. Devoret, R., Blanco, M. 1970. *Mol. Gen. Genet.* 107:272–80

26a. Donch, J., Greenberg, J., Green, M. H. L. 1970. *Genet. Res.* 15:87–97

27. Echols, H., Gingery, R. 1968. *J. Mol. Biol* 34:239–49
28. Echols, H., Gingery, R., Moore, L. 1968. *J. Mol. Biol.* 34:251–60
29. Eisenstark, A., Eisenstark, R., Van Dillewijn, J., Rörsch, A. 1969. *Mutation Res.* 8:497–504
30. Emmerson, P. T. 1968. *Genetics* 60: 19–30
31. Emmerson, P. T., Howard-Flanders, P. 1967. *J. Bacteriol.* 93:1729–31
32. Ephrussi-Taylor, H., Gray, T. C. 1966. *J. Gen. Physiol.* 49 (Suppl.):211–31
33. Epstein, R. H., et al. 1963. *Cold Spring Harbor Symp. Quant. Biol.* 28:375–96
34. Ganesan, A. Personal communication
35. Ganesan, A. K., Smith, K. C. 1969. *J. Bacteriol.* 97:1129–33
36. Goldmark, P., Linn, S. Personal communication
37. Goldmark, P., Linn, S. 1970. *Proc. Nat. Acad. Sci. USA* 67:434–41
38. Gorini, L., Beckwith, J. R. 1966. *Ann. Rev. Microbiol.* 20:401–22
39. Gray, T. C., Ephrussi-Taylor, H. 1967. *Genetics* 57:125–53
40. Gross, J. Personal communication
41. Guarneros, G., Echols, H. 1970. *J. Mol. Biol.* 47:565–74
42. Guarneros, G., Echols, H., Jordan, E. *Proc. Int. Congr. Microbiol., X* In press
43. Haefner, K. 1968. *J. Bacteriol.* 96: 652–59
44. Hegyi, J., Clark, A. J. Unpublished results

44a. Hertman, J. M. 1969. *Genet. Res.* 14:291–307

45. Hertman, I., Luria, S. E. 1967. *J. Mol. Biol.* 23:117–33
46. Hoekstra, W. P. M. Personal communication

47. Holland, I. B., Threlfall, E. J., Holland, E. M., Darby, V., Samson, A. C. R. 1970. *J. Gen. Microbiol.* 62:371–82
48. Holliday, R. 1964. *Genet. Res.* 5: 282–304
49. Holloway, B. W. 1969. *Bacteriol. Rev.* 33:419–43
50. Horii, Z. I., Clark, A. J. Unpublished results
51. Horii, Z. I., Suzuki, K. 1970. *Photochem. Photobiol.* 11:99–107
52. Hotchkiss, R. D. 1971. *Advan. Genet.* 16. In press
53. Hotchkiss, R. D., Gabor, M. 1970. *Ann. Rev. Genet* 4:193–224
53a. Hout, A., van de Putte, P., de Jonge, A. J. R., Schuite, A., Oosterbaan, R. A. 1970. *Biochim. Biophys. Acta* 224:285–87
54. Howard-Flanders, P. 1968. *Ann. Rev. Biochem.* 37:175–200
55. Howard-Flanders, P. Personal communication
56. Howard-Flanders, P., Boyce, R. P. 1966. *Radiat. Res. Suppl.* 6:156–84
57. Howard-Flanders, P., Rupp, W. D., Wilde, C., Reno, D. *Proc. Int. Congr. Microbiol., X* In press
58. Howard-Flanders, P., Rupp, W. D., Wilkins, B. M., Cole, R. S. 1968. *Cold Spring Harbor Symp. Quant. Biol.* 33:195–207
59. Howard-Flanders, P., Theriot, L. 1966. *Genetics* 53:1137–50
60. Howard-Flanders, P., Theriot, L., Stedeford, J. B. 1969. *J. Bacteriol.* 97:1134–41
61. Jordan, E., Meselson, M. 1965. *Genetics* 51:77–86
62. Kellenberger-Gujer, G. 1971. In *The Bacteriophage Lambda,* ed. A. D. Hershey. Long Island: Cold Spring Harbor Laboratory. In press
63. Kellenberger, G., Zichichi, M. L., Weigle, J. J. 1961. *Proc. Nat. Acad. Sci. USA* 47:869–78
64. Kirby, E. P., Jacob, F., Goldthwaite, D. A. 1967. *Proc. Nat. Acad. Sci. USA* 58:1903–10
65. Klenow, H., Overgaard-Hansen, K. 1970. *FEBS Lett.* 6:25–27
66. Koerner, J. F. 1970. *Ann. Rev. Biochem.* 39:291–322
67. Kushner, S. R., Clark, A. J. Unpublished results
68. Kushner, S. R., Nagaishi, H., Templin, A., Clark, A. J. 1971. *Proc. Nat. Acad. Sci. USA* 68:824–27
69. Kutter, E. M., Wiberg, J. S. 1969. *J. Virol.* 4:439–53
70. Lacks, S. 1966. *Genetics* 53:207–35
71. Lee, C. S., Davis, R. W., Davidson, N. 1970. *J. Mol. Biol.* 48:1–22
72. Lehman, I. R., Nussbaum, A. L. 1964. *J. Biol. Chem.* 239:2628–36
73. Lindahl, G., Sironi, G., Bialy, H., Calendar, R. 1970. *Proc. Nat. Acad. Sci. USA* 66:587–94
74. Little, J. W. 1967. *J. Biol. Chem.* 242:679–86
75. Low, B. 1968. *Proc. Nat. Acad. Sci. USA* 60:160–67
75a. Low, B. Personal communication
76. MacPhee, D. G. 1970. *J. Bacteriol.* 104:345–50
77. Marinus, M. G., Adelberg, E. A. 1970. *J. Bacteriol.* 104:1266–72
78. McCarthy, B. J., Church, R. B. 1970. *Ann. Rev. Biochem.* 39:131–50
79. Meselson, M. 1964. *J. Mol. Biol.* 9: 734–45
80. Meselson, M., Weigle, J. J. 1961. *Proc. Nat. Acad. Sci. USA* 47: 857–68
81. Miura, A., Tomizawa, J. 1968. *Mol.* Gen. Genet. 103:1–10
82. Moody, E. E. M., Lukin, A. 1970. *J. Mol. Biol.* 48:209–17
83. Morgan, T. H. 1911. *J. Exp. Zool.* 11:365–414
84. Mosig, G. 1970. *Advan. Genet.* In press
85. Mount, D. M. Personal communication
86. Norkin, L. C. 1970. *J. Mol. Biol.* 51: 633–55
87. Ogawa, H., Shimada, K., Tomizawa, J. 1968. *Mol. Gen. Genet.* 101: 227–44
88. Oishi, M. 1969. *Proc. Nat. Acad. Sci. USA* 64:1292–99
89. Oppenheim, A. B., Riley, M. 1967. *J. Mol. Biol.* 28:503–11
90. Oppenheim, A. B., Riley, M. 1966. *J. Mol. Biol.* 20:331–57
91. van de Putte, P., Zwenk, H., Rörsch, A. 1966. *Mutation Res.* 3:381–92
92. Radding, C. M. 1970. *J. Mol. Biol.* 52:491–99
93. Radding, C. M., Carter, D. M. *J. Biol. Chem.* 246:2513–18
94. Riley, M., Mailing, B., Chamberlin, M. J. 1966. *J. Mol. Biol.* 20:359–89
95. Rörsch, A. Personal communication
96. Scarsella, A., Clark, A. J. Unpublished data
96a. Sgaramella, V., van de Sande, J. H., Khorana, H. G. 1970. *Proc.*

Nat. Acad. Sci. USA 67:1468–75
97. Shulman, M. J., Hallick, L. M., Echols, H., Signer, E. R. 1970. *J. Mol. Biol.* 52:501–20
98. Signer, E. 1971. In *The Bacteriophage Lambda,* ed. A. D. Hershey. Long Island: Cold Spring Harbor Laboratory. In press
99. Signer, E. R., Weil, J. 1968. *Cold Spring Harbor Symp. Quant. Biol.* 33:715–19
100. Signer, E. R., Weil, J. 1968. *J. Mol. Biol.* 34:261–71
101. Sironi, G., Bialy, H., Lozeron, H., Calendar, R. 1971. Submitted for publication
102. Smith, K. C. 1971. *Photophysiology* 6. In press
103. Stahl, F. W. 1969. *Genetics* (Suppl. 1) 61:1–13
104. Stahl, F. W., Stahl, M. 1971. In *The Bacteriophage Lambda,* ed. A. D. Hershey. Long Island: Cold Spring Harbor Laboratory. In press
105. Storm, P. K. Personal communication
105a. Taylor, A. L. 1970. *Bacteriol. Rev.* 34:155–75
106. Tessman, E. S. 1966. *J. Mol. Biol.* 17:218–36
107. Tessman, I. 1968. *Science* 161:481–82
108. Thomas, C. A. 1966. *Prog. Nucl. Acid Res. Mol. Biol.* 5:315–37
109. Thomas, C. A. 1967. In *The Neurosciences: A Study Program,* ed. G. C. Quarton, T. Melnechuk, F. O. Schmitt, 162–82. New York: Rockefeller University Press
110. Threlfall, E. J., Holland, I. B. 1970. *J. Gen. Microbiol.* 62:383–98
111. Tomizawa, J., Anraku, N. 1964. *J. Mol. Biol.* 8:516–40
112. Tomizawa, J., Ogawa, H. 1968. *Cold Spring Harbor Symp. Quant. Biol.* 33:243–51
114. Unger, R., Echols, H., Clark, A. J. Unpublished results
115. Weil, J., Signer, E. R. 1968. *J. Mol. Biol.* 34:273–79
116. Weisberg, R. A., Kellenberger-Gujer, G. 1971. In *The Bacteriophage Lambda,* ed. A. D. Hershey. Long Island: Cold Spring Harbor Laboratory. In press
117. Whitehouse, H. L. K. 1970. *Biol. Rev.* 45:265–315
118. Willetts, N. S. Personal communication
119. Willetts, N. S. Unpublished data
120. Willetts, N. S., Clark, A. J. 1969. *J. Bacteriol.* 100:231–39
120a. Willetts, N. S., Mount, D. W. 1969. *J. Bacteriol.* 100:923–34
121. Willetts, N. S., Clark, A. J., Low, B. 1969. *J. Bacteriol.* 97:244–49
122. Wing, J. P., Levine, M., Smith, H. O. 1968. *J. Bacteriol.* 95:1828–34
123. Witkin, E. M. Personal communication
124. Witkin, E. M. 1969. *Ann. Rev. Genet.* 3:525–52
125. Witkin, E. M. 1969. *Mutation Res.* 8:9–14
125a. Wright, M., Buttin, G. 1969. *Bull. Soc. Chim. Biol.* 51:1373–83
126. Yajko, D., Clark, A. J. Unpublished results
127. Yamagami, H., Yamamoto, N. 1970. *J. Mol. Biol.* 53:281–85
128. Zissler, J., Signer, E. R. 1971. In *The Bacteriophage Lambda,* ed. A. D. Hershey. Long Island: Cold Spring Harbor Laboratory. In press

EMERGING DISEASES OF MAN AND ANIMALS

1579

DAVID J. SENCER, M.D., AND STAFF OF LABORATORY DIVISION AND EPIDEMIOLOGY PROGRAM

Center for Disease Control, Public Health Service,
U. S. Department of Health, Education and Welfare,
Atlanta, Georgia

CONTENTS

INTRODUCTION

Infectious diseases of man and animals currently emerging as public health problems include some old acquaintances and some that are new in respect to identity or concept.

The resurgence of old diseases is often intertwined with other major social and economic problems of our times. In areas where food supplies, sanitary facilities, and health services have never been adequate, existing miseries are compounded by the population explosion, which further diminishes the amount of goods and services available per person while increasing the number of susceptible individuals living under adverse conditions. The spread of cholera through the Middle East and of dysenteric shigella in Central America are manifestations of such problems. In the United States in recent years we have experienced sporadic outbreaks of preventable diseases such as diphtheria, polio, and measles. These have occurred most frequently in communities where there has been a letdown in immunizations among migrants and others in the low socioeconomic class. The growing mobility of people of all classes and the speed with which they travel make it essential, as never before, to recognize and cope with disease at its source.

Of the newly recognized agents of disease, the Marburg and Lassa viruses come immediately to mind. There are other analogous diseases such as amoebic meningitis, which will not be dealt with in this review. But there are also many familiar organisms formerly considered nonpathogenic that are now associated with nosocomial infections, use of artificial kidneys, and the acceptance or rejection of organ transplants, for example.

And so infectious disease, one of man's oldest enemies, survives as an adversary that calls forth our best efforts.

BACTERIAL DISEASES

Anaerobe Infections

Exogenous clostridia have long been recognized as important sources of human disease, including botulism, tetanus, cellulitis, and gas gangrene. More recently, with proper anaerobic culture methods, endogenous anaerobic bacteria, particularly those common to the gastrointestinal tract, the genitourinary tract, or the oral cavity, are increasingly isolated from persons with a variety of acute and chronic infections. The anaerobic bacteria involved are clostridia, nonsporulating Gram-negative and Gram-positive bacilli, and Gram-negative and Gram-positive cocci. Infections caused by nonspore-forming bacteria such as *Bacteroides, Fusobacterium, Propionibacterium,* and *Peptostreptococcus* (anaerobic streptococci) outnumber the clostridial infections.

The host-parasite interactions which predispose to endogenous infections are poorly understood, but susceptibility increases when the host defenses diminish. Such diverse factors as antimicrobial agents, adrenocortical steroids, and immunosuppressive agents have been implicated. Patients with underlying debilitating diseases are particularly susceptible to infection by anaerobic as well as by aerobic organisms. The rising proportion of elderly, chronically debilitated patients, the increasing number of surgical procedures performed, and the emergence of drug-resistant microorganisms all contribute to the problem. Types of infection include septicemias, deep abscesses, traumatic and postoperative wound infections, and bacterial endocarditis.

Clostridium septicum has long been recognized as an agent of traumatic gas gangrene, especially during World War II. Recent reports of clostridial bacteremias not associated with obvious, grossly contaminated wounds in humans, however, have emphasized its endogenous origin and its striking association with malignancy (4). A bowel lesion, possibly due to underlying disease or to antimetabolite or steroid therapy, could enable this organism to enter the bloodstream.

Bacteroides and *Fusobacterium* species are commonly associated with endogenous bacteremias (15, 44, 105). Recently, Felner & Dowell (35) analyzed clinical and bacteriologic data related to 250 patients with "bacteroides" bacteremia and found that all probably had an endogenous source of infection. Prior surgery, malignant neoplasms, diabetes mellitus, and steroid, immunosuppressive, or cytotoxic therapy were commonly associated factors; thrombophlebitis, septic embolization, and metastatic abscess formation were the hallmarks of their disease.

The same authors recently reported a study of clinical and bacteriologic data pertaining to 33 patients with anaerobic bacterial endocarditis (34). *Bacteroides fragilis* was the most common anaerobe involved.

Clearly, diagnostic laboratories should employ anaerobic as well as aerobic techniques in examining blood, abscess fluid, and intra-abdominal tissue specimens, especially from patients with malignancies or intestinal lesions. The clinician also must increase his awareness of the likelihood of an aner obe being totally or partially responsible for a patient's illness and request the laboratory to employ anaerobic techniques.

Cholera

During the 1960's, cholera, a scourge of antiquity, once again emerged from obscurity. Since 1961, the disease has spread from its focus in the Celebes in Indonesia through most of Asia and the Middle East into Eastern Europe and Africa (Fig. 1). The current pandemic, the seventh, is caused by the El Tor biotype. Ominous as it may seem, the risk to American travelers and to the United States is actually quite small.

Cholera is unique among enteric diseases in that diarrhea can be so severe as to cause cardiovascular collapse and death within one or two days. Nevertheless, the disease is no longer dreaded as it was in the past because even severe cases respond dramatically to intravenous fluid and electrolyte replacement therapy (18), and milder cases can be treated with either intravenous or oral fluid therapy and tetracycline (22). In good treatment centers, even in the developing countries, case-fatality rates no longer exceed 1 percent.

Most infections are mild or subclinical. There may be 100 or more persons with mild or asymptomatic infections for each one that is clinically ill (8). This accounts for the futility of quarantine measures in preventing spread of this disease.

Although millions of Americans have traveled and lived in cholera-infected areas during the past decade, only four documented cases have oc-

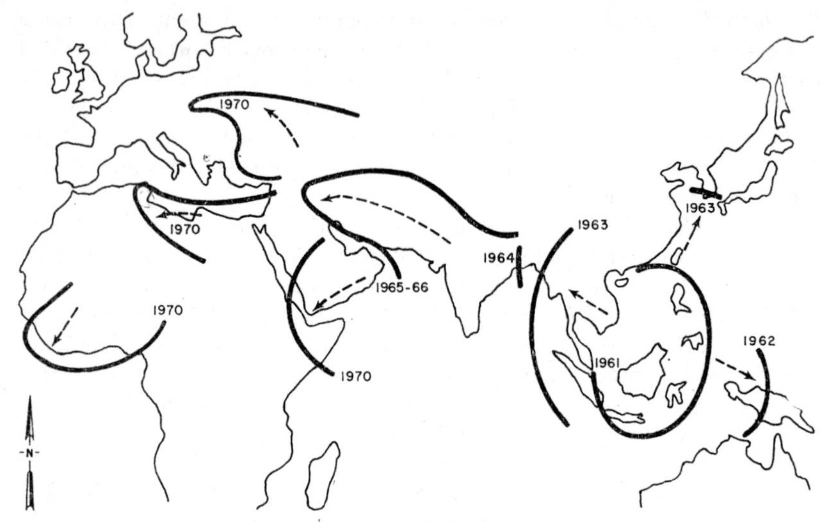

* SOURCE: PRINCIPLES AND PRACTICE OF CHOLERA CONTROL, PUBLIC HEALTH PAPERS NO. 40, WHO, GENEVA, 1970. REVISED ACCORDING TO WEEKLY EPIDEMIOLOGICAL RECORD, VOL. 45, NOS. 1-52, 1970

FIGURE 1. Extension of El Tor cholera, 1961–1970.

curred in them (1, 68, 114, 115), all nonfatal. During the rapid geographic spread which occurred in 1970, only one case was recorded in a traveler from a western nation, England (33). The risk is obviously so small that no one should need to cancel or alter any travel plans. Most American travelers avoid questionable water and food supplies by frequenting tourist accommodations which maintain relatively high sanitary standards. Added to such nonspecific defenses as gastric acid, intestinal motility, and the normal intestinal flora, these factors constitute a formidable barrier against the disease.

There may be sporadic importations into the United States but further spread is unlikely. The major mode of transmission is contaminated water; in most communities water supplies are treated and people ill with diarrhea use modern sanitary facilities.

Physicians should consider cholera in any patient who has diarrhea within five days after returning from an infected area. A rectal swab or stool culture should be taken immediately to a laboratory or transported in Cary-Blair (19) or Sea-salt (110) media or, if these are not available, a piece of blotting paper dipped into the stool and inserted into a leak-proof plastic bag will do (9). The organism grows readily on most available media, such as nutrient or blood agar, or on MacConkey's agar, but not on inhibitory media such as SS. A highly selective medium, thiosulphate citrate bile salts sucrose agar (TCBS) has been found very useful (40). The diagnosis should not be made until the organism has been definitely identified.

Serologic tests are available, but are not valid for persons recently vaccinated.

Although nonagglutinating or noncholera vibrios have been recognized as a cause of enteric illness, their importance as a source of diarrheal disease in travelers has not been established.

By international agreement, all cases of cholera must be reported to the World Health Organization; therefore, any proven or suspected case of cholera should be reported promptly to local and state health authorities and in turn to the Center for Disease Control.

Diphtheria

Diphtheria is an example of a disease that is recrudescent after reaching very low levels in the United States. Even before diphtheria toxoid became widely used, the disease was becoming a decreasing problem here, and this trend accelerated as diphtheria immunization became commonplace. From an estimated 200 cases per 100,000 population in the early 1920's, the incidence plummeted to 40 cases per 100,000 in 1933, and reached a nadir of 0.086 cases per 100,000 (164 reported cases) in 1965. Preliminary data for 1970 show that more cases (439 reported, provisionally) occurred in that year than in any since 1962 (27).

The increase is largely accounted for by urban epidemics that involve as few as 6 or 8 cases or as many as 200 or more. Some outbreaks have continued for more than a year, as in Austin, Tex., and Phoenix, Ariz. They are occurring most frequently in southern states and primarily affecting unimmunized children who are members of minority groups living in low income areas.

The public health implications are quite clear. Immunizations and other preventive services must be provided in those areas shown to be at greatest risk if a resurgence of diphtheria and other preventable diseases is to be forestalled in this country.

Epidemic Shiga Dysentery

Shigella dysenteriae type 1, the classic Shiga bacillus, is an organism of unusual virulence. During the late 19th and early 20th centuries, it caused numerous outbreaks of dysentery with a high mortality rate on all continents. About 1920, for reasons unknown, it virtually disappeared from the world scene only to re-emerge as an epidemic threat in Central America in 1968.

The organism is spread predominantly by water or food contaminated by the feces of an infected person. It causes an illness characterized by painful, bloody, mucoid diarrhea, and occasionally fever. If untreated, the disease may be fatal in 5 to 15 percent of infected persons. With proper treatment, the case-fatality ratio was reduced from this high level (41) to about 1 percent in most Central American treatment centers in 1970.

The current pandemic began late in 1968 in southwestern Guatemala and spread rapidly during 1969 and 1970 through most of Middle America

(24, 26, 80, 82, 84–86, 88). During 1969 there were more than 112,000 reported cases and 13,500 deaths in Guatemala, and uncounted thousands of persons were affected in other Middle American countries.

The epidemic derived from a long-standing endemic base dating back to the early part of the century (73). Among the factors contributing to the exacerbation of this endemic focus were the population explosion which increased the number of susceptible persons, who were also often crowded together in inadequate housing; improved roads and public transportation which increased the mobility of the population; extensive environmental pollution of water supplies by human excrement; primitive sanitation that has remained essentially unchanged for decades; severe climatic conditions in 1969; and the known virulence and communicability of the Shiga bacillus. The chronic nature of the socioeconomic problems suggests that the epidemic will persist for years before a dormant, endemic situation is again resumed.

The danger of spread of this pandemic into contiguous areas to the north and south is real. Migrants and Indians could introduce the disease into poverty areas, especially in the southwestern part of the United States, and so create epidemic foci. However, Shiga dysentery is not likely to become a serious problem in this country where sanitary standards tend to be generally adequate.

Importations to other countries, including the United States, by air travelers has already been documented. There have been two deaths among American tourists who became ill after visiting Middle American countries (24–26, 80, 82, 84–86, 88). With a few exceptions, indigenous transmission has not occurred in the United States.

Although the epidemic strains are resistant to many commonly used chemotherapeutic agents such as chloramphenicol, tetracycline, and the sulfa drugs, they have been found to be susceptible to moderately large doses of ampicillin and penicillin. Work is in progress to develop an oral attenuated vaccine which, after appropriate testing, might be useful in the control or eradication of the disease.

Gonorrhea

Despite the fact that we have had a specific cure for gonorrhea for two decades, the disease ranks first among the reportable bacterial infections in the United States (Fig. 2).

Current estimates put gonorrheal incidence at 5000 cases or more a day; and reported morbidity of 573,200 cases for the fiscal year ending June 1970 (28) was 15 percent over the previous year, a trend that is continuing (Fig. 3).

Clearly, gonorrhea is out of control and must be considered a national epidemic of special significance. Among 164,000 women screened routinely in a wide variety of clinics over the past year, one in ten was found to be infected by the gonococcus.

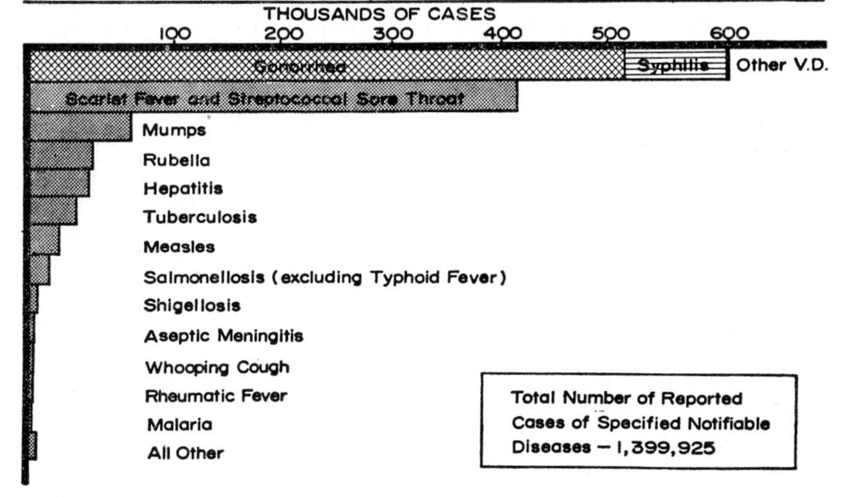

FIGURE 2. Communicable diseases in the United States—reported cases, 1969.

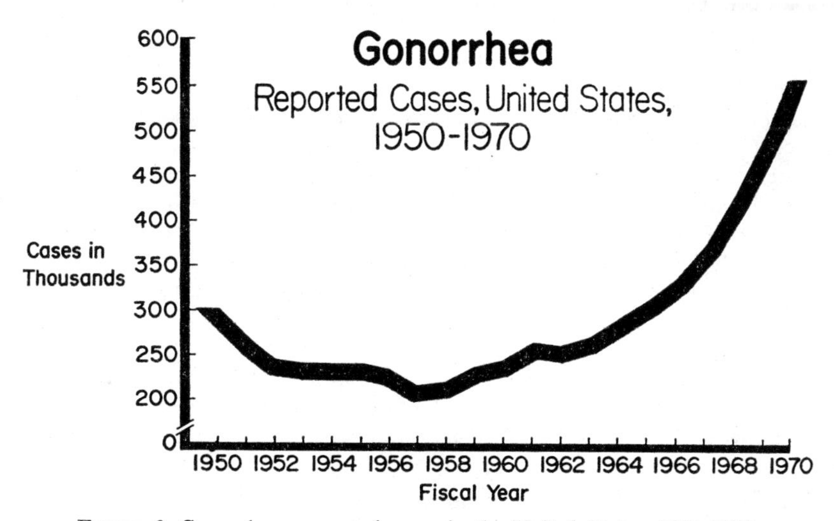

FIGURE 3. Gonorrhea—reported cases in the United States, 1950–1970.

Contributing to the growing problem of gonorrhea is the organism's increasing resistance to antibiotics. The CDC's Venereal Disease Research Laboratory has conducted in vitro sensitivity studies of *Neisseria gonorrhoeae* over some 15 years (70). Prior to 1958, most of the gonococcal strains which they studied were inhibited by penicillin concentrations of 0.2

units/ml or less. Today, there are strains from almost every state that require minimal inhibitory concentrations of 0.5 units/ml or more.

This tendency towards resistance on the part of the gonococcus extends to alternate antibiotics as well. Wide attention has been given recently to treatment with tetracyclines, but resistance to these drugs has grown much more rapidly than it ever did to penicillin (99). In 1965, concentrations of less than 0.5 mcg/ml inhibited 77 percent of cultures studied; only 1 percent required concentrations greater than 1.0 mcg/ml. By 1970, concentrations of less than 0.5 mcg/ml inhibited only 16 percent, and 27 percent required concentrations greater than 1.0 mcg/ml. In fact, one West Coast strain studied in 1970 continued to grow after exposure not only to tetracycline in concentration of 3.0 mcg/ml, but also to doxycycline, another widely used tetracycline analog.

Properly administered combinations of ampicillin and probenecid still cure almost all gonorrhea cases, but no case may be taken for granted; and follow-up cultures—urethral, endocervical, and rectal—are *sine qua non* in gonorrhea management. This should no longer be a problem for the private practitioner, regardless of his location. A new transport medium (Transgrow) will allow cultures to be taken in a doctor's office and safely transported for a period of 48 hours or longer en route to a central laboratory (71).

Melioidosis

Melioidosis has emerged as a problem in military personnel in Indochina where the causative organism, *Pseudomonas pseudomallei,* finds a natural habitat in soil and water. In their 1966 review, Redfearn, Palleroni & Stanier (95) found that all reported isolations of the organism occurred between the latitudes 20°N and 20°S, or were from patients who had been in such localities. Its invasive powers are not well defined but the organism is capable of infecting man and several other mammalian hosts. Transmission is probably from contaminated soil and water to wounds or to the respiratory tract. Person-to-person transmission apparently has not been described. Monkey-to-man transmission has been suspected and one serious laboratory infection has been reported (47).

Clinical forms of infection vary from those detectable only by serological methods to rapidly fatal systemic disease. The first described human infections with *P. pseudomallei* were reported from Rangoon in 1912 by Whitmore & Krishnaswami (113). Most infections have been acquired in the Far East. Many American military personnel contracted meliodidosis while on active duty in the Far East during World War II and more recently in Vietnam (48, 55, 89, 92).

Because melioidosis occurs in chronic forms that may not become manifest in military personnel until after their discharge, civilian as well as military hospitals should be prepared to diagnose the disease. In a case reported by Nelson & Albright, symptoms of infection did not appear until four years after the patient had left an endemic area.

Again, the clinician and the microbiologist must maintain close liaison when melioidosis is suspected. The organism is not especially difficult to grow, but it may go undetected if the microbiologist is not alerted to the possibility.

Nosocomial Infections

A nosocomial infection is one that develops in a patient after his admission to a hospital and as a result of in-hospital exposure to the causative agent, whether endogenous or exogenous. As defined, nosocomial infections include both preventable and nonpreventable infections.

In the United States, about 30 million patients are admitted to acute-care hospitals each year. Of this number, about 5 percent, or approximately one and one-half million persons, develop nosocomial infections, and about 1 percent of the patients thus infected, or 15 thousand people, die as a direct consequence of these hospital-associated infections.

These infections impose a considerable economic burden on our society. Research has demonstrated that the average patient with a nosocomial infection spends about 3.4 excess days in the hospital. Since the present per-day cost of hospitalization is about $100.00, a minimal estimate of the resultant excess cost is more than $500 million per year. This estimate does not include the additional expenses involved in the diagnosis and therapy of an infection; the extra expenses incurred in isolation of infected patients; the time lost by patients from productive employment because of the additional hospitalization; the inconvenience and monetary costs of lawsuits deriving from such infection; nor the costs, direct and indirect, of deaths resulting from nosocomial infections.

The total annual cost of surveillance and control programs in all short-stay hospitals in the United States has been estimated to be $31 million. From economic considerations alone, therefore, such surveillance and control programs would need to be capable of reducing the currently reported infection rate of 5.0 percent to 4.7 percent, or by a mere 0.3 percent, in order to produce a net economic benefit to the country as a whole. However, a survey conducted as recently as 1968 found that only 19 percent of a sample of 489 general hospitals had appointed nurse surveillance officers and only 20 percent had appointed hospital epidemiologists.[1]

Attack rates for nosocomial infections vary with the nature of the hospital. Community hospitals with less than 300 beds report the lowest rates, whereas the highest rates are reported from hospitals for chronic diseases. The services most affected in acute-care hospitals are, in descending order, gynecology, surgery, medicine, obstetrics, and pediatric-newborn services (12). Urinary-tract infections account for approximately one-half of all no-

[1] Himmelsbach, C. K., Chiazze, L., Dawson, J. M., Wise, D. L. Evaluation of Conference on Infection Control in Hospitals—A Report from Department of Community Medicine and International Health, Georgetown University School of Medicine, Washington, D. C. (unpublished).

socomial infections, surgical wounds for one-fourth, and infections of the respiratory tract for one-eighth of the total.

Escherichia coli is the pathogen most frequently isolated from nosocomial infections, accounting for approximately one-fourth of all isolates, whereas *Staphylococcus aureus* as an agent has been declining for many years and presently comprises about one-eighth of all isolates from nosocomial infections.

The pathogens that usually cause endemic nosocomial infections differ substantially from those responsible for hospital outbreaks. For example, *Salmonella* accounts for only about 0.1 percent of all isolates reported from nosocomial infections, yet it produces about twice as many recognized outbreaks as any other pathogen (12).

The bulk of nosocomial infections represent endemic and not epidemic problems. Such problems are often hospital-wide, are caused by organisms with multiple resistance to antimicrobial agents, and are difficult to control. In all types of hospitals, there is an increasing realization of the need for hospital-wide control measures to prevent cross-infections and those infections resulting from endogenous microorganisms. Effective isolation procedures, adequate handwashing facilities and practices, proper disinfection of respiratory equipment (51, 94), and judicious handling of urinary (36, 69) and intravenous catheters (14, 116) are among the measures being instituted and stressed in all types of hospitals.

Common-source nosocomial infections have recently attracted attention as a potentially significant problem among hospitals in the United States. Experience with contaminated products and equipment distributed on a nationwide basis, such as prepackaged catheter kits (49), hand lotions (77), and carmine dye (65), has highlighted the importance of infections deriving from such sources. This experience has re-emphasized the desirability of a surveillance system capable of bringing information on contaminated products, as well as on all newly discovered vehicles of infection, to the immediate attention of all hospitals.

Both endemic and epidemic nosocomial infections may become more of a problem in the future. The emergence of multiply resistant Gram-negative organisms with the ability to transfer these resistances between genera have made infections more difficult to treat. Although multiply resistant organisms have most often been associated with the persistent problem of endemic nosocomial infections, epidemics in hospitals, due to such organisms, have recently been recognized (42, 75).

Increasing resistance in agents of disease is not limited to the Gram-negative organisms. For example, methicillin and the semisynthetic penicillins have been the mainstay of therapy for *S. aureus* infections caused by penicillin-resistant staphylococci, which emerged as a significant problem in the 1950's. But for the past several years, *S. aureus* strains resistant to the semi-synthetic penicillins have been recognized as a major nosocomial infection problem in other countries, although to date only a few small outbreaks of methicillin-resistant *S. aureus* have been reported in the United States (7, 17, 90).

The increasing susceptibility of hospitalized patients to these infectious agents has compounded the nosocomial infections problem. The advancing age of our population and the availability of medical care to the elderly have resulted in an advance in the average age of the hospitalized patient. With the progress of medicine has come the development of new diagnostic procedures which, in many instances, create a portal of entry for pathogens or greatly increase the chance of exposure, or both. Also, new therapeutic procedures capable of prolonging life, including chronic renal dialysis, cancer chemotherapy, and steroid therapy, may greatly reduce the resistance of the patient to infections. Finally, transmission of infections is facilitated by patient overloads that result in crowding, and by a shortage of hospital personnel. This dual problem is becoming increasingly common in many present-day hospitals.

All of these factors will operate to increase the risk of infections among hospitalized patients. Thus, effective and persistent surveillance and control of nosocomial infections will become even more important in an effort to combat the morbidity and mortality associated with this burgeoning public health problem.

MYCOSES

Cryptococcosis

Before 1965, cryptococcosis was considered an uncommon disease. With the introduction of specific and sensitive serological tests for *Cryptococcus neoformans* antigens or antibodies, awareness of the increasing prevalence of this disease is now possible. The development of serologic tests is fortunate because cryptococcosis does not always present clear-cut symptoms and may not be recognized clinically until postmortem (2, 3, 5, 13, 59).

Between 1965 and 1968, the number of specimens submitted to the Center for Disease Control increased from 99 to 702, or by 609 percent. In 1969 clinical specimens from 478 patients were tested; 85 patients were serologically positive and 44 proved to have cryptococcosis by culture or histology (Figs. 4–5). The continuing increase in cases indicates that this disease is of greater public health importance than previously realized.

Opportunistic Mycoses

In recent years, a new spectrum of infectious diseases has developed that causes deep concern to the medical profession. The greatly expanded use of antibiotics, corticosteroids, cytotoxic drugs, immunosuppressives, prosthetic devices such as heart valves, and radiation therapy, has led to the development of secondary mycotic infections in patients with lowered resistance to disease (21, 39, 50). These infections create diagnostic and therapeutic problems for the physician and frequently are the actual cause of death of the patient.

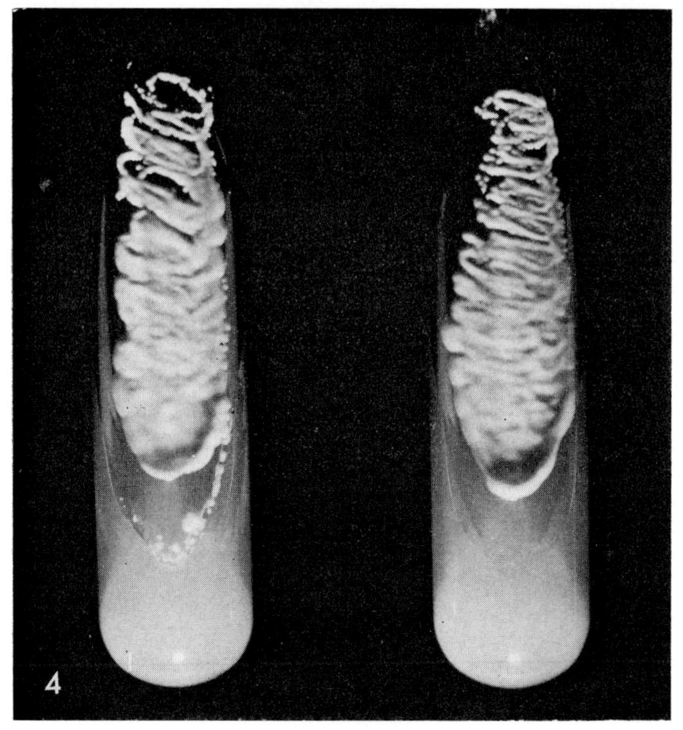

FIGURE 4. Yeastlike appearance of *Cryptococcus neoformans* in culture.

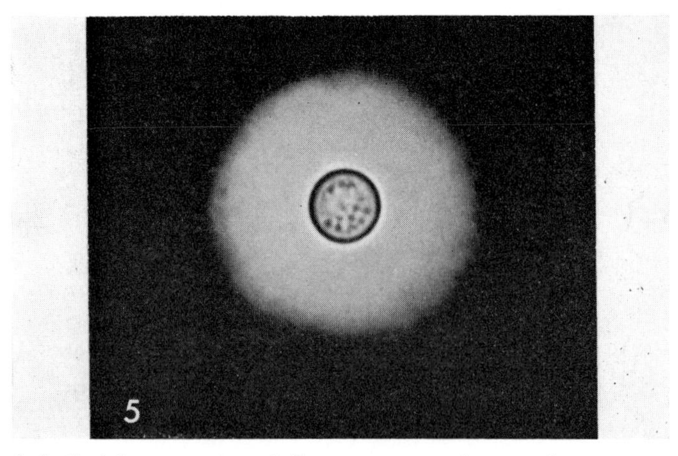

FIGURE 5. India ink preparation of *Cryptococcus neoformans* demonstrates capsule surrounding yeast cell.

A steadily increasing number of papers report such opportunistic fungus infection caused by a large variety of yeasts and filamentous fungi. The species involved frequently are ones formerly considered to be harmless and insignificant (67, 91). In many cases, no one bothered to identify them. Even when they were identified, the clinician did not attach any causal relationship to the patient's condition. In addition to these saprophytic fungi, recognized pathogenic fungi in increasing numbers are infecting compromised hosts, endangering their lives and complicating their treatment.

The frequency of these new mycotic diseases points to the need for providing competent medical mycological diagnostic services on a local, state, and national basis. Unless this problem of imposed fungus infections is resolved, it will jeopardize the use of many of the newer life-saving treatments (organ transplants, deep radiation, steroids, etc.) in many cases.

PARASITIC DISEASES

Babesiosis

Babesiosis, or piroplasmosis, is a disease caused by protozoan parasites that inhabit the red blood cells of a variety of wild and domestic animals (dogs, horses, cattle, sheep, rodents, etc.). Although the infection is often subclinical, it can produce a febrile, hemolytic disease in livestock, of considerable economic importance in countries where it is endemic. The organisms undergo development in and are transmitted by several species of hard ticks which are the only known arthropod vectors.

Only four human cases have been described in the literature (37, 98, 106, 111); of these, three have been recorded within the past three years and two were in the United States. Two of the cases were fatal. For various reasons, all but one of the infected persons had been splenectomized. The syndrome in the infected humans has included fever, chills, hemolytic anemia, jaundice, and renal failure.

How frequently the disease masquerades as falciparum malaria in Vietnam veterans and in the populations of malarious areas where babesiosis of animals also is common merits further study. The most definitive diagnosis is based upon the morphologic appearance of the red blood cell parasite.

Malaria

By the mid-1950's, malaria had been eradicated as an endemic disease in this country, although it was recognized that a small problem could arise because of widespread international travel of civilians and military personnel. The upsurge in reported cases since 1966 to more than 4000 per year is attributable almost entirely to the return of infected troops from Vietnam. Failure to complete the recommended courses of chemosuppression following return to this country is the primary cause of military cases occurring here.

The frequency of imported malaria infections in returning servicemen, combined with common intravenous heroin use in servicemen and civilians, has resulted in transmission of malaria from servicemen to civilian groups by means of shared needles and syringes. Several large malaria epidemics among heroin users have occurred since July 1970 in the United States, and all of the outbreaks were related to servicemen who shared injection equipment after their return.

Re-establishment of malaria transmission in the United States appears unlikely, but continuing surveillance will be required for some time to come. In the few instances in which mosquitoes carried the parasite, probable sources of infection were veterans who had served in Vietnam, and the outbreaks have been readily controlled.

Pneumocystosis

Pneumocystosis is an emerging parasitic disease seen almost exclusively in premature infants, debilitated infants and children, and adults with some defect in the immune response (93, 102). It has been described as a "plasma cell" pneumonia. Kagan et al (58) have discussed its epidemiological and clinical manifestations in some detail.

The parasite *Pneumocystis carinii,* believed to be protozoan, is found in the lungs of man and animals. Evidence suggests that it may be ubiquitous but of low pathogenicity except when the host defenses are impaired. Because the organism does not stain with routine stains used in pathology, it may be overlooked during microscopic examination of infected lung tissue. The capsule of the parasite is outlined well with Giemsa stain and silver impregnation technics.

Until the discovery that pentamidine isethionate was a useful therapeutic agent against this organism (54), pneumocystosis in infants was usually fatal. With the increased use of steroids and immunosuppressants (e.g., in preparation for organ transplants), *P. carinii* infections in adults are being reported with increasing frequency. From 1968 through 1970, the Parasitic Disease Drug Service of the Center for Disease Control had more than 400 calls for this drug from all parts of the country. The use of pentamidine has been documented recently by Western, Perera & Schultz (112), in regard to both efficacy and toxicity.

VIRAL DISEASES

Arboviral Diseases

Although several arboviral diseases, notably eastern, western, and St. Louis encephalitis, have long been recognized in the United States, at least two additional ones are presently emerging as actual or potential public health problems. They are, respectively, California encephalitis (29, 52) and Venezuelan equine encephalomyelitis.

California encephalitis.—California encephalitis (CE), an acute central nervous system (CNS) disease, occurs spottily throughout the United States but has been most widely recognized in the Midwest. Most cases involve children under 15 years of age. Symptoms vary from a mild aseptic meningitis to a severe encephalitis. Early signs consist of fever and headache, commonly followed by vomiting, lethargy, and convulsions.

This disease is caused by one or more of at least seven closely related arboviruses comprising the California virus complex. These viruses are transmitted mostly by *Aedes* mosquitoes in woodland areas, and various wild rodents serve as their natural vertebrate hosts. Humans are frequently bitten as they work or seek outdoor recreation in these areas.

Although California encephalitis was first recognized in 1943, its true medical importance is only now becoming apparent. Suddenly in the 1960's, cases began to be noted—not because the virus was only recently introduced, but because of clinicians' increased awareness of its role. Newer knowledge of the ecology of the causative virus complex and improved diagnostic capabilities in state and federal laboratories led to more frequent identification of the disease. Two hundred and eighty-five cases were recognized between 1963 and 1968 in 12 central and eastern states; this number, although significant, probably represents only a fraction of the actual number of cases that occurred.

As clinical acuity continues to increase, undoubtedly a still greater proportion of cases will be correctly diagnosed. And as our population grows and makes greater use of woodland recreational areas, human exposure will also increase.

Venezuelan equine encephalomyelitis.—Since 1938, Venezuelan equine encephalomyelitis (VEE) has been recognized as a devastating mosquito-transmitted CNS disease of horses in northern South America, and it has gained the reputation of being one of the most easily acquired laboratory infections of man (60). Until 1962, however, it was rarely known as a naturally acquired human disease.

Between 1962 and 1964, a huge outbreak occurred in Venezuela, sweeping eastward from the Colombian Guajira peninsula. Reported human cases numbered 31,966, with 190 deaths. Innumerable horses also died (101). Estimates of the actual number of human cases ran as high as 100,000. Symptoms ranged from a relatively mild influenza-like illness to a severe, fulminating encephalitis. Death rates were high among those individuals (particularly children) who showed the more severe CNS involvement, and disabling neurological sequellae were common in survivors of this group.

During 1967, other more restricted outbreaks took place in Colombia in horses and man.[2] In Ecuador between January and June of 1969, an estimated 20,000 human cases with about 200 deaths[3] were reported, as well as 30,000 to 60,000 equine deaths.[4] In the summer and fall of 1969, more equine outbreaks were recognized in Guatemala, Honduras, El Salvador, and Nica-

[2] Personal communication, 1967. Dr. Carlos Sanmartin, Universidad del Valle, Cali, Colombia.

[3] Personal communication, 1969. Dr. Thomas P. Monath, Virology Section, CDC.

[4] Personal communication, 1969. Dr. Roy W. Chamberlain, Virology Section, CDC.

ragua,[5] again with associated human cases. Outbreaks continued in the summer and fall of 1970, affecting Costa Rica to the south and Chiapas state of Mexico to the north.[6]

The recent pattern of progression appears to threaten the southern United States, particularly Texas. Certainly, with the extensive activity of VEE since 1962, which has accounted for hundreds of thousands of equine deaths and many thousands of human cases, serious thought must be given to control of this disease.

A less virulent strain of the virus has been reported from Florida. Possibly this strain could be used to develop a safe and effective VEE vaccine for use in horses. If so, its use in areas where virulent strains exist would indirectly protect humans, because horses serve as a source of infection for species of mosquitoes that also feed on man.

Arenoviral Diseases

An emerging or newly recognized complex of severe hemorrhagic diseases is caused by a number of lymphocytic choriomeningitis-related viruses for which the name "arenoviruses" has been proposed (97). The term arenovirus reflects the characteristic fine, sandlike granules seen in the virion in ultrathin sections.

Argentinian hemorrhagic fever.—The first of these diseases to receive attention was Argentinian hemorrhagic fever, caused by Junin virus. Although not recognized until 1955, in retrospect it must have occurred much earlier. Details of the ecology of Junin virus are still uncertain; small wild rodents in corn-growing districts near Buenos Aires apparently become chronically infected and excrete the virus in their urine. Man acquires the infection while working in the cornfields at harvest time, presumably through contact with the rodents' infected urine. Around 300 to 1000 human cases occur annually. Mortality rates as high as 20 percent have been reported by Shelokov (103) and by Casals & Clarke (20).

Bolivian hemorrhagic fever.—In 1963, a severe outbreak of a similar hemorrhagic disease, with an even higher death rate, occurred for the first time in the town of San Joaquin, Bolivia. Investigations showed that this disease, Bolvian hemorrhagic fever, is caused by a virus related to Junin, subsequently named Machupo virus (103), which is excreted in the urine of certain chronically infected wild rats, *Calomys callosus* (57). Apparently, this species of rodent had encroached upon human habitations following a gradual decimation of the local domestic cat population, a die-off due to a

[5] Hinman, Alan, et al. Paper on VEE in Guatemala, 1969, in preparation.

[6] Personal communication, 1969 and 1970, Dr. Robert W. McKinney, Walter Reed Army Medical Unit, U. S. Army Biological Laboratories, Ft. Detrick, Maryland.

neurological disorder circumstantially attributed to widespread use of DDT within the town (56). Successful control of the rodents and of the Bolivian hemorrhagic disease outbreak was accomplished by a well-organized rat-trapping program (64). However, more recently a second outbreak has occurred in another Bolivian community, warning us that the threat still remains.[7]

Lassa fever.—In January and February of 1969, Lassa fever, a hitherto unknown disease, caused the death of two missionary nurses in Nigeria and a severe illness in a third (38). The initial infection was contracted in Lassa, a town in northern Nigeria. The second and third cases contracted their disease through exposure to the first case. The illnesses were characterized by marked malaise, generalized myalgia, headache, severe sore throat, and small ulcerations in the pharyngeal and buccal mucosa. Convulsions, macular and petechial rash, and some hemorrhagic manifestations were also noted.

In the laboratory studies that followed, a virus was isolated from the blood of all three cases and was shown to be related to Junin and Machupo viruses (16). During the course of subsequent laboratory studies in the United States, a medical scientist and a laboratory technician in turn became infected. The latter died; the former survived but had virus in the urine until the 32nd day (66).

Activity of the disease was again apparent in Nigeria in January and February of 1970, when 26 persons with suspected cases were hospitalized at Jos. Symptoms ranged from mild to very severe; 10 of the 26 died. Epidemiological observations indicated that transmission was through contact with infected individuals. Additional cases were also seen at another hospital 13 miles from Jos (109).

The ecology of Lassa virus is still a mystery, but since it is related to Junin and Machupo viruses, a rodent host is suspected. In view of its extreme infectiousness and virulence, it appears to pose a considerable public health threat and is therefore being studied further to elucidate its distribution, its natural hosts, and the factors that permit its spread to man.

Marburg Disease: African Green Monkey or Vervet Disease

In the late summer of 1967, an outbreak of a hitherto unknown hemorrhagic disease, now called Marburg disease, occurred in laboratory workers in Germany and Yugoslavia engaged in removal and processing of African green monkey kidneys for cell cultures (53). The monkeys involved were all from the same supplier in Uganda and had been shipped in two lots by air with only an overnight stop in London. Twenty-five primary human cases stemmed from direct contact with blood or tissues of monkeys from one or both shipments, and seven of these cases were fatal.

[7] Johnson, Karl M., 1970, oral report, Annual Meeting of American Society of Tropical Medicine and Hygiene, San Francisco.

Five of six secondary cases resulted from contact with tissues or blood of patients or from accidental pricking with hypodermic needles used on these patients. One secondary case was venereally acquired by the wife of a patient 87 days after onset of his illness. Virus was recovered from his semen, which suggested a method of spread among monkeys in nature.

The incubation period in man apparently was four to seven days. Onset was abrupt and generally biphasic. Early symptoms included high fever, headache, vomiting, diarrhea, and conjunctivitis. The second stage was characterized by an exanthem and hemorrhagic signs that included gastrointestinal and nasal bleeding. Bronchopneumonia, hepatitis, and pancreatitis were commonly seen; less commonly, meningitis, encephalitis, orchitis, and myocarditis (72, 108). In fatal cases, the liver, the kidneys, and lymphoid tissues were most severely affected (11, 43).

The causative viral agent was isolated in guinea pigs from tissues obtained at autopsy and from acute-phase blood specimens (61, 74, 100, 107). Electron microscopic examination showed this virus to be vastly different from any other virus particle known. It was cylindrical in shape, sometimes branched, 90–100 nm in diameter and 130–2600 nm in length; genetic material was RNA (61).

Serological studies of African green monkeys at the Lake Kyoga region of Uganda, the area of origin of the suspect animals, indicated that Marburg virus infection had occurred there. Evidence at hand thus far suggests that monkeys, probably African greens, are the natural hosts and usually undergo only asymptomatic infections. However, the ability of this bizarre virus to produce fatal human disease was forcefully demonstrated in Germany and Yugoslavia.

Much biomedical research and vaccine production is dependent upon constant supplies of African green monkeys—about 12,000 a year are imported into the United States for this purpose. In view of the potential of this virus to cause disastrous focal outbreaks among biomedical personnel, Marburg disease must be considered a definite threat. Research must be continued to make clear which precautionary measures—quarantine and other—must be enforced to prevent further human infections.

Viral Hepatitis

The ecology of serum hepatitis is currently changing in subtle and challenging ways, leading to increased incidence and clinical problems, and to new epidemiological directions.

In 1952, a national system of viral hepatitis surveillance was established by state and territorial epidemiologists. Tabulation of cases received by this surveillance network reveals peak hepatitis incidence in 1953–54 and again in 1960–61. During the past four years there has been a steady increase in reported cases (23).

Before 1967, hepatitis reports maintained a definite seasonal pattern, which was thought to be a reflection of seasonal variations in the occurrence of infectious hepatitis. Recently, however, there has been a change in this seasonal pattern that very likely indicates the emerging epidemiologic importance of serum hepatitis.

The rates of reported serum hepatitis are relatively low when compared with the total hepatitis reported, but large numbers of cases of serum hepatitis go unreported or are misclassified. Even so, the rates of reported serum hepatitis during the past four years have increased dramatically. During the same period, the peak age and sex distribution for viral hepatitis has shifted from children aged 5–19 to teenage and young adult males (23, 32). Clinical experience and recent epidemiologic reports (31, 81, 83, 96) suggest that these changing age and sex characteristics reflect a growing epidemic of parenteral drug use in the United States.

While epidemiologists have linked recent increases in the incidence of serum hepatitis with epidemic parenteral drug abuse, laboratory workers have discovered a serum antigen that is closely related to the etiologic agent of serum hepatitis. This antigen, called hepatitis-associated antigen (HAA, Australia antigen) has physical, chemical, and structural characteristics that suggest a virus or a viral remnant (6, 10); however, tests for nucleotides (DNA and RNA) (76) and attempts at culturing HAA have thus far been negative.

Epidemiologic observations suggest that HAA is related to the infective agent of serum hepatitis. In prospective studies of patients with diagnosed serum hepatitis in whom serial bloods were collected, greater than 90 percent have demonstrable HAA during their acute illness (45, 104). However, in experimental or well-defined outbreaks of infectious hepatitis, HAA has not been detected except in rare cases (30, 45, 78). In patients with acute serum hepatitis, the period of antigenemia usually precedes clinical symptoms by approximately two weeks and persists for approximately four weeks. However, occasionally HAA may persist indefinitely (62).

In the United States, approximately 0.1 percent of apparently healthy individuals have been found to harbor HAA; when blood from these HAA-positive individuals is transfused, a significant percent of recipients will develop hepatitis (46). Therefore, in an effort to reduce the incidence of transfusion-associated hepatitis, several blood banks are now screening and eliminating units that contain HAA.

In addition to the practical application of HAA screening in blood banking, HAA has also helped to redefine the epidemiology of serum hepatitis. It is now well accepted that serum hepatitis is spread by an oral as well as a parenteral route in unique experimental and epidemic settings (63, 79, 87). Using HAA as a marker for serum hepatitis, investigators have uncovered more HAA-positive individuals that can be accounted for by parenteral transmission alone. It is possible than many of these patients contracted infection by nonparenteral routes, and that orally acquired infection is more commonplace than previously acknowledged.

SUMMARY

Man is a member of the biological universe in which no species is independent of others or immune to changes in the environment. Changes, great or small, natural or man-made, set off a chain reaction as nature seeks to re-establish a balance, but not necessarily to restore the old.

The microbiological system is closely allied with man; changes in the environment alter his relationship with organisms whether they be beneficial, symbiotic, or pathogenic. Man's way of life, his human behavior, his technological advances, his mere existence foster the conquest of some disease organisms, the emergence of others, and his introduction to unfamiliar ones.

The infectious disease picture, therefore, is as subject to change as life itself.

LITERATURE CITED

1. Abrutyn, E., Gangarosa, E., Forrest, J., Mosley, W. H. 1971. *Ann. Intern. Med.* 74:228–31
2. Ajello, L. 1967. *Bacteriol. Rev.* 31: 6–24
3. Ajello, L. 1970. Pan American Health Organization, *Proc. Int. Symp. on Mycoses.* Sci. Publ. 205
4. Alpern, R. J., Dowell, V. R., Jr. 1969. *J. Am. Med. Assoc.* 209: 385–88
5. Bardana, E. J., Kaufman, L., Benner, E. J. 1968. *Arch. Intern. Med.* 122:517–20
6. Barker, L. F., Smith, K. O., Gehle, W. D. 1969. *J. Immunol.* 102: 1529–32
7. Barrett, F. F., McGehee, R. F., Finland, M. 1968. *N Engl. J. Med.* 279:441–48
8. Bart, K. J., Huq, Z., Khan, M., Mosley, W. H. 1970. *J. Infect. Dis.* 121:S17–24
9. Barua, D. 1970. *World Health Organ. Pub. Health Papers No. 40,* 47–52. World Health Organization: Geneva.
10. Bayer, M. E., Blumberg, B. S., Werner, B. 1968. *Nature (London)* 218:1057–59
11. Bechtelsheimer, H., Jacob, H., Solcher, H. 1968. *Deut. Med. Wochenschr.* 93:602–4
12. Bennett, J. V., Scheckler, W. E., Maki, D. G., Brachman, P. S. 1970. *Current Patterns of Nosocomial Infections in the United States.* Presented at Int. Conf. Nosocomial Infections, Center for Disease Control, Atlanta, Ga. 30333
13. Bennington, J. L., Haber, S. L., Morgensten, N. L. 1964. *Dis. Chest* 45:262–63
14. Bentley, D. W., Lepper, M. H. 1968. *J. Am. Med. Assoc.* 206:1749–52
15. Bodner, S. J., Koenig, M. G., Goodman, J. S. 1970. *Ann Intern. Med.* 73:537–44
16. Buckley, S. M., Casals, J. 1970. *Am. J. Trop. Med. Hyg.* 19:680–91
17. Bulger, R. J. 1967. *Ann. Intern. Med.* 57:81–89
18. Carpenter, C. C. J., et al. 1966. *Bull. Johns Hopkins Hosp.* 118:165–68
19. Cary, S. G., Blair, E. B. 1964. *J. Bacteriol.* 88:96–98
20. Casals, J., Clarke, D. H. 1965. *Viral and Rickettsial Infections of Man,* 4th ed., 659–84. Ed. F. L. Horsfall, Jr., and I. Tamm. Philadelphia, Pa.: J. B. Lippincott Co.
21. Casazza, A. R., Duvall, C. P., Carbone, P. P. 1966. *J. Am. Med. Assoc.* 197:118–24
22. Cash, R. A., et al. 1970. *Am. J. Trop. Med. Hyg.* 19:653–56
23. Center for Disease Control. 1970. *Hepatitis Surveillance Rep. No. 32*
24. Center for Disease Control. 1969. *Morbid. Mortal. Ann. Suppl.–* Summary. 18
25. Center for Disease Control. 1970. *Morbid. Mortal. Weekly Rep.* 19: 269–70
26. Center for Disease Control. 1970. *Morbid. Mortal Weekly Rep.* 19: 381–82
27. Center for Disease Control. 1971. *Morbid. Mortal. Weekly Rep.* 19: 495
28. Center for Disease Control. 1970. *VD Fact Sheet* (27th ed.)
29. Chamberlain, R. W. 1971. *Viruses Affecting Man and Animals. Int. Symp. Med. & Appl. Virol., 3rd.*

Compiled and edited by M. Sanders, M. Schaeffer, Chap. 21, 309–26. St. Louis, Mo.: Warren H. Green, Inc.
30. Chang, L. W., O'Brien, T. F. 1970. *Lancet* 2:59
31. Cherubin, C. E., Hargrove, R. L., Prince, A. M. 1970. *Am. J. Epidemiol.* 91:510–17
32. Dismukes, W. E., Karchmer, A. W., Johnson, R. F., Dougherty, W. J. 1968. *J. Am. Med. Assoc.* 206: 1048–52
33. Editorial and letter. 1970. *Brit. Med. J.* 4:2–3
34. Felner, J. M., Dowell, V. R., Jr. 1970. *N. Engl. J. Med.* 283:1188–92
35. Felner, J. M., Dowell, V. R., Jr. 1971. *Am. J. Med.* 50:787–96
36. Finkelberg, Z., Kunin, C. M. 1969. *J. Am. Med. Assoc.* 207:1657–62
37. Fitzpatrick, J. E. P., et al. 1968. *Nature (London)* 217:861–62
38. Frame, J. D., Baldwin, J. M., Jr., Gocke, D. J., Troup, J. M. 1970. *Am. J. Trop. Med. Hyg.* 19:670–76
39. Frenkel, J. K. 1962. *Lab. Invest.* 11: 1192–1208
40. Gangarosa, E. J., Dewitt, W. E., Huq, I., Zarifi, A. 1968. *Trans. Royal Soc. Trop. Med. Hyg.* 62: 693–99
41. Gangarosa, E. J., et al. 1970. *J. Infect. Dis.* 122:181–90
42. Gardner, P., Smith, D. H. 1969. *Ann. Intern. Med.* 71:1–9
43. Gedigh, P., Bechtelsheimer, H., Korb, G. 1968. *Deut. Med. Wochenschr.* 93:590–601
44. Gelb, A. F., Seligman, S. J. 1970. *J. Am. Med. Assoc.* 212:1038–41
45. Giles, J. P., et al. 1969. *N. Engl. J. Med.* 281:119–22
46. Gocke, D. J., Greenberg, H. B., Kavey, N. B. 1969. *Lancet* 2:248–49
47. Green, R. N., Tuffnell, P. G. 1968. *Am. J. Med.* 44:599–605
48. Greenawald, K. A., Nash, G., Foley, F. D. 1969. *Am. J. Clin. Pathol.* 52:188–98
49. Hardy, P. C., Ederer, G. M., Matsen, J. M. 1970. *N. Engl. J. Med.* 282:33–35
50. Hart, P. D., Russell, E., Jr., Remington, J. S. 1969. *J. Infect. Dis.* 120:169–91
51. Hellewell, J., Jeanes, A. L., Watkin, R. R., Gibbs, F. J. 1967. *Anaesthesia* 22:497–503
52. Henderson, B. E., Coleman, P. H. *Progr. Med. Virol.* In press
53. Henessen, W., Bonin, O., Mauler, R. 1968. *Deut. Med. Wochenschr.* 93:582–89
54. Ivady, G., Paldy, L. 1958. *Kinderheilk* 106:10
55. James, A. E., Dixon, G. D., Johnson, H. F. 1967. *Radiology* 89: 230–35
56. Johnson, K. M. 1965. *Am. J. Trop. Med. Hyg.* 14:816–18
57. Johnson, K. M., Mackenzie, R. B., Webb, P. A., Kuns, M. L. 1965. *Science* 150:1618–19
58. Kagan, I. G., Fox, H. A., Walls, K. W., Healy, G. R. 1967. *Clin. Pediat.* 6:641–54
59. Kaufman, L., Blumer, S. 1968. *Appl. Microbiol.* 16:1907–12
60. Kissling. R. E., Chamberlain, R. W. 1967. *Advan. Vet. Sci.* 11:65–84
61. Kissling, R. E., Robinson, R. Q., Murphy, F. A., Whitfield, S. G. 1968. *Science* 160:888–90
62. Krugman, S., Giles, J. P. 1970. *J. Am. Am. Med. Assoc.* 212:1019–29
63. Krugman, S., Giles, J. P., Hammond, J. 1967. *J. Am. Med. Assoc.* 200:365–73
64. Kuns, M. L. 1965. *Am. J. Trop. Med. Hyg.* 14:813–16
65. Lang, D. J., Kunz, L. F., Martin, A. R., Schroeder, S. A., Thomason, L. A. 1967. *N. Engl. J. Med.* 276: 829–32
66. Leifer, E., Gocke, D. J., Bourne, H. 1970. *Am. J. Trop. Med. Hyg.* 19: 677–79
67. Louria, D. B., Blevins, A., Armstrong, D., Burdick, R., Lieberman, P. 1967. *Arch. Intern. Med.* 119:247–52
68. Marr, W. L., Gaines, S. 1964. *Mil. Med.* 129:1061–63
69. Martin, C. M., Bookrajian, E. N. 1962. *Arch. Intern. Med.* 110: 703–11
70. Martin, J. E., Jr., Lester, A., Price, E. V., Schmale, J. D. 1970. *J. Infect. Dis.* 122:459–61
71. Martin, J. E., Jr., Lester, A. 1971. *HSMHA Health Rep.* 86:30–33
72. Martini, G. A., Knauff, H. G., Schmidt, H. A., Mayer, G., Baltzar, G. 1968. *Deut. Med. Wochenschr.* 93:559–71
73. Mata, L. J., Gangarosa, E. J., Caceres, A., Perera, D. R., Mejicanos, M. L. 1970. *J. Infect. Dis.* 122:170–80
74. May, G., Knothe, H. 1968. *Deut. Med. Wochenschr.* 93:620–22
75. Medeiros, A. A., O'Brien, T. F.

1969. *Antimicrob. Ag. Chemother.—1968 (Proc. Intersci. Conf. Antimicrob. Ag. Chemother., 8th)*, 30–35. Bethesda, Md.: Am. Soc. Microbiol.
76. Millman, I., Loeb, L. A., Bayer, M. E., Blumberg, B. S. 1970. *J. Exp. Med.* 131:1190–99
77. Morse, L. J., Schonbeck, L. E. 1968. *N. Engl. J. Med.* 278:376–78
78. Mosley, J. W., Barker, L. F., Shulman, N. R., Hatch, M. H. 1970. *Nature (London)* 225:953–55
79. National Communicable Disease Center. 1970. *Hepatitis Surveillance Rep. 31*
80. National Communicable Disease Center. 1969. *Morbid. Mortal. Weekly Rep.* 18:367
81. National Communicable Disease Center. 1969. *Morbid. Mortal. Weekly Rep.* 18:433–34
82. National Communicable Disease Center. 1969. *Morbid. Mortal Weekly Rep.* 18:441
83. National Communicable Disease Center. 1970. *Morbid. Mortal. Weekly Rep.* 19:30
84. National Communicable Disease Center. 1970. *Morbid. Mortal. Weekly Rep.* 19:69–70
85. National Communicable Disease Center. 1970. *Morbid. Mortal. Weekly Rep.* 19:104–5
86. National Communicable Disease Center. 1970. *Morbid. Mortal. Weekly Rep.* 19:172–73
87. National Communicable Disease Center. 1970. *Morbid. Mortal. Weekly Rep.* 19:198–99
88. National Communicable Disease Center. 1970. *Morbid. Mortal. Weekly Rep.* 19:207
89. Nelson, R. N., Albright, C. R. 1967. *Oral Surg., Oral Med., Oral Pathol.* 24:128–36
90. O'Toole, R. D., Drew, L. W., Dahlgren, B. J., Beaty, H. N. 1970. *J. Am. Med. Assoc.* 213:257–63
91. Parkhurst, G. F., Vlahides, G. D. 1967. *J. Am. Med. Assoc.* 202:279–81
92. Patterson, M. C., Darling, C. L., Blumenthal, J. B. 1967. *J. Am. Med. Assoc.* 200:447–51
93. Patterson, J. H., Lindsey, I. L., Edwards, E. S., Logan, W. D. 1967. *Pediatrics* 38:388–97
94. Pierce, A. K., Sanford, J. P., Thomas, G. D., Leonard, J. S. 1970. *N. Engl. J. Med.* 282:528–31
95. Redfearn, M. S., Palleroni, N. J., Stanier, R. Y. 1966. *J. Gen. Microbiol.* 43:293–313
96. Richards, L. G., Carroll, E. E. 1970. *Publ. Health Rep.* 85:1035–41
97. Rowe, W. P., et al. 1970. *J. Virol.* 5:651–52
98. Scholtens, R. G., Braff, E. H., Healy, G. R., Gleason, N. N. 1968. *Am. J. Trop. Med. Hyg.* 17:810–13
99. Schroeter, A. L., Pazin, G. J. 1970. *Ann. Intern. Med.* 72:553–59
100. Siegert, R., Shu, H. L., Slenczka, W., Peters, D., Müller, G. 1967. *Deut. Med. Wochenschr.* 92:2341–43
101. Sellers, R. F., Bergold, G. H., Suarez, O. M., Morales, A. 1965. *Am. J. Trop. Med. Hyg.* 14:460–69
102. Sheldon, W. H. 1962. *J. Pediat.* 61:780–91
103. Shelokov, A. 1965. *Am. J. Trop. Med. Hyg.* 14:790–92
104. Shulman, N. R., Hirschman, R. J., Barker, L. F. 1970. *Ann. Intern. Med.* 72:257–69
105. Sinkovics, J. G., Smith, J. P. 1970. *Cancer* 25:663–71
106. Škrabalo, Z., Deanović, Ž. 1957. *Doc. Med. Geogr. Trop.* 9:11–16
107. Smith, C. E. G., Simpson, D. I. H., Bowen, E. T. W., Zlotnik, I. 1967. *Lancet* 2:1119–21
108. Stille, W., Böhle, E., Helm, E., Van Rey, W., Siede, W. 1968. *Deut. Med. Wochenschr.* 93:572–82
109. Troup, J. M., White, H. A., Fom, A. L., Carey, D. E. 1970. *Am. J. Trop. Med. Hyg.* 19:695–96
110. Venkatraman, K. V., Ramakrishman, C. S. 1941. *Indian J. Med. Res.* 29:681–84
111. Western, K. A., Benson, G. D., Gleason, N. N., Healy, G. R., Schultz, M. G. 1970. *N. Engl. J. Med.* 283:854–56
112. Western, K. A., Perera, D. R., Schultz, M. G. 1970. *Ann. Intern. Med.* 73:695–702
113. Whitmore, A., Krishnaswami, C. 1912. *Indian M. Gaz.* 47:262–67 (cited by A. L. Prevatt, J. S. Hunt. 1957. *Am. J. Med.* 23:810–23)
114. *WHO Weekly Epidemiologic Report.* 1963. 38:570
115. *WHO Weekly Epidemiologic Report.* 1967. 42:364
116. Zinner, S. H., et al. 1969. *J. Infect. Dis.* 120:616–19

INHIBITORS OF RIBOSOME FUNCTIONS[1]

1580

SIDNEY PESTKA[2]

Roche Institute of Molecular Biology, Nutley, New Jersey

CONTENTS

[1] Abbreviations used: AA-tRNA: (aminoacyl-tRNA); Ac-Phe-tRNA: (N-acetyl-phenylalanyl-tRNA); fMet-tRNA: (formylmethionyl-tRNA); Ac-Phe-puromycin: (acetyl-phenylalanyl-puromycin); fMet-puromycin: (formylmethionyl-puromycin); GDPCP: (guanylyl-5′-methylenediphosphonate); ATA: (aurintricarboxylic acid); Spc: (spectinomycin); Sm: (streptomycin); TC: (tetracycline); CTC: (chlortetracycline); CM: (chloramphenicol); ERY: (erythromycin).

[2] Many of my colleagues have kindly sent me preprints of copies of their results prior to publication. I very much thank them for helping to keep this review current.

INTRODUCTION

The biosynthesis of a protein is a product of a large number of highly coordinated events. A summary of the process as it occurs in *Escherichia coli* is given in Figure 1. A detailed review of protein biosynthesis is presented by Lucas-Lenard & Lipmann (1) and by Lengyel & Söll (2). As illustrated in Figure 1, proteins are synthesized on the ribosome which serves as the focus for reactions involving many ribosomal and supernatant enzymic factors and cofactors. During several steps there is transient association between the supernatant factors and the ribosome. Inhibitors of either the supernatant factors or ribosomes may thus influence reactions of one another. The present review will concentrate on those inhibitors which ap-

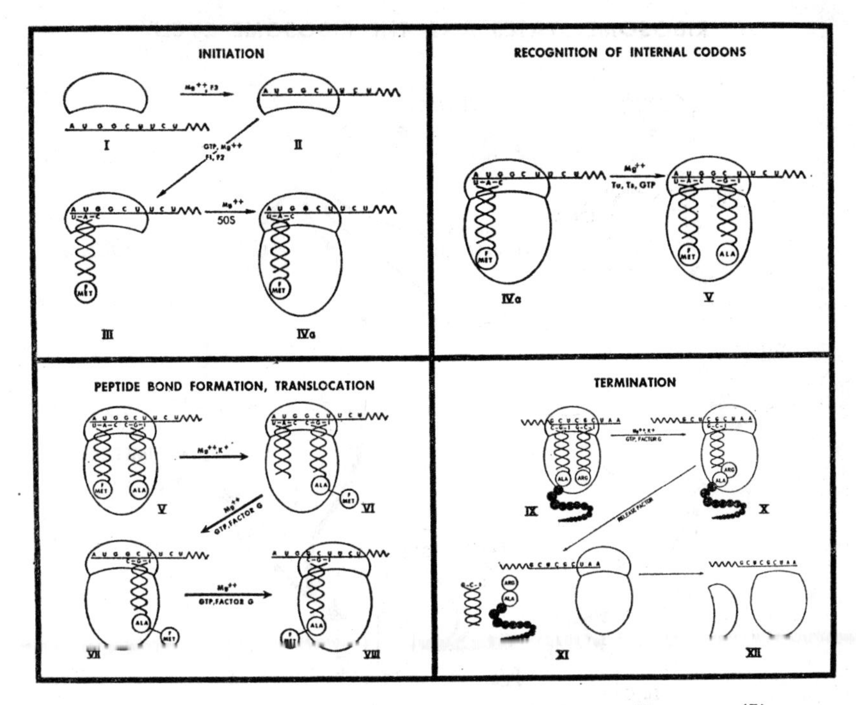

FIGURE 1. Schematic summary of protein synthesis. The semilunar cap (I) represents the free 30S subunit. Initiation of protein synthesis involves attachment of mRNA to a 30S subunit (I) to form complex II; this process requires Mg^{++} as well as initiation factor F3. Subsequent attachment of fMet-tRNA in response to initiation codon AUG to form complex III requires GTP and initiation factors F1 and F2. Junction of the 50S subunit to complex III produces complex IVa; it is probable that prior to complex IVa an intermediate state IV exists where fMet-tRNA is in the "A" or nonpuromycin reactive site; the transition from IV to IVa possibly occurs on hydrolysis of GTP. Enzymic recognition of internal codons involves factors Tu, Ts, and GTP: the Tu:GTP:Ala-tRNA complex binds to the ribosomes in response to the GCU codon to form complex V. Peptide bond formation occurs by transfer of the fMet (peptidyl) group to form fMet-Ala (VI); peptidyltransfer requires only ribosomes and K^+. Translocation involves several coordinate processes: release of deacylated $tRNA^{fMet}$ to form state VII; one codon movement of mRNA and ribosome with respect to each other, precisely positioning the next codon UCU into position for translation and coordinate movement of peptidyl-tRNA (fMet-Ala) from the "A" to "P" site resulting in state VIII. By repetition of the codon recognition step Ser-tRNA would enter the A-site in response to the codon UCU. Complex IX represents a peptidyl-tRNA with a polypeptide almost completed. Transpeptidation and translocation produces complex X with a completed protein still attached to tRNA and a termination codon, UAA, in the next recognition site. In response to a release factor the completed protein is released and perhaps also tRNA after translocation (XI). Provided no further cistrons are to be translated, the ribosome may be dissociated into 30S and 50S subunits with release of mRNA (XII). Alternatively, mRNA may be degraded prior to this stage. See reviews (1, 2) for details.

RIBOSOME STATES AND THE RIBOSOME CYCLE

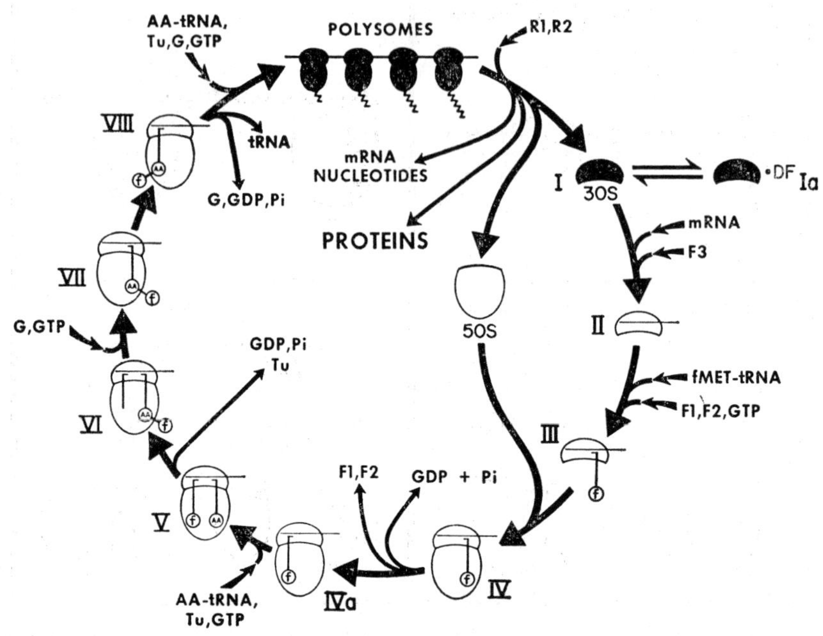

FIGURE 2. Ribosomal states and the ribosomal cycle. The individual complexes I to VIII have been described in the legend to Figure 1. The initiation sequence comprises stages I to IVa. For every addition of a single amino acid, elongation, the ribosome state cycles from V to VIII. State IVa is formally equivalent to VIII. Thus, the elongation epicycle (V to VIII) is superimposed on the overall ribosome cycle and repeats itself on each amino acid addition. The termination sequence is not given in detail for it has not been precisely delineated. Dissociation factor (DF) bound to 30S subunits (Ia) may be identical to one of the initiation factors and may possibly be involved in dissociation of 70S ribosomes following termination. Discussion of the ribosome cycle is presented in several reports (3–6). The symbol "f" represents fMet.

pear to act chiefly through inhibition of some ribosomal function rather than through inhibition of the supernatant factors, themselves.

RIBOSOME STRUCTURE AND FUNCTION

Throughout protein biosynthesis, ribosomes and subunits continually interact with one another; and subunits recycle (3–6) through the pool of free monomers and subunits (Fig. 2). Of all the enzymic reactions in protein synthesis (Figs. 1 and 2), only the formation of the peptide bond can be clearly designated as a ribosomal function not requiring other nonribosomal protein factors (7). For the majority of translational operations, the ribosome provides a scaffold on which appropriate apposition of mRNA codons and tRNA can occur while coordinately peptidyl- and aminoacyl-tRNA are oriented into position for peptide bond formation. To accomplish these tasks ribosomes provide this precise coordination for relatively distant molecular events such as those of codon and peptidyl-tRNA translocation (VI-VIII). The ribosome, therefore, is a multi-macromolecular complex containing numerous sites for binding mRNA, tRNA, and various factors and cofactors, which interact with ribosomes in a precise temporal and spacial sequence as shown in Figures 1 and 2. However, ribosomes are far from being inert scaffolds upon which protein synthesis occurs; for they appear to participate intimately and to modulate the reactions while they themselves undergo conformational modifications as protein synthesis progresses. During the various stages, ribosomes and subunits probably occupy discrete states related to their position in protein synthesis; also, each state, I to XI, may represent a discrete conformation and may have a characteristic affinity for an antibiotic or other factor. In addition, it is possible that the presence of a factor or tRNA on a ribosome may sterically impede or enhance the attachment of a particular antibiotic.

Examination of polysome patterns in cells treated with an antibiotic provides a convenient way of determining where in the ribosome cycle an antibiotic acts, for an antibiotic which may find ribosomes accessible during some stages of the ribosome cycle, may find ribosomes inaccessible to antibiotics during other stages. Thus, a specific inhibitor of initiation would be expected to produce polysome breakdown; an inhibitor of translocation (viz, movement of mRNA and ribosomes with respect to each other), stabilization of the polysome pattern; and an inhibitor of termination should produce polysomes of larger size than normal as ribosomes pile up along mRNA (8). It should be kept in mind that inhibition of protein synthesis may reflect inhibition of any of the individual steps comprising initiation, elongation, or termination; and that analysis of the ribosomal cycle can help define the area of inhibition.

Antibiotics, as will be readily seen later on in this review, which affect ribosomes produce numerous pleiotropic effects. Thus, antibiotics which predominantly affect 30S subunits may also influence 50S function. Conversely, antibiotics which predominantly influence 50S subunit function can also produce effects on 30S subunits. Some of these pleiotropic effects are demonstrable in intact cells as well as in cell-free extracts. In other cases the effects observed in cell-free extracts may simply be pleiotropic manifestations of ribosomal topology, but be only remotely related to antibiotic action in intact cells. No doubt, the marked interdependence of structural ribosomal components gives rise to the observed pleiotropy.

In order to correlate results in cell-free assays with antibiotic action in intact cells, it is useful to correlate the concentration at which these effects are observed in cell-free assays with that necessary for inhibition of growth or of protein synthesis in intact cells. To this end, it is useful to know the intracellular antibiotic concentration. In fact, many antibiotics which affect Gram-positive but not Gram-negative organisms simply are not able to penetrate Gram-negative organisms readily (antibiotics such as thiostrepton, streptogramin A, fusidic acid, and many others). For many antibiotics it is frequently possible to demonstrate many effects over a wide concentration range for, at high concentrations, charged antibiotics may act as polyanions or -cations. Nevertheless, it is noteworthy that specific effects of an antibiotic often occur at a ratio of one bound antibiotic molecule per ribosome.

At present the purification of ribosomal proteins and RNA and their reconstitution to active subunits permits an insight into ribosomal function and structure not previously amenable to analysis (9–15). Ultimately, it is hoped the three-dimensional structure of the ribosome should be known. Thus, for each antibiotic, factor and cofactor there should be a precise physical site on the ribosome for its interaction and precise temporal positions during protein synthesis when interaction can occur. At present we can at best localize antibiotics to a particular subunit and, in some cases, to a specific protein. Yet, so far we have little idea as to their exact position in or on the intact ribosome. However, antibiotics and other molecules which bind with great specificity to ribosomes can be used to help map the ribosome with respect to function and mutual interaction.

To localize antibiotic action to a particular subunit or site, numerous criteria may be used (see review, 16): binding of the radioactive antibiotic to a particular subunit; inteference with binding of a radioactive antibiotic known to interact with a particular subunit; interference with a function ascribed to a particular subunit; use of 70S ribosomes reassociated in all four combinations from subunits obtained from antibiotic-resistant and -sensitive bacterial strains; use of reconstituted subunits lacking a specific protein or containing a substitution for a specific protein, replacing that protein from an antibiotic-sensitive strain with that from a strain resistant to the antibiotic. In this review as in other reports (16, 17) inhibitors of ribosome function will be considered in two major classes: 30S (40S) and 50S (60S) inhibitors of bacterial ribosomal functions. Each group is further subdivided according to antibiotic chemical structure. A summary of inhibitors of protein synthesis and their major site of action is given in Table I.

30S INHIBITORS

Aurintricarboxylic Acid

Grollman & Stewart (18) showed that aurintricarboxylic (ATA) acid prevents the attachment of f2 or Q_β mRNA to *E. coli* ribosomes. Complete inhibition of protein synthesis directed by viral RNA occurred at 1×10^{-5} *M* ATA, and a 50% inhibition at 3×10^{-6} *M* concentration. There is a much smaller effect of ATA on endogenous protein synthesis where initiation and attachment of mRNA to ribosomes are minimal: 50% inhibition

TABLE 1. Inhibitors of Protein Synthesis

SUPERNATANT	30S (40S)	50S (60S)	
FOLIC ACID ANTAGONISTS	AMINOGLYCOSIDES Streptomycin Dihydrostreptomycin Paromomycin Neomycin Kanamycin Gentamycin Bluensomycin Spectinomycin Kasugamycin TETRACYCLINES Chlortetracycline EDEINE Edeine A Edeine B	CHLORAMPHENICOL MACROLIDES Niddamycin Carbomycin Spiramycin III Tylosin Leukomycin Erythromycin Chalcomycin Oleandomycin Lankamycin Methymycin LINCOMYCIN STREPTOGRAMIN A GROUP Ostreogrycin A Synergistin A (PA114A) Streptogramin A Vernamycin A Mikamycin A STREPTOGRAMIN B GROUP Ostreogrycin B Synergistin B (PA114B), etc. Viridogrisein (etamycin) THIOSTREPTON GROUP Thiostrepton (bryamycin, thiactin) Siomycin A (sporangiomycin, A-59) Micrococcin	PROCARYOTES
AMINOALKYL ADENYLATES GUANYLYL-5'-METHYLENE DIPHOSPHONATE FUSIDIC ACID	PACTAMYCIN AURINTRICARBOXYLIC ACID	PUROMYCIN 4-AMINOHEXOSE PYRIMIDINE NUCLEOSIDES Gougerotin Amicetin Blasticidin S Plicacetin Bamicetin SPARSOMYCIN	BOTH
DIPHTHERIA TOXIN		GLUTARIMIDES Cycloheximide (actidione) Acetoxycycloheximide Streptovitacin A IPECAC ALKALOIDS Emetine	EUCARYOTES

occurs at about 3×10^{-5} *M*, ten times the concentration necessary for similar inhibition of f2 mRNA initiated protein synthesis. Two other triphenylmethane dyes, gallein and pyrocatechol violet, were approximately equal to ATA in inhibitory capacity (19). ATA interfered with initiation in extracts from wheat embryos (20); and inhibited binding of poly U (21) and globin mRNA-ribonucleoprotein complex (22) to reticulocyte ribosomes. Also, binding of f2 RNA to *E. coli* 30S subunits (18) and globin mRNA to rabbit reticulocyte 40S subunits was inhibited (22). Recently, Grollman (personal communication) has shown that [^{3}H]ATA binds to 30S, but not to derived 50S subunits. If phage Q_β RNA is incubated with ribosomes, fMet-tRNA, GTP, and initiation factors to form an initiation complex, neither Q_β RNA nor fMet-tRNA is displaced by later addition of ATA (18). Thus, ATA inhibits the first step of initiation, namely, the binding of mRNA to ribosomes from both pro- and eucaryotes. When ATA is added after protein synthesis has already begun, peptides are released on completion and ribosomal subunits accumulate at the expense of polysomes (23).

Since f2 mRNA and ATA do not compete for binding sites on ribosomes, it is likely that ATA inhibits binding of mRNA allosterically (18). The increased sedimentation constant (76S) of 70S *E. coli* ribosomes (18) and the decreased sedimentation constant (60S) of 80S reticulocyte ribosomes (22) may reflect conformational changes produced by ATA.

Since endogenous protein synthesis, chain propagation, and chain termination, have been shown to be inhibited by ATA at much higher concentrations than necessary to inhibit viral mRNA attachment to ribosomes (18, 22, 24), it is probable that at high concentrations ATA produces effects on protein synthesis other than those on initiation. Recent reports suggest that ATA may inhibit charging of tRNA (25), GTP hydrolysis (25), ribonuclease V activity (25, 26) and transfer factor Ts (27). However, these effects are apparently minimal at low ATA concentrations during translation of natural mRNA since chain elongation is unaffected by ATA concentrations inhibitory to initiation (23, 24). The ability of ATA to bind to various proteins (28) may account for some nonspecific inhibitory effects produced at high concentrations.

Grollman (19) suggested that ATA and similar compounds may be used as models for viral inhibitors. Although ATA does not enter a variety of mammalian and bacterial cells tested, modification of ATA may permit it to enter cells without altering its inhibitory effects on viral attachment.

Summary.—ATA prevents attachment of mRNA to ribosomes and subunits of pro- and eukaryotes. Accordingly, on inhibition of protein synthesis, subunits accumulate at the expense of polysomes. At high ATA concentrations, additional effects are observed.

Streptomycin and Other Aminoglycoside Antibiotics

It was established by Fitzgerald, Bernheim & Fitzgerald (29) that streptomycin (Sm) inhibited protein biosynthesis in bacteria sensitive to the antibiotic. In Sm-dependent *E. coli* strains, Spotts (30) demonstrated that the synthesis of various enzymes was inhibited from 0–95% when these strains were deprived of antibiotic. Inhibition of protein synthesis was not seen in cells resistant to the antibiotic.

Ribosomal localization of streptomycin sensitivity, resistance, and dependence.—Using all possible combinations of supernatant and ribosomes from Sm-sensitive and -resistant *E. coli* strains, several laboratories (31–33) localized the Sm-sensitive site to the ribosome. By using various combinations of 50 and 30S subunits from Sm-sensitive and -resistant *E. coli* strains, Davies (34) and Cox, White & Flaks (35) showed that Sm-sensitivity resided in the 30S subunit. Sm did not interfere with the binding of labeled poly U to Sm-sensitive ribosomes (34). In other studies, Kaji, Suzuka & Kaji (36) and Pestka (37) showed directly that recognition of AA-tRNA by isolated 30S subunits prepared from Sm-sensitive *E. coli* was inhibited by Sm.

Miscoding during protein synthesis.—During studies of conditional Sm-dependence, Gorini & Kataja (38) suggested that Sm could influence the accuracy of translation of the genetic code. Using cell-free extracts from Sm-sensitive *E. coli* strains, Davies, Gilbert & Gorini (39) and Van Knippenberg et al (40) showed that Sm and other aminoglycoside antibiotics can stimulate miscoding. Specifically, in the presence of poly U, Sm stimulates leucine, isoleucine, serine, and tyrosine incorporation. Miscoding was small or negligible with extracts prepared from Sm-resistant strains. In addition, Likover & Kurland (41) indicated that enhancement of isoleucine and serine incorporation in response to poly U requires the presence of denatured DNA in the incubation mixture as a cofactor as well as Sm. Several groups (42–45) presented evidence that in the presence of aminoglycoside antibiotics miscoded amino acids were incorporated into protein as substitutions for the proper amino acids; thus, Bissel (44) has predictably found that in the presence of neomycin or Sm, cells produced some altered β-galactosidase.

Davies, Gorini & Davis (46) surveyed the effect of seven aminoglycoside antibiotics on the coding specificity of the homopolymers, poly U, poly A, poly C, and poly I. The spectrum of misreading varied with the antibiotic as well as with the polynucleotide template. Using synthetic polynucleotide templates of alternating nucleotide sequence, Davies, Jones, & Khorana (47) found that the coding changes produced by Sm are more limited than those produced by neomycin. They concluded that Sm stimulates the misreading of the 5′ terminal and internal positions of a trinucleotide codon, of only one base at a time, and of pyrimidines more frequently than purines. Neomycin, however, produces the Sm coding errors as well as misreading of the 3′-terminal nucleoside and sometimes of two bases of a codon at one time. Moreover, it appears that reading frame shifts need not be invoked to explain any of the antibiotic-induced misreading phenomena. In the case of alternating copolymers, poly $(AC)_n$ and $(AG)_n$, Sm produced no miscoding but only inhibition of amino acid incorporation stimulated by these templates.

Kanamycin produces substantial miscoding with extracts from kanamycin-sensitive *E. coli* (48). Miscoding was reduced in cell-free extracts from a kanamycin-resistant strain.

Pestka, Marshall & Nirenberg (49), Kaji & Kaji (50), and Pestka (51) showed that Ile-, Leu-, and Ser-tRNA binding to ribosomes was stimulated by Sm when ribosomes from Sm-sensitive *E. coli* were used, but was affected only slightly when ribosomes were prepared from Sm-resistant *E. coli.* The effects on AA-tRNA recognition occurred at low levels of Sm where the Sm: ribosome ratio was approximately 1. It appears that only in certain states during protein synthesis are ribosomes sensitive to Sm action. White & White (52) and Yamaki & Tanaka (53) noted that puromycin, which produces polysome breakdown, enhances killing by Sm. Pestka (51) and Kaji (54) reported that in the presence of Sm, ribosomes which were never dissociated miscoded more than ribosomes which had been dissociated. In contrast, Sm hardly influenced the Phe-tRNA binding to nondissociated ribosomes, but inhibited the binding of Phe-tRNA in response to poly U to reassociated ribosomes from Sm-sensitive *E. coli* approximately 50% (51). Sm-resistant bacteria were resistant to both these effects. Studying intact cells of *E. coli,* Kogut & Harris (55) concluded that the interaction of intracellular Sm with ribosomes in vivo also must be restricted to a certain state of the cycle of ribosome function, for growth inhibitory effects were dependent on growth rate.

Aminoglycoside antibiotics vary widely in degree of misreading produced as a function of their concentration (56). Thus, neomycin, kanamycin, and gentamicin show a 3- to 5-fold increase in level of misreading as drug concentration was raised from 10^{-6} to 10^{-4} *M*. This suggested that each of these drugs may act on more than one site and is consistent with the absence of one-step mutants with high level resistance to these antibiotics (56). In contrast, little increase in misreading over the same concentration range was seen with Sm and paromomycin for which one-step high level resistant mutants are known. These latter drugs may act only on a single site.

Polynucleotides which ordinarily do not serve as templates for protein synthesis can function as templates in the presence of some aminoglycoside antibiotics, particularly neomycin and kanamycin. For example, McCarthy and Holland (57, 58) showed that in the presence of neomycin denatured DNA, rRNA, and tRNA could serve as direct templates for protein synthesis in extracts from *E. coli.* Masukawa & Tanaka (59) also showed that the incorporation of amino acids into protein in the presence of DNA was markedly increased by kanamycin and neomycin. Less stimulation was observed with Sm; and kasugamycin did not significantly stimulate amino acid incorporation by DNA. Ribosomes prepared from aminoglycoside-resistant strains were resistant to these effects. Price & Rottman (60) showed that neomycin and Sm enhanced the template activity of oligonucleotides containing 2′-0-methyladenosine in binding Lys-tRNA to ribosomes. Dunlap & Rottman (61), furthermore, showed that amino acid incorporation dependent on polynucleotides containing 2′-0-methyl analogs was stimulated by neomycin. The homopolymers containing 2′-0-methyl analogs (poly A_m, poly C_m, and poly U_m) were inactive as templates; however, neomycin was able to induce template activity of poly U_m although poly A_m and poly C_m remained inactive. Thus, some aminoglycoside antibiotics clearly alter the structural requirements for the template in protein synthesis and cause substantial perturbation of the ribosome-template interaction, permitting as templates structures that ordinarily do not serve this function.

In addition, Hatfield (62) and Hatfield & Nirenberg (personal communication quoted in ref. 37) showed that Sm not only inhibited the binding of Phe-tRNA to ribosomes stimulated by [^{3}H]UUC and [^{3}H]UUUC, but also inhibited the binding of these templates. Thus, it is conceivable that Sm actually distorts the template interaction on the ribosome in a very localized way. In this case, instability of the binding of small templates to ribosomes was evident.

Physical effects of bound streptomycin on ribosomes.—Some direct effects of Sm on physical properties of ribosomes have been demonstrated. Thus, Leon & Brock (63) showed that Sm and neomycin protect the ribosomes of a Sm-sensitive, neomycin-sensitive *E. coli* strain against thermal denaturation, whereas ribosomes from a Sm-resistant, neomycin-sensitive strain are only protected by neomycin. Wolfe & Hahn (64) also showed that the ribosomes from Sm-sensitive, but not from Sm-resistant, *E. coli* were stabilized to heat in the presence of Sm. Sherman & Simpson (65) reported that Sm produces changes in the hydrogen exchange pattern of ribosomes: at low Sm to ribosome ratios there is apparent loosening of the ribosomal structure; at high Sm to ribosome ratios, there appears to be tightening of the ribosomes structure. None of these effects is observed with Sm-resistant ribosomes.

Bindings of radioactive streptomycin to ribosomes.—Kaji & Tanaka (66) have shown that approximately one molecule of dihydrostreptomycin binds to each 30S ribosomal subunit at 24°C; there was no detectable binding to 50S subunits. Binding required magnesium and was dependent on the presence of uridine or cytidine containing polynucleotides. Little or no binding of dihydrostreptomycin to 30S subunits derived from Sm-resistant *E. coli* strains was observed. Vogel et al (67) showed that 30S subunits bind radioactive dihydrostreptomycin in proportion to their state of activity: active 30S subunits bind the antibiotic, but inactive subunits do not. Inactive 30S subunits can be converted to active subunits by incubation in the presence of ammonium or potassium ions. They also noted that poly U has no direct effect on binding dihydrostreptomycin to 30S subunits, but accelerates the conversion of inactive to active 30S subunits. Similarly, Wolfgang & Lawrence (68) showed binding of [^{14}C]streptomycin and [^{3}H]dihydrostreptomycin to ribosomes from Sm-sensitive ***Bacillus megaterium.***

Other effects on protein synthesis.—Streptomycin did not inhibit the formation of polylysyl-tRNA:ribosome complexes in response to poly A at 0.01 *M* magnesium (69). The formation of Ac-Phe-tRNA complexes with ribosomes was inhibited only about 30% by 10^{-5} *M* Sm and not any more by higher concentrations. Neomycin did not influence the binding of Ac-Phe-tRNA to ribosomes in response to poly U. Although neomycin and Sm did not inhibit the binding of polylysyl-tRNA to ribosomes, they did inhibit release of polylysine from tRNA by puromycin. Neomycin produced 50% inhibition of release at approximately 3×10^{-6} *M* and Sm, at 10^{-3} *M* (69). Moreover, at high concentrations, Sm (3×10^{-4} *M*) can inhibit termination reactions of protein synthesis (70). It is unlikely, however, that Sm inhibits transpeptidation in cells, for 10^{-4} *M* Sm did not interfere with the rate of peptidyl-[^{3}H] puromycin synthesis in cell-free *E. coli* extracts containing active polyribosomes (477); 10^{-4} *M* neomycin inhibited peptidyl-[^{3}H] puromycin synthesis 50% (477).

In addition, Igarashi, Ishitsuka & Kaji (71) have suggested that Sm inhibits translocation as well as binding of AA-tRNA. They found that Sm does not inhibit peptide bond formation from Phe-tRNA bound in the donor site; at high magnesium concentrations when Phe-tRNA is bound both in the donor and acceptor sites, maximal formation of phenylalanyl-puromycin is dependent on GTP and factor G translocation and this reaction is inhibited by Sm; also, release of $tRNA^{Phi}$ from ribosome: poly U complexes is inhibited by Sm. It is possible, however, that these in vitro assays starting with artificial complexes of ribosomes and poly U do not behave similarly to in vivo translocation reactions where the donor or acceptor sites (or both), are generally filled. Under these conditions the ribosome may not be accessible to Sm.

Ribosomes prepared from Sm-resistant *E. coli* strains, although relatively insensitive to inhibitions and stimulations of protein synthesis and codon recognition produced by Sm, show reproducible slight effects in the presence of Sm (39, 40, 49, 51). In fact, Sm stimulated miscoding with ribosomes prepared from Sm-resistant *E. coli* strains (72); and the ability of the ribosomes to miscode in the presence of Sm in vitro correlated with the ability of Sm to correct or substitute for an arginine requirement (ornithine transcarbamylase) in vivo.

Regarding the effects of kanamycin on protein synthesis, Suzuki, Kunimoto & Hori (73) showed that kanamycin, like chloramphenicol, prevents ribosomes from moving along mRNA through inhibition of peptide bond synthesis: 4×10^{-5} *M* kanamycin produced 44% inhibition of the puromycin reaction with polyphenylalanyl-tRNA.

Mammalian cells are relatively insensitive to Sm (74). Neomycin and paromomycin affected yeast growth and inhibited amino acid incorporation by mitochondrial and cytoplasmic ribosomes (75); in contrast, Sm and kanamycin had little effect on either. The mitochondrial system was more sensitive than the cytoplasmic ribosomes to neomycin and paromomycin in cell-free extracts. Protein synthesis by mitochondria prepared from rat liver, however, was insensitive to neomycin and paromomycin.

Structure activity relationships.—Tanaka and others (76, 77) have provided a structural basis for miscoding activity. Both deoxystreptamine and streptamine, although lacking in antimicrobial activity, cause miscoding in cell-free extracts. Deoxystreptamine or streptamine is contained in all aminoglycosides which cause miscoding such as kanamycin, neomycin, paromomycin, gentamicin, hygromycin B, and Sm; but not in kasugamycin and spectinomycin, which do not produce miscoding. The aminoglycoside moieties of kanamycin may enhance or modify the activity of deoxystreptamine.

The streptomycin protein and other aminoglycoside 30S proteins—Using the procedures for the reconstitution of ribosomal subunits, several laboratories (78–81) have shown that the distinction between 30S ribosomes from Sm-sensitive and Sm-resistant bacteria resides in the core proteins, and not in the RNA or the split proteins of the 30S particles. Ozaki, Mizushima & Nomura (82) have identified a specific protein constituent of the 30S particle to be controlled by the *Sm* locus in *E. coli*. Their results indicated that 30S ribosomal particles reconstituted with protein P_{10} from Sm-resistant strains and all other proteins from the Sm-sensitive strain were resistant to Sm. Conversely, particles reconstituted using P_{10} from the Sm-sensitive strain and all other proteins from the Sm-resistant strain were sensitive to the antibiotic. Particles containing P_{10} protein from the Sm-resistant strain or those particles not containing any P_{10} protein bound little or no dihydrostreptomycin. Protein P_{10} itself, however, did not bind dihydrostreptomycin. Thus, it appears that some structure involving both protein P_{10} and other ribosomal components is essential for binding of dihydrostreptomycin, and the mutation or alteration in P_{10} which accompanies mutation from sensitivity to resistance abolishes the binding ability of this structure.

It is noteworthy that reconstituted particles lacking P_{10} retain some functions of protein synthesis. In fact, poly U-directed polyphenylalanine synthesis was about 50–80% as active with P_{10}-deficient particles as with control particles and Sm had no effect on the polyphenylalanine synthesis with P_{10}-deficient particles no matter what the source of the protein used in the reconstitution of the 30S particles. However, in phage f2 RNA-directed valine incorporation, P_{10}-deficient particles were only about 20% as active as control particles and residual activity of the P_{10}-deficient particles was partially inhibited by Sm whether the proteins were obtained from sensitive or resistant strains. P_{10} deficient particles were only 6–17% as active as control reconstituted particles containing all the 30S proteins in binding fMet-tRNA to ribosomes in response to AUG, 50S subunits, and initiation factors. Thus, P_{10}-deficient particles are deficient in chain initiation. A temperature-sensitive 30S mutant isolated by Kang (83, 84) has properties similar to P_{10}-deficient particles. P_{10}-deficient particles exhibit drastic decrease of Sm-stimulated misreading; and they are also resistant to ambiguity induced by other aminoglycoside antibiotics as paromomycin, neomycin B, kanamycin A, and even ethyl alcohol. It is probable that the failure to observe deletions in the *Sm* locus (85) may reflect the indispensability of the P_{10} protein. Since P_{10}-deficient 30S particles are able to translate synthetic mRNA more accurately than complete 30S particles, the presence of translational ambiguity is an inherent property of ribosomes. Although P_{10} is not essential for the codon-anticodon interaction, it was able to modulate and make this interaction susceptible to external ambiguity-inducing factors, such as Sm. The genetic data on the *Sm* locus described below suggests that there are many possible alterations in this modulation as a result of mutational changes in P_{10} or related structures.

Birge & Kurland (86) identified a single protein that is functionally altered in ribosomes obtained from Sm-dependent *E. coli.* This protein is the same protein that is altered in ribosomes resistant to Sm. Apirion et al (87) also have observed that reversion from Sm dependence appears to involve mutation in a specific 30S protein.

By similar techniques as described above, ribosomal proteins corresponding to other aminoglycoside antibiotics have been identified. Kanamycin sensitivity has been localized to the 30S P_{10} protein (88, 89). Bollen and co-workers (90), Bollen & Herzog (91), and Dekio & Takata (92) identified a single protein of the 30S particle which is responsible for spectinomycin sensitivity and resistance. This protein P_4 clearly differs from protein P_{10}.

Genetic studies on streptomycin and the aminoglycoside antibiotics.—Among auxotrophic mutants of *E. coli,* a class of mutants is found for which Sm can partially correct the defect. This correction disappears if Sm is removed from the medium. For instance, an arginine auxotroph can grow in the absence of arginine provided that Sm is present in the growth medium (93). However, growth occurs at a slow rate. Also, defective bacteriophages may produce mature particles at a low efficiency in a nonpermissive host if Sm is present during infection (94–96). Thus, bacterial mutants selected for their resistance to Sm (so that they are Sm-resistant or -dependent) frequently manifest phenotypes which have no obvious relation to the Sm-resistant character itself (38, 93–104). Since there was evidence that mutations of the *Sm* locus influenced ribosome structure (31–33), Gorini & Kataja (38) suggested that Sm could influence the accuracy of reading of the genetic code. This interpretation was consistent with the observed effects of Sm in suppressing apparently unrelated genetic defects, including suppression of amber suppressor mutations. Observations indicate that the allele borne at the *Sm* locus, and therefore the structure of the ribosome, influences not only suppression by Sm, but also genetic suppression (95, 97, 105–108). Accordingly, several classes of Sm-resistant alleles have been distinguished in *E. coli* on the basis of their patterns of phenotypic suppression (109–111).

Rosset & Gorini (112) analyzed a ribosomal ambiguity mutation (*ram*) for a 30S ribosomal component controlling translational ambiguity. In vivo, the *ram* mutation confers on the cell the ability to suppress all three nonsense codons. Sm induces a similar generalized ambiguity phenotypically both in vitro and in vivo. The effects of *ram* and *Sm* mutations are additive. Wild-type *ram*$^+$ *Sm* A$^+$ and *ram* 1 *Sm* A$^+$ are equally sensitive to the bacteriocidal action of Sm; however, the growth of *ram* mutants (such as *ram* 1 *Sm* A$^+$) is inhibited by sublethal concentrations of Sm (2–5 μg/ml), while no bacteriostatic action is demonstrable with *ram*$^+$ *Sm* A$^+$ wild type at sublethal Sm concentrations. Oversuppressed mutants (*ram* 1 *Sm* A1) do not grow in the presence of high Sm concentrations, but Sm does not kill these cells (113). Thus, inhibition of growth (protein synthesis) by Sm and killing are independent phenomena controlled by the same locus; and miscoding, which is responsible for oversuppression, and growth inhibition are distinct from its bacteriocidal action. Also, some revertants from Sm-dependent strains (106, 107) are unable to grow in the presence of Sm, but Sm was bacteriostatic despite in vitro misreading; the revertants appeared to be in a locus close to but not identical to the *Sm* locus.

The ribosome consists of interdependent components and thus a single change in one component can modify its relationship with others as well as modifying the entire structure. Thus, it is not surprising that ribosomal mutations show extensive pleiotropic effects; and that ribosomes often show alterations in their response to antibiotics when selection has been carried out by unrelated antibiotics. In fact, it is possible that a mutation in one subunit can change the response to an antibiotic that exerts its primary effect through the other ribosomal subunit. In this regard, Brownstein & Lewandowski (106, 107) have noted ribosomal mutations outside the *Sm* locus which affect Sm sensitivity. Tanaka et al (48) have reported that kanamycin produced greater miscoding in cell-free extracts from a Sm-resistant, kanamycin-sensitive *E. coli* strain than in extracts from the parent strain sensitive to both antibiotics; namely, the mutation to Sm-resistance has made the ribosomes more sensitive to kanamycin. Apirion & Schlessinger (114) have found that erythromycin mutants often show an increased sensitivity to spectinomycin while neomycin mutants show an increased resistance to Sm and spectinomycin. In studies of strains resistant to neomycin and kanamaycin it appears that the neomycin and kanamycin markers are very close or identical and the ribosomes are changed from those of the parental strain (115). The neomycin-, kanamycin-resistant mutants are closely linked to Sm and spectinomycin markers. As in the case of Sm, these mutants were recessive to the sensitive wild-type allele. Neomycin-kanamycin-resistant mutants exhibited increased resistance to low levels of Sm and spectinomycin, but decreased resistance to chloramphenicol. The high pleiotropy of Sm mutations has been evident in numerous other studies (96, 98–104).

Mechanism of inhibition of protein synthesis with natural templates and in intact cells; relation to bacteriocidal action.—Although Sm causes many effects on protein synthesis in cell-free extracts as shown above, these effects may not be related to inhibition of protein synthesis in intact cells or to the lethal action of the antibiotic. Modolell & Davis (116–118) have examined the effects of Sm on various aspects of f2 RNA-directed polypeptide synthesis in cell-free extracts. With f2 RNA as a template, Sm (3.4×10^{-5} *M*) rapidly inhibits chain extension before two amino acids are incorporated per active ribosome. Sm also rapidly inhibits endogenous amino acid incorporation with natural *E. coli* mRNA (118). After addition of Sm to extracts actively incorporating amino acids, polysomes appear stabilized at first, but then break down after incubation for 4 min and longer. Sm probably halts peptidyl-tRNA in the P-site because puromycin releases all bound peptide from peptidyl-tRNA within 15 sec at the same rate in the presence or absence of Sm. This indicates that 3.4×10^{-5} *M* does not inhibit peptidyl transfer. Polysome breakdown in the presence of Sm was inhibited by chlortetracycline, chloramphenicol, sparsomycin, and fusidic acid. Modolell & Davis (119) have examined the effect of Sm on fMet-tRNA binding to ribosomes. They found that formation of the fMet-tRNA:poly AUG:30S complex is not inhibited by Sm; however, after complexes form in the presence of 50S subunits, Sm stimulates dissociation of the complexes so that fMet-tRNA is released. Sm releases the fMet-tRNA from the P-site, for at least 75% of the fMet is released by puromycin. Perhaps, initiation factors and GTP present in the 30S:fMet-tRNA:template complex stabilize it; but, after hydrolysis of the GTP upon addition of 50S subunits and release of initiation factors, the complexes may be destabilized by Sm. Predictably, then, when the analog GDPCP was used, Sm could not dissociate the complexes and the fMet-tRNA does not become puromycin reactive. Similar effects of Sm on destabilization of fMet-tRNA complexes have been reported by LeLong and Gros (personal communication). It has been suggested by Hishizawa, Lessard & Pestka (120) that the 50S portion of the P-site which interacts with the fMet end of fMet-tRNA does not produce a stable configuration with the ribosome; in fact, it is possible that there is very little affinity of the 50S subunit for the peptidyl end of peptidyl-tRNA. Thus, as soon as fMet-tRNA is released from the A-site to which it is bound by initiation factors and GTP, on hydrolysis of the GTP and release of the initiation factors, destabilization of the tRNA molecule by Sm on the 30S subunit becomes evident. The 70S monosomes which accumulate in

Sm-killed cells can be dissociated by the ribosome dissociation factor and 0.05 *M* NaCl, conditions which dissociate runoff ribosomes, but not ribosomes carrying peptidyl-tRNA and mRNA (4). These studies suggest that Sm rapidly inhibits chain elongation, probably through inhibition of AA-tRNA binding. Sm bound to 30S subunits distorts ribosomes so that peptidyl-tRNA is slowly released with concomitant release of ribosomes from polysomes.

Luzzatto, Apirion & Schlessinger (5, 121–123) have investigated polyribosome metabolism in cultures treated with Sm. After addition of Sm to intact cells, protein synthesis was rapidly inhibited with subsequent changes in ribosome distribution: decrease in large polysomes and in free 30 and 50S subunits; and concomitant accumulation of 70S monomers. Pulse-labeled RNA continues to be associated with the remaining polyribosomes and 70S monomers; this mRNA is comparable in size to mRNA extracted from untreated cells, suggesting that it is stabilized from degradation on Sm monosomes. They propose that Sm alters normal interaction of 30 and 50S subunits on mRNA, and that this leads to appearance of Sm monosomes which they consider to be ribosomes blocked in aberrant initiation complexes.

Studies of ribosomal subunit exchange (3) are in accord with these results: Sm inhibits chain elongation, but permits ribosomes to leave mRNA; upon release ribosomes dissociate into subunits which reassociate to form irreversibly inactivated complexes, incapable of further subunit exchange.

If the modified complexes cannot properly initiate and translocate along the messenger, addition of further ribosomes is inhibited and the normal cycle of ribosome function is interrupted. Sm monosomes which accumulate are incapable of protein synthesis either in vivo or in vitro; they do not arise during cell lysis in Sm; they do not form in Sm-resistant cells; nor are they nonspecific consequences of growth arrest, for glucose starvation does not produce them (5, 121–123). Rate of loss of viability is well correlated with increase in Sm monosomes. Sm monosomes are very likely identical to the unusually stable 70S ribosomes previously observed by Herzog (124): 70S ribosomes of Sm-sensitive *E. coli* treated with Sm did not dissociate readily at low Mg^{++} concentration, unlike normal 70S ribosomes or those from Sm-resistant bacteria treated with Sm.

In extracts, Sm blocks polypeptide synthesis on natural mRNA templates (121) in contrast to variable inhibition when synthetic templates such as poly U are used (31, 32, 46); and Sm inhibits binding to ribosomes of fMet- and Ala-tRNA which correspond to the first and second amino acids incorporated into the nascent phage coat protein directed by R17 RNA (121). Extracts of cells killed by Sm can no longer respond to natural mRNA such as R17 RNA; however, polyphenylalanine synthesis in response to poly U can still occur at 50% of the level of untreated cells. Initiation and protein synthesis with natural mRNA, unlike that directed by poly U, is selectively blocked by Sm. Thus, 30S particles with bound Sm resemble P_{10}-deficient particles. Irreversibility of Sm action is correlated with its very tight binding to ribosomes. Even after prolonged dialysis, Sm monosomes show no capacity to respond to natural mRNA.

It has been stated that protein synthesis is required for Sm to kill cells, for chloramphenicol prevents Sm killing (125–127); in contrast, however, Stern and Cohen (128, 129) have shown that Sm killing can occur in amino acid auxotrophs deprived of a required amino acid under conditions in which protein synthesis is negligible. Chloramphenicol prevents the destabilization of polysomes by Sm. It is possible that this effect of chloramphenicol, rather than its inhibition of protein synthesis, is responsible for preventing Sm lethality, for amino acid starvation possibly may not prevent polysome disaggregation produced by Sm. Moreover, it has been noted that anaerobiosis protects cells from the lethal action of Sm (129). The nature of this protection is uncertain.

Dominance of the bacteriocidal effect of aminoglycosides has been reported (130, 130a); however, Breckenridge & Gorini (130b) found that the immediate effect of Sm on sensitive/resistant heterodiploids of *E. coli* B is only bacteriostatic, not bacteriocidal. Aberrant Sm monosomes will form from the sensitive ribosomes (5, 116–119, 121–123). Since each aberrant Sm monosome contains a mRNA template and two tRNAs (131), it was suggested that these monosomes could trap available mRNA. Conceivably, these monosomes may trap tRNA also, particularly fMet-tRNA, making fMet-tRNA unavailable for initiation by Sm-resistant ribosomes. Such events may contribute to the immediate bacteriostasis. After dilution into Sm-free medium, sensitive/resistant heterodiploids can produce colonies; this suggests that the inactive Sm monosomes irreversibly inactivated by Sm do not interfere with the function of the resistant ribosomes. In the case of the bacteriocidal neomycin, Sm-like monosomes accumulate; however, with the bacteriostatic spectinomycin, they do not (131).

Regarding Sm lethality, several other observations deserve comment. Actinomycin which inhibits mRNA synthesis prevents Sm killing (132); hydroxyurea which inhibits DNA synthesis does not (133). Starvation of cells for a carbon source as well as anaerobiosis blocks Sm killing. Membrane damage is observed in Sm-killed cells (134), but since one-step mutations to resistance simultaneously alter ribosomes and eliminate membrane damage (132), the membrane damage has been considered a consequence of ribosome-mediated effects. However, mutation to Sm resistance or dependence simultaneously alters bacterial cell walls (135) as well as ribosomes. Can the P_{10} protein be a constituent of the cell wall or membrane as well as part of the ribosomes?

Spectinomycin and kasugamycin.—Unlike most of the other aminoglycoside antibiotics, spectinomycin (Spc) is not bacteriocidal and its inhibitory effects can be reversed by washing exposed cells (136, 137). One-step mutants with highly resistant 30S subunits are readily isolated, suggesting that a single site for Spc action exists. Sensitivity and resistance to Spc are properties of the 30S subunit (138) and protein P_4 of that subunit (90–92).

Binding of [^{3}H]dihydrospectinomycin to ribosomes was examined (139). At low antibiotic concentrations, ribosomes and 30S subunits from Spc-sensitive *E. coli* bind more drug than those particles from Spc-resistant strains. Neomycin, kanamycin, or Sm do not affect its binding to ribosomes. In contrast to Sm, binding of dihydrospectinomycin is reversible: the antibiotic can be removed from ribosomes completely by dialysis for 2 hr.

Inhibition of protein synthesis by Spc is maximal at 10^{-6} *M* and depends on nucleotide composition of synthetic templates: inhibition is absent with poly U or poly A; increases with C and G content of polynucleotides; is complete with poly I (138). Effects are independent of Mg^{++}. Spectinomycin does not interfere with puromycin release or with binding of AA-tRNA to ribosomes. Although there is a striking requirement for the presence of C or G in templates for inhibition, this is unrelated to any specific AA-tRNA for the inclusion of G in a template equally sensitizes to inhibition amino acids whose codons contain G as those with codons containing no G. Spc does not produce miscoding. Mutants dependent on Spc and phenotypic suppression by Spc have not been found. Its inhibitory effect does not depend on initiation; however, initiation reactions were not directly examined (138). It was suggested that Spc inhibits some translocation-related event since other steps of protein synthesis were not influenced and inhibition was directly related to nucleotide composition (138). In merodiploids prepared from Spc-sensitive and -resistant *E. coli,* sensitivity was dominant over resistance (130). Dominance of sensitivity is consistent with a block in translocation. Inhibition of ribosomal subunit exchange by Spc (3) is also in accord with a block in translocation.

Although there was some accumulation of subunits and 70S ribosomes in cells treated with Spc, this accumulation was comparable to that occurring on chloramphenicol treatment (131). The increase in cellular mRNA content during treatment with chloramphenicol or Spc so that there is more mRNA per ribosome than in control cells may explain the accumulation of small polysomes and monosomes in the presence of these antibiotics. Lack of accumulation of polysome aggregates indicates that termination is not specifically inhibited.

Kasugamycin is an aminoglycoside which inhibits protein synthesis (140). Neither spectinomycin nor kasugamycin contains deoxystreptamine or streptamine and thus produces no miscoding (140–142). Kasugamycin inhibits Phe-tRNA binding to ribosomes in response to poly U and is presumably an inhibitor of the recognition process. Sparling (143) demonstrated that mutations to kasugamycin resistance alter 30S subunits of *E. coli.* Although all other known 30S mutations are located in a cluster near the Sm region, kasugamycin resistance is located at a distance from this region near the leucine marker.

Summary.—Streptomycin inhibits chain elongation by interfering with AA-tRNA binding to the A-site of ribosomes. Binding of streptomycin occurs at a single site on the 30S subunit containing a structure including protein P_{10}. Some time after polypeptide synthesis has ceased, peptidyl-tRNA is slowly released from ribosomes, which themselves are released from mRNA as subunits or 70S monomers. After release, the subunits reassociate into 70S monomers which cannot dissociate. These streptomycin monosomes contain mRNA and are 70S ribosomes which are irreversibly inactivated by streptomycin; they are inactive in both cells and extracts, and may be aberrant initation complexes. Streptomycin also stimulates miscoding, which occurs in cells and cell-free extracts. At high concentrations, streptomycin can more directly interfere with translocation and peptide bond formation. The precise causes of lethality are still uncertain, but are related to its irreversible binding to ribosomes.

Neomycin and kanamycin also interact with 30S subunits, probably at several sites. They strongly inhibit peptide bond formation. Inactive 70S monosomes form in the presence of neomycin. They produce extensive miscoding due to the deoxystreptamine moiety. In the case of kanamycin, and probably neomycin, the P_{10} protein is involved in its inhibitory action.

Spectinomycin and kasugamycin are bacteriostatic, not bacteriocidal as the other aminoglycosides; their inhibitions are reversible. They both inhibit protein synthesis through interaction with the 30S subunit and in the case of spectinomycin with protein P_4 as well. Inhibition of protein synthesis by spectinomycin requires the presence of cytidylic or guanylic acid residues in templates; it interferes with mRNA:30S subunit interactions occurring during translocation. It does not inhibit peptide bond formation, AA-tRNA binding, initiation, or termination. Kasugamycin interferes specifically with AA-tRNA binding to ribosomes. Neither antibiotic produces miscoding.

Pactamycin

Pactamycin is an antitumor antibiotic which is a potent inhibitor of protein synthesis in cells and cell-free extracts of pro- and eukaryotes (144–151). The complete structure of pactamycin has recently been reported (152).

Felicetti, Colombo & Baglioni (148) demonstrated that the ribosome contained the site sensitive to pactamycin on examining protein synthesis in extracts obtained from reticulocytes incubated with and without pactamycin. Ribosomes from cells incubated with pactamycin showed little amino acid incorporation, no matter what the source of the supernatant factor; in contrast, ribosomes from control cells were active with supernatant fraction from both control cells or those incubated with pactamycin. Single ribosomes replace polyribosomes in reticulocytes treated with pactamycin (147, 148).

Pactamycin does not interfere with the puromycin-stimulated release of nascent peptides from reticulocyte polyribosomes (148). Thus, pactamycin does not inhibit peptide bond formation itself as also shown by others (151, 153, 154) in bacterial systems. Pactamycin (10^{-4} *M*) may interfere only slightly (15–20%), if at all with AA-tRNA formation (148).

Macdonald & Goldberg (155) have examined the effect of pactamycin on the initiation of protein synthesis in reticulocytes. Low levels of pactamycin (10^{-6} *M*) stop polypeptide synthesis by intact reticulocytes and their lysates after a lag of about 2 min, at which time inhibition is complete. This effect is similar to that of poly A, a known inhibitor of polypeptide chain initiation (156). Cells pretreated with NaF, which converts polysomes to monosomes and subunits (157), are more sensitive to pactamycin than are untreated cells. A high concentration of pactamycin ($\pm$ 10^{-5} *M*) also inhibits chain elongation in addition to initiation, for it decreases the rate of protein synthesis immediately without lag and prevents ribosomal subunit exchange (3). Inhibition of chain elongation may result from an effect on translocation since peptide bond formation is unaffected. Of note is that streptomycin also can inhibit translocation at high concentration. They demonstrated (155) that [^{3}H]pactamycin binds to 80S ribosomes as well as 40S subunits at 0°C; monosomes produced by sodium fluoride treatment bind [^{3}H]pactamycin very well, whereas polysomes as well as monosomes produced by RNase treatment of polyribosomes do not bind pactamycin effectively. However, at 37°C polyribosomes bind pactamycin. The ribosomes accumulated by treatment of reticulocytes with NaF presumably lack peptidyl-tRNA, AA-tRNA, and protein factors, but probably not mRNA (158). On the other hand, RNase-produced monosomes carry a fragment of mRNA as well as the other components. Therefore, increase in binding of [^{3}H]pactamycin to ribosomes produced by NaF treatment is due to the availability of a binding site present on 40S subunits which is less available on RNase-produced monosomes; once the initiation complex is formed and chain elongation is in progress, the binding site for pactamycin is less accessible to or has a decreased affinity for the antibiotic so that a higher concentration of pactamycin is necessary in order to produce inhibition of chain elongation. Alternatively, at high concentrations pactamycin may influence additional sites.

Pactamycin (10^{-4} *M*) at 0.007 *M* Mg^{++} inhibits poly A-dependent binding of polylysyl-tRNA to *E. coli* ribosomes about 60%; the same concentration of chlortetracycline stimulates binding slightly (151). Pactamycin does not interfere with the binding of poly A to ribosomes. Maximal inhibition (50–60%) of peptidyl-tRNA binding to ribosomes is reached at 3×10^{-6} *M* pactamycin, at which concentration polylysine synthesis is almost completely inhibited. Apparently, only half of the polylysyl-tRNA binds to a pactamycin-sensitive site. At 0.007 *M* Mg^{++}, pactamycin (10^{-4} *M*) produces about a 30% inhibition of Lys-tRNA binding to ribosomes. In addition to interfering with the binding of polylysyl-tRNA to ribosomes, pactamycin also stimulates release of polylysyl-tRNA already bound to ribosomes without the cleavage of polylysine from the tRNA. At 3×10^{-6} M pactamycin, the extent of release levels off at about 60% of that initially bound. At higher Mg^{++} concentrations (0.01–0.014 M) the pactamycin inhibition or dissolution of the complex decreases. Since complete inhibition of polylysine synthesis by pactamycin (10^{-4} M) is observed at 0.014 M Mg^{++}, where the effect on binding of both polylysyl- and Lys-tRNA to ribosomes is relatively small, it is not clear how these actions of the antibiotic are related. It is possible that at higher Mg^{++} concentrations Mg^{++} masks the destabilization of Lys- and polylysyl-tRNA binding to ribosomes, but that pactamycin still inhibits their functional attachment.

Cohen, Herner and Goldberg (153, 159) have shown that 8×10^{-4} M pactamycin inhibits polyphenylalanine synthesis by cell-free extracts from *E. coli* at low Mg^{++} concentrations where initiation by Ac-Phe-tRNA is rate-limiting. Inhibition by pactamycin is decreased by prior incubation of the extracts with Ac-Phe-tRNA, however, prior incubation with pactamycin results in increased inhibition of polypeptide synthesis; the delayed addition of pactamycin until completion of initiation results in a lack of effect on chain elongation. Furthermore, it has been shown that endogenous protein synthesis in which there is mainly completion of existing peptide chains is more resistant to pactamycin inhibition than is that promoted by exogenous phage MS2 RNA. At low Mg^{++} concentrations, the binding of Ac-Phe-tRNA to ribosomes which is dependent upon initiation factors and GTP is inhibited by 10^{-6} M pactamycin. At high Mg^{++} (0.02 M) concentration or 0.01 M spermidine, inhibition produced by pactamycin is lost. Similar to streptomycin, pactamycin stimulates the release of prebound Ac-Phe-tRNA, but does not directly inhibit formation of Ac-Phe-puromycin. In summary, pactamycin alters the structure of the initiation complex causing destabilization and dissociation.

Also, pactamycin inhibits the formation of the 30S initiation complex (160): it also inhibits binding of Ac-Phe-tRNA to 30S ribosomal subunits dependent on poly U, initiation factors, and GTP. The inhibition is decreased by high Mg^{++} and increased by high NH_4^+ levels. Pactamycin causes release of prebound Ac-Phe-tRNA from this complex. [^3H] pactamycin binds to 30S, but not to 50S subunits. One molecule of pactamycin was bound for every three 30S subunits. By analogy with streptomycin, it is possible that protein P_{10} or a neighboring structure of the 30S subunit may be involved in binding pactamycin. However, pactamycin does not produce misreading of mRNA. Kersten et al (149, 150) have examined the mode of action of pactamycin in *Bacillus subtilis*. Their results indicate that also in *B. subtilis* pactamycin interferes with the synthesis of protein at the ribosomal level and interferes with methylation of rRNA and tRNA.

Summary.—Pactamycin inhibits initiation in pro- and eukaryotes by interfering with firm attachment of initiator tRNA to the initiation complex.

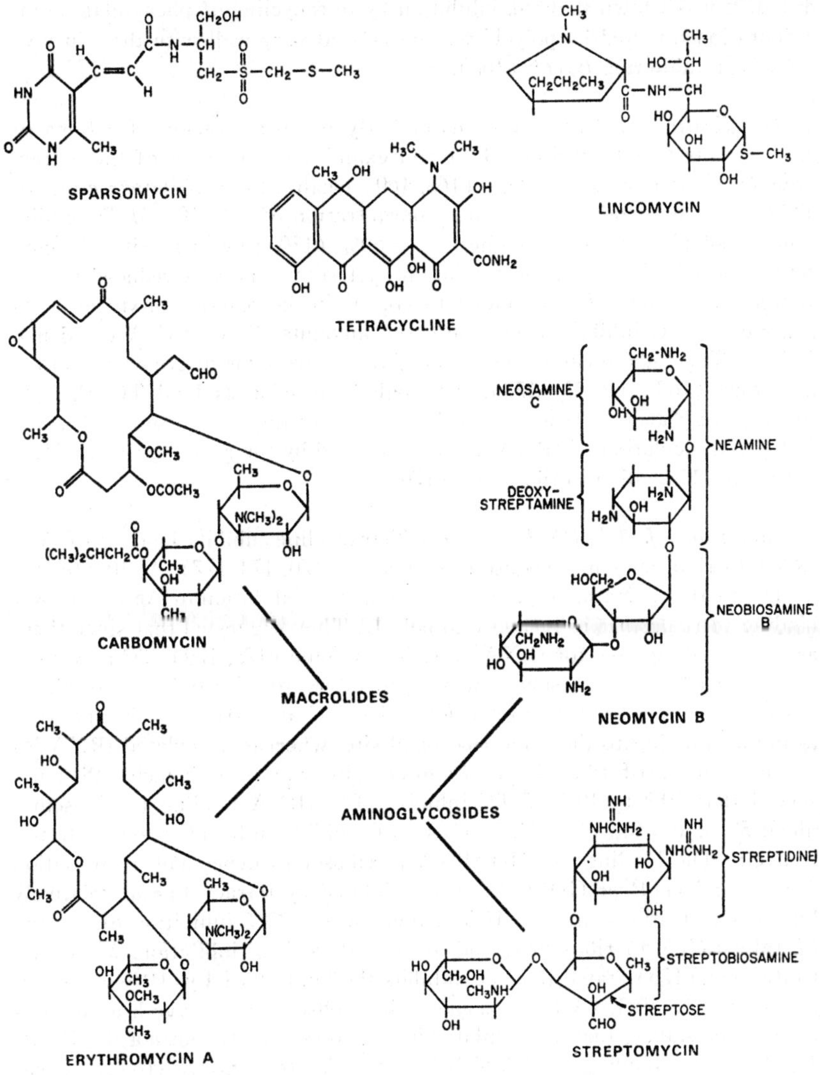

FIGURE 3. Structure of several antibiotics (references are given in the text).

It alters the structure of the initiation complex causing destabilization and dissociation. At higher concentrations, inhibition of translocation occurs.

THE TETRACYCLINES

The tetracyclines (Fig. 3) have been shown to inhibit protein synthesis in cell-free extracts of both mammalian and bacterial cells (161–166). In addition, it was shown that the inhibition by tetracycline of phenylalanine incorporation directed by poly U was correlated very well with their in vivo activity in inhibiting *E. coli* (167).

Localization of tetracycline action.—By exposing various fractions of cell extracts to tetracycline (TC) and examining the effect of these fractions on protein synthesis, Day (168, 169) localized the inhibitory effect of TC to the ribosomal fraction. At a concentration of 5×10^{-6} *M* TC inhibition of polyphenylalanine synthesis was 60% (170); by increasing the concentration of ribosomes, the inhibition by tetracycline was reduced to 6%. Increased amounts of ribosomes plus poly U in the reaction mixtures could reverse the TC inhibition, but increased amounts of AA-tRNA could not. Labeled TC binds to ribosomes, although there are some discrepancies in the literature (168, 169, 171–173). Although both subunits bind TC, the 30S subunit binds at least twice as much TC as does the 50S subunit (168, 171, 172, 173). Acylation of tRNA is not influenced by the antibiotic (164, 165), nor is poly U binding to ribosomes (170).

Inhibition of AA-tRNA binding.—Tetracycline inhibits binding of AA-tRNA to ribosomes in response to templates (170, 174–177). At 10^{-5} to 10^{-4} *M* TC, Suarez & Nathans (170) reported a 50% inhibition of Ac-Phe-tRNA binding to ribosomes in response to poly U. They suggested that since there are two binding sites for tRNA on each ribosome (178, 179), TC may combine with ribosomes in such a way as to interfere with only one of them. Seeds, Retsma & Conway (180) noted that TC and streptomycin appeared to inhibit binding to the same ribosomal site, whereas deacylated tRNA inhibited binding of Phe-tRNA to another site. Sarkar & Thach (181) observed that 10^{-5} to 10^{-4} *M* TC inhibited Phe-tRNA binding to ribosomes about 50%, and the Phe-tRNA resistant to inhibition by TC reacts with puromycin. The binding of fMet-tRNA to ribosomes dependent on initiation factors and GTP or GDPCP is also inhibited by tetracycline at relatively high concentrations; at 10^{-4} *M* TC, there was a 30% inhibition of binding of fMet-tRNA to ribosomes, and at 10^{-3} *M*, a 70% inhibition. Moreover, Gottesman (176) stated that TC inhibits the binding of Lys-tRNA, but not that of polylysyl-tRNA to ribosomes. These observations have been used as a basis for stating that TC inhibits binding to the A-site, but not the P-site. By definition, binding of AA-tRNA to the A-site is nonreactive, whereas binding to the P-site is reactive to puromycin.

In other studies, poly A-dependent binding of polylysyl-tRNA to ribosomes was inhibited about 50% by 5×10^{-4} *M* TC or chlortetracycline (CTC) at 0.01 *M* Mg^{++} in the absence of initiation factors or GTP (182). Also, Ac-Phe-tRNA binding to ribosomes was inhibited about 70% by CTC. There was little or no inhibition if TC was added after formation of the complex. It was previously reported (176, 183–185) that TC had little or no inhibitory effect on polylysyl-tRNA binding to ribosomes; however, in these studies either TC was added after the ternary complex was formed or concentrations of TC 10^{-4} *M* or less were used. Thus, to higher concentrations than necessary for inhibition of AA-tRNA binding to ribosomes, binding of polylysyl-tRNA can be inhibited.

The results of Bodley & Zieve (186) indicate that TC inhibition of Phe-tRNA binding to ribosomes is a function of magnesium concentration. At low magnesium concentrations (<0.01 *M*), TC has little or no inhibition of Phe-tRNA binding to ribosomes. At higher magnesium concentrations, inhibition of Phe-tRNA binding to ribosomes by TC is substantial. This is consistent with the report (187) that at low Mg^{++} (0.005 *M*), Phe-tRNA is primarily bound to the P-site, but at high Mg^{++} (>0.013 *M*), to both sites. These observations may help explain some of the discrepancies.

Binding of fMet-tRNA to ribosomes with AUG or poly AUG is inhibited by TC (50% inhibition at 3×10^{-4} *M*); however, R17 mRNA-dependent binding of fMet-tRNA in the presence of initiation factors and GTP is stimulated by concentrations of TC less than 4×10^{-4} *M*, but inhibited by concentrations greater than 5×10^{-4} *M* (188). The results were similar with GTP or GDPCP. When 1.5×10^{-3} *M* TC was added to the fMet-RNA: ribosome complex containing AUG or poly AUG, no breakdown of the complex was observed. However, 1×10^{-3} *M* TC could destabilize the initiation complex with R17-RNA; destabilization of the complex occurred only when the complex was formed in the presence of GTP but not when formed in the presence of GDPCP; fMet-puromycin formation was rapid and similar in the presence or absence of TC. Thus, at high TC concentrations and only with natural mRNA, TC destabilizes fMet-tRNA from the P-site. However, TC does not destabilize peptidyl-tRNA from polysomes formed from R17-RNA.

Furthermore, Suzuka, Kaji & Kaji (189) and Pestka & Nirenberg (177) reported that binding of AA-tRNA to 30S subunits in response to templates is inhibited by TC. Tetracycline (6×10^{-4} *M*) inhibits the binding of oligolys-tRNA to 30S subunits, but stimulates the binding to isolated 50S subunits both in the presence or absence of poly A (190).

Effect of tetracycline on peptide bond formation.—Published reports are contradictory regarding the effect of TC on peptide bond formation. Cerna et al (182) showed that 10^{-4} *M* TC inhibits the puromycin reaction with oligolys-tRNA 50% as does neomycin (50% inhibition at 10^{-6} *M*) and streptomycin (50% inhibition at 10^{-3} *M*). They report that TC inhibits the puromycin reaction with Ac-Phe-tRNA or Ac-Leu-tRNA just as effectively as it inhibited polylysyl-puromycin formation at about 5×10^{-4} *M*. Traut & Monro (191) observed that the puromycin reaction with polyphenylalanine-charged ribosomes was inhibited 59% by a CTC concentration of 1×10^{-3} *M*; however, 50% inhibition of polyphenylalanine synthesis occurred at 5×10^{-6} *M* CTC. In addition, high concentrations of TC can inhibit the formation of fMet-puromycin by *E. coli* ribosomes from C-A-A-C-C-A(f-

Met) and puromycin: 10^{-4} and 10^{-3} *M* TC produced 13% and 75% inhibition, respectively (192); and high TC concentrations (10^{-4} *M*) can inhibit the binding of C-A-C-C-A(Phe) to ribosomes (193).

The above reports are in apparent contradiction to the results of Sarkar & Thach (181) and Vogel et al (194) who observed no inhibition of fMet-puromycin formation by 10^{-3} *M* and 3×10^{-4} *M* TC, respectively; and to the results of Gottesman (176), using polylysyl-tRNA and *E. coli* ribosomes, and Cundliffe & McQuillen (154) using protoplasts of *Bacillus megaterium*, who found no inhibition of the puromycin reaction by TC. Although the reports are inconsistent, it appears that at high concentrations ($>10^{-4}$ *M*) TC may inhibit peptide bond formation to some extent, depending on the particular assay conditions. This may be related to reports that radioactive tetracycline binds to 50S subunits about half as well as to 30S subunits. It is unlikely, however, that TC inhibits transpeptidation in the intact cell, for 10^{-4} *M* TC did not interfere with the rate of peptidyl-[^{3}H] puromycin synthesis in extracts containing active polyribosomes (477).

Genetics and mutants.—At present no proven ribosome mutants of TC-resistant bacterial strains are known although currently some are under study (195). One TC-resistant mutant of *E. coli* analyzed by Laskin & Chan (167) proved to have cell-free extracts sensitive to the antibiotic. Conceivably, however, 30S ribosomal mutants may be obtained from some TC-resistant organisms, unless the mutation to TC resistance is lethal. Kirshmann & Davis (196) have shown that at slightly inhibitory concentrations, TC can antagonize the growth inhibitory effect of low concentrations of streptomycin on Sm-sensitive strains; and can even replace Sm as a phenotypic suppressor, supporting the growth of conditionally Sm-dependent strains.

Termination.—Tetracycline is a good inhibitor of termination of protein synthesis as measured by release of fMet from fMet-tRNA bound to ribosomes (67, 194) or by release of amino-terminal coat protein fragment of R17-RNA (197). At 5×10^{-4} *M* TC, there is greater than 90% inhibition of termination.

Polyribosome metabolism and effects on intact cells.—Cundliffe (198) examined the effects of CTC on polyribosomes of exponentially growing protoplasts of *B. megaterium*. Low concentrations of CTC (2×10^{-6} to 10^{-5} *M*) produced rapid polysome breakdown within 1 to 6 min. Polysome breakdown occurred only in intact protoplasts, but not in control lysates to which CTC had been added prior to analysis. The products of polyribosome breakdown in low CTC concentrations were ribosome subunits and not 70S ribosomes (199) as had been previously indicated (198). In the presence of high CTC concentrations (3×10^{-4} *M*), polysome breakdown was prevented. Also, after breakdown of polysomes in media with a low CTC concentration, addition of CTC to high concentrations produced aggregation of 70S ribosomes which was prevented by actinomycin D, indicating that new mRNA synthesis was necessary for the aggregation process. A substantial amount of aggregated ribosomes was found in the pellet compared to control reactions. Thus, these aggregates are on the average heavier than normal polysomes. At high concentrations of CTC (1×10^{-4} *M*) protein synthesis was immediately shut off, whereas at the low concentrations protein synthesis was gradually totally inhibited over a 5-min period, that period necessary for breakdown of polysomes. In explanation, it was suggested that low concentrations of CTC inhibit binding of coded AA-tRNA to ribosomes present on polyribosomes. This would prevent peptide bond formation and subsequent translocation. Those ribosomes on the 3′ side of the affected ribosome would be free to continue and complete their chains; those on the 5′ side would be impeded. Increasing the CTC concentration would increase the number of antibiotic interactions per polysome, thus allowing fewer 70S ribosomes to be released. Additionally, it is possible that inhibition of termination might explain the formation of aggregates.

Gurgo et al (131) reported that TC (2×10^{-4} *M*) produced some breakdown of polysomes, but new pulse-labeled mRNA continued to enter polysomes in the presence of this TC concentration even though protein synthesis was completely inhibited. Possibly, ribosomal movement and peptide bond formation may be uncoupled under the influence of TC and other antibiotics as they suggest. At 2×10^{-4} *M* TC, analysis of ribosomes indicated that there were two tRNA molecules per ribosome; at 8×10^{-4} *M* TC, tRNA binding to ribosomes was inhibited in vivo, polyribosome formation was blocked, and 30 and 50S subunits accumulated. It is conceivable that in the presence of ribosomes distorted by antibiotics, mRNA can translocate, be released or degraded, and new mRNA enter polyribosomes; however, entry of newly pulse-labeled mRNA into polysomes may not be a valid criterion for translocation, for distorted ribosomes may contain more than one mRNA per ribosome.

Summary.—Tetracycline inhibits binding of AA-tRNA to the ribosome A-site in vitro and in vivo. At low tetracycline concentrations the amount of polysomes is diminished. At higher tetracycline concentrations, polysomes appear increased in size; this may be a consequence of the ability of the tetracyclines to inhibit termination as well as AA-tRNA binding to ribosomes. At further higher concentrations of the antibiotic, it is sometimes possible to demonstrate other effects on various steps of protein synthesis: inhibition of initiation and of peptide bond formation. Tetracycline binds preferentially to the 30S subunits.

Edeine

Bacillus brevis Vm4 produces at least two closely related basic polypeptide antibiotics, edeine A and B. What has been termed edeine in previous studies (*200–203*) is edeine A (Kurylo-Borowska, personal communication). It is likely that edeine B will have similar properties as edeine A, for they are similar in structure. Several recent papers have elucidated the structures of the edeine antibiotics (*204–206*).

Edeine inhibits the synthesis of bacterial DNA, but not RNA in intact bacteria (*200, 201*). In vivo, the inhibition of protein synthesis did not appear to be as significant as the inhibition of DNA synthesis. Nevertheless, Hierowski and Kurylo-Borowska (*202, 203*) demonstrated that edeine has a potent inhibitory effect on polypeptide synthesis in cell-free extracts from *E. coli,* but had no effect on the transfer of amino acids to tRNA. At a concentration as low as 2×10^{-7} *M* edeine almost completely inhibits polyphenylalanine synthesis directed by poly U. The inhibition of polypeptide synthesis directed by synthetic polynucleotides is not dependent on the composition of the polynucleotide although amino acid incorporation is inhibited less with natural than with synthetic templates.

Edeine inhibits the binding of AA-tRNA to ribosomes in response to template codons. The antibiotic does not affect the binding of poly U to ribosomes. When the complex of ribosomes, poly U, and Phe-tRNA is formed prior to addition to edeine, the antibiotic has only a very slight inhibitory effect on subsequent polymerization (*202*). Thus, when the ternary functional complex is formed prior to the addition of edeine, it appears the antibiotic is ineffective. Although edeine is a basic antibiotic like streptomycin, it does not alter the coding properties of poly U as does streptomycin (*203*). Edeine (10^{-6} *M*) does not inhibit peptide bond formation as measured by fMet-puromycin synthesis (*192*).

Although edeine is discussed under 30S inhibitors, edeine may be a 30S and 50S inhibitor, for [^{14}C]edeine binds to 70S, 30S, and 50S particles (*203*). At 0.01 *M* Mg^{++}, *the* molar ratio of [^{14}C]edeine bound is 2–3 and 1–2 to 30 and 50S subunits, respectively (*203, 207*). Binding of edeine to ribosomes does not depend on ammonium chloride and can occur at 1×10^{-4} *M* Mg^{++}. In addition, 50S subunits treated with edeine show loss of ability to bind fMet-tRNA in response to AUG and MS2 RNA when added to reaction mixtures containing untreated 30S subunits (W. Szer and Z. Kurylo-Borowska, personal communication). Therefore, edeine may, in fact, be both a 30S and 50S inhibitor.

Ribosomes which are exposed to edeine and then dialyzed in 1×10^{-4} *M* Mg^{++} do not dissociate into subunits as do control ribosomes not exposed to the antibiotic (*203*). Also, 30 and 50S subunits reassociated in the presence of the antibiotic even at 1×10^{-4} *M* Mg^{++}, at which concentration control subunits do not reassociate. Perhaps, the edeine binds to sites normally occupied by peptidyl- and AA-tRNA and, as bound peptidyl- and AA-tRNA, prevents the component subunits from dissociating.

Szer & Kurylo-Borowska (*207*) have shown that edeine preferentially inhibits Phe- and fMet-tRNA binding to the P-site, although at low Mg^{++} (0.003–0.005 *M*) both sites are vulnerable to edeine. At 0.01–0.02 *M* Mg^{++}

the A-site is resistant to edeine inhibition while the P-site is sensitive: edeine ($3–5 \times 10^{-6}$ *M*) inhibits Phe-tRNA binding to ribosomes about 40% and the Phe-tRNA resistant to inhibition does not react with puromycin; 10^{-4} *M* tetracycline inhibits binding about 50%, while the remaining Phe-tRNA is puromycin-reactive; and edeine plus tetracycline produce greater than 90% inhibition of Phe-tRNA binding to ribosomes. In addition, edeine (5×10^{-6} *M*) inhibits the Phe- and fMet-tRNA binding to 30S subunits almost completely at low and high Mg^{++}; however, it is difficult to extrapolate the effects on 30S subunits to effects on A- and P-sites since the operational definition of these sites requires 50S subunits. Furthermore, edeine does not destabilize fMet- or Phe-tRNA already bound to ribosomes in response to natural or synthetic polynucleotide templates. Enzymic binding of fMet-tRNA to 30S subunits is more sensitive to edeine than nonenzymic binding, whereas opposite sensitivity to tetracycline prevails. The current definition of A- and P-sites and their antibiotic sensitivity requires modification for many inconsistencies prevail: for example, the 30S subunit is considered to have only one binding site (A or P?) (189); yet, tetracycline and edeine, which inhibit binding to different sites, each inhibit binding of AA-tRNA and fMet-tRNA to 30S subunits. Perhaps, there are two tRNA binding sites on 30S subunits as suggested by Pestka & Nirenberg (208).

Although poly U-directed polyphenylalanine synthesis is inhibited about 90% at 1×10^{-7} *M* edeine in cell-free *E. coli* extracts, only 10% inhibition occurs at this concentration of edeine using mammalian cell-free extracts. Even at 10^{-4} *M* edeine mammalian cell-free extracts show only a 30% inhibition (209). Thus, mammalian cell-free extracts appear to be less sensitive to inhibition by edeine than extracts from bacteria. To date there are no published reports of the effect of edeine on polyribosomes and many other steps of protein synthesis. Since edeine inhibits binding to 30S subunits, it may be a good inhibitor of initiation in cell-free bacterial extracts.

Summary.—Edeine is a potent inhibitor of protein synthesis in bacterial cell-free extracts. It inhibits AA-tRNA and fMet-tRNA binding to the P-site at 0.01-0.02 *M* Mg^{++}, but to both sites at lower Mg^{++}. Thus, at high Mg^{++}, edeine and tetracycline are complementary in that they inhibit P- and A-sites, respectively. Edeine inhibits binding of AA-tRNA to 30S subunits almost completely at low and high Mg^{++}. The antibiotic stabilizes the association of subunits in 70S ribosomes and binds to both 30 and 50S subunits.

50S INHIBITORS

Puromycin

Puromycin has been shown to be an inhibitor of protein synthesis in intact cells as well as cell-free extracts from diverse organisms (210). Its chemical structure (211–213), crystal conformation (214), and its conformation in solution (215, 216) have been reported. Yarmolinsky & de la Haba (217) pointed out the structural resemblance between puromycin and

N(CH₃)₂ — DIMETHYL ADENIDE — PUROMYCIN AMINONUCLEOSIDE — p-METHOXY-PHENYLALANINE

PUROMYCIN

R' = REST OF tRNA CHAIN
R = AMINO ACID SIDE CHAIN

AMINOACYL ADENOSINE END OF AMINOACYL-tRNA

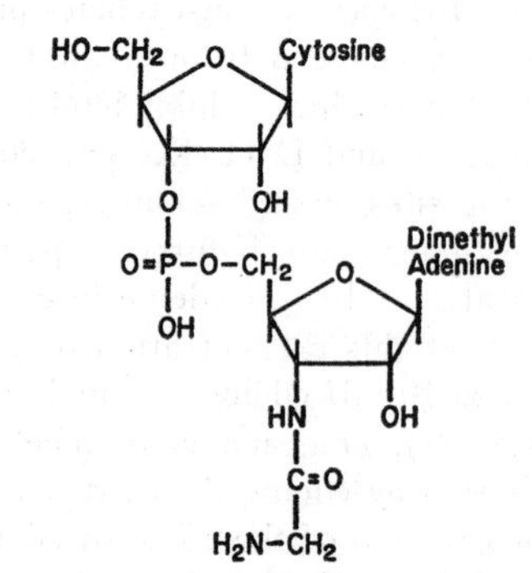

3'-N-GLYCYLPUROMYCIN AMINONUCLEOSIDE

CYTIDINE-3'-PHOSPHORYL-5'-O-(3'-N-GLYCYL) PUROMYCIN AMINONUCLEOSIDE

FIGURE 4. Structures of puromycin, aminoacyl-adenosine end of AA-tRNA and puromycin derivatives (references in text).

the aminoacyl-adenylyl terminus of AA-tRNA (Fig. 4). This provided a basis for understanding the mechanism of puromycin action. Puromycin, like AA-tRNA, can serve as an acceptor of the nascent peptide chain of ribosome-bound peptidyl-tRNA (218, 219), and is incorporated into the prematurely released nascent chains (220, 221). The entire puromycin molecule is linked to the released peptides through the amino group of the side chain by a peptide bond through the carboxy-terminal end of the polypeptides in a reaction similar to the transfer of peptide from peptidyl-tRNA to the next AA-tRNA (Fig. 1). On analysis of lysine peptides released by puromycin, Smith et al (222) also concluded that one puromycin molecule was bound to each lysine peptide. They also noted that although dilysyl-, trilysyl-, and tetralysyl-puromycin were found, no lysyl-puromycin was detectable. In short, puromycin inhibits protein synthesis by substituting for the incoming coded AA-tRNA, in effect competing with the aminoacyl-adenylyl terminus of AA-tRNA. Puromycin does not interfere with AA-tRNA formation (217), or with mRNA or AA-tRNA binding to ribosomes (223).

The following conclusions could be drawn concerning the structural requirements for puromycin inhibition of protein synthesis (224, 225): both the aminonucleoside and the amino acid side chain are necessary; for maximal inhibition the amino acid should be in the L-configuration; the amino group must be unsubstituted although at high concentrations N-acetyl-puromycin was inhibitory to termination (194); for maximal activity the amino acid side chain should contain an aromatic residue. Since puromycin acts as an analog of AA-tRNA, it was paradoxical that the aliphatic derivatives of puromycin such as L-leucyl-, L-alanyl-, L-prolyl-, and L-methionyl-puromycin aminonucleosides are relatively inactive. Nevertheless, several independent groups have indicated that a single aromatic substitution in aminoacyl-nucleosides permits them to exhibit maximal puromycin-like activity (224–227). Although alanine, proline, and glycine substitutions were inactive, leucyl-adenosine (226) and 3′-N-leucyl-puromycin aminonucleoside (225) had about 50% and 10%, respectively, of the puromycin activity. The S-benzyl-L-cysteine derivative of puromycin was almost as active as puromycin itself (225). It is possible, therefore, that the aromatic amino acids can participate in some interactions which may enhance their binding to ribosomes compared to nonaromatic aminoacyl-nucleosides. Examination of molecular models of A(Phe), CpA(Phe), and puromycin suggested that the benzene ring of L-phenylalanine (but not that of D-phenylalanine), of A(Phe) and puromycin can occupy spacial positions which may be occupied by the cytosine moiety of CpA(Phe). Thus, interaction of the phenyl group of phenylalanine with a ribosomal region ordinarily interacting with the cytosine ring may explain the increased activity of the aromatic analogs of puromycin and the aminoacyl-adenosines (225, 228). As noted above, the models also suggest an explanation for the observation that the D-phenylalanine derivative of puromycin is inactive.

Other compounds have puromycin-like activity. Takanami (229) reported that digests of unfractionated AA-tRNA, containing aminoacyl-oligonucleotide fragments ranging in length from 4 to 9 nucleotides, were considerably more effective than pancreatic ribonuclease digests, containing aminoacyl-adenosines, in stimulating the release of nascent peptide chains. In supporting this conclusion, Rychlik et al (227) have shown that aminoacyl-nucleosides are most inhibitory to protein synthesis when adenosine comprises the terminal base. Phenylalanyl-adenosine, A(Phe), was almost as active as puromycin; A(Phe), I(Phe), and C(Phe) had about 70, 15, and 3% of the acceptor activity of puromycin in transferring the peptide residue from polylysyl-tRNA to the acceptor substrates. The aminoacyl-nucleosides, U(Phe) and G(Phe), were inactive as acceptors. The addition of a cytidine 3′-phosphoryl group to the inactive A(Gly) (230) or 3′-N-glycyl-puromycin aminonucleoside (225) converted these inactive compounds to molecules with 80 to 90% of the puromycin activity.

A secondary effect caused by the action of puromycin is the breakdown of polysomes. When intact reticulocytes were incubated with a high puromycin concentration ($> 5 \times 10^{-4}$ *M*) there was a marked fall in the number of large polysomes and an increase in the number of 80S monosomes (231, 232). Similar observations have been made with rat liver (233), *E. coli* (234), and in cell-free preparations from pro- and eukaryotes (116, 232, 235, 236). Direct demonstration of substitution by puromycin for AA-tRNA in intact reticulocytes has been made (237). Tryptophan occurs only near the amino-terminal ends of rabbit hemoglobin, in position 14 of the α-chain and positions 15 and 17 of the β-chain. The selective omission of tryptophan from an otherwise complete medium for optimum hemoglobin synthesis brings about rapid breakdown of polysomes. Polysome breakdown occurs because ribosomes, which are beyond the tryptophan codons, continue to complete the globin chains, but those prior to the point of the tryptophan codons are halted because of a lack of Trp-tRNA. On the other hand, if puromycin could substitute for Trp-tRNA (as it should if it were a true analog of AA-tRNA), then the block could be overcome and ribosomes may continue beyond it. By using a low concentration of puromycin (2.5×10^{-5} *M*) which by itself did not produce polyribosome breakdown, the polyribosome pattern was restored to that of the control. The results suggest that puromycin can substitute for Trp-tRNA in vivo and that ribosomes losing their nascent peptides can continue to synthesize the remaining C-terminal polypeptide.

Since AA-tRNA is intimately involved in coordinate functioning of the ribosome (238), replacement of AA-tRNA and release of peptidyl-tRNA by puromycin may loosen the junction of the subunits so that ribosomes fall off mRNA. In this regard it has been shown that puromycin almost instantaneously stimulates ribosomal subunit exchange (3). Thus, when peptidyl-tRNA is released, 50S subunits probably dissociate and new 50S subunits reassociate with 30S subunits remaining attached to mRNA. The random distribution of N-terminal amino acids of puromycin peptides (236) suggests that 30S subunits can remain attached to mRNA. Breakdown of polysomes but not release of peptides by puromycin requires energy (236) and results in 70S ribosomes rather than subunits (231, 232, 239). Perhaps, also occasionally 70S ribosomes or 30S subunits containing tRNA or AA-tRNA, or both, are released from mRNA subsequent to release of peptides from peptidyl-tRNA by puromycin. The presence of tRNA or AA-tRNA on ribosomes can stabilize them as 70S monosomes, which may account for the formation of monomers rather than subunits after polysome breakdown produced by puromycin. In addition, the energy-dependent step of polysome breakdown (probably translocational events) can be inhibited by cycloheximide while peptide release is unaffected (236).

Immediately after release of nascent peptides by puromycin, there may be neither a peptidyl-tRNA nor an AA-tRNA on the ribosome. In order to continue, the ribosome must therefore reinitiate protein synthesis at a normally internal codon. The cellular mechanism for such internal reinitiation should be of interest. Also, despite its causing premature release of nascent proteins, puromycin may inhibit the normal termination mechanism, for both hydroxy-puromycin and acetyl-puromycin can inhibit chain termination (194).

Puromycin can directly inhibit the binding of the aminoacyl end of AA-tRNA to isolated ribosomes (193, 240, 241). Binding of C-A-C-C-A(Phe), the aminocyl-oligonucleotide terminus of Phe-tRNA, to ribosomes is inhibited by puromycin. This localized inhibition of binding of the AA-tRNA thus may be another action of puromycin distinct from, but related to, its stimulation of release of nascent protein chains from ribosomes.

Since puromycin becomes linked to the released peptides by means of a peptide bond, the reaction has been used extensively as a model of peptide bond formation in protein synthesis (176, 191, 242–245). Formation of peptidyl-puromycin from peptidyl-tRNA attached to ribosomes does not require GTP or supernatant enzymes (220) although the total amount released can be stimulated approximately twofold by GTP and supernatant enzymes (191). Such studies have suggested a two-site model for peptidyl- and AA-tRNA attachment in protein synthesis and have provided an operational definition for the donor (peptidyl- or P-site) and acceptor (A-site) sites: AA-tRNA or N-blocked derivatives such as acetyl-AA-tRNA or various peptidyl-tRNA species which react with puromycin are considered to be in the P-site. Those unable to react with puromycin are considered to be in the A-site. Others have suggested a similar two-site model (178, 179, 246). Donors for peptidyl-puromycin synthesis include polyphenylalanyl-tRNA (191, 242), polylysyl-tRNA (247), fMet-tRNA (244, 245), Ac-Phe-tRNA (248–250) as well as natural peptidyl-tRNA (220). In addition, terminal fragments of fMet-tRNA, Ac-Phe-tRNA, or Ac-Leu-tRNA have served as peptidyl donors. Thus, C-A-A-C-C-A(fMet) was shown (251) to react with puromycin in the presence of washed ribosomes; the reaction required ethanol as well as K^+. Monro (7) further demonstrated that isolated 50S subunits could catalyze such transpeptidation. The puromycin reaction in various forms has been used as an assay for peptide bond formation to evaluate the activity of various antibiotics on peptidyl transfer. It must be kept in mind, however, that reactions with isolated components may not reflect conditions in cells; and that antibiotics which inhibit the puromycin reaction with isolated components may not inhibit normal peptide bond formation during cellular protein synthesis.

Summary.—Puromycin, through its resemblance to the aminoacyl-adenylyl end of AA-tRNA, inhibits protein synthesis by competing with AA-tRNA for this site on 50S subunits of ribosomes. In place of AA-tRNA it accepts nascent peptides, causing premature release of incomplete polypeptide chains. As a result of these actions, puromycin produces breakdown of polysomes.

FIVE AMINOHEXOSE PYRIMIDINE NUCLEOSIDE ANTIBIOTICS

The five aminohexose pyrimidine nucleoside antibiotics consist of gougerotin, amicetin, blasticidin S, bamicetin, and plicacetin. The effect of the first three on protein synthesis has been studied; however, the latter two have not been evaluated although they (plicacetin and bamicetin) are quite similar to amicetin in structure. A review of these and other nucleoside antibiotics has been compiled by Suhadolnik (252). The structures of gougerotin, blasticidin S, and amicetin are given in Figure 5.

Gougerotin.—Clark & Gunther (253) showed that gougerotin was an inhibitor of protein synthesis in cell-free extracts from *E. coli*. Gougerotin inhibited polyphenylalanine synthesis dependent on poly U. The antibiotic did not inhibit Phe-tRNA formation. Subsequently, gougerotin has been shown to be an inhibitor of protein synthesis in cell-free extracts obtained from mouse liver (254) and rabbit reticulocytes (255, 256). Clark & Chang (255) reported that gougerotin inhibits the puromycin-induced release of nascent protein chains from reticulocyte ribosomes. However, gougerotin itself could not stimulate the release of nascent chains from reticulocyte ribosomes as does puromycin (255, 256). They suggested that gougerotin inhibited translocation also for it inhibited the release of free ribosomes from peptidyl-tRNA:mRNA:ribosome complexes. On the other hand, gougerotin did not inhibit GTP-dependent release of completed globin chains from prelabeled rabbit reticulocyte ribosomes. In contrast to puromycin which accelerates the breakdown of polysomes (236), polysome structures are maintained by gougerotin (256) and amicetin (Ennis, personal communication). Coutsogeorgopoulos (257) also found that gougerotin, as well as blasticidin S and amicetin, inhibited polyphenylalanine synthesis directed by poly U. He was not able to reverse the inhibition by increasing amounts of Phe-tRNA using an S-30 extract for protein synthesis.

Goldberg & Mitsugi (185) examined the effect of gougerotin on the puromycin-stimulated release of polylysine from polylysyl-tRNA. Gougerotin (10^{-4} *M*) inhibited this puromycin reaction approximately 50%. Although Casjens & Morris (256) suggested that gougerotin was a competitive inhibitor of puromycin, Goldberg & Mitsugi (185) and Pestka (249) concluded that gougerotin was a mixed type inhibitor of the puromycin reaction. In addition, gougerotin was a more effective inhibitor of proline incorporation stimulated by a poly C or poly UC template than it was of phenylalanine incorporation stimulated by poly U or lysine incorporation stimulated by poly A (258). Phenylalanine incorporation was inhibited 33% while lysine incorporation was inhibited only 22% by 4×10^{-5} *M* gougerotin. Under similar conditions proline incorporation was inhibited 70–80%. This contrasts

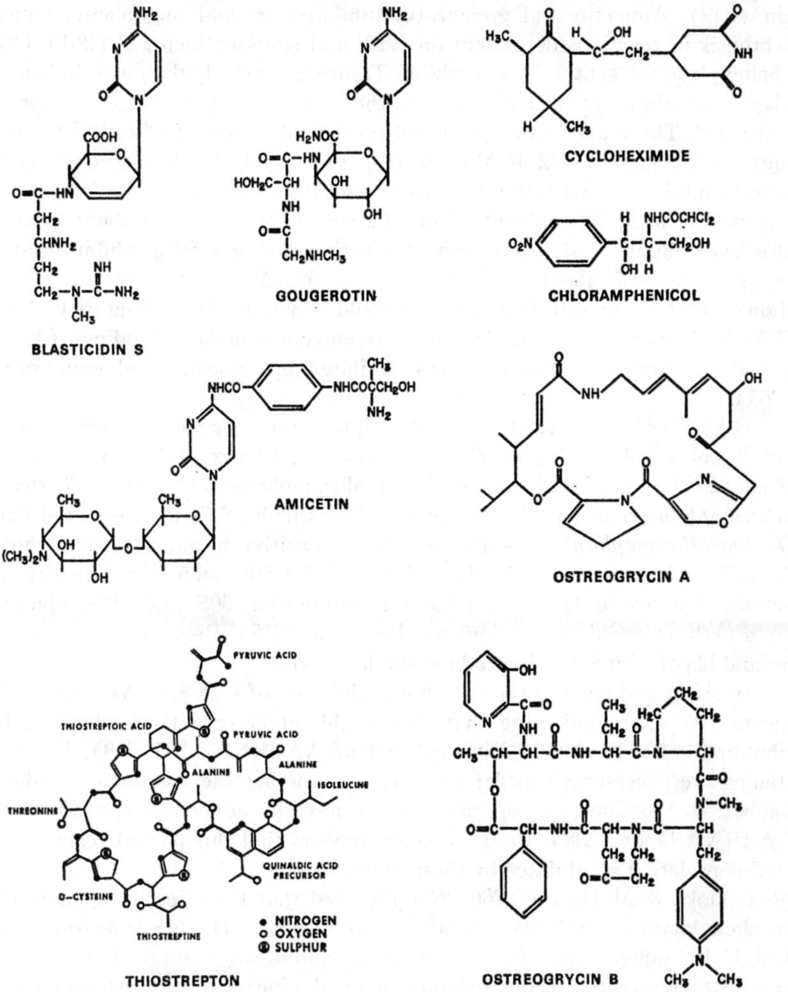

FIGURE 5. Structures of several antibiotics (references in text).

with chloramphenicol which inhibits both lysine and proline incorporation much better than it does phenylalanine incorporation. Herner, Goldberg & Cohen (259) described the stimulation of Ac-Phe-tRNA binding to ribosomes by sparsomycin; gougerotin could interfere with this sparsomycin-stimulated binding of Ac-Phe-tRNA to ribosomes. At high concentrations (10^{-4} *M* and above) gougerotin itself could produce stimulation of Ac-Phe-tRNA binding to ribosomes (259, 260).

Gougerotin and amicetin (10^{-4} *M* and above) were effective in inhibiting Ac-Phe-puromycin synthesis from Ac-Phe-tRNA, ribosomes, and puromycin (249). Amicetin and gougerotin inhibited di- and oligophenylalanine synthesis in reactions dependent on GTP and transfer factor G (261). Oligophenylalanine synthesis was inhibited more extensively than was diphenylalanine synthesis at 0.02 *M* Mg^{++} in the absence or presence of factor G and GTP. These antibiotics produced no net inhibition of Phe-tRNA binding to ribosomes at 0.02 *M* Mg^{++} in response to poly U. Amicetin and gougerotin inhibited fMet-puromycin formation from C-A-A-C-C-A(fMet) and puromycin with 50S subunits from *E. coli* (192). Both antibiotics were effective inhibitors of this fragment reaction giving a 50% inhibition at a concentration of about 3×10^{-5} *M*. Celma, Monro & Vazquez (262) showed that gougerotin and amicetin could stimulate the binding of C-A-C-C-A(Ac-Leu) to ribosomes, but the sparsomycin stimulated binding of C-A-C-C-A(Ac-Leu) to ribosomes was inhibited by amicetin and gougerotin (263).

Although Vazquez (264) showed no interference by amicetin with [^{14}C]-chloramphenicol binding to *E. coli* ribosomes, Chang, Siddhikol & Weisblum (265) showed inhibition of [^{14}C]chloramphenicol binding to *B. stearothermophilus* ribosomes by amicetin and blasticidin S. They suggested that *B. stearothermophilus* ribosomes are more sensitive to amicetin than those from *E. coli*. Independently they identified the 50S subunit as the site of action of amicetin by using ribosomes containing 30S and 50S subunits from both organisms in various combinations (265). Presumably, gougerotin and blasticidin S would produce similar results.

Amicetin and gougerotin were both inhibitors of C-A-C-C-A(Phe) binding to ribosomes indicating that they could interfere with the binding to ribosomes of an isolated aminoacyl end of AA-tRNA (193, 240). Further studies are necessary in order to ascertain whether the fragment is indeed binding to the same site on ribosomes where the aminoacyl end of intact AA-tRNA binds (228). In any case, it appears that this partial reaction of transpeptidation is inhibited by these antibiotics.

Yukioka & Morisawa (266–268) reported that the effect of gougerotin on phenylalanine synthesis depends on the K^+ and NH_4^+ concentrations. At 0.05 *M* K^+, gougerotin inhibited polyphenylalanine formation. At 0.2 *M* K^+ or NH_4^+ gougerotin stimulated polyphenylalanine synthesis. Although the number of peptide chains synthesized increased in the presence of gougerotin, the average chain length decreased. When ribosomes were incubated with gougerotin before additions of the complete reaction mixture, stimulation produced by gougerotin was even greater. Gougerotin did not stimulate release of polyphenylalanine peptides and polyphenylalanine was still bound to ribosomes. Since there is apparently one binding site for peptidyl-tRNA per ribosome, the increased number of polyphenylalanine peptides suggests there is an increase in the number of active ribosomes which are participating in peptide bond formation. Stimulation of polyphenylalanine synthesis by gougerotin appeared to be related to the amount of T factor present in the high speed supernatant. Thus, further investigation revealed gougerotin as well as blasticidin S and puromycin enhanced T factor-dependent enzymatic binding of Phe-tRNA to ribosomes, but had no effect on nonenzymic binding to ribosomes (268). The antibiotics had no effect on the GTP binding activity of T factor. This gougerotin-dependent stimulation of enzymic binding is responsible for the increased number of polyphenylalanine peptide chains produced and the overall stimulation of polyphenylalanine synthesis. Puromycin-dependent stimulation of enzymic binding of AA-tRNA to ribosomes was observed previously in cell-free extracts from rat liver (269). Gougerotin is an inhibitor of transpeptidation through its action on the 50S subunit. Through this same site gougerotin stimulates enzymic binding of AA-tRNA by stabilizing the 50S subunit in a conformation that makes 30S-50S couples more likely and enhances formation of the AA-tRNA: factor T: GTP complexes with 70S ribosomes. These antibiotics interact with ribosomes rather weakly and, although they stabilize the ribosome in a certain conformation, they can be easily displaced by the more stable binding of the aminoacyl end of AA-tRNA when this reaction occurs enzymatically. Experiments have indicated that these antibiotics can directly interfere with binding of isolated fragments of aminoacyl ends of AA-tRNA, such as C-A-C-C-A(Phe) to ribosomes (193).

Neth et al (270) have examined the effect of a number of antibiotics on the puromycin reaction using C-A-A-C-C-A(fMet) and puromycin as substrates with human tonsil ribosomes. Their results indicate that amicetin and gougerotin can inhibit the puromycin reaction of human tonsil ribosomes although concentrations at which they produce a 50% inhibition of the reaction are quite high being 10^{-4} to 10^{-3} *M*. Also, C-A-C-C-A(Phe) can accept polypeptides from peptidyl-tRNA of human placental ribosomes; this is inhibited by amicetin, gougerotin, puromycin, and blasticidin S (209).

Capecchi & Klein (197) have indicated that 5×10^{-4} *M* gougerotin can inhibit peptide chain termination; and Vogel et al (194) have found inhibition of peptidyl transfer and peptide chain termination by amicetin to be correlated very well as a function of antibiotic concentration.

Amicetin.—The results with amicetin are similar to those with gougerotin and have been described for the most part above. Brock (271) found that amicetin inhibited the growth of *Streptococcus pyogenes* and synthesis of M protein. Bloch & Coutsogeorgopoulos (272) showed that amicetin could inhibit poly U-directed phenylalanine incorporation in cell-free extracts from *E. coli*. In addition, amicetin inhibited growth as well as protein synthesis in intact *E. coli* at concentrations where RNA and DNA synthesis continued. Phe-tRNA formation was not inhibited by amicetin.

Blasticidin S.—Effects of blasticidin S on protein synthesis are similar to amicetin and gougerotin as described above. Yamaguchi, Yamamoto & Tanaka (273) demonstrated that blasticidin S inhibited amino acid incorporation directed by native as well as synthetic mRNAs. Poly U-directed polyphenylalanine incorporation was very sensitive to blasticidin S especially if ribosomes were pretreated with the antibiotic (274). Blasticidin S inhibits the puromycin-dependent release of polyphenylalanine from ribosomes containing polyphenylalanyl-tRNA. Inhibition of the puromycin reaction was decreased by increasing puromycin concentration, suggesting that puromycin and blasticidin S may be competing for sites on the ribosomes. Increased amounts of AA-tRNA significantly reversed the blasticidin S inhibition of polyphenylalanine formation. By using various derivatives of and compounds similar to blasticidin S, Yamaguchi & Tanaka (274) were able to show that removal of the amino group from the cytosine moiety or removal of the cytosine moiety reduced or eliminated the activity of the antibiotic.

Blasticidin S inhibits interaction of C-A-C-C-A(Phe) with ribosomes from *E. coli* or human placenta (209, 275). Mammalian ribosomes seem to be more sensitive to blasticidin S than bacterial ribosomes; whereas bacterial ribosomes appear more sensitive to amicetin and gougerotin than mammalian ribosomes.

Kinoshita, Tanaka & Umezawa (276) examined the binding of blasticidin S to ribosomes using equilibrium dialysis. There is a single binding site for each ribosome; the association constant for the binding to *E. coli* B ribosomes is $5 \times 10^5 M^{-1}$. The binding site is localized to the 50S ribosomal subunit and binding is reversible so that the complex with ribosomes cannot be isolated by gel filtration. At $2 \times 10^{-6} M$ blasticidin, ribosomes are half-saturated with antibiotic. This agrees well with the concentration needed to give 50% inhibition of polypeptide synthesis in cell-free extracts of *E. coli* and suggests close correlation between the binding of blasticidin S to ribosomes and inhibition of polypeptide synthesis. The free energy change of binding is calculated to be −7.3 kcal/mole which is equivalent to approximately two hydrogen bonds. In this regard, it is interesting to note that the N-methylguanido group and the amino group in the cytosine moiety of blasticidin S have been reported to play an important role in inhibition of polypeptide synthesis (17). Binding of blasticidin S to ribosomes was inhibited by gougerotin, but was not affected by chloramphenicol, lincomycin, erythromycin, or puromycin. Mikamycin A stimulates the binding of blasticidin S to ribosomes; mikamycin A is of the streptogramin A group of antibiotics. In accordance with the lack of influence of puromycin on blasticidin S binding, Lessard & Pestka (241) have found no effect of $10^{-4} M$ amicetin or gougerotin on puromycin binding to ribosomes.

Nucleocidin.—Nucleocidin is an adenine nucleoside (252). It is included here since an individual section is not warranted for the limited amount of work which has been reported. It appears, however, that protein synthesis is inhibited by nucleocidin in both intact animals and in cell-free extracts obtained from rat liver (277). Growth of Gram-positive and Gram-negative bacteria is inhibited by nucleocidin. Nucleocidin did not inhibit the binding of C-A-C-C-A(Phe) to *E. coli* ribosomes under conditions in which amicetin, gougerotin and blasticidin S were inhibitors of this reaction (Pestka, unpublished experiments).

Summary.—Amicetin, gougerotin, and blasticidin S inhibit transpeptidation reactions. They can inhibit binding of C-A-C-C-A(Phe) and stabilize binding of C-A-C-C-A(Ac-Leu). They may also inhibit some translocation-related events for polysomes are stabilized by gougerotin and amicetin. Binding of blasticidin S to *E. coli* ribosomes is localized to the 50S subunit and is not inhibited by puromycin.

CHLORAMPHENICOL

Chloramphenicol is chiefly a bacteriostatic agent. It is a broad spectrum antibiotic, inhibiting the growth of Gram-positive and Gram-negative bacteria; eukaryotic cells are generally resistant to the antibiotic (278). The chemical structure of CM is one of the simplest of the known antibiotics (279–281) as shown in Figure 5. There are 4 stereoisomers, only one of which is active, D(—)-threo-chloramphenicol (282). A variety of electronegative groups can be substituted for the aromatic nitro group without major loss of antimicrobial activity (278).

Chloramphenicol inhibits protein synthesis (161, 164, 283, 284); it does not inhibit amino acid-dependent pyrophosphate exchange (285) or formation of aminoacyl-tRNA (164, 286); nor does it inhibit binding of poly A or poly UC to ribosomes (287, 288), in contrast to the report that CM inhibited poly U binding to *E. coli* ribosomes (289). Chloramphenicol was found to inhibit incorporation of amino acids into protein in cell-free extracts from *E. coli* (164, 290, 291), but not in those from eukaryotic cells (292–296). The resistance of protein synthesis by a combination of yeast ribosomes and *E. coli* supernatant, but not by *E. coli* ribosomes and yeast supernatant to CM, suggested that the site of action of CM is on the ribosome (296).

Chloramphenicol inhibits peptide bond formation as measured by various assays. In intact bacteria, CM inhibits peptidyl-puromycin formation (297, 298). Traut & Monro (191) showed that CM inhibits polyphenylalanyl-puromycin formation. Synthesis of polylysyl-puromycin (176, 185, 299, 300), fMet-puromycin (7, 251), and acetyl-phenylalanyl-puromycin (248, 249) was also inhibited by CM. Ribosomes catalyze the reaction of C-A-A-C-C-A(fMet) with puromycin; CM inhibited this reaction (251). In studying the effect of CM on acetyl-phenylalanyl-puromycin synthesis, Pestka (249) showed that CM acted as a competitive inhibitor of puromycin. In fact, puromycin can inhibit binding of CM to ribosomes (241). Since puromycin is an analog of the aminoacyl end of AA-tRNA, then CM also may be. A similar suggestion was made by Coutsogeorgopoulos (301).

Isolated aminoacyl-oligonucleotide fragments of AA-tRNA are bound well to ribosomes in the absence of templates (193, 240). Binding of these aminoacyl-oligonucleotides such as C-A-C-C-A(Phe) to ribosomes was inhibited by CM (193, 228, 240, 241). Only the bacteriostatically active isomer significantly inhibited the binding of these aminoacyl-oligonucleotides to ribosomes (241). These results demonstrated directly that CM inhibits the binding of the aminoacyl end of AA-tRNA to ribosomes. Inhibition of functional attachment of the aminoacyl end of AA-tRNA could account for inhibition of peptide bond formation. Celma et al (262) showed that CM slightly stimulated binding of C-A-C-C-A(Ac-Leu) to ribosomes. Weber & DeMoss (302), on the other hand, have concluded that CM does not directly block transpeptidation, but inhibits the conversion of peptidyl-tRNA into the puromycin-susceptible donor state.

Radioactive CM is bound to purified ribosomes (264, 265, 303–306). Binding is relatively weak so that CM can be removed by washing; binding to intact *E. coli* was also weak (307). Adherence of CM to ribosomes requires K^+ or NH_4^+ and is localized to the 50S subunit (264). Many antibiotics interfere with [^{14}C]chloramphenicol binding to ribosomes; those that do are presumed to be 50S inhibitors. Streptogramin A, lincomycin, and most macrolide antibiotics inhibit CM binding to ribosomes (264). Aminoglycosides and tetracyclines do not. Studies of this sort have led to a relatively consistent picture of the subunit localization of various inhibitors.

Effect of CM on polynucleotide-directed protein synthesis was a function of template used (287, 288, 308). Polypeptide synthesis directed by poly U and poly UA templates was substantially more resistant to the action of CM than peptide synthesis directed by poly A and poly UC templates. It should be noted that protein synthesis directed by natural messengers such as phage f2 RNA is markedly inhibited by CM. The apparent paradox partially may be a reflection of the differential solubility and precipitability of the various polypeptides formed under the direction of different templates. Small phenylalanine peptides of chain length 4 are substantially precipitable by cold trichloroacetic acid (309). Lysine peptides or those containing a high proportion of lysine tend to be soluble in trichloroacetic acid, the usual methods of assaying protein synthesis in cell-free extracts (310). Julian (311, 312) examined in detail the distribution of the lysine peptides formed in the presence and absence of CM under the direction of a poly A template. In the presence of CM the distribution of peptides was shifted from long-chain material to chains of shorter length. Although 2×10^{-4} *M* CM caused a 60% inhibition of lysine peptides precipitable by tungstate, it produced only a 23% inhibition of total peptide bond formation and no inhibition of the total number of lysine peptide chains formed. Irvin & Julian (313) reported that, in contrast to lysine peptides, CM inhibited all sizes of proline peptides synthesized in the presence of a poly C template although large peptides were inhibited more than small ones. With regard to diphenylalanine and oligophenylalanine (phenylalanines of chain length 3 and greater) synthesis, Pestka (240, 314) found that, although the rate of oligophenylalanine synthesis was inhibited by CM, extent of formation of diphenylalanine was either unchanged or stimulated. The formation of increased amounts of shorter chains is consistent with inhibition of peptide bond formation by CM, for chain-propagating systems in general show greater quantities of shorter chains in inhibited than in the uninhibited systems (315). Another possible contribution to the small inhibition of polyphenylalanine synthesis by CM may be related to interaction of the amino acid with ribosomes. As indicated above in the discussion of puromycin, puromycin analogs possess maximum puromycin-like activity when they contain constituent aromatic residues (224–227). The aminoacyl end of Phe-tRNA may have greater affinity for the ribosome than nonaromatic AA-tRNAs; and thus CM may inhibit peptide bond formation less in the case of aromatic than in the case of aliphatic AA-tRNA.

Chloramphenicol has been shown to prevent polysome breakdown when protein synthesis is inhibited by it for short times of 5 min or less (306, 316–318); it can prevent breakdown of polysomes caused by puromycin or actinomycin (317) and can block degradation of mRNA observed in the presence of actinomycin (319, 320). Although CM did not inhibit attachment of ribosomes to mRNA (306, 321), Dresden & Hoagland (321) reported that CM inhibited formation of polysomes in cells recovering from glucose starvation; this suggests that CM inhibits ribosome movement along mRNA. On longer exposure of cells to CM, polysome breakdown can occur (317). Nevertheless, Gurgo, Apirion & Schlessinger (322) have reported the continued entrance of mRNA into polysomes in the presence of 3×10^{-4} *M* CM for 30 min at almost a normal rate when protein synthesis is inhibited 95%; from this they inferred that movement of ribosomes on mRNA continues in cells treated with CM and that transpeptidation is uncoupled from translocation. In the presence of fusidic acid, polyribosomes are fixed so that little mRNA enters polysomes (322). However, continual association of newly synthesized mRNA with ribosomes distorted by antibiotics may not imply translocation. It is probable that ribosomal movement along mRNA occurs at a slow rate in the presence of CM; however, this rate of movement is much slower than that occurring during protein synthesis and thus the apparent protection of polysomes during short incubations with CM. On longer exposure of cells to CM, polysome dissolution is observed (317). Kaempfer (3, 324) has shown that in the presence of 6×10^{-4} *M* CM, ribosome subunit exchange and protein synthesis in cell-free extracts can occur at about one-fifth the rate without antibiotic. Slow rate of protein synthesis may account for some of the entry of newly synthesized mRNA into polysomes. The results of Weber & DeMoss (316) and Cameron & Julian (323) indicate that peptidyl-tRNA remains attached to ribosomes on binding of CM, for the polysome-specific activity does not decrease shortly after CM is added to growing cultures of *E. coli,* at least for several minutes. They (316, 323) have also indicated that CM stimulated polysome formation; this was shown to be due to nonfunctional polysomes formed from the combination of 30S and 50S subunits with newly synthesized mRNA. Perhaps, the incorporation of mRNA into polysomes seen by Gurgo et al

(322) is similarly due to stimulation of the formation of nonfunctional polysomal complexes.

Chloroplast ribosomes from tobacco leaves show the same stereospecific inhibition of protein synthesis by CM as do bacterial ribosomes (325); under the same conditions cytoplasmic ribosomes are unaffected by CM. Protein synthesis by isolated mitochondria from rabbit liver (326, 327) and yeast (328) is inhibited by CM. It appears that all ribosomes of the 70S class containing 16S and 23S RNA are inhibited by CM.

Chloramphenicol has been shown to inhibit peptide synthesis with reticulocyte ribosomes when exogenous templates are used (329–331), particularly poly U. The effect is variable and strongly dependent on Mg^{++} concentration. If extracts are preincubated with poly U for 5 min prior to addition of CM, the antibiotic has no inhibitory effect. Although this effect of CM is observed at concentrations of 10^{-4} *M* or less, it is probably unrelated to the effects described on procaryotic ribosomes, and may be a reaction of CM with exogenous templates or with ribosomal sites to which these exogenous templates bind. The effect of the other three enantiomers of chloramphenicol in this system has not been reported. The resemblance of CM to uridylic acid (332) may possibly enable CM to compete with uridylic residues on template codons for ribosomal sites.

Although mammalian tissues are relatively insensitive to CM, antibody synthesis appears to be an exception since it is inhibited by concentrations of CM which are bacteriostatic (333–336). In spleen, ribosomes bound to endoplasmic reticulum were sensitive to inhibition by 10^{-5} to 10^{-4} *M* CM, whereas free cytoplasmic ribosomes were relatively insensitive (337). Protein synthesis in intact reticulocytes was also inhibited by CM at high concentrations (3×10^{-3} *M*). However, the inhibition occurred equally well with the bacteriostatically inactive L(+)threo enantiomer and RNA synthesis as well as protein synthesis was inhibited (338).

Summary.—Chloramphenicol binds specifically to 50S subunits. It inhibits functional attachment of the aminoacyl end of AA-tRNA to the 50S subunit thus inhibiting transpeptidation, but may also have a direct inhibitory effect on the peptidyl transferase. The antibiotic decreases the rate of polysome breakdown, but ribosomes can slowly progress along mRNA in the absence of peptide bond formation. Mitochondrial protein synthesis as well as bacterial protein synthesis is inhibited by chloramphenicol.

SPARSOMYCIN

The structure of sparsomycin has recently been determined by Wiley & MacKellar (339) and is shown in Figure 3. The antibiotic has a broad spectrum of activity and is toxic to both prokaryotic and eukaryotic cells (340–342). In studies on the mode of action of sparsomycin in *E. coli* B, Slechta (341) showed that protein synthesis rather than DNA or RNA synthesis was the primary site of inhibition. Using cell-free extracts from rabbit reticulocytes, Colombo, Felicetti & Baglioni (147) showed that 10^{-5} *M* sparsomycin completely inhibited protein synthesis and prevented breakdown of polysomes. Protein synthesis of intact reticulocytes was not affected. However, protein synthesis in L cells grown in tissue culture was markedly inhibited by sparsomycin (342): RNA synthesis is inhibited 25%, protein synthesis is inhibited 95%, and DNA synthesis inhibited to an intermediate extent.

Sparsomycin does not influence the formation of aminoacyl-tRNA (258, 342), but prevents protein synthesis in cell-free extracts from *E. coli*. Inhibition of polypeptide synthesis increased with decreasing amounts of uridylic acid in the polynucleotide templates (342). The antibiotic did not inhibit binding of Phe-tRNA or poly U to ribosomes. Sparsomycin blocked the puromycin-induced release of polyphenylalanine from ribosomes containing polyphenylalanyl-tRNA (185, 258, 342) and puromycin release of polylysine from polylysyl-tRNA in the presence of poly A and ribosomes. Sparsomycin was a competitive inhibitor of puromycin in this reaction. Also, the addition of one lysine residue to polylysyl-tRNA was inhibited 50% by 10^{-7} *M* sparsomycin (343). Furthermore, sparsomycin was a competitive inhibitor of puromycin for the formation of acetyl-phenylalanyl-puromycin from Ac-Phe-tRNA and puromycin (240, 249, 261); and the antibiotic inhibited diphenylalanine synthesis.

Herner, Goldberg & Cohen (259) demonstrated that the binding of Ac-Phe-tRNA to ribosomes is stabilized by low concentrations of sparsomycin (half maximal effect at 10^{-7} *M*). The initial rate is not affected, but the extent of complex formation is stimulated. Sparsomycin also blocks the pactamycin-induced dissociation of initiation complexes. Stabilization of binding by sparsomycin requires addition of 50S to 30S subunits. Antibiotics such as gougerotin and chloramphenicol exhibit qualitatively similar actions, but at much higher concentrations, and they compete with sparsomycin for this effect.

Trakatellis (344) showed that sparsomycin is a potent inhibitor of protein synthesis in cell-free extracts as well as in intact mouse liver. Ribosomes were prepared from the livers of mice after intraperitoneal injections of sparsomycin as well as from mice not injected with the antibiotic. The ribosomal preparations isolated from sparsomycin-injected mice incorporated less amino acid than did ribosome preparations isolated from normal animals. These studies localized the action of sparsomycin to the ribosome. Sparsomycin not only prevents polysome breakdown (147, 344), but also inhibits ribosomal subunit exchange (3). It also markedly inhibits termination at a concentration of less than 10^{-6} *M* (70, 194, 197).

Sparsomycin inhibited peptidyl transferase activity of 50S ribosomal subunits (192). In the presence of K^+, Mg^{++}, and ethanol, 50S subunits from *E. coli* catalyzed the reaction of various N′-acyl-aminoacyl-oligonucleotide fragments with puromycin to give the corresponding N-acyl-aminoacyl-puromycin products. Sparsomycin appears to produce an inert complex between the N-blocked aminoacyl-oligonucleotide fragments and the 50S ribosomal subunit (263). Specifically, sparsomycin stimulated formation of a ribosomal complex with C-A-C-C-A(Ac-Leu) and C-C-A(Ac-Leu). Hishizawa, Lessard & Pestka (120) confirmed that sparsomycin stimulates the formation of complexes between C-A-C-C-A(Ac-Phe) and ribosomes. In addition, sparsomycin inhibited binding of C-A-C-C-A(Phe) to ribosomes at low concentrations (193, 241). Jimenez, Monro & Vazquez (260, 345) also showed that increased amounts of Ac-Phe-tRNA binds to ribosomes in the presence of sparsomycin as well as in the presence of amicetin and gougerotin; and that Ac-Phe-tRNA bound to 70S ribosomes was resistant to hydroxylamine action compared to free Ac-Phe-tRNA. Although Goldberg & Mitsugi (258) indicated that sparsomycin did not interfere with [^{14}C]chloramphenicol binding to ribosomes of *E. coli,* Weisblum & Davies (16) and Lessard & Pestka (241) reported that sparsomycin inhibits binding of [^{14}C]chloramphenicol to ribosomes from *B. stearothermophilus* and *E. coli,* respectively.

Summary.—Sparsomycin is an effective inhibitor of peptide bond formation with 70S and 80S ribosomes. It inhibits binding of the aminoacyl end of AA-tRNA to ribosomes and simultaneously stimulates the formation of inert complexes between the peptidyl end of peptidyl-tRNA and ribosomes. Sparsomycin prevents breakdown of polysomes.

Anisomycin

Anisomycin specifically inhibits 80S ribosomal systems (346, 347). The effects of anisomycin are completely reversible and the antibiotic stabilizes the polysome pattern in HeLa cells (346). Vazquez et al (348) have shown that anisomycin can inhibit the formation of fMet-puromycin with yeast ribosomes. Anisomycin produces complete inhibition of the puromycin reaction at 1×10^{-4} *M* and 50% inhibition at 10^{-6} *M*. Similar results were obtained by Neth et al (270), using human tonsil ribosomes. In addition, anisomycin is a potent inhibitor of transpeptidation by human placental ribosomes when C-A-C-C-A(Phe) is used as an acceptor (209).

Summary.—Anisomycin inhibits peptide bond synthesis in eukaryotes only. The antibiotic stabilizes polyribosomes.

Macrolides

The macrolide antibiotics consist of a group of structurally related compounds produced by various species of streptomyces. All the macrolide antibiotics contain a large lactone ring of 12–22 atoms which contains few or no double bonds and no nitrogen atoms (349); in addition, they have one or more sugars attached to the lactone ring (see the structures of erythromycin (ERY) and carbomycin in Figure 3). Celmer (350) has discussed the biogenetic and stereochemical relationships within the macrolide antibiotics.

As a group, macrolide antibiotics are active against Gram-positive bacteria and often show cross-resistance within the macrolide group. Gram-negative bacteria are often resistant or at least relatively resistant to most macrolides; yeasts and protozoa are also resistant as well as animal cells. In general, macrolides are bacteriostatic at concentrations near minimum growth inhibitory concentrations; however, at concentrations tenfold higher, they appear bacteriocidal. Erythromycin, chalcomycin, spiramycin, and probably other macrolides inhibit bacterial growth by blocking protein synthesis (349, 351–354).

Although the macrolides are structurally related, they do not have identical effects on protein synthesis. In fact, they exert a spectrum of inhibitions: each particular macrolide can be characterized by the steps inhibited within this spectrum. Macrolides with large lactone rings tend to inhibit more reactions than those with small rings.

Binding of antibiotics to ribosomes.—Spiramycin, oleandomycin, carbomycin, ERY, and angolamycin inhibit the binding of chloramphenicol to bacterial ribosomes as shown by Vazquez (264, 304) and Wolfe & Hahn (305). Methymycin and lankamycin also inhibit chloramphenicol binding to ribosomes; however, the effect of these two antibiotics is small and is readily reversible (264). However, Oleinick, Wilhelm & Corcoran (355) reported that ERY did not displace chloramphenicol from *B. subtilis* ribosomes; nor did chloramphenicol displace ERY. The inhibition of both antibiotics on polylysine synthesis was additive.

Many studies of macrolide binding to ribosomes have been reported (355–372). Shimizu et al (356) and Vazquez (357) reported binding of [^{14}C]spiramycin to ribosomes. Ribosomes from antibiotic-sensitive *Staphylococcus aureus* bound spiramycin 5–10 times better than ribosomes from spiramycin-resistant strains (356). Spiramycin was bound to 70S and 50S particles, but not to 30S subunits; NH_4^+ is required for binding (357). Binding of spiramycin to ribosomes was inhibited by most of the macrolide antibiotics, but little or none by a chloramphenicol concentration 300 times that of the [^{14}C]spiramycin.

Binding of [^{14}C]erythromycin to ribosomes from different bacteria has been reported (358–367). Chloramphenicol does not inhibit ERY binding (355, 361). Ribosome-bound ERY is readily displaced by unbound antibiotic (359, 360). Mao and Putterman (362, 363) and Corcoran and co-workers (364–367) have studied binding of [^{14}C]erythromycin to ribosomes from *S. aureus* and *B. subtilis,* respectively. They reported binding of the antibiotic to 50S subunits. Binding required K^+ or NH_4^+, for which Na^+ could not substitute. The association constant for the ERY : ribosome complex is 3×10^5 M^{-1} (362). Displacement of [^{14}C]erythromycin from ribosomes by unlabeled derivatives correlates well with their antibacterial activity. Binding of one molecule of ERY to each sensitive ribosome occurs at 4×10^{-7} *M;* at concentrations above 10^{-5} *M* ERY multiple interactions occur, which are not confined to 50S subunits and do not require K^+ or NH_4^+ (366). Although one molecule of ERY binds to each ribosome from resistant bacteria at 10^{-5} *M* ERY and below, ERY-resistant ribosomes have less affinity for ERY than ribosomes from ERY-sensitive bacteria: association constants for the ERY:ribosome complex are 2.6×10^7 and 8×10^5 M^{-1} for ribosomes from ERY-sensitive and -resistant *B. subtilis,* respectively. At high concentration, binding of ERY to RNA, poly U, poly C, and poly A was reported. Lincomycin competes with ERY binding as do other macrolide antibiotics; only at high concentrations do the small macrolide antibiotics compete; streptomycin has a slight inhibitory effect and puromycin stimulates ERY binding (367). By using various derivatives of ERY to compete with [^{3}H]erythromycin binding to ribosomes, they were able to estimate which structural features are essential for activity (367). Thus, most compounds possessing both sugars inhibited ERY binding as well as protein synthesis. Derivatives which have altered ribose structures do not have significantly reduced activity, but derivatives with only one or none of the sugars lost most activity as measured by competition with ERY for binding to ribosomes and inhibition of cell-free polypeptide synthesis.

Teraoka (368) reported that ribosomes which do not bind ERY at 0°C can be reactivated to bind it if incubated in high NH_4^+ or K^+ concentrations at 37°C. These results suggest that a favorable conformation of the ribosomes is necessary for binding ERY.

Tanaka et al (369, 370) reported that ERY combines with 50S ribosomal subunits from ERY-sensitive *E. coli* (strain Q13), while ribosomes from an ERY-resistant mutant of this strain have little affinity for the antibiotic. They identified a protein component of 50S subunits of the mutant strain which is distinct from that of the parent Q13 strain; and suggest that this 50S protein is involved in binding ERY. Ribosomes from ERY-resistant mutants have lower activity for peptide bond synthesis in cell-free extracts (370); however, if these ribosomes are incubated in high K^+ or NH_4^+ concentrations, then ERY can be bound to them and peptidyl transferase activity increased.

Although ERY has been found to bind to many varieties of bacterial ribosomes, rat liver or rabbit reticulocyte ribosomes do not bind the antibiotic. It is thus probable that the inability of mammalian ribosomes to bind macrolides is responsible for the selective action of these antibiotics on bacterial protein synthesis (371).

Effect on polypeptide synthesis.—Erythromycin inhibits incorporation of phenylalanine into polyphenylalanine directed by poly U in a crude 30,000 $\times$ *g* supernatant extract (354, 373); 50% inhibition of polyphenylalanine synthesis was observed at approximately 10^{-6} *M* ERY. In more purified systems with ribosomes, 100,000 $\times$ *g* supernatant, and Phe-tRNA, inhibition of polyphenylalanine synthesis by ERY could not be demonstrated in *E. coli* cell-free extracts (Pestka, unpublished studies). Tanaka & Teraoka (374) have shown that ERY can inhibit polylysine formation directed by poly A in cell-free extracts of *E. coli* containing 100,000 $\times$ *g* supernatant and ribosomes; 2$\times$ 10^{-6} *M* ERY produced 90% inhibition. On the other hand, di- and trilysine formation were not inhibited although the higher polymers of lysine were. Nonenzymic binding of Lys-tRNA and Phe-tRNA to ribosomes in response to their respective templates was not influenced by ERY (358).

Ahmed (*372, 375*) has shown that spiramycin inhibits incorporation of amino acids into protein without affecting RNA synthesis in *E. coli* cells. In extracts both rate and extent of phenylalanine incorporation into polyphenylalanine directed by poly U was inhibited approximately 50% by 4 $\times$ 10^{-6} *M* spiramycin. AA-tRNA formation is unaffected. Spiramycin does not produce miscoding, premature release of peptide chains, or disaggregation of polyribosomes; nor does it interfere with poly U binding to ribosomes. Poly U-dependent binding of Phe-tRNA to ribosomes and 30S subunits of both spiramycin-sensitive and -resistant cells is inhibited by high concentrations of spiramycin (3 $\times$ 10^{-3} *M*); this effect at extremely high antibiotic concentrations therefore is unrelated to its interference with polypeptide synthesis which was inhibited at 4 $\times$ 10^{-6} *M* spiramycin in extracts from spiramycin-sensitive cells. No appreciable inhibition of Phe-tRNA binding to ribosomes was detectable at 4 $\times$ 10^{-6} *M* spiramycin. At very high macrolide concentrations, Mao & Wiegand (*376*) also reported inhibition of AA-tRNA binding to ribosomes. Another macrolide antibiotic, tylosin, produced no inhibition of Phe-tRNA binding to ribosomes (261). In addition, spiramycin inhibited proline, lysine, and phenylalanine incorporation directed by poly C, poly A, and poly U, respectively (357). Methymycin and lankamycin inhibit proline and lysine incorporation directed by poly C and poly A, but do not significantly inhibit polyphenylalanine formation directed by poly U (308).

In comparing the concentration of macrolides necessary for inhibition of growth with that for inhibition of protein synthesis in cell-free extracts, it appears that some macrolides such as ERY, oleandomycin, chalcomycin, and lankamycin produced only 50% inhibition of protein synthesis at 1000-fold the concentration necessary for inhibition of growth of *S. aureus.* On the other hand, spiramycin, tylosin, carbomycin, and niddamycin inhibited protein synthesis and growth at comparable concentrations (*376*). The discrepancy between the concentrations necessary for inhibition of growth and protein synthesis may be explained partially by the observation that Gram-positive bacteria can accumulate ERY by 100-fold or more over the external concentration (*377*). Results indicate that ERY and oleandomycin (*376*) inhibit protein synthesis best at high pH. This suggests that the active molecule of ERY is nonprotonated. Since peptidyl-tRNA or fMet-tRNA also are uncharged at their aminoacyl ends, this is consistent with inhibition of peptidyl-tRNA function by ERY.

Endogenous protein synthesis in cell-free extracts is insensitive to ERY under conditions where polylysine synthesis directed by poly A is sensitive to low concentrations of ERY (3×10^{-6} M) (377, 378). By removing peptidyl-tRNA by two methods, puromycin and dialysis of ribosomes against low magnesium concentrations, [^{14}C]erythromycin binding to ribosomes increased (378). Also, ERY had no inhibitory effect on formation of Ac-Phe-tRNA complexes with 30S subunits in the presence of poly U and initiation factors at low magnesium; nor on additional binding of Phe-tRNA to this initiation complex when 50S subunits were added. In the absence of GTP and transfer factors, the combination of Ac-Phe-tRNA and Phe-tRNA on the ribosomes primarily synthesized acetylated dipeptides; ERY had no effect on this dipeptide synthesis (378). On addition of GTP and transfer factors, there is a large increase in polymeric material initiated with the acetyl-phenylalanine. It is this material whose formation is inhibited by ERY particularly when 50S subunits are first treated with ERY. When ERY is added later, the inhibition is significantly reduced. These results suggest that peptide bond formation itself may not be sensitive to ERY; and that ERY may inhibit translocation related steps, but can do so best when it binds to 50S ribosomal subunits prior to completion of the active complex.

Mao & Robishaw (379) have examined the effect of various macrolides on the extent and kinetics of lysine peptide synthesis. It appears that ERY stimulates rate and extent of di- and trilysine synthesis, but inhibits formation of the larger lysine peptides stimulated by poly A templates (379; also 380). Spiramycin inhibits the rate, but ultimately stimulates extent of formation of dilysine; formation of trilysine and larger lysine peptides is markedly reduced by spiramycin. Niddamycin inhibits rate and extent of formation of all lysine peptides. In these studies the macrolides appear to divide into three groups: ERY and oleandomycin; spiramycin and tylosin; and niddamycin and carbomycin. In examining dilysine synthesis in the absence of translocation, it was observed that ERY has either no effect or is slightly stimulatory to dilysine synthesis; niddamycin strongly inhibits; spiramycin produces small inhibition of dilysine synthesis. Similar studies of the effect of antibiotics on lysine peptide synthesis were carried out by Pulkrabek, Cerna & Rychlik (381). Their results, slightly at variance with those of Mao & Robishaw (379), also permitted division of antibiotics into three groups according to their effects on lysine peptide synthesis: antibiotics which inhibit the first peptide bond include neomycin and spiramycin; antibiotics which do not inhibit dilysine synthesis, but do prevent trilysine synthesis are carbomycin and chlortetracycline; antibiotics which do not inhibit small peptides, but interfere with synthesis of large peptides include ERY, oleandomycin, and chloramphenicol. Perhaps differences between the studies may relate to differences in antibiotic potency or purity. They did not observe inhibition of fMet-tRNA binding to ribosomes by the macrolides. Macrolides showed no significant effect on G-dependent GTP hydrolysis, formation of [^{3}H]GTP : factor T : Lys-tRNA complex, T-dependent GTP hydrolysis, and Phe-tRNA binding to ribosomes (261, 379). Factor G can reduce the binding of ERY to ribosomes; and factor G and peptidyl-tRNA may compete with ERY for binding to ribosomes (379). They did not, however, detect any inhibition of ERY binding to ribosomes by polylysyl-tRNA.

Effects on peptide bond formation and translocation.—Peptide bond formation can be inhibited by ERY; however, inhibition is markedly dependent on the assay for transpeptidation. Thus, 10^{-6} *M* ERY can inhibit release of polylysine from polysyl-tRNA by puromycin (69) and prevent the addition of a single lysine unit to polylysine-tRNA (343). Yet, even higher concentrations of ERY do not inhibit acetyl-phenylalanyl-puromycin synthesis (69). In contrast, spiramycin, tylosin, carbomycin, and niddamycin are inhibitory to peptide bond formation in most of the assays (69, 192, 249, 379); however, in the case of tylosin and spiramycin, inhibition of peptide bond formation reached a plateau of 50–75% at concentrations of antibiotics greater than 10^{-5} *M* (192, 249). It is possible that peptide bond formation may not be the primary site of action of spiramycin and tylosin, for under certain conditions they stimulate dilysine and diphenylalanine formation (261, 379).

The action of niddamycin and carbomycin differs from that of ERY, oleandomycin, spiramycin, and tylosin. Niddamycin and carbomycin are strong inhibitors of peptide bond formation in most assays for transpeptidation (69, 379), and inhibit dipeptide synthesis under all conditions during complete protein synthesis. Carbomycin inhibits the puromycin reaction with fMet-hexanucleotide (192), with polylysyl-tRNA and with Ac-Phe-tRNA (69). However, in investigations of the effect of carbomycin, spiramycin, tylosin, and ERY on peptidyl-[^{3}H] puromycin formation by *E. coli* extracts containing active polyribosomes, neither carbomycin (10^{-4} *M*), tylosin (10^{-4} *M*), or spiramycin (10^{-4} *M*) interfered with the rate of transfer of polypeptides from polysomes to [^{3}H] puromycin (477). ERY, however, inhibited the rate of this transpeptidation about 75% (477). Such extracts containing active polyribosomes probably more nearly represent conditions existing in intact cells than the usual assays for peptide bond synthesis with ribosomes washed in high salt. It is, therefore, apparent that, although ERY can inhibit such peptidyl-puromycin formation, carbomycin, spiramycin, and tylosin cannot. In whole cells these latter macrolides may inhibit initiation of polypeptide chains.

Cerna & Rychlik (382) have examined release of polylysine from polylysyl-tRNA by puromycin using ribosomes obtained from ERY-sensitive and -resistant *E. coli* B. Ribosomes from the resistant strain were resistant to inhibition of the puromycin reaction by ERY and, in addition, were also more resistant to inhibition by spiramycin and chloramphenicol. It should be noted, however, that ribosomes from the ERY-resistant strain were less active in stimulating polylysyl-puromycin formation than those from the sensitive strain. The results of Teraoka, Tamaki & Tanaka (383) have also indicated that ribosomes obtained from *E. coli* strains resistant to ERY have reduced ERY binding ability as well as lower peptidyl transfer activity. Ribosomes from the ERY-resistant strain could be activated by high K^+ so that they were as active as ribosomes from the parent strain.

Cerna, Rychlik & Pulkrabek (69) examined the effect of ERY, spiramycin, and carbomycin on polylysyl- and Ac-Phe-tRNA binding to ribosomes. They observed no effect of these macrolides on polylysyl-tRNA binding to ribosomes. ERY had no effect on Ac-Phe-tRNA binding to ribosomes (69, 259); however, spiramycin (10^{-3} *M*) and carbomycin (10^{-3} *M*) produced 76% and 47% stimulation of the binding of Ac-Phe-tRNA to ribosomes. All macrolide antibiotics examined (oleandomycin, carbomycin, ERY, and spiramycin) were markedly inhibitory to release of polylysine from polylysyl-tRNA by puromycin: at 10^{-7} *M* inhibition of this reaction was evident and was complete for most macrolides at 10^{-6} *M*. In contrast, transfer of Ac-Phe from Ac-Phe-tRNA to puromycin is inhibited by spiramycin and carbomycin at tenfold higher concentrations than required to inhibit transfer from polylysyl-tRNA; however, whereas 5×10^{-6} *M* ERY completely inhibited transfer of peptides from polylysyl-tRNA, it stimulated the puromycin reaction with Ac-Phe-tRNA even at concentrations 100 times higher.

Monro & Vazquez (192) have examined the effect of a number of macrolides on fMet-puromycin formation, using 50S subunits, C-A-A-C-C-A(f-Met), and puromycin. Several macrolides (angolamycin, carbomycin, spiramycin III, and lankamycin) inhibited the fragment reaction. Carbomycin was most inhibitory: 10^{-3} *M* angolamycin, spiramycin III, and lankamycin produced 48%, 72%, and 25% inhibition, respectively; carbomycin at 10^{-4} *M* produced 99% inhibition. In addition, 10^{-4} *M* spiramycin and carbomycin inhibited binding of C-A-C-C-A(Ac-Leu) to 50S subunits 88% (262). In contrast, 10^{-3} *M* oleandomycin and ERY each produced 25% and 56% stimulation of this binding, respectively. ERY does not inhibit C-A-A-C-C-A(Phe) binding to ribosomes (193, 240); but tylosin, spiramycin III, and carbomycin inhibit the binding (241). Thus, some macrolides can inhibit binding of the aminoacyl-oligonucleotide terminus of AA-tRNA to ribosomes although ERY and the smaller macrolides cannot. Inhibition of C-A-C-C-A(Phe) binding to ribosomes by antibiotics correlates well with their effects on transpeptidation as measured by fMet- or Ac-Phe-puromycin formation. The binding of these fragments to ribosomes is simply a partial reaction of transpeptidation.

Erythromycin inhibited puromycin release of nascent peptides in protoplasts of *B. megaterium;* however, in the presence of chlortetracycline, ERY permitted much release (154). Assuming that chlortetracycline inhibits AA-tRNA binding to the ribosomal A-site, peptidyl-tRNA then would remain in the P-site in the presence of chlortetracyline; since ERY did not inhibit release when peptidyl-tRNA resided in the P-site, they suggested that ERY inhibits translocation: specifically, ERY prevents peptidyl-tRNA from entering the P-site consequently inhibiting the puromycin reaction.

Also ERY (10^{-5} M) inhibited release of $tRNA^{Phe}$ dependent on translocation 50% (71).

Polysome metabolism.—Cundliffe (384) examined the effect of antibiotics on polyribosome metabolism in *B. megaterium.* Lincomycin, streptogramin A, and spiramycin III, each at a concentration of 5×10^{-4} M, inhibited protein synthesis in protoplasts rapidly and essentially completely. Under conditions in which protein synthesis was inhibited, breakdown of polyribosomes rapidly occurred into 30S and 50S subunits, not as 70S monomers. Breakdown was rapid so that by 40 sec after addition of antibiotic, 60% of the polyribosomes were converted to subunits. In control experiments, actinomycin D led to complete breakdown of polyribosomes into 30S and 50S subunits. Thus, the residual polyribosome-like material present after breakdown with these antibiotics may represent inactive multi-initiation complexes. These antibiotics may cause premature detachment of ribosomes from mRNA or, alternatively, they may selectively inhibit initiation. In the presence of ERY, slow subunit ribosomal exchange is seen (3). Extensive studies of polyribosome metabolism with macrolides have not been reported.

Inducible resistance and other effects.—When cells are exposed to a subinhibitory concentration of ERY (10^{-8} to 10^{-7} M) in an otherwise complete medium for growth, cells become resistant to macrolides, to lincosamides, and to streptogramin-type B antibiotics (356, 385, 386). Concentrations substantially greater than 10^{-7} M block induction due to inhibition of protein synthesis. Induced cells begin to appear early and by about one generation time nearly the entire culture is resistant. If ERY is removed, the induced cells revert to sensitivity within one to two generations. The process of induction is inhibited by chloramphenicol and inhibitors of RNA synthesis. The expression of induced resistance involves alteration of ribosomal function and structure so that the ribosome binds various antibiotics with decreased affinity. Thus, cells which are induced to macrolide resistance have ribosomes with much less affinity for spiramycin than comparable ribosomes obtained from uninduced sensitive cells (356). The mechanism by which induced resistance occurs is not clear.

At slightly inhibitory concentrations, several reversible inhibitors of the ribosome, such as chloramphenicol, tetracycline, ERY, and spectinomycin can antagonize the inhibitory effect of low concentrations of streptomycin on the growth of Sm-sensitive strains of *E. coli* (196). In addition, these compounds can also replace Sm as phenotypic suppressors, supporting the growth of conditional Sm-dependent mutants. Thus, ERY, which is a 50S inhibitor can apparently alter the Sm binding site, producing both a decrease in affinity for Sm and a stimulation of misreading. Presumably, ERY can stimulate substantial conformational changes throughout the ribosome.

Although the macrolide antibiotics are similar in structures and compete for a single major site on the 50S unit, the individual antibiotics have distinct actions. It is likely that the spectrum of effects they manifest represents different interactions of each antibiotic with this site. It is quite probable that some larger macrolides (carbomycin and niddamycin) inhibit peptide bond formation by both inhibiting binding of acceptor as well as donor substrates. ERY and oleandomycin also bind to this same center, but do not prevent the acceptor substrates from attaching to the ribosome (193, 240, 262) and probably have no effect on the small donor substrates such as fMet- or Ac-Phe-tRNA. However, when the donor substrate is large (such as polylysyl-tRNA), ERY does interfere with proper situation of peptidyl-tRNA at which time ERY becomes a potent inhibitor of protein synthesis (154, 379). Accordingly, small macrolides such as methymycin produce less interference with protein synthesis. With such a model most of the enigmatic observations can be explained, particularly the differential sensitivity of the first, second, and later peptide bonds to inhibition by macrolides. Since macrolides bind to isolated 50S subunits, it is possible that they may be effective inhibitors of initiation as well. It should be mentioned that the precise obstruction produced by a macrolide is not merely a function of its size but is related to its chemical structure. In addition, although it is not necessary to invoke allosteric changes, these probably also occur (196).

Since ERY inhibits proper positioning of large peptidyl-donors, it becomes a matter of semantics to give this inhibitory effect a term, for it is equally an inhibitor of transpeptidation and translocation.

Summary.—Macrolides bind to a common site on 50S subunits. In relation to their size and structure, they produce conformational changes in the ribosome as well as direct effects on adjacent areas. Thus, some macrolides (carbomycin, niddamycin) inhibit functional attachment of both donor and acceptance substrates to ribosomes (and are strong inhibitors of transpeptidation); spiramycin III and tylosin inhibit binding to both donor and acceptor sites to a smaller degree (and are moderate inhibitors of transpeptidation); erythromycin and oleandomycin produce little or no effect on binding of the acceptor substrates and interfere with binding of the donor substrates only when these contain large peptide moieties (thus are good inhibitors of transpeptidation only with large peptidyl-donors). In whole cells, however, it is likely that many macrolides (carbomycin, tylosin, and spiramycin, for example) cannot inhibit transpeptidation on polyribosomes, although erythromycin can through direct action or an indirect effect on translocation. Direct effects of the macrolides on initiation through interaction with free 50S subunits have not been excluded as a major mode of action.

Lincomycin

The structure of lincomycin is given in Figure 2 (387). Josten & Allen (388) indicated that lincomycin inhibits protein synthesis in Gram-positive, but not Gram-negative bacteria, whereas there was little interference with

DNA or RNA synthesis. Lincomycin has little or no effect on protein synthesis in reticulocyte extracts (389). The antibacterial properties of chemically modified lincomycin indicate that activity is dependent on substitution of the ring nitrogen for the N-dimethyl-lincomycin is only 10% as active as lincomycin, whereas N-ethyl-lincomycin is as active as lincomycin (390). In addition, substitution of the sulfur is also essential for S-ethyl-lincomycin is 130% as active as lincomycin while S-dimethyl-lincomycin is inactive.

Studies by Chang, Sih & Weisblum (391) indicated that cell-free extracts and ribosomes from *E. coli* appear to be resistant to inhibition by lincomycin at concentrations of 10^{-4} *M*. In contrast, phenylalanine incorporation directed by poly U was sensitive to these concentrations of lincomycin in extracts from *B. stearothermophilus*. In experiments using reconstituted ribosomes with mixed subunits from *E. coli* and *B. stearothermophilus,* they concluded that the 50S subunit contained the lincomycin-sensitive site. Lincomycin did not interfere with poly U binding to ribosomes.

Although they reported that lincomycin inhibited Phe-tRNA binding to *B. stearothermophilus* ribosomes in response to poly U, inhibition of diphenylalanine formation was not excluded. [^{14}C]Lincomycin binds to 70S and 50S particles of *B. stearothermophilus,* but not 30S subunits (392); binding requires K^+ or NH_4^+ and Mg^{++}. Binding to 70S ribosomes of *E. coli* was not demonstrated. On 100-fold dilution of [^{14}C]lincomycin: ribosome complexes, there is rapid dissociation of about half the complexes within several minutes. Unlabeled lincomycin and erythromycin can displace [^{14}C]lincomycin from its association with ribosomes; however, on additcon of a 15-fold concentration of unlabeled lincomycin or erythromycin, only slightly more than half of the complexes are dissociated. It is therefore possible that lincomycin is bound irreversibly to about half of the lincomycin: ribosome complexes. In erythromycin-resistant *S. aureus,* Griffith (393) reported that erythromycin can act as an antagonist of lincomycin. It may be possible to account for the antagonism observed in intact organisms by their mutual antagonism in binding to ribosomes. Vazquez (264) reported that lincomycin inhibited binding of [^{14}C]chloramphenicol to ribosomes of *B. megaterium.* He also showed that lincomycin (10^{-4} *M*) inhibits poly A-directed lysine and poly C-directed proline incorporation in cell-free *E. coli* extracts. Thus, lincomycin can interact with *E. coli* ribosomes.

Lincomycin inhibits N-formyl-methionyl-puromycin formation from N-formyl-methionyl-tRNA and puromycin under conditions in which erythromycin was not inhibitory (394). At low concentrations of erythromycin (from 10^{-7} to 10^{-5} *M*) the inhibition of this reaction by lincomycin could be abolished. Lincomycin and erythromycin both inhibit polylysine synthesis directed by poly A; in the presence of 10^{-5} *M* lincomycin, increasing concentrations of erythromycin did not appear to further increase the inhibition of polylysine synthesis, suggesting that the antibiotic effects are not additive. Erythromycin, lincomycin, and chloramphenicol all inhibited polylysine incorporation dependent on poly A in cell-free *E. coli* systems (380).

Lincomycin and chloramphenicol required much higher concentrations than ERY for inhibition; ERY appeared to be active at a concentration of 1 molecule per ribosome. ERY thus has a higher affinity than chloramphenicol or lincomycin for binding to *E. coli* ribosomes. Chloramphenicol and lincomycin do not readily interfere with [^{14}C]erythromycin binding to ribosomes (358, 380). Erythromycin, lincomycin, and chloramphenicol seem to bind to ribosomes in a mutually exclusive way.

Lincomycin can effectively inhibit the formation of fMet-puromycin from fMet-hexanucleotide and puromycin (192): at 10^{-4} *M* lincomycin there was 82% inhibition of the reaction. Lincomycin also inhibits binding of C-A-C-C-A(Ac-Leu) to 50S subunits (262). Lincomycin inhibited C-A-C-C-A(Phe) binding to *E. coli* ribosomes (193). These results suggest localization of lincomycin action, at least in part, to inhibition of binding of the aminoacyl end of AA-tRNA to ribosomes. The mechanism of this inhibition may be direct or indirect by producing a conformational change in the ribosome. Lincomycin inhibits peptide bond formation as assayed by the puromycin reaction with Phe-tRNA, but does not inhibit the release of deacylated tRNA from the ribosome (71). Lincomycin therefore inhibits peptide bond formation without interfering with this step of translocation. In comparing fMet-puromycin formation from fMet-tRNA with fMet release by termination factors it was noted (194) that lincomycin inhibited both peptide bond synthesis and termination comparably.

Krembel and Apirion (395, 396) have isolated erythromycin-lincomycin mutants which are more sensitive or resistant to these antibiotics than the wild-type *E. coli*. By comparison of acrylamide gel electrophoresis patterns isolated from 50S subunits of wild type and mutant strains, they have detected at least one altered protein in these mutant strains.

Cundliffe (384) showed that lincomycin (as well as streptogramin A and spiramycin) rapidly caused redistribution of ribosomes or polyribosomes into 30S and 50S subunits. Polyribosome breakdown was not complete so that 20–30% of the polysomes still remained. These results indicate that lincomycin either destabilizes ribosomes so that they fall off mRNA or prevents initiation. Pestka (477) reported that lincomycin does not interfere with transpeptidation (peptidyl-puromycin formation) in extracts containing active polyribosomes. Thus, it is possible that in whole cells lincomycin may inhibit initiation. The effect of lincomycin on initiation reactions has not been reported.

Summary.—Lincomycin inhibits peptide bond synthesis by binding to 50S subunits. With washed ribosomes the bound lincomycin interferes with proper positioning of the acceptor (and possibly donor) substrates, namely, peptidyl- and AA-tRNA, so that peptide bond synthesis is interdicted. In the intact cell, however, lincomycin probably does not interfere with transpeptidation. The breakdown of polysomes by lincomycin indicates that ribosomes with bound lincomycin cannot maintain attachment to mRNA and/or that lincomycin inhibits initiation.

Streptogramin Group

Almost all of the streptogramin antibiotics are produced as a complex of at least two antibiotics which are highly synergistic and fall into two structural classes. Each component alone can inhibit bacterial growth, but together they are markedly synergistic (397). Some members of the group are streptogramins, ostreogrycins, vernamycins, synergistins (PA114), and mikamycins; each of these antibiotic complexes contains at least one A and one B group compound. The probable structure of ostreogrycin A (Fig. 5) has been reported by Delpierre et al (398). From physical and chemical data it appears that all the antibiotics of the A group, nonpeptide macrolides, are probably very similar in structure although their complete structures have not yet been determined (399). All the compounds in the B group contain a macrocyclic lactone ring consisting of a cyclic peptide of 6 or 7 amino acids closed through the hydroxyl of threonine. In general, the antibiotics are inhibitory to the growth of Gram-positive, but not Gram-negative organisms (397, 417). Separately and in combination, this group of antibiotics inhibits protein synthesis in intact Gram-positive bacteria and in cell-free extracts from Gram-positive or Gram-negative bacteria (400–411). The natural resistance of Gram-negative bacteria to the antibiotic is due to the inability of the antibiotic to enter these cells (412). It has been shown that these antibiotics do not influence tRNA acylation or mRNA binding to ribosomes (308, 400, 412–414).

Streptogramin A group.—Streptogramin A antibiotics seem to inhibit slightly AA-tRNA binding to ribosomes in response to templates (261, 411, 413). They do not inhibit poly U binding to ribosomes or produce miscoding (410). They are strong inhibitors of protein synthesis in cell-free extracts dependent on synthetic polynucleotide templates; endogenous polypeptide synthesis is more resistant to inhibition by the streptogramin A compounds (410). In fact, there is less inhibition of protein synthesis if these antibiotics are added after protein synthesis has begun than if the antibiotic was present initially (411, 413). It was, therefore, suggested that the presence of peptidyl-tRNA inhibits binding of these antibiotics to ribosomes.

Inhibition of protein synthesis by streptogramin A group antibiotics was localized to the 50S subunit. Ennis (411, 412) and Vazquez (415) showed that 50S, but not 30S, subunits once treated with PA114A and streptogramin A, respectively, were inactive in protein synthesis even after extensive dialysis. [^{3}H]Vernamycin A binds to 50S and 70S particles, but not to 30S subunits (416); the larger macrolide antibiotics such as erythromycin, carbomycin, and spiramycin III inhibit binding of vernamycin A to ribosomes, but smaller macrolides such as methymycin, lankamycin, or oleandomycin do not. No other 50S inhibitors interfered with vernamycin A binding to ribosomes: thus, chloramphenicol, lincomycin, vernamycin Bα, amicetin, sparsomycin, puromycin, and thiostrepton did not. Inhibitors of the 30S subunit were also ineffective in inhibiting vernamycin A binding. Binding of streptogamin A antibiotics to ribosomes requires K^+ or NH_4^+ (412, 415, 416). Although chloramphenicol cannot prevent vernamycin A binding, ostreogrycin A and streptogramin A inhibit binding of chloramphenicol to ribosomes (16, 264, 303). The finding that some macrolides can antagonize vernamycin A binding to ribosomes may indicate a similar or overlapping site for binding to ribosomes and may account for the observation of cross-resistance between macrolide antibiotics (erythromycin and oleandomycin) and the streptogramin group (409, 416, 444). The binding of streptogramin A antibiotics to ribosomes is virtually irreversible as long as K^+ is present in the medium. The dissociation constant for the vernamycin A: ribosome complex is 1×10^{-8} *M* (416).

Examination of peptide bond formation dissociated from other steps of protein synthesis has indicated that streptogramin A antibiotics can inhibit transpeptidation effectively. Thus, polylysyl-puromycin (412), fMet-puromycin (192), and acetyl-phenylalanyl-puromycin (249) are inhibited by group A compounds at low concentrations. Streptogramin A antibiotics are also good inhibitors of binding of C-A-C-C-A(Phe) (193, 240) and of C-A-C-C-A(Ac-Leu) to ribosomes (262). The streptogramin A antibiotics are thus very effective inhibitors of transpeptidation with isolated ribosomes, for they inhibit substrate binding for peptidyl-transfer (see discussion under other 50S inhibitors above). However, similar to some of the macrolides (carbomycin, spiramycin, and tylosin) and lincomycin, streptogramin A antibiotics do not prevent peptidyl-puromycin formation in extracts containing active polyribosomes (477). Therefore, in the intact cell these antibiotics probably do not interfere with transpeptidation.

Polyribosomes rapidly disappear in cells treated with streptogramin A antibiotics, with concomitant appearance of 30S and 50S subunits (384) or 70S ribosomes (412). Thus, these antibiotics either destabilize ribosomes from mRNA or inhibit initiation of protein synthesis (see similar discussion under macrolides). They may prevent normal functional attachment of free 50S subunits to fMet-tRNA :30S initiation complexes. Streptogramin A antibiotics effectively inhibit fMet-tRNA binding to 70S ribosomes so that it is possible that loss of polysomes is due to inhibition of initiation or of peptide bond synthesis shortly thereafter (H. Ennis, personal communication). Also, once protein synthesis has commenced ribosomal sites for binding streptogramin A antibiotics may no longer be available or have a reduced affinity for these antibiotics. Perhaps, then, only free 50S subunits or 70S ribosomes readily bind these antibiotics.

Streptogramin B group.—In contrast to A group compounds, inhibition of protein synthesis by streptogramin B antibiotics was greater the longer protein synthesis was permitted to proceed prior to addition of the antibiotic (413). This group of antibiotics may stimulate AA-tRNA binding to ribosomes (17). They do not inhibit transpeptidation, but stimulate acetyl-phenylalanyl-puromycin formation (249).

In addition, PA114B stimulates the binding of C-A-C-C-A(Phe) to ribosomes (193, 240). Viridogrisein (another group B antibiotic) stimulates the binding of C-A-C-C-A(Ac-Leu) to ribosomes (262). Also streptogramin B compounds enhance the binding of PA114A and [^{3}H]vernamycin A to ribosomes and prevent the dissociation of ribosome-bound PA114A from ribosomes in the absence of K^+ or NH_4^+ (412, 416). In this regard, it would be useful to determine if the streptogramin B antibiotics could enhance the binding of erythromycin and other macrolides (as well as lincomycin) to ribosomes since the macrolides structurally resemble the streptogramin A group of antibiotics. However, streptogramin B compounds have been reported to inhibit spiramycin (357) binding to *E. coli* ribosomes and chloramphenicol (265) binding to *B. stearothermophilus* ribosomes. There remain some discrepancies in the literature, for it has also been reported in later studies that streptogramin B compounds do not inhibit chloramphenicol binding to ribosomes from *B. megaterium* (264).

As do streptogramin A compounds, PA114B produces loss of polysomes in protoplasts of *B. subtilis* (412). They may thus destabilize ribosomes on mRNA or inhibit initiation or other early events. However, there are no reports of the influence of these antibiotics on initiation. Since streptogramin B antibiotics stimulate binding of AA-tRNA, C-A-C-C-A(Phe) and C-A-C-C-A(Ac-Leu) to ribosomes, these antibiotics may interfere with movement of tRNAs during translocational events by fixing peptidyl- or AA-tRNA attachment to functional sites for peptidyl transfer.

Summary.—Streptogramin A antibiotics bind tightly to 50S subunits; their binding is enhanced by streptogramin B antibiotics. With isolated ribosomes streptogramin A compounds inhibit transpeptidation through inhibition of binding of the aminoacyl end of AA-tRNA to ribosomes, but they do not inhibit protein synthesis readily after protein synthesis has well begun nor do they inhibit peptidyl-puromycin synthesis on polyribosomes. They produce a rapid decrease in number of polysomes with concomitant increase in number of free subunits and ribosomes. It is likely they inhibit initiation or early events of protein synthesis. Streptogramin B antibiotics may inhibit protein synthesis by fixing peptidyl- or AA-tRNA ends onto 50S subunits, thus interfering with the necessary coordinated movements of these moieties on the ribosome.

Thiostrepton Group

To date many compounds of this group probably have not been identified for the complete structures of most are unkown and relatively few studies of their biological characteristics have been reported. They consist of sulfur-containing cyclic polypeptides. The structure of thiostrepton has been reported and is given in Figure 5 (418). Compounds in this group include thiostrepton (bryamycin, thiactin), siomycin A (A-59, sporangiomycin), micrococcin, and perhaps other sulfur-containing cyclic polypeptides such as multhiomycin and althiomycin (419–423). In general, these antibiotics inhibit growth of Gram-positive, but not Gram-negative bacteria. Protein synthesis in intact Gram-positive (424–426), and in extracts from Gram-positive and -negative microorganisms (424–428) is effectively inhibited by these antibiotics. They do not inhibit protein synthesis in cell-free extracts from mammalian cells (209, 424).

The results of Weisblum & DeMohn (425) and Tanaka et al (426) showed that thiostrepton and siomycin A, respectively, inhibit protein synthesis through inhibition of 50S subunit function. The antibiotic binds very tightly to 50S subunits, for subunits which have been exposed to thiostrepton and subsequently dialyzed remain inactive (425). In addition, inhibition of protein synthesis by these antibiotics can be reversed by excess ribosomes (425–428), and particularly 50S subunits (425, 427).

These antibiotics inhibit translocation. Enhancement of acetyl-phenylalanyl-puromycin synthesis by G factor and GTP was suppressed by siomycin A (426). Oligophenylalanine synthesis dependent on G factor and GTP was prevented by thiostrepton (427). In addition, thiostrepton prevented G factor and GTP-independent oligophenylalanine synthesis; this result indicated that a ribosomal function of translocation was inhibited by this antibiotic (427). Furthermore, G factor-dependent hydrolysis of GTP was also inhibited by thiostrepton (427, 429) and by siomycin A (430). In the presence of thiostrepton and siomycin A, GDP or GTP:G factor:ribosomes complexes are very much reduced in the presence or absence of fusidic acid (429–431). This indicates that these two antibiotics may inhibit attachment of G to ribosomes or may prevent ribosome-bound factor G from binding nucleotides. These two antibiotics completely inhibit protein synthesis and translocational events at 10^{-6} *M* (425–427, 430, 431). At 100-fold higher concentrations they are ineffective in inhibiting AA-tRNA binding to ribosomes or transpeptidation as measured by acetyl-phenylalanyl-puromycin formation or similar assays (426, 427, 430, 431). Thiostrepton did not inhibit vernamycin A binding to ribosomes (416) nor did siomycin A prevent chloramphenicol, erythromycin, or lincomycin binding to ribosomes (430).

Micrococcin and micrococcin P, like thiostrepton and siomycin A, are potent inhibitors of protein synthesis and of translocation both dependent and independent of G factor; also, like thiostrepton and siomycin A, they do not inhibit peptide bond formation nor AA-tRNA binding to ribosomes; however, whereas they are good inhibitors of translocation, they do not inhibit factor G-ribosome-dependent hydrolysis of GTP nor binding of factor G:GTP complexes to ribosomes (431). Thiostrepton, siomycin A, micrococcin and micrococcin P stabilize polyribosomes and inhibit puromycin breakdown of polyribosomes (431). Effects of these antibiotics on initiation, termination, and many other reactions related to protein synthesis have not been reported. Fujimoto et al (432) reported that althiomycin inhibits protein synthesis in intact cells and in cell-free extracts of *E. coli* B; that althiomycin inhibits acetyl-phenylalanyl-puromycin synthesis. Althiomycin does not appear to be a specific inhibitor of translocation (431), but inhibits transpeptidation.

Since the antibiotics of the thiostrepton group do not interfere with peptide bond synthesis, but strongly interfere with translocation through an effect on a 50S subunit function, it is likely they will be useful in genetic and biochemical analyses of 50S subunit proteins. Presumably, most of the antibiotics of this group will have close genetic linkage and, very probably, antibiotic cross-resistance patterns. In *B. subtilis,* thiostrepton and micrococcin markers are closely linked to each other and to other markers for antibiotics which affect ribosomal function (433). These antibiotics are highly insoluble in aqueous media and thus have found little therapeutic use clinically in humans.

Summary.—Thiostrepton and siomycin A (and perhaps micrococcin and micrococcin P) bind to 50S ribosomal subunits. All are strong inhibitors of translocation through inhibition of ribosomal functions which are an integral part of the translocational events. Thus, they all stabilize polysomes. Both thiostrepton and siomycin A, but not micrococcin P and micrococcin, inhibit ribosome-factor G-dependent GTP hydrolysis as well as factor G: GTP complexes with ribosomes; micrococcin P and micrococcin thus permit GTP hydrolysis as they uncouple this hydrolysis from translocation. None of these antibiotics inhibits transpeptidation, but thiostrepton and siomycin A inhibit Phe-tRNA binding to ribosomes.

Cycloheximide and Other Glutarimide Antibiotics

Many glutarimide antibiotics have been isolated among which are cycloheximide, acetoxycycloheximide, isocycloheximide, streptimidone, inactone, and the streptovitacins (434, 435). From the large number of active and inactive glutarimide antibiotics isolated and synthesized, some basic structure-activity relationships have been established (Fig. 5). The ketone-carbonyl, the hydroxyl, or the imide-nitrogen groups cannot be removed or altered without substantial loss of activity (434, 436). Cycloheximide (actidione) has been the best well-studied of these antibiotics so that this discussion will primarily concern it; however, the general conclusions apply to most of these active glutarimide antibiotics. Glutarimide antibiotics are active against eukaryotic organisms and essentially inactive in inhibiting intact and cell-free extracts from prokaryotes. Although cycloheximide has been shown to inhibit DNA synthesis in intact cells, DNA formation in cell-free extracts is unaffected by it and effects of cycloheximide on RNA synthesis and other metabolic processes are probably secondary to its effect on protein synthesis (411, 434–438).

Cycloheximide has been reported by many groups to inhibit protein synthesis in yeasts (439, 440) and mammalian cells (411, 437, 438, 441–443) as well as in cell-free extracts from eukaryotes (411, 437, 438, 444–446). Cycloheximide does not interfere with amino acid activation or transfer to tRNA (447). Acetoxycycloheximide and streptovitacin A had a smaller effect on inhibition of protein synthesis in intact cells than cycloheximide, but had a significantly greater effect than cycloheximide in cell-free extracts (434, 435, 438).

Grollman (448) has suggested a structural analogy between the ipecac alkaloids and the glutarimide antibiotics, which is the basis for their inhibition of protein biosynthesis. The alkaloid tubulosine appears to contain the active portion of the molecule of the glutarimide antibiotics as well as possibly the ipecac alkaloids as pointed out by Grollman (449). In contrast to cycloheximide, however, inhibition could not be reversed by washing and resuspending the cells in fresh medium in the absence of tubulosine. Cell-free extracts from eukaryotic organisms were less sensitive to inhibition by tubulosine than intact cells, and *E. coli* extracts were unaffected. Like glutarimide antibiotics (436, 438, 449) tubulosine action has the following characteristics: (*a*) it is active against most eukaryotic organisms, but inactive against *E. coli;* (*b*) it is structurally specific; (*c*) it is selective on protein synthesis.

Hemoglobin synthesis by rabbit reticulocytes was inhibited 96% 30 sec after the addition of 1.4×10^{-5} *M* cycloheximide (450). Protein synthesis by cell-free extracts from rabbit reticulocytes was also inhibited by cycloheximide: 2×10^{-5} *M* cycloheximide produced a 60–70% inhibition of protein synthesis. It is noteworthy that with rabbit reticulocytes as well as some of the studies with yeasts, intact cells are more sensitive to cycloheximide than cell-free extracts.

Cycloheximide did not influence the polyribosome pattern as compared to that found in reticulocytes not treated with the antibiotic, and nascent protein chains were not removed from polyribosomes (449). After NaF disaggregation of polyribosomes, cycloheximide prevented reformation of polysomes after the removal of NaF. In addition, in the presence of cycloheximide the rate of removal of the nascent chains from polyribosomes was substantially retarded. Cycloheximide and streptovitacin A, which are structurally related, have a similar mechanism of action (147). However, inhibition by cycloheximide is reversible while that produced by streptovitacin A is irreversible in intact reticulocytes. Breakdown and reassembly of polysomes is inhibited by cycloheximide and streptovitacin A (147, 148, 449).

Using yeast extracts, Siegel & Sisler (447) showed that poly U-directed phenylalanine incorporation was as sentitive to cycloheximide as endogenous amino acid incorporation in the absence of added templates. Extreme differences in sensitivity to cycloheximide have been observed in *Saccharomyces* species. For example, *S. fragilis* can grow in a cycloheximide concentration of 1000 μg/ml, whereas *S. pastorianus* does not grow at all in 0.17 μg/ml of cycloheximide (451). Thus, by using ribosomes and supernatant from both these species, Siegel & Sisler (452) found that inhibition of cell-free extracts by cycloheximide was determined by the source of the ribosomes, not by the source of the supernatant enzymes. Thus, these antibiotics act at ribosomal sites.

Strains of *Saccharomyces cerevisiae* carrying a recessive resistance gene yielded extracts resistant to cycloheximide (453). Resistance was clearly associated with the ribosomes. The heterozygous diploid carrying the resistance marker and its wild-type allele, while phenotypically sensitive, yielded ribosomes with an intermediate resistance in vitro. Thus, the diploid cells apparently contain a mixed complement of both sensitive and resistant ribosomes which give rise to mixed polysomes. With both resistant and sensitive ribosomes using the same mRNA simultaneously, movement of resistant ribosomes along mRNA may be blocked by sensitive ribosomes whose movement is inhibited by the antibiotic (453). In addition, with the use of cycloheximide-resistant and -sensitive strains of *Saccharomyces,* Rao & Grollman (454) indicated that resistance and sensitivity to cycloheximide are properties associated with 60S ribosomal subunits.

Several groups have examined the in vitro and in vivo incorporation of amino acids into mitochondria and mitochondrial protein synthesis products. It was shown that amino acid incorporation by rat liver mitochondria in vitro was insensitive to cycloheximide concentrations to 10^{-3} *M* (455–457). In addition, in vivo cycloheximide-resistant amino acid incorporation into mitochondrial proteins has also been demonstrated (458, 459).

Jondorf and co-workers (460, 461) have described an in vivo effect of cycloheximide and ipecac alkaloids: the administration of cycloheximide to rats increases the amino acid incorporation in vitro in a rat liver microsomal system. A single injection of cycloheximide stimulates the in vitro system within the first hour. It is possible that the stabilization of polysomes by these antibiotics may produce fractions which are more active in protein synthesis.

Cycloheximide prevents GTP-dependent polyribosome breakdown which is coupled to protein synthesis (446). Polyribosome breakdown was inhibited to approximately the same extent as was protein synthesis. Breakdown of liver polyribosomes following the administration of actinomycin or ethionine was inhibited by injection of cycloheximide; and reassembly of polyribosomes produced by reversing the effect of ethionine was inhibited by cycloheximide (462). The incorporation capacity of ribosomal preparations from normal animals was inhibited when the supernatant used in the assay for protein synthesis was obtained from cycloheximide-injected animals. A similar effect was seen by Fellicetti et al (148) using intact reticulocytes and extracts prepared from reticulocytes exposed to streptovitacin A.

Godchaux, Adamson & Herbert (463) reported that 4×10^{-6} *M* cycloheximide partially inhibited protein synthesis in reticulocytes, and caused a slight increase in the polyribosome content of cells. When reticulocytes were incubated with NaF or without added iron salts, polyribosomes were converted to single ribosomes. This conversion to single ribosomes could be prevented by the addition of 4×10^{-6} *M* cycloheximide. When cycloheximide was added after polysome breakdown had taken place, polyribosomes were reformed. At 10^{-4} *M*, cycloheximide caused breakdown of polyribosomes during incubation. However, polyribosome breakdown did not take place when the reticulocytes were incubated with 1.5×10^{-2} *M* cycloheximide. This concentration of cycloheximide (1.5×10^{-2} *M*) prevented breakdown of polyribosomes in cells incubated with NaF, but did not permit polyribosome reformation when added after breakdown had taken place. At these high concentrations of cycloheximide there was no detectable incorporation of amino acids into polypeptide. These results are consistent with the hypothesis that cycloheximide slows down the movement and detachment of ribosomes from polyribosomes at 4×10^{-6} *M* and that the process is strongly inhibited by 1.5×10^{-2} *M* cycloheximide.

Several groups have indicated that cycloheximide inhibits translocation in cell-free extracts. Cycloheximide can inhibit polysome aggregation in much smaller concentrations than are needed to prevent chain elongation (464, 465). Also action of cycloheximide on chain elongation could be partially reversed by high concentrations of glutathione (464). It was concluded that inhibitory action of cycloheximide involves inactivation of transferase II; and that the inhibition of polysome aggregation (initiation?) which occurs at lower cycloheximide concentrations may be distinct from its inhibition of chain extension. High cycloheximide concentration (1 mg/ml) inhibited translocation of peptidyl-tRNA from A-site to the P-site, a reaction which requires transferase II (466), for prior incubation of extracts with cycloheximide prevented subsequent formation of peptidyl-puromycin. Both NaF and cycloheximide were found to inhibit the initiation of new chains on ribosomes in intact rabbit reticulocytes as well as in extracts (467). Studies of the rate of amino acid incorporation indicated that cycloheximide also inhibits incorporation into nascent chains that was initiated in intact cells and remains attached to ribosomes during their isolation. NaF on the other hand was found to have no apparent effect on the rate of incorporation into these previously initiated chains.

Cycloheximide inhibits chain elongation as well as chain initiation (468, 469). Cycloheximide and related glutarimide antibiotics inhibit binding, transferase II-dependent movement, and release of tRNA from the donor site of reticulocyte ribosomes. They inhibit binding of deacylated $tRNA^{Phe}$ to ribosomes in the absence of transfer enzymes at concentrations which inhibit initiation ($\leq 10^{-4}$ *M*). Somewhat higher concentrations are required for comparable inhibition of peptide chain extension and transferase II-dependent translocation of peptidyl-tRNA from the acceptor to donor ribosomal sites. This translocation reaction is dependent upon transferase II and GTP hydrolysis; however, cycloheximide has no effect on the ribosome-dependent GTPase activity of transferase II. Furthermore, peptidyl transfer by which peptides on tRNA in the donor ribosomal site are transferred to an amino acid on AA-tRNA in the acceptor site is not inhibited by cycloheximide (466, 469).

Cycloheximide has no effect on the reaction with puromycin of various N-acyl-aminoacyl-oligonucleotide fragments of tRNA to give N-acyl-aminoacyl-puromycin products with 60S ribosomal subunits from yeast (270). Also, the glutarimide antibiotics have no effect on transpeptidation with human ribosomes (209).

Summary.—The glutarimide antibiotics inhibit chain initiation as well as chain elongation by interaction with 60S, but not 50S, ribosomal subunits. Initiation is blocked at lower concentrations than is chain extension. They interfere with chain elongation by inhibiting entry of peptidyl-tRNA from A- to P-site through interference with release of deacylated tRNA from the donor site. They may interfere with initiation by inhibiting the binding of initiator tRNA to ribosomes or junction of 60S subunits to initiation complexes.

Emetine and Other Alkaloids

According to Grollman (448), the active structure of the glutarimide antibiotics is similar to that of the ipecac alkaloids. The indole alkaloid tubulosine contains this same stereochemical structure and is a good inhibitor of eukaryotic protein synthesis (449). Emetine is an effective inhibitor of protein synthesis in suspension cultures of HeLa cells, inhibiting protein synthesis by 50% at a concentration of 4×10^{-8} M in the medium since it is rapidly concentrated by cells (470). In contrast to effects of cycloheximide on protein synthesis which are reversible, effects of emetine on protein synthesis are irreversible. Synthesis of cellular RNA compared to protein synthesis is essentially unaffected by emetine. Emetine, like the glutarimide antibiotics, prevents the breakdown of polyribosomes produced by puromycin, but does not in itself affect the puromycin-stimulated release of nascent polypeptides. Binding [^{3}H]emetine to ribosomes was not detected (470).

Prior incubation of polysomes with cycloheximide or fusidic acid inhibits subsequent puromycin release of nascent chains; however, prior incubation with emetine does not inhibit peptidyl-puromycin formation although emetine preserves the polyribosome pattern even in the presence of puromycin (466). It was concluded that cycloheximide and fusidic acid inhibit peptidyl-tRNA transfer to the donor site, but emetine has no effect on this transfer. Furthermore, emetine has no effect on transpeptidation with human ribosomes (209). These studies indicate emetine (*a*) preserves polysome patterns even in the presence of puromycin; (*b*) does not inhibit transpeptidation; (*c*) does not inhibit translocation of peptidyl-tRNA from the acceptor to donor sites, translocational events. Emetine clearly has an affect on 80S ribosomes, for the rapid puromycin breakdown of polysomes is entirely prevented by the alkaloid. What effect, then, does emetine have on ribosomes? It stabilizes 40's–60S couples so that that they no longer can move along mRNA or fall off mRNA after release of peptides by puromycin. Perhaps also these 80S : emetine ribosomes on polysomes can no longer bind AA-tRNA. Answers must await further data, but the similarity of action of glutarimide antibiotics and ipecac alkaloids is not complete. Differences may be related to the huge size of the alkaloids compared to the antibiotics. However, judging from their similarities, it is likely that the site of action of the active alkaloids such as emetine is on the 60S ribosomal subunit. It is also possible that some functions of transferase II in translocation are blocked through inhibition of a ribosomal site involved in the translocation processes as shown for cycloheximide. Discovering what these functions are may elucidate further the translocational events in protein synthesis.

Jondorf and his collaborators (471–473) found that prior treatment with emetine in vivo stimulates the in vitro amino acid incorporating activity of liver microsomal fractions subsequently prepared from treated animals. The stimulation is associated predominantly with the microsomal fraction and appears related to stabilization of the activity of the in vitro system and of endogenous mRNA activity, in particular. The actual incorporation of labeled amino acids into rat liver protein in vivo, however, is inhibited by treatment of rats with emetine.

Summary.—Emetine and similar active alkaloids are effective inhibitors of protein synthesis in eukaryotes. Emetine does not prevent transpeptidation or translocation of peptidyl-tRNA from acceptor to donor sites; however, it does stabilize ribosomes on polysomes, and, perhaps, the resultant 80S:emetine complexes can no longer carry out AA-tRNA binding or some translocational functions.

CONCLUDING REMARKS

The antibiotics and inhibitors discussed in this review inhibit ribosome functions in many ways. Their binding to ribosomes is generally highly specific both spacially and temporally. Thus, an antibiotic binds not only to a specific ribosomal site, but the availability of this site varies with the state of the ribosome (see Figs. 1 and 2). The sensitivity of various ribosomal states to binding by antibiotics is summarized in Table II. The effects of antibiotic binding to ribosomes is analogous to effects resulting from the toss of a jamming agent into complex machinery—almost anything can happen. Accordingly, often an antibiotic produces innumerable effects apparently far removed from its major site of action. For example, most antibiotics which inhibit chain extension can be expected to inhibit polysome breakdown and mRNase (474–476); but as jams they may also uncouple functions of this complex machinery. It would thus not be surprising to discover that some ribosome inhibitors inhibit chain elongation, while stimulating mRNase activity. Uncoupling of peptide bond formation and even the tightly coordinated internal events of translocation has been observed: thus,

TABLE 2. Sensitivity of Ribosomal States to Antibiotic Binding

ANTIBIOTIC	RIBOSOMAL STATES									
	I	II	III	IV	IVa	V	VI	VII	VIII	IX
AURINTRICARBOXYLIC ACID	++	—	—	—	—	—	—	—	—	—
STREPTOMYCIN	++	++	++	—	++	—	—	—	++	—
NEOMYCIN, KANAMYCIN	++	++	++	—	++	—	—	—	++	—
SPECTINOMYCIN	++	++	++	++	++	++	++	++	++	++
KASUGAMYCIN	++	++	—	—	++	—	—	—	++	—
PACTAMYCIN	++	++	++	+	+	+	+	+	+	+
TETRACYCLINE	++	++	—	—	++	—	—	—	++	—
EDEINE	++	++	++	++	—	—	—	++	—	—
	Free 50S (60S)									
PUROMYCIN	++			—	++	—	—	—	++	—
GOUGEROTIN, AMICETIN	++			—	++	—	—	—	++	—
CHLORAMPHENICOL	++			—	++	—	—	—	++	—
SPARSOMYCIN	++			—	++	—	—	—	++	—
ANISOMYCIN	++			—	++	—	—	—	++	—
NIDDAMYCIN, CARBOMYCIN	++			++	—	—	—	—	—	—
SPIRAMYCIN III	++			++	—	—	—	—	—	—
ERYTHROMYCIN	++			++	—	—	++	++	—	—
LINCOMYCIN	++			—	—	—	—	—	—	—
STREPTOGRAMIN A	++			—	—	—	—	—	—	—
THIOSTREPTON	++			++	++	++	++	++	++	++
CYCLOHEXIMIDE	++			++	—	—	+	+	—	—
EMETINE	++			++	—	—	+	+	—	—

The ribosomal states indicated in the table are described in detail in Figure 1 and 2. The sensitivity of a given ribosome state to binding by the antibiotics is estimated from studies described in detail in the text. —— No binding. + Little binding or good binding at high antibiotic concentration. ++ Good binding.

emetine inhibits mRNA:ribosomal movement, but does not inhibit transfer of peptidyl-tRNA from the A- to P-site, nor does it inhibit transpeptidation (466); and micrococcin inhibits translocational events, but not ribosome-factor G-dependent GTP hydrolysis which are normally coupled (431). A summary of the effects of these inhibitors on various steps of protein synthesis is presented in Table 3. As our knowledge increases, the gaps in this picture can be filled.

TABLE 3. Effect of Antibiotics on the Steps of Protein Synthesis

Antibiotic	Initiation			Codon Recognition		Transpeptidation			Translocation							Termination
	mRNA binding	$tRNA^{fMet}$ binding	Subunit junction	AA-tRNA binding	Miscoding	Catalytic center	C-C-A (AA) binding	C-C-A (PEP) binding	Abnormal 30S-50S 40S–60S couples	tRNA release	PEP-tRNA A→P	Rib Subunit-Exchange	mRNA: RIB movement	GTP hydrolysis	mRNase	Termination
Aurintricarboxylic acid	●	—	—	—	—	—	—	—		—	—	—	—	—	—	—
Streptomycin	—	●	●	●	●	○	—	—	●	○	—	○	—	—	●	○
Neomycin,Kanamycin	—		●		●	○	○	—	●	○	—	○	—	—	●	○
Spectinomycin	—	—	—	—	—	—	—	—	—			○	●			
Kasugamycin	—			●	—				—							
Pactamycin	—	●		○	—	—	—	—				○				
Tetracycline	—	○		●	—	○	○					●	—		—	●
Edeine	—	●	○	○	—	—	—	—	●							
Puromycin	—	—		—	—	—	●		○			●				○
Gougerotin, Amicetin	—	—		—	—	●	●	○					○	—		○
Chloramphenicol	—	—	—	—	—	●	●	—				○	○	—	●	○
Sparsomycin	—	—	—	—	—	●	●	●				●	○	—		○
Anisomycin						●	●									
Niddamycin, Carbomycin		—		—	—	○	○	○		○	○			—		
Spiramycin III, Tylosin		—		—	—	○	○	○		○	○			—		
Erythromycin		—		—	—	○	—	●		○	●			—		
Lincomycin				—	—	○	○			—		○		—		○
Streptogramin A	—	●	●	○	—	○	○	○				○		—		
Thiostrepton				○	—	—	—	—					●	●		
Micrococcin				—	—	—	—	—					●	—		
Cycloheximide		●				—	—	—		●	●		●	—		
Emetine						—	—	—	●		—		●			

Effect of antibiotics on various steps has been estimated from studies described in detail in the text. In some cases it was conjectured from studies with closely related antibiotics or from genetic evidence. For example, inhibition of mRNA:ribosome movement is taken from stabilization of polyribosome patterns by an antibiotic or from studies of entry of pulse-labeled mRNA into polysomes. In many cases there was no experimental data reported and it was felt a reasonable guess could not be made: these represent gaps in our knowledge and appear as empty places in the table. Obviously, adequate interpretation of the table requires reference to the appropriate section in the text.

Most of the column headings are self-explanatory except, perhaps, for the following: $tRNA^{fMet}$ binding refers to the binding of initiator tRNA to ribosomes; in bacteria this is fMet-$tRNA^{fMet}$; in mammalian cells, Met-$tRNA^{fMet}$ or possibly even other species of initiator tRNA. Interference with binding by an antibiotic does not imply interference with rate of formation of the complex, for it may equally be due to destabilization of the complex after its formation. Binding of C-C-A(AA) and C-C-A(PEP) refer to interaction of the aminoacyl-termini of peptidyl- and AA-tRNA with the substrate binding sites of the peptidyl transferase. Release of tRNA refers to release of deacylated tRNA species after transfer of its polypeptide to AA-tRNA. Pep-tRNA (A→P) refers to the translocation of peptidyl-tRNA from the acceptor (A) to donor (P) sites. The column headed mRNase refers to ribonuclease V, whose exonuclease activity is ordinarily coupled to translocation (474–476). — No effect. ○ Small effect; or large effect at high antibiotic concentration; or effect may not be relevant to antibiotic action in vivo. ● Major effect.

ACKNOWLEDGMENT

I would like to thank Elena Smith and Karin Klein for their excellent assistance in the preparation of this manuscript.

LITERATURE CITED

1. Lucas-Lenard, J., Lipmann, F. 1971. *Ann. Rev. Biochem.* In press
2. Lengyel, P., Söll, D. 1969. *Bacteriol. Rev.* 33:264–301
3. Kaempfer, R., Meselson, M. 1969. *Cold Spring Harbor Symp. Quant. Biol.* 34:209–20
4. Subramanian, A. R., Davis, B. D., Beller, R. J. 1969. *Cold Spring Harbor Symp. Quant. Biol.* 34: 223–30
5. Schlessinger, D., Gurgo, C., Luzzatto, L., Apirion, D. 1969. *Cold Spring Harbor Symp. Quant. Biol.* 34:231–42
6. Phillips, L. A., Franklin, R. M. 1969. *Cold Spring Harbor Symp. Quant. Biol.* 34:243–53
7. Monro, R. E. 1967. *J. Mol. Biol.* 26: 147–51
8. Hori, M., Rabinovitz, M. 1968. *Proc. Nat. Acad. Sci. USA* 59:1349–55
9. Kurland, C. G. The Proteins of the Bacterial Ribosome. *Protein Synthesis: A Series of Advances,* Vol. 1, ed. E. McConkey, New York: Marcel Dekker, 1970 pp.
10. Traut, R. R., et al. 1969. *Cold Spring Harbor Symp. Quant. Biol.* 34: 25–38
11. Nomura, M., Mizushima, S., Ozaki, M., Traub, P., Lowry, C. V. 1969. *Cold Spring Harbor Symp. Quant. Biol.* 34:49–61
12. Traub, P., Nomura, M. 1969. *Cold Spring Harbor Symp. Quant. Biol.* 34:63–67
13. Traub, P., Nomura, M. 1968. *Proc. Nat. Acad. Sci. USA* 59:777–84
14. Traub, P., Nomura, M. 1969. *J. Mol. Biol.* 40:391–413
15. Nashimoto, H., Nomura, M. 1970. *Proc. Nat. Acad. Sci. USA* 67: 1440–47 (1970).
16. Weisblum, B., Davies, J. 1968. *Bacteriol. Rev.* 32:493–528
17. Vazquez, D., Monro, R. E. 1967. *Biochim. Biophys. Acta* 142:155–73
18. Grollman, A. P., Stewart, M. L. 1968. *Proc. Nat. Acad. Sci. USA* 61:719–25
19. Grollman, A. P. 1969. *Antimicrob. Ag. Chemother. 1968.* 36–40
20. Marcus, A., Bewley, J. D., Weeks, D. P. 1970. *Science* 167:1735–36
21. Grollman, A. P., Huang, M. T. 1970. *Fed. Proc.* 29:1624
22. Lebleu, B., Marbaix, G., Werenne, J., Burny, A., Huez, G. 1970. *Biochem. Biophys. Res. Commun.* 40:731–39
23. Stewart, M. L., Grollman, A. P., Huang, M. T. 1971. *Proc. Nat. Acad Sci. USA* 68:97–101
24. Webster, R. E., Zinder, N. D. 1969. *J. Mol. Biol.* 42:425–39
25. Siegelman, F., Apirion, D. 1970. *Bacteriol. Proc.* 149
26. Kuwano, M., Schlessinger, D., Apirion, D. 1970. *Nature (London)* 226:514–16
27. Miller, D. L., Weissbach, H. 1970. *Biochem. Biophys. Res. Commun.* 38:1016–22
28. Lindenbaum, A., Schubert, J. 1956. *J. Phys. Chem.* 60:1663–65
29. Fitzgerald, R. J., Bernheim, F., Fitzgerald, D. B. 1948. *J. Biol. Chem.* 175:195–200
30. Spotts, C. R. 1962. *J. Gen. Microbiol.* 28:347–65
31. Speyer, J. F., Lengyel, P., Basilio, C. 1962. *Proc. Nat. Acad. Sci. USA* 48:684–86
32. Flaks, J. G., Cox, E. C., Witting, M. L., White, J. R. 1962. *Biochem. Biophys. Res. Commun.* 7:390–93
33. Mager, J. Benedict, M., Artman, M. 1962. *Biochim. Biophys. Acta* 62: 202–4
34. Davies, J. E. 1964. *Proc. Nat. Acad. Sci. USA* 51:659–64
35. Cox, E. C., White, J. R., Flaks, J. G. 1964. *Proc. Nat. Acad. Sci. USA* 51:703–9
36. Kaji, A., Suzuka, I., Kaji, H. 1966. *J. Biol. Chem.* 241:1251–56
37. Pestka, S. 1967. *Bull. N. Y. Acad. Med.* 43:126–48
38. Gorini, L., Kataja, E. 1964. *Proc. Nat. Acad. Sci. USA* 51:487–93
39. Davies, J., Gilbert, W., Gorini, L. 1964. *Proc. Nat. Acad. Sci. USA* 51:883–90
40. Van Knippenberg, P. H., Claasen, J. C., Van Ravenswaay, Grijm-Vos, M., Veldstra, H., Bosch, L. 1965. *Biochim. Biophys. Acta* 95: 461–73

41. Likover, T. E., Kurland, C. G. 1967. *Proc. Nat. Acad. Sci. USA* 58: 2385–92
42. Schwartz, J. H. 1965. *Proc. Nat. Acad. Sci. USA* 53:1133–40
43. Old, D., Gorini, L. 1965. *Science* 150:1290–92
44. Bissel, D. M. 1965. *J. Mol. Biol.* 14: 619–22
45. Bodley, J. W., Davie, E. W. 1966. *J. Mol. Biol.* 18:344–55
46. Davies, J., Gorini, L., Davis, B. D. 1965. *Mol. Pharmacol.* 1:93–106
47. Davies, J., Jones, D. S., Khorana, H. G. 1966. *J. Mol. Biol.* 18:48–57
48. Tanaka, N., Sashikata, K., Umezawa, H. 1967. *J. Antibiot.* 20: 115–19
49. Pestka, S., Marshall, R., Nirenberg, M. 1965. *Proc. Nat. Acad. Sci. USA* 53:639–46
50. Kaji, H., Kaji, A. 1965. *Proc. Nat. Acad. Sci. USA* 54:213–19
51. Pestka, S. 1966. *J. Biol. Chem.* 241: 367–72
52. White, J. R., White, H. L. 1964. *Science* 146:772–74
53. Yamaki, H., Tanaka, N. 1963. *J. Antibiot.* A16:222–26
54. Kaji, H. 1967. *Biochim. Biophys. Acta* 134:134–42
55. Kogut, M., Harris, M. 1969. *Eur. J. Biochem.* 9:42–49
56. Davies, J., Davis, B. D. 1968. *J. Biol. Chem.* 243:3312–16
57. McCarthy, B. J., Holland, J. J. 1965. *Proc. Nat. Acad. Sci. USA* 54: 880–86
58. Holland, J. J., Buck, C. A., McCarthy, B. J. 1966. *Biochemistry* 5: 358–65
59. Masukawa, H., Tanaka, N. 1967. *J. Biochem. (Tokyo)* 62:202–9
60. Price, A. R., Rottman, F. 1970. *Biochemistry* 9:4524–29
61. Dunlap, B. E., Rottman, F. *Biochemistry* Submitted for publication
62. Hatfield, D. 1966. *Cold Spring Harbor Symp. Quant. Biol.* 34:619–22
63. Leon, S. A., Brock, T. D. 1967. *J. Mol. Biol.* 24:391–404
64. Wolfe, A. D., Hahn, F. E. 1968. *Biochem. Biophys. Res. Commun.* 31:945–49
65. Sherman, M. I., Simpson, M. V. 1969. *Proc. Nat. Acad. Sci. USA* 64:1388–95
66. Kaji, H., Tanaka, Y. 1968. *J. Mol. Biol.* 32:221–30
67. Vogel, Z., Vogel, T., Zamir, A., Elson, D. 1970. *J. Mol. Biol.* 54:379–86
68. Wolfgang, R. W., Lawrence, N. L. 1967. *J. Mol. Biol.* 29:531–35
69. Cerna, J., Rychlik, I., Pulkrabek, P. 1969. *Eur. J. Biochem.* 9:27–35
70. Scolnick, E., Tompkins, R., Caskey, T., Nirenberg, M. 1968. *Proc. Nat. Acad. Sci. USA* 61:768–74
71. Igarashi, K., Ishitsuka, H., Kaji, A. 1969. *Biochem. Biophys. Res. Commun.* 37:499–504
72. Anderson, W. F., Gorini, L., Breckenridge, L. 1965. *Proc. Nat. Acad. Sci. USA* 54:1076–83
73. Suzuki, J., Kunimoto, T., Hori, M. 1970. *J. Antibiot.* 23:99–101
74. Moskowitz, M., Kelker, N. E. 1963. *Science* 141:647–48
75. Davey, P. J., Haslam, J. M., Linnane, A. W. 1969. *Arch. Biochem. Biophys.* 136:54–64
76. Tanaka, N., Masukawa, H., Umezawa, H. 1967. *Biochem. Biophys. Res. Commun.* 26:544–49
77. Masukawa, H., Tanaka, N. 1968. *J. Antibiot.* 21:70–72
78. Tanaka, Y., Kaji, H. 1968. *Biochem. Biophys. Res. Commun.* 32:313–19
79. Traub, P., Hosokawa, K., Nomura, M. 1966. *J. Mol. Biol.* 19:211–14
80. Staehelin, T., Meselson, M. 1966. *J. Mol. Biol.* 19:207–10
81. Traub, P., Nomura, M. 1968. *Science* 160:198–99
82. Ozaki, M., Mizushima, S., Nomura, M. 1969. *Nature (London)* 222: 333–39
83. Kang, S. S. 1970. *Nature (London)* 225:1132–33
84. Kang, S. S. 1970. *Proc. Nat. Acad. Sci. USA* 65:544–50
85. Silengo, L., Schlessinger, D., Mangiarotti, G., Apirion, D. 1967. *Mutation Res.* 4:701–3
86. Birge, E. A., Kurland, C. G. 1969. *Science* 166:1282–84
87. Apirion, D. L., Schlessinger, D., Phillips, S., Sypherd, P. 1969. *J. Mol. Biol.* 43:327–29
88. Masukawa, H., Tanaka, N., Umezawa, H. 1968. *J. Antibiot.* 21: 517–18
89. Masakuwa, H. 1969. *J. Antibiot.* 22: 612–23
90. Bollen, A., Davies, J., Ozaki, M., Mizushima, S. 1969. *Science* 165: 85–86
91. Bollen A., Herzog, A. 1970. *FEBS Lett.* 6:69–72

92. Dekio, S., Takata, R. 1969. *Mol. Gen. Genet.* 105:219–24
93. Gorini, L., Gundersen, W., Burger, M. 1961. *Cold Spring Harbor Symp. Quant. Biol.* 26:173–82
94. Valentine, R. C., Zinder, N. D. 1964. *Science* 144:1458–59
95. Gartner, T. K., Orias, E. 1966. *J. Bacteriol.* 91:1021–28
96. Gorini, L. 1966. *Bull. N. Y. Acad. Med.* 42:633–37
97. Lederberg, E. M., Cavalli-Sforza, L., Lederberg, J. 1964. *Proc. Nat. Acad. Sci. USA* 51:678–82
98. Couturier, M., Desmet, L., Thomas, R. 1964. *Biochem. Biophys. Res. Commun.* 16:244–48
99. Rosenkranz, H. S., Bendich, A., Carr, H. S. 1964. *Biochim. Biophys. Acta* 82:110–17
100. Gorini, L., Rosset, R., Zimmermann, R. A. 1967. *Science* 157:1314–17
101. Otsuji, N., Aono, H. 1968. *J. Bacteriol.* 96:43–50
102. Kuwano, M., Ishizawa, M., Endo, H. 1968. *J. Mol. Biol.* 33:513–16
103. Kuwano, M., Endo. H. 1969. *J. Virol.* 4:252–55
104. Scafati, A. R. 1967. *Virology* 32: 543–52
105. Gorini, L., Beckwith, J. R. 1966. *Ann. Rev. Microbiol.* 20:401–22
106. Brownstein, B. L., Lewandowski, L. J. 1967. *J. Mol. Biol.* 25:99–109
107. Lewandowski, L. J., Brownstein, B. L. 1966. *Biochem. Biophys. Res. Commun.* 25:554–61
108. Apirion, D., Schlessinger, D. 1968. *J. Bacteriol.* 96:768–76
109. Gorini, L., Jacoby, G. A., Breckenridge, L. 1966. *Cold Spring Harbor Symp. Quant. Biol.* 31:657–64
110. Breckenridge, L., Gorini, L. 1970. *Genetics* 65:9–25
111. Gorini, L. 1969. *Cold Spring Harbor Symp. Quant. Biol.* 34:101–11
112. Rosset, R., Gorini, L. 1969. *J. Mol. Biol.* 39:95–112
113. Gorini, L., Kataja, E. 1964. *Proc. Nat. Acad. Sci. USA* 51:995–1001
114. Apirion, D., Schlessinger, D. 1969. *Proc. Int. Congr. Genet., 12th* 2: 51–52
115. Apirion, D., Schlessinger, D. 1968. *J. Bacteriol.* 96:768–76
116. Modolell, J., Davis, B. D. 1969. *Nature (London)* 224:345–48
117. Modolell, J., Davis, B. D. 1969. *Cold Spring Harbor Symp. Quant, Biol.* 34:113–18
118. Modolell, J., Davis, B. D. 1968. *Proc. Nat. Acad. Sci. USA* 61: 1279–86
119. Modolell, J., Davis, B. D. 1970. *Proc. Nat. Acad. Sci. USA* 67: 1148–55
120. Hishizawa, T., Lessard, J. L., Pestka, S. 1970. *Proc. Nat. Acad. Sci. USA* 66:523–30
121. Luzzatto, L., Apirion, D., Schlessinger, D. 1968. *Proc. Nat. Acad. Sci. USA* 60:873–80
122. Luzzatto, L., Apirion, D., Schlessinger, D. 1969. *J. Mol. Biol.* 42: 315–35
123. Apirion, D., Phillips, S. L., Schlessinger, D. 1969. *Cold Spring Harbor Symp. Quant. Biol.* 34:117–28
124. Herzog, A. 1964. *Biochem. Biophys. Res. Commun.* 15:172–76
125. Jawetz, E., Gunnison, J. B., Speck, R. S. 1951. *Am. J. Med. Sci.* 222: 404–12
126. Plotz, P. H., Davis, B. D. 1962. *J. Bacteriol.* 83:802–5
127. Hurwitz, C., Rosano, C. L. 1960. *Biochim. Biophys. Acta* 41:162–63
128. Stern, J. L., Cohen, S. S. 1964. *Proc. Nat. Acad. Sci. USA,* 51: 859–65
129. Stern, J. L., Barner, H. D., Cohen, S. S. 1966. *J. Mol. Biol.* 17:188–217
130. Sparling, P. F., Modolell, J., Takeda, Y., Davis, B. D. 1968. *J. Mol. Biol.* 37:407–21
130a. Lederberg, J. 1951. *J. Bacteriol.* 61: 549–50
130b. Breckenridge, L., Gorini, L. 1969. *Proc. Nat. Acad. Sci. USA* 62: 979–85
131. Gurgo, C., Apirion, D., Schlessinger, D. 1969. *J. Mol. Biol.* 45:205–20
132. Davis, B. D. 1968. *Asian Med. J.* 11:78–85
133. Rosenkranz, H. S., Garro, A. J., Levy, J. A., Carr, H. S. 1966. *Biochim. Biophys. Acta* 114:501–15
134. Anand, N., Davis, B. D. 1960. *Nature (London)* 185:22–23
135. Lawrence, N. L. 1967. *Biochem. Biophys. Res. Commun.* 26:284–89
136. Davies, J., Anderson, P., Davis, B. D. 1965. *Science* 149:1096–98
137. Anderson, P. 1969. *J. Bacteriol.* 100:939–47
138. Anderson, P., Davies, J., Davis, B. D. 1967. *J. Mol. Biol.* 29:203–15
139. Bollen, A., Helser, T., Yamada, T.,

Davies, J. 1969. *Cold Spring Harbor Symp. Quant. Biol.* 34:95–100
140. Umezawa, H., et al. 1966. *Antimicrob. Ag. Chemother. 1965* 753–57
141. Tanaka, N., Yoshida, Y., Sashikata, K., Yamaguchi, H., Umezawa, H. 1966. *J. Antibiot.* 19:65–68
142. Tanaka, N., Yamaguchi, H., Umezawa, H. 1966. *J. Biochem. (Tokyo)* 60:429–34
143. Sparling, P. F. 1970. *Science* 167: 56–58
144. Bhuyan, B. K. 1967. *Antibiotics, I. Mechanism of Action,* 170–72. Ed. D. Gottlieb, P. D. Shaw. New York: Springer-Verlag, 785 pp.
145. Bhuyan, B. K. 1967. *Biochem. Pharmacol.* 16:1411–20
146. Young, C. W. 1966. *Mol. Pharmacol.* 2:50–55
147. Colombo, B., Felicetti, L., Baglioni, C. 1966. *Biochim. Biophys. Acta* 119:109–19
148. Felicetti, L., Colombo, B., Baglioni, C. 1966. *Biochim. Biophys. Acta* 119:120–29
149. Kersten, H., Chandra, P., Tanck, W., Wiedemhöver, W., Kersten, W. 1968. *Hoppe-Seyler's Z. Physiol. Chem.* 349:659–63
150. Kersten, H., Kersten, W., Emmerich, B., Chandra, P. 1967. *Hoppe-Seyler's Z. Physiol. Chem.* 348: 1424–30
151. Cohen, L. B., Goldberg, I. H. 1967. *Biochem. Biophys. Res. Commun.* 29:617–22
152. Wiley, P. F., Jahnke, H. K., MacKellar, F. A., Kelly, R. B., Argoudelis, A. D. 1970. *J. Org. Chem.* 35:1420–25
153. Cohen, L. B., Herner, A. E., Goldberg, I. H. 1969. *Biochemistry* 8: 1312–26
154. Cundliffe, E., McQuillen, K. 1967. *J. Mol. Biol.* 30:137–46
155. Macdonald, J. S., Goldberg, I. H. 1970. *Biochem. Biophys. Res. Commun.* 41:1–8
156. Hardesty, B., Miller, R., Schweet, R. 1963. *Proc. Nat. Acad. Sci. USA* 50:924–31
157. Marks, P. A., Burka, E. R., Conconi, F. M., Perl, W., Rifkind, R. A. 1965. *Proc. Nat. Acad. Sci. USA* 53:1437–43
158. Hoerz, W., McCarty, K. S. 1969. *Proc. Nat. Acad. Sci. USA* 63: 1206–13
159. Herner, A. E., Cohen, L. B., Goldberg, I. H. 1968. *Fed. Proc.,* 27: 771
160. Cohen, L. B., Goldberg, I. H., Herner, A. E. 1969. *Biochemistry* 8: 1327–35
161. Gale, E. F., Folkes, J. P. 1953. *Biochem. J.* 53:493–98
162. Nikolov, T. K., Ilkov, A. I. 1961. Abstracts, *Int. Congr. Biochem., 5th, Moscow, 1961,* Sect. 2, p. 44, Abstract No. 2, 114
163. Hash, J. H., Wishnick, M., Miller, P. A. 1964. *J. Biol. Chem.* 239: 2070–78
164. Rendi, R., Ochoa, S. 1967. *J. Biol. Chem* 237:3711–13
165. Franklin, T. J. 1963 *Biochem. J.* 87: 449–53
166. Franklin, T. J. 1964. *Biochem. J.* 90:624–28
167. Laskin, A. I., Chan, W. M. 1964. *Biochem. Biophys. Res. Commun.* 14:137–42
168. Day, L. E. 1966. *J. Bacteriol.* 91: 1917–23
169. Day, L. E. 1966. *J. Bacteriol.* 92: 197–203
170. Suarez, G., Nathans, D. 1965. *Biochem. Biophys. Res. Commun.* 18:743–50
171. Connamacher, R. E., Mandel, H. G. 1965. *Biochem. Biophys. Res. Commun.* 20:98–103
172. Connamacher, R. E., Mandel, H. G. 1968 *Biochim. Biophys. Acta* 166:475–86
173. Maxwell, I. H. 1968. *Mol. Pharmacol.* 4:25–37
174. Hierowski, M. 1965. *Proc. Nat. Acad. Sci. USA* 53:594–99
175. Suzuka, I., Kaji, H., Kaji, A. 1966. *Proc. Nat. Acad. Sci. USA* 55: 1483–90
176. Gottesman, M. E. 1967. *J. Biol. Chem.* 242:5564–71
177. Pestka, S., Nirenberg, M. 1966. *Cold Spring Harbor Symp. Quant. Biol.* 31:641–56
178. Warner, J. R., Rich, A. 1964. *Proc. Nat. Acad. Sci. USA* 51:1134–41
179. Arlinghaus, R., Shaefer, J., Schweet, R. 1964. *Proc. Nat. Acad. Sci. USA* 51:1291–99
180. Seeds, N. W., Retsma, J. A., Conway, T. W. 1967. *J. Mol. Biol.* 27:421–30
181. Sarkar, S., Thach, R. E. 1968. *Proc. Nat. Acad. Sci. USA* 60:1479–86
182. Cerna, J., Rychlik, I., Pulkrabek, P. 1969. *Eur. J. Biochem.* 9:27–35
183. Rychlik, I. 1966. *Biochim. Biophys. Acta* 114:425–27
184. Rychlik, I. 1966. *Collect. Czech. Chem. Commun.* 31:2583–95
185. Goldberg, I. H., Mitsugi, K. 1967.

Biochemistry 6:383–91
186. Bodley, J. W., Zieve, P. J. 1969. *Biochem. Biophys. Res. Commun.* 36:463–68
187. Kaji, A., Igarashi, K., Ishitsuka, H. 1969. *Cold Spring Harbor Symp. Quant. Biol.* 34:167–77
188. Modolell, J. 1970. *Int. Congr. Biochem., 8th, Switzerland, 1970*
189. Suzuka, I., Kaji, H., Kaji, A. 1966. *Proc. Nat. Acad. Sci. USA* 55: 1483–90
190. Jonak, J., Rychlik, I. 1970. *Biochim. Biophys. Acta* 199:421–34
191. Traut, R. R., Monro, R. E. 1964. *J. Mol. Biol.* 10:63–72
192. Monro, R. E., Vazquez, D. 1967. *J. Mol. Biol.* 28:161–65
193. Pestka, S. 1969. *Proc. Nat. Acad. Sci USA* 64:709–14
194. Vogel, Z., Zamir, A., Elson, D. 1969. *Biochemistry* 8:5161–68
195. Craven, G. R., Gavin, R., Fanning, T. 1969. *Cold Spring Harbor Symp. Quant. Biol.* 34:129–37
196. Kirschmann, C., Davis, B. D. 1969. *J. Bacteriol.* 98:152–59
197. Capecchi, M. R., Klein, H. A. 1969. *Cold Spring Harbor, Symp. Quant. Biol.* 34:469–77
198. Cundliffe, E. 1967. *Mol. Pharmacol.* 3:401–11
199. Cundliffe, E. 1968. *Biochem. Biophys. Res. Commun.* 33:247–52
200. Kurylo-Borowska, Z. 1962. *Biochim. Biophys. Acta* 61:897–902
201. Kurylo-Borowska, Z. 1964. *Biochim. Biophys. Acta* 87:305–13
202. Hierowski, M., Kurylo-Borowska, Z. 1965. *Biochim. Biophys. Acta* 95: 578–89
203. Kurylo-Borowska, Z., Hierowski, M. 1965. *Biochim. Biophys. Acta* 95: 590–97
204. Hettinger, T. P., Craig, L. C. 1968. *Biochemistry* 7:4147–52
205. Hettinger, T. P., Kurylo-Borowska, Z., Craig, L. C. 1968. *Biochemistry* 7:4153–60
206. Hettinger, T. P., Craig, L. C. 1970. *Biochemistry* 9:1224–32
207. Szer, W., Kurylo-Borowska, Z. 1970. *Biochim. Biophys. Acta* 224:477–86
208. Pestka, S., Nirenberg, M. 1966. *J. Mol. Biol.* 21:145–71
209. Pestka, S. *Arch. Biochem. Biophys.* In press
210. Nathans, D. 1967. Puromycin. *Antibiotics, I. Mechanism of Action.* 259–77. Ed. D. Gottlieb, P. D. Shaw. New York: Springer-Verlag, 785 pp.
211. Waller, C. W., Fryth, P. W., Hutchings, B. L., Williams, J. H. 1953. *J. Am. Chem. Soc.* 75:2025
212. Baker, B. R., Schaub, R. E., Williams, J. H. 1955. *J. Am. Chem. Soc.* 77:7–12
213. Baker, B. R., Schaub, R. E., Joseph, J. P., Williams, J. H. 1954. *J. Am. Chem. Soc.* 76:4044–45
214. Sundaralingam, M., Arora, S. K. 1969. *Proc. Nat. Acad. Sci. USA* 64:1021–26
215. Jardetzky, O. 1963. *J. Am. Chem. Soc.* 85:1823–25
216. Johnson, L. F., Bhacca, N. S. 1963. *J. Am. Chem. Soc.* 85:3700–1
217. Yarmolinsky, M. B., de la Haba, G. L. 1959. *Proc. Nat. Acad. Sci. USA* 45:1721–29
218. Rabinovitz, M., Fisher, J. M. 1962. *J. Biol. Chem.* 237:477–81
219. Morris, A. J., Schweet, R. S. 1961. *Biochim. Biophys. Acta* 47:415–16
220. Allen, D. W., Zamecnik, P. C. 1962. *Biochim. Biophys. Acta* 55:865–74
221. Nathans, D. 1964. *Proc. Nat. Acad. Sci. USA* 51:585–92
222. Smith, J. D., Traut, R. R., Blackburn, G. M., Monro, R. E. 1965. *J. Mol. Biol.* 13:617–28
223. Spyrides, G. J. 1964. *Proc. Nat. Acad. Sci USA* 51:1220–26
224. Nathans, D., Neidle, A. 1963. *Nature (London)* 197:1076–77
225. Symons, R. H., Harris, R. J., Clarke, L. P., Wheldrake, J. F., Elliott, W. H. 1969. *Biochim. Biophys. Acta* 179:248–50
226. Waller, J. P., Erdös, T., LeMoine, F., Guttmann, S., Sandrin, E. 1966. *Biochim. Biophys. Acta* 119:566–80
227. Rychlik, I., Cerna, J., Chladek, S., Zemlicka, J., Haladova, Z. 1969. *J. Mol. Biol.* 43:13–24
228. Pestka, S., Hishizawa, T., Lessard, J. L. 1970. *J. Biol. Chem.* 245: 6208–19
229. Takanami, M. 1964. *Proc. Nat. Acad. Sci. USA* 52:1271–76
230. Rychlik, I., Chladek, S., Zemlicka, J. 1967. *Biochim. Biophys. Acta* 138:640–42
231. Hardesty, B., Miller, R., Schweet, R. 1963. *Proc. Nat. Acad. Sci. USA* 50:924–31
232. Burka, E. R., Marks, P. A. 1964. *J. Mol. Biol.* 9:439–51
233. Villa-Trevino, S., Farber, E., Staehelin, T., Wettstein, F. O.,

Noll, H. 1964. *J. Biol. Chem.* 239:3826–33
234. Ron, E. Z., Kohler, R. E., Davis, B. D. 1966. *Proc. Nat. Acad. Sci. USA* 56:471–75
235. Noll, H., Staehelin, T., Wettstein, F. O. 1963. *Nature (London)* 198: 632–38
236. Williamson, A. R., Schweet, R. 1965. *J. Mol. Biol.* 11:358–72
237. Freedman, M. L., Fisher, J. M., Rabinovitz, M. 1968. *J. Mol. Biol.* 33:315–18
238. Spirin, A. S. 1969. *Cold Spring Harbor Symp. Quant. Biol.* 34:197–207
239. Kohler, R., Ron, E., Davis, B. D. 1968. *J. Mol. Biol.* 36:71–82
240. Pestka, S. 1969. *Cold Spring Harbor Symp. Quant. Biol.* 34:395–410
241. Lessard, J. L., Pestka, S. In preparation
242. Gilbert, W. 1963. *J. Mol. Biol.* 6: 389–403
243. Heintz, R. L., Salas, M. L., Schweet, R. S. 1968. *Arch. Biochem. Biophys.* 125:488–96
244. Bretscher, M. S., Marcker, K. A. 1966. *Nature (London)* 211:380–84
245. Zamir, A., Leder, P., Elson, D. 1966. *Proc. Nat. Acad. Sci. USA* 56:1794–1801
246. Watson, J. D. 1964. *Bull. Soc. Chem. Biol.* 46:1399–1425
247. Rychlik, I. 1966. *Biochim. Biophys. Acta* 114:425–27
248. Weissbach, H., Redfield, B., Brot, N. 1968. *Arch. Biochem. Biophys.* 127:705–10
249. Pestka, S. 1970. *Arch. Biochem. Biophys.* 136:80–88
250. Lucas-Lenard, J., Lipmann, F. 1967. *Proc. Nat. Acad. Sci. USA* 57: 1050–57
251. Monro, R. E., Marcker, K. A. 1967. *J. Mol. Biol.* 25:347–50
252. Suhadolnik, R. J. 1970. *Nucleoside Antibiotics.* New York: Wiley-Interscience, John Wiley & Sons, Inc.
253. Clark, J. M., Gunther, J. K. 1963. *Biochim. Biophys. Acta* 76:636–38
254. Sinohara, H., Sky-Peck, H. H. 1965. *Biochem. Biophys. Res. Commun.* 18:98–102
255. Clark, J. M., Chang, A. Y. 1965. *J. Biol. Chem.* 240:4734–39
256. Casjens, S. R., Morris, A. J. 1965. *Biochim. Biophys. Acta* 108:677–86
257. Coutsogeorgopoulos, C. 1967. *Biochemistry* 6:1704–11
258. Goldberg, I. H., Mitsugi, K. 1967. *Biochemistry* 6:372–83
259. Herner, A. E., Goldberg, I. H., Cohen, L. B. 1969. *Biochemistry* 8: 1335–44
260. Jimenez, A., Monro, R. E., Vazquez, D. 1970. *FEBS Lett.* 7:103–8
261. Pestka, S. 1970. *Arch. Biochem. Biophys.* 136:89–96
262. Celma, M. L., Monro, R. E., Vazquez, D. 1970. *FEBS Lett.* 6: 273–77
263. Monro, R. E., Celma, M. L., Vazquez, D. 1969. *Nature (London)* 222:356–58
264. Vazquez, D. 1965. *Biochim. Biophys. Acta* 114:277–88
265. Chang, F. N., Siddhikol, C., Weisblum, B. 1969. *Biochim. Biophys. Acta* 186:396–98
266. Yukioka, M., Morisawa, S. 1969. *J. Biochem. Tokyo* 66:225–32
267. Yukioka, M., Morisawa, S. 1969. *J. Biochem. Tokyo* 66:233–39
268. Yukioka, M., Morisawa, S. 1969. *J. Biochem. Tokyo* 66:241–47
269. Ibuki, F., Moldave, K. 1968. *J. Biol. Chem.* 243:791–98
270. Neth, R., Monro, R. E., Heller, G., Battaner, E., Vazquez, D. 1970. *FEBS Lett.* 6:198–202
271. Brock, T. D. 1963. *J. Bacteriol.* 85: 527–31
272. Bloch, A., Coutsogeorgopoulos, C. 1966. *Biochemistry* 5:3345–51
273. Yamaguchi, H., Yamamoto, C., Tanaka, N. 1965. *J. Biochem. Tokyo* 57:667–77
274. Yamaguchi, H., Tanaka, N. 1966. *J. Biochem. Tokyo* 60:632–42
275. Lessard, J. L., Pestka, S. In press
276. Kinoshita, T., Tanaka, N., Umezawa, H. 1970. *J. Antibiot.* 23: 288–90
277. Florini, J. R., Bird, H. H., Bell, P. H. 1966. *J. Biol. Chem.* 241: 1091–98
278. Hahn, F. E. 1967. *Antibiotics, I, Mechanism of Action,* 308–30. Ed. D. Gottlieb, P. D. Shaw. New York: Springer-Verlag, 785 pp.
279. Rebstock, M. C., Crooks, H. M., Controulis, J., Bartz, Q. R. 1949. *J. Am. Chem. Soc.* 71:2458–62
280. Controulis, J., Rebstock, M. C., Crooks, H. M. 1949. *J. Am. Chem. Soc.* 71:2463–68
281. Dunitz, J. D. 1952. *J. Am. Chem. Soc.* 74:995–99
282. Maxwell, R. E., Nickel, V. S. 1954. *Antibiot. Chemother. (Washington, D.C.)* 4:289–95
283. Hahn, F. E., Wisseman, C. L. 1951.

Proc. Soc. Exp. Biol. Med. 76: 533–35
284. Wisseman, C. L., Smadel, J. E., Hahn, F. E., Hopps, H. E. 1954. *J. Bacteriol.* 67:662–73
285. DeMoss, J. A., Novelli, G. D. 1956. *Biochim. Biophys. Acta* 22:49–61
286. Lacks, S., Gros, F. 1959. *J. Mol. Biol.* 1:301–20
287. Speyer, J. F., et al. 1963. *Cold Spring Harbor Symp. Quant. Biol.* 28:559–67
288. Kucan, Z., Lipmann, F. 1964. *J. Biol. Chem.* 239:516–20
289. Jardetzky, O., Julian, G. R. 1964. *Nature (London)* 201:397–98
290. Tissieres, A., Schlessinger, D., Gros, F. 1960. *Proc. Nat. Acad. Sci. USA* 46:1450–63
291. Lamborg, M. R., Zamecnik, P. C. 1960. *Biochim. Biophys. Acta* 42: 206–11
292. Rendi, R. 1959. *Exp. Cell Res.* 18: 187–89
293. Von Ehrenstein, G., Lipmann, F. 1961. *Proc. Nat. Acad. Sci. USA* 47:941–50
294. Allen, E. H., Schweet, R. S. 1962. *J. Biol. Chem.* 237:760–67
295. Borsock, H., Fischer, E. H., Keighley, G. 1957. *J. Biol. Chem.* 299: 1059–70
296. So. A. G., Davie, E. W. 1963. *Biochemistry* 2:132–36
297. Nathans, D., von Ehrenstein, G., Monro, R., Lipmann, F. 1962. *Fed. Proc.* 21:127–35
298. Nathans, D. 1964. *Proc. Nat. Acad. Sci. USA* 51:585–92
299. Rychlik, I. 1966. *Biochim. Biophys. Acta* 114:425–27
300. Coutsogeorgopoulos, C. 1967. *Biochem. Biophys. Res. Commun.* 27:46–52
301. Coutsogeorgopoulos, C. 1966. *Biochim. Biophys. Acta* 129:214–17
302. Weber, M. J., DeMoss, J. A. 1969. *J. Bacteriol.* 97:1099–1105
303. Vazquez, D. 1963. *Biochem. Biophys. Res. Commun.* 12:409–13
304. Vazquez, D. 1964. *Biochem. Biophys. Res. Commun.* 15:464–68
305. Wolfe, A. D., Hahn, F. E. 1965. *Biochim. Biophys. Acta* 95:146–55
306. Das, H. K., Goldstein, A., Kanner, L. C. 1966. *Mol. Pharmacol.* 2: 158–70
307. Hurwitz, C., Braun, C. B. 1967. *J. Bacteriol.* 93:1671–76
308. Vazquez, D. 1966. *Biochim. Biophys. Acta* 114:289–95
309. Pestka, S., Heck, B. H., Scolnick, E. M. 1969. *Anal. Biochem.* 28:376–84
310. Gardner, R. S. et al. 1962. *Proc. Nat. Acad. Sci. USA* 48:2087–94
311. Julian, G. R. 1965. *J. Mol. Biol.* 12: 9–16
312. Julian, G. R. 1966. *Antimicrob. Ag. Chemother. 1965.* 992:1000
313. Irvin, J. D., Julian, G. R. 1970. *FEBS Lett.* 8:129–32
314. Pestka, S. 1968. *Proc. Nat. Acad. Sci. USA* 61:726–33
315. Frost, A. A., Pearson, R. G. 1953. *Kinetics and Mechanisms.* New York: Wiley & Sons, 233 pp.
316. Weber, M. J., DeMoss, J. A. 1966. *Proc. Nat. Acad. Sci. USA* 55: 1224–30
317. Flessel, C. P. 1968. *Biochem. Biophys. Res. Commun.* 32:438–46
318. Dresden, M. H., Hoagland, M. B. 1967. *J. Biol. Chem.* 242:1065–68
319. Levinthal, C., Fan, D. P., Higa, A., Zimmerman, R. A. 1964. *Cold Spring Harbor. Symp. Quant. Biol.* 28:183–90
320. Gros, F. et al. 1964. *Cold Spring Harbor Symp. Quant. Biol.* 28: 299–313
321. Dresden, M. H., Hoagland, M. B. 1967. *J. Biol. Chem.* 242:1069–73
322. Gurgo, C., Apirion, D., Schlessinger, D. 1969. *FEBS Lett.* 3:34–36
323. Cameron, H. J., Julian, G. R. 1968. *Biochim. Biophys. Acta* 169:373–80
324. Kaempfer, R. 1968. *Proc. Nat. Acad. Sci. USA* 61:106–13
325. Ellis, R. J. 1969. *Science* 163:477–78
326. Kroon, A. M. 1963. *Biochim. Biophys. Acta* 72:391–402
327. Ashwell, M. A., Work, T. S. 1968. *Biochem. Biophys. Res. Commun.* 32:1006–12
328. Wintersberger, E. 1965. *Biochem. Z.* 341:409–19
329. Weisberger, A. S., Armentrout, S., Wolfe, S. 1963. *Proc. Nat. Acad. Sci. USA* 50:86–93
330. Armentrout, S. A., Weisberger, A. S. 1967. *Biochem. Biophys. Res. Commun.* 26:712–16
331. Armentrout, S. A., Weisberger, A. S. 1968. *Biochim. Biophys. Acta* 161:180–87
332. Jardetzky, O. 1963. *J. Biol. Chem.* 238:2498–2508
333. Ambrose, C. T., Coons, A. H. 1963. *J. Exp. Med.* 117:1075–88
334. Weisberger, A. S., Daniel, T. M., Hoffman, A. 1964. *J. Exp. Med.* 120:183–96

335. Svehag, S. 1964. *Science* 146:659–61
336. Cruchaud, A., Coons, A. H. 1964. *J. Exp. Med.* 120:1061–74
337. Talal, N., Exum, E. D. 1966. *Proc. Nat. Acad. Sci. USA* 55:1288–95
338. Godchaux, W., III, Herbert, E. 1966. *J. Mol. Biol.* 21:537–53
339. Wiley, P. F., MacKellar, F. A. 1970. *J. Am. Chem. Soc.* 92:417–18
340. Slechta, L. 1967. *Antibiotics,* I, *Mechanism of Action,* 410–14. Ed. D. Gottlieb, P. D. Shaw. New York: Springer-Verlag, 785 pp.
341. Slechta, L. 1966. *Antimicrob. Ag. Chemother. 1965* 326–33
342. Goldberg, I. H., Mitsugi, K. 1966. *Biochem. Biophys. Res. Commun.* 23:453–59
343. Jayaraman, J., Goldberg, I. H. 1968. *Biochemistry* 7:418–21
344. Trakatellis, A. C. 1968. *Proc. Nat. Acad. Sci. USA* 59:854–60
345. Jimenez, A., Monro, R. E., Vazquez, D. 1970. *FEBS Lett.* 7:109–11
346. Grollman, A. P. 1967. *J. Biol. Chem.* 242:3226–33
347. Monro, R. E., et al. 1969. *Int. Congr. Chemother. 6th, Tokyo, 1969* (Proc. 2:473–81, Univ. of Tokyo Press, 1970)
348. Vazquez, D., Battaner, E., Neth, R., Heller, G., Monro, R. E. 1969. *Cold Spring Harbor Symp. Quant. Biol.* 34:369–75
349. Vazquez, D. 1967. *Antibiotics,* I, *Mechanism of Action,* 366–77. Ed. D. Gottlieb, P. D. Shaw. New York: Springer-Verlag, 785 pp.
350. Celmer, W. D. 1966. *Antimicrob. Ag. Chemother. 1965* 144–56
351. Brock, T. D., Brock, M. L. 1959. *Biochim. Biophys. Acta* 33:274–75
352. Jordan, D. C. 1963. *Can. J. Microbiol.* 9:129–32
353. Taubman, S. B., Young, F. E., Corcoran, J. W. 1963. *Proc. Nat. Acad. Sci. USA* 50:955–62
354. Wolfe, A. D., Hahn, F. E. 1964. *Science* 143:1445–46
355. Oleinick, N. L., Wilhelm, J. M., Corcoran, J. W. 1968. *Biochim. Biophys. Acta* 155:290–92
356. Shimizu, M., Saito, T., Hashimoto, H., Mitsuhashi, S. 1970. *J. Antibiot.* 23:63–67
357. Vazquez, D. 1967. *Life Sci.* 6:845–53
358. Tanaka, K., Teraoka, H., Nagira, T., Tamaki, M. 1966. *Biochim. Biophys. Acta* 123:435–37
359. Tanaka, K., Teraoka, H. 1966. *Biochim. Biophys. Acta* 114:204–6
360. Tanaka, K., Teraoka, H., Nagira, T., Tamaki, M. 1966. *J. Biochem. Tokyo* 59:632–34
361. Taubman, S. B., Jones, N. R., Young, F. E., Corcoran, J. W. 1966. *Biochim. Biophys. Acta* 123:438–40
362. Mao, J. C.-H., Putterman, M. 1969. *J. Mol. Biol.* 44:347–61
363. Mao, J. C.-H. 1967. *Biochem. Pharmacol.* 16:2441–43
364. Wilhelm, J. M., Corcoran, J. W. 1967. *Biochemistry* 6:2578–85
365. Taubman, S. B., Jones, N. R., Young, F. E., Corcoran, J. W. 1966. *Biochim. Biophys. Acta* 123:438–40
366. Oleinick, N. L., Corcoran, J. W. 1969. *J. Biol. Chem.* 244:727–35
367. Wilhelm, J. M., Oleinick, N. L., Corcoran, J. W. 1968. *Antimicrob. Ag. Chemother. 1967* 236–50
368. Teraoka, H. 1970. *J. Mol. Biol.* 48:511–15
369. Tanaka, K., Teraoka, H., Tamaki, M., Otaka, E., Osawa, S. 1968. *Science* 162:576–78
370. Otaka, E., Teraoka, H., Tamaki, M., Tanaka, K., Osawa, S. 1970. *J. Mol. Biol.* 48:499–510
371. Mao, J. C.-H., Putterman, M., Wiegand, R. G. 1970. *Biochem. Pharmacol.* 19:391–99
372. Ahmed, A. 1968. *Biochim. Biophys. Acta* 166:205–17
373. Taubman, S. B., So, A. G., Young, F. E., Davie, E. W., Corcoran, J. W. 1964. *Antimicrob. Ag. Chemother. 1963* 395–401
374. Tanaka, K., Teraoka, H. 1968. *J. Biochem. Tokyo* 64:635–48
375. Ahmed, A. 1968. *Biochim. Biophys. Acta* 166:218–28
376. Mao, J. C.-H., Wiegand, R. G. 1968. *Biochim. Biophys. Acta* 157:404–13
377. Mao, J. C.-H., Putterman, M. 1968. *J. Bacteriol.* 95:1111–17
378. Oleinick, N. L., Corcoran, J. W. 1969. *Int. Congr. Chemother. 6th, Tokyo, 1969* (Proc. 1:202–8, Baltimore, Md., University Park, 1970)
379. Mao, J. C.-H., Robishaw, E. E. *J. Mol. Biol.* In press
380. Teraoka, H., Tanaka, K., Tamaki, M. 1969. *Biochim. Biophys. Acta* 174:776–78
381. Pulkrabek, P., Cerna, J., Rychlik, I. 1970. *Collect. Czech. Chem. Comm.* 35:2973–82
382. Cerna, J., Rychlik, I. 1968. *Biochim. Biophys. Acta* 157:436–38
383. Teraoka, H., Tamaki, M., Tanaka,

K. 1970. *Biochem. Biophys. Res. Commun.* 38:328–32
384. Cundliffe, E. 1969. *Biochemistry* 8: 2063–66
385. Weisblum, B. 1969. *Fed. Proc.* 28: 466
386. Weisblum, B., Demohn, V. 1969. *J. Bacteriol.* 98:447–52
387. Hoeksema, H., et al. 1964. *J. Am. Chem. Soc.* 86:4223–24
388. Josten, J. J., Allen, P. M. 1964. *Biochem. Biophys. Res. Commun.* 14:241–44
389. Baglioni, C. 1966. *Biochim. Biophys. Acta* 129:642–45
390. Mason, D. J., Lewis, C. 1964. *Antimicrob. Ag. Chemother. 1963* 7–12
391. Chang, F. N., Sih, C. J., Weisblum, B. 1966. *Proc. Nat. Acad. Sci. USA* 55:431–38
392. Chang, F. N., Weisblum, B. 1967. *Biochemistry* 6:836–43
393. Griffith, L. J., Ostrander, W. E., Mullins, C. G., Beswick, D. E. 1965. *Science* 147:746–47
394. Chang, F. N. 1968. *Studies on Antibiotic Inhibitors of the 50S Subunit of Bacterial Ribosomes.* Thesis submitted to the University of Wisconsin in partial fulfillment of the doctoral degree
395. Krembel, J., Apirion, D. 1968. *J. Mol. Biol.* 33:363–68
396. Apirion, D. 1967. *J. Mol. Biol.* 30: 255–75
397. Vazquez, D. 1967. The Streptogramin Family of Antibodies. In *Antibodies I. Mechanism of Action,* 387–403. Ed. D. Gottlieb, P. D. Shaw. New York: Springer-Verlag, 785 pp.
398. Delpierre, G. R., et al. 1966. *Tetrahedron Lett.* 4:369–72
399. Smith, L. E. 1963. *J. Gen. Microbiol.* 33:111
400. Laskin, A. I., Chan, W. M. 1965. *Antimicrob. Ag. Chemother.-1964* 485–88
401. Cocito, C. 1969. *J. Gen. Microbiol.* 57:179–94
402. Tanaka, N., Yamaguchi, H., Umezawa, H. 1961. *J. Antibiot.* 14:60
403. Laskin, A. I., Chan, W. M. 1966. *Antimicrob. Ag. Chemother.-1965*:321–25
404. Yamaguchi, H. 1961. *J. Antibiot.* 14:313–23
405. Vazquez, D. 1962. *Biochim. Biophys. Acta* 61:849–51
406. Yamaguchi, H. 1963. *J. Antibiot.* 16:92–96
407. Yamaguchi, H. 1963. *J. Antibiot.* 16:97–106
408. Yamaguchi, H., Tanaka, N. 1964. *Nature (London)* 201:499–501
409. Ennis, H. L. 1965. *J. Bacteriol.* 90: 1102–8
410. Ennis, H. L. 1965. *J. Bacteriol.* 90: 1109–19
411. Ennis, H. L. 1966. *Mol. Pharmacol.* 2:543–57
412. Ennis, H. L. 1970. *Proc. Int. Congr. Chemotherapy, 6th, Tokyo,* 2: 489–98. Univ. of Tokyo Press
413. Yamaguchi, H., Tanaka, N. 1967. *J. Biochem. (Tokyo)* 61:18–25
414. Yamaguchi, H., Yoshida, Y., Tanaka, N. 1966. *J. Biochem. Tokyo* 60:246–55
415. Vazquez, D. 1967. *Life Sci.* 6:381–86
416. Ennis, H. L. 1971. *Biochemistry* 10: 1265–70
417. Garrod, L. P., Waterworth, P. M. 1956. *Brit. Med. J.* 2:61–65
418. Anderson, B., Hodgkin, D. C., Viswamitra, M. A. 1970. *Nature (London)* 225:233–35
419. Pagano, J. F., Weinstein, M. J., Stout, H. A., Donovick, R. 1955–1956. *Antibiot. Ann.* 554–59
420. Ebata, M., Miyazaki, K., Otsuka, H. 1969. *J. Antibiot.* 22:364–68
421. Yamaguchi, H. et al. 1957. *J. Antibiot.* 10:195–200
422. Kondo, S. et al. 1961. *J. Antibiot.* 14: 194–98
423. Tanaka, T. et al. 1970. *J. Antibiot.* 23: 231–37
424. Tanaka, K., Watanabe, S., Tamaki, M. 1970. *J. Antibiot.* 23:13–19
425. Weisblum, B., Demohn, V. 1970. *J. Bacteriol.* 101:1073–75
426. Tanaka, K., Watanabe, S., Teraoka, H., Tamaki, M. 1970. *Biochem. Biophys. Res. Commun.* 39:1189–93
427. Pestka, S. 1970. *Biochem. Biophys. Res. Commun.* 40:667–74
428. Tanaka, T., Sakaguchi, K., Yonehara, H. 1970. *J. Antibiot.* 23: 401–6
429. Weisblum, B., Demohn, V. 1970. *FEBS Lett.* 11:149–52
430. Modolell, J., Vazquez, D., Monro, R. E. 1971. *Nature New Biol.* 230: 109–12
431. Pestka, S., Brot, N. In preparation
432. Fujimoto, H., Kinoshita, T., Suzuki, H., Umezawa, H. 1970. *J. Antibiot.* 23:271–75
433. Dubnau, D., Goldthwaite, C., Smith, I., Marmur, J. 1967. *J. Mol. Biol.* 27:163–85
434. Sisler, H. D., Siegel, M. R. 1967. *Antibiotics, I, Mechanism of Action* 283–307. Ed. D. Gottlieb, P.

D. Shaw. New York: Springer-Verlag, 785 pp.
435. Sisler, H. D. 1969. *Ann. Rev. Phytopathol.* 7:311–30
436. Siegel, M. R., Sisler, H. D., Johnson, F. 1966. *Biochem. Pharmacol.* 15:1213–23
437. Ennis, H. L., Lubin, M. 1964. *Science* 146:1474–76
438. Ennis, H. L. 1968. *Biochem. Pharmacol.* 17:1197–1206
439. Kerridge, D. 1958. *J. Gen. Microbiol.* 19:497–506
440. Shepherd, C. J. 1958. *J. Gen. Microbiol.* 18:iv–v
441. Gorski, J., Axman, M. C. 1964. *Arch. Biochem. Biophys.* 105:517–20
442. Young, C. W., Robinson, P. F., Sacktor, B. 1963. *Biochem. Pharmacol.* 12:855–65
443. Bennett, L. L., Ward, V. L., Brockman, R. W. 1965. *Biochim. Biophys. Acta* 103:478–85
444. Bennett, L. L., Smithers, D., Ward, C. T. 1964. *Biochim. Biophys. Acta* 87:60–69
445. Siegel, M. R., Sisler, H. D. 1963. *Nature (London)* 200:675–76
446. Wettstein, F. O., Noll, H., Penman, S. 1964. *Biochim. Biophys. Acta* 87:525–28
447. Siegel, M. R., Sisler, H. D. 1964. *Biochim. Biophys. Acta* 87:83–89
448. Grollman, A. P. 1966. *Proc. Nat. Acad. Sci. USA* 56:1867–74
449. Grollman, A. P. 1967. *Science* 157:84–85
450. Colombo, B., Felicetti, L., Baglioni, C. 1965. *Biochem. Biophys. Res. Commun.* 18:389–95
451. Whiffin, A. J. 1948. *J. Bacteriol.* 56:283–91
452. Siegel, M. R., Sisler, H. D. 1965. *Biochim. Biophys. Acta* 103:558–67
453. Cooper, D., Banthorpe, D. V., Wilkie, D. 1967. *J. Mol. Biol.* 26:347–50
454. Rao, S. S., Grollman, A. P. 1967. *Biochem. Biophys. Res. Commun.* 29:696–704
455. Beattie, D. S., Basford, R. E., Koritz, S. B. 1967. *Biochemistry* 6:3099–3106
456. Borst, P., Kroon, A. M., Ruttenberg, G. J. C. N. 1967. *Genetic Elements, Properties and Function*, ed. D. Shugar. London: Academic Press, and Warsaw: Polish Sci. Publishers, 81 pp.
457. Sebald, W., Höfstotter, T., Hacker, D., Bücher, T. 1969. *FEBS Lett.* 2:177–80
458. Sebald, W., Schwab, A. J., Bücher, T. 1969. *FEBS Lett.* 4:243–46
459. Beattie, D. S. 1970. *FEBS Lett.* 9:232–34
460. Jondorf, W. R., Simon, D. C., Avnimelech, M. 1966. *Mol. Pharmacol.* 2:506–17
461. Jondorf, W. R. 1968. *Arch. Biochem. Biophys.* 126:194–205
462. Trakatellis, A. C., Montjar, M., Axelrod, A. E. 1965. *Biochemistry* 4:2065–71
463. Godchaux, W., III, Adamson, S. D., Herbert, E. 1967. *J. Mol. Biol.* 27:57–72
464. Munro, H. N., Baliga, B. S., Pronczuk, A. W. 1968. *Nature (London)* 219:944–46
465. Baliga, B. S., Pronczuk, A. W., Munro, H. N. 1969. *J. Biol. Chem.* 244:4480–89
466. Baliga, B. S., Cohen, S. A., Munro, H. N. 1970. *FEBS Lett.* 8:249–52
467. Lin, S.-Y., Mosteller, R. D., Hardesty, B. 1966. *J. Mol. Biol.* 21:51–69
468. McKeehan, W., Hardesty, B. 1969. *Biochem. Biophys. Res. Commun.* 36:625–30
469. Obrig, T. G., Culp, W., McKeehan, W., Hardesty, B. 1971. *J. Biol. Chem.* 246:174–81
470. Grollman, A. P. 1968. *J. Biol. Chem.* 243:4089–94
471. Jondorf, W. R., Szapary, D. 1968. *Arch. Biochem. Biophys.* 126:892–904
472. Jondorf, W. R., Johnson, R. K., Donahue, J. D. 1969. *Arch. Biochem. Biophys.* 134:233–41
473. Jondorf, W. R., Drassner, J. D., Johnson, R. K., Miller, H. H. 1969. *Arch. Biochem. Biophys.* 131:163–69
474. Kuwano, M., Kwan, C. N., Apirion, D., Schlessinger, D. 1969. *Proc. Nat. Acad. Sci. USA* 64:693–700
475. Kuwano, M., Schlessinger, D., Apirion, D. 1970. *J. Mol. Biol.* 51:75–82
476. Kuwano, M., Apirion, D., Schlessinger, D. 1970. *J. Mol. Biol.* 51:453–57
477. Pestka, S. 1971. *Proc. Symp. Molecular Mechanisms of Antibiotic Action on Protein Biosynthesis and Membranes. 1971, Granada, Spain.* Ed. E. Munoz, F. Ferrandiz, D. Vazquez. New York: Springer-Verlag. In press

MICROBIAL CRITERIA OF ENVIRONMENT QUALITIES

1581

E. FJERDINGSTAD
Institute of Hygiene, University of Copenhagen,
DK 2100 Copenhagen, Denmark

CONTENTS

INTRODUCTION

Clean water low in nutrients is described as being oligotrophic, while the term, eutrophic, characterizes water rich in nutrients. It is a natural development that, in course of time, an oligotrophic water will gradually become more rich in nutrients in order to arrive finally at a stage in which it is quite overgrown. This is a process which takes a long time, however. The gradual change in the fertility of a lake has a marked influence on plants as well as on animals. The total number of living organisms increases, and some species are replaced by other species. While an oligotrophic lake has a fairly low production of plankton, preferably desmids as, e.g., *Staurastrum anatinum* and *Arthrodesmus* species; with increasing fertility the desmids will be replaced by diatoms, later by flagellates and green algae, and finally by blue-green algae such as *Aphanizomenon flos-aquae, Microcystis aeruginosa, Microcystis flos-aquae, Anabaena flos-aquae, Oscillatoria rubescens, Oscillatoria agardhii,* and others. When the lake has reached the stage of blue-green algae, the production of these algae in summer may be sufficiently high to cause a green coloration of the water (water bloom). At the same time there is an increase in the number of small animals, preferably crustacea.

The development from oligotrophic to eutrophic is violently accelerated if sewage is introduced into the lake, even if the sewage has passed through a sewage treatment plant since the effluents from most of the plants used at present are very high in nutrients, particularly phosphate and nitrogen, which are necessary for a high production of algae.

Berg et al (1) have studied a large lake, Furesøen, which is situated north of Copenhagen and considerably influenced by human activities. In this study, calculations showed that purified sewage introduces 24 tons of nitrogen and 4 tons of phosphate phosphorus annually into the lake. This corresponds to about 12 percent of the amounts of nitrogen and phosphate which are consumed by the plankton. To this it should be added that the nitrogen and phosphate of plankton are again available as food for organisms and thus take part in a metabolic cycle an unknown number of times annually. With the outflow from the lake about 3.5 tons of nitrogen, but only about 87 kg of phosphate phosphorus are removed from the lake.

When eutrophication has occurred, a lake will continue as eutrophic for a very long time after the introduction of nutrients has ceased. As early as 1947 Hassler (30) published a survey of 37 lakes in Europe and the United States which had undergone eutrophication because of introduction of domestic sewage. In most cases the surface areas of the lakes were large and their depths considerable. Eutrophication started quite suddenly. Hasler mentions, for instance, that the diatom, *Tabularia fenestrala,* showed explosive growth in Zürichsee in 1896, and two years later the blue-green alga *Oscillatoria rubescens* produced water bloom. This blue-green alga had not previously been found in the lake.

Eutrophication of a lake will make it unattractive and less useful for recreative purposes. Bathing in a lake of this type may cause irritation of the skin or, in particularly unfortunate cases on accidental ingestion of water, serious disease due to toxic products resulting from the metabolism of the blue-green algae. There are, moreover, many reports of death of birds, cattle, etc., which had drunk water from a lake with water bloom caused by blue-green algae. A number of experimental studies also confirm the toxic properties of the algae. Such toxic products cannot be removed from the water either by cooling or heating, or by filtration through water-works filters, and it is thus—at any rate during the bloom period—impossible to use the water from such lakes as stand-by for drinking water supplies.

In deeper layers, near the bottom of the lake, oxygen deficiency will occur in eutrophic lakes because of the bacterial degradation of the large number of dead organisms which settle on the bottom.

Streams are even more exposed to eutrophication than lakes, because the increasing intensification of agriculture results in an increased application of artificial fertilizers of which some is washed with rain water into streams, but the sewage which, from towns, is allowed to flow into rivers is of even greater significance. Also rain water may contain phosphates. Contents of 49 and 80 μg/liter have been reported (cf. Keup, 37).

PRIMARY PRODUCTION AND BIOTESTS

Recent ecological studies have been concerned in particular with questions such as primary production and biotests. Since these concepts are significant in connection with Caspers & Karbe's (12) saprobic system, a brief survey will be given in the following.

The production is defined as the energy which is annually transferred from the sun to a biotope. The primary production is determined by means of C^{14} isotopes and is chiefly a measure of the production of phytoplankton. Several investigations of the primary production in lakes have been made, and in 1956 Odum (44) applied the method to flowing waters. While the method is readily applicable to lakes and oceans, its application to water courses is somewhat doubtful. Hynes (33) has rightly pointed out that at any rate a very great part of the production in water courses is based on photosynthesis which takes place elsewhere, leaves from trees growing on land fall, for instance, in the water. Hynes writes further: "without these sources of allochthonous organic matter the inherent stability of running water and its consequent inability to grow much in the way of plants, except in places specially favoured, would cause it to be almost a desert habitat."

There is, of course, some primary production in water courses. Odum has thus shown that rivers which are recovering from organic pollution may be among the most productive areas on the earth.

In a recent publication Steemann Nielsen & Wium Andersen (54) have drawn attention to the significance of copper ions in lake and sea water. They write that a content of copper ions even as low as 1–2 μg/liter is poisonous for phytosynthesis and growth of unicellular algae. When ordinarily copper in natural waters does not affect algae this is because it forms complexes with various organic substances whereby the toxicity is lost. The authors further point out that some manufacturers of C^{14} ampules use ordinary distilled water which may frequently contain about 250 μg/liter Cu, and that it is consequently possible that the toxic effect may be introduced in this way.

Biotests

Bringmann (2) and Bringmann & Kühn (3–6) have developed a biological pollution assessment method which depends on physiological measurements. They assume that in case of an increase in the pollution of a receiving water because of introduction of domestic sewage or industrial wastes of a corresponding chemical composition, the content of organic nitrogen in the receiving water will increase proportionally to the degree of pollution.

As a test organism *Escherichia coli* in pure culture is used. It is introduced into the material to be tested after filtration, sterilization, and addition of glucose, KH_2PO_4, and $MgSO_4$ to the latter. The rate of multiplication of the test organism, i.e., optimum biomass produced, is considered to be an integrated biological measure of the content of biologically active organic nitrogen in the water and thus a measure of the saprobicity.

For the determination of the degree of trophicity a pure culture of the green alga *Scenedesmus quadricauda* is inoculated into the sample, and the resulting optimum population density (biomass) is considered to be a measure of the trophicity of the water with respect to photo-autotrophic organisms. Bearing in mind the complexity of waste waters and their frequent contents of toxic substances or substances which may either inhibit or favor growth, it is doubtful whether the method is useful. Thus, Felföldy (17) adopts a critical attitude towards the method and writes that the results are greatly dependent on the presence of suitable physiological characters in the test organisms.

Stange-Bursche (53) has made a number of experiments using either such green algae as *Scenedesmus quadricauda* and *Ankistrodesmus falcatus,* or diatoms such as *Asterionella formosa* and *Diatoma elongatum* as test organisms, and using lake water with or without the addition of nutrients. She points out that the amount of algae present does not give any indication of the concentration of a single nutrient, or of special loads of sewage, but represents merely a summary reaction to the combination of growth-favoring and growth-inhibiting components in the water. The reaction of the algae to the sewage load depends *inter alia* on their state of nutrition, and thus this reaction is not species-specific. Stange-Bursche finally states that an algal test of this type can in no way replace the methods hitherto used for pollution assessments, and that it will be of value as a supplement in special cases only.

Incidentally, Bursche (9) has shown the chlorophyll content and cell volume in plankton algae to vary with varying growth conditions.

Elster (16) mentions that in all experimental ecological tests and biotests, natural populations taken from the water to be examined, will be forced to live under unnatural conditions. We do not know what—under such altered conditions—will happen with regard to loss and multiplication of organisms, or even whether the organisms do change from what they would be under natural conditions.

The present author is entirely in agreement with the mentioned criticism of the applicability of the method. However, when Elster, after his abovementioned critical comments, nevertheless says that a further development of this biomass method should prove valuable, e.g., for re-examining other test organisms according to ecological requirements, the present author is unable to agree with him.

Bringmann & Kühn themselves draw attention to the fact that the results obtained are greatly dependent on the temperature of the sampled water and on the magnitude of the rate of flow. It should be added, moreover, that in the case of a water course the time at which the sample is collected will also influence the result of the test, see, e.g., Figures 8–9, Fjerdingstad (25). Just as for the chemical and bacteriological methods used at present, the test will, at best, provide only information as to conditions at a certain moment, quite apart from the fact that we cannot as a matter of course draw conclusions from such a laboratory test with regard to conditions in a water course. The test is somewhat similar to the BOD test which frequently yields rather misleading results because it is extremely sensitive to toxic substances and in particular to metallic poisons.

Hobbie & Wright (31) have likewise suggested a bioassay procedure using a pure culture of a suitable bacterium. However, Jannasch (35) quite rightly points out the difficulties involved in the method: "1. that the uptake constants theoretically will be affected by the conditions introduced by the sample of unknown composition, and 2. that obtaining substratefree water as a diluting medium with unchanged biological qualities presents a considerable problem."

Caspers (10) has described a method for the measurement of the momentary biological activity in sediments in which he adds 10 ml of fresh bottom sediment to 1 litre of tap water and measures the oxygen content at once and again when the sample has been left for 48 hours in the dark at 20°C. To develop maximum reactivity the sample is shaken several times so as to distribute the sediment finely and to prevent stratification. For the last 2 hours the sample is not shaken so that the sediment settles. The oxygen content may be determined according to Ohle's iodine difference method. According to Caspers, the values obtained represent the metabolic activity of the aerobic bacteria. The use of water for dilution no doubt introduces many sources of error, and the true values of assimilation and disassimilation are presumably obtained only when allowance is made for the respiratory quotient, CO_2/O_2 (cf. also Wetzel, 57).

Bucksteeg & Thiele (8) determine the activity in the sediment by means of 2,3,5-triphenyl-tetrazolium chloride (TTC) which is water-soluble and colorless. On exposure to the action of enzymes, living cells assume a reddish-violet color because TTC is converted into formazan which is insoluble in water. The reaction depends on the nature of the cells and may be observed also in bacteria which turn red. The color is resistant to air, but is light-sensitive, it may be extracted with ethyl alcohol and determined photometrically.

In an extremely sensitive test, Holm-Hansen & Brooth (32) measure the content of adenosine triphosphate (ATP) which is a biochemically essential substance of low molecular weight present in all living matter. However, two important problems remain to be solved—the turnover time (i.e., ratio of production to mean standing crop) and amount of ATP per cell under varying physiological conditions, while it must also be shown that there is no ATP in the detritus.

With regard to the value of these biotests the present author must agree with Jannasch (35) who has written in a publication in 1969: "More important than biomass determination are in case of microorganisms direct measurements of in situ activities. Here again, what can easily be done in more or less well-supplemented bacteriological cultures is extremely difficult to reproduce under natural conditions. Most attempts to assess natural transformation rates deal with respiration or with the degradation of certain important energy sources for microbial growth."

In 1968, Knöpp (38) described new metabolic-dynamic methods which might serve as supplement to methods already used (e.g., BOD) for the examination of waters and waste waters. The first stage of self-purification is considered to be a biochemical oxygenation of organic matter, the next step a photo-reassimilation of the mineralization products attended by liberation of oxygen (biogen Belüftung).

The methods comprise: 1. determination of BOD 7hr ("zusätzliche Zehrung, ZZ); 2. determination of the biogenous rate of oxygenation ("biogen Belüftungrate," BBR); 3. determination of oxygen production potential (SPP); and 4. assimilation test ("Assimilations-Zehrungstest, A-Z). ZZ is the measure of the amount and activity of heterotrophic bacteria (microbial degradation). BBR is a determination of the biogenous rate of oxygenation of phytoplankton. SPP is a relative measure of the activity of phytoplankton under standard conditions, and A-Z a toxicological and metabolic-dynamic evaluation of sewage.

The determinations are made at a temperature of 20–22°C. For further information reference should be made to Knöpp's detailed description.

It is Knöpp's view that these metabolic-dynamic model analyses might serve to bridge the gap between purely analyticochemical and sewage engineering considerations on the one hand, and limnological considerations on the other. Actually, the methods must be said to represent a further development of the BOD determination. Whether they will be able to yield very much more than the latter may be seen when results from a great number of localities are available, but it must be reasonable to point out that with regard to an energy estimate it will be of little value to measure the respiration rate at a temperature to which the organism is not exposed in nature (cf. also Cummins, 15).

BIOLOGICAL ASSESSMENT OF EUTROPHICATION AND POLLUTION

Even before 1850, the first attempts were made in Europe to use the fauna and flora in the evaluation of water—the observation having been made that water possessed a certain ability to purify itself. In 1908, Kolkwitz & Marsson (41) published a saprobic system according to which it was possible by means of indicator species to divide waters into four zones according to decreasing pollution, viz., polysaprobic, α-mesosaprobic, β-mesosaprobic, and oligosaprobic, a terminology which is still used on the European continent. In England, the terms used are foul pollution, pollution, mild pollution, repurified water, while in the United States such terms as septic, contaminated, recovery zone, clean water zone are often used. As Kolkwitz' (39) β-mesosaprobicity represents mineralized pollution, it thus corresponds to the concept of eutrophication.

Brinley (7) found from investigations of algal and protozoon plankton and fishes in the Ohio River that instead of applying the usual chemical classification to waters in terms as septic, polluted, contaminated, and clean water, they might be defined in terms of biological population and dissolved oxygen levels. The organisms were divided into three classes:

Class I. Organisms which are able to live in a medium low inorganic material, but which are also able to tolerate a more concentrated medium. To this class the genera *Chrysococcus, Cryptomonas, Dinobryon, Chromulina* are referred together with various diatom genera.

Class II. Organisms which prefer a rich medium or which feed on bacteria or solid particles. To this group most of the species of genera *Euglena, Lepocinclis, Phacus, Synura, Anabaena,* and bacteria-eating ciliates as *Paramaecium, Colpidium,* and *Vorticella* are referred.

Class III. The largest class contains those organisms which are able to develop under nutritive conditions between the two extremes. These organisms are considered as intermediates and belong mostly to the Chlorophyseae.

It is assumed that there is a close relation between the amount of nutrients present and the plankton population. The criteria are said to show quite satisfactory correlation with the BOD test and with the bacterial count (*coli-aerogenes* group).

Five distinct zones in a polluted stream are characterized as follows:

Zone I. Zone of active bacterial decomposition. A low oxygen concentration (not over 3 ppm and often zero), a high BOD, and high bacterial count. This zone is biologically dominated by bacteria-eating ciliates such as *Paramaecium, Colpidium* (Class II), and large numbers of stalked ciliates (*Vorticella* and *Carchesium*), a few flagellates such as *Chlamydomonas, Chrysococcus,* and *Cryptomonas* (Class I).

Zone II. Zone of intermediate decomposition. Dissolved oxygen between 3 and 5 ppm during the daytime, but drops slightly at night. Plankton production volume is higher than in Zone I, but the organisms are still largely of Class II with an increase in the number of chlorophyll-bearing phytoplankton. *Oscillatoria* and other blue-green algae are commonly found along the margins and on the bottom of the stream.

Zone III. Fertile zone. Dissolved oxygen above 5 ppm and often supersaturation in the daytime; subject to diurnal variation. Plankton over 1 ppm, usually several parts per million, largely Chlorophycea of the intermediate group accompanied by a decrease in the ciliated protozoa (Class II) and a slight increase in Class I forms.

Zone IV. Game fish zone. Dissolved oxygen above 5 ppm and approximating saturation. Plankton between 0.3 and 1 ppm, Class I forms persistent, Class II forms scarce.

Zone V. Biologically poor. Dissolved oxygen near saturation. Plankton less than 1 ppm and consisting almost entirely of *Chrysococcus, Cryptomonas,* and diatoms (Class I). Fish are scarce compared with the previous zone, but mainly game fish.

With regard to zonation this system shows great resemblance to Kolkwitz & Marsson's system (41), and the values given by Brinley for oxygen content correspond on the whole to the values given by Kolkwitz & Marsson.

However, the volume of plankton organisms without any precise definition of the species concerned must be said to be a very loose foundation on which to build a biological system. The fact that a few genera are mentioned as characteristic of the zones does not help, because within most genera there are species which occur in strongly polluted areas as well as species which occur in the cleanest waters, and in addition species which are more or less indifferent and which may be found within the whole range of pollution. *Vorticella microstoma* is polysaprobic; *V. convallaria*, α-mesosaprobic; *V. campanula*, β-mesosaprobic, and *V. nebulifera* var. *similis*, oligosaprobic; *Colpidium colpoda*, polysaprobic to α-mesosaprobic; *Chlamydomonas* spp. may be polysaprobic to oligosaprobic; *Chromulina rasonoffi*, oligosaprobic; *Paramaecium putrinum*, polysaprobic to α-mesosaprobic; *P. caudatum*, α-mesosaprobic to polysaprobic; while *P. bursaria* is β-mesosaprobic. Unfortunately, it is difficult to identify protozoa down to species, and accurate identification down to species is necessary (cf. Lackey, 42), since different species within the same genus react quite differently. To this it must be added—as pointed out by Šrámek-Husek (52)—that numerical abundance of the various species is as important as their occurrence, and that no biological investigation which does not take this into account is reliable.

Another difficulty is provided by the fact that the soil is full of protozoa which are continuously washed into rivers (cf. Gray, 29) who has studied this phenomenon and found the number to vary with weather conditions. Following heavy showers the numbers in the stream increased. As Hynes (33) writes: "it will be appreciated that it is impossible at this stage in our knowledge to give satisfactory account of the effects of organic pollution on microscopical animals."

On the basis of investigations made in the United States, Patrick (45) developed, in 1949, a system of biological water examination, in which she classes both animals and plants in seven groups. On the basis of their occurrence in water courses, she divides the water courses in the following zones:

Zone I. Healthy, with great abundance of species belonging to all seven groups.

Zone 2. Semi-healthy, with some groups overabundant and other groups reduced in numbers.

Zone 3. Polluted, with intensive development of worms, Cyanophyceae and rotifers, and great reduction in numbers of insects and fish species (one or both of these groups may be eliminated).

Zone 4. Very polluted, with the majority of groups eliminated.

Zone 5. Septic conditions.

The organisms are grouped in such a way that group I comprises blue-green algae (*Oscillatoria* sp., *Rivularia* sp., *Nostoc* sp.), certain green algae (*Spirogyra, Stigeocloneum* sp., *Tribonema* sp.), and certain rotifers plus *Cephalodella megalocephala* and *Proales decipiens;* Group II, Oligochaetes, leeches, and pulmonate snails; group III, protozoa (*Euglena* sp.); group IV, diatoms, red algae, and most of the green algae (*Ulothrix, Zygnema* sp.); group V all rotifers not included in group I, clams, gill-breathing snails, and tricladid flatworms; group VI all insects and crustacea; group VII, all fish.

The results are plotted in a histogram, and it is assumed that stations differing markedly from the controls will show biological imbalance in that the columns will be of very unequal heights. With respect to this grouping of organisms it should be pointed out that the extent of the pollution cannot with any reasonable certainty be assessed on this basis even if it should be possible to demonstrate a variation in the numbers within the organism groups. It should further be borne in mind that blue-green algae as well as flagellates comprise both indicators of pollution and markedly indifferent or even katharobic species. It has also been pointed out by Lackey (42) that the genus *Euglena* cannot rightly be regarded as an indicator of pollution since it includes species which are not associated with polluted localities. In a discussion of Patrick's system (cf. Patrick, 46), Butcher has rightly pointed out that it is unwise to regard plants and animals as a unit.

In a subsequent work in 1954, Patrick, Hohn & Wallace (47) described a new method for determining the pattern of the diatom flora. They submerged slides in the water for specified periods and then counted and identified the diatoms deposited on the slides. The results were plotted as number of species per interval against number of specimens per species on a logarithmic scale. It appears from the histogram that in a polluted stream there are few species and few of these are present in great numbers.

In his study of the river Mølleå in Denmark, Fjerdingstad (19) found that diatoms are relatively indifferent to pollution, and they are consequently not suitable indicators (even such a clean water form as *Meridion circulare* has subsequently been found to be able to survive in almost unchanged numbers when a locality is suddenly heavily loaded with drainage from silage).

In 1961, Sládeček (49, 50) summarized experience from Czechoslovakia (authors: Šrámek-Husek, Kreda, Bulich, Cyrus, Sierp) and extended and modified Kolkwitz & Marsson's (41) saprobic system. In addition to domestic sewage, the system also covers industrial wastes, ground water, and drinking water. Sládeček distinguishes among four main groups:

I. Katharobic, i.e., nonpolluted waters, for example, ground water and drinking water.

II. Limnosaprobic, i.e., surface water and ground water which do not display the greatest degree of purity.

III. Eusaprobic, i.e., wastes which are readily or difficultly decomposed by microorganisms.

IV. Trans-saprobic, i.e., wastes which cannot be decomposed by microorganisms (which contain organic, inorganic, or radioactive substances).

These four main groups are subdivided altogether into eleven stages:

I. KATHAROBIC

Stage 1: Katharobic

II. LIMNOSAPROBIC

Stage 2: Oligosaprobic
Stage 3: β-mesosaprobic
Stage 4: α-mesosaprobic
Stage 5: Polysaprobic (*Sphaerotilus trinet*). Organisms: *Sphaerotilus natans, Zoogloea, Paramaecium caudatum, Euglena viridis, Chlamydomonas reinhardii,* etc., *Chlorogonium elongatum,* etc., *Carteria multifiliis,* etc BOD_5: 10–40–(80) mg/liter; coliform count, 200–20,000/cm^3.

III. EUSAPROBIC

Stage 6: Isosaprobic (the ciliate stages of wastes). Organisms: *Paramaecium caudatum, Colpidium colpoda, Glaucoma scientillans, Tetrahymena pyriformis, Dexiotricha centralis, Vorticella microstoma, Polytoma uvella, Zoogloea, Streptococcus.* Anaerobic type, raw domestic sewage. BOD_5 40–400–(600): coliform count, 20,000–3,000,000/cm^3.
Stage 7: Metasaprobic (the flagellate stage of wastes). Organisms: *Cercobodo longicauda, Bodo putrinus, Oicomonas mutabilis, Trepomonas compressa, Tetramitus pyriformis, Hexamitus* (several species), *Spirillum, Lamprocystis, Peloploca,* and other indicators of the presence of H_2S. Anaerobic type: putrefying domestic sewage. BOD_5:200–700 mg/liter; H_2S, 10 mg/liter or more; coliform count, up to 10,000,000/cm^3.
Stage 8: Hypersaprobic (the bacterium and mycophyte stage of wastes). Organisms: bacteria, mycophyta, true saprobiontic species. Anaerobic type; putrefying sludge. BOD_5: 500–1500–2000 mg/liter; H_2S present; coliform count, 1000/cm^3.
Stage 9: Ultrans-saprobic. No organisms; azoic stage before decomposition begins. Anaerobic. BOD_5: 1000–60,000 mg/liter; H_2S generally not present.

IV. TRANS-SAPROBIC

Stage 10: Antisaprobic (no life, no decomposition because toxicity is outside the saprobic system). Aerobic or anaerobic type; industrial wastes containing poisons. BOD_5: 0; coliform count: 0.
Stage 11: Radiosaprobic. Organisms absorb radioactive substances, cryptolethal; other indications vary.

In a later work, Sládeček (49) gave a more detailed account of his reasons for drawing up this system of abiotic zones in water and wastes. In this work he points out that he considers katharobic a designation which applies to completely clean water containing no organic substances, even cleaner that that indicated according to Kolkwitz & Marsson's system (41), and he refers to the views held by Zelinka & Marvan (58) who write that organisms in springs and small streams should not be described as katharobic but as xenosaprobic (previously β-oligosaprobic).

Waters influenced by poisonous substances to such an extent that all organisms die are described as antisaprobic, but lack of organisms may also be due to physical reasons, e.g., the presence of coal dust, minerals, etc. which influences all objects and organisms and prevents decomposition. Such waters should be described as cryptosaprobic.

The ultrans-saprobic stage is characterized by being abiotic, but not toxic (bacteria: 0–10, mycophyta: 0–10, BOD_5: 1000–60,000 mg/liter), but since there is no decomposition and no community or organisms, this stage does not seem to form part of the system. It is probably wastes from sugar factories which Sládeček has had in mind. The postulate that abiotic water is nontoxic appears unrealistic.

In a still later publication (51), Sládeček acknowledged that katharobic, in the sense in which he uses the term, as well as trans-saprobic and radiosaprobic do not belong in a saprobic system. For this reason he now replaces Kolkwitz' term, katharobic, by xenosaprobic, designating clean water with organisms.

Regarding trans-saprobicity, Sládeček writes "since at all trans-saprobic degrees interference with saprobicity often occurs, it is possible to replace the scheme by developing three (or four) independent systems, i.e.: system of saprobicity, system of toxicity, system of radioactivity.

It seems inadvisable to specify numbers of coliform bacteria, and as the figures given show wide variations, they are actually of no value with reference to the system. The BOD values given for each individual stage also show great differences between minimum and maximum values. The BOD values may be said to be no more than vaguely related to the complex of biological processes which take place in the water. This may possibly be due to the fact that the BOD value is greatly dependent on toxic substances in the water. Even in low concentrations such substances as lead, copper, mercury, or chromium present as chromate are toxic to bacteria, and zinc, nickel, and cadmium cause a marked reduction in the rate of oxygen uptake. However, Purdy & Butterfield's hypothesis should also be taken into consideration. They write that "the presence of bacteria-consuming protozoa exerts a great influence on oxygen consumption. Protozoa, by feeding on bacteria, reduce their number and stimulate their further development. Thus, the bacterial metabolism is kept at a high level and the development of protozoa may greatly influence the oxygen consumption during the BOD test." In 1963, Javornický & Prokesová (36) were able to confirm this hypothesis by laboratory experiments. As early as 1947 Weston (56) demonstrated the fact that at least eighteen variables may have an important influence on the BOD test.

Sládeček's saprobic system can only be considered to be an extension of Kolkwitz & Marsson's original system.

Caspers & Schulz (13) have discussed the saprobic system on the basis of studies of the Isebaek canal near Hamburg. They found that using the list of indicator species given by Kolkwitz, Liebmann and Šrámek-Husek the results of the biological assessment did not agree with the actual conditions. According to the present author's opinion, this is due, however, to the fact that the last-mentioned authors have attached excessive importance to phytoplankton organisms, which are chiefly indicators of eutrophication, and to protozoa which—as mentioned in the above—are unsuitable as indicators of pollution. In a work published in 1962, Caspers & Schulz (14) tried to use the chemical assessment method depending on the oxygen consumption during 48 hours which Richter (48) has developed; this method was also found to be unsatisfactory. Caspers & Karbe (11) then tried to consider trophicity and saprobicity as a metabolic-dynamic complex. This point of view was discussed at a symposium at Prague where the trophicity concept gained ground, and where the following formulation of the term saprobicity was agreed upon: "Within the bioactivity of a body of water, saprobicity is the sum total of all those metabolic processes which are the antithesis of primary production. It is therefore the sum total of all those processes which are accompanied by loss of potential energy" (Part I, Prague Convention).

Caspers & Karbe (12) have proposed a saprobiological classification of waters. In this system, saprobicity is considered to be a common phenomenon within the framework of the metabolic processes in the waters, thus, it is not considered to apply merely to the conversion of anthropogenous material and, secondly, unlike the trophic components in the metabolic-dynamic complex, the saprobic components comprise all the phenomena connected with breakdown of organic substances. While Elster (16) has defined trophism as the intensity of the primary production, Caspers & Karbe (11) defines saprobicity as the intensity of the decomposition of dead matter. By combining metabolic-dynamic and biocoenotic points of view with regard to the evaluation of waters we arrive at the second part of the Prague convention: "From the saprobicity appears by combination of the biogenous and the physical supply of oxygen the saprobic degree of the water. The latter may be determined both by metabolic-dynamic measurements and by analysis of the communities."

Caspers & Karbe (12) give the following criteria for the delimitation of oxygen content in the saprobic stages: 1. β-oligosaprobic: rate of assimilation and respiration, low O_2 content independent of this, determined by hydrographic factors. 2. α-oligosaprobic: oxygen content depends equally on assimilation and respiration activity of the organisms and on hydrographic factors. 3. β-mesosaprobic: the diurnal curve representing the oxygen content is clearly seen to depend on the activity of the organisms, during the daytime frequently supersaturation with oxygen, in the night obviously oxygen deficit. 4. α-mesosaprobic: oxygen content generally below saturation, during the night conditions are frequently anaerobic; in addition to the activity of the organisms variations in the supply of putrifying material are observed. 5. β-polysaprobic: even during daytime chiefly anaerobic conditions, and it is difficult to observe any influence due to assimilation. The diurnal oxygen curve is characterized by putrefying material. 6. α-polysaprobic: constantly anaerobic conditions.

Table 1 gives Caspers & Karbe's (12) delimitation of the saprobic levels according to bioactivity components, and Table 2 gives the delimitation of the saprobic levels according to the trophic structure of the organism communities.

Saprobicity derives either from an autosaprobic or an allosaprobic component, where autosaprobicity equals the intensity of the decomposition of organic matter of autochthonous origin, while allosaprobicity equals the intensity of the decomposition of introduced organic matter, i.e., of allochthonous origin (cf. Caspers & Karbe, 11).

In that the oxygen content is assumed to be the only causal factor, the system actually approaches Kolkwitz' system in which the saprobic zones are characterized by the oxygen saturation: polysaprobic = 0, α-mesosaprobic < 50 percent, β-mesosaprobic > 50 percent, which may perhaps appear reasonable when animal organisms are concerned but certainly not when assessment is based on phytomicroorganisms. Much of the criticism directed against Kolkwitz' system has centered on the fact that he has based it too one-sidedly on oxygen saturation. We know now that there is a series of factors on which the presence of organisms depends: H_2S, CH_4, etc.; a great number of inorganic trace substances, vitamins, hormones, amino acids, and to this should be added the metabolic products of the individual organisms which may either favor or inhibit growth or be toxic to other organisms (cf. Fjerdingstad, 24).

Even though studies of metabolic dynamics are of importance for the understanding of the complex of processes occurring in the water, these methods have not as yet reached their final development. It must be taken into consideration that it has been shown that the conversion of carbon dioxide into sugar is the result of enzymatic action. A number of studies of algae and purple bacteria have shown that radiant energy absorbed by chlorophyll may be utilized in different ways, not necessarily in absorbing carbon dioxide and liberating oxygen, but, for example, in the conversion of acetic acid (which may result from bacterial degradation of organic matter). Under certain conditions many algae are able to assimilate carbon dioxide without developing oxygen, or they may store radiant energy in the form of phosphate rich in energy (Gaffron, 28). Myers (43) states that the ratio CO_2/O_2 varies markedly with the nitrogen source. To base biological assessment of waters (saprobic system) on metabolic-dynamic points of view will

TABLE 1. Saprobicity Levels According to the Trophic Structure of the Communities of Organisms

Saprobicity level	Structure of the communities of organisms
I. β-oligosaprobic	Balanced relationship between producers, consumers, and destroyers; the communities of organisms are poor in individuals but there is a moderate variety of species, small biomass, and low bioactivity.
II. α-oligosaprobic	Balanced relationship between producers, consumers, and destroyers; the communities of organisms are rich in individuals and species with a large biomass and high bioactivity.
III. β-mesosaprobic	Substantially balanced relationship between producers, consumers, and destroyers; a relative increase in the abundance of destroyers and, accordingly, of the consumers living on them; communities of organisms are rich in individuals and species with large biomass and high bioactivity.
IV. α-mesosaprobic	Producers decline as compared with an increase in consumers and destroyers; mixotrophic and amphitrophic forms predominate among the producers; communities and organisms rich in individuals but poor in species with large biomass and extremely high bioactivity; still only few species of macroorganisms; mass development of bacteria and bacteria-eating ciliates.
V. β-polysaprobic	Producers drastically decline; communities of organisms are extremely rich in individuals but poor in species with a large biomass and high bioactivity; macrofauna represented only by a few species of tubificids and chironomids; as in IV, these are in great abundance; mass development of bacteria and bacteria-eating ciliates.
VI. α-polysaprobic	Producers are absent; the total biomass is formed practically solely by anaerobic bacteria and fungi; macroorganisms are absent; flagellates outnumber ciliates among the protozoa.

hardly serve to promote the use of biological methods in practice and would consequently be highly regrettable. Incidentally, the system does not provide any information about the organisms which belong in the different zones.

In a number of publications Fjerdingstad (18–27) has discussed biological assessment of pollution and has suggested a new saprobic system on the

TABLE 2. Delimitation of Saprobicity Levels According to Bioactivity Components (According to Caspers & Karbe)

Saprobicity level	Supply of organic matter	Primary production	Respiration rate	Organic drift and rate of sedimentation	P/R quotient
	Supply of allochothonous potential energy	Autochthonous conversion of kinetic into potential energy	Conversion of potential energy and liberation as kinetic energy	nonutilized potential energy	
I β-oligo	negligible	small	small	negligible	about 1
Ia saprobic	small	small[a]	small	small	>1
II α-oligo	small	medium	medium	small	about 1
IIa saprobic	small	high	medium	medium	>1
III β-meso	high	medium	high	medium	<1
IIIa saprobic	high	high[a]	high	high	$1->1$
IV α-meso	very high	high	very high	high	$\ll 1$
IVa saprobic	very high	high[a]	very high	high[a]	<1
V β-polysaprobic	extremely high	small	extremely high	extremely high	$\ll<1$
VI α-polysaprobic	extremely high	about 0	extremely high	extremely high	about 0

[a] denotes an increase of the rate above the normal type

basis of microorganism communities (algae, bacteria, etc.). In agreement with Thienemann (55), he first divides the relevant organisms into four groups (autecology): 1. coenobiontic species that are characteristic of the community and that mostly occur in fairly large numbers of individuals; 2. coenophilous species which may occur also in other communities, that is, they are to a certain extent indifferent; 3. coenoxenous species, species occurring incidentally and present mostly in very small numbers (introduced from outside); 4. saprophobous species which cannot exist in polluted water. The number of coenobiontic species mentioned is 67; of saprophilous species, 66 are given; of saproxenous, 34; and of saprophobous, 17 species. Next, the communities are considered (synecology) in which one or more species are predominant and are considered to form communities and give their names to the communities. The system is built on the communities and consists of nine zones with subzones (see Table 3).

The species which usually occur as associates in the individual communities are mentioned, and the individual zones are characterized as follows:

Zone I. The coprozoic zone. This zone comprises the undiluted brown fecal water in which formation of H_2S has not as yet begun, and domestic sewage before it enters the purification plant. The domestic sewage is grayish with a stale odor; if H_2S formation has started it must be fairly insignificant as otherwise the grayish sludge particles would turn black. The BOD is high and so is the total nitrogen content, while contents of NH_4 and NO_3 may be fairly small.

TABLE 3. SURVEY OF THE SAPROBIC ZONES AND THE CORRESPONDING COMMUNITIES

Zone I	Coprozoic zone a. bacterium community; b. Bodo community, c. both communities
Zone II	α-polysaprobic zone 1. *Euglena* community, 2. Rhodo- and thio-bacterium community 3. pure *Chlorobacterium* community
Zone III	β-polysaprobic zone 1. *Beggiatoa* community, 2. *Thiothrix nivea* community, 3. *Euglena* community
Zone IV	γ-polysaprobic zone 1. *Oscillatoria chlorina* community, 2. *Sphaerotilus natans* community
Zone V	α-mesosaprobic zone a. *Ulothrix zonata* community, b. *Oscillatoria benthonicum* community, c. *Stigeoclonium tenue* community
Zone VI	β-mesosaprobic zone a. *Cladophora fracta* community, b. *Phormidium* community
Zone VII	γ-mesosaprobic zone a. Rhodophyce community (*Bactrachospermum moniliforme* or *Lemanea fluviatilis*) b. Chlorophyce community (*Cladophora glomerata* or *Ulothrix zonata* (clean-water type)).
Zone VIII	Oligosaprobic zone a. Chlorophyce community (*Draparanaldia glomerata*), b. pure *Meridion circulare* community, c. Rhodophyce community (*Lemanea annulata*, *Batrachospermum vagum*, or *Hildenbrandia rivularis*), d. *Vaucheria sessilis* community, e. *Phormidium inundatum* community.
Zone IX	Katharobic zone a. Chlorophyce community (*Chlorotylium cataractum* and *Draparnaldia plumosa*), b. Rhodophyce community (*Hildenbrandia rivularis*), c. lime-encrusting algal communities (*Chamaesiphon polonius* and various *Calothrix* species).

a, b, c,—as alternatives.
1, 2, 3,—as differences in degree.

Zone II. The α-polysaprobic zone. This is the H_2S environment; O_2 is either absent or present in negligible quantities only but the decomposition of organic matter is in full progress, manifesting itself by a high NH_4 content.

Zone III. The β-polysaprobic zone. A low O_2 content is essential for the communities belonging to this zone, but H_2S is still present. The phosphate content is high in this zone and there may be a very considerable content of NH_4.

Zone IV. The γ-polysaprobic zone. Chemically, this zone is characterized by the presence of small amounts of H_2S, a low O_2 saturation percentage, and decreasing NH_4 content.

Zone V. The α-mesosaprobic zone. Chemically, the zone is characterized by a high content of amino acids; by $H_2S = O$; by $O_2 < 50$ percent saturation; and by BOD > 10 mg/liter (generally).

Zone VI. The β-mesosaprobic zone. Chemically, this zone is characterized by continued oxidation and mineralization of organic matter. Usually $O_2 > 50$ percent saturation; BOD < 10 mg/liter; and $NO^-_3 > NO^-_2 > NH^+_4$.

Zone VII. The γ-mesosaprobic zone. Chemically, the characteristics of this zone are almost complete decomposition of organic matter, increasing oxygen saturation, and BOD = 3–6 mg/liter.

Zone VIII. The oligosaprobic zone. In this zone the mineralization of organic matter has been completed and BOD < 3 mg/liter.

Zone IX. The katharobic zone. This zone comprises waters that have not been exposed to pollution.

The system is intended for receiving waters into which domestic sewage with a certain admixture of industrial wastes is introduced. It is the author's opinion that actual industrial wastes, the nature and composition of which vary so much, do not belong in a saprobic system. However, with regard to industrial waste he distinguishes between two zones, a chemotoxic zone in which there are no living organisms and a chemobiontic zone in which there are organisms which are resistant to the toxic substances. The chemobiotic zone is illustrated by examples of various types of industrial wastes and the species occurring in them.

In a subsequent work (26), Fjerdingstad has given an evaluation of the saprobic valence of 182 phytomicroorganisms on the basis of his own experience and information in the literature with regard to occurrence and ecology, and in this way he has further developed and motivated the system.

Since 1964 the system has been used successfully, for example, for the annual assessment of the pollution of the river Odenseå (Denmark) and its tributaries. In an examination which was made in 1968–1969 of Denmark's largest river, Gudenå, in addition to the assessment which was made according to the above system, sediment samples for bacteriological examination were collected throughout 27 km of the stream. The bacteriological examination comprised counts of colonies per gram on meat-peptone gelatine (21°C), on meat-peptone agar (37°C), counts of *E. coli* types I (45°C), number of *Clostridium perfringens,* total number of sulfite-reducing and sulfate-reducing bacteria, and number of sulfur-oxidizing bacteria of the genus *Thiobacillus.* Cultures for the purpose of counting cellulose-decomposing bacteria (aerobic + anaerobic at 30°C) and *Pseudomonas* sp. (21–30–37–42°C) were also made. It appeared that there was no correlation between the degree of pollution and the number of colonies on gelatine or agar media. The number of *Pseudomonas* sp. decreased considerably beyond the outlet from the three water treatment plants situated within this reach of the river. The same applied to the cellulose-decomposing bacteria, while the number of sulfur bacteria (sulfite-reducing, sulfate-reducing and *Thiobacillus*) corresponds fairly well to the increases and decrease in the degree of pollution. A metabolic-dynamic criterion may thus be said to apply also to this system, although it is not based on the oxygen consumption, but on the highly significant sulfur reactions.

COMMENTS

It is remarkable that most of the authors who, in the course of time, have criticized the biological assessment of pollution have always maintained that the method was not sufficiently substantiated, or that certain indicator organisms were not correctly placed in the system while, nevertheless, they conclude by saying that the biological method is indispensable. After Fjerdingstad's work on the saprobic valence of 182 species, such criticism seems no longer justified.

Hynes (34), who is greatly in favor of biological water assessment, writes: "Pollution is essentially a biological phenomenon, and it is a source of constant surprise to me that it is not generally studied from a biological standpoint. . . . It is even odder when one appreciates the enormous amount of labour which is undertaken every day by men in white coats measuring DO, OA, BOD, SS and many other letters on the alphabet on hundreds of bottles of variously dirty water . . . a biologist on the staff could save them from this, and he could do the job more quickly, more cheaply and very often more reliably." How right Hynes is!

If we look toward the future, it must be said that it is not sufficient to fight the increasing eutrophication and pollution of our waters by stopping the introduction of nutrients, active efforts must be made toward restoration of the lakes and rivers which have already been spoiled. Endeavors in this direction have been started in a number of countries, e.g., Sweden, the United States, and Switzerland, using methods such as removal of sludge from the bottom and aeration of the bottom layer of water in order thereby to accelerate the oxidation of organic matter. In Sweden, a great scheme has moreover been started for the purpose of restoring an overgrown lake to its original state in order thereby to create a natural park.

LITERATURE CITED[1]

1. Berg, Kaj et al. 1958. *Folia Limnol. Scand.* 10:1–189
2. Bringmann, G. 1960. *Vom Wasser* 27:86–98
3. Bringmann, G., Kühn, R. 1956. *Gesundh. Ing.* 77:374–81
4. Ibid. 1958. 79:329–33
5. Ibid. 1958. 79:50–54
6. Bringmann, G., Kühn, R. 1962. *Int. Rev. gestamten Hydrobiol.* 47:123–45
7. Brinley, F. J. 1942. *Sewage Works J.* 14:147–59
8. Bucksteeg, W. H., Thiele, H. 1959. *Gas.-Wasserfach.* 100:912
9. Bursche, E. M. 1961. *Int. Rev. Gesamten. Hydrobiol.* 46:610:52
10. Caspers, H. 1962. *Int. Rev. Gesamten. Hydrobiol.* 47:581–86
11. Caspers, H., Karbe, A. 1966. *Arch. Hydrobiol.* 61:453–70
12. Caspers, H., Karbe, A. 1967. *Int. Rev. Gesamten. Hydrobiol.* 52:145–62
13. Caspers, H., Schulz, H. 1960. *Int. Rev. Gesamten. Hydrobiol.* 45:535–65
14. Caspers, H., Schulz, H. 1962. *Int. Rev. Gesamten. Hydrobiol.* 47:100–17
15. Cummins, K. W. 1969. *Stream Ecosystem, Tech. Rep.* 7:31–37
16. Elster, H. J. 1966. *Verh. Int. Ver. Limnol.* 16:759–85
17. Felföldy, L. J. M. 1962. *Ann. Biol. Tihanny.* 29:85–93
18. Fjerdingstad, E. 1950. *Dan. Bot. Ark.* 14:1–44
19. Fjerdingstad, E. 1950. *Folia. Limnol. Scand.* 5:1–123
20. Fjerdingstad, E. 1954. *Hydrobiologia* 6:328–30
21. Fjerdingstad, E. 1957. *Arch. Hydrobiol.* 53:240–49
22. Fjerdingstad, E. 1960. *Nord. Hyg. Tidskr.* 41:149–96
23. Fjerdingstad, E. 1965. Biol. Problems in Water Pollution, 3rd Seminar 1962. *U.S. Pub. Health Serv.* 232–33
24. Fjerdingstad, E. 1964. *Symp. Pollution Marine Microorg. Monaco,* 209–16

[1] For supplementary bibliographic material (15 pages), order NAPS Document 01357 from ASIS National Auxiliary Publications Service, c/o CCM Information Corporation, 909 Third Avenue, New York, N.Y. 10022.

25. Fjerdingstad, E. 1964. *Int. Rev. Gesamten Hydrobiol.* 49:63–131
26. Fjerdingstad, E. 1965. *Int. Rev. Gesamten Hydrobiol.* 50:475–604
27. Fjerdingstad, E., Hvid-Hansen, N. 1951. *Nord. Hyg. Tidskr.* 32:159–80
28. Gaffron, H. 1962. *Beitr. Physiol. Morphol. Algen. Stuttgart.* 1–12
29. Gray, E. A. 1956. *Verh. Int. Ver. Limnol.* 12:814–17
30. Hassler, A. 1947. *Ecology* 28:383–95
31. Hobbie, J. E., Wright, R. T. 1965. *Limnol. Oceanogr.* 10:47–474
32. Holm-Hansen, V., Brooth, C. R. 1966. *Limnol. Oceanogr.* 11:510–51
33. Hynes, H. B. N. 1960. *The Biology of Polluted Waters.* Liverpool: Liverpool Univ. Press, 202 pp.
34. Hynes, H. B. N. 1961. *River Board Ass. Animal Conf. Brighton,* 3–8
35. Jannasch, H. W. 1969. *Verh. Int. Ver. Limnol.* 17:25–39
36. Javornický, P., Prokesová, V. 1963. *Int. Rev. Gesamten. Hydrobiol.* 14:335–50
37. Keup, L. E. 1968. *Water Res.* 2:373–86
38. Knöpp, H. 1968. *Int. Rev. Gesamten. Hydrobiol.* 53:409–11
39. Kolkwitz, R. 1935. *Pflanzenphysiologie,* 3rd ed. Jena. 310 pp.
40. Kolkwitz, R. 1950. *Ver. Wass.-Boden-Lufthyg.* 4:1–64
41. Kolkwitz, R., Marsson, M. 1908. *Ber. Deut. Bot. Ges.* 26:505–19
42. Lackey, J. B. 1938. *Pub. Health Rep.* 53:1499–1507
43. Myers, J. 1962. *Beitr. Physiol. Morphol. Algen. Stuttgart* 13–19
44. Odum, H. T. 1957. *Ecol. Monogr.* 27: 55–112
45. Patrick, R. 1949. *Proc. Nat. Acad. Sci. Phila.* 101:271–344
46. Patrick, R. 1951. *Verh. Int. Ver. Limnol.* 11:299–307
47. Patrick, R., Hohn, M. H., Wallace, J. H. 1954. *Proc. Nat. Acad. Sci. Phila.* 259:1–12
48. Richter, K. 1959. *Vom Wasser.* 26: 56–67
49. Sládeček, V. 1961. *Arch. Hydrobiol.* 58:103–21
50. Sládeček, V. 1964. *Arch. Hydrobiol.* 60:241–43
51. Sládeček, V. 1966. *Verh. Int. Ver. Limnol.* 16:809–16

52. Šrámek-Husek, R. 1956. *Arch. Hydrobiol.* 51:376–90
53. Stange-Bursche, E. M. 1964. *Int. Rev. Gesamten. Hydrobiol.* 49:361–74
54. Steemann Nielsen, E., Wium-Andersen, S. 1970. *Int. J. Life Oceans Coast. Wat.* 6:93–97
55. Thienemann, A. 1920. *Festschr. Zchokke, Basel* 4:1–14
56. Weston, R. F. 1947. *Sewage Works J.* 19:871–74
57. Wezel, A. 1969. *Technische Hydrobiologie, Trink-Brauch-Abwasser. Akad. Verl. Leipzig* 407 pp.
58. Zelinka, M., Marvan, P. 1961. *Arch. Hydrobiol.* 57:389–407

1582

THE MICROBIOLOGY OF BREWING

JOHN KLEYN AND JAMES HOUGH

Department of Biology, University of Puget Sound, Tacoma, Washington, and British School of Malting and Brewing, Department of Biochemistry, University of Birmingham, Birmingham, England

CONTENTS

HISTORICAL

Making beer is an ancient craft, but fermentation has been understood for only some 130 years. Yeast was previously regarded as undesirable scum and, with a few enlightened exceptions, brewers initiated fermentation fortuitously from yeast either clinging to badly-cleaned equipment or present in unsterilized wort. About 1836, sugar fermentation was ascribed to vital activity of yeast (33) and the fungal nature of yeast was recognized (201). At this time bottom-fermentation yeast was used only in Bavaria, but its use spread rapidly in Europe and then to the United States, because of emigration of many German brewmasters. Pasteur began his microbiological research, developed a reasonable theory of fermentation (182) and, in 1876, reported on beer spoilage by bacteria (183). Pasteur's work was extended by Hansen who developed methods for isolating single yeast cells and, from a selected cell, propagating a clone sufficient for commercial-scale fermentation (89). Modern developments in brewing microbiology relate to maintaining yeast free of bacteria and to rapid methods of fermentation.

ORGANIZATION OF THE BREWING INDUSTRY

The industry is well organized into national and international technical associations. Of particular note are 1. the American Society of Brewing Chemists, 2. the European Brewery Convention, 3. the Institute of Brewing (Great Britain and Australia), and 4. the Master Brewers' Association of America. Except for 3 international congresses are organized regularly. Collaborative research and the development of analytical methods feature in the work of 1, 2, and 3. At the present time, these organizations hope to establish universally accepted methods of analysis. Such developments are undoubtedly encouraged by the international sharing of technical knowledge, by companies building overseas breweries, and by amalgamation of brewery companies. The size of the industry can be judged by the estimated world production figure for beer or circa 5×10^{10} liters (231).

LITERATURE OF BREWING

There are many specialized journals devoted to brewing science and technology including: *Brauwelt, Brauwissenschaft, Bulletin of Brewing Science* (Tokyo), *Communications of the Master Brewers' Association of America, Journal of the Institute of Brewing, Proceedings of American Society of Brewing Chemists, Proceedings of the European Brewery Convention, Wallerstein Laboratory Communications.* Additional microbiological information relating to breweries appears in several international journals devoted solely to microbiology.

TEACHING AND RESEARCH IN THE INDUSTRY

Particularly in Europe, the brewing industry in individual countries has helped to organize centers of teaching and research in brewing science, usually in universities (130). In addition, Brewing Research Institutes have been set up in certain countries by national brewery associations and supplement the work carried out by the laboratories in individual companies.

BACTERIA ENCOUNTERED IN BREWERIES

The number of bacterial genera encountered in breweries is small. Gram-positive genera comprise *Lactobacillus* (frequent in top-fermentation breweries) and *Pediococcus* (more common in bottom-fermentation breweries). The gram-negative genera are virtually confined to *Aerobacter (Klebsiella), Acetobacter, Acetomonas, Obesumbacterium* and *Zymomonas.* Table 1 shows the stage of production at which they occur (3).

TABLE 1. Brewery Spoilage Organisms and the Stage of Production at Which They Occur (3)

Stage	Bacteria encountered
I Mashing and sweet wort	Thermophilic lactic acid bacteria (rare)
II Cooling of wort to pitching with yeast	Acetic acid bacteria (rare) Lactic acid bacteria (rare) *Obesumbacterium* (rare)
III Fermentation	*Obesumbacterium* Acetic acid bacteria Lactic acid bacteria
IV After-fermentation stages	Acetic acid bacteria Lactic acid bacteria *Zymomonas* (rare)

Ault, R. G. 1965. *J. Inst. Brew.* 71:376–91.

Lactic Acid Bacteria

Many species of *Lactobacillus* and *Pediococcus* have been claimed to be associated with brewing (12, 59, 63, 236) but it is possible that all of them may well be varieties of (*a*) *L. brevis* and *L. pastorianus* which are heterofermentative species with long rod-shaped cells, and (*b*) *P. damnosus*, a homofermentative species with coccal-shaped cells (42, 81, 230). Carbohydrate is degraded by the heterofermentative species by the phosphoketolase pathway (189). In contrast to homofermenters, the glycolytic pathway is inactive because aldolase and hexose isomerase are missing (250). The end products of metabolism include lactic acid, ethanol, glycerol, acetic acid, and carbon dioxide. Homofermenters produce lactic acid but some strains yield a small amount of diacetyl (9) which, at levels as low as 0.2 ppm, can spoil beer (a condition called sarcina sickness). A wide variety of carbon sources will serve, especially maltose (249) but some strains require in addition carbon dioxide (195).

Both homofermentative and heterofermentative strains need a wide range of amino acids, nitrogenous bases, and vitamins (162), and therefore growth in beer depends on incomplete uptake of these materials by the yeast. Growth of isolated strains is best in the pH range 5.0–6.5 but in the breweries the same strains may grow at pH values below 4.5. Some strains are introduced into the mash in Continental breweries in order to produce lactic acid and lower the pH of the wort: thermophilic strains such as *Lactobacillus delbrückii* are most suitable (229). Certain strains produce an extracellular or thixotrophic slime which is a heteropolymer containing glucose, mannose, nucleic acid, and sometimes protein (53, 246). This slime, or rope, as it is called, spoils beer. Other undesirable effects of the bacteria upon beer include turbidity, acidity, and off-flavors (229). The bacteria may reinfect by persisting in the pitching yeast or in nonsterile equipment. Certain strains are capable of flocculating yeast (160).

Coliform Bacteria

These facultative anaerobes comprise strains of *Escherichia coli* and *Aerobacter (Klebsiella) aerogenes* that grow in wort rapidly, but in beer only when the pH is above 4.3 (229, 232). They produce a wide range of products of metabolism which impart flavors and odors to the wort that may be sweet and fruity, or resemble the smell of cooked cabbage. Coliforms may be introduced into the wort from water used for cooling or washing; they may be transmitted to the next fermentation via the pitching yeast but this is probably unusual. When wort is stored for continuous fermentation, coliforms may cause serious spoilage (3).

Acetic Acid Bacteria

When motile, *Acetobacter* has peritrichous flagella in contrast to the polar flagella of motile strains of *Acetomonas* (135, 207). The former genus has stronger powers of oxidation, the latter being scarcely able to oxidize ethanol further than acetic acid (190). *Acetobacter* strains may metabolize glucose via the hexose monophosphate pathway and the tricarboxylic acid cycle, but in some glucose is oxidized rather than phosphorylated (48, 242). In *Acetomonas* strains, the HMP pathway is normally used but the TCA cycle fails to operate due to lack of activity of isocitrate dehydrogenase and possibly other enzymes (245). There is therefore a requirement for certain amino acids or the corresponding oxo-acids. In contrast, *Acetobacter* strains can synthesize their entire complement of nitrogenous compounds from ammonia and suitable carbon fragments (25, 40). The simple nutritional requirements help to make them almost ubiquitous in some breweries and so become the most frequent cause of acidity, off-flavors, and turbidity (3). Certain strains are capable of causing yeast cells to die (71, 118). Acetic acid bacteria may also produce a dextranous "rope" in substantial quantities in beer (87). Frequently, the acetic acid bacteria grow as a greasy pellicle in order to increase exposure to atmospheric oxygen. Growth is very restricted in the absence of oxygen.

The variability of species of acetic acid bacteria is so marked that classification at this level may be of little value (206). A continuous spectrum of strains may exist with neighboring strains differing only in their ability to produce one or two enzymes.

Obesumbacterium proteus

This nonmotile species with short fat rod-like cells has the unique ability of growing in competition with actively multiplying yeast cells (209). It is a facultative anaerobe with a pH optimum of about 6.0 which grows with difficulty at pH levels below 4.5 (208). The glycolytic pathway, the HMP pathway and TCA cycle are active but terminal oxidation via the cytochromes is weak (221). Some strains at least require a spectrum of nitrogenous bases and amino acids for growth (214). It is not necessary for vitamins to be present in the growth medium and a large variety of sources of carbon will serve (214). The growth of the organism during a brewery fermentation depends on the strain of yeast used and, conversely, the rate of fermentation by the yeast is influenced by this bacterium (215). A characteristic odor of parsnips (204), possibly dimethyl sulfide (221), is transmitted to the beer by the metabolism of the bacterium. The following factors are important in the development of the organism in the early stages of fermentation: choice of yeast strain, seeding rate of bacteria, the pH of the wort, and the rate at which it falls during fermentation (35). If the numbers of *O. proteus* within the pitching yeast are high, the final pH of the beer will be greater than normal and the bacteria will be harvested in still larger numbers in the yeast crop (35).

Zymomonas anaerobia

The strains are usually highly motile with rod-like cells bearing 1–5 polar flagella (205). The species is a strict anaerobe and can grow over a wide pH range (3.5–7.5). Glucose and fructose, sometimes sucrose present in beer, but not maltose, are utilized as carbon sources (11). Energy is provided by the Entner-Doudoroff pathway, and the end products are ethanol and carbon dioxide (155). Minor products such as hydrogen sulfide and acetaldehyde give a highly objectionable odor (205). The organism grows in beer sweetened with sucrose or invert sugar and is transmitted by nonsterile equipment (3).

WILD YEASTS

Wild yeasts are those strains that are present in wort, beer, or other brewery materials which, by their action, do not enhance, and often spoil, the final products. Many wild yeasts are strains of *Saccharomyces cerevisiae* or *S. carlsbergensis* and cause off-flavors, fermentation of dextrins, and turbidity of beers (244). They are difficult to distinguish from culture yeasts on morphological and physiological grounds but very recent work with selective media has been encouraging (85). Serological methods have been used routinely to identify them in pitching yeasts (193). The current methods are based on antigenic relationships within the genus *Saccharomyces* (34, 199), and in the case of top yeasts, the antiserum to *S. pastorianus* is obtained from rabbits. This is absorbed with the culture yeast used in the brewery and then used for treating samples of pitching yeast under test. The wild yeasts may be present in proportions as low as a few cells per million cells of culture yeast (192), but the antiserum may be located around the wild yeasts by using an antirabbit serum from goats which has been coupled to a fluorescent dye such as fluorescein. A suitable microscope with ultraviolet illumination permits an operator to distinguish the fluorescing cells of the wild yeast (194).

Surveys of wild yeasts have been carried out in Britain on raw materials (244), pitching yeasts (18), and on draught beers (98). Apart from wild yeasts of the genus *Saccharomyces,* representatives from other genera such as *Candida, Debaryomyces, Hanseniaspora, Hansenula, Kloeckera, Pichia, Rhodotorula,* and *Torulopsis* have been encountered. Species of *Saccharomyces* fail to grow on a medium in which lysine is the sole source of nitrogen, while those of other yeast genera usually do, and therefore a lysine test enables the brewer to identify certain wild yeasts in low numbers in the presence of large numbers of culture yeasts (163).

OUTLINE OF TRADITIONAL BREWING PROCESSES

Traditional brewing processes include (*a*) malting, a process whereby barley is germinated, forming malt, thereby increasing the array of enzymes necessary for converting various substances in malt and malt adjuncts to forms capable of assimilation by fermenting yeast; (*b*) brewing of wort includes cooking, mashing, wort filtration, kettle boiling, and wort cooling; (*c*) fermentation of wort; (*d*) aging of fermented beer; (*e*) finishing and filtration; and (*f*) beer packaging.

Due primarily to space, equipment, and technical know-how, brewers, except for some of the largest ones, depend on the maltster for finished malt. As a prelude to successful malting, a pure variety of barley must be selected to ensure evenness of germination during malting (46). The major malting steps include grain storage to insure proper dormancy, steeping of the grain to acquire the necessary moisture content for malting, malting of the grain to prepare the malt, and kilning of the malt which serves to arrest germination and modification and provides malt with its characteristic flavor. Kilning of the malt is divided into two distinct phases: (*a*) drying which is the last step in germination, and (*b*) curing which is essentially a physicochemical reaction between different malt constituents. The manner of controlling these two phases helps determine whether the final malt will be a pale or dark malt (46). In all of the above malting steps, careful regulation of temperature, moisture, and aeration are important to produce properly modified malt. Related microbiological problems are discussed under the microbiology of brewing materials.

The malt as delivered to the brewery is next ground for efficient release and extraction of nutrients therein during mashing. Most American beers, in addition to malt, contain a starchy adjunct, usually corn or rice, which must by hydrolyzed to simpler sugars via amylolytic enzymes present in malt. In contrast, adjuncts are not permitted in German beer. The adjunct is first gelatinized by boiling in a cooker with a small amount of ground malt. Following gelatinization, the cooker contents are transferred to a mash tun which contains the bulk of the ground malt mixed with water. The primary objective of mashing is enzymatic conversion of the majority of starches, proteins, lipids, and organic phosphates into simple water-soluble forms. The resultant extract is named wort. The mashing temperature(s) and pH are dependent on the particular mashing process. The major mashing processes are infusion, decoction, and mixed mashing which is a combination of the first two. Infusion mashing is used for ale, whereas decoction mashing is used for lager beer and mixed mashing for production of Lambic and other old beer types (46). A primary difference between infusion and decoction mashing is boiling—in infusion mashing the temperature is held constant without boiling, whereas in decoction mashing a part of the mash is withdrawn, boiled and returned to the mash tun to raise the temperature of the whole mash. Boiling is not necessary for infusion mashing since highly modified malts are used which are already partly peptonized. Furthermore, a low proportion of unmalted cereal adjuncts are used and these are often precooked flakes.

Following mashing, the wort is separated from the spent grains and boiled in a kettle. Wort separation is essentially a filtration process which can be accomplished by two different methods. For ale production the mash tun is used for both mashing and filtration, whereas in lager beer production, filtration is almost always carried out in a separate vessel, either a lauter tub or a mash filter.

Major reasons for wort boiling are wort sterilization, enzyme destruction, coagulation of unstable colloidal protein named trub, and extraction of bittering substances and essential oils from hops. Boiled wort does contain viable *Bacillus* and *Clostridium* spores which, due to the acid environment, cannot propagate therein (3). In addition, hop resins contain substances toxic to many Gram-positive rod-shaped bacteria. According to Wackerbauer & Emeis (239), all types of lactobacilli can grow in unhopped beer but only a few can grow in the presence of hop resins. Unfortunately, the hopping rate in many countries is too low (e.g. ca 0.3 lb per bbl in the United States) to be of bactericidal value.

After boiling, spent hops are extracted by passage of the wort through a strainer known as the hop jack. The wort is next pumped to a closed hot wort tank followed by centrifugation to remove one form of proteinaceous particle called "hot trub" which coagulates in hot wort. In Britain, whirlpool-type hot wort tanks are often used for removing hot trub, thereby obviating the need for centrifuges. Wort also contains protein particles called "cold trub" which coagulate only in cold wort. The latter particles are probably structurally related to another proteinaceous material called "chill haze" which often forms in previously chilled warm beer. In former years an open tank, a coolship, was used for wort cooling, thereby enabling precipitation of cold, as well as hot trub particles. The wort is next cooled to pitching temperature by passage through a cooler, where air is simultaneously injected to help reduce the lag phase of yeast growth. Some breweries even use a limited additional aeration upon initiation of fermentation. One danger in the latter procedure is that of inducing excess diacetyl production (186). In a modern brewing operation all of the above processes from kettle boil to fermenter delivery occur in a closed system thereby helping to assure wort sterility at the onset of fermentation.

The two major beer types are lager and ale which are fermented with bottom and top yeasts, respectively. Most bottom yeasts are *Saccharomyces carlsbergensis* species whereas most top yeasts are *S. cerevisiae* species. Bottom yeasts flocculate and settle out to the fermenter bottom near the end of fermentation, whereas top yeasts remain powdery and form a yeasty head on the wort surface. As a consequence thereof, top yeasts are collected for repitching from the surface layers of the beer, whereas bottom yeasts are collected from the fermenter bottom. Lager beers are produced throughout the world, whereas ale is restricted primarily to countries of British origin. Some American breweries producing both lager and ale use only one yeast, a bottom yeast, for both beer types, thereby eliminating the problems involved in propagating and keeping two yeast cultures separate (188). Such ales produced with bottom yeast are designated "bastard ales." Conversely, in England, a "bastard lager" is produced from infusion wort plus bottom yeast. A lager yeast used to ferment ale is usually not repitched because of enhanced autolysis due primarily to the higher fermentation temperatures (15–20°C) of ale as compared to lager beer (6–11°C). An additional consequence of the higher fermentation temperature is a shorter fermentation interval for ale (4–5 days) as compared to lager beer (6–7 days). Melibiose fermentation is a diagnostic test used to separate the two yeasts. *S. carlsbergensis* is melibiose-positive, whereas *S. cerevisiae* is usually negative (191). In addition, *S. cerevisiae* usually sporulates more readily then *S. carlsbergensis.*

Following primary fermentation, the beer is subject to various aging (lagering) and filtration procedures depending upon the beer type and available facilities. In most instances, the beer is cooled to approximately 0°C near the end of primary fermentation and kept at that temperature until packaged. The low temperature helps retain carbonation and prevent microbiological contamination. Some major objectives in transferring beer to aging tanks following primary fermentation are: (*a*) beer clarification will hopefully occur through precipitation of most of the remaining yeast and cold trub. (*b*) The low storage temperature also helps precipitate a proteinaceous chill haze complex so as to obtain a more brilliant beer, as well as prevent subsequent haze formation when the beer is chilled after bottling. (*c*) Aging helps improve beer flavor by various, as yet little understood, chemical changes. (*d*) Some brewers use the aging period as a means of saturating the beer with carbon dioxide through secondary fermentation. Other brewers subject the aged beer to a primary filtration and then saturate the beer artifically with carbon dioxide in finishing tanks. Purging with CO_2 also helps wash undesirable volatiles out of the beer.

After finishing, the beer is subject to a secondary filtration followed by transfer to bottling tanks. After filling, package beers are often pasteurized in conventional tunnel pasteurizers.

OUTLINE OF ADVANCED BREWING PROCESSES

Dry milling of malt requires complex precision equipment to achieve uniform trituration of the endosperm to coarse flour and, at the same time, preserve the husks almost intact (46). In several breweries, the husk is rendered less brittle by either brief steaming or steeping. Usually steeping is followed by wet milling to produce a mash and therefore a separate mash-mixer is not required (132). In order to reduce the requirement for mashing equipment capacity, syrups resembling concentrated sweet wort are used by breweries in several countries (149). The syrups may be derived from many unmalted cereals, but most commonly from maize and barley, and the degradation is achieved by industrial amylolytic enzymes (142). In the case of maize, acid hydrolysis may precede enzymatic degradation (46). Hops are bulky and are usually stored in cold-rooms; they are replaced in many breweries by a vacuum-packed powder produced by milling the hops (14) and selecting a fraction rich in lupulin glands, the site of bitter resins and essential oils (30). Alternatively, the resins may be extracted by an organic solvent to yield a semisolid that may be canned (159). In the copper (kettle) the principal bitter resins are isomerized to isohumulones (212). Similarly, the extract may be isomerized by boiling with dilute alkali (107). This preparation is usually added after fermentation in order to obviate adsorption of the isohumulones to precipitated protein in the copper and yeast in the fermenter (50).

Enclosed fermentation vessels are replacing open ones because the former are more easily spray-cleaned, have reduced susceptibility to air-borne microbial infection, are cooled more readily and facilitate collection of evolved carbon dioxide (233). In cylindro-conical vessels, which are a popular type of enclosed fermenters, the yeast separates from the beer at the end of fermentation into the cone from which yeast slurry can easily be removed (238). Under slight pressure, the carbon dioxide produced from fermentation will readily dissolve in the beer. Large open-air vessels, now used in some breweries for fermentation and beer storage, may be built and maintained cheaply for their capacity (220). The speed at which yeast is naturally separated from beer has, in the past, determined to some degree the duration of the beer in fermenters, but centrifuges are now commonly used to achieve the separation (37).

Continuous fermenters recently introduced have included stirred vessels coupled in series which are superior in performance to a single stirred vessel (15, 105). Yeast recycled, from the emerging beer, to the fermenters increases the overall fermentation rate but at high yeast concentrations the yeast cells divide slowly and the quality of the beer produced is altered (104). Unstirred continuous fermenters of tower form permit dwell-times in the order of 6–9 hr and are used commercially (4, 198). Semicontinuous or accelerated batch systems have been used successfully on at least pilot scale (83).

MICROBIOLOGY OF BREWING MATERIALS

The prime ingredients in beer are water, malt, hops, and yeast. Some European beers are all malt beers, whereas most American beers contain ca 50 per cent malt and 50 per cent adjunct, usually corn or rice.

Production of suitable brewing water is usually not a major problem since it can be treated to either add or remove various metal ions and salts (39, 46). It is an almost universal practice to harden or mineralize the mash water or mash with calcium and magnesium sulfates, thereby lowering the pH of the mash. A high pH is unfavorable for a number of important reactions in the brewing process, e.g., at mashing saccharification will not proceed smoothly, the coagulation of protein (trub) at boiling is incomplete, and the resultant beers with a high pH are biologically unstable and liable to infection with lactic acid bacteria (46). Furthermore, the damaging effect of nitrates on fermentation is diminished (235). Other effects of inorganic ions, e.g. zinc, are discussed in the section on growth and fermentation kinetics.

Also of importance is the requirement that the water supply be relatively free of decaying vegetation, e.g., algae, and industrial wastes such as phenolics (241), since these can contribute off-tastes and odors that are difficult to remove (157).

Production of suitable malt begins on the farm where a pure variety of barley is necessary to obtain evenness of germination during malting (46). In moist growing seasons the barley may become infected with *Fusarium* spores that grow during the steeping process, producing substances which cause gushing in beer (73). Growth of foetid bacteria on the husks during steeping also occurs and can be eliminated in part by frequent changing of the steep liquor (46). Mold growth during steeping is reduced by raising the pH of the steep liquor with lime (46). Unfortunately, this process encourages growth of coliform-type bacteria (203). Another microbiological problem relates to the use of gibberellic acid to speed up germination during malting. The malts produced thereby have higher soluble N content which encourages growth of thermophilic lactobacilli during mashing (189). In certain cases, this might be desirable, particularly in countries such as Germany where the use of acids is prohibited in brewing liquor (171). Recent studies (61, 252) emphasize the importance of having the proper amount of amino acids in wort for achieving a satisfactory fermentation. The composition thereof is dependent both on the quality of the raw materials and mashing procedure (61, 184).

Hops, in addition to antiseptic value (239), contribute to the foam-stabilizing properties of wort and are important for achieving satisfactory yeast head production in ale-type beers (51). Current interest in fermenting unhopped wort with the subsequent addition of a preisomerized hop extract may complicate this problem (50).

SELECTION AND PROPAGATION OF BREWERS' YEAST

It is usual to select strains of yeast for brewing from yeasts already in commercial use. While the application of genetic principles to the production of new strains of bakers' yeast has been successful (65), there have been few instances of induced hybridization for commercial brewing (112). Mutation and transformation (178) have also been suggested for producing brewing strains with new properties but there has been no commercial exploitation. Desirable features in a brewing yeast include (*a*) the capacity to produce a beer of good flavor and aroma; (*b*) the ability to ferment wort rapidly until fructose, glucose, sucrose, maltose, and maltotriose have been used; and (*c*) the propensity to grow in wort rapidly until, in normal batch fermentation, the total concentration of yeast is in the range 2.5–5.5g dry weight/liter. In batch fermentation, it is desirable that the yeast separates readily from the beer at the conclusion of fermentation although less necessary if centrifuges are used for separation (37). Selection is normally based on the results of small-scale fermentations (213). For continuous fermentation using unstirred towers, it is necessary to have a yeast which is strongly sedimentary throughout the fermentation in order to maintain a yeast plug at the base of the tower (4, 32, 198).

Some breweries isolate, select and maintain their yeast strains but others engage specialist laboratories to provide this service. The entire yeast within a brewery may be derived from a single cell, from several isolated cells, from a single yeast colony or from several colonies (32). Again, some breweries choose to have two or more strains that may be employed in mixture or in separate fermentation vessels. Proportions of strains in a mixture may, however, change because of alterations in materials or procedure, and individual strains may be eliminated (103). Nevertheless, a yeast of several strains may adapt more successfully than a single clone. Cultures may be maintained at 10°C on wort-agar slopes or at 4°C in carbohydrate media such as 10 per cent sucrose, wort, or Wickerham's malt extract medium (243). Subculturing is carried out at regular intervals (24), preferably at less than three-month intervals. Lyophilized cultures have not been used extensively because there is a high mortality of cells during freeze-drying, and thus mutants and variants may be selected (251).

Many yeast propagators are based on the pioneer work of Hansen and Kühle and operate either semicontinuously or on a batch basis (88). Sterile wort is run into a vessel that has been presterilized by steam and cooled. Sterile air or oxygen is perfused through the wort, and the culture of yeast from the laboratory is inoculated. Aeration or oxygenation may be supplied continuously but, because of foam formation, it is more usually intermittent (43). In a modern example, the propagator is charged with 23 hl of wort of specific gravity 1.040 and is pitched with 91 g of pressed yeast;

aeration is provided for 1 min in every 5 at 0–11 m^3/ minute. At approximately 18°C exponential growth occurs for 48 hr when some 54 kg of pressed yeast is available in partly fermented beer of specific gravity 1.016. When the entire contents of the propagator are discharged into 250 hl of fresh wort there is no lag phase (5, 222). With some strains of yeast, the pH levels of beers produced in the propagator are low and the cells are elongated, but these effects are lost when the yeast is used normally (43). Modern cylindro-conical fermenters may be used as yeast propagators and stirred-tank continuous fermenters are particularly good (62).

Propagation of brewers' yeast enables a brewery to replace the entire stock of yeast on a predetermined basis. Frequently, a batch of yeast is used only about 12 times before it is discarded. There are, however, breweries claiming that their yeast has not been changed for 50 years or more (139). The changes in a yeast that persuade brewers to discard them relate either to infection with bacteria or wild yeast, poor settling near the end of fermentation if a bottom yeast, or partial loss of ability to grow, ferment, and produce the expected quality of beer.

YEAST MANAGEMENT

In the average brewery, a large inoculum of cells is used (ca 5–15 million cells/ml of wort). In each fermentation the number of cells increases three- to fourfold. Therefore, one-third to one-fourth of the yeast crop of each fermentation is used for inoculation of the next batch. If the brewery is of sufficient size (ca 1 million bbls of beer or more annually in the United States), drying of the remaining yeast for use as an animal food supplement becomes economically feasible. Alternatively, the yeast is used for manufacture of yeast extract or for fermentation in grain distilleries.

Yeast collected for repitching is usually mixed with 2–3 volumes of chilled water and passed through a vibrating screen to help remove bitter cold trub particles (196). In a modern brewing operation, the screened yeast passes directly into a scale hopper thereby providing the required amount of yeast for repitching (Editorial 1959, *Brewers Digest* 24:11). One danger in washing with water is a change in metabolic activity from fermentation to respiration (31), thereby increasing susceptibility to autolysis (116). Conversely, storage under chilled water is believed to hold autolysis to a minimum (100). Yeast to be stored for a prolonged period of time is best left in the fermenter under beer (38). One danger of prolonged storage is incomplete ability to ferment upon reuse (197). A minimum 24-hr rest period is believed necessary before reusing a yeast (197), but present practice in Britain with top and bottom yeasts in cylindro-conical vessels belies this belief.

Some suggestions for reducing yeast autolysis include iron enrichment and maintenance of a high C to N ratio (117), and the addition of unsaturated fatty acids to wort (223, 224). An important index of yeast autolysis is increased proteolytic activity (10).

Yeast contaminated with beer spoilage bacteria may either be replaced with a pure culture or washed with acids such as phosphoric acid (45), ammonium persulfate (27), or a combination thereof (7), thereby eliminating the necessity for replacement. Yeast replacement or acid washing can affect beer flavor since it usually requires several fermentations for fresh yeast to become acclimatized to the brewery (16). Related information on yeast replacement and acid washing is found in sections Selection and Propagation of Brewers' Yeast and Microbiological Control in Brewing, Fermentation, and Packaging Including Sanitation.

GROWTH AND METABOLISM

Brewers' wort (145) commonly has 8–14 per cent total solids, of which 90–92 per cent are carbohydrates. The major carbohydrate components of wort are glucose, fructose, maltose, sucrose, maltotriose, and a group of linear and perhaps also branched polymers of glucose containing four or more units. Brewers' yeast uses the sugars up to maltotriose but not the larger molecules (91). More fermentable worts are produced if the malts used are rich in amylolytic enzymes; unkilned malts are particularly rich. Lowering the mashing temperatures increases fermentability (86). Raising the proportion of unmalted cereal or the temperature of mashing diminishes wort fermentability (13, 110). Similarly, the concentration of nitrogenous material in the wort is influenced by the malt and other materials used in wortmaking and by mashing and wort boiling conditions (109, 200). Commercial worts commonly have 70–110 mg N/100 ml, and the nitrogenous constituents include ammonia, simple amines, amino acids, purines, and simple peptides to complex proteins (145). The most important source of nitrogen is the amino acids. Proline, an imino acid, is abundant but is scarcely used (113). Biotin, inositol, pantothenic acid, pyridoxine, and thiamine are present in wort and utilized by brewers' yeast. The total ash content of wort represents about 2 per cent of the wort solids; phosphates, chlorides, sulfates and other anions are present with the cations Na, K, Ca, Mg, Fe, Cu, and Zn. Phosphate content is in the range 60–120 mg/100 ml (64), and sulfate content in the region of 400 mg/liter (125). Dissolved oxygen content varies from about 4–14 mg/liter (154).

The growth and metabolism of brewers' yeast have recently been reviewed (191). Yeast cells readily take up monosaccharides by facilitated diffusion (120) but di- and trisaccharides enter the cell by means of a permease system (92, 93) which is inducible in some strains, constitutitive in others. Maltotriose is the last fermentable carbohydrate to be taken up. There is also a sequence of uptake of amino acids (Table 2) probably because of competition at the permease sites between the various acids (113, 114). The yeast is able to synthesize certain amino acids more easily than others. Thus, lysine, histidine, arginine, and leucine yield oxo-acids which are not furnished to any extent from carbohydrate metabolism and therefore changes in their concentration may affect the general metabolism of the yeast and hence the quality of the final beer. Nitrogen nutrition is complicated, however, by the ability of yeasts to release amino acids and nucleo-

TABLE 2. Order of Absorption of Amino Acids from Wort by Brewers' Yeast (113)

Group A	Group B	Group C	Group D
Immediately absorbed	Absorbed gradually during fermentation	Absorbed after a lag	Only slowly absorbed after 60 hr
Arginine	Histidine	α-Alanine	Proline
Asparagine	Isoleucine	Ammonia	
Aspartate	Leucine	Glycine	
Glutamate	Methionine	Phenylalanine	
Glutamine	Valine	Tryptophan	
Lysine		Tyrosine	
Serine			
Threonine			

From Jones, M., Pierce, J. S. 1964. *J. Inst. Brew.* 70:307–15.

tides into the medium especially when changing the medium, thereby causing alterations in membrane permeability (49, 136).

When yeast is pitched into aerated or oxygenated wort, there is at first a lag period when the cells actively take up materials from the wort, including the dissolved oxygen. It is not certain why the oxygen is important for the growth of the yeast but it may well permit synthesis of unsaturated lipids (2, 23) and influence mitochondrial function (36). The level of oxygen (about 4–14 mg/liter) is insufficient for any significant aerobic respiration and indeed the high levels of fermentable sugar ensure by the Crabtree effect (47, 211) that the metabolism is anaerobic. The major energy-yielding pathway is the glycolytic Embden-Meyerhoff-Parnas (EMP) one, but the hexose monophosphate shunt mechanism operates to a limited extent, mainly for the synthesis of pentoses (102). Pyruvic acid, the product of the EMP pathway, undergoes enzymic decarboxylation and reduction to ethanol and carbon dioxide. While this is the outstanding feature of yeast metabolism during beer production, special flavors and aromas of beers may arise from minor biochemical reactions (Table 3), notably those stemming from pyruvic acid. For instance, esters arise from an intracellular reaction involving acyl-CoA compounds, alcohols, and ATP (175). Ethyl acetate is thus produced from acetyl-CoA and ethanol, both products of pyruvic acid metabolism. The various fatty acids available within the cell compete in ester synthesis, except that propionic, isobutyric, and isovaleric acids do not furnish ethyl esters. Leakage of acyl CoA-compounds from the synthesis of higher fatty acids may also contribute to the level of esters, for instance, ethyl caprylate (174).

Esters other than ethyl esters utilize fusel alcohols which arise from either carbohydrate or amino acid metabolism, giving a range of oxo-acids (6, 106, 111, 115). Oxo-acids in excess of the requirements of the yeast may

TABLE 3. Taste Thresholds of Some Beer Constituents (μg/ml)

	In water	In lager beer	In degassed beer
Methanol	36.9	—	100
Ethanol	8.20	—	—
Propanol	6.08	—	50
Isopropanol	6.01	—	100
2-Methylpropanol	0.565	—	100
2-Methylbutanol	4.15	—	50
3-Methylbutanol	0.291	—	50
β-phenylethanol	0.00317	47.9	50
Ethyl acetate	0.257	93.5	5
Butyl acetate	0.043	2.63	—
Isobutyl acetate	0.073	—	1
Amyl acetate	0.009	3.44	—
Isoamyl acetate	0.019	2.30	1
Diacetyl	0.00261	0.162	0.005

From the results given in References 96 and 202.

be enzymically decarboxylated to the corresponding aldehyde which is then reduced to yield the fusel alcohols (Table 4). Thus, the uptake of isoleucine, leucine, valine, and phenylalanine from wort results in production by the yeast of 2- and 3-methyl butanol, iso-butanol and phenethyl alcohol. The choice of yeast strain, conditions of fermentation, and wort composition each affect fusel alcohol formation, thereby modifying beer flavor and aroma and providing material for ester synthesis.

Acetoin, diacetyl, and 2,3-pentanedione are normal beer constituents but in excess they spoil the beer by their musty, buttery, and honey flavors, respectively. The threshold of tolerance for vicinal diketones is in the order of 0.2–0.5 μg/mg (52, 240). Acetoin is produced from "active acetaldehyde" (hydroxyethyl-2-thiamine pyrophosphate) and free acetaldehyde in the presence of a carboligase. Yeast does not oxidize acetoin to diacetyl but instead tends to reduce diacetyl; thus, yeast is often added to filtered beer if the level of vicinal diketones is too high. Active acetaldehyde will react with pyruvic and oxo-butyric acids to yield acetolactic and acetohydroxy butyric acids, respectively, and it is believed that these acids (which may be precursors of valine and isoleucine) diffuse to some extent from the yeast cells into the beer. By decarboxylation and oxidation within the beer, the vicinal diketones are produced (218). Strains of *Pediococcus* and respiratory-deficient mutants of brewers' yeasts are sometimes responsible for high levels of vicinal diketones (44).

Yeast requires sulfur for the production of proteins, coenzymes, vitamins, etc., and takes up organic sulfur from wort, chiefly as methionine, and inorganic sulfur in the form of sulfate (152). Hydrogen sulfide is generated during yeast metabolism and depends, in brewery fermentations, on the

TABLE 4. Alcohols, Aldehydes, Oxo Acids, and Amino Acids Identified in Yeast (217)

Alcohols	Aldehydes	Oxo acids	Amino acids
Ethanol	Acetaldehyde	Pyruvic acid	Alanine +
Glycol	Glyoxal	Hydroxypyruvic acid	Serine
Propanol	Propionaldehyde	α-Oxobutyric acid	α-Aminobutyric acid
Isopropanol	—	—	—
Butanol	Butyraldehyde	—	—
Isobutanol	Isobutyraldehyde	α-Oxoisovaleric acid	Valine
Sec. butanol	—	—	—
Tert. butanol	—	—	—
Isoamyl alcohol	Isovaleraldehyde	α-Oxoisocaproic acid	Leucine
Act. amyl alcohol	Act. valeraldehyde	α-Oxo-β-methyl valeric acid	Isoleucine
Hexanol	Hexanal	—	—
Heptanol	Heptanal	—	—
—	—	Oxalacetic acid	Aspartic acid
—	—	α-Oxoglutaric acid	Glutamic acid
Phenethyl alcohol	—	Phenylpyruvic acid	Phenylalanine
Tyrosol	—	Hydroxyphenyl pyruvic acid	Tyrosine
Tryptophol	—	—	Tryptophan

Suomalainen, H. 1968. *Aspects of Yeast Metabolism*, 17, ed. A. K. Mills, H. Krebs. Oxford: Blackwell.

yeast strain used, the temperature, and the wort composition (123). The gas, unpleasant over certain threshold levels, arises either from leakage of sulfide ions during the enzymic reduction of sulfate or more likely by the action of cysteine desulfhydrase on cysteine (133). Mercaptans, sulfides, and thicarboxyls have been implicated in the flavor of beer (227). Nevertheless, growth of yeast in synthetic media and wort gives rise to no significant levels of volatile organic sulfur compounds (97, 176). These compounds arise from nonenzymic reactions in the beer (170) and from the metabolism of spoilage bacteria (1).

GROWTH AND FERMENTATION KINETICS

Brewery fermentations are characterized by the use of a complex medium and a large inoculum. In batch fermentations, the pitching rate is in the order of 0.2–0.4 mg dry wt/ml and the final harvest is 5–10 times this amount, depending on the yeast strain used, the composition of the wort and the process conditions that apply. Variations in pitching rate strongly influence the time to achieve fermentations (237) and in the ability of the yeast to utilize maltose (79). The small number of cell divisions normally occurring usually means that a true exponential growth phase is absent and is replaced by an almost linear increase in cell mass. Arrest of growth often occurs because there is an insufficiency of assimilable carbohydrate but may also occur if the yeast has come out of suspension due to premature sedimentation (bottom yeast) or premature yeast-head formation (top yeast). Rates of utilization of maltose and maltotriose by individual yeast strains appear to be dependent on the malto-permease system, and not on the overall maltase activity (79). Levels of pH and various cations, including K^+, Zn^{2+}, Mg^{2+}, and NH_4^+ appear to be important in utilization of maltose and maltotriose (234). The importance of Zn^{2+} levels in wort in influencing the quantity of yeast crop is recognized (147).

The role of oxygen has been mentioned earlier in connection with the Crabtree effect. With growth and fermentation kinetics, the continued use of wort containing 0.5 ppm dissolved oxygen immediately before pitching (6 per cent of air saturation level) leads to progressively poorer fermentations (108). In the range 0–20 per cent saturation of wort by oxygen before pitching, dissolved oxygen concentration is directly proportional to yeast crop and to fermentation rate (150).

YEAST CELL WALL

The brewing yeast cell wall is fairly rigid, layered, 100–200 nm thick, comprising glucans (40 per cent), mannans (40 per cent), protein (8 per cent), lipid (7 per cent), inorganic material (3 per cent), hexosamine and polymers (2 per cent). These proportions are approximate and vary according to yeast strain, cell age, and growth conditions (80, 156). The wall has associated with it, in free and bound forms, invertase, acid phosphatase, catalase, proteases and, in the case of *S. carlsbergensis,* melibiase (129, 146). Glucamylase is present in the wall of *S. diastaticus,* a wild yeast (101) and hydrolyzes beer dextrins. Other extracellular enzymes in the wall include glucanases (148), mannanases (148), protein disulphide reductases (173) which, with the proteases, are probably responsible for hydrolyzing the wall (161) and thus influence the budding process and the final shape of the bud (22). In top fermenting yeasts, certain strains are characterized by buds failing to detach so that chains of cells are formed (26).

The cells of other brewing yeasts may clump together to form flocs, especially in the absence of sugars and in the presence of divalent cations (54). The mechanism of flocculation probably involves the creation of salt bridges between superficial phospho-mannan-proteins of adjacent cells (141), and hydrogen bonding stabilizes the bridging (95, 158). The creation of flocs is important in beer clarification by sedimentation or centrifugation. Most strains of brewing yeast have a strong negative charge because of superficial phosphate and carboxyl groupings (55) and will therefore react with positively charged fining agents such as collagens (134) and certain gelatins. The ability of top fermentation yeasts to form a yeast-head (51) is also a function of the cell wall surface but the substances responsible for holding the cells at a gas/liquid interface have not been identified.

BREWING YEAST GENETICS

Recent reviews by Matile, Moor & Robinow (151), Fowell (67), and Mortimer & Hawthorne (165) provide timely information on three interrelated subjects: yeast cytology, sporulation, and genetics, respectively. Brewing yeast genetics, although moving at a steady pace, has been retarded somewhat because of the following: (*a*) One of the most important but least understood variables, the effect of yeast on beer flavor, is not yet amenable to genetic control (247). Some factors which are, include flocculation (69, 225), fermentation rate and limit (72, 228), maltose (78, 185, 228), maltotriose (228), and dextrin fermentation (68, 228), and total yeast cell count and cell size (69, 123). (*b*) The brewing characteristics of laboratory yeast hybrids are not always the same when evaluated in different breweries, indicating thereby that the environment plays a role (60). Also desirable for brewing is the development of new strains for use in continuous brewing and fermentation (143). (*c*) Most industrial yeasts are polyploids which sporulate poorly (57, 248). Associated therewith is low ascospore viability (56). This approach for altering the genotype can be obviated somewhat by induction of mitotic haploidization (58). (*d*) Mutants produced artificially have not produced suitable brewing yeasts (228). This is not surprising since most mutations are deleterious. Other methods for altering the genotype which may prove more beneficial include mitotic crossing-over, gene conversion, and nondisjunction (67). These occur at a higher frequency than mutation and can sometimes be induced by chemicals such as *p*-fluorophenylalanine. Another promising approach is transformation (178, 179), although to date it has not proven to be successful with yeast (94, 131, 164).

MICROBIOLOGICAL CONTROL IN BREWING, FERMENTATION, AND PACKAGING INCLUDING SANITATION

In general, methods for microbiological control have not changed significantly during the past 30 years except where dictated by a change in beer processing. Major changes in beer processing include the use of closed as compared to open vessels for brewing, fermentation, aging and packaging, attendant therewith is the adoption of in-place cleaning systems, the production of nonpasteurized package beer, and continuous brewing and fermentation. The first two processing changes help reduce bacteriological contamination, thereby lessening the need to examine more and more samples. Two additional parameters worthy of consideration in choosing a microbiological quality control program are: (*a*) Use of standards whenever possible, e.g., in a fermenting beer what constitutes an acceptable level of contamination? This will vary depending upon the overall design and construction of the brewery. Thus, open fermenters in older breweries with wooden surfaces are difficult to clean and will no doubt have a considerably higher level of contamination than modern breweries with closed fermenters with easy-to-clean rounded surfaces such as stainless steel. (*b*) Common sense realism is necessary, e.g., a beer to be pasteurized or a draft beer to be consumed within a short period of time may contain a small number of organisms, whereas a beer subject to aseptic filling can withstand very few, if any, contaminant organisms.

Some of the more recent survey papers on microbiological control are listed herein (17, 41, 66, 82, 128). The three major control areas are monitoring of brewers yeast quality; detection, enumeration, and control of microbial contaminants; and sanitation of the brewing environment.

Yeast quality can be determined by periodic measurement of the following parameters during the course of fermentation: fermentation rate (226), flocculation characteristics (41), degree of wort attenuation (210), total number of cells and percentage of budding and viable cells (122), chain length of cells (26), effect on beer flavor, and the presence or absence of foreign microorganisms. Some of the more recent innovations include the use of fluorescent dyes (74) in place of methylene blue or rhodamin B (177) to determine yeast cell viability, and the use of a Coulter counter in place of a hemocytometer to determine cell numbers (144).

The most reliable method for detecting contaminant organisms is plating on a differential growth medium (75, 119, 127, 169). Direct microscopic examination is often of little or no value since the contamination level may be too low to detect in this manner (20).

Recent years have seen marked improvement in formulation of media for detection of microbial contaminants due largely to a better understanding of their exact nutritional requirements (21). An excellent example is Nakagawa's medium (169) for detection of beer sarcinae. It contains 1. mannose, an energy source not used by lactobacilli, 2. ascorbic acid for inhibition of aerobic bacteria, and 3. actidione to inhibit brewers' yeast. A low pH serves to inhibit acid-sensitive bacteria.

Two additional measures used by many breweries to offset yeast contamination are acid washing of harvested yeast (see section on Yeast Management) and periodic replacement of yeast with a fresh culture (see section on Yeast Propagation). Fresh cultures are also employed to help correct inherent yeast differences such as a reduced rate of fermentation which may be the result of a genetic change in the yeast population.

One of the greatest deterrants to microbial contamination of brewers yeast is in-place cleaning (Editorial. 1969. *Intern. Brewers J.* 105:1251:59–67) which, in recent years, has helped revolutionize brewing quality control. As indicated earlier, the success of such an operation is dependent on having a closed system from kettle boil through fermentation. By definition, in-place cleaning is the process of circulating cleaning solutions through process equipment and removing the soiling material by chemical action. Two types of circuits are in use, a closed circuit for heat exchangers and pipe lines and an open circuit for brewhouse vessels, fermentors, and the like. Automation is possible and seepage of cleaning solution past a damaged valve seating is eliminated by arranging for the product to maintain a higher pressure than the cleaning solution. A wide choice of cleaning and sanitizing agents is available, the proper choice depending in part on the nature of the material to be cleaned (180). Some of the more commonly used sanitizing agents include various halogens (76), quaternary ammonium compounds (90), and miscellaneous compounds such as formaldehyde (167) and chloramine T (8).

Sanitation of the brewery environment is achieved in part by the following: (*a*) the use of ultraviolet light for sterilization of water used to rinse equipment and lines. The ultraviolet light source may be incorporated directly in the city water line. In order to achieve 100 per cent sterilization, it is important that the water be free of foreign particles such as iron deposits which can shield bacteria from the light source. (Kleyn, J. 1963. Personal observation). (*b*) Ultraviolet light may similarly be used for air sterilization (126). However, it is not thought to be in wide use with the possible exception of aseptic filling room sterilization. (*c*) Dehumidification of various cellars and packaging areas, thereby inhibiting mold growth which, in turn, will help eliminate musty odors as well as extend paint life (153). Passage of air through a dehumidification system also helps reduce the number of microorganisms present in the air. (*d*) Clean room techniques such as the use of positive air pressure in an aseptic filling room. Many of these are an offshoot of the space research program. (*e*) Incorporation of fungicides into grouting cement used in fermentation cellars.

ASEPTIC PACKAGING

Aseptic packaging is a brewing process whereby a biologically stable package beer is produced by means other than conventional tunnel pasteurization. Current commercial practices include flash pasteurization, membrane filtration, and chemical additives. Other proposed methods include ionizing radiations and high-frequency electrical fields (138). A major advantage relates to producing beer with an improved draft beer-like flavor. Conversely, there are brewers who will not adopt this process since tunnel pasteurization is believed to make a positive contribution to the flavor of their beer. A second advantage in certain aseptic packaging methods is a reduced cost of packaging as compared to conventional tunnel pasteurization (19). A third advantage relates to a saving of space through the elimination of conventional tunnel pasteurizers which require considerable floor area. For a recent review on aseptic packaging consult Portno (187).

One problem posed by flash pasteurization and membrane filtration and, to some extent, by chemical additives is the necessity of having an essentially sterile packaging area coupled with brewery workers trained in microbiological techniques necessary for aseptic packaging. Additional requirements to facilitate aseptic packaging include: (*a*) production of a beer with little fermentable sugar present. Absence thereof will inhibit reproduction of most contaminant microorganisms. The desired attenuation can be achieved through use of the proper yeast strain coupled with good brewing and fermentation practices. (*b*) Production of beer with a low oxygen content. Oxygen stimulates both microbial growth and oxidative rancidity and should therefore be reduced to a minimum. Methods proposed for reducing the air content of beer include the use of nitrogen in various stages of the bottling process (99). (*c*) Production of beer with a low number of organisms able to grow in packaged beer. A major requirement herein is selection of a brewers yeast strain which is unable to grow in well-attenuated beer. If microbial contaminants are present in high numbers before filtration, one will not be able to eliminate them with filtration (172). High numbers also pose a problem with chemical additives since their activity is dependent in part on the concentration of contaminant microorganisms present in the beer.

One of the major contaminants found in spoiled package beer is *Saccharomyces diastaticus* (77, 124, 219), a yeast able to utilize dextrins left in beer which regular brewers yeasts, *S. carlsbergensis* and *S. cerevisiae* cannot degrade (70). Greenspan (77) determined that as few as four *S. diastaticus* cells could infect a package beer of any size. Brumsted & Glenister (28) determined that a second major contaminant of package beer, a *Lactobacillus,* could infect with as few as one cell per 12 oz. bottle of beer. Both of the above organisms grow well in package beer in that they can tolerate reduced oxygen concentrations. Other organisms found occasionally in spoiled package beer include members of the genus *Pediococcus, Obesumbacterium,* and *Brettanomyces.*

Major problems of flash pasteurization include defects in beer flavor due in part to uneven heating of the beer and the requirement for aseptic filling. The latter requirement is also necessary when using membrane filtration but is avoided to some extent by chemical additives. Mulvany (168) and Posada & Galindo (181) compare merits and demerits of the first two mentioned processes. Chemical additives currently used in beer include *n*-heptyl *p*-hydroxybenzoate (216) and octyl gallate (140). Both compounds, although effective, have their limitations, thereby providing opportunity for new and better chemical additives. Some other preservatives evaluated but not in use commercially include diethyl pyrocarbonate (166), hydrogen peroxide (29), and salts of ethylene diamine tetraacetic acid (121).

The mechanism of action of *n*-heptyl *p*-hydroxybenzoate is thought to involve destruction of the cytoplasmic membrane (137), whereas EDTA is thought to function by binding metal ions such as Mg^{++} essential for yeast growth (84). Interestingly enough, the activity of *n*-heptyl-*p*-hydroxybenzoate is enhanced by the addition of Ca^{++} or Mg^{++} (137), whereas the reverse occurs with EDTA (121).

LITERATURE CITED

1. Anderson, R. W. 1970. Unpublished results.
2. Andreasen, A. A., Stier, T. J. B. 1953. *J. Cell. Comp. Physiol.* 41: 23–36
3. Ault, R. G. 1965. *J. Inst. Brew.* 71:376–91
4. Ault, R. G., Hampton, A. N., New-

ton, R., Roberts, R. H. 1969. *J. Inst. Brew. London* 75:260–77
5. Ault, R. G., Woodward, J. D. 1965. *Brew. Guardian* 94:17–30
6. Äyräpää, T. 1963. *Eur. Brew. Conv. Proc. Congr. (Nice)* 276–87
7. Bah, S., McKun, W. E. 1965. *Can. J. Microbiol.* 11:309–18
8. Barrett, M. A. Nov. 1969. *Process Biochem.* 4:23–24
9. Bärwald, G., Kesselschläger, J., Dellweg, H. 1969. *Naturwissenschaften* 56:285
10. Bergander, E., Bahrmann, K. 1957. *Nahrung* 1:74–87
11. Bexon, J., Dawes, E. A. 1970. *J. Gen. Microbiol.* 60:421–23
12. Bhandari, R. R., Walker, T. K. 1953. *J. Gen. Microbiol.* 8:330–32
13. Birtwistle, S. E., Hudson, J. R., MacWilliam, I. C. 1963. *J. Inst. Brew.* 69:239
14. Bishop, L. R. 1966. *Proc. Am. Soc. Brew. Chem.* 64–65
15. Bishop, L. R. 1970. *J. Inst. Brew.* 76:172–81
16. Bockelmann, J. B. 1954. *Proc. A.M. Master Brew. Assoc. Am.* 63–68
17. Bourgeois, C. 1968. *Biotechnique* 3: 2–39
18. Brady, B. L. 1958. *J. Inst. Brew.* 64:304–7
19. Brenner, M. W., Iffland, H. 1966. *Tech. Quart. Master Brew. Assoc. Am.* 3:193–99
20. Brenner, M. 1970. *Tech. Quart. Master Brew. Assoc. Am.* 7:43–49
21. Bridson, E. Y. 1969. *Lab. Equip. Digest.* Abs. from *Int. Brew. J.* Sept. 1969, 42–44
22. Brown, C. M., Hough, J. S. 1965. *Eur. Brew. Conv. Proc. Congr. (Stockholm)* 223–37
23. Brown, C. M., Johnson, B. J. 1970. *J. Gen. Microbiol.* 64:279–87
24. Brown, C. M., Wight, R. L., 1968. *Process Biochem.* 3:21–22, 28
25. Brown, G. D., Rainbow, C. 1956. *J. Gen. Microbiol* 15:61–69
26. Brown, M. L. 1970. *J. Inst. Brew.* 76:61–65
27. Bruch, C. W., Hoffman, A., Gosine, R. M., Brenner, M. W. 1964. *J. Inst. Brew.* 70:242–46
28. Brumsted, D. D., Glenister, P. R. 1963. *Proc. A.M. Am. Soc. Brew. Chem.* 12–15
29. Bulgakov, N. I., Zubenko, A. P., Antonova, I. I. 1954. *Tr. Vses. Nauch. Issled. Inst. Pivovar. Prom.* 4:38–40 Abs. from *Chem. Abs.* 1958 52:19012f
30. Burgess, A. H., 1964. *Hops, Botany, Cultivation and Utilization.* London: Leonard Hill. 300 pp.
31. Burns, J. A. Apr. 1957. *Brew. Guardian* 17
32. Burns, J. A. 1964. *Brew. Guild. J.* 50:589–98
33. Cagniard-Latour, C. 1838. *Ann. de Chimie et de Physique* 68:206–22
34. Campbell, I. 1967. *Eur. Brew. Conv. Proc. Congr. (Madrid)* 145–54
35. Case, A. C. 1965. *J. Inst. Brew.* 71: 250–56
36. Chapman, C., Bartley, W. 1968. *Biochem. J.* 107:455–65
37. Clark, D. F. 1967. *Process Biochem.* 2:41–42
38. Comrie, A. A. D. 1958. *Brew. Guild J.* 44:68–86
39. Comrie, A. A. D. 1967. *J. Inst. Brew.* 73:335–41
40. Cooksey, K. E., Rainbow, C. 1962. *J. Gen. Microbiol.* 27:135–42
41. Corran, H. S. 1965. *Tech. Quart. Master Brew. Assoc. Am.* 2:210–13
42. Coster, E., White, H. R. 1956. *J. Gen. Microbiol.* 37:15–31
43. Curtis, N. S., Clark, A. G. 1957. *Eur. Brew. Conv. Proc. Congr. (Copenhagen)* 249–262
44. Czarnecki, H. T., Van Engel, E. L. 1959. *Brew. Dig.* 52–56
45. Dean, B. T. 1957. *J. Inst. Brew.* 63: 36–43
46. de Clerck, J. 1957. *A Textbook of Brewing, Vol. 1,* trans. K. Barton-Wright. London: Chapman and Hall Ltd. 587 pp.
47. de Deken, R. H. 1966. *J. Gen. Microbiol.* 44:149–56
48. de Ley, J. 1961. *J. Gen. Microbiol.* 24:31–50
49. Delisle, A. L., Phaff, H. J. 1961, *Proc. Am. Soc. Brew. Chem.* 103–18
50. Dixon, I. J. 1967. *J. Inst. Brew.* 73: 488–93
51. Dixon, I. J., Kirsop, B. H. 1969. *J. Inst. Brew.* 75:200–4
52. Drews, B., Specht, H., Oelscher, H. J., Thürauf, F. M. 1962. *Monatsschr. Brau.* 15:109–13
53. Dunican, L. K., Seeley, H. W. 1965. *J. Gen. Microbiol.* 40:297–308
54. Eddy, A. A. 1958. *J. Inst. Brew.* 64: 143–50
55. Eddy, A. A., Rudin, A. D. 1958. *Proc. Roy. Soc. Ser. B. (London)* 148:419

56. Emeis, C. C. 1958. *Braueri wiss. Beil.* 11:160–63
57. Emeis, C. C. 1966. *Monatsschr. Brau.* 191–19
58. Emeis, C. C. 1966. *Z. Naturforsch.* 216:816–17
59. Emeis, C. C. 1969. *Monatsschr. Brau.* 22:8–11
60. Emeis, C. C. 1970. *Brew. Dig.* 45: 66–70
61. Enari, T. M., Linko, M., Lossa, M., Makinen, V. 1970. *Tech. Quart. Master Brew. Assoc. Am.* 7:237–40
62. Erda, A. R., Maresca, L., Laufer, S. 1961. *Eur. Brew. Conv. Proc. Congr. (Vienna)* 235–45
63. Eschenbecher, F. 1969. *Brauwissenschaft* 22:14–28
64. Essery, R. E. 1951. *J. Inst. Brew.* 57:170–74
65. Fowell, R. R. 1966. *Process Biochem.* 1:25–28, 60
66. Fowell, R. R. 1967. *Process Biochem.* 2:11–15
67. Fowell, R. R. 1970. *Sporulation and Hybridization of Yeasts.* In *The Yeasts,* Vol. 1, ed. A. H. Rose, J. S. Harrison, 303–83. London and New York: Academic. 508 pp.
68. Gilliland, R. B. 1953. *Eur. Brew. Conv. Proc. Congr. (Nice)* 121–34
69. Gilliland, R. B. 1965. *J. Inst. Brew.* 71:287–88
70. Gilliland, R. B. 1966. *J. Inst. Brew.* 72:271–75
71. Gilliland, R. B., Lacey, J. P. 1966. *J. Inst. Brew.* 72:291–303
72. Gilliland, R. B. 1969. *Eur. Brew. Conv. Proc. Congr. (Interlaken)* 303–10
73. Gjertsen, P., Trolle, B., Anderson, K. 1965. *Eur. Brew. Conv. Proc. Congr. (Stockholm)* 428–38
74. Graham, R. K. 1970. *J. Inst. Brew.* 76:16–20
75. Green, S. R. 1955. *Wallerstein Lab. Commun.* 18:239–51
76. Greenfield, J. R. 1965. *Proc. Am. Soc. Brew. Chem.* 72–83
77. Greenspan, R. P. 1966. *Proc. Am. Soc. Brew. Chem.* 109–12
78. Griffin, S. R. 1969. *J. Inst. Brew.* 75:342–46
79. Griffin, S. R. 1970. *J. Inst. Brew.* 76:41–45, 357–61
80. Griffin, S. R., MacWilliam, I. C. 1969. *J. Inst. Brew.* 75:355–58
81. Gunther, H. L. 1959. *Nature (London)* 183:903–4
82. Haas, G. J. 1960. *Advan. Appl. Microbiol.* 2:113–62
83. Haboucha, J., Jenard, H., Devreux, A., Masschelein, C. A. 1969. *Eur. Brew. Conv. Proc. Congr. (Interlaken)* 241–51
84. Hagler, A., Kleyn, J. G. 1968. Unpublished results
85. Hall, J. F. 1970. *J. Inst. Brew.* 76: 522–23
86. Hall, R. D. 1957. *Eur. Brew. Conv. Proc. Congr. (Copenhagen)* 314–26
87. Hampshire, P. 1922. *Bur. Biotechnol. Bull. No. 6,* 179
88. Hansen, A. 1948. *Jorgensen's Microorganisms and Fermentation.* London: Griffin. 550 pp. 37 plates
89. Hansen, E. C. 1888. *Medd. Carlsberg Lab.* 2:257–322
90. Harris, C. 1967. *Process Biochem.* 2:28–31
91. Harris, G., Barton-Wright, E. C., Curtis, N. S. 1951. *J. Inst. Brew.* 57:264–84
92. Harris, G., Millin, D. J. 1963. *Eur. Brew. Conv. Proc. Congr. (Brussels)* 400–11
93. Harris, G., Thompson, C. C. 1960. *J. Inst. Brew.* 66:213–17, 293–97
94. Harris, G., Thompson, C. C. 1960. *Nature (London)* 188:1212–13
95. Harris, J. O. 1959. *J. Inst. Brew.* 65:5–6
96. Harrison, G. A. F. 1963. *Eur. Brew. Conv. Proc. Congr. (Brussels)* 247–51
97. Hashimoto, N., Kuroiwa, Y. 1967. *Rep. Res. Lab. Kirin Brew. Co.* 10:51–60
98. Hemmons, L. M. 1954. *J. Inst. Brew.* 60:288–91
99. Hill, A. M. 1967. *Brew. Guild J.* 53: 509–13
100. Holahan, R. C. 1954. *Proc. Master Brew. Assoc. Am.* 57–59
101. Hopkins, R. H. 1955. *Eur. Brew. Conv. Proc. Congr. (Baden-Baden)* 52–54
102. Horecker, B. L., et. al. 1968. *Aspects of Yeast Metabolism,* ed. A. K. Mills, H. Krebs, 71–103. Oxford:Blackwell
103. Hough, J. S. 1957. *J. Inst. Brew.* 63: 483–87
104. Hough, J. S. 1961. *Eur. Brew. Conv. Proc. Congr. (Vienna)* 160–71
105. Hough, J. S., Rudin, A. D. 1958. *J. Inst. Brew.* 64:404–10
106. Hough, J. S., Stevens, R. 1961. *J. Inst. Brew.* 67:488–94
107. Howard, G. A. 1959. *J. Inst. Brew.* 65:414–18

108. Hudson, J. R. 1967. *Eur. Brew. Conv. Proc. Congr. (Madrid)* 187–95
109. Hudson, J. R., Birtwistle, S. E. 1966. *J. Inst. Brew.* 72:46–50
110. Hudson, J. R., MacWilliam, I. C., Birtwistle, S. E. 1963. *J. Inst. Brew.* 69:308
111. Ingraham, J. L. 1965. *Tech. Quart. Masters Brew. Assoc. Am.* 2:85–87
112. Johnson, J. R. 1963. *Eur. Brew. Conv. Proc. Congr. (Brussels)* 412–21
113. Jones, M., Pierce, J. S. 1964. *J. Inst. Brew.* 70:307–15
114. Jones, M., Pierce, J. S. 1969. *Eur. Brew. Conv. Proc. Congr. (Interlaken)* 151–60
115. Jones, M., Power, D. M., Pierce, J. S. 1965. *Eur. Brew Conv. Proc. Congr. (Stockholm)* 182–94
116. Joslyn, M. A. 1955. *Proc. Master Brew. Assoc. Am.* 49–56
117. Joslyn, M. A. 1955. *Wallerstein Lab. Commun.* 18:107–19
118. Kaneko, T., Yamamoto, Y. 1966. *Rep. Res. Lab. Kirin Brewery Co.* 37–50
119. Kato, S. 1967. *Bull. Brew. Sci.* 13:19
120. Kleinzeller, A., Kotyk, A. 1968. *Aspects of Yeast Metabolism,* ed. A. K. Mills, H. Krebs, 33–95. Oxford–Blackwell. Sols, A. in the above, 47–66
121. Kleyn, J. G. 1969. *Antonie van Leeuwenhoek J. Microbiol. Serol.* Suppl. 35:41
122. Kleyn, J., Mildner, R., Riggs, W. 1962. *Brew. Dig.* 37:42–46
123. Kleyn, J. G., Vacano, N. L. 1966. *Brew. Dig.* 41:65–68
124. Kleyn, J. G., Vacano, N. L., Kain, N. A. 1965. *Wallerstein Lab. Commun.* 28:35–49
125. Kolbach, P., Rinke, W. 1964. *Monatsschr. Brau.* 17:51
126. Kornboch, K. 1955. *Brauerei* 9:667–69
127. Kozulis, J. A., Page, H. E. 1968. *Proc. Am. Soc. Brew. Chem.* 52–58
128. Lakios, G. C. 1965. *Proc. Am. Soc. Brew. Chem.* 60–66
129. Lampen, J. O. 1968. *Antonie van Leeuwenhoek J. Microbiol. Serol.* 34:1–18
130. Laneau, R. 1963. *Eur. Brew. Conv. Proc. Congr. (Brussels)* 11–36
131. Laskowski, W., Lochmann, E. R. 1961. *Naturwissenschaften* 48:225
132. Lauridsen, P. B. 1966. *Tech. Quart. Master Brew. Assoc. Am.* 3:34–42
133. Lawrence, W. C., Cole, E. R. 1968. *Wallerstein Lab. Commun.* 31:95–115
134. Leach, A. A. 1967. *J. Inst. Brew.* 73:8–16
135. Leifson, E. 1954. *Antonie van Leeuwenhoek J. Microbiol. Serol.* 20:102–10
136. Lewis, M. J., Stephanopoulos, D. 1967. *J. Bacteriol.* 93:976–84
137. Lewis, M. J. 1968. *Proc. Am. Soc. Brew. Chem.* 5–9
138. Linke, W. 1967. *Monatsschr. Brau.* 353–60
139. Lloyd Hind, H. 1948. *Brewing Science and Practice,* Vol. 2, 507–1020. London: Chapman & Hall
140. Loncin, M., Kozulis, J. A., Bayne, P. D. 1970. *Proc. Am. Soc. Brew. Chem.* In press
141. Lyons, T. P., Hough, J. S. 1970. *J. Inst. Brew.* 76:564–71
142. Macey, A., Stowell, K. C., White, H. B. 1967. *Eur. Brew. Conv. Proc. Congr. (Madrid)* 283–94
143. MacKay, C. F. 1965. *J. Inst. Brew.* 71:530–35
144. Macrae, R. M. 1963. *Eur. Brew. Conv. Proc. Congr. (Brussels)* 510–12
145. MacWilliam, I. C. 1968. *J. Inst. Brew.* 74:38–54
146. Maddox, I. S., Hough, J. S. 1970. *Biochem. J.* 117:843–52
147. Maddox, I. S., Hough, J .S. 1970. *J. Inst. Brew.* 76:262–64
148. Maddox, I. S., Hough, J. S. 1971. *J. Inst. Brew.* 77 In press
149. Maiden, A. M. 1962. *Brew. Guild J.* 48:Part 1 565–78; Part 2 630–42
150. Markham, E. 1969. *Wallerstein Lab. Commun.* 32:5–14
151. Matile, Ph., Moor, H., Robinow, C. F. 1970. Yeast Cytology. In *The Yeasts,* Vol. 1, ed. A. H. Rose, J. S. Harrison, 219–302. London and New York: Academic. 508 pp.
152. Maw, G. A. 1965. *Wallerstein Lab. Commun.* 28:49–70
153. McCraine, T. 1968. *Tech. Quart. Master Brew. Assoc. Am.* 5:148–51
154. McCully, R., Laufer, L., Stewart, E. D., Brenner, M. W. 1949. *Proc. Am. Soc. Brew. Chem.* 116–26
155. McGill, D. J., Ribbons, D. W., Dawes, E. A. 1965. *Biochem. J.* 97:44

156. McMurrough, I., Rose, A. H. 1967. *Biochem. J.* 105:189–203
157. Middlebrooks, E. J. 1966. *Wallerstein Lab. Commun.* 29:117–26
158. Mill, P. J. 1964. *J. Gen. Microbiol.* 35:53–68
159. Mitchell, W. M. 1970. *Brew. Guardian* 99(3):51–58
160. Momose, M., Iwano, K., Tonoike, R. 1969. *J. Appl. Microbiol. (Tokyo)* 15:19–26
161. Moor, H. 1967. *Arch. Microbiol.* 57: 135–46
162. Moore, W. B., Rainbow, C. 1955. *J. Gen. Microbiol.* 13:190–97
163. Morris, E. O., Eddy, A. A. 1957. *J. Inst. Brew.* 63:34–35
164. Mortimer, R. K., Hawthorne, D. C. 1966. *Ann. Rev. Microbiol.* 20: 151–68
165. Mortimer, R. K., Hawthorne, D. C. 1970. Yeast Genetics. In *The Yeasts,* Vol. 1, ed. A. H. Rose, J. S. Harrison, 385–460. London and New York: Academic. 508 pp.
166. Mouch, G. 1961. *Brauwissenschaft* 14:257
167. Mrozek, H. 1969. *Mittder. Versuchssta. Gaerwungsgewerbe Wien* 23:81–86
168. Mulvany, J. 1966. *Process Biochem.* 1:470–73
169. Nakagawa, A. 1964. *Bull. Brew. Sci.* 10:7–10
170. Nakayama, T. O. M., Fly, W. H. 1968. *Proc. Am. Soc. Brew. Chem.* 198–202
171. Narziss, L. 1967. *J. Inst. Brew.* 72: 13–24
172. Nelson, R. H., Siebold, R. H. 1967. *Proc. Am. Soc. Brew. Chem.* 181–86
173. Nickerson, W. J. 1963. *Bacteriol. Rev.* 27:305–24
174. Nordstrom, K. 1964. *J. Inst. Brew.* 70:233–42
175. Nordstrom, K. 1964. *J. Inst. Brew.* 70:328–36
176. Okada, K. 1967. *Rep. Res. Lab. Kirin Brewery Co.* 10:9–18
177. Oppenoorth, W. F. F. 1958. *Brauwissenschaft* 11:113
178. Oppenoorth, W. F. F. 1961. *Eur. Brew. Conv. Proc. Congr. (Vienna)* 172–204
179. Oppenoorth, W. F. F. 1962. *Nature (London)* 193:706
180. Parry, D. 1969. *Brew. Guardian* 98: 46–50
181. Posada, J., Garcia Galindo, J. 1967. *Eur. Brew. Conv. Proc. Congr. (Madrid)* 353–64
182. Pasteur, L. 1860. *Ann. Chim. Phys.* 58:323–426
183. Pasteur, L. 1876. *Etudes sur la biere.* Paris: Gauthier-Villars. 387 pp.
184. Piendl, A. 1968. *Brauwelt* 108:418–25
185. Piendl, A. 1969. *Brauwissenshaft* 21:346–53
186. Portno, A. D. 1966. *J. Inst. Brew.* 72:458–61
187. Portno, A. D. 1968. *J. Inst. Brew.* 74:291–300
188. Prechtl, C. 1968. *Tech. Quart. Master Brew. Assoc. Am.* 3:177–82
189. Rainbow, C. 1965. *Brew. Dig.* 40: 50–52
190. Rainbow, C. 1966. *Wallerstein Lab. Commun.* 29:5–16
191. Rainbow, C. 1970. Brewers' Yeasts. In *The Yeasts* Vol. 3, ed. A. H. Rose, J. J. Harrison, 147–224. London & New York: Academic. 590 pp.
192. Richards, M., Cowland, T. W. 1967. *J. Inst. Brew.* 72:552–58
193. Richards, M. 1968. *J. Inst. Brew.* 74:433–35
194. Richards, M. 1969. *J. Inst. Brew.* 75:476–80
195. Ritter, W. 1953. Die Ernharung der Milch-Saurebaketerien, *Int. Congr. Microbiol., 6th, 1953,* 157. Symp. Nutrition and Growth Factors
196. Roessler, J. G. 1968. *Brew. Dig.* 4: 94, 96, 98, 102, 115
197. Roth, M. 1954. *Proc. Master Brew. Assoc. Am.* 59–62
198. Royston, M. G. 1966. *Process Biochem.* 1:215–21
199. Sandula, J., Kocková-Kratochvilová, A., Zamecnikova, A. 1964. *Brauwissenschaft* 17:130–37
200. Schuster, K., Hager, H. 1963. *Brauwissenschaft* 16:109–17
201. Schwann, T. 1837. *Ann. Phys. Chem.* 41:184–93
202. Sega, G. M., Lewis, M. J., Woskow, M. M. 1967. *Proc. Am. Soc. Brew. Chem.* 156–64
203. Sheneman, J. M., Hollenbeck, C. M. 1960. *Proc. Am. Soc. Brew. Chem.* 22–27
204. Shimwell, J. L. 1936. *J. Inst. Brew.* 42:119–27
205. Shimwell, J. L. 1950. *J. Inst. Brew.* 56:179–82
206. Shimwell, J. L. 1959. *Antonie van Leeuwenhoek J. Microbiol. Serol.* 25:49
207. Shimwell, J. L. 1960. *Brew. J.* 515–17, 563–65

208. Shimwell, J. L. 1964. *Brew. J.* 99: 759–60
209. Shimwell, J. L., 1964. *J. Inst. Brew.* 70:247–48
210. Silbereisen, K. 1938. *Wochschr. Brau.* 55:217
211. Slonimski, P. 1958. *Soc. Chem. Ind. Monogr.* No. 3:7–20
212. Stevens, R. 1967. *Chem. Rev.* 67: 19–71
213. Stevens, T. J. 1966. *J. Inst. Brew.* 72:369–73
214. Strandskov, F. B., Bockelmann, J. B. 1955. *Proc. Am. Soc. Brew. Chem.* 36–42
215. Strandskov, F. B., Bockelmann, J. B. 1957. *Proc. Am. Soc. Brew. Chem.* 94–97
216. Strandskov, F. B., Ziliotto, H. L., Brescia, J. A., Bockelmann, J. B. 1965. *Proc. Am. Soc. Brew. Chem.* 129–34
217. Suomalainen, H. 1968. *Aspects of Yeast Metabolism,* ed. A. K. Mills, H. Krebs, 17. Oxford: Blackwell
218. Suomalainen, H., Ronkainen, P. 1968. *Nature (London)* 220:792–93
219. Takahashi, T. 1966. *Bull. Brew. Sci.* 12:9–14
220. Takayanagi, S., Harada, T. 1967. *Eur. Brew. Conv. Proc. Congr. (Madrid)* 473–88
221. Thomas, M. 1970. *A study of* Obesumbacterium proteus, *a common brewery bacterium.* PhD thesis, Univ. Birmingham, England
222. Thompson, C. C. 1964. *Brew. Guild J.* 50:599–605
223. Thompson, C. C., Ralph, D. J. 1967. *Eur. Brew. Conv. Proc. Congr. (Madrid)* 177–86
224. Thompson, C. C. 1968. *Brew. Dig.* 43:56–64
225. Thorne, R. S. W. 1951. *Eur. Brew. Conv. Proc. Congr. (Brighton)* 21–34
226. Thorne, R. S. W. 1954. *J. Inst. Brew.* 60:238–48
227. Thorne, R. S. W. 1966. *Master Brew. Assoc. Am. Tech. Quart.* 3:160–68
228. Thorne, R. S. W. 1968. *J. Inst. Brew.* 74:516–24
229. Thorwest, A. 1965. *Brauwelt* 105: 845–51
230. Uli, A., Kühbeck, C. 1969. *Brauwissenschaft* 22:121–29, 199–208, 248–54
231. United Nations Statistical Year Book 1967. 244
232. Urbanek, J. 1967. *Kvasny Prum.* 13: 127–31
233. Vermeylen, V. 1962. *Traité de la fabrication du malt et de la biere.* Assoc. Roy. des. Anciens Eleves de l'Institut Sup. des. Fermentations Gand., Belgium (2 vols) 1624 pp.
234. Visuri, K., Kirsop, B. H. 1970. *J. Inst. Brew.* 76:362–66
235. Vogl, K., Schumann, G., Proepsting, W. 1967. *Monatsschr. Brau.* 20: 116
236. Wackerbauer, K. 1968. *Monatsschr. Brau.* 21(10):288–93; 21(11):328–33
237. Wackerbauer, K. 1969. *Monatsschr. Brau.* 22:211–16
238. Wackerbauer, K. 1969. *Eur. Brew. Conv. Proc. Congr. (Interlaken)* 523–37
239. Wackerbauer, K., Emeis, C. C. 1968. *Monatsschr. Brau.* 21:328–33
240. West, D. B., Lautenbach, A. L., Becker, K. 1952. *Proc. Am. Soc. Brew. Chem.* 81–88
241. West, D. B., Lautenbach, A. L., Brumsted, D. D. 1963. *Proc. Am. Soc. Brew. Chem.* 194–99
242. White, G. A., Wang, C. H. 1964 *Biochem. J.* 90:408–33
243. Wickerham, L. J. 1951. *Tech. Bull. U. S. Dept. Agr. No 1029*
244. Wiles, A. E. 1953. *J. Inst. Brew.* 59: 265–84
245. Williams, P. T. C. B., Rainbow, C. 1964. *J. Gen. Microbiol.* 35:237–47
246. Williamson, D. H. 1959. *J. Appl. Bacteriol.* 22:392–402
247. Windisch, S. 1965. *Monatsschr. Brau.* 18:274–81
248. Winge, O. 1944. *C. R. Trav. Lab. Carlsberg Ser. Physiol.* 24:79–95
249. Wood, B. J. A., Rainbow, C. 1961. *Biochem. J.* 78:204–9
250. Wood, W. A. 1961. Fermentation of Carbohydrates and Related Compounds, 59–151. In *The Bacteria,* ed. I. C. Gunsalus, R. Y. Stanier, New York: Academic, 572 pp.
251. Wynants, J. 1962. *J. Inst. Brew.* 68: 350–54
252. Yoshida, T. 1968. *Rep. Res. Lab. Kirin Brewing Co.* 11:77–86

MECHANISM OF CELL TRANSFORMATION BY RNA TUMOR VIRUSES

1583

Howard M. Temin

McArdle Laboratory, University of Wisconsin, Madison, Wisconsin

Contents

INTRODUCTION

The mechanisms of cell transformation and viral replication by RNA tumor viruses have been studied for many years by only a relatively small group of investigators. The members of this group, after long years of apprenticeship, were able to cope with some of the complexities of these viruses. Recently, the RNA tumor viruses have become of interest to a much larger number of people. This review will attempt to provide them with a comprehensible picture of my view of the present status of work on mechanisms of cell transformation and viral replication by RNA tumor viruses. [The chapters on RNA tumor viruses in the book edited by Tooze (336) represent a somewhat parallel attempt and should also be consulted.] The present review is written from the point of view of the DNA provirus hypothesis and will cover primarily work with cell cultures. It will also often be illustrated with work on avian viruses, both because I know more about them and because more is known about them.

GENERAL PROPERTIES OF RNA TUMOR VIRUSES

There are not many comprehensive reviews of the RNA tumor viruses. Vogt (351) published an extensive one in 1965, and Fenner in his book, *The Biology of Animal Viruses* (93), has also thoroughly covered these viruses. Two excellent new reviews have recently appeared (297a, 349). In addition, there are reviews on separate groups of RNA tumor viruses or on more limited topics. When these latter reviews are relevant (and I am aware of them), I shall refer to them later.

Classification

Even the proper name for this class of viruses is a subject for disagreement. "RNA tumor viruses" is not in the proper form for the name of a class of viruses. In addition, all of the viruses which appear to share the properties of this class are not related to tumors. "Oncornaviruses" is of the proper form for the name of a class of viruses, but is still subject to the criticism about relationship to tumors. "C-type RNA viruses" depends upon a morphological distinction of unknown structural basis. It also puts the mammary tumor viruses in a different class. "Leukovirus" or "Rousvirus," as a general name for RNA tumor viruses and their relatives, seems preferable to me, but these names are not widely used. "Thylaxoviridae" has also been proposed. Here, the more combersome designation "RNA tumor viruses and their close relatives" or "RNA tumor viruses" will be used.

The major groups of RNA tumor viruses are the avian, the murine, and the mouse mammary tumor viruses. (The murine group includes feline viruses, and so the name should be broadened.) Additional groups probably will be defined soon, that is, amphibian, primate and, perhaps, bovine. Visna virus and Simian Foamy virus may be representatives of related types of viruses. This review will concern mainly the avian and murine groups.

The groups are distinguished on the basis of the species in which they were originally isolated, host range, internal (group specific) antigens, and types of tumors produced. Each group contains several different viruses. However, the classification and naming of the viruses within each group is also a subject for disagreement.

The avian RNA tumor virus group has been divided into subgroups in two ways: 1. on the basis of the three correlated properties of neutralizing antigen, efficiency of plating on different chicken cells, and interference with other avian RNA tumor viruses (86, 355). These properties probably depend upon a property of one or more of the glycoproteins of the envelope. 2. On the basis of the type of neoplasia induced (30). The individual virus strains are named in 1 by a name and a number or letter, for example, Schmidt-Ruppin virus-A (SRV-A) or Rous Associated virus-49 (RAV-49). In 2, the viruses are designated by a disease and a number or letter, for example, BAI strain A, avian myeloblastosis virus. Mixed names, for example, avian myeloblastosis virus-2, also exist.

The murine RNA tumor viruses are subdivided into mouse, rat, hamster, and feline viruses on the basis of species in which they originally were isolated and of internal group-specific antigens. One of these antigens is common to all of the murine RNA tumor viruses, one is distinct for each subgroup (102). Few rat, hamster, and feline RNA tumor viruses are known. There are many mouse RNA tumor viruses. The mouse viruses have been classified by two different sets of criteria as have been the avian viruses, but classification is less well worked out (137, 187). Furthermore, at present no correlation of serotype and host range appears to exist. Other RNA tumor viruses have been reported in reptiles, monkey, and possibly humans (58, 111, 121, 160, 335).

RNA tumor viruses cause primarily leukemias and sarcomas, but some appear to be non-neoplastic. Mammary tumor virus induces adenocarcinoma, and several avian leukemia viruses cause liver, kidney, and ovarian carcinomas (30, 50, 53, 54). Therefore, RNA tumor viruses are interesting both for their unique mode of replication and for their possible relationship to "spontaneous" neoplasia.

Another source of confusion is the naming of viruses produced in experimental work. Phenotypic mixing is very common among RNA tumor viruses. A virion can have a genome and proteins coded by different virus genomes, for example, Rous Sarcoma Virus in a Rous-Associated Virus envelope. Such phenotypically mixed virus has been called a pseudotype, and is designated "genome (envelope)," for example, "RSV(RAV)." With the discovery that the virion RNA-directed DNA polymerase can also be involved in phenotypic mixing, the situation becomes more complicated. New naming systems will have to be developed.

The naming of infected cells is equally confused. The convention we use is "cell (virus[es])," for example, "*Ch* (RSV, RAV)."

VIRIONS

Structure.—Since virions of RNA tumor viruses have been recognized in the electron microscope and have been available in a relatively pure form for years, the general morphology and chemical composition have long been known. Furthermore, electron microscopists can distinguish between A, B, and C-type particles, and even between avian and murine C-type viruses (Gelderblom, Bauer & Frank, 109). However, we still have no clear idea of the organization of the virion components. There is an outer protein envelope with two or more glycoproteins, and lipid. Inside this envelope there may be an inner membrane, and there is a core or nucleoid. This inner core contains the large viral RNA (see later), some smaller proteins with group-specific antigenicity, and the virion DNA polymerase system. Several groups seem to favor the idea that the core is a helically symmetric ribonucleoprotein strand coiled in some, perhaps cubically symmetric, fashion (Bader, Steck & Kakefuda, 17; Lacour et al, 180; Sarker & Moore, 263). However, the evidence for this particular form of organization is not very convincing. Other structures have also been proposed (73).

The envelope contains the viral subgroup- and type-specific antigens. It also may contain host cell-specified material (71). Avian myeloblastosis virus grown in myeloblasts contains a membrane ATPase, whereas the same virus grown in fibroblasts does not (De Thé, 70). The RNase in the envelope (252) is also probably host material.

Disruption of the envelope by treatment with nonionic detergents or ether leads to activation of the virion DNA polymerase system, the accessibility of the group-specific antigen to antibody, and the accessibility of the RNA to RNase (223, 324). A ribonucleoprotein particle can be isolated from preparations of disrupted virions (16, 64, 73, 81, 96, 110, 180). This particle has a higher density in equilibrium sucrose density gradient centrifugation than the virion, that is, 1.22–1.27 g/cc for the core versus 1.14–1.17 g/cc for the virion. The core contains the large viral RNA, the group-specific antigens, and the virion DNA polymerase enzyme system. It is not clear whether or not the structure of the core remains after the RNA is digested by RNase, as is the case for the ribonucleoprotein of influenza virus (Duesberg 80; Kingsbury & Webster, 169). It has been reported that infectious cores and infectious nucleic acids can be obtained (34, 133, 207, 276, 277). It would be useful for experimental work if these observations were reproducible.

The number of proteins in the virion, their size, location, antigenicity, and enzyme activity are being studied in many laboratories (5, 5a, 43a, 83, 101, 152a, 200, 222, 236, 250, 270, 339, and others who have not yet published). There are certainly more than five polypeptide chains in the virions, but until more work is done, it would be premature to draw detailed conclusions about their number, location, and sizes. Already it is known that there must be molecules with type-, subgroup-, and group-specific antigenic,

DNA polynumerase, nuclease, and ligase activities (see later).

Nucleic acids.—There are several sizes of nucleic acids in the virion: single-stranded RNAs of 70S, 28S, 18S, 7S, and 4S; and DNA of small size (see Bishop et al, 36, 37).

70S RNA.—It has generally been found that extraction of nucleic acids by standard techniques from relatively clean preparations of virions gives a major component of single-stranded RNA with a sedimentation constant near 70S, larger than that for any other RNA virus (Duesberg, 81; Robinson, Pitkanen & Rubin, 249). The base composition of this RNA is not unusual. It is also generally agreed that exposing this RNA to denaturing conditions, such as heat or dimethylsulfoxide, results in RNA of a lower sedimentation coefficient, that is near 35S (Duesberg, 79; Erikson, 90; Montagnier, Goldé & Vigier, 209). Two major explanations can be proposed for this change: a conformational change or a decrease in molecular weight. Because the mobility of the RNA in polyacrylamide gel electrophoresis increases after denaturation, the hypothesis of a decrease in molecular weight seems more likely (17, 81). The experiments of Duesberg & Vogt (85) suggest that the 35S material may be resolvable into more than one component, and that the ratio of the components may be different in different preparations or strains of virus. Sarker & Moore (264) report that the RNA isolated from mammary tumor virus has a length consistent with a maximum molecular weight of 4×10^6 daltons. But Kakefuda & Bader (162), looking at electron micrographs of RNA from murine leukemia virus, do not report such a clear result.

At present, a reasonable model is that the major RNA of the virion is a relatively specific aggregate of 3 or 4 pieces of 3 or 4×10^6 daltons each. If this model is correct, numerous questions arise: are these pieces different? Are they all made from the same type of intermediate with the same enzymes? Are the intermediates located at the same cellular site? How does the aggregate get formed, and what is its structure? These questions are similar to those posed by the fragmented genomes of influenza and reoviruses.

4S RNA.—Although it is clear that breakdown must play a part in formation of the smaller RNAs in the virion, there appears to be another origin for some of them. The labeling kinetics of the 70S and the smaller RNA molecules may be different (Bader, 14; Bauer, 27). The 4S RNA contains transfer RNA molecules with acceptor activity. The amino acid-accepting activity of the virion transfer RNAs of Avian Myeloblastosis Virus is not a random sample of the amino acid-accepting capacity of the transfer RNA from myeloblasts, the cell which produces Avian Myeloblastosis Virus (Carnegie et al, 52; Erikson & Erikson, 92; Travnicek & Riman, 341). These virion 4S RNAs have the hybridization properties expected of cellular tRNA (Baluda & Nayak, 25; Wollman & Kirsten, 363; Yoshikawa-Fukada

& Ebert, 366). Bishop et al (37) report that the base composition of 4S RNAs from virions of Avian Myeloblastosis Virus and Bryan High Titer Rous Sarcoma Virus are different. The role of these tRNAs in the viral life cycle is unclear. Since no protein synthesis appears to be needed by the virus before formation of the provirus, there is no necessity for the virion to have tRNAs, even if they are viral-specific. These tRNAs may reflect a contaminant picked up during formation of virions, or they may represent a relic of some ancestor of the RNA tumor viruses. It would be interesting to know what tRNAs are in fibroblast-grown Avian Myeloblast Virus or in RNA sarcoma viruses. The relative amounts of 4S RNA differ in Avian Myeloblastosis Virus and Rous Sarcoma Virus (Erikson, personal communication; Hanafusa, personal communication).

7S, 18S, 28S RNAs and DNA.—These RNAs appear to be real (Bishop et al, 36, 37). The 18S and 28S are the same size as ribosomal RNA. The 7S RNA is different from the known small cellular RNAs. The role of these RNAs is completely unknown. As with the 4S RNA, they may just "happen" to be included in the virion. It would be valuable to know whether these small RNAs are located in the virion core with the 70S RNA and whether the ribosomal-like RNAs are in ribosomes. It would also be interesting to know whether there is a constant number of each of these molecules in every virion. The small adenine-rich RNA of reovirus presents a somewhat similar problem (31).

Several groups have also reported the presence of a small DNA in the virion (39, 91, 185, 248, 251). This DNA may fall into two density classes. Its function and relationship to viral RNA or host DNA is unknown. It probably is not necessary for viral infectivity since virus produced in the presence of inhibitors of DNA synthesis is fully infectious.

Life Cycles

A study of the life cycles of RNA tumor viruses is complicated by a number of variables, which are not present with most other animal viruses. These variables relate to infection, virus production, transformation, genetic variation, and phenotypic mixing. They do not involve cell killing which does not result from infection with RNA tumor viruses.

Whether an RNA tumor virus infects a cell appears to depend upon the genome of the virus, the nature of the proteins in the virion, the genome of the cell, whether it is already infected with an RNA tumor virus, and the differentiated state of the cell. Infection is usually most efficient in cells from the species of origin, but genetically resistant individuals exist in the species of origin. For chickens, sensitivity to avian tumor viruses is controlled by a single dominant cellular gene; for mice, sensitivity to murine leukemia viruses is controlled by recessive genes. RNA tumor viruses can often infect cells from other species. Usually the infection is more efficient the more closely the species are related. Chicken viruses often infect turkeys, ducks, and quail very efficiently; mammals much less so. Mouse viruses infect mice, rats, and hamsters very efficiently (mice more efficiently than rats), but chickens not at all. In addition to cellular genetic resistance to a particular virus, there is virus-specified interference in a cell already infected with another RNA tumor virus of the same subgroup.

The relationship between transformation and virus production is affected by the genome of the host cell. A mouse cell infected by a murine sarcoma virus will become transformed and produce infectious virus. A chicken cell infected by an avian sarcoma virus will also become transformed and produce virus. However, a mouse cell infected by an avian sarcoma virus may become transformed, but no virus production will occur. Certain variant avian tumor viruses will be produced only by chicken cells carrying specific genetic factors. In the absence of these factors, the variant avian tumor viruses will infect and transform the chicken cells, but no infectious virus will be produced.

Often an infected cell has many of its properties altered, for example, alterations in shape, orientation, rate of multiplication, and transplantability in animals. Usually all of these alterations appear together. Whether alterations occur or not depends on the virus genome and on the differentiated state of the infected cell. For example, Avian Myeloblastosis Virus infects chicken fibroblasts, virus is produced, but no transformation appears. On the other hand, Avian Myeloblastosis Virus infects chicken hematopoietic cells and also causes transformation.

Genetic variation occurs by mutation or, perhaps, recombination between different RNA tumor viruses or between RNA tumor viruses and cells. Genetic variation appears to occur frequently. The RNA → DNA and DNA → RNA transfers of information, as well as integration with the cell genome, give sources of variation not present for other viruses.

Phenotypic mixing occurs when two genetically different RNA tumor viruses infect the same cell. The mixing can involve the envelope proteins and the virion DNA polymerase(s). (As other virion proteins are identified, they, too, probably will be found to be involved in phenotypical mixing.) The resulting viruses are affected only for steps (*a*), initial events, and (*b*), formation of provirus (see below). As a result of the widespread occurrence of mixed infections, phenotypic mixing seems to occur more frequently with RNA tumor viruses than with most other viruses.

RNA sarcoma virus in sensitive cells.—Following is a hypothetical life cycle of an RNA sarcoma virus replicating in sensitive host cells (for example, Schmidt-Ruppin-D strain of avian sarcoma virus infecting C/A chicken cells): (It must be remembered that many of the specific mechanisms are not now supported by experimental evidence.) Steps 1, 2, and 3 occur sequentially. Steps 4, 5, 6, 7, and 8 occur concurrently and for as long as the transformed cell lives.

Step 1. Initial events: A virion attaches to and enters a cell. It is uncoated so that virion core enzymes are exposed and the core is transported to the cell nucleus. These processes are controlled by the presence and specificities of specific proteins in the virion and specific receptors in the cell.

Step 2. Formation of provirus: DNA is synthesized using the viral RNA as a template. The DNA is integrated at a specific site in the host chromosome to form an integrated DNA provirus. These processes are controlled by the presence and specificities of enzymes in the virion, and by the presence of specific regions of the host cell chromosomes.

Step 3. Activation of virus-specified RNA and protein synthesis: RNA synthesis using the provirus DNA as template begins. This virus-specified RNA is processed and transported to the cytoplasm. Virus-specified, RNA-directed protein synthesis begins. These processes are controlled by cell enzymes, cell division, and by the sequences of the provirus DNA and virus RNA.

Step 4. Synthesis of virus-specified RNAs: A modified cellular polymerase transcribes RNA from the provirus DNA. This RNA is processed and transported to the cytoplasm by modified cellular machinery. Step 4 involves modified cellular machinery, whereas step 3 involves unmodified cellular machinery.

Step 5. Synthesis of virus-specific proteins: Virus-specified proteins are made from virus-specified RNA by cellular machinery. Viral glycoproteins are made by cellular enzymes which add carbohydrate to viral-specified proteins.

Step 6. Formation of virions: Membranes of the infected cell are modified by virion envelope components. Virion cores are formed; virions are formed and released by budding from the cell plasma membrane.

Step 7. Replication of provirus: The provirus is replicated together with the rest of the cellular chromosomes and is passed to daughter cells at mitosis. Viral-modified machinery is not involved in this step.

Step 8. Neoplastic transformation: Properties of the infected cell are altered by virus-specified proteins which are not essential for virus replication and are not structural components of the virion.

RNA sarcoma virus in resistant cells of class of origin.—This life cycle is modified when an RNA sarcoma virus is inoculated on resistant cells of the class from which the virus was originally isolated (for example, Schmidt-Ruppin-D strain of avian sarcoma virus inoculated on C/D chicken cells or chicken cells infected with RAV-50):

Step 1. Initial events: Initial events are prevented, so no virus replication occurs. However, if the specificity controls of step 1 are circumvented, for example, by use of inactivated Sendai virus, the life cycle of an RNA sarcoma virus in these resistant cells continues as it did in sensitive cells.

RNA sarcoma virus in cells of a class other than that of origin.—If an RNA sarcoma virus is inoculated on cells of a class other than that of origin (for example, Schmidt-Ruppin strain of avian sarcoma virus inoculated on rat or other mammalian cells), the life cycle is further modified. Step 1, initial events, occurs at a low efficiency and does not involve the specific virion proteins and cell receptors. Step 2, formation of provirus, occurs at a lower efficiency than in sensitive cells because of the absence in mammalian cells of a region of the chromosomes specific for integration of avian sarcoma viruses. Step 3, activation of virus-specified RNA and protein synthesis is the same as in sensitive cells. Steps 4, synthesis of virus-specified RNA, 5, synthesis of virus-specified proteins, and 6, formation of virions, each takes place with lower efficiencies. The cumulative result of these lowered efficiencies is that few or no virions are produced. Steps 7, replication of provirus, and 8, neoplastic transformation, are the same as in sensitive cells.

RNA leukemia virus in sensitive cells.—The life cycle of an RNA leukemia virus in sensitive cells may differ from that of an RNA sarcoma virus in sensitive cells only at step 8, neoplastic transformation. RNA leukemia viruses cause neoplastic transformation only of cells in a special differentiated state. RNA leukemia viruses may also differ from RNA sarcoma viruses in steps 2, formation of provirus, and 7, replication of provirus, as a result of the nonchromosomal location of their provirus.

Comparison of Life Cycles With Those of Other Animal Viruses

These steps in the hypothetical life cycles of RNA tumor viruses will now be compared with what has been found for other animal viruses.

Step 1. Initial events. Attachment, entrance, uncoating, and transport to some site in the cell are processes common to all animal viruses. The RNA tumor viruses seem to differ from other aimal viruses only in the extent of genetic resistance in the species of origin and in subgroup-specific interference. Except possibly for influenza viruses, the RNA tumor viruses may be the only RNA viruses transported to the cell nucleus.

Step 2. Formation of provirus. Synthesis of DNA from viral RNA is unique to RNA tumor viruses and their relatives, as is a DNA intermediate for the replication of an RNA virus. Chromosomal integration may also occur with the oncogenic papova, adeno- and herpesviruses. The presence of polymerases in the virion is fairly common among animal viruses: for example, virions of pox-, diplorna- and rhabdoviruses contain various types of RNA polymerases. However, these polymerases seem to be involved in formation of messenger RNA rather than in genome replication.

Step 3. Activation of RNA and protein synthesis. A requirement for cell division-controlled activation of virus production seems unique to the RNA tumor viruses. Papova viruses require one cell division for fixation of transformation and further divisions for expression of transformation.

Steps 4. Synthesis of virus-specified RNAs, 5, synthesis of virus-specified proteins, and 6, formation of virions do not appear to involve processes different from those found with other animal viruses.

Step 7. Replication of provirus. The replication of provirus by cellular machinery may also occur in cells transformed by oncogenic DNA viruses. It also may be related to the apparent replication of picodnavirus DNA by cellular machinery.

Step 8. Neoplastic transformation may involve a mechanism unique for RNA tumor viruses. Although DNA tumor viruses cause neoplastic transformation, they probably use another mechanism. All viruses, whether they are tumor viruses or not, transform some properties of the infected cells. Since the infected cells are usually killed, there cannot then be neoplastic transformation. Cell killing by viruses seems to require interaction of some viral product(s) with specific cell functions. Similarly, interaction of some viral product(s) with specific cell functions appears to be required for neoplastic transformation by RNA tumor viruses. Some viruses which are not members of the RNA Tumor Virus group also do not cause cell killing in some infected cells; however, this moderate infection does not lead to neoplastic transformation.

Assays

Infectious virus.—The only really definitive assay for a virus is the production of new infectious virus. Although animals can be used for such an assay, cell culture is preferable when it is possible. We shall discuss only cell culture assays for infectious virus. (Sarma et al, 268, gives a good summary of assay methods and references.) The first requirement for replication and assay of virus is sensitive cells. This sensitivity may be hard to predict. For example, virus produced from single foci induced by Bryan High Titer RSV often does not infect cells of the type it came from (Hanafusa, 126), and virus from New Zealand mice appears to infect rat or hamster cells, but not mouse cells (Levy & Pincus, 186).

The sensitive cells must be able to divide after infection. The first division or two is necessary for activation of virus-specified RNA and protein synthesis (step 3 above). Further divisions may be necessary to give transformed cells a selective advantage in multiplication or to make them distinguishable from uninfected cells (see later). Environmental conditions, that is, temperature, pH, nutrients, and macromolecular factors in the medium, must be permissive for transformation. For example, it has been reported in some cases that excess fetal bovine serum or other proteins in the medium, or the absence of tryptose phosphate broth prevents expression of transformation (Prince, 235; Rubin, 255; Temin, 306, 308; Vigier, 347). Infection by *Mycoplasma* also prevents expression of transformation (Ponten & Macpherson, 234; Somerson & Cook, 283).

If the infecting virus is a sarcoma virus or a transforming leukemia virus, for example, MC29, these conditions lead to either foci of transformed cells or rapid transformation of an entire culture of cells (43, 125, 305). If the virus is nontransforming, tests for virions must then be made. The simplest of these tests is "induction" or "rescue" of sarcoma virus from converted cells not producing virus (Hanafusa, 126; Levy & Pincus, 186; Temin, 300). Then, focus formation by this induced or rescued virus can be assayed (95, 172, 247a, 333).

Virions.—Assays for virions, in addition to infectivity, depend on a test for virion RNA and proteins, the latter usually as antigens. One of the simplest tests is for production of a particle containing radioactive uridine and of density of about 1.15 g/cc in equilibrium sucrose density gradient centrifugation. Refinements on this test are the demonstration that the particle has a 70S RNA that changes to 35S after denaturation and has an RNA-directed DNA polymerase. Electron microscopy is a useful complement to this analysis. In addition, infectious messenger activity of RNA from avian myeloblastosis virus has been reported (Hlozanek et al, 147).

Since the virion has both internal and envelope proteins these can be used for assays. The first of these assays to be developed was the RIF-test (Rubin, 256) based on interference between a nontransforming virus and a transforming virus of the same subgroup. Other subgroup- or type-specific tests are serum-blocking, membrane fluorescent antibody, and immune adherence (217). Tests for internal antigens include complement-fixation, precipitation, and fluorescent antibody tests. Syncytium formation with a line of rat-Rous cells appears to be a specific test for murine leukemia virus production (Klement et al, 173). A plaque test for murine leukemia viruses based on this observation has been developed (Rowe et al, 254). Only an unadapted L-cell virus has been negative in this plaque assay.

Labeled specific DNA, the product of the virion RNA-directed DNA polymerases, is a useful additional reagent for hybridization with cellular or virion RNA (107a, 114a). An additional type of assay is for the presence of specific polymerases or polymerase-containing particles (64, 106).

These tests can all be applied when noninfectious virus is produced or when no virions are produced. All of these tests, of course, require appropriate controls. Unfortunately, these are not always in evidence.

Virus-Cell States

One of the key questions in defining the mechanism of cell transformation by RNA tumor viruses is that of the relationship of the viral genome to the cell genome in infected cells long after infection. Although this problem has been the subject of an enormous amount of work, it still is not resolved in several crucial aspects. As discussed above, infected cells can be classified as to whether they are producing virus and whether they are transformed. These determinations are straightforward.

Cells producing virus.—Generally it has been assumed that the viral genome in all virus-producing cells is in the same state, independent of whether or not the cells are transformed. What that state is thought to be differs with different investigators. However, there is certain evidence which suggests that the simplifying assumption of a single type of virus-cell state in virus-producing cells may not be valid. This evidence is the segregation of noninfected cells from infected virus-producing cells. For example, chickens congenitally infected with avian leukosis virus appear to produce uninfected eggs and sperm (Dougherty & Di Stefano, 77). Some clones of Balb/3T3 cells infected with MSV-MuLV segregate cells that still produce MuLV but appear to have lost MSV (Todaro & Aaronson, 334). Segregants which no longer produce virus have been seen in MuLV-induced leukemias and myelomas (Cohn, 65; Ferrer & Gibbs, 94). Segregation of noninfected cells from cells exposed to virus just before mitosis has also been reported (Trager & Rubin, 340). Here, the segregation probably occurred before the formation of provirus or the first replication of provirus.

In contrast is all other work with virus-producing cells. There, segregants were not seen (for instance, see Temin, 299). This disparity may represent a difference between proviruses located in the chromosomes and proviruses in other locations in the cell.

However, there is no firm evidence as to the location of the provirus (but see Payne & Chubb, 227). My guess is that it is usually integrated in a special site in a chromosome. Hybridization of viral RNA with DNA from various cell fractions might be one way of answering this question. Nuclear or chromosomal transplants would be a more elegant method.

In addition to cells producing one type of virus, there are cells producing two types of virus. The second virus is introduced either by simultaneous or superinfection or by mutation. No immunity of the kind seen in bacterial lysogeny appears to exist. (Of course, the viral subgroup-specific interference and host cell resistances still operate in infected cells.) Further, there appears to be no complementation between viruses for steps 1, early events, 2, formation of provirus, or 3, activation of virus-specified RNA and protein synthesis (84, 316, 348). In superinfection experiments, when both the first and the superinfecting viruses were sarcoma viruses (transforming), the resulting superinfected cells produced either both viruses or only the superinfecting one. When the first virus was a sarcoma virus and the superinfecting virus was a leukemia virus (nontransforming), all superinfected cells produced both viruses, that is, no cells were found which produced only the superinfecting leukemia virus. This contrast may be explained by a difference in site of integration of the provirus for leukemia and sarcoma viruses. After virus production begins, it is not known whether there are controls on the rate of production. In doubly infected cells, both viruses can be produced at the same rate, or there can be a large excess of one type of virus (Hanafusa, 123; Temin, 299).

The Murine Sarcoma Virus-Murine Leukemia Virus system is a confusing form of virus-cell interaction because of the complex kinetics involved in focus formation and the different properties of virus produced at different times (97, 120, 135, 282). But this complexity hides a combination of the factors discussed before, that is, there are transforming and nontransforming viruses, phenotypic mixing, and subgroup-specific interference, plus new factors of reduced multiplication of some transformed cells and formation of virus aggregates (2, 3, 219, 224, 279). The same factors are probably operative in the formation of tumors by Murine Sarcoma Virus (Berman & Allison, 33, Siegler, 278).

Cells not producing virus.—Transformed cells which do not produce virus have been studied extensively. It is possible that the difference between these cells and the virus-producing cells discussed in the previous section is only a matter of quantity. No evidence suggests that there are systematic differences in integration or location of proviruses, for example. As with virus-producing cells, segregation of uninfected cells has been found (Macpherson, 195), but this segregation is a rare exception. Two major cases of transformed cells which do not produce virus exist: avian sarcoma virus in mammalian cells, and Murine Sarcoma Virus (MSV) in the absence of leukemia viruses.

Cells infected only with MSV are tumor cells. They do not produce infectious virus or virions and do not contain detectable virion antigens. The MSV can be "rescued" or recovered by superinfection with murine leukemia viruses, resulting in phenotypically mixed virions containing the MSV genome and the MuLV envelope (3, 69, 95, 172, 333). So far, no viral function except that of undergoing phenotypic mixing with leukemia viruses has been specifically assigned to MSV. MSV may represent genes for neoplastic transformation which, as a mRNA, can aggregate with leukemia virus RNA and therefore become incorporated in a virion and be a template and substrate for the leukemia virus polymerase and integration systems.

Mammalian cells can be transformed to neoplastic cells by infection with avian sarcoma viruses. These transformed cells usually produce no infectious virus, although occasionally very small amounts have been found (Klement & Vesely, 171; Svec and colleagues, 294, 295; Svoboda & Klement, 296). Virus particles have been seen in these cells by electron microscopy (190, 344). These particles probably represent contaminating murine leukemia viruses (Gelderblom, Bauer & Frank, 109). Avian tumor virus group-specific antigen is also found in these transformed mammalian cells which do not produce virus. There is a large range in the amounts of group-specific antigen in these cells. Although most transformed cells are positive for antigen, some are negative but cause production of antibody to the group-specific antigen after injection into test animals (26, 134, 329). In most cases, complete infectious avian sarcoma virus can be recovered from these mammalian cells after fusion with uninfected chicken cells. The efficiency of this recovery varies. Up to 100 percent of heterokaryons were reported to produce virus by Machala, Donner & Svoboda (191), but lower values were reported by Svoboda & Dourmashkin (297), Yamaguchi, Takeuchi & Yamamoto (365), and Altaner & Temin (7). There may be a correlation between efficiency of virus recovery and amount of avian tumor virus group-specific antigens in the mammalian cells.

The state of avian sarcoma viruses in mammalian cells may therefore be fundamentally different from that of MSV in the absence of murine leukemia virus. The avian sarcoma viruses in mammalian cells seem to have a quantitative block to virus production. Perhaps step 4, synthesis of virus-specified RNAs by modified cellular polymerase, processing, and transport machinery, is less efficient for avian viruses in mammalian cells than in avian cells. The integration machinery of avian viruses (step 2, formation of provirus) also seems to be less efficient in mammalian cells (see later).

Recent work has shown that RSV(O) does not form converted, nonvirus-producing cells, but does form cells that produce viruses of unique and narrow host ranges (Hanafusa, 126; Vogt, 352; Weiss, 359). However, with Harris RSV and with Avian Myeloblastosis Virus, foci which do not produce virus are reported (Moscovici & Zanetti, 213; Reamur & Okazaki, 241). Goldé (112), Hanafusa (126), and Toyoshima, Friis & Vogt (337) have reported that mutagenizing stocks of avian sarcoma viruses gave rise to variants that induce foci which do not produce virus. Virus was not recovered from these variant avian sarcoma virus-induced foci or from some of the Avian Myeloblastosis Virus-induced foci by exposure to avian leukemia viruses. If the leukemia virus was able to superinfect the cells and if no transforming virus could be recovered by cell fusion, this type of transformed cell which did not produce virus would be different from the ones already discussed. If so, the piece of viral RNA responsible for transformation must have lost its ability to aggregate with other avian RNA tumor virus RNA, and to undergo phenotypic mixing with avian RNA tumor virus proteins.

There are also reports, for mammals injected with RSV, of tumors that contain no RSV group-specific antigen and from which virus cannot be recovered by fusion (134, 329). At present it is not clear whether the RSV genome is present but makes little or no virus-specified products, or whether these tumors were not induced by RSV.

Another type of state is maintained by vertical transmission in animals and does not involve recent infection nor transformation. This state is represented by the genes for Mammary Tumor Virus in mice and for avian RNA tumor virus group-specific antigen in chickens which are inherited as dominant Mendelian genes (Bentvelzen et al, 32; Payne & Chubb, 227). Weiss (360) and Hanafusa (126) have furthered the investigation of chickens having genes for the avian tumor virus group-specific antigen by showing that they also have linked genes for envelope antigens of avian RNA tumor viruses. The mRNA produced from these genes can be transferred to another cell by phenotypic mixing with an avian tumor virus (Hanafusa, Miyamoto & Hanafusa, 129). This process appears to be similar to that proposed above for MSV. The Mendelian inheritance of the genes controlling these viral antigens is the best evidence for chromosomal integration of proviruses of RNA tumor virus. A puzzling feature of these cells is that the envelope antigens which are products of the genes do not cause the expected type of subgroup-specific interference.

Human cells can be infected with avian or murine RNA tumor viruses (4, 98, 158, 194, 269, 364). The details of the virus-cell interactions have not yet been described.

VIRUS-VIRUS INTERACTIONS

The common host cells for RNA tumor viruses, chicken and mouse fibroblasts, come from animals which frequently are already infected with RNA tumor viruses by transmission from their parents, that is, by vertical transmission. Therefore, it is difficult to maintain an unmixed population of virus, and virus-virus interactions are much more common than for most other animal viruses (see also next section). One might remark that all animal viruses grown in these hosts are probably contaminated with RNA tumor viruses. The long latent period of RNA tumor viruses and their sensitivity to inhibitors prevent their being a major problem in most studies of other mouse- or chicken-grown animal viruses.

Three types of virus-virus interactions have already been mentioned: subgroup-specific interference, phentoypic mixing, and formation of aggregates. A fourth, recombination, has been suggested. Hanafusa (124) has given a good general review of this subject, so we will not go into much detail here.

Rubin (256) first reported that infection of fibroblasts with a nontransforming virus prevents later infection with a transforming virus of the same subgroup, and he developed the RIF-test for leukemia viruses. The detailed mechanism of this interference phenomenon is still unclear. It involves step 1, early events. Complexities in patterns of interference have been found by Duff & Vogt (86). They reported interference between viruses of subgroups B and D; and no interference in C/B cells between two viruses of subgroup D. These complexities may relate to the possibility that there are two determinants of virus type (Bauer & Graf, 29), or they may relate to a requirement for some threshold amount of a viral product to set up interference. Perhaps viruses which are produced at low rates do not set up strong interference. [For instance, it might be interesting to determine if *s-C/O* (RSVα) cells are sensitive to RSVβ(0).]

In contrast, preinfection with virus of one subgroup can sometimes increase the efficiency of infection with virus of a second subgroup (Hanafusa & Hanafusa, 127; Ishizaki & Shimizu, 156). This enhancement also appears to involve step 1, early events. It is probably not a result of complementation by products of the first viral genome, but may be related to the alteration of cell properties by the first virus. The effects of DEAE-dextran on increasing the efficiency of infection may be related (78, 281, 354).

Phenotypic mixing can involve the virion DNA polymerases as well as the envelope (Hanafusa, Hanafusa & Miyamoto, 131). Phenotypically mixed virus stocks are common. Phenotypic mixing can occur between two leukemia viruses, as well as between sarcoma and leukemia viruses (Chung & Hinz, 61; Vogt, 353). Phenotypic mixing between avian and murine RNA tumor viruses has not been reported thus far. However, artificial aggregates can be made between murine leukemia virus and influenza virus (O'Connor & Rauscher, 220).

Stocks of Murine Sarcoma Virus-Murine Leukemia Virus often contain particles with both genomes (Guillemain, Hampe & Boiron, 118; O'Connor & Fischinger, 219). These particles appear to be simple aggregates, rather than the heterozygotes seen with Newcastle disease virus (Kingsbury & Granoff, 168). It is not clear whether a cell produces agregates or whether they are formed in solution or during purification.

One case of a stable change in viral properties has been reported after mixed infections: the change of RSVα to RSVβ after mixed infection of *Ch*(RSVα) with RAV-1 (Hanafusa, 124). However, in the other cases of mixed infection which are common with these RNA tumor viruses, production of such stable variants has not occurred after mixed infection. These results may be explained by the hypothesis that the viral recombination process involved in the change from RSVα to RSVβ is a reassortment of separate pieces of the 70S RNA aggregates. One piece of 35S RNA may contain the information for the major virus structural proteins and for transformation, and another piece may contain the information for the virion polymerase system. Recombination may occur only between markers on separate RNA molecules. Therefore, since all viral markers, except the polymerase system, are on the same molecule, only RSVα (which does not have a polymerase), could be observed to recombine. As more markers are available for other viral functions, this hypothesis will be easy to test.

Variation of Viruses

A major difficulty observed in work with RNA tumor viruses is the nature and extent of variation of viruses. Because of the widespread natural occurrence of RNA tumor viruses, and the resultant phenotypic mixing, cloning of a virus stock may result in a virus with properties entirely different from those of the mixture. For example, viruses of two different subgroups, neither of which caused myeloblastosis, were isolated by cloning of virus from the BAI strain of Avian Myeloblastosis Virus (Smith & Moscovici, 280). Even after a virus is cloned its properties may change as a result of introduction of a contaminant virus from the cells.

In addition, cloned virus grown on a noninfected cell frequently appears to vary. The frequency of genetic change of morphologic types of foci, antigenic type, and ability to grow at high temperatures appears to be high (60, 299, 321). In the virus life cycle, genetic variation could occur during steps 2, formation of provirus, 4, synthesis of viral RNA, and 7, replication of provirus. Two of these steps involve an unusual type of transfer of genetic information to progeny molecules. As yet there is no information on the error frequency of these transfers. It may be high.

In addition, if the aggregate model for the virion RNA is correct, a phenomenon related to that of incomplete myxovirus I should occur. Either virus passage at high multiplicities of infection or long-term passage of clones of virus-infected cells, which would involve repeated step 7, provirus transfers, might lead to an accumulation of incomplete or defective viruses. The formation of aleukemogenic murine leukemia virus after long-term culture of virus-infected cells may represent this phenomenon (Yoshikura et al, 369), as might the gradual decrease in efficiency of recovery of RSV from hamster tumors induced by RSV with increased passage level of the tumor (Shevliaghyn et al, 275).

These types of variation have ignored possible interactions with the host cell genome. Three types of such interaction have been described. Weiss (360) and Hanafusa, Miyamoto & Hanafusa (128) reported a host-dependent change in the envelope properties of a strain of RSV on passage in two different types of host cells. This change, from RSVβ to RSVβ' and from RSVβ' to RSVβ, [using the terminology of (128)] was completely reversible, and it appeared to be the result of phenotypic mixing of an RSV with the product(s) of a chicken cell gene(s). This phenotypic mixing appears to be the same process as that discussed earlier between two different RNA tumor viruses. The only difference is that in this case the genetic information controlling the proteins of one of the elements involved in the phenotypic mixing is cellular, rather than viral.

After passage of avian sarcoma virus through rat or mouse cells, the virus which is recovered and grown in chicken cells has altered properties. These include increased plating efficiency on the mammalian host, and sometimes altered envelope antigens (Altaner & Temin, 7; Kryukova, Obukh & Tot 177; Obukh & Kryukova, 218). This change may represent recombination of the avian sarcoma virus with genetic elements in the mammalian cell.

An apparently different change is seen when Murine Sarcoma Virus of mouse origin, that is, MSV or FBJ virus, is injected into hamsters or rats. A tumor appears which at first contains no infectious virus. Upon passage of the tumor a virus may appear with a host range very different from that of the virus used for the original infection. The recovered virus may be a mixture of a nontransforming virus and a sarcoma virus, but this is not always apparent (167, 230, 245, 246, 267, 331, 332). The appearance of RAV-60 in preparations of RSV(0) may be the result of a related process (130).

Several alternative hypotheses can be offered to explain this apparent change in host range of MSV: (*a*) a new passenger virus infects the tumor and by phenotypic mixing transfers the MSV genome with a new host range determined by the envelope of the passenger virus; (*b*) there has been recombination between the infecting virus and genetic elements in the host cell, plus selection of an altered virus with a new host range; and (c) there has been mutation of the MSV, plus selection of a mutant with a new host range. The isolation of endogenous nontransforming viruses from uninfected rats and hamsters (Stenback, Van Hoosier & Trentin, 289, 290) supports hypothesis (*a*). Probably, fingerprinting of nucleic acids by the methods of Sanger, Brownlee & Barrell (262) will be the best way of deciding among these hypotheses.

MECHANISM OF TRANSFORMATION

Introduction

The question, what is the mechanism of neoplastic transformation by RNA tumor viruses can be broken into several sub-questions: on a cellular level does the transformation result from information brought into a cell by the infecting virus? What form does the added information take in the cell? On a molecular level, how does the information bring about conversion of the properties of the infected cell?

Mechanisms of transformation can be separated into direct and indirect mechanisms. Direct mechanisms can be separated into hit-and-run, quantitative, site-specific, and product-specific. In direct mechanisms, the cell which is transformed has been infected by the virus. In hit-and-run mechanisms, the infecting virus switches on the transformation, and its later presence is irrelevant to the transformation. In site-specific mechanisms, the presence of the viral genome in some special cellular site causes the transformation. In quantitative and product-specific mechanisms, viral products cause the transformation either by their amount or by their special property(s). [Elsewhere (317) I have presented further discussion of this topic and comparisons among tumor viruses.]

RNA tumor viruses transform cells by a direct mechanism, and the viral presence is relevant to the transformation. However, it appears that no one of the simply studied parameters of viral replication controls transformation.

In order to produce transformation (see earlier discussion of assays), infection must be with a transforming virus in a cell permissive for transformation and able to divide in a proper environment. Whether virus is produced or not, and whether the provirus is stably integrated or not does not appear to be important for the occurrence of transformation.

Furthermore, it may be unrealistic to talk about a single mechanism for transformation by RNA tumor viruses. Infection with RSV can lead to chromosome breakage (Nichols et al, 216). This breakage may be important for neoplastic transformation under circumstances not considered here.

The most dramatic evidence of the separation of transformation and virus replication is the existence of transforming and nontransforming viruses as determined by infection of fibroblasts. No general differences between these two types of RNA tumor viruses have been found in their mode or rates of replication. Furthermore, the virion proteins seem to be interchangeable, as judged by the extensive occurrence of phenotypic mixing. In addition, the two types of viruses seem to be closely related since there can be transitions between them (see later). Therefore, we conclude that some specific virus-specified product(s) which is not a structural component of the virion causes transformation. This product(s) may act directly, by altering the site of integration of the provirus, or in other fashions.

In order to get neoplastic transformation, the viral information must interact with some particular expressed host information. Electron microscopy shows virus replicating in most organs of an animal infected with leukemia viruses (*72, 77*). However, the rate of neoplastic transformation per cell is very low in these animals. The work of Baluda, Moscovici & Goetz (23) showed that avian myeloblastosis virus could replicate in many cells, but could transform only precursors of hematopoetic cells (see also 212). The work of Meier, Myers & Huebner (202) showed that a single dominant gene in the mouse made a great difference in the incidence of leukemia, but had no effect on the incidence of leukemia virus replication.

Furthermore, it appears that exposure of RNA sarcoma viruses to irradiation leads to the appearance of nontransforming RNA tumor viruses (Goldé, 112; Graf & Bauer, 113; Toyoshima, Friis & Vogt, 337). The possible presence in the stocks of RNA sarcoma virus of contaminating nontransforming viruses makes it impossible to prove absolutely that the nontransforming viruses were derived from the transforming viruses, but that is a likely explanation. Again, nucleic acid fingerprinting might help to resolve this problem.

The mutants of avian sarcoma viruses which are temperature-sensitive for maintenance of transformation present clearer evidence for the close relation between transforming and nontransforming viruses. Transformed cells of these mutants, when shifted from permissive to nonpermissive temperatures, lose their transformed phenotype (Martin, 198; Temin, 318; Vigier, 350; Vogt, 356). These experiments appear to establish definitively that some viral product is required for maintenance of transformation. Since, at least for some of these viruses, cells continue to produce infectious virus at the temperature which is nonpermissive for transformation, a situation analogous to that of leukemia viruses which are genetically unable to transform fibroblasts, exists for sarcoma viruses. Earlier studies of virus variants which gave rise to different cellular morphologies led to the same conclusion about the requirement for a continuing viral product for transformation (Temin 298, 299).

This result and the existence of pairs of nontransforming and transforming viruses (for example, an RSV and RAV of subgroup B, so that the virions appear to be the same, although only the RSV transforms fibroblasts) support the hypothesis that transformation is mediated by some virus-specified product(s) which is a nonstructural component of the virion. Furthermore, this component(s) appears not to be essential for virus replication. The information for such a product might have appeared originally from viral or from cellular information (Temin, 322).

DNA Provirus

Having concluded that viral information is necessary to cause transformation, we can ask about the form of this information in converted cells. Most of the experimental evidence is relevant to the question of the form of the viral information in converted cells. We assume that the information for transformation has the same form.

The first indication that the intermediate for replication of RNA tumor viruses might be different from the replicative intermediate of other RNA viruses was the observation of regular inheritance of this intermediate at cell division. This regular inheritance, and the evidence that there were only one or two genetic copies of the intermediate, led, in the early 1960's, to the concept of the provirus (Temin, 298–300). The concept of a chromosomal location for the provirus is supported by the genetic studies of Payne & Chubb (227) and Bentvelzen et al (32), discussed previously.

The first indication that the provirus might be DNA came from the sensitivity of viral replication to the antibiotic actinomycin D (Temin, 301). This sensitivity of viral replication to actinomycin D is a general property of all RNA tumor viruses tested. It appears to be a direct result of blocking of viral RNA synthesis by the actinomycin (Bader, 14; Baluda & Nayak, 24). Further, in most cells infected with RNA tumor viruses, treatment with actinomycin D appeared to block all RNA synthesis. Double-stranded RNA or RNA strands complementary to the viral RNA have not been found in most cases (see section following).

The next indication that the provirus might be DNA came from the requirement for new DNA synthesis for infection (Bader, 11; Temin, 302). These early experiments were complicated by the necessity for cell division for activation (step 3), which also was blocked by preventing DNA synthesis (see below). However, later experiments with cells in a non-DNA synthesizing phase of the cell cycle, G1, showed a requirement for DNA synthesis for infection (step 2, formation of provirus) in addition to the requirement for DNA synthesis for cell division which, in turn, was required for activation (step 3) (Murray & Temin, 215; Temin, 309, 311).

Hybridization between labeled RNA from virions and DNA from uninfected and infected cells has given two types of results. The early work of Temin (303, 304) reporting increased hybridization of viral RNA with DNA from infected cells was not confirmed by later work (132, 240, 362, 363, 366), with some exceptions (13, 88). The discrepancy in results appears to come from the high background of hybridization of RNA tumor virus with DNA from uninfected cells, that is, avian viruses with chicken cells, but not with mouse or rat cells. This high background hybridization of viral nucleic acid with DNA from uninfected cells was also observed by Aloni et al (6) with SV40. The uninfected cell DNA may contain proviruses of vertically transmitted viruses, defective proviruses, or protoviruses. Another problem in the hybridization studies with RNA tumor viruses appears to have been the hybridization of the tRNA in the virion.

The most recent and careful experiments of Baluda & Nayak (25) and Baluda (22) show the expected increased hybridization of viral RNA with DNA of transformed cells. The high melting temperature of these hybrids suggests that they are specific. The base composition of the hybridized RNA is the same as that of the input viral RNA. This work should be confirmed in other laboratories. The virion DNA polymerase system might supply good reagents for such studies.

Formation of the provirus (step 2).—At present the strongest evidence for the existence of the DNA provirus comes from study of the formation of the provirus. Although there have been many studies involving inhibitors, few have separated the effects on formation of the provirus (step 2, involving viral DNA synthesis) and effects on activation of virus-specified RNA and protein synthesis (step 3, involving cell division which requires cell DNA synthesis). (See Vigier, 349, for review of inhibitor studies.) Formation of the provirus is defined as formation, in a form resistant to inhibition of DNA synthesis, of a stable nonvirion structure containing viral information.

Temin (309) first used synchronized cultures to separate the two steps of provirus formation and activation. Later, we found stationary cultures were more useful than synchronized cultures (215, 311, 316). Inhibitors of DNA synthesis prevented infection if they were applied soon after inoculation of virus. This effect was independent of any effects on cell S-phase DNA synthesis or cell division.

Further support for the DNA provirus hypothesis came from the BUDR-labeling and light inactivation studies of Boettiger & Temin (42) and Balduzzi & Morgan (19). They found that the newly synthesized provirus could be labeled with BUDR and then inactivated with visible light under conditions in which there was no cell killing. Boettiger (42) further showed that an increased multiplicity of infection led to increased resistance to the light inactivation, suggesting that more proviruses were made after the multiple infection.

However, as yet there has been no isolation and chemical characterization of this putative newly synthesized DNA. Several groups have reported some stimulation of DNA synthesis after infection of cells with RNA tumor viruses (Hirschman, Fischinger & O'Connor, 146; Kara, 165; Lacour, 179; Lee, Kaighn & Ebert, 182; Macieira-Coelho & Ponten, 193). Most of this stimulation seems to be either a result of some other component in the virus inoculum (Macieira-Coelho, Hiu & Garcia-Giralt, 192) or to occur too late in time for formation of provirus DNA to occur. (Sundelin, 292, finds cells with over 2N amounts of DNA soon after transformation, days after infection.) Boettiger (41) has found some increased DNA synthesis after inoculation of multiplicities of infection of over 100 focus-forming units of RSV per stationary chicken cell. Density labeling and CsCl analysis established that this was synthesis of cellular DNA. Since this effect was not seen at lower multiplicities of infection, it appears not to be related to infection per se, but to some aspect of the stimulation of cell DNA synthesis by specific stimuli (see Temin, Pierson & Dulak, 327). Clearly, it is necessary to find some newly synthesized provirus DNA to establish the DNA provirus hypothesis.

It is not yet entirely clear whether any new RNA synthesis is required for provirus formation. The experiments of Buck & Bather (47) suggested a transient early actinomycin D sensitivity which was not explainable by cell killing. We (204) have been unable to confirm this observation.

At first, it seemed protein synthesis was required for provirus formation. It was reported that interferon treatment inhibited focus formation by RNA sarcoma viruses (Bader, 10; Peries et al, 229; Rhim et al, 243–245; Sarma et al 265, 266; Traub & Morgan, 342). The report by Bader (12) of resistance of early infection to puromycin treatment was ignored. Later we rechecked the interferon studies using stationary cells and controls for effects on cell division and found no effect of interferon on infection (Mizutani quoted in 321). These differences in results were explained by the work of Gresser et al (115) which showed that interferon could inhibit cell division and therefore block infection by preventing step 3, activation.

Our studies of provirus formation in stationary cells showed that it was resistant to cycloheximide treatment (Mizutani quoted in 318, 321). (We also found that puromycin killed these stationary cells at concentrations which were inhibitory to most amino acid incorporation.) Therefore, we looked in the virion of RSV for an RNA-directed DNA polymerase (Temin & Mizutani, 324). Baltimore (20), following up the discovery of an RNA-dependent RNA polymerase in virions of vesicular stomatitis virus (Baltimore, Huang & Stampfer, 21), did the same.

DNA polymerase system in the virion.—Both groups found such an enzyme. This discovery is now confirmed for all RNA tumor viruses and is extended to Visna and Simian Foamy virus (20, 82, 105, 110, 114, 199, 205, 206, 248, 251, 271, 272, 284–286, 324).

Because of the rapid development of this particular area of research, it would not be worthwhile to go into too much detail writing in December, 1970 for publication at the end of 1971. What appear to be the most significant and solid findings at the time of writing will be reported.

In a test tube, the disrupted virion enzyme system uses an RNA template to make small pieces of RNA-DNA hybrid, and single- and double-stranded DNAs. The virion enzyme system is located in the virion core. When freed from its endogenous template, it can synthesize DNA from a variety of natural and synthetic nucleic acid templates; that is, single- and double-stranded RNAs, RNA-DNA hybrids, and single- and double-stranded DNAs. The DNAs made are homologous to the templates. RNA is not synthesized from any of these templates. In addition, the virions contain DNA endo- and exonuclease and DNA ligase activity. The details of the template preferences of the enzyme system and the nature of the intermediates are disputed.

Several questions, in addition to the obvious molecular ones, are raised by these discoveries. Why does the virion have an enzyme system? How do these enzymes act in the cell? What is the significance of these enzymes as markers for hidden RNA tumor viruses?

Why does the virion have an enzyme system? Vaccinia, reo- and vesicular stomatitis viruses, the other animal viruses with polymerases in their virions (Baltimore, Huang & Stampfer, 21; Borsa & Graham, 45; Kates & McAuslan, 166; Munyon, Paoletti & Grace, 214; Shatkin & Sipe, 274) require novel enzymes to make mRNA from the virion genome nucleic acids, that is, DNA-directed RNA polymerases or RNA-directed RNA polymerases in the cytoplasm of animal cells. There does not appear to be any compelling reason why RNA tumor viruses could not make the necessary DNA polymerase system after infection before provirus formation. Since they apparently do not, we must assume either that the virion RNA is not a messenger for the DNA polymerase system, that is, a complementary strand or a cell messenger RNA is required, or that the topological problems of cytoplasmic synthesis and nuclear integration made postinfection synthesis of the enzymes too inefficient. No strands of RNA complementary to viral RNA have been found consistently in infected cells. This result bears against the first alternative. An alternative explanation is that the enzymes in the virion reflect the origin of RNA tumor viruses.

The presence of these enzymes in the virion may explain the greater ultraviolet resistance of RNA tumor viruses than of Newcastle disease virus (Levinson & Rubin, 184; Rubin & Temin, 261). The ultraviolet target for inactivation of focus formation would not include genes for these enzymes or other virion components (Coffin, personal communication).

How do these enzymes act in the cell? We have hypothesized that the virion enzyme system is sufficient both for formation of DNA from the viral RNA template and for integration of this DNA into its specific cellular site (205, 206). Where these processes take place and how many copies of DNA are made from each RNA is not known. Payne, Solomon & Purchase (228) reported virus antigens in the nucleus soon after infection. Boettiger's inactivation curves (42) suggest only one new DNA per infecting virus. The existence in the virion of integration machinery which reconizes specific sites in the host cell could explain the inefficiency of cross-class infections. In a cross-class infection, some of the virion machinery does not appear to be operative (Altaner & Temin, 7), which could also explain possible abortive infections in mammals exposed to avian RNA tumor viruses (Bauer et al, 28; Kotler, 175).

What is the significance of these enzymes as markers for RNA tumor viruses? A number of possible explanations can be given for the presence of RNA-dependent DNA polymerase activity in a tumor (328): it is a normal cellular DNA polymerase which can use RNA as template (Cavalieri & Carroll, 55); it is an unusual cell polymerase for differentiation (the protovirus hypothesis) (Temin, 317, 321, 322); it is an RNA-dependent DNA polymerase of a nononcogenic virus (Lin & Thormar, 189; Scolnick et al, 272); it is the polymerase of a passenger RNA tumor virus; it is the polymerase of an RNA tumor virus activated by the neoplastic change but not related to it; it is the polymerase of an RNA tumor virus which caused the neoplastic transformation, but the polymerase is not essential to the persistence of the neoplastic transformation; or, it is the polymerase of an RNA tumor virus which caused the neoplastic transformation and the polymerase is essential to the persistence of the neoplastic transformation. Each polymerase will have to be studied carefully to determine its biological role. It is obvious that the significance of RNA-directed DNA polymerase activity in leukemia cells from man (106) may be quite different from the significance of polymerase activity in rat tumor cells transformed by avian sarcoma virus (64).

Activation.—When a provirus is formed in stationary cells, no cell transformation or virus production occurs. Cell division is required for initiation of virus production and cell transformation. This process is called activation of virus-specified RNA and protein syntheses (step 3).

The requirement for cell division for activation was foreshadowed in the early findings that the capacity of chicken cells to produce RSV was very sensitive to inactivation by X ray or ultraviolet light (Rubin & Temin, 261). The capacity became much more resistant to irradiation soon after infection (Temin & Rubin, 326; Yoshikura, 368). [The later belief by Rubin (257) that the sensitivity of the capacity was an artifact has been resolved by the demonstration (Coffin, 62) that irradiated chicken cells still undergo one or two divisions.] Other inhibitors of cell division, for example, mitomycin, actinomycin, inhibitors of DNA synthesis, and colchicine, were also found to block initiation of virus production.

The use of cultures of synchronized cells demonstrated more clearly the requirement for cell division by showing that the viral latent period varied with the phase of the cell cycle at the time of virus inoculation (148, 308, 330, 368). Finally, cultures of stationary cells were used (146, 215, 311, 316).

It is not known whether cell division is required for activation of a provirus formed in a virus-producing cell. The superinfection experiments which could have tested this point were done with multiplying cells and therefore did not separate the effects on provirus formation from effects on activation (84, 348).

As yet, the molecular details of activation are not known. Coffin & Hilgers (63) observed in preliminary experiments that stationary chicken cells infected with RSV did not contain viral antigens, but became positive for these antigens during the first mitosis. If these results are confirmed, it could mean that the first viral RNAs synthesized after infection are unable to escape from the nucleus until mitosis occurs. Then, after viral-specified protein synthesis begins, a system capable of transporting viral RNA through the nuclear membrane is established.

It is not clear whether one or two initial divisions are required for transformation. If transformation involves a viral product activating a cellular gene, a second division might be required to allow the cell messenger RNA to get out of the nucleus. (A cell division requirement for shifts in normal differentiation is not uncommon, 149, 201.) This hypothesis might explain the stage-specific effect of rifampicin on focus formation, but not on virus production (Diggelmann & Weissmann, 74). The requirement for a second division to initiate conversion might also explain the finding of Yoshikura (367) that MSV caused conversion in wounded monolayers of mouse cells only after exposure to virus at particular times after wounding. Weiss (361), however, reports evidence suggesting that some aspects of transformation appear after one division.

Synthesis of viral RNA and protein (steps 4 and 5).—Details of these steps are lacking. A major question is whether all of the viral RNA is made from the DNA provirus or whether some is made from an RNA intermediate. Two lines of evidence have been put forward in favor of the latter hypothesis. Two groups have reported double-stranded RNA in cells chronically infected with Murine Sarcoma Virus—Murine Leukemia Virus, as well as hybridization between viral RNA and cell RNA (Biswal & Benyesh-Melnick, 38; Van Griensven, Emanoil-Ravicovitch & Boiron, 345). However, these authors may be looking at RNA-DNA interactions, normally occurring double-stranded RNAs (Colby & Duesberg, 66), contaminating myxoviruses, or DNA labeled by a label from uridine (Reddi, 242). The second line of evidence is the assertion of Spiegelman and his colleagues that an RNA replicase exists in leukemia virus-infected cells (138, 358). However, these studies did not have preparations from parallel control cells. Therefore, it is not possible to conclude that the enzyme activity is other than a normal cellular polymerase.

Against the hypothesis of some viral RNA being synthesized with an RNA template are the results of experiments with actinomycin D (discussed previously). Bader & Bader (15) have also presented evidence that RSV-transformed cells with BUDR-substituted DNA make more noninfectious virions than do infected cells with normal DNA. They suggest that this result demonstrates that information for virion proteins comes from DNA.

Viral RNA enters into virions within about 1 hr after synthesis (14, 24, 221, 238). The molecules which enter the virion and those which act as messengers may be different. Infected cell nuclei have large amounts of nucleolar RNA (291), and there is increased nucleolar labeling in actinomycin D-treated cells soon after exposure to RNA tumor viruses (293).

There may be extra-large-sized polysomes in infected cells (48). Viral protein enters virions soon after synthesis (24, 203). There may be particulate intermediates before final virion synthesis. Haguenau, Michelson-Fiske & Rabotti (119), using electron microscopic autoradioraphy, found precursor particles. Fleissner (99) found several group-specific antigen-containing particles in the cytoplasm of transformed cells which were not producing virus. Kirsten (170) found cores in the cytoplasm of virus-producing cells, and suggested that cleavage of viral peptides accompanies formation of virions. Coffin & Temin (64) found a variety of ribonucleoprotein particles containing RNA-dependent DNA polymerase(s) in virus-producing cells and in transformed cells which did not produce virus. More detailed studies are needed before the role of these particles in virus production can be established.

Calculations of the amount of labeled uridine incorporated into cells and that released in virions suggests that 0.1 percent of cell RNA synthesis may be viral (62, 247, 302). If this figure is correct, provirus DNA is transcribed more frequently than most genes except those for ribosomal and transfer RNAs. Therefore, we might expect a modified DNA-directed RNA polymerase to be engaged in this transcription. Although there are reports of increased polymerase activity in virus-induced leukemias (Lin & Rich, 188; Staines & Yamada, 287), we cannot conclude this increase is in any way related to viral RNA synthesis. The difference in effect of toyocamycin on viral RNA synthesis and cell RNA synthesis may be relevant to the hypothesis of an altered RNA polymerase (Brdar, Rifkin & Reich, 46; but see 295a). We do not know the number of viral messengers, whether they are polycistronic, or their relation to the virion RNAs.

Replication of provirus.—When a cell divides the provirus goes to each daughter cell. Since the provirus usually appears to be integrated with cellular DNA, it probably is replicated and separated to progeny cells in the same way as the cellular DNA. The provirus could be a separate replicon or part of a neighboring replicon. It probably is replicated by cellular enzymes even though most RNA tumor viruses carry a DNA polymerase. RSVα, which appears not to have a DNA polymerase in the virion, is still regularly inherited as a provirus (Hanafusa, Hanafusa & Miyamoto, 131).

RNA tumor viruses do not have the means or the need to reverse provirus integration. However, the formation of viral RNA from provirus DNA means there is continual transfer of information from the integrated provirus without excision. This process is probably different from that of integrated DNA viruses, where excision appears to be necessary to get viral information out of the chromosome.

Conversion

Provirus formation and activation, steps 2 and 3, appear necessary before conversion can occur. Results with the viral mutants for focus morphology and for temperature-sensitive conversion suggest there must be continued viral-specific synthesis, presumably of RNA and protein, to maintain conversion. The previous discussion suggested that the required products are not structural components of the virion and not essential for viral replication, and that complex interactions between viral products, cellular products, and the environment are necessary for conversion. As yet there is no information enabling us to connect a particular viral-specified product with conversion. Since a very subtle type of interaction may be involved, attention has been directed to the properties of the converted cell, in the hope of describing those features which are specifically different from those of untransformed cells. Perhaps, in the near future the study of converted cellular properties and of virus-specified products will merge.

It should be added that this discussion is concerned only with the requirements for conversion at a cellular level. This conversion is a necessary precursor to tumor formation. However, it is clear that there are many other factors which determine whether or not a converted cell will give rise to a tumor; for example, see (75).

RNA tumor viruses provide good material for making comparisons between normal and tumor cell properties. Since the replication and the virions of transforming and nontransforming RNA tumor viruses appear to be the same in fibroblasts, a nontransformed control cell to compare to tumor cells is available. Of course, such a normal control is not always possible for RNA tumor virus-transformed cells, for example, in the study of tumors in mammals induced by RSV.

Surface changes.—There are many alterations in the surfaces of cells infected with RNA tumor viruses: changes in morphology and orientation, the presence of virion envelope components and tumor-specific antigens, increased hyaluronic acid synthesis and glucose transport, etc.

The discovery by Manaker & Groupé (196) of alterations in the morphology of chicken embryo fibroblasts exposed to RSV started the tissue culture study of RNA tumor viruses. Alteration in morphology is a constant correlate of conversion. Different viruses make different changes in morphology. The most extreme alternatives are the round and fusiform cells induced by different mutants of RSV or Fujinami virus (Temin, 298). Concomitant with the altered morphology of the converted cell is an altered orientation to other cells. Fusiform cells are more ordered than are normal fibroblasts; round cells are less so. Other RNA tumor viruses, for example, MSV and Schmidt-Ruppin strains of RSV make a crisscross type of focus. The molecular basis for this (or these) alterations is not known. Martin (198) noted that some of his temperature-sensitive mutants of RSV induce foci of round cells at permissive temperatures, foci of fusiform cells at suprapermissive temperatures, and no foci at nonpermissive temperatures. We have noticed that many viruses selected as negative for focus formation at high temperature are really making fusiform foci (321).

These morphological changes seem to be controlled somewhat independently of the other changes, that is, both round cells and fusiform cells otherwise have the same properties of converted cells. Furthermore, Toyoshima, Friis & Vogt (337) report that two avian leukosis viruses, RAV-50 and CZAV, alter the morphology of chicken embryo fibroblasts in vitro. The alterations are an epithelioid appearance of the cultures and a tendency of the cells to pile up in several layers. MC29 virus, which induces myelocytomatosis in chickens, also alters the appearance of fibroblasts (Bolognesi et al, 43). If these viruses do not induce sarcomas in chickens, these results would separate the two processes of morphological alteration and neoplastic transformation (see further discussion at the end of this section).

Since virions are produced by budding from the outer cell membrane, it is not surprising that infected cells have new viral-specific surface antigens. These antigens may appear within 6 hr after infection (253). However, transformed cells which do not produce virus also have new surface antigens (178, 181). And, Pasternak & Pasternak (225) and Aoki et al (9) demonstrated by serological means that virus-producing cells also have new antigens which appear to be different from those of the virions. These antigens are reportedly no different in Murine Sarcoma Virus- and Murine Leukemia Virus-infected cells (59). Vaage (343) reports that each mouse mammary tumor induced by Mammary Tumor Virus has a different nonvirion-associated, tumor-specific transplantation antigen.

The increased amount of hyaluronic acid in tumors induced in chickens by Rous sarcoma virus is matched by increased production of hyaluronic acid by chicken cells converted in cell culture by RSV. This increase is caused by an increase in the amount of the enzyme hyaluronic acid synthetase. This enzyme is probably located in the plasma membrane of the cells (Ishimoto, Temin & Strominger, 155). However, although increased acid mucopolysaccharide production has been found in some other RNA tumor virus-infected cells (Hamerman et al, 122; Van Tuyen, Maunoury & Febora, 346), this increased production is not a constant correlate of conversion by RNA tumor viruses (Rakusanova, 239; Temin, 312).

Hatanaka, in a series of papers (139–144) has shown that an increased rate of glucose uptake accompanies conversion by RNA sarcoma viruses. This increase appears to be the result of an altered membrane transport system.

A number of other miscellaneous changes in the surfaces of converted cells have been found. Montagnier (208) reports that acid mucopolysaccharides inhibit BHK cells, but do not inhibit multiplication of RSV-transformed BHK cells in suspension. Klement et al (173) have shown that a line of rat-RSV cells form syncytia, when grown in culture with mouse MuLV-infected cells. Rifkin & Reich (247) have shown that some avian tumor virus-infected cells are more sensitive to lysis by dibucaine than are uninfected chicken cells. Moore (210) has reported that rat-RSV cells differ from uninfected rat cells in their adhesion to solid substrates in the absence of serum.

A number of surface changes widely found to be concomitants of neoplastic transformation have not been found in cells converted by RNA tumor viruses. Patinkin, Zaritsky & Doljanski (226) found that conversion of chicken cells by RSV does not alter the surface charge, pH mobility relationship, calcium-binding, or effect of neuriminidase on the cell surfaces. Moore & Temin (211) found that conversion of chicken, mouse, or rat cells by RSV or MSV did not usually alter the agglutination of these cells by wheat germ agglutinin or concanavalin A.

Therefore, it appears that there is no simple correlation between surface changes and neoplastic transformation with RNA tumor viruses. Certainly, when cells are infected and converted by RNA tumor viruses there are alterations in some surface properties. (It is probably true that any alteration in a cell state is accompanied by changes in certain surface properties.) However, we cannot determine now whether these surface changes are responsible for the neoplastic properties of the cell, or whether these surface changes are only reflections of alterations elsewhere in the cell. The actual picture may be more complex; for example, alterations in invasiveness and ease of metastasis might reflect specific surface changes; alterations in multiplication control might not.

Multiplication control.—Elsewhere this topic has been discussed in more detail (319, 327). Cells converted by infection with RNA sarcoma viruses do not appear to have an intrinsically faster multiplication rate than untransformed cells. Only under conditions of some limitation on the rate of cell multiplication do the converted cells multiply more than uninfected cells (Colby & Rubin, 67; Hanafusa, 125; Temin, 305). This difference in multiplication results from a decreased requirement of the converted cells for specific multiplication-stimulating factors in the medium (Jainchill & Todaro, 157; Temin, 307, 310, 312). Untransformed cells producing RNA tumor viruses do not have this decreased requirement for the specific multiplication-stimulating factors (Biquard, 35).

In their decreased requirement for serum, which is the usual source of the specific multiplication-stimulating factors, cells converted by RNA tumor viruses are similar to other transformed cells, for example, see (176). Although the correlation of this change in serum requirement with tumorigenicity is good, there are apparent exceptions in other transformation systems (154, 159).

A detailed study has been carried out to determine the factors affecting the multiplication of RNA tumor virus-converted chicken and rat cells, as compared with untransformed chicken and rat cells (Temin, 312, 314, 315, 319, 320). The following factors were found to be important: cells produced multiplication-stimulating activity, potentiators of multiplication-stimulating activity, and toxins or inhibitors of multiplication-stimulating activity. The cells removed multiplication-stimulating activity from the medium. Converted cells produced more of the toxic material, but otherwise seemed similar in these respects to the untransformed cells. The time and mode of action of the multiplication-stimulating factors appeared similar for uninfected and converted cells. Converted cells appeared to differ from uninfected cells in their efficiency of utilization of multiplication-stimulating activity. The most striking manifestation of this increased efficiency was their ability to continue to multiply in medium after removal of multiplication-stimulating activity.

As multiplication-stimulating factors become available in pure form (87, 231), and as their mode of action is better defined (323), we might hope to find a molecular basis for these differences in efficiency of utilization between uninfected and converted cells.

In addition, Rubin (258–260) has emphasized the importance of proteases released, presumably, from dying converted cells in affecting the growth of the cells.

The alteration in control of cell multiplication in cells transformed by RNA tumor viruses may be somewhat different from that of the same cells transformed by DNA tumor viruses. Ponten & Macintyre (233) reported that bovine cells transformed by RSV multiplied on a layer of untransformed bovine cells, but polyoma-transformed bovine cells did not.

Other cell properties altered by RNA tumor viruses.—A variety of other cellular properties have been reported to be altered after infection with RNA tumor viruses. Their general significance and relation to the properties already discussed is not clear.

Glycolysis is altered in converted cells (Steck, Kaufman & Bader, 288; Temin, 312, 313). The converted cells glycolyze more than the untransformed cells. Since glycolysis and cell multiplication appear to be controlled by serum in the same manner (Temin, 313), the previous discussion of multiplication could be repeated here. Cytochemical tests of foci of converted cells shows no decrease in amounts of respiratory enzymes (100).

When serum is withdrawn from the medium, converted cells at first multiply more and glycolyze more than uninfected ones (Temin, 313, 314). However, soon more of them die than do uninfected cells. This death may relate to the toxic factors produced at a higher rate by transformed cells than by uninfected cells (Kotler, 174; Rubin, 258; Temin, 305, 314). It also may relate to the proteases released in cultures of converted cells (Rubin, 259, 260).

Infection and conversion may also affect other aspects of a cell's differentiation. Pigmented iris cells no longer can make pigment after conversion (Ephrussi & Temin, 89; Temin, 305). This effect may be secondary to the changes in control of cell multiplication. Leukemia virus-infected animals have a depressed immune response (44, 104, 273). Cells transformed by MSV are reported to be resistant to the protective effects of interferon and to contain a substance which antagonizes interferon action in other cells (51, 56, 57); but these results are not confirmed by Freeman et al (103). The observations that multiplication of fowl plague virus is less sensitive to inhibition by actinomycin D or ultraviolet irradiation in RNA tumor virus-infected cells may be related (Zavada, 370).

In spite of their efficient transforming action, RNA sarcoma viruses do not seem to have much direct effect on the chromosomes of cells infected by these viruses. Cells studied soon after infection and conversion appear diploid. Later, chromosomal alterations may occur (76, 197, 232, 325, 357). However, with some other cell systems, usually involving heterologous systems with no conversion, chromosome breaks do occur (Nichols et al, 216).

One strain of RSV appears to cause nuclear fragmentation in chicken fibroblasts (Levinson 183).

It is probable that a great many other differences could be found between converted and uninfected cells. Furthermore, these changes might be regular concomitants of conversion. However, correlational changes are not meaningful until they are related either to some important aspect of the neoplastic phenotype or to a direct action of a provirus product. Since a cell is a complex interconnected network, any specific changes in a cell brought about by an RNA tumor virus will lead to a series of related changes in other areas and aspects of the cell. These other, secondary changes may merely be a reflection of the primary changes. Alternatively, the primary changes may of themselves be trivial, and it may be only some propagated tertiary changes that are responsible for the neoplastic transformation. At present, it is a somewhat vain hope that we shall be able to understand the changes leading to neoplastic transformation without understanding too much of the rest of the cell matrix in which the change takes place.

ORIGINS

One of the most intriguing problems related to the RNA tumor viruses is their origin. Of course, this is a general problem for all viruses. However, with the RNA tumor viruses, we may be able to study this problem.

In the absence of experimental infection, RNA tumor viruses seem usually to be "naturally" transmitted by vertical transmission, that is, from parent to offspring. Usually this is transmission of virions (40, 49, 116, 164). However, in addition, there appears at times to be vertical transmission in the absence of virions. Bentvelzen et al (32) and Payne & Chubb (227) have reported Mendelian inheritance in the absence of virions for the genetic determinants for the mammary tumor virus and the avian tumor virus group-specific antigen and O envelope. Similarly, Aaronson, Hartley & Todaro (1) report the appearance of murine leukemia viruses in long-term cultures of mouse cells when no virions appeared to be present in the primary cultures.

Another property of RNA tumor viruses is their widespread occurrence in different classes and individuals (8, 136, 163). Of course, the almost ubiquitous existence of virions of these viruses makes it extremely difficult for the experiments on vertical transmission, discussed in the last paragraph, to rule out contamination with infectious virus.

In addition, Huebner and co-workers have reported that animals are tolerant to the group-specific antigen of the RNA tumor viruses of their group, that is, chickens are tolerant to the avian tumor virus group-specific antigen, etc. (151). They also report that murine leukemia virus or antigens appear in most or all murine tumors induced by chemicals (153); that murine leukemia virus antigens are present in all rapidly dividing tissues of mice (150); and that there is a synergism between infection with murine leukemia viruses and chemical carcinogens in cell culture (102). The status of these reports is unclear. All have been challenged. Rabotti & Blackham (237) report antibody to avian tumor virus group-specific antigen in chickens; and Geering, Aoki & Old (108) report antibody to murine tumor virus group-specific antigen in the rat. Hilgers, Nowinski & Old (145), using a sensitive immunofluorescence assay and sera absorbed in vivo, do not find such a widespread incidence of occurrence of murine leukemia virus group-specific antigens. We have been unable to reproduce the interaction of diethylnitrosamine and MuLV in rat cell culture. More work by other groups will be necessary to resolve these conflicts.

Two hypotheses have been proposed as a result of observations on the widespread occurrence of RNA tumor viruses: the oncogene hypothesis and the protovirus hypothesis. In the terms used in this review, the oncogene theory states that the germ-line of all vertebrates contains the DNA provirus for an RNA tumor virus. In development or carcinogenesis, this provirus is partially or entirely activated to make viral-specific products and virions (Huebner & Todaro, 152). In contrast, the protovirus hypothesis states that the germ-line of vertebrates contains regions of DNA which can evolve in various directions through DNA to RNA to DNA transfers in the somatic cells. This type of evolution is normally part of embryonic differentiation, but abnormal evolution might lead to the formation of the genome of an RNA tumor virus (Temin, 317, 321, 322). Of course, an alternative to both of these hypotheses is that RNA tumor viruses or related genetic elements have no special relationship to the genomes of uninfected cells or play no special part in most oncogenesis.

Clearly, these ideas have implications beyond the special topic of RNA tumor viruses. Particularly, the protovirus hypothesis with its DNA → RNA → DNA transfer provides a useful mechanism for the somatic translocation of antibody genes postulated by Gally & Edelman (107). The further ramifications of this hypothesis might also give a further connection between histocompatibility genes and antibodies, as suggested by Jerne (161), and help to explain the high rate of mutation in histocompatability genes (18) and the close linkage of genes in chickens for susceptibility to avian tumor viruses and an erythrocyte isoantigen (68). Gruneberg (117) discusses the high mutation rate for genes affecting skeletal differences in mice. Again, this protovirus DNA to RNA to DNA type of information transfer might be involved.

A more detailed future form of the protovirus hypothesis might also explain the RNA tumor virus virion with its collection of polymerases, nucleases, and ligases, and its collection of different classes of nucleic acids.

ACKNOWLEDGEMENTS

The work from my laboratory was supported by Public Health Research Grant CA 07175 from the National Cancer Institute. I hold Research Career Development Award 10K 3-CA8182 from the same Institute. D. Boettiger, J. Coffin, E. Humphries, and I. Riegel made helpful comments on the manuscript.

LITERATURE CITED

1. Aaronson, S. A., Hartley, J. W., Todaro, G. J. 1969. *Proc. Nat. Acad. Sci. USA* 64:87–94
2. Aaronson, S. A., Jainchill, J. L., Todaro, G. J. 1970. *Proc. Nat. Acad. Sci. USA* 66:1236–43
3. Aaronson, S. A., Rowe, W. P. 1970. *Virology* 42:9–19
4. Aaronson, S. A., Todaro, G. J. 1970. *Nature (London)* 225: 458–59
5. Allen, D. W. 1969. *Virology* 38:32–41
5a. Allen, D. W., Sarma, P. S., Niall, H. D., Sauer, R. 1970. *Proc. Nat. Acad. Sci. USA* 69:837–42
6. Aloni, Y., Winocour, E., Sachs, L., Torten, J. 1969. *J. Mol. Biol.* 44: 333–45
7. Altaner, C., Temin, H. M. 1970. *Virology* 40:118–34
8. Aoki, T., Boyse, E. A., Old, L. J. 1968. *J. Nat. Cancer Inst.* 41:89–96
9. Aoki, T. et al. 1970. *Proc. Nat. Acad. Sci. USA* 65:569–76
10. Bader, J. P. 1962. *Virology* 16:436–43
11. Bader, J. P. 1965. *Virology* 26:253–61
12. Bader, J. P. 1966. *Virology* 29:444–51
13. Bader, J. P. 1966. in *Subviral Carcinog. Int. Symp. Tumor Viruses, 1st, 1966*:144–155
14. Bader, J. P. 1970. *Virology* 40:494–504
15. Bader, J. P., Bader, A. V. 1970. *Proc. Nat. Acad. Sci. USA* 67: 843–50
16. Bader, J. P., Brown, N. R., Bader, A. V. 1970. *Virology* 41:718–28
17. Bader, J. P., Steck, T. H., Kakefuda, T. 1970. *Curr. Top. Microbiol. Immunol.* 51:106–13
18. Bailey, D. W. 1966. *Transplantation* 4:482–88
19. Balduzzi, P., Morgan, H. R. 1970. *J. Virol.* 5:470–77
20. Baltimore, D. 1970. *Nature (London)* 226:1209–11
21. Baltimore, D., Huang, A. S., Stampfer, M. 1970. *Proc. Nat. Acad. Sci. USA* 66:572–76
22. Baluda, M. A., Nayak, D. P. 1971. *Nature New Biol. (London)* 231: May 19
23. Baluda, M. A., Moscovici, I. E., Goetz, I. E. 1964. *Nat. Cancer Inst. Monogr.* 17:449–58
24. Baluda, M. A., Nayak, D. P. 1969. *J. Virol.* 4:554–66
25. Baluda, M. A., Nayak, D. P. 1970. *Proc. Nat. Acad. Sci. USA* 66: 329–36
26. Bataillon, G. 1969. *C. R. Acad. Sci. Paris* 269:2156–58
27. Bauer, H. 1966. *Z. Naturforsch.* 216: 453–460
28. Bauer, H., Bubenik, J., Graf, T., Allgaier, C. 1969. *Virology* 39: 482–90
29. Bauer, H., Graf, T. 1969. *Virology* 37:157–61
30. Beard, D., Chabot, J. F., Langlois, A. J., Hillman, E. A., Beard, J. W. 1970. *Arch. Geschwulstforsch.* 35:315–25
31. Bellamy, A. R., Joklik, W. K., 1967. *Proc. Nat. Acad. Sci. USA* 58: 1389–96
32. Bentvelzen, P., Daams, J. H., Hageman, P., Calafat, J. 1970. *Proc. Nat. Acad. Sci. USA* 67:377–84
33. Berman, L. D., Allison, A. C. 1969. *Int. J. Cancer* 4:820–36
34. Bielka, H., Graffi, A., Yen, C. Y. 1963. *Acta Biol. Med. Ger.* 10: 63–78
35. Biquard, J. M. 1970. *C. R. Acad. Sci. Paris* 270:440–43
36. Bishop, J. M. et al. 1970. *Virology* 42:182–95

37. Bishop, J. M. et al. 1970. *Virology* 42:927–37
38. Biswal, N., Benyesh-Melnick, M. 1969. *Proc. Nat. Acad. Sci. USA* 64:1372–79
39. Biswal, N., McCain, B., Benyesh-Melnick, M. 1971. *Proc. Lepetit Colloq., 2nd,* 221–31. Amsterdam: North-Holland
40. Bittner, J. J. 1942. *Science* 95:462–63
41. Boettiger, D. 1970. Unpublished observations
42. Boettiger, D., Temin, H. M. 1970. *Nature (London)* 288: 662–64
43. Bolognesi, D. P., Langlois, A. J., Sverak, L., Bonar, R. A., Beard, J. W. 1968. *J. Virol.* 2:576–86
43a. Bolognesi, D. P., Bauer, H. 1970. *Virology* 42:1097–1112
44. Borella, L. 1969. *J. Immunol.* 103: 185–95
45. Borsa, J., Graham, A. F. 1968. *Biochem. Biophys. Res. Commun.* 33:895–901
46. Brdar, B., Rifkin, D., Reich, E. 1971. *Proc. Nat. Acad. Sci. USA* In press
47. Buck, B. M., Bather, R. 1969. *J. Gen. Virol.* 4:457–60
48. Burghouts, J. T. M. 1970. *The Influence of the infection with an RNA tumor virus on polyribosome profiles.* PhD thesis, Univ. of Nijmegen, Netherlands
49. Burmester, B. R. 1962. *Cold Spring Harbor Symp. Quant. Biol.* 27: 471–77
50. Burmester, B. R., Walter, W. G., Gross, M. A., Fontes, A. K. 1959. *J. Nat. Cancer Inst.* 23:277–91
51. Canivet, M., Peries, J., Boiron, M. 1969. *C. R. Acad. Sci. Paris* 268: 2527–29
52. Carnegie, J. W., Deaney, A. O. C., Olson, K. C., Beaudreau, G. S. 1969. *Biochim. Biophys. Acta* 190: 274–84
53. Carr, J. G. 1956. *Brit. J. Cancer* 10: 379–83
54. Carr, J. G. 1960. *Brit. J. Cancer* 14: 77–82
55. Cavalieri, L. F., Carroll, E. 1970. *Biochem. Biophys. Res. Commun.* 41:1055–60
56. Chany, C., Gregoire, A., Lemaitre, J. 1969. *C. R. Acad. Sci. Paris* 269: 1236–37
57. Chany, C., Robbe-Maridor, F. 1969. *Proc. Soc. Exp. Biol. Med.* 131: 30–35
58. Chopra, H. C., Mason, M. M. 1970. *Cancer Res.* 30:2081–86
59. Chuat, J. C., Berman, L., Gunvén, P., Klein, E. 1969. *Int. J. Cancer* 4:465–79
60. Chubb, R. C., Biggs, P. M. 1968. *J. Gen. Virol.* 3:87–96
61. Chung, M., Hinz, R. W. 1970. *Proc. Soc. Exp. Biol. Med.* 133:20–24
62. Coffin, J. 1970. Unpublished observations
63. Coffin, J., Hilgers, J. 1970. Unpublished observations
64. Coffin, J., Temin, H. M. 1971. *J. Virol.* 7:625–34
65. Cohn, M. 1967 *Cold Spring Harbor Symp. Quant. Biol.* 32:211–21
66. Colby, C., Duesberg, P. H. 1969. *Nature (London)* 222:940–44
67. Colby, C., Rubin, H. 1969. *J. Nat. Cancer Inst.* 43:437–44
68. Crittenden, L. B., Briles, W. E., Stone, H. A. 1970. *Science* 169: 1324–25
69. De Petris, S., Harvey, J. J. 1969. *J. Gen. Virol.* 5:561–64
70. De Thé, G. 1964. *Nat. Cancer Inst. Monogr.* 17:651–71
71. De Thé, G. 1966. *Int. J. Cancer* 1: 119–38
72. De Thé, G. 1968. *Norm. Rev. Franc. Hematol.* 8:769–92
73. De Thé, G., O'Connor, T. E. 1966. *Virology* 28:713–28
74. Diggelmann, H., Weissmann, C. 1969. *Nature (London)* 224: 1277–79
75. Dinowitz, M., Rubin, H. 1970. *Int. J. Cancer* 6:160–71
76. Donner, L., Turano, A., Bubenik, J. 1969. *Folia Biol. (Praha)* 15: 226–28
77. Dougherty, R. M., Di Stefano, H. S. 1969. *Progr. Med. Virol.* 11:154–84
78. Duc-Nguyen, H. 1968. *J. Virol.* 2: 643–44
79. Duesberg, P. H. 1968. *Proc. Nat. Acad. Sci. USA* 60:1511–18
80. Duesberg, P. H. 1969. *J. Mol. Biol.* 42:485–599
81. Duesberg, P. H. 1970. *Curr. Top. Microbiol. Immunol.* 51:79–104
82. Duesberg, P. H., Canaani, E. 1970. *Virology* 42:783–88
83. Duesberg, P. H., Martin, G. S., Vogt, P. K. 1970. *Virology* 41: 631–46
84. Duesberg, P. H., Vogt, P. K. 1969. *Proc. Nat. Acad. Sci. USA* 64: 939–46
85. Duesberg, P. H., Vogt, P. K. 1970.

Proc. Nat. Acad. Sci. USA 67: 1673–80

86. Duff, R. G., Vogt, P. K. 1969. *Virology* 39:18–30
87. Dulak, N., Temin, H. M. 1971. In preparation
88. Emanoil-Ravicovitch, R., Baudelarie, M. F., Borion, M. 1969. *C. R. Acad. Sci. Paris* 269:1903–5
89. Ephrussi, B., Temin, H. M. 1960. *Virology* 11:547–52
90. Erikson, R. L. 1969. *Virology* 37:124–31
91. Erikson, R. 1970. Personal communication
92. Erikson, E., Erikson, R. L. 1970. *J. Mol. Biol.* 52:387–90
93. Fenner, F. 1968. *The Biology of Animal Viruses.* New York: Academic
94. Ferrer, J. F., Gibbs, F. A., Jr. 1969. *J. Nat. Cancer Inst.* 43:1317–30
95. Fieldsteel, A. H., Kurahara, C., Dawoni, P. J. 1969. *Nature (London)* 223:1274
96. Fink, M. A., Sibal, L. R., Wivel, N. A., Cowles, C. A., O'Connor, T. E. 1969. *Virology* 37:605–14
97. Fischinger, P. J., Moore, C. O., O'Connor, T. E. 1969. *J. Nat. Cancer Inst.* 42:605–22
98. Fischinger, P. J., O'Connor, T. E. 1970. *J. Nat. Cancer Inst.* 44: 429–38
99. Fleissner, E. 1970. *J. Virol.* 5:14–21
100. Francois, D. 1970. *J. Gen. Virol.* 6: 187–99
101. Franker, C. K., Gruca, M. 1969. *Virology* 37:489–92
102. Freeman, A. E. et al. 1970. *J. Nat. Cancer Inst.* 44:65–78
103. Freeman, A. E., Uhlendorf, C. P., Younkers, P. E., Baron, S. 1970. *J. Cell Physiol.* 76:365–72
104. Friedman, H., Ceglowski, W. S. 1968. *Nature (London)* 218:1232–34
105. Fujinaga, K., Parsons, J. T., Beard, J. W., Beard, D., Green, M. 1970. *Proc. Nat. Acad. Sci. USA* 67: 1432–39
106. Gallo, R., Yang, S. S., Ting, R. C. 1970. *Nature (London)* 228:927–29
107. Gally, J. A., Edelman, G. M. 1970. *Nature (London)* 227:341–48

107a. Garapin, A. C., Leong, J., Fanshier, L., Levinson, W. E., Bishop, J. M. 1971. *Biochem. Biophys. Res. Commun.* 42:919–25

108. Geering, G., Aoki, T., Old, L. J. 1970. *Nature (London)* 226:265–66
109. Gelderblom, H., Bauer, H., Frank, H. 1970. *J. Gen. Virol.* 7:33–45
110. Gerwin, B. I., Todaro, G. J., Zeve, V., Scolnick, E. M., Aaronson, S. A. 1970. *Nature (London)* 228: 435–38
111. Gilden, R. V., Lee, Y. K., Oroszlan, S., Walker, J. L., Huebner, R. J. 1970. *Virology* 11:187–90
112. Goldé, A. 1970. *Virology* 40:1022–29
113. Graf, T., Bauer, H. 1970. Personal communication
114. Green, M., Rokutanda, M., Fujinaga, K., Ray, R. K., Rokutanda, H., Gurgo, C. 1970. *Proc. Nat. Acad. Sci. USA* 67:385–93

114a. Green, M., Rokutanda, H., Rokutanda, M. 1971. *Nature New Biol. (London)* 230:229–32

115. Gresser, I., Bronty-Boye, D., Thomas, M. T., Macieira-Coelho, A. 1970. *Proc. Nat. Acad. Sci. USA* 66:1052–58
116. Gross, L. 1951. *Proc. Soc. Exp. Biol. Med.* 76:27–32
117. Gruneberg, H. 1970. *Nature (London)* 225:139–41
118. Guillemain, B., Hampe, A., Boiron, M. 1969. *C. R. Acad. Sci. Paris* 269:2283–86
119. Haguenau, F., Michelson-Fiske, S., Rabotti, G. 1970. *C. R. Acad. Sci. Paris* 270:1954–57
120. Hahn, G. M., Decleve, A., Lieberman, M., Kaplan, H. S. 1970. *J. Virol.* 5:432–36
121. Hall, W. T., Morton, D. L., Malmgren, R. A. 1970. *J. Nat. Cancer Inst.* 44:507–13
122. Hamerman, D., Barski, G., Youn, J. K., Green, H. 1968. *Rev. Fr. Etud. Clin. Biol.* 13:800–1
123. Hanafusa, H. 1964. *Nat. Cancer Inst. Monogr.* 17:543–56
124. Hanafusa, H. 1969. *Advan. Cancer Res.* 12:137–65
125. Hanafusa, H. 1969. *Proc. Nat. Acad. Sci. USA* 63:318–25
126. Hanafusa, H. 1970. *Curr. Top. Microbiol. Immunol.* 51:114–23
127. Hanafusa, T., Hanafusa, H. 1967. *Proc. Nat. Acad. Sci. USA* 58: 818–25
128. Hanafusa, T., Miyamoto, T., Hanafusa, H. 1970. *Virology* 40:55–64
129. Hanafusa, H., Miyamoto, T., Hanafusa, T. 1970. *Proc. Nat. Acad. Sci. USA* 66:314–21
130. Hanafusa, T., Hanafusa, H., Mi-

yamoto, T. 1970. *Proc. Nat. Acad. Sci. USA* 67:1797–1803

131. Hanafusa, H., Hanafusa, T., Miyamoto, T. 1971. *Proc. Lepetit Colloq., 2nd,* 170–75. Amsterdam: North-Holland
132. Harel, L., Harel, J., Lacour, F., Huppert, J. 1966. *C. R. Acad. Sci. Paris* 263:616–19
133. Harel, J., Huppert, J., Lacour, F., Lacour, J. 1959. *Bull. Assoc. Fr. Etude Cancer* 46:75–91
134. Harris, R. J. C., Chesterman, F. C., McClelland, R. M. 1969. *Int. J. Cancer* 4:31–41
135. Hartley, J. W., Rowe, W. P. 1966. *Proc. Nat. Acad. Sci. USA* 55: 780–86
136. Hartley, J. W., Rowe, W. P., Copps, W. I., Huebner, R. J. 1969. *J. Virol.* 3:126–32
137. Hartley, J. W., Rowe, W. P., Huebner, R. J. 1970. *J. Virol.* 5: 221–25
138. Haruna, I., Watanabe, I., Ohno, T. 1970. Abstracts *Proc. Int. Cancer Congr., 10th, Houston,* 389
139. Hatanaka, M., Augl, C., Gilden, R. V. 1970. *J. Biol. Chem.* 245:714–17
140. Hatanaka, M., Gilden, R. V. 1970. *J. Nat. Cancer Inst.* 45:87–89
141. Hatanaka, M., Hanafusa, H. 1970. *Virology* 41:647–52
142. Hatanaka, M., Huebner, R. J., Gilden, R. V. 1969. *J. Nat. Cancer Inst.* 43:1091–96
143. Hatanaka, M., Huebner, R. J., Gilden, R. V. 1970. *Proc. Nat. Acad. Sci USA* 67:143–47
144. Hatanaka, M., Todaro, G. J., Gilden, R. V. 1970. *Int. J. Cancer* 5:224–28
145. Hilgers, J., Nowinski, R., Old, L. 1970. Personal communication
146. Hirschman, S. Z., Fischinger, P. J., O'Connor, T. E. 1970. *J. Nat. Cancer Inst.* 44:107–16
147. Hlozanek, I., Sovova, V., Riman, J., Veprek, L. 1970. *J. Gen. Virol.* 6: 163–68
148. Hobom-Schnegg, B., Robinson, H. L., Robinson, W. S. 1970. *J. Gen. Virol.* 7:85–93
149. Holtzer, H., Abbott, J. 1968. *Results Prob. Cell Diff.* 1:1–16
150. Huebner, R. J. et al. 1970. *Proc. Nat. Acad. Sci. USA* 67:366–76
151. Huebner, R. J. et al. 1970. *Ann. N. Y. Acad. Sci.* In press
152. Huebner, R. J., Todaro, G. J. 1969. *Proc. Nat. Acad. Sci. USA* 64: 1087–94

152a. Hung, P. P., Robinson, H. L., Robinson, W. S. 1971. *Virology* 43:251–66

153. Igel, H. J., Huebner, R. J., Turner, H. C., Kotin, P., Falk, H. L. 1969. *Science* 166: 1624–66
154. Inbar, M., Rabinowitz, Z., Sachs, L. 1969. *Int. J. Cancer* 4:690–96
155. Ishimoto, N., Temin, H. M., Strominger, J. L. 1966. *J. Biol. Chem.* 241:2052–57
156. Ishizaki, R., Shimizu, T. 1970. *Virology* 40:415–17
157. Jainchill, J. L., Todaro, G. J. 1970. *Exp. Cell Res.* 59:137–146
158. Jarrett, O., Laid, H. M., Hay, D. 1969. *Nature (London)* 244: 1208–9
159. Jarrett, O., Macpherson, I. 1968. *Int. J. Cancer* 3:654–62
160. Jensen, E. M., Zelljadt, I., Chopra, H. C., Mason, M. M. 1970. *Cancer Res.* 30:2388–91
161. Jerne, N. K. 1970. In *Immune Surveillance,* ed. R. T. Smith, M. Landy, 345. New York:
162. Kakefuda, T., Bader, J. P. 1969. *J. Virol.* 4:460–74
163. Kajima, J., Pollard, M. 1968. *Nature (London)* 218:188–89
164. Kaplan, H. S. 1967. *Cancer Res.* 27: 1325–40
165. Kara, J. 1968. *Biochem. Biophys. Res. Commun.* 32:817–24
166. Kates, J. R., McAuslan, B. R. 1967. *Proc. Nat. Acad. Sci. USA* 57: 314–20
167. Kelloff, G., Huebner, R. J., Lee, Y. K., Toni, R., Gilden, R. 1970. *Proc. Nat. Acad. Sci. USA* 65: 310–17
168. Kingsbury, D. W., Granoff, A. 1970. *Virology* 42:262–65
169. Kingsbury, D. W., Webster, R. G. 1970. *J. Virol.* 4:219–25
170. Kirsten, W. 1970. *Proc. Int. Cancer Congr., 10th, Houston*
171. Klement, V., Vesely, P. 1965. *Neoplasma* 12:147–53
172. Klement, V., Hartley, J. W., Rowe, W. P., Huebner, R. J. 1969. *J. Nat. Cancer Inst.* 43:925–34
173. Klement, V., Rowe, W. P., Hartley, J. W., Pugh, W. E. 1969. *Proc. Nat. Acad. Sci. USA* 63:753–58
174. Kotler, M. 1970. *Cancer Res.* 30: 2493–96
175. Kotler, M. 1971. *J. Gen. Virol.* In press
176. Kruse, P. F., Jr., Whittle, W., Mie-

dema, E. 1969. *J. Cell Biol.* 42: 113–21
177. Kriukova, I. N., Obukh, I. B., Tot, F. 1970. *J. Nat. Cancer Inst.* 45: 49–57
178. Kumanishi, T., Yamamoto, T. 1970. *Jap. J. Exp. Med.* 40:79–86
179. Lacour, F., Fourcade, A., Huynh, T. 1970. In *Biology of Large RNA Viruses*, ed. R. D. Barry, B. W. J. Mahy, 125–220. New York: Academic
180. Lacour, F., Fourcade, A., Verger, C., Delain, E. 1970. *J. Gen. Virol.* 9 (part I):89–92
181. Law, L. W., Ting, R. C. 1970. *J. Nat. Cancer Inst.* 44:615–21
182. Lee, H. H., Kaighn, M. E., Ebert, J. D. 1968. *Int. J. Cancer* 3:126–36
183. Levinson, W. 1970. *J. Nat. Cancer Inst.* 44:151–58
184. Levinson, W., Rubin, H. 1966. *Virology* 28:533–42
185. Levinson, W., Bishop, J. M., Quintrell, N., Jackson, J. 1970. *Nature (London)* 227:1023–25
186. Levy, J. A., Pincus, T. 1970. *Science* 170: 326–27
187. Levy, J. A., Varet, B., Oppenheim, E., Laduc, J. C. 1969. *Nature (London)* 224:606–8
188. Lin, F. H., Rich, M. A. 1968. *Biochim. Biophys. Acta* 157:310–21
189. Lin, F. H., Thormar, H. 1970. *J. Virol.* 6:702–4
190. Lindberg, L. G. 1968. *Acta Pathol. Microbiol. Scand.* 74:189–98
191. Machala, O., Donner, L., Svoboda, J. 1970. *J. Gen. Virol.* 8:219–29
192. Macieira-Coelho, A., Hiu, I. J., Garcia-Giralt, E. 1969. *Nature (London)* 222:1172
193. Macieira-Coelho, A., Ponten, J. 1967. *Biochem. Biophys. Res. Commun.* 29:316–21
194. Macintyre, E. H., Grimes, R. A., Vatter, A. E. 1969. *J. Cell Sci.* 5: 583–602
195. Macpherson, I. 1965. *Science* 148: 1731–33
196. Manaker, R., Groupé, V. 1956. *Virology* 2:838–40
197. Mark, J. 1969. *Cancer* 5:307–15
198. Martin, G. S. 1970. *Nature (London)* 227:1021–23
199. McDonnell, J. P. et al. 1970. *Nature (London)* 288:433–35
200. McDugald, L. V., Panem, S., Kirsten, W. H. 1970. *Int. J. Cancer* 5:64–71
201. McGarry, M. P., Vanable, J. W., Jr. 1969. *Develop. Biol.* 20:291–303
202. Meier, H., Myers, D. D., Huebner, R. J. 1969. *Proc. Nat. Acad. Sci. USA* 63:759–66
203. Michelson-Fiske, S., Haguenau, F., Rabotti, G. F. 1969. *C. R. Acad. Sci. Paris* 269:2475–78
204. Mizutani, S. 1969. Unpublished observations
205. Mizutani, S., Boettiger, D., Temin, H. M. 1970. *Nature (London)* 288:424–27
206. Mizutani, S., Temin, H. M., Kodama, M., Wells, R. D. 1971. *Nature New Biol. (London)* 230:232–35
207. Moloney, J. B. 1963. *Acta Unio Int. Contra Cancrum* 19:250
208. Montagnier, L. 1969. *C. R. Acad. Sci. Paris* 263:2218–21
209. Montagnier, L., Goldé, A., Vigier, P. 1969. *J. Gen. Virol.* 4:449–52
210. Moore, E. M. 1970. Unpublished observations
211. Moore, E. M., Temin, H. M. 1971. *Nature (London)* 231:117–18
212. Moscovici, C., Moscovici, M. G., Zanetti, M. 1969. *J. Cell. Physiol.* 73:105–8
213. Moscovici, C., Zanetti, M. 1970. *Virology* 42:61–67
214. Munyon, W., Paoletti, E., Grace, J. T. 1967. *Proc. Nat. Acad. Sci. USA* 58:2280–87
215. Murray, R. K., Temin, H. M. 1970. *Int. J. Cancer* 5:320–26
216. Nichols, W. W., Levan, A., Coriell, L. L., Goldner, H., Ahlstrom, C. G. 1964. *Science* 146:248–50
217. Nordenskjold, B. A., Klein, E., Tachibana, T., Fenyo, E. M. 1970. *J. Nat. Cancer Inst.* 44:403–12
218. Obukh, I. B., Kryukova, I. N. 1969. *Int. J. Cancer* 4:809–12
219. O'Connor, T., Fischinger, P. J. 1969. *J. Nat. Cancer Inst.* 43:487–97
220. O'Connor, T. E., Rauscher, F. J. 1964. *Science* 146:787–90
221. Okano, H., Rich, M. A. 1969. *Nature (London)* 224:77–78
222. Oroszlan, S., Fisher, C. L., Stanley, T. B., Gilden, R. V. 1970. *J. Gen. Virol.* 8:1–10
223. Oroszlan, S., Gilden, R. V. 1970. *Science* 168:1478–80
224. Parkman, R., Levy, J. A., Ting, R. C. 1970. *Science* 168:387–89
225. Pasternak, L., Pasternak, G. 1968. *Arch. Geschwultforsch.* 32:301–8
226. Patinkin, D., Zaritsky, A., Doljanski, F. 1970. *Cancer Res.* 30:498–503
227. Payne, L. N., Chubb, R. C. 1968. *J. Gen. Virol.* 3:379–91

228. Payne, F. E., Solomon, J. J., Purchase, H. G. 1966. *Proc. Nat. Acad. Sci. USA* 55:341–49
229. Peries, J., Canivet, M., Guillemain, B., Boiron, M. 1968. *J. Gen. Virol.* 3:465–68
230. Perk, K., Violo, M. V., Smith, K. L., Wivel, N. A., Moloney, J. B. 1969. *Cancer Res.* 29:1089–1102
231. Pierson, R. W., Jr., Temin, H. M. 1971. Submitted for publication
232. Ponten, J., Lithner, F. 1966. *Int. J. Cancer* 1:589–98
233. Ponten, J., Macintyre, E. H. 1968. *J. Cell Sci.* 3:603–13
234. Ponten, J., Macpherson, I. 1966. *Ann. Med. Exp. Biol. Fenn.* 44: 260–64
235. Prince, A. M. 1962. *Virology* 18: 524–34
236. Rabotti, G. F. 1969. *C. R. Acad. Sci. Paris* 268:456–58
237. Rabotti, G. F., Blackham, E. 1970. *J. Nat. Cancer Inst.* 44:985–91
238. Rabotti, G. F., Michelson-Fiske, S., Haguenau, F. 1969. *C. R. Acad. Sci. Paris* 269:2291–94
239. Rakusanova, T. 1969. *Folia Biol. (Praha)* 15:87–95
240. Rakusanova, T. 1969. *Folia Biol. (Praha)* 15:96–103
241. Reamur, R. H., Okazaki, W. 1970. *J. Nat. Cancer Inst.* 44:763–67
242. Reddi, K. K. 1971. *Nature New Biol. (London)* 229:25–27
243. Rhim, J. S., Greenawalt, C., Huebner, R. J. 1969. *Nature (London)* 222:1166–68
244. Rhim, J. S., Huebner, R. J., Gisim, S. 1969. *Proc. Soc. Exp. Biol. Med.* 130:181–87
245. Rhim, J. S., Huebner, R. J., Lane, W. T., Turner, H. C., Rubstein, L. 1969. *Proc. Soc. Exp. Biol. Med.* 132:1091–98
246. Rhim, J. S., Huebner, R. J., Ting, R. C., Wivel, N., Vass, W. 1970. *Int. J. Cancer* 5:28–38
247. Rifkin, D., Reich, E. 1971. In press
247a. Rispens, B. H., Long, P. A., Okazaki, W., Burmester, B. R. 1970. *Avian Dis.* 14:738–51
248. Riman, J., Beaudreau, G. S. 1970. *Nature (London)* 228:427–30
249. Robinson, W. S., Pitkanen, A., Rubin, H. 1965. *Proc. Nat. Acad. Sci. USA* 54:137–44
250. Robinson, W. S., Hung, P., Robinson, H. L., Ralph, D. D. 1970. *J. Virol.* 6:695–98
251. Rokutanda, M. et al. 1970. *Nature (London)* 227:1026–28
252. Rosenbergova, M., Lacour, F., Huppert, J. 1965. *C. R. Acad. Sci. Paris* 260:5145–48
253. Rossi, G. B., Cudkowicz, G., Friend, C. 1970. *J. Exp. Med.* 131:765–81
254. Rowe, W. P., Pugh, W. E., Hartley, J. W. 1970. *Virology* 43:1136–39
255. Rubin, H. 1960. *Virology* 10:29–49
256. Rubin, H. 1960. *Proc. Nat. Acad. Sci. USA* 46:1105–19
257. Rubin, H. 1960. *Virology* 11:28–47
258. Rubin, H. 1966. *Exp. Cell Res.* 41: 149–61
259. Rubin, H. 1970. *Proc. Int. Symp. Tumor Viruses 2nd,* 183:11–17
260. Rubin, H. 1970. *Science* 167:1271–72
261. Rubin, H., Temin, H. M. 1959. *Virology* 7:75–91
262. Sanger, F., Brownlee, G. G., Barrell, B. G. 1965. *J. Mol. Biol.* 13:373–98
263. Sarkar, N. H., Moore, D. H. 1968. *J. Microsc. Paris* 7:539–48
264. Sarkar, N. H., Moore, D. H. 1970. *J. Virol.* 5:230–36
265. Sarma, P. S., Baron, S., Huebner, R. J., Shia, G. 1969. *Nature (London)* 224:604–5
266. Sarma, P. S., Shia, G., Baron, S., Huebner, R. J. 1969. *Nature (London)* 223:845–46
267. Sarma, P. S., Huebner, R. J., Barber, J. F., Bernon, L. Gilden, R. V. 1970. *Science* 168:1098–1100
268. Sarma, P. S. et al. 1970. *Virology* 41:377–81
269. Sarma, P. S., Log, T., Gilden, R. V. 1970. *Proc. Soc. Exp. Biol. Med.* 133:718–22
270. Schafer, W., Anderer, F. A., Bauer, H., Pister, L. 1969. *Virology* 38: 387–94
271. Scolnick, E. M., Aaronson, S. A., Todaro, G. J. 1970. *Proc. Nat. Acad. Sci. USA* 67:1034–41
272. Scolnick, E., Rands, E., Aaronson, S. A., Todaro, G. J. 1970. *Proc. Nat. Acad. Sci. USA* 67:1789–96
273. Seidel, H. J., Lavenstein, K. 1969. *Z. Krebsforsch.* 72:219–25
274. Shatkin, A. J., Sipe, J. D. 1968. *Proc. Nat. Acad. Sci. USA* 61: 1462–69
275. Shevliaghyn, V. J., Biryulina, T. I., Tikhonova, Z. N., Karazas, N. V. 1969. *Int. J. Cancer* 4:42–46
276. Shibley, G. P., Deur, F. E., Shidlovsky, G., Wright, B. S., Schmitter, R. 1967. *Science* 156:1610–13
277. Shibley, G. P. et al. 1969. *Cancer Res.* 29:905–11

278. Siegler, R. 1970. *J. Nat. Cancer Inst.* 45:135–47
279. Simons, P., Pepper, S., Baker, R. 1969. *Proc. Soc. Exp. Biol. Med.* 131:454–56
280. Smith, R. E., Moscovici, C. 1969. *Cancer Res.* 29:1356–66
281. Somers, K. D., Kirsten, W. H. 1968. *Virology* 36:155–57
282. Somers, K. D., Kirsten, W. H. 1969. *Int. J. Cancer* 4:697–704
283. Somerson, N. L., Cook, M. K. 1965. *J. Bacteriol.* 90:534–40
284. Spiegelman, S. et al. 1970. *Nature (London)* 228:430–32
285. Spiegelman, S. et al. 1970. *Nature (London)* 227:1029–31
286. Spiegelman, S. et al. 1970. *Nature (London)* 227:563–67
287. Staines, R. L., Yamada, E. W. 1970. *Biochim. Biophys. Acta* 209: 75–85
288. Steck, T. L., Kaufman, S., Bader, J. P. 1968. *Cancer Res* 28:1611–19
289. Stenback, W. A., Van Hoosier, G. L., Trentin, J. J. 1966. *Proc. Soc. Exp. Biol. Med.* 122:1219–23
290. Stenback, W. A., Van Hoosier, G. L., Trentin, J. J. 1968. *J. Virol.* 2:1115–21
291. Sundelin, P. 1967. *Exp. Cell Res.* 46:581–92
292. Sundelin, P. 1968. *Exp. Cell Res.* 50:233–38
293. Suskind, R. G., Pry, T. W., Rabotti, G. F. 1969. *Cancer Res.* 29:1598–1605
294. Svec, F., Altaner, C., Hlavay, E. 1966. *J. Nat. Cancer Inst.* 36: 389–404
295. Svec, J., Svec, F., Simkovic, D., Thurzo, V. 1970. *J. Nat. Cancer Inst.* 44:521–32
295a. Sverak, L., Bonar, R. A., Langlois, A. J., Beard, J. W. 1970. *Biochim. Biophys. Acta* 224:441–50
296. Svoboda, J., Klement, V. 1963. *Folia Biol. (Praha)* 9:403–11
297. Svoboda, J., Dourmashkin, R. 1969. *J. Gen. Virol.* 4:523–29
297a. Svoboda, J., Hlozanek, I. 1970. *Advan. Cancer Res.* 13:217–69
298. Temin, H. M. 1960. *Virology* 10: 182–97
299. Temin, H. M., 1961. *Virology* 13: 158–63
300. Temin, H. M. 1962. *Cold Spring Harbor Symp. Quant. Biol.* 27: 405–14
301. Temin, H. M. 1963. *Virology* 20: 577–82
302. Temin, H. M. 1964. *Virology* 23: 486–94
303. Temin, H. M. 1964. *Nat. Cancer Inst. Monogr.* 17:557–70
304. Temin, H. M. 1964. *Proc. Nat. Acad. Sci. USA* 52:323–29
305. Temin, H. M. 1965. *J. Nat. Cancer Inst.* 35:679–93
306. Temin, H. M. 1966. *Cancer Res.* 26: 212–216
307. Temin, H. M. 1966. *J. Nat. Cancer Inst.* 35:167–75
308. Temin, H. M. 1967. In *Mol. Biol. Viruses,* J. S. Colter, W. Paranchych, 709–15. New York: Academic
309. Temin, H. M. 1967. *J. Cell Physiol.* 69:53–63
310. Temin, H. M. 1967. *J. Cell Physiol.* 69:377–84
311. Temin, H. M. 1968. *Cancer Res.* 28: 1835–38
312. Temin, H. M. 1968. *Int. J. Cancer* 3:491–503
313. Temin, H. M. 1968. *Int. J. Cancer* 3:273–82
314. Temin, H. M. 1968. *Int. J. Cancer* 3:771–787
315. Temin, H. M. 1969. *J. Cell Physiol.* 74:9–15
316. Temin, H. M. 1970. In *Biology of Large RNA Viruses,* ed. R. D. Barry, B. W. J. Mahy, 233–49. New York: Academic
317. Temin, H. M. 1970. *Perspect. Biol. Med.* 14:11–26
318. Temin, H. M. 1970. *Proc. Int. Cancer Congr., 10th, Houston*
319. Temin, H. M. 1970. In *Biology of Large RNA Viruses,* ed. R. D. Barry, B. W. J. Mahy, 697–711. New York: Academic
320. Temin, H. M. 1970. *J. Cell Physiol.* 75:107–20
321. Temin, H. M. 1971. *Proc. Lepetit Colloq., 2nd,* 176–87. Amsterdam: North-Holland
322. Temin, H. M. 1971. *J. Nat. Cancer Inst.* 46:III–VI
323. Temin, H. M. 1971. *J. Cell Physiol.* In press
324. Temin, H. M., Mizutani, S. 1970. *Nature* 266:1211–13
325. Temin, H. M., Rubin, H. 1968. *Virology* 6:669–88
326. Temin, H. M., Rubin, H. 1969. *Virology* 8:209–22
327. Temin, H. M., Pierson, R., Dulak, N. 1971. In *Nutrition and Metabolism of Cells in Culture,* ed. V. J. Cristofalo, G. H. Rothblat. New York: Academic

328. Temin, H. M., Mizutani, S., Coffin, J. 1971. *Proc. Miami Winter Symposia,* 3rd. Amsterdam: North-Holland
329. Thurzo, V., Simkovičová, M., Simkovič, D. 1969. *Int. J. Cancer* 4: 852–58
330. Ting, R. C. 1964. *Virology* 22:568–74
331. Ting, R. C. 1968. *J. Virol.* 2:865–68
332. Ting, R. C., Bader, A. V. 1969. *Virology* 39:194–204
333. Ting, R. C., Law, L. W. 1970. *Int. J. Cancer* 5:202–10
334. Todaro, G. J., Aaronson, S. A. 1969. *Virology* 38:174–79
335. Todaro, G. J., Zeve, V., Aaronson, S. A. 1970. *Nature* 226:1047–49
336. Tooze, J., Ed. 1971. *Molecular Biology of Animal Tumor Viruses.* NewYork: Cold Spring Harbor Laboratory
337. Toyoshima, K., Friis, R. R., Vogt, P. K. 1970. *Virology* 42:163–70
338. Toyoshima, K., Vogt, P. K. 1969. *Virology* 39:930–31
339. Tozawa, H., Bauer, H., Graf, T., Gelerblom, H. 1970. *Virology* 40:530–39
340. Trager, G. W., Rubin, H. 1966. *Virology* 30:266–74
341. Travnicek, M., Riman, J. 1970. *Biochim. Biophys. Acta* 199:283–85
342. Traub, W. H., Morgan, H. R. 1967. *Arch. Ges. Virusforsch.* 20:1–10
343. Vaage, J. 1968. *Cancer Res.* 28: 2477–83
344. Valentine, A. F., Bader, J. P. 1968. *J. Virol.* 2:224–37
345. Van Griensven, L., Emanoil-Ravicovitch, R., Boiron, M. 1970. *C. R. Acad. Sci. Paris* 270:1723–26
346. Van Tuyen, Vu, Maunoury, R., Febvre, H. 1967. *C. R. Acad. Sci. Paris* 265:1345–48
347. Vigier, P. 1970. *Virology* 40:179–92
348. Vigier, P. 1970. *C. R. Acad. Sci. Paris* 270:1192–95
349. Vigier, P. 1970. *Progr. Med. Virol.* 12:240–83
350. Vigier, P., Biquard, J. M. 1971. *Proc. Lepetit Colloq., 2nd,* 326–30. Amsterdam: North-Holland
351. Vogt, P. K. 1965. *Advan. Virus Res.* 11:293–385
352. Vogt, P. K. 1967. *Proc. Nat. Acad. Sci. USA* 58:801–8
353. Vogt, P. K. 1967. *Virology* 32:708–17
354. Vogt, P. K. 1967. *Virology* 33:175–77
355. Vogt, P. K. 1970. *Int. Symp. Comp. Leukemia Res., 3rd,* Basel: Karger
356. Vogt, P. K., Friis, R. R., Toyoshima, K. 1971. *Proc. Lepetit Colloq., 2nd,* 313–16. Amsterdam: North-Holland
357. Vrba, M., Donner, L. 1964. *Folia Biol. (Praha)* 10:373–80
358. Watson, K. F., Beaudreau, G. S. 1969. *Biochem. Biophys. Res. Commun.* 37:925–32
359. Weiss, R. A. 1967. *Virology* 12: 719–23
360. Weiss, R. A. 1969. *J. Gen. Virol.* 5: 511–28
361. Weiss, R. A. 1970. *Int. J. Cancer* 6: 333–45
362. Wilson, D. E., Bauer, H. 1967. *Virology* 33:754–57
363. Wollmann, R. L., Kirsten, W. H. 1968. *J. Virol.* 3:1241–48
364. Wright, B. S., Korol, W. 1969. *Cancer Res.* 29:1886–88
365. Yamaguchi, N., Takeuchi, M., Yamamoto, T. 1969. *Int. J. Cancer* 4: 678–89
366. Yoshikawa-Fukada, M., Ebert, J. D. 1969. *Proc. Nat. Acad. Sci. USA* 64:870–77
367. Yoshikura, H. 1970. *J. Gen. Virol.* 6:183–85
368. Yoshikura, H. 1970. *J. Gen. Virol.* 8:113–20
369. Yoshikura, H., Hirokawa, Y., Ikawa, Y., Sugano, H. 1969. *Int. J. Cancer* 4:636–40
370. Zavada, J. 1969. *J. Gen. Virol.* 4: 571–76

THE BDELLOVIBRIOS

1584

MORTIMER P. STARR AND RAMON J. SEIDLER
Department of Bacteriology, University of California, Davis, California 95616
Department of Microbiology, Oregon State University, Corvallis, Oregon 97331

CONTENTS

INTRODUCTION

A variety of well-documented (21, 95, 96) bacteriolytic phenomena is known in which microbes or their products cause the lysis or dissolution of living or dead bacteria and cell wall preparations. These biological bacteriolyses are generally mediated by antibiotics or enzymes that are excreted into the environment. None of these, however, acts in a manner comparable to the intimate and antagonistic symbiotic relationship exhibited by bacteria of the genus *Bdellovibrio*. These intriguing predatory, endoparasitic, and bacteriolytic bacteria have captured the attention of an international spectrum of microbiologists, for whom *Bdellovibrio* provides an experimentally feasible model for studying the basic principles of intracellular parasitism (6, 33, 54, 55, 78, 100). It is the goal of this review to present and to discuss the developments pertaining to the parasitic interaction, life-cycle, structure, ecology, and taxonomy of the bdellovibrios. A current review by Starr & Huang (81) emphasizes the physiological and biochemical properties of the bdellovibrios. Admirable treatments by Shilo (73, 74) and Stolp (85, 88, 89, 90, 91) have helped to put into perspective the burgeoning literature on the bdellovibrios.

THE DISCOVERY

While attempting to enrich for bacteriophage, Heinz Stolp—then (1962) in Berlin-Dahlem—accidentally discovered a hitherto unknown sort of microbe which had lytic activity against *Pseudomonas* bacteria (92). Stolp had inoculated a soil filtrate onto a lawn of *Pseudomonas phaseolicola* in a Petri plate. The lawn showed no phage plaques following a 24 hr incubation period. For some unmentioned reason, Stolp did not discard the plate, but re-examined it 24 hr later. The delayed examination was crucial for, at this time, cleared zones of lysis (plaques) were evident which continued to increase in size for about a week. Because of the delay in starting plaque formation and the continuing increase in plaque size, Stolp concluded that conventional phages were probably not the cause of the lytic reaction.

Microscopic observations of material removed from these growing plaques demonstrated a great number of rapidly moving tiny microbes in addition to a few cells of the pseudomonad. These small organisms collided with the much larger pseudomonad cells, adhered to their surfaces, and seemed to cause them to lyse. It was confirmed that the lytic principle was not a phage: the agent did not pass through a 200 nm filter, nor did plaques form on a streptomycin-containing medium with a lawn of streptomycin-resistant pseudomonad cells, whereas phage does induce plaques under these conditions.

In 1963, Stolp & Starr (94) published a comprehensive study on these unusual microbes. After isolating eleven additional strains from soil and sewage in California and characterizing them and Stolp's original isolate, they concluded that these microbes are indeed unique bacteria. Based on the parasitic properties and small cell size of these bacteria, Stolp & Starr (94) established a new genus, *Bdellovibrio,* with the specific epithet, *bacteriovorus,* for the single species then recognized. The generic name describes the organism with respect to its morphology and mode of attachment to its host (*"Bdello-"* is derived from the Greek word for leech), while *"bacteriovorus"* refers to the fact that *Bdellovibrio* devours its host. As will be seen below in the section on Taxonomy, it is likely that several species can now be delineated within the genus *Bdellovibrio;* for this reason, we use the formal designation *"Bdellovibrio bacteriovorus"* sparingly in this essay and only for the nomenclatural type of this species (strain H-D *Bd.* 100 of Stolp & Starr, 94) as well as for those strains which we now believe are closely related to H-D *Bd.* 100. It seems prudent—in the present state of ignorance about the precise limits of each species—to use (as we do here) the trivial name "bdellovibrio" and to refer to a particular strain by its informal laboratory designation including, where necessary for clarity, the prefixes "H-D" (for host-dependent) and "H-I" (for host-independent).

Structural Features of Host-Dependent Bdellovibrios

The two most striking morphological features of host-dependent (H-D; i.e., "parasitic") bdellovibrios, as observed by phase-contrast microscopy, are their small cell size (actually, their narrow width; see the later section on Small size) and their rapid motility. In the extracellular stage, the bdellovibrios typically measure 1 to 2 μm in length and only about 0.35 μm in width. Most strains are usually vibrioid to nearly rod-shaped during the extracellular stages (73, 94). They move several times as fast as a pseudomonad and have been estimated to travel at the astonishing speed of 2000 cell-lengths per min (76) or 100 cell-lengths per sec (89)!

Flagella.—One of the distinctive morphological features is the unusually thick polar flagellum which was originally reported (on the basis of shadowed electron micrographs) to measure 55 nm in width, or three to four times the thickness of typical bacterial flagella (92–94). Preliminary abstracts (40, 56) indicated that the flagellum is sheathed. These observations were subsequently confirmed and extended (1, 2, 10, 11, 67, 68). Negative stains of the *Bdellovibrio* flagellum showed it to be of uniform width: 28 nm, or about twice the thickness of typical bacterial flagella. An inner core measures 13 nm while the sheath, consisting of a three-layered unit, measures 7.5 nm. Seidler & Starr (68) demonstrated that this flagellum consists of two distinct major chemical components, by showing the differential disruption of the sheath following exposure of intact flagella to 6 M urea. The flagellar sheath appears to be an extension of the cell wall (2, 10, 11, 67, 68). Sheathed flagella have been demonstrated in a variety of bacteria (44, 52). One organism, *Vibrio metschnikovii,* was thought to be related to *Bdellovibrio* because of the similarity in DNA base composition, cell shape, and sheathed flagellum. However, attempts to demonstrate further relationships have proved unsuccessful (66, 71).

Anterior end.—Some investigators have described a morphological specialization found at the aflagellated (anterior) end of some bdellovibrio isolates. This specialization has been referred to as a holdfast or infection cushion (2, 10, 11, 47, 63). The cell tip is slightly swollen and appears convoluted in thin section. A more recent comment by one group suggests that this distortion may be an artifact of preparative techniques and may merely represent an incompletely developed region of new cell wall material (12). It is the opinion of Abram & Davis (1) that the entire cell surface of *Bdellovibrio* is susceptible to similar disorganizations. Within the anterior region, a lamellar mesosome is usually apparent (11, 63, 79). It has been postulated—with due caution—that the mesosome may be crucial in enzymatic activities involved in the penetration or later stages of host cell dissolution.

An additional organelle at the anterior cell tip consists of fine filaments (fibers) measuring 45–55 Å in diameter and up to 0.8 μm in length (1, 2, 73). In addition to two to three (rarely up to six) such fibers emerging from a small number of cells, some 6 to 12 electron-dense circular structures have been revealed with certain negative stains (1). These ring-like structures are associated with the membrane and wall and occasionally have fibers emerging from them. The functions of both fibers and ring-like structures are unknown. It has been proposed that the fibers play a direct role in boring a hole through the host wall (89) or that the fibers may function to connect and hold the parasite to its host (1). Since comparatively few of the unattached bdellovibrio cells have such fibers, it is assumed that fiber formation occurs only after attachment (1). It seems not to have been suggested before that these fibers may be pili which happen to form sporadically and only at the cell anterior, and which may or may not have a direct function in the parasitic process.

Encysted resting cells.—Burger, Drews & Ladwig (9) isolated an unusual bdellovibrio strain (*Bd.* W) which is said to have the capacity to parasitize two kinds of Gram-positive bacteria and also to produce occasionally an "encysted resting cell" within its *Rhodospirillum rubrum* host. Hoeniger, Moor & Ladwig (36) have studied in more detail the development and fine structure of the encysted resting cells which are formed by *Bd.* W grown in *R. rubrum* (but not when grown in *Escherichia coli* B). The development of *Bd.* W encysted resting cells begins some 3 hr following infection of *R. rubrum* in broth, with the noticeable deposition of a small amount of amorphous matter around the periphery of the *Bd.* W cell. One hour later, the outer layer of the parasite becomes more electron-dense and the encysted resting cell matures. The encysted resting cell is some two times the thickness of the free-living parasite cell. This encysted resting cell is structurally more simple than are endospores, *Azotobacter* cysts, and myxobacterial microcysts. Maximum production of encysted resting cells occurs after two days of incubation in shaken broth cultures; they are not formed in lawns of host cells on plates (36).

Some of the details of encysted resting cell formation are not yet clear. For example, when infection occurs at low multiplicities-of-infection in batch cultures, it is not specified whether the parasites go through the normal growth cycle and proceed to reinfect until host cells become limiting. A further uncertainty arises concerning whether or not the encysted resting cell is formed directly from an infecting parasite before it undergoes intracellular multiplication. We are unable to rationalize the bases for any possible selective advantage in the immediate formation of a resting structure shortly after the parasite enters its host and before propagation of its progeny begins. An attempt was made (36) to determine whether another *Bdellovibrio* strain (*Bd.* 109) could also form encysted resting cells, but *Bd.* 109 did not infect *R. rubrum* in these trials. In other experiments, *Bd.* W was allowed to infect *E. coli* B, but encysted resting cells of *Bd.* W were not produced in this host. It is unfortunate that a possible influence of medium composition on encysted resting cell formation cannot be ruled out in these trials, since infection of *E. coli* by *Bd.* W was studied in a chemically different medium from that used for *R. rubrum.*

Cell wall and membrane.—Abram & Davis (1) report that the walls of bdellovibrio cells show freeze-fracturing tendencies rather different from those of other bacteria. The fractured wall usually shows two smooth surfaces which correspond to two layers; the boundaries of the innermost wall layer are serrated and clearly separated from the outermost layer. The outer surface of the *Bdellovibrio* cytoplasmic membrane is covered with scattered particles (diameter: 6–10 nm) and often shows patches which are probably fragments of the innermost wall layer. Although Abram & Davis (1) state that plasmolysis (retraction of the protoplast from the cell wall) has never been observed by them in *Bdellovibrio,* micrographs published by others (11, 79) certainly do show such plasmolytic retraction. The cell walls of H-D *Bd.* 109 are reported (99) to contain muramic acid, glucosamine, and 13 amino acids—all similar to the wall components found in other bacteria.

Cytoplasm and nucleoplasm.—In thin-section preparations, the cytoplasm of *Bdellovibrio* appears identical to that of other bacteria with respect to its centrally located nucleoplasm surrounded by densely packed particles which are assumed to be ribosomes. Ribosomes as well as ribosomal RNA have recently been isolated from several host-independent (H-I; i.e., "saprophytic") *Bdellovibrio* strains using standard techniques (66). These particles appear identical to the corresponding components in *E. coli* with respect to their co-sedimentation properties in sucrose gradients. The stabilization of 70S ribosomes from *Bdellovibrio,* however, requires a higher Mg^{++} concentration than do those from *E. coli* (Seidler & Mandel, unpublished observations).

There is a rather large volume of deoxyribonucleic acid relative to the small cell size. The nucleoplasm appears to occupy about two-thirds of the total cytoplasmic area in the thin-sections illustrated by Starr & Baigent (79). In addition, one or more densely staining inclusion bodies occupies a portion of the nucleoplasm (11, 79). It has recently been demonstrated that, at least in one *Bdellovibrio* isolate, a rich medium induces the formation of prominent inclusion bodies which are visible by phase microscopy (61). The chemical composition and function of these inclusions are unknown.

Parasitic Interaction

Early stages.—The initial interaction of *Bdellovibrio* with its host might be described in terms of a possible "recognition of prey" quite likely mediated by a chemotaxis (presently under study by us), followed by a violent collision of the highly motile bdellovibrio cells with a prospective host cell. The bdellovibrio, which has been estimated (89) to swim at the astonishing speed of 100 cell-lengths per sec, collides so violently with its host that it may occasionally carry a host cell with 10 times the mass of the bdellovibrio across a 1000× oil-immersion microscopic field. The attachment may be reversible in the earliest stages, i.e., the attachment may be aborted and the parasite swims away to another host cell.

Attachment and penetration.—When an active preparation of H-D bdellovibrios is added to a suspension of susceptible host cells, attachments by the parasites begin immediately. We have observed that the early stages of contact may be nonspecific. In a mixed population of susceptible and nonsusceptible cells, some H-D bdellovibrios collide with and momentarily attach to Gram-positive bacteria such as *Bacillus megaterium*. Bdellovibrios are also sometimes observed attached to the coverslip. It is not known whether these sorts of attachments have any relationship to the parasitic connection with a host cell.

According to Stolp's (86, 87) high speed cinematographic observations, the parasite, following attachment, undergoes a fast rotation around its own long axis with speeds up to 100 revolutions per sec. This interpretation is difficult to reconcile with the evidence of the tight bond formed between the bdellovibrio and host cell (79) and the absence of a rapid rotation of the host cell which would be expected under the motive power of a firmly attached bdellovibrio. On the other hand, other interpretations appear more plausible. Starr & Baigent (79) described the rotation relative to the attachment point as suggestive of an arm-in-socket motion. Burnham, Hashimoto & Conti (11) observed that the drilling action does not appear to involve a spinning about the long axis, but rather a swiveling of the posterior end of the parasite with no rotation at the attachment tip.

The next discernible stage involves the preparation for penetration through the host cell wall. There is now both morphological and biochemical evidence to indicate that at least some of these processes are facilitated or mediated by enzymatic action. Shortly after attachment the original smooth inner murein layer of the host cell becomes irregular and convoluted just prior to spheroplast formation (79). More recent studies of serial sections show that one of the earliest detectable events involves a localized bulging of the host cell wall in the area of attachment (11). Some workers believe that the pore which is formed in the host cell wall might result from physical motion of the parasite, while most agree that enzymatic decomposition of the periplasmic matrix is suggested. Some of these differences in interpretation may stem from the use of different strains or, indeed, different species of *Bdellovibrio,* since independent isolates were used in the studies. [Recent investigations do indeed confirm substantial differences among these strains, which are probably representatives of different species of *Bdellovibrio* (66; see Taxonomy below).] The most convincing biochemical evidence for early enzyme production by H-D *Bdellovibrio* comes from the studies of Varon & Shilo (104), who demonstrated that antibiotics which inhibited protein synthesis also inhibited penetration but not attachment. Thus, it is likely that such enzymes are inducible, possibly as a consequence of the physical interaction of *Bdellovibrio* with its host.

Following the establishment of a successful contact, the next noticeable phase of the interaction involves the formation of a pore in the host cell wall, through which the parasite enters. One group has been able to demonstrate that local damage occurs on the host cell surface even though *Bdellovibrio* may have subsequently become detached (2, 73). Their negatively stained preparations of host cells which have been attacked by *Bdellovibrio* revealed holes and pitted areas in the surface of the host wall. It is unknown whether host autolytic processes (95) play a direct role in this pore formation. The actual penetration process of the parasite is probably facilitated by its swiveling or drilling motion. The penetration pore is smaller than the diameter of the bdellovibrio cell, which becomes constricted as penetration proceeds (10, 11, 79).

With many *Bdellovibrio*-host combinations, the attacked cell is converted into a spherical body (a spheroplast?) at a very early stage following attachment and before penetration is microscopically apparent. When the host cells are short rods, the entire cell is converted into a spherical body (63, 70, 79). In longer pseudomonad hosts and at low multiplicities-of-infection, the "spheroplasting" is more localized and a kind of multiple "ballooning" is often seen (79). There is some apparent disagreement as to when, if ever, spheroplast formation is induced. For example, Burnham, Hashimoto & Conti (11)—contrary to the aforementioned observations—illustrate that spheroplast formation in *E. coli* becomes prevalent only in later stages (presumably following penetration) of infection by *Bd.* UKi2. We wish to emphasize that *Bd.* UKi2 differs uniquely in several physiological and molecular aspects from other strains, and we believe that these differences might be reflected also in variations in parasitic mechanisms (see Taxonomy).

Several kinetic studies have been reported dealing with the attachment and multiplication of *Bdellovibrio* in host cells. Varon & Shilo (104) devised a method by which to study the kinetics of attachment. The technique relies on the differential filtration of parasites and host cells through a 1.2 μm membrane filter. About 98 percent of *E. coli* cells are retained on the filter, whereas 90–95 percent of the unattached parasites pass into the filtrate. This procedure made it possible to examine a variety of parameters affecting attachment. Under optimum conditions, 65–90 percent of the total number of parasites were capable of attachment within 15–20 min. Nutrients and cations facilitated, but were not essential for, establishing an infection in this particular system. Substantial attachment occurred in the temperature range of 20–40°C and pH range of 6.0 to 9.0. Various inhibitors of attachment (10^{-5} M H^+, NaCl, EDTA, citrate) also prevented motility, thus confirming early comments (93, 94) on the importance of motility in establishing an infection.

An agitation device causes detachment of *Bdellovibrio* during a 4- to 5-min period following the addition of host cells (104), and thus provides a procedure for following the penetration of parasites into host cells. After 20 to 30 min from the beginning of attachment, agitation no longer releases *Bdellovibrio,* making it possible to show that streptomycin and chloramphenicol inhibit penetration but not attachment. It should be pointed out that the kinetic data revealed that the saturation level for penetration was two-thirds that achieved in attachment (104). This would imply that, of the total number of parasites attaching in this particular system, only two-thirds possessed the capacity for penetration.

Host viability is not always necessary for attachment of the bdellovibrios. There are, however, certain restrictions to the manner in which host cells can be killed (74). For example, Varon & Shilo (104, 106) demonstrated that lethal exposure of *E. coli* to ultraviolet irradiation had no effect on *Bdellovibrio* attachment and multiplication and that certain antibiotics may inhibit penetration but not attachment. When *E. coli* was killed by heat (15 min, 70°C), attachment was reduced to 7–8 percent of the control level. Of those parasites capable of attachment, further development proceeded at a slightly reduced rate and the burst size was somewhat reduced (106). *Bd.* 6–5-S could multiply, at least by a factor of 10^5, in autoclaved (5 min, 121°C) cell suspensions of *E. coli* ML35 and *Spirillum serpens* VHL provided Ca^{++}, Mg^{++}, or both, were present (14). Cells of *Bd.* 109 could elongate, but not divide—in the absence of living host cells—in yeast-peptone broth, host cell extract, and host cell wall extract (4, 61); an increase in the number of *Bd.* 109 cells was noted microscopically in host-parasite lysates, but it was not determined whether these are host-dependent or host-independent forms, nor whether any living host cells were present (4). *Bd.* 6–5-S, *Bd.* 100, *Bd.* 109, and *Bd.* A3.12 could multiply in heat-killed (10 min, 70°C; 15 min, 121°C) cells of *Spirillum serpens* VHL, provided Ca^{++} were added (41). Similarly, *Bd.* W was shown to be capable of multiplication in heat-killed (20 min, 80°C) *Rhodospirillum rubrum* (9). In these cases, the fraction of parasites capable of growth was not always indicated. The depressed attachment in the case of heat-killed *E. coli* (104) might be explained by the irreversible denaturation and disruption of the wall surface structures (receptors?) which might be necessary for successful attachment.

The influence of the chemical composition of the host cell wall in permitting attachment by *Bdellovibrio* has been studied using a series of mutants of *E. coli* and *Salmonella* spp. (105). By following the kinetics of attachment to host cells lacking the O-specific side chain but containing the complete rough core (chemotype Ra), such mutants were shown to be more susceptible to attachment than the smooth wild-type parent strains. When portions of the remaining rough core were removed (cf glucosamine, chemotype Rb), attachment was depressed compared to the chemotype Ra control. Additional deficiencies reduced attachment further, implicating the R-antigen as a possible receptor site. Contrary to expectations, the plaquing efficiencies on various host mutants did not vary, indicating that the activities of attachment which occur in 20–40 min are not necessarily indicative of the events taking place during four to six days of incubation in a host lawn.

Intracellular growth stages.—As *Bdellovibrio* continues its penetration, the host cell membrane is pushed away from the cell wall and the parasite lodges between the host membrane and wall (9–11, 70, 79). It must be emphasized that the host cell protoplast is not breached and that the *Bdellovibrio* initially occupies an intraintegumental position; on the basis of this locus, Stanier (78) has argued most eloquently to the effect that bdellovibrios are not endosymbionts. There are certain conceptual advantages stemming from this clarification and, while we continue in this essay to use the terms "endoparasite" and "ectoparasite" in the currently conventional ways, the facts must not be obscured by etymology: the bdellovibrio is not inside the protoplast although it is within a space between the cell wall and the cytoplasmic membrane of its host cell. Once inside its host, the small vibrioid parasite begins its growth phase and elongates into a C- or spiral-shaped cell; the spiral may be 5 to 10 times, or more, longer than the infecting cell (2, 63, 70, 79). During this intracellular growth phase, there is a progressive disorganization and dissolution of host cell cytoplasm, which is also accompanied by the release of ultraviolet-absorbing material into the lysate. The spiral finally undergoes constrictions prior to its segmentation into daughter cells (2, 63, 70, 79).

The nature of the mechanism(s) by which the progeny parasites leave the ghosted host cell is not entirely clear. The juvenile parasites are rapidly motile within the ghosted host cell, and it has been suggested that they are able to break down host cell remnants by a physical mechanism and depart to reinitiate infection. In some instances, the destruction of the host cell is so extensive that no apparent physical barrier remains to retard departure of the progeny (79).

These observations bring into focus various questions concerning the extent and nature of the damage to the host cell membrane. If the parasite is to assimilate host cell components successfully, the nature of the damage should allow controlled passage of metabolites across the membrane. This might also include *Bdellovibrio* digestive enzymes such as protease and lysozyme (42, 69, 71, 75). Electron microscopy has clearly revealed that extensive deterioration of the membrane and host cell cytoplasmic contents takes place at rather early stages of intracellular infection. It is not entirely clear whether the parasite induces all of the membrane damage directly, whether disruption is osmotically facilitated prior to and during penetration, whether autolytic processes in the host cell are induced, or whether a combination of these or other factors is important. It is also quite possible that some artifactitious damage is induced during the fixation and sectioning processes.

The first studies involving a biochemical approach to some of the above questions have recently been reported (62, 102). In a rather novel approach, the effect of *Bdellovibrio* infection on host permeability was followed by the unmasking of β-galactosidase in a cryptic (permeaseless) strain of *E. coli*. It was discovered that—even at very early stages of infection (before complete penetration)—the host membrane is so damaged as to permit the free passage of small molecules across the membrane of the host. The rate of membrane damage (as measured by the unmasking of β-galastosidase activity) was shown to be directly related to the multiplicity-of-infection (62).

Three kinetic studies have dealt with the multiplication of *Bdellovibrio* and factors affecting intracellular growth. The initial studies followed the plaque-forming unit (PFU) titer of *Bdellovibrio* and the viable host cell count; a lack of proportionality was reported between the number of host cells killed and the number of *Bdellovibrio* cells produced in a given time interval (93, 94). In addition, after 26 hr of incubation, when the death rate of host cells decreased considerably, there still ensued an increase in the PFU titer. From these observations, it was concluded that the parasites develop at the expense of nutrients released from lysed host cells (94). At the time of that report (1963), the intracellular multiplication of *Bdellovibrio* was unknown and, in retrospect, the lack of correlation between host death and parasite production can be predicted from the intracellular nature of *Bdellovibrio* growth. Thus, it can be assumed that host cells are killed shortly after attachment, but that the new rate of PFU titer increase would not manifest itself until the life cycle is completed and a new round of bursts begin. The same argument can be used to explain the final increase in PFU titer at a time when few additional host cells are being killed.

Following the report of Varon & Shilo (104) on attachment kinetics, studies dealing with the later stages of intracellular growth were reported

(70, 106). A procedure was described by Seidler & Starr (70) for initiating synchronous infections which lead to a single round of intracellular growth. These experiments demonstrated that, under optimum conditions, the average burst size of *Bd.* 109 growing in *E. coli* is 5.7. The burst was always preceded by a lag or latent period during which the parasites were proceeding through attachment, penetration, and intracellular multiplication. It must, therefore, be assumed that multiplication of H-D bdellovibrios takes place only within a host cell. Attachment and intracellular growth in various media had little effect on the average burst size; a ten-fold reduction in nutrient reduced the average burst size by only 15 percent. In addition, 75 percent of the control's burst size was observed in Tris buffer (70). These observations must indicate that H-D bdellovibrios have a major dependence upon host cells for their supply of carbon and energy.

Effects of cations.—Divalent cations are often added to the culture media in which the *Bdellovibrio* and its host are being grown. The very first strain of *Bdellovibrio* isolated was cultivated in a nutrient medium supplemented with Fe^{++}, Ca^{++}, and Mn^{++} (92). An abstract (40) mentions that a successful *Bdellovibrio*-host relationship is established only in the presence of added Ca^{++} or Mg^{++}, or both, and another abstract (38) notes that the growth of *Bd.* 6–5-S in nonsynchronous cultures of *S. serpens* VHL is, within limits, proportional to the concentrations of divalent cations and that the delayed addition of the cations to a host-parasite culture results in an immediate rise in the PFU titer. Still other abstracts (14, 15) record similar observations; no explanations were provided to account for the stimulatory effect(s) of the added cations.

Cations seem to function in the bdellovibrio-host interactions in a number of ways, including stabilization of the living host protoplast—which latter function is clearly the case in the *Bd.* 109-*E. coli* system studied by Seidler & Starr (70) who used several approaches to study the effect of cations on the host-parasite relationship. Attachment experiments in the control medium (NB/10 supplemented with 3 mM Mg^{++} and 2 mM Ca^{++}) resulted in more than 70 percent attachment, as was previously reported also by Varon & Shilo (104). When no cations were added to the NB/10 medium, attachment was reduced to 50 percent of the control; chemical analysis for the Ca^{++} and Mg^{++} content of NB/10 revealed a total concentration of about 0.07 mM (unpublished data). On the other hand, attachment in Tris buffer (no detectible Ca^{++} or Mg^{++}) was reduced to 20–30 percent of the control (70). The attachment in Tris buffer containing *E. coli* B cells grown in complex medium (NB/10) probably indicates the presence of some bound cations in or on the surface of host cells. The cations may function in the neutralization of the host cell surface charges which permit a higher percentage of successful attachments. However, the possibility of cations serving as cofactors for enzyme(s) concerned with attachment, penetration, or lysis cannot be ruled out. Indeed, the most recent evidence for the mechanism of host lysis by *Bdellovibrio* implicates protease, muramidase, and lipase; the addition of Ca^{++} and Mg^{++} to the reaction mixture activated the protease (42, 81).

In addition to facilitating early infection, cations are also important in the later stages of infection of *E. coli* by some *Bdellovibrio* isolates (70). In the absence of added cations, an increase in the latent period and a depression in the average burst size is recorded during growth in NB/10 broth. Under these conditions, Seidler & Starr (70) showed by means of phase contrast microscopy that premature lysis of *E. coli* sometimes occurred, that the prematurely lysed host cells usually did not form into spherical bodies (spheroplasts?), and that the bdellovibrio cells often did not continue their development and eventually were themselves lysed. These observations correlate with the observed 20 percent decrease in PFU titer during the latent period, and it was concluded (70) that cations, in addition to facilitating attachment, also serve to stabilize the infected host protoplast so as to maintain its integrity during most of the infection period.

Apparent discrepancies in the literature concerning the necessity of added cations for the *Bd.* 109 life cycle have been traced to the use of two somewhat different strains, both designated *Bd.* 109. Thus, Rittenberg & Shilo (62) have demonstrated that, at the very early states of infection of *E. coli* by their strain of *Bd.* 109, host membrane damage permits nonspecific passage of small molecules such as lactose. However, no leakage of β-galactosidase out of the cells was detected during the 60-min sampling period. These observations are not compatible with those of Seidler & Starr (70) described in the foregoing paragraph. The recent studies of Huang & Starr (41), as well as comments from Rittenberg, Shilo, Conti, and Varon (personal communications), have pointed to a series of differences (morphology, cation requirements, phage-typing patterns, etc.) between the two "*Bd.* 109" strains used in these studies. Whatever the origin of the differences, it is now clear that the *Bd.* 109 (Davis strain) isolated by Stolp & Starr (94), known descendants of which were used in our subsequent work (41, 42, 65–72, 79, 93, 94), is indeed different from the strain bearing the same designation [*Bd.* 109 (Jerusalem strain)] and used by Shilo and his co-workers (61, 62, 102–106).

Huang & Starr (41) have studied the effects of cations on the multiplication of bdellovibrios in thrice-washed H-D *Bdellovibrio*-host suspensions in Tris buffer. Growth was scored by an increase in the PFU titer. Using *E. coli* B cells harvested from a low-Ca^{++} minimal medium, a normal rise in PFU titer requires the addition of Ca^{++}. When the host cells are grown in complex media, the requirement for Ca^{++} is sometimes obscured. For example, the multiplication of *Bd.* 100, *Bd.* 109 (Davis), *Bd.* 109 (Jerusalem), *Bd.* A3.12, and *Bd.* 6–5-S in *E. coli* B cells harvested from the low-Ca^{++} minimal medium requires the addition of Ca^{++}. *Bd.* 109 (Jerusalem) developed after a lag period of about two days, without added Ca^{++}, in *E. coli* B cells which had been grown in a complex medium; however—in agreement with an earlier report (70)—*Bd.* 109 (Davis) did not multiply at all under these conditions unless Ca^{++} was added. It thus appears that *Bd.* 109 (Jerusalem) requires a relatively low level of Ca^{++} which can be satisfied under some conditions by host cell Ca^{++}, and that *Bd.* 109 (Davis) might require a different form or a higher level of Ca^{++} than does *Bd.* 109 (Jerusalem).

In summary, added cations have multiple effects on the host-*Bdellovibrio* interaction. These effects are complicated by the fact that a requirement for added cations will depend upon the particular *Bdellovibrio*-host system under study and the basal levels of the cations in and around the host cells.

Bacteriolytic enzymes.—Evidence has been obtained for extracellular lytic enzyme production in the *Bd.* 6–5-S and *S. serpens* VHL system (19, 39, 42). Multiplication of *Bd.* 6–5-S within *S. serpens* (including heat-killed cells) results in the release of submicroscopic particles containing amino sugars, and of soluble muramic acid (42). Enzyme preparations, precipitated with ammonium sulfate or cold acetone, solubilize isolated ^{14}C-labeled peptidoglycan (42). Using radiochemical analysis, as well as procedures involving collagen or casein as substrates, a lysozyme-like enzyme and proteolytic enzymes have been isolated by means of DEAE-cellulose or Sephadex (G-100) gel filtration (42). An abstract (19) records the isolation of an enzyme, from a *Bd.* 6–5-S and *S. serpens* VHL culture, which degrades purified mucopeptide isolated from *S. serpens.*

Host-dependent *Bd.* 6–5-S, *Bd.* 100, *Bd.* 109, and *Bd.* A3.12 all produce protease in cell suspensions of *S. serpens* VHL (42). Protease production by the corresponding H-I *Bdellovibrio* cultures was reported earlier (71).

A lipase capable of hydrolyzing tributyrin incorporated in agar was detected in the cold acetone precipitate of *Bd.* 6–5-S lysates (42). This enzyme might facilitate host cell wall degradation, in addition to weakening the cell membrane to allow the parasite access to the host cell cytoplasmic contents. We might refer, in passing, to two abstracts (83, 84) in which the first studies on the unique phospholipids of *Bd.* UKi2 are sketched.

Much remains to be done in clarifying the relationships of these enzymes to the parasitic interaction of the *Bdellovibrio* and host cells. However, it does seem reasonable to suppose that production by *Bdellovibrio* of muramidase, protease, and lipase might be relevant to penetration and digestion of the host cell by the parasite. Thus, such studies may ultimately provide an understanding of the enzymatic basis of the *Bdellovibrio* parasitism.

Ecological Studies on Host-Dependent Bdellovibrios

Isolation and enumeration procedures.—All procedures for the isolation of *Bdellovibrio* from natural habitats rely on the property of the relatively small cell size (narrow width; low mass) of the parasites. The technique used by Stolp & Starr (94) involves a series of differential filtrations through membranes of decreasing porosity. Portions of filtrates from 0.45

μm-, 0.65 μm-, or 0.8 μm-membranes are mixed with prospective host cells and plated as for phage using the double-agar layer technique. Phage plaques are marked after 24 hr and *Bdellovibrio* plaques appear after one to six days of additional incubation. The major difficulty with this technique lies in the small recovery of the original number of parasites present; less than 0.1 percent according to Shilo (74).

Several more recent techniques have proven to be of value for the quantitative enumeration of *Bdellovibrio*. Klein & Casida (46) developed a direct plating procedure for enumerating soil bdellovibrios. In their dilution-to-extinction technique, decimal serial dilutions of soil suspensions were mixed with suspensions of *E. coli* and incubated three days at 26°C. Samples were plated on host lawns and examined for plaque development. For direct plating, decimal dilutions of soil suspensions were directly added to double-layer plates. A most significant contribution of this study was the demonstration of the effect of centrifugation of the soil suspension on the recovery of parasites. A one-log decrease was noted in the PFU titer following a 30 min centrifugation at 500 *g*. Centrifugation, which had been used in previous isolation procedures (16, 93), was thus shown to make quantitative recoveries uncertain. One group (16), nevertheless, maintains that three centrifugations still give a higher recovery rate than a single filtration through a 0.45 μm filter.

Two additional quantitative isolation techniques are known (74, 103, 104). One procedure (104) involves a single filtration of the material through a 1.2 μm filter. In this procedure, 90–95 percent of the host bacteria are retained on the filter, while nearly complete recovery of *Bdellovibrio* is obtained from the filtrate. The second technique (103) involves the differential centrifugation of the parasites and the larger host bacteria through a linear gradient of Ficoll. Under optimum conditions, about 80 percent of the bdellovibrios are recovered from the upper portion of the gradient, while 70–90 percent of the larger bacteria move further toward the bottom. To the best of our knowledge, neither technique has been applied in practice for quantitative ecological studies.

Habitat and host range.—Bdellovibrios have been reported from a great variety of habitats including soil and sewage (16, 45, 46, 50, 57, 92–94, 109), rivers (25–29, 31), and marine environments (51, 73). We are not certain that all of the works cited pertain to members of the genus *Bdellovibrio;* see the section below on "*Bdellovibrio*-like" bacteria.

We know only two studies dealing primarily with the quantitative enumeration of *Bdellovibrio* from natural habitats. The studies of Dias & Bhat (16) involved the enumeration of phage and *Bdellovibrio* in raw sewage. On different sampling days, the *Bdellovibrio* titer ranged from 0 to more than 800 PFU/ml. The titer also varied drastically depending on the host employed. *Serratia marcescens* and *Pseudomonas aeruginosa* were poor hosts, i.e., very few bdellovibrios were active against them. *Pseudomonas fluorescens, Salmonella typhosa, S. paratyphi,* and *Aerobacter aerogenes* provided the highest counts of *Bdellovibrio*. These investigators were the first to illustrate that *Bdellovibrio* is capable of parasitizing enteric pathogens, a subject later worked on by Guélin and co-workers (26–29, 31) and Gillis & Nakamura (22). Since the largest titers of *Bdellovibrio* were found on lawns of these enteric pathogens (*S. typhosa*), it may be assumed—if natural numbers permit physical contact—that the bdellovibrios play a significant role in their removal from raw sewage. It must be pointed out that the isolation technique used for enumeration of *Bdellovibrio* in these studies (16) involved three centrifugations (each, 30 min at 2000 *g*), which quite likely reduced the *Bdellovibrio* titer at least tenfold.

Klein & Casida (46) used a direct plating procedure and were able to demonstrate soil bdellovibrios active against *E. coli* in all 23 samples collected from a variety of habitats in the central and eastern United States. It was also demonstrated that the parasites were capable of parasitizing indigenous soil bacteria. The parasite titers on *E. coli* ranged from 1×10^3 to 7×10^4 PFU per g of soil. The highest titers were obtained from soils known to have been contaminated with sewage plant effluents.

In addition to these two quantitative studies, the qualitative significance of *Bdellovibrio* as a natural parasite has been discussed. Based on preliminary investigations of the biological bacteriostasis of sea water, Mitchell et al (51) suggested that *Bdellovibrio* plays a possible role in the elimination of *E. coli* from sea water. In somewhat more detailed studies, Guélin and collaborators (25, 26, 28, 29, 31, 47) have concluded that bdellovibrios are important in the removal of *Salmonella* from polluted rivers.

In their first work on *Bdellovibrio,* Stolp & Starr (93, 94) isolated 12 strains and determined their host spectra against 34 bacterial strains by the plaque assay technique. Activity of all *Bdellovibrio* strains was limited to Gram-negative host species. However, not all tested Gram-negative bacteria were susceptible to these bdellovibrio strains; for example, none of their bdellovibrios lysed cells of *Rhizobium* and *Agrobacterium.* Postgate (60) attempted to isolate bdellovibrios active against *Azotobacter* cultures; although plaques were indeed formed when soil extracts were placed on *Azotobacter* lawns, the plaques were induced not by bdellovibrios but by bacteriolytic pseudomonads which seemed to exert their lytic action only when closely apposed to the surface of the *Azotobacter* cells. Sullivan & Casida (97) have since attempted to isolate, directly from soil, *Bdellovibrio* capable of parasitizing *Azotobacter* and *Rhizobium* species, but also without success. They were able, however, to demonstrate plaque formation on *Azotobacter chroococcum,* but not on *A. vinelandii* or *Rhizobium* spp., by a *Bdellovibrio* originally isolated on *E. coli.* Recently, Parker & Grove (57) were able to isolate, directly from soil, bdellovibrios parasitic on *Rhizobium meliloti, R. trifolii, Agrobacterium tumefaciens,* and *A. radiobacter.* The ecological significance, if any, of the difficulty in isolating soil bdellovibrios active against these indigenous soil bacteria is unknown.

There have been reports concerning *Bdellovibrio* strains which are claimed to be capable of parasitizing Gram-positive hosts (9, 30, 32). One group (30, 32) reports the lysis of *Clostridium perfringens* and *C. histolyticum* by bacterial parasite Xpfr. However, Xpfr is significantly different from the bdellovibrios in that Xpfr cells are pleomorphic, seemingly aflagellated, long rods which, moreover, have not been shown to enter the clostridia but rather to be located outside the cell wall. Guélin (personal communication) has taken note of the differences between *Bdellovibrio* and Xpfr and proposes to call the latter organism *Bdellobacter*. Xpfr may be related to other ectoparasitic (ectocommensal?) bacteria such as those described by Wood & Hirsch (109) on *Hyphomicrobium* and by Postgate (60) on *Azotobacter* (see the section below on "*Bdellovibrio*-like" bacteria).

The other report (9) deals with *Bd.* W, originally isolated on *Rhodospirillum rubrum*. *Bd.* W is alleged to form plaques on enterobacteria, *Streptococcus faecalis*, and *Lactobacillus plantarum*, but not on *Pseudomonas aeruginosa* or *Spirillum serpens*. Our experience with *Bd.* W has been that it does not parasitize or lyse available strains of one of the Gram-positive alleged hosts, *Streptococcus faecalis* (we have had no opportunity to determine what effect, if any, it might have on lactobacilli), and that it does indeed parasitize and lyse *S. serpens* VHL. The reason for the discrepancy is presently unknown. In addition to the alleged activity on some Gram-positive host bacteria, *Bd.* W is interesting because of its reported encysted resting cells (see Structural Features) and its variations in plaquing efficiency on different hosts. Thus, if the relative titer on *R. rubrum* was taken as 1.0, the plaquing efficiency ranged from 10^{-4} to 10^{-5} for *Serratia marcescens* and 10^{1} for *Proteus vulgaris*. At least one other example of variation in plaquing efficiency has been reported (74). The nature of the variable plaquing efficiency is unknown, but may well result from differences in the accessibility of receptor sites of the different hosts. It is possible that access to receptor sites (105) may be impeded when hosts are embedded in agar, for the host range of at least one *Bdellovibrio* strain is known to be broader when determined in two-membered broth cultures as compared with plaque formation on agar lawns (75).

Host-Independent Bdellovibrios

Isolation procedures.—The experiments of Stolp & Petzold (92) clearly demonstrated that, practically speaking, the host-dependent (H-D) *Bd.* 321 which they studied is an obligate parasite. Thus, upon plating lysates containing 10^{8} or fewer *Bdellovibrio* cells per ml onto a variety of nutrient media, no development of *Bdellovibrio* colonies was observed. However, when lysates were concentrated to $> 10^{9}$ PFU/ml and plated onto a yeast extract-peptone medium, a few yellow colonies of host-independent (H-I) *Bd.* 321 appeared after five to six days. When a colony of H-I *Bd.* 321 was replated on a host lawn, there was either no activity or plaques with a central colony were formed (92).

Stolp & Starr (94) isolated H-I derivatives of several of the other H-D *Bdellovibrio* strains by inoculation of concentrated suspensions of the H-D strains onto yeast extract-peptone agar. The H-I strains so isolated were strongly proteolytic and catalase-negative. Shilo & Bruff (75) isolated H-I strains from H-D *Bd.* A3.12. The isolation was accomplished by a single differential filtration of the lysate through a 0.45 μm filter followed by plating onto a nutrient medium. This isolation procedure proved adequate only for isolating the aforementioned H-I strains from H-D *Bd.* A3.12 but not from other *Bdellovibrio* strains including *Bd.* 109 (8).

More recently, another procedure has been described for the routine isolation of H-I bdellovibrios (69, 71). This technique relies on a preliminary stage involving the isolation of spontaneous streptomycin-resistant mutants of H-D bdellovibrios, followed by their propagation on a streptomycin-sensitive host. In this manner, large numbers of streptomycin-resistant H-D bdellovibrio cells can be transferred to a selection medium containing streptomycin. The antibiotic prevents the growth of residual host cells in the lysate and this counter-selects H-D bdellovibrio cells. In this manner, it has been possible to isolate H-I derivatives from every one of 16 different H-D strains studied. Since the 16 H-D cultures had been isolated from different geographical regions and habitats, it was concluded that bdellovibrios in general have the potentiality to give rise to H-I derivatives.

Physiology and biochemistry.—An exocellular enzyme produced by H-I and H-D strains of *Bd.* A3.12, was reported to induce lysis and digestion of damaged host cells (75). This enzyme, characterized as a protease, digested a commercial preparation of collagen (Azocoll), had no effect on normal host cells, but would dissolve both Gram-positive and Gram-negative heat-killed cells. When host cells were exposed to the enzyme immediately after attachment by parasites, an increase in the lysis of the culture was observed. These observations led to the speculation that lysis of host cells proceeds in a two-step fashion: attachment first induces localized damage to the outermost cell components which, secondly, exposes other components to the action of the enzyme. This mechanism could not be generalized for all *Bdellovibrio* strains, since the enzyme was not demonstrable in cultures of two other H-D isolates (75).

In view of the importance ascribed to the *Bd* A3.12 protease, 12 H-I strains examined in another study (71) were assayed for this enzyme. All cultures showed detectable protease activity, thus lending further credence to the earlier proposals (75). Some of the digestion by bdellovibrios of heat-killed host strains (attributed to protease) may indeed be due partially to lysozyme-like enzymes or perhaps lipases, as well as to proteases (19, 39, 42, 81).

A variety of physiological properties of H-I strains has been studied by Seidler & Starr (71). All H-I cultures examined could liquefy gelatin and produce ammonia from peptone, all are oxidase-positive, and all grow in the temperature range of 23°C to 37°C but not at 12°C and 42°C. None of the cultures used in that study produces indole from tryptophan nor reduces nitrate. However, at least two *Bdellovibrio* strains are known that can reduce nitrate, namely H-I *Bd.* 321 (Sp. 19; 94) and H-I *Bd.* UKi2 (17). Many H-I cultures are catalase-positive upon initial isolation, but some cultures lose this enzyme activity upon repeated transfer through host-free media. It would be interesting to determine if this enzyme activity is regained in the presence of host cells when *Bdellovibrio* is subcultured back to the H-D form.

None of the 9 H-I strains so examined could utilize a variety of carbohydrates and organic acids as sources of carbon and energy (71). This might be expected of an organism that, during H-D growth, is suspected of utilizing only host cell protein (70). Gelatin liquefaction and ammonia production by all isolates, taken together with the absence or possibly limited ability to utilize sugars (71), would seem to indicate metabolic specialization to the extent that *Bdellovibrio* may be limited to the utilization of only proteins, peptides, and amino acids.

Results consistent with these observations have come from two biochemical studies. The comparative studies of Seidler, Mandel & Baptist (66) have extended the one made on *Bd.* 6–5-S (76) to an additional seven H-I cultures which are representative of the three presently known species of *Bdellovibrio*. Both studies agree that bdellovibrios contain a functional TCA cycle as well as alanine and glutamic dehydrogenases. Assays were negative for glucose-6-phosphate dehydrogenase and 6-phosphogluconate dehydrogenase. Contrary to the earlier report (76) on *Bd.* 6–5-S, assays for aldolase and glyceraldehyde-3-phosphate dehydrogenase in seven other strains were negative even when cultures were grown in the presence of 1.0 percent glucose (66).

Attempts have been made to ascertain the vitamin requirements of H-I *Bd.* A3.12 growing in a basal salts medium containing 1 percent peptone as the carbon and nitrogen source (75). The results demonstrate that only thiamine could partially replace the stimulatory effect of 0.1 percent yeast extract. Since H-I *Bd.* A3.12 is biochemically quite different from most, if not all, other known strains of *Bdellovibrio* (see Taxonomy), it is unlikely that other strains would have identical nutritional requirements.

Structure.—Microscopically, H-I cells appear heterogeneous in length and can be pleomorphic in structure (12, 71, 91). Within the same exponentially growing culture, the population consists of uniform vibrio-shaped cells to long spiral- or bizarrely shaped cells, with many gradations between. Cells are usually of uniform width, 0.3 μm to 0.4 μm, similar to the cells of H-D cultures.

Upon primary isolation, most H-I cells in a culture are motile. However, the percentage of motile cells is dependent upon the growth stage, the number of transfers through host-free media, and the particular *Bdellovibrio* strain. Flagellated H-I cells have a single, sheathed, polar flagellum identical in size and structure to that of the H-D strains [12, 68, 71; but see Stolp & Petzold (92) for remarks about multiple flagella in H-I *Bd.* 321].

A detailed structural study of one particular and unusual H-I bdellovibrio strain has recently been reported by Burnham, Hashimoto & Conti (12). By a combination of continuous phase contrast observations and electron microscopy, H-I *Bd.* UKi2—when growing as a saprophyte, i.e., apart from the host—was shown to go through the same morphological sequences as the intracellular H-D strains. The vibrio-shaped H-I cell elongates into a spiral, the flagellum degenerating in the process and, following the sequential fragmentation of the spiral, motile daughter cells are produced.

A unique kind of morphological specialization seems to occur with respect to flagella development. Spirals formed by H-I cells often generate flagella before all daughter cells separate from the filament; this development parallels that of the H-D strains (2, 11, 12). Thus, flagellar development in the daughter cells along the spiral can occur randomly; anterior to anterior, anterior to posterior, and posterior to posterior. However, it was observed that the anterior-posterior positions were most prevalent (12).

We would like to emphasize that the parallel in cell development between a H-I strain and its H-D counterpart is known only in the case of *Bd.* UKi2. Continuous phase microscopic observations have not been made on other isolates. Constrictions in the terminal and central positions of developing H-I spirals have been illustrated in electron micrographs of other strains (71); these might be analogous to several stages of the sequential fragmention processes discussed earlier.

A peculiar structural abnormality in the cell shape of H-I and H-D cells has been recorded by three independent groups (12, Fig. 5; 61, Figs. 8–11; 71, Fig. 6). Under normal circumstances, the *Bdellovibrio* cell assumes a rather open coiled or spiral shape. Rarely, a much more tightly coiled cell is observed. One group, which has reported limited growth of a H-D strain in cell extracts (61), indicated that the tightly coiled form is induced under low ionic conditions. This report implies that the osmolarity may influence *Bdellovibrio* morphology and cellular development.

Bacteriophages.—The use of H-I *Bdellovibrio* strains has facilitated the isolation of bacteriophages. Hashimoto, Diedrich & Conti (34) have isolated from sewage a phage, designated H-D C-1, which is active only on H-I *Bd.* UKi2. The phage is hexagonal in shape, contains two distinct capsomeres, measures 60 to 70 nm, has no tail, and contains single-stranded DNA. Recently, 12 additional phages have been isolated on other H-I bdellovibrios and a preliminary abstract concerning their properties is available (3). Three morphological types have been isolated: the H-D C-1 type described above, another group which has round heads and long tails, and a third group which has polyhedral heads and short tails. Host-range studies on 18 H-I *Bdellovibrio* strains resulted in five phage groups. Some of the phages infected H-D bdellovibrios in three-membered (bdellovibrio, host, phage) systems.

Virulence of H-I cultures.—In an attempt to determine the physiological potential for virulence of several H-I isolates, Seidler & Starr (71) examined in a semi-quantitative manner the ability of these strains to grow on living host cells. H-I cells were removed from log-phase cultures, added to host cell suspensions, and examined periodically for three days. At that time, in every H-I strain investigated, bdellovibrios had been reselected possessing parasitic and bacteriolytic abilities seemingly identical to the original H-D cultures. On the other hand, isolated plaque formation by log phase H-I cells on double-layer plates was observed only with the "facultatively" parasitic H-I A3.12 strain. The basic characteristics of H-I *Bd.* UKi2 (17) appear similar to those of the majority of H-I strains isolated previously (71) with one important exception, namely, the "facultative" parasitism. In this respect, H-I *Bd.* UKi2 appears to be similar to the unusual H-I *Bd.* A3.12 isolate of Bruff (8) which has also been set off from the others by its "facultative" nature (75) and its high virulence compared to other H-I strains (71).

We would like to characterize the "facultatively" parasitic (F-P) situation in *Bdellovibrio* as one in which a population of a H-I strain grows with equal ability in the presence or absence of host cells. It appears from the descriptions in the literature that H-I *Bd.* UKi2 does have some discrete, but important, differences from H-I *Bd.* A3.12 with respect to its F-P nature. Thus, H-I *Bd.* A3.12 is capable of forming colonies on media or plaques on a host lawn in a 1:1 ratio of colonies and plaques (75). In addition, when H-I *Bd.* A3.12 is grown in broth with host cells, the ratio of plaques to colonies is 1:1 under optimum conditions (75). On the other hand, under what is presently thought to be optimum conditions for H-I growth, not more than 65 percent of the H-I *Bd.* UKi2 cells were capable of parasitizing host cells (17).

The virulence of H-I *Bd.* A3.12 was shown by Shilo & Bruff (75) to be gradually lost (as were its F-P capabilities) with successive transfers through host-free media. The number of transfers of H-I *Bd.* UKi2 that had been made before virulence was tested was not indicated (17); on the other hand, it was mentioned that H-I *Bd.* UKi2 is stable with respect to its F-P potential since its endoparasitic ability was retained after 250 transfers through host-free media. However, it is impossible to evaluate the precise significance of this statement since the proportion of H-I *Bd.* UKi2 cells retaining virulence was not reported. The implication is that a large proportion of the H-I *Bd.* UKi2 cells retains F-P capabilities after 250 transfers. Our own experiences with this strain indicate that—like all other H-I derivatives isolated by us in 1967 (71) and since tested for reselection to the H-D forms—it is quite difficult (but, in our hands, invariably possible) to reisolate endoparasitic H-D forms from an active H-I *Bd.*

UKi2 culture. This is also true for H-I *Bd.* A3.12 which, upon its original isolation and for some 25 to 30 subcultures afterward, retained a high percentage of F-P cells (75). The point we would like to make is that, like many other parasitic organisms, all H-I bdellovibrios lose virulence when cultivated away from living hosts, making it increasingly difficult to reselect for H-D forms. Thus, it becomes apparent that only some bdellovibrios (at present, only *Bd.* A3.12 and *Bd.* UKi2 are known) may be F-P, but that even with these strains the F-P property is always a transient character for which there is no positive selection upon repeated growth in host-free media.

Taxonomy

Relationships of bdellovibrios to other bacteria.—The isolation of H-I bdellovibrio strains has been an important aid in providing a satisfactory experimental tool for physiological studies as well as for studying the systematics of bdellovibrios (66, 71, 72, 75, 94). Many experiments are feasible with H-I bdellovibrios grown in axenic cultures which are made difficult to impossible when bdellovibrios are grown in two-membered systems with host cells. Moreover, the cell yield of *Bdellovibrio* is increased at least four- to six-fold when cultivated in the H-I mode as compared to the yield from H-D lysates. Host-independent strains therefore make it more convenient to do biochemical or nucleic acid studies where a greater cell mass is essential.

The characterizations of some 16 H-I strains based on a variety of diagnostic, physiological, and morphological studies have revealed homogeneity among all these strains except for H-I *Bd.* A3.12 (71). Significant differential features separating H-I *Bd.* A3.12 from other bdellovibrios included resistance to a vibriostat, greater virulence in reversion to the H-D mode, differences in the cytochrome spectra, and variations in the amounts of proteolytic activity (71). All of the H-I cultures included in that study (71) are nonfermentative, do not use carbohydrates as an energy source, do not grow anaerobically with nitrate, and are oxidase-positive. It might appear then, from these traits, that *Bdellovibrio* could be placed in the family *Spirillaceae* as described by Véron (107, 108).

The DNA base composition of 19 separate isolates of *Bdellovibrio* has been determined from the T_m and buoyant density in CsCl (71, 72). The GC content of 17 of these isolates was 50 to 51 percent, while two isolates (*Bd.* 321 and *Bd.* A3.12) have 43 percent GC. Unfortunately, *Bd.* 321 was not available for the later comparative studies which further demonstrated the uniqueness of *Bd.* A3.12 (66, 71). Recently, another isolate, *Bd.* UKi2 (12, 17), was found to have 42 percent GC (66).

On the basis of nutritional characteristics, cell shape, and seeming inertness toward carbohydrates, *Bdellovibrio* might be thought to be related to members of the *Campylobacter* group (107). However, since campylobacters are not proteolytic, do reduce nitrate, and have a 35 percent GC base composition (64, 107), we believe that there are no close taxonomic affinities of campylobacters with bdellovibrios.

Bdellovibrio shares common properties with some bacteria of the genus *Vibrio*. These include mutual antibiotic sensitivities, cytochrome content and absorption peaks, as well as cell shape and DNA base composition [about 40 to 50 percent GC for *Vibrio* (107)] (13, 71, 72). In addition, some species of *Vibrio (V. metschnikovii; V. cholerae)* have a sheathed flagellum morphologically identical to that of *Bdellovibrio* (20, 23, 44, 52, 68). However, there are some equally important properties of *Vibrio* which would exclude *Bdellovibrio;* these include the anaerobic or facultatively anaerobic metabolism of *Vibrio* as well as its ability to utilize glucose (13, 107, 108).

We concede that many problems arise from drawing taxonomic conclusions based on the kinds of comparative studies already made on *Bdellovibrio*. For example, little is known concerning the taxonomic significance of similarities in cytochrome absorption peaks, antibiotic sensitivities, sheathed flagella, etc. However, it is our view that, taken collectively, these and other similarities cannot be dismissed as coincidences, and we believe that they are indicative of some relationship of *Bdellovibrio* to *Vibrio*.

Molecular heterogeneity among bdellovibrio isolates.—Recently, a study has been made of the molecular relationships among several different *Bdellovibrio* isolates (66). The goal was to assess, in an objective and quantitative manner, the variation in molecular attributes of the bdellovibrios. In a sense, the logic of using the predation/parasitism/bacteriolysis complex as an exclusive (i.e., monothetic) defining trait has been challenged with the demonstration that such a criterion—used alone—has brought together a heterogeneous assemblage of organisms. This conclusion is based on a study involving DNA/DNA and DNA/RNA reassociations, estimates of genome sizes, and zone electrophoresis of several soluble enzymes.

The zymogram technique, which involves gel electrophoresis and enzyme-specific staining, has been shown to be a valuable and sensitive technique in bacterial taxonomy (5). The same enzyme from different strains of *E. coli* and *Shigella* spp. will usually migrate at the same rate, whereas the same enzyme from *Salmonella* species migrates at different rates (5). Based on an examination of eight different enzymes, H-I bdellovibrios fell into three groups. The largest group contained five isolates showing very little dissimilarities in enzyme migrations. All these strains have a DNA base composition of 50–51 percent GC. Host-independent *Bd.* A3.12 and H-I *Bd.* UKi2 not only differed from each other in enzyme migration, but also differed from all other *Bdellovibrio* isolates examined (66).

Studies of the genetic relatedness among groups of organisms have been extensively investigated by the techniques of reassociation of their nucleic acids (7, 37, 49). DNA/ribosomal RNA reassociations have consistently demonstrated a high degree of conservation of this small segment of the genome, and reactions are readily observed in both interspecific and intergeneric renaturations (18, 53). Such reassociations were examined among several bdellovibrios, two *Vibrio* spp., and *E. coli,* using the DNA of H-I

Bd. 109 as the reference DNA. The results obtained by Seidler, Mandel & Baptist (66) demonstrate that RNA from H-I *Bd.* B, H-I *Bd.* 110, and H-I *Bd.* 2484Se-2 competed as well as the homologous H-I *Bd.* 109 control, thus indicating identity in these nucleotide sequences. RNA from H-I *Bd.* A3.12 as well as from the two *Vibrio* spp. demonstrated a much smaller degree of competition (20–30 percent), while *E. coli* RNA showed essentially no competition. It is possible to bring these data into perspective by citing the results of analogous experiments with other bacteria. Duplexes comparable to the levels observed for *Bd.* 109/*Bd.* A3.12 (11–23 percent) have been reported between the nucleic acids of a number of pairs of taxonomically divergent bacteria such as *B. subtilis* versus *E. coli* or *P. aeruginosa,* and *Moraxella lwoffi* versus *Rhodomicrobium vanielii* (43, 98).

DNA/DNA reassociation reactions were studied among the various molecular types of *Bdellovibrio* by an optical technique (65). Using the nomenclatural type strain of *Bdellovibrio bacteriovorus* (*Bd.* 100; 94) as the reference DNA, approximately complete homology was demonstrated with the DNAs of isolates H-I *Bd.* 109, H-I *Bd.* 114, and H-I *Bd.* 118. This is in agreement with all other types of taxonomic data (71, 72). As would have been expected on the basis of base composition differences alone, however, no detectible DNA/DNA homology was observed with the H-I *Bd.* 100/H-I *Bd.* A3.12 pair or with the H-I *Bd.* 100/H-I *Bd.* UKi2 pair. The reassociation between the DNAs from H-I *Bd.* A3.12 and H-I *Bd.* UKi2 was only 16 percent and is comparable in level to the *E. coli*/*S. typhimurium* reaction (7). The results of all the *Bdellovibrio* DNA/DNA reassociation reactions are consistent with the observations on genome sizes, DNA base composition, DNA/RNA reassociation, and enzyme electrophoresis (66, 72).

Species of the genus Bdellovibrio.—Thus, it has become apparent that there is an array of molecular-genetic types of bdellovibrios. In a recent report, Seidler, Mandel & Baptist (66) have formalized the above evidence and have designated two new species within the genus *Bdellovibrio*: *Bdellovibrio stolpii* (with H-D *Bd.* UKi2 as the nomenclatural type culture) and *Bdellovibrio starrii* (with H-D *Bd.* A3.12 as the nomenclatural type culture). *Bdellovibrio bacteriovorus* remains as the type species, with H-D *Bd.* 100 as the nomenclatural type culture. This species delineation is considered by these workers to be strongly supported and quite justifiable on the basis of complete uniformity of evidence in the various molecular approaches used.

We have indicated on several occasions in this essay that there are discrepant reports about parasitic behavior, fine structure, and physiology in the different bdellovibrio strains used by various investigators. In the light of the foregoing molecular studies, these apparent discrepancies may be well founded and thus represent manifestations of molecular differences in structure and function exhibited by the several species of *Bdellovibrio*.

A Concept of *Bdellovibrio*

Parasitism and predation.—The genus *Bdellovibrio* was based initially (92, 94) on two properties, namely, the predatory/ectoparasitic/bacteriolytic complex (as seen in the microscope and on host lawns), and the small size (as detected by microscopy and passage through filters). The concept of *Bdellovibrio* was broadened by the finding (79) that bdellovibrios are endoparasites or, at least, intraintegumental parasites of other bacteria [rather than ectoparasites, as originally (75, 92, 94) and, inexplicably, also currently (101) believed]. Moreover, the lingering doubt that they might not be prokaryotic protista was surely removed with the publication of the first ultrastructure studies (63, 79). The finding of host-independent derivatives and, later, of facultatively parasitic strains raised the issue of whether there might be permanently nonparasitic bdellovibrios but, as related in the section on Virulence of Host-Independent Isolates, we have found that every H-I and F-P *Bdellovibrio* strain which we have examined (and this includes most of the known strains) has the potential of becoming an endoparasitic H-D *Bdellovibrio*. Based on this experience, and the inability of any student of *Bdellovibrio* to locate as yet a similar potential in non-*Bdellovibrio* vibrios and spirilla, Heise & Starr (in preparation) conclude that actual or potential intracellular (or intraintegumental) parasitism of other bacteria is a necessary and sufficient condition of a small, aerobic, vibrioid bacterium (which has all or most of the other properties listed in the section on Toward a Definition of *Bdellovibrio*) being a bdellovibrio. By "potential parasitism" they mean the genetic capacity to shift toward or to pass on to its progeny the actual endoparasitic mode. This trait, capacity to endoparasitize other bacteria, has an "epistemological primacy," by which term Heise & Starr have in mind two related matters: (*a*) an experienced student of *Bdellovibrio* would early direct his efforts to determining whether this trait is present in the course of deciding whether or not an unidentified organism is a bdellovibrio; and (*b*) he would be more reluctant to accept as a bdellovibrio an organism which lacks the endoparasitic capacity than he would if it lacked some other trait, say, the vibrio shape.

Small size.—The cells of bdellovibrios are indeed small: about one-tenth the mass of the cells of *E. coli.* Actually, they may be 1 to 2 μm in length, much longer in the cases of the intracellular predivision C-shaped and spirillar "snakes" of H-D bdellovibrios and certain life forms of some H-I and F-P strains. However, they are unusually narrow; somewhere around 0.35 μm would be the mean width of bdellovibrio cells. This small size (narrow width; low mass) may have several consequences: a very rapid and very active motility (which may have a bearing on the force and the outcome of the initial collision with a host cell); a possible limitation in space for the organelles needed by a free-living organism (59); the ability to pass through filters which retain ordinary bacteria; and ecological aspects of the increased resistance to gravity (easier flotation in aquatic habitats; possibility of isolation from nature by differential centrifugation). The possibility that the cell volume might be inadequate for all the organelles (including genome) needed by a free-living prokaryote has been dismissed by the findings that the genome size (66) is within 60 percent of that of *E. coli* (despite the 1:10 ratio in cell volume); that all 20 ordinary amino acids are coded for (making a triplet code, rather than a duplex code, mandatory) by the four conventional bases (72); that an astonishing array of internal and external structures is indeed present in and upon the bdellovibrio cell (1, 11, 12, 79); and, moreover, that free-living, self-nourishing H-I forms of bdellovibrios are known.

"Bdellovibrio-*like" bacteria.*—As with other unusual bacteria (82) which are newly uncovered, there has been some confusion with organisms that are only superficially related to *Bdellovibrio*. For example, if one uses only the criterion of slowly-forming plaques on a bacterial lawn as presumptive evidence for a bdellovibrio, there are apt to be mistakes stemming from such agents as: (*a*) antibiotic- and colicin-producing organisms; (*b*) bacteriolytic microorganisms which are not bacteria; and (*c*) commensal, parasitic, or predatory bacteria which are not vibrioid endoparasites of other bacteria. Detailed consideration of these undeniably interesting "*Bdellovibrio*-like" organisms might well provide a separate essay entitled "What *Bdellovibrio* is Not"! However, brief mention of certain cases here might be instructive in our groping for a concept of *Bdellovibrio*.

Confusion with antibiotics and colicins is readily eliminated by showing the nonorganismic nature of the plaque-producing agent. The early confusion of *Bdellovibrio* with *Caulobacter* (92) was soon sorted out (82, 94) on a number of bases including the ectocommensal and nonparasitic nature of the relationship of *Caulobacter* to other bacteria. Bacteriolytic myxobacteria are not endoparasitic and differ in many other ways from *Bdellovibrio*. The fascinating predatory bacteria (*Dictyobacter, Cyclobacter, Trigonobacter, Streptobacter, Desmobacter,* and *Teratobacter*) described by Perfil'ev & Gabe (58) seem not to be endoparasites. We have already commented on the interesting bacteriolytic pseudomonads which are active against azotobacters (60); these may be similar in action to the "heterobacteriolytic" bacteria described some forty years earlier by Söhngen (77) and to the bacteriolytic bacteria reported in an abstract (109), which have been further characterized (Hirsch, personal communication) as not entering the "host" cell and differing in other ways (e.g., GC content of about 60 percent or higher) from *Bdellovibrio*. The "parasite" of *Clostridium perfringens*, strain Xpfr of Guélin and co-workers (30, 33), also does not enter the "host" cell. And, finally, the exciting bacterial parasite of *Chlorella*, reported by Mamkaeva (48) and Gromov & Mamkaeva (24), is now considered by these workers (Gromov, personal communication) to be a new species of *Bdellovibrio* for which they are proposing the name *Bdellovibrio chlorellavorus* in a forthcoming paper in *Tsitologiya*. It would be very interesting, after the necessary side-by-side comparisons are made and the taxonomic placement is indeed confirmed, to have a genuine bdellovibrio in a eukaryotic host. However, from what we presently know, this organism does not enter the *Chlorella* cells and it has conventional (i.e., not sheathed) flagella.

Toward a definition of Bdellovibrio.—It is only just that an exclusionary dictum on the theme of "What *Bdellovibrio* is Not" be followed by an attempt at an affirmative definition. Definitions provide very tempting targets for the barbs of critics; we will try to minimize our vulnerability.

The epistemological roles of the type specimen (nomenifer) in defining the name of a taxon have been laid out (35, 80). Among these roles, we find that the specific nomenifer can serve as an ostensive definition (i.e., definition by example) of the name of a species. From this it follows that *Bdellovibrio bacteriovorus* can be defined ostensively in terms of an actual culture of H-D *Bd.* 100 (the type specimen of that species designated by Stolp & Starr, 94). Similarly, *Bdellovibrio stolpii* is defined ostensively by a culture of H-D *Bd.* UKi2 and *Bdellovibrio starrii* by a culture of H-D *Bd.* A3.12 (in accordance with the designations made by Seidler, Mandel & Baptist, 66). Although it is not the conventional view, one may look on the name of a genus as being ostensively defined by the specific nomenifers of all included species (35, 80). The genus *Bdellovibrio* would thus on this view (presently) be circumscribed by H-D *Bd.* 100, H-D *Bd.* A3.12, and H-D *Bd.* UKi2. In other words, the genus *Bdellovibrio* (as presently conceived) would have to contain H-D *Bd.* 100, H-D *Bd.* A3.12, and H-D *Bd.* UKi2; and these strains serve as an ostensive definition of the name of the genus. The genus *Bdellovibrio* may, of course, contain other organisms which are reasonably similar to these nomenifers.

In addition to ostensive definition, bacteriologists usually use intensional definitions (definitions in terms of words expressing properties—usually the "key" properties, possession of most of which is considered essential for inclusion in the class). An intensional definition of the genus *Bdellovibrio* might be laid out in the following terms: *Bdellovibrio* is the genus of bacterial prokaryotes which are: (*a*) generally vibrioid, sometimes spirillar; (*b*) unusually narrow (about 0.3–0.45 μm in width); (*c*) polarly flagellated (very rarely with more than one flagellum but the flagellum is always sheathed and the insertion is always polar); (*d*) Gram-negative; (*e*) strictly aerobic; (*f*) chemo-organotrophic; (*g*) endowed with DNA which is highly homogeneous and contains 42–51 percent GC; (*h*) markedly and invariably proteolytic; (*i*) able to metabolize through the TCA cycle; (*j*) generally not capable of using carbohydrates; (*k*) fitted with a dimorphic developmental cycle: alternation of flagellated, predatory, vibrioid swarmers (capable of attaching by the aflagellated tip to a host cell) and nonflagellated spirillar (usually intracellular) vegetative stages; and (*l*) actually capable of (or have the genetic potential for) entering the cells of certain other bacteria and developing and multiplying therein. We have already indicated that the last-named trait is considered to have an epistemological primacy.

Thus, Heinz Stolp's curiosity about the unusual lytic zones on his Petri plates was highly rewarded. An entirely new type of biological interaction was thereby and subsequently uncovered: the entrance of an unusual sort of bacterium into other bacteria, with propagation of the intruder and lysis of the penetrated cell the outcome of the interaction. But why had *Bdellovibrio* not been recognized before Heinz Stolp examined microscopically the material from the slowly-developing plaques? Given the ubiquity of the bdellovibrios, the probability is very high that some earlier scientists had "seen-without-seeing" these tiny microbes—perhaps because they had no a priori concept of such creatures. Moral: the open-minded shall inherit the earth—or at least discover its contents!

ACKNOWLEDGMENTS

Supported, in part, by a grant-in-aid from American Bioculture, Inc., and by research grant AI-08426 from the National Institute of Allergy and Infectious Diseases, U.S. Public Health Service. We are greatly obliged to our colleagues the world over who very generously shared their latest findings and thoughts with us prior to publication. We are grateful to H. R. Heise, M. Mandel, and J. C.-C. Huang for critical reviews of the manuscript. We appreciate very much the editorial and bibliographic aid provided by B. V. Daniel and P. B. Starr.

LITERATURE CITED

1. Abram, D., Davis, B. K. 1970. *J. Bacteriol.* 104:948–65
2. Abram, D., Shilo, M. 1967. *Bacteriol. Proc.* 41–42
3. Althauser, M., Conti, S. F. 1971. *Bacteriol. Proc.* 173
4. Amemiya, K., Kim, J. 1970. *Bacteriol. Proc.* 65
5. Baptist, J. N., Shaw, C. R., Mandel, M. 1969. *J. Bacteriol.* 99:180–88
6. Barnett, H. L. 1963. *Ann. Rev. Microbiol.* 17:1–14
7. Brenner, D. J., Fanning, G. R., Johnson, K. E., Citarella, R. V., Falkow, S. 1969. *J. Bacteriol.* 98: 637–50
8. Bruff, B. S. 1964. *Studies on a predacious vibrio.* MA thesis, University of California, Berkeley
9. Burger, A., Drews, G., Ladwig, R. 1968. *Arch. Mikrobiol.* 61:261–79
10. Burnham, J. C., Hashimoto, T., Conti, S. F. 1968. *Bacteriol. Proc.* 22
11. Burnham, J. C., Hashimoto, T., Conti, S. F. 1968. *J. Bacteriol.* 96:1366–81
12. Burnham, J. C., Hashimoto, T., Conti, S. F. 1970. *J. Bacteriol.* 101:997–1004
13. Colwell, R. R. 1970. *J. Bacteriol.* 104:410–33
14. Crothers, S. F., Fackrell, H. B., Robinson, J. 1970. *Abstracts, Can. Soc. Microbiol., 20th Ann. Meet.* 88
15. Crothers, S. F., Robinson, J. 1970. *Abstracts, Can. Soc. Microbiol., 20th Ann. Meet.* 83
16. Dias, F. F., Bhat, J. V. 1965. *Appl. Microbiol.* 13:257–61
17. Diedrich, D. L., Denny, C. F., Hashimoto, T., Conti, S. F. 1970. *J. Bacteriol.* 101:989–96
18. Doi, R. H., Igarashi, R. T. 1965. *J. Bacteriol.* 92:88–96
19. Fackrell, H. B., Campbell, G. K., Robinson, J. 1970. *Abstracts, Can. Soc. Microbiol., 20th Ann. Meet.* 82
20. Follett, E. A. C., Gordon, J. 1963. *J. Gen. Microbiol.* 32:235–39
21. Ghuysen, J.-M. 1968. *Bacteriol. Rev.* 32:425–64
22. Gillis, J. R., Nakamura, M. 1970. *Infect. Immunity* 2:340–41
23. Glauert, A. M., Kerridge, D., Horne, R. W. 1963. *J. Cell Biol.* 18:327–36
24. Gromov, B. V., Mamkeva, K. A. 1966. *Microbiology* [USSR; English Transl.] 35:893–98
25. Guélin, A., Cabioch, L. 1970. *C. R. Acad. Sci. Paris* 271:137–40
26. Guélin, A., Lamblin, D. 1966. *Bull. Acad. Nat. Med. Paris* 150:526–32
27. Guélin. A., Lépine, P., Lamblin, D. 1967. *Ann. Inst. Pasteur* 113:660–65
28. Guélin, A., Lépine, P., Lamblin, D. 1968. *Rev. Int. Oceanogr. Med.* 10:221–27
29. Guélin. A., Lépine, P., Lamblin, D. 1969. *Verhandl. Int. Verein. Limnol.* 17:744–46
30. Guélin, A., Lépine, P., Lamblin, D. 1969. *C. R. Acad. Sci. Paris* 268: 2828–30
31. Guélin, A., Lépine, P., Lamblin, D., Petitprez, A. 1969. *Bull. Franc. de Piscicult.* 233:101–7
32. Guélin, A., Lépine, P., Lamblin, D., Sisman, J. 1968. *C. R. Acad. Sci. Paris* 266:2508–9
33. Hanks, J. J. 1966. *Bacteriol. Rev.* 30:114–35
34. Hashimoto, T., Diedrich, D. L., Conti, S. F. 1970. *J. Virol.* 5:97–98
35. Heise, H., Starr, M. P. 1968. *System. Zool.* 17:458–67
36. Hoeniger, J. F. M., Moor, H., Ladwig, R. 1971. In preparation
37. Hoyer, B. H., McCarthy, B. J., Bolton, E. T. 1964. *Science* 144: 959–67
38. Huang, J. C. C. 1968. *Bacteriol. Proc.* 23
39. Huang, J. C.-C., Robinson, J. 1969. *Bacteriol. Proc.* 41
40. Huang, J. C.-C., Robinson, J., Murray, R. G. E. 1966. *Abstracts, Can. Soc. Microbiol., 16th Ann. Meet.* 45
41. Huang, J. C.-C., Starr, M. P. 1971. In preparation
42. Huang, J. C.-C., Starr, M. P. 1971. In preparation
43. Johnson, J. L., Anderson, R. S., Ordal, E. J. 1970. *J. Bacteriol.* 101: 568–73
44. Joys, T. M. 1968. *Antonie van Leeuwenhoek J. Microbiol. Serol.* 34: 205–25
45. Klein, D. A., Dennis, S. C., Casida, L. E., Jr. 1966. *Bacteriol Proc.* 2
46. Klein, D. A., Casida, L. E., Jr. 1967.

Can. J. Microbiol. 13:1235–41
47. Lépine, P., Guélin, A., Sisman, J., Lamblin, D. 1967. *C. R. Acad. Sci. Paris* 264:2957–60
48. Mamkaeva, K. A. 1966. *Microbiology* [USSR; English Transl.] 35:724–28
49. McCarthy, B. J., Bolton, E. T. 1963. *Proc. Nat. Acad. Sci., U.S.A.* 50:156–64
50. Mishustin, E. N., Nikitina, E. S. 1970. *Bull. Acad. Sci. USSR, Biol. Ser.* 3:423–6
51. Mitchell, R., Yankofsky, S., Jannasch, H. W. 1967. *Nature (London)* 215:891–93
52. Miwatani, T., Shinoda, S., Fujino, T. 1970. *Biken J.* 13:149–55
53. Moore, R. L., McCarthy, B. J. 1967. *J. Bacteriol.* 94:1066–74
54. Moulder, J. W. 1962. *The biochemistry of intracellular parasitism.* Chicago: University of Chicago Press, 171 pp.
55. Moulder, J. W. 1966. *Ann. Rev. Microbiol.* 20:107–30
56. Murray, R. G. E. 1964. *Abstracts, Can. Soc. Microbiol., 14th Ann. Meet.* 52
57. Parker, C. A., Grove, P. L. 1970. *J. Appl. Bacteriol.* 33:253–55
58. Perfil'ev, B. V., Gabe, D. R. 1969. *Capillary methods of investigating microorganisms.* Edinburgh: Oliver and Boyd, 627 pp.
59. Pirie, N. W. 1969. Possible limitations to the size of free-living organisms. In *The Mycoplasmatales and the L-phase of bacteria,* ed. L. Hayflick, 3–14. New York: Appleton-Century-Crofts, 731 pp.
60. Postgate, J. R. 1967. *Antonie van Leeuwenhoek J. Microbiol. Serol.* 33:113–20
61. Reiner, A. M., Shilo, M. 1969. *J. Gen. Microbiol.* 59:401–10
62. Rittenberg, S. C., Shilo, M. 1970. *J. Bacteriol.* 102:149–60
63. Scherff, R. H., De Vay, J. E., Carroll, T. W. 1966. *Phytopathology* 56:627–32
64. Sebald, M., Véron, M. 1963. *Ann. Inst. Pasteur* 105:897–910
65. Seidler, R. J., Mandel, M. 1971. *J. Bacteriol.* 106:608–14
66. Seidler, R. J., Mandel, M., Baptist, J. N. 1971. *J. Bacteriol.* In press
67. Seidler, R. J., Starr, M. P. 1967. *Bacteriol. Proc.* 42
68. Seidler, R. J., Starr, M. P. 1968. *J. Bacteriol.* 95:1952–55
69. Seidler, R. J., Starr, M. P. 1968. *Bacteriol. Proc.* 23
70. Seidler, R. J., Starr, M. P. 1969. *J. Bacteriol.* 97:912–23
71. Seidler, R. J., Starr, M. P. 1969. *J. Bacteriol.* 100:769–85
72. Seidler, R. J., Starr, M. P., Mandel, M. 1969. *J. Bacteriol.* 100:786–90
73. Shilo, M. 1966. *Sci. J.* 2:33–37
74. Shilo, M. 1969. *Curr. Top. Microbiol. Immunol.* 50:174–204
75. Shilo, M., Bruff, B. 1965. *J. Gen. Microbiol.* 40:317–28
76. Simpson, F. J., Robinson, J. 1968. *Can. J. Biochem.* 46:865–73
77. Söhngen, N. L. 1927. *Kon. Akad. Wet. Amsterdam* 36:1–6
78. Stanier, R. Y. 1970. *Symp. Soc. Gen. Microbiol.* 20:1–38
79. Starr, M. P., Baigent, N. L. 1966. *J. Bacteriol.* 91:2006–17
80. Starr, M. P., Heise, H. 1969. *Int. J. System. Bacteriol.* 19:173–81
81. Starr, M. P., Huang, J. C.-C. 1972. *Advan. Microbial Physiol.* 8. In press
82. Starr, M. P., Skerman, V. B. D. 1965. *Ann. Rev. Microbiol.* 19:407–54
83. Steiner, S., Conti, S. F. 1970. *Bacteriol. Proc.* 74
84. Steiner, S., Lester, R. L., Conti, S. F. 1971. *Bacteriol. Proc.* 143
85. Stolp, H. 1965. *Zentralbl. Bakteriol. Parasitenk., Abt. I,* Supplementheft 1:52–6
86. Stolp, H. 1967. Bdellovibrio bacteriovorus *(Pseudomonadaceae). Parasitischer Befall und Lysis von* Spirillum serpens. Film E-1314. Göttingen: Institut für den wissenschaftlichen Film
87. Stolp, H. 1967. *Lysis von Bakterien durch den Parasiten* Bdellovibrio bacteriovorus. Film C-972. Göttingen: Institut für den wissenschaftlichen Film
88. Stolp, H. 1967. *Umschau* 2:58–59
89. Stolp, H. 1968. *Naturwissenschaften* 55:57–63
90. Stolp, H. 1969. *Image, Medical Photo Reports Roche* 30:2–5
91. Stolp, H. 1969. *Ärztl. Praxis* 21:2273, 2293–97
92. Stolp, H., Petzold, H. 1962. *Phytopathol. Z.* 45:364–90
93. Stolp, H., Starr, M. P. 1963. *Bacteriol. Proc.* 47
94. Stolp, H., Starr, M. P. 1963. *Antonie van Leeuwenhoek J. Microbiol. Serol.* 29:217–48

95. Stolp, H., Starr, M. P. 1965. *Ann. Rev. Microbiol.* 19:79–104
96. Strominger, J. L., Ghuysen, J.-M. 1967. *Science* 156:213–21
97. Sullivan, C. W., Casida, L. E., Jr. 1968. *Antonie van Leeuwenhoek J. Microbiol. Serol.* 34:188–96
98. Takahashi, M., Saito, M., Ikeda, Y. 1967. *Biochim. Biophys. Acta* 134:124–33
99. Tinelli, R., Shilo, M., Laurent, M., Ghuysen, J.-M. 1970. *C. R. Acad. Sci. Paris* 270:2600–2
100. Trager, W. 1960. Intra-cellular parasitism and symbiosis. In *The Cell; Biochemistry, Physiology, Morphology. IV. Specialized cells: Part 1,* ed. J. Brachet, A. E. Mirsky, 150–213. New York: Academic Press, 511 pp.
101. Uematsu, T., Wakimoto, S. 1970. *Ann. Phytopathol. Soc. Japan* 36: 48–55
102. Varon, M., Drucker, I., Shilo, M 1969. *Biochem. Biophys. Res. Commun.* 37:518–25
103. Varon, M., Shilo, M. 1966. *Isr. J. Med. Sci.* 2:654
104. Varon, M., Shilo, M. 1968. *J. Bacteriol.* 95:744–53
105. Varon, M., Shilo, M. 1969. *J. Bacteriol.* 97:977–79
106. Varon, M., Shilo, M. 1969. *J. Bacteriol.* 99:136–41
107. Véron, M. 1965. *C. R. Acad. Sci. Paris* 261:5243–46
108. Véron, M. 1966. *Ann. Inst. Pasteur* 111:671–709
109. Wood, O. L., Hirsch, P. 1966. *Bacteriol. Proc.* 25

HISTOCOMPATIBILITY ANTIGENS

1585

Dean L. Mann and John L. Fahey[1]
Immunology Branch, National Cancer Institute,
National Institutes of Health, Bethesda, Maryland

Contents

INTRODUCTION

One of the major biologic features of mammalian cell surface membranes (in contrast to bacteria, for example) is the presence of histocompa-

[1] Present address: Department of Medical Microbiology and Immunology, UCLA School of Medicine, Los Angeles, California.

tibility antigens. The histocompatibility antigen systems are notable (*a*) for the extensive polymorphism seen in every species, and (*b*) for the ruthless immune rejection excited in normal individuals following engraftment of tissues from another individual of the same species bearing other histocompatibility antigens of that species.

Every species examined closely has been found to have many transplantation antigen systems. At least one of these is termed a major histocompatibility system, and it excites an immune response which causes prompt and vigorous rejection of tissue grafts. The immune response elicited by major histocompatibility antigens, furthermore, is extremely difficult to suppress.

There are also minor histocompatibility systems which, individually, cause less strenuous immune rejection reactions, although combined minor system differences may cause relatively rapid tissue rejection. The rejection reactions are easier to suppress. In the mouse, for example, there is one major histocompatibility system, termed H-2, and 14 minor histocompatibility systems that have been identified and as many as 25 postulated to exist (17).

Each major histocompatibility antigen system contains many different antigenic specificities. These are grouped into related or allelic series so that an individual member of a species has only a few of the antigens. In man, for example, over 20 specificities are known in the HL-A system but an individual has only 2 to 4 of these specificities.

The major histocompatibility antigens are viewed here as physical configurations determined by the primary structure (probably amino acid sequence) of cell membrane glycoproteins. Furthermore, these antigens are believed to represent a relatively small part of a larger molecule, other portions of which are responsible for association with other membrane components that assure firm attachment of the antigenic molecule to the plasma (outer) surface of the cell membrane.

The histocompatibility antigens or their genes are associated with a wide variety of important biological systems. Although the transplantation antigen systems now receive principal attention for their role in tissue graft rejection, it is possible that they developed to meet other, more common biologic needs. Such needs would include effective distinction between fetal and maternal tissue, and protection against engraftment of cannibalized tissue. Since they are located on the surface membranes of most, if not all, cells, they may assure protection against viruses and other external microorganisms, as well as reaction against mutation and neoplastic transformation. Further, they may, as suggested by Jerne (74), provide an effective stimulus to development and diversity of the normal immune response. Continued progress in defining the histocompatibility antigens, their genes and associated functions, will help to illuminate the biologic and disease significance of the histocompatibility systems.

The major histocompatibility antigen systems have been best defined in mouse and man. This review will deal primarily with the detection, purification, and characteristics of these antigens, the genes controlling them as well as their possible biological significance in these species.

DETECTION OF HISTOCOMPATIBILITY ANTIGENS

There are three general test systems for detection of transplantation antigens. Allograft immunity demonstrates the ability of the antigen to accelerate the rejection of grafted tissues or organs. Serologic assays have proven to be important in the in vitro identification of antigen determinants on cells (tissue typing) and for assay of subcellular materials containing transplantation antigens. The more indirect tests such as lymphocyte blast transformation and delayed hypersensitivity reactions are much less sensitive than the above and lack the ability to detect allotypic specificity.

One of the first definitive demonstrations of specific graft immunity was reported by Gibson & Medawar (58). Skin grafts were rejected in an accelerated fashion after prior sensitization of the host animal with tissue from the same donor. They concluded that this was an immunologic phenomenon in that the reaction demonstrated specificity and the response of rejection was more vigorous after initial sensitization. The in vivo demonstration of accelerated rejection is, by definition, the sole criterion for detecting transplantation antigens.

The serologic assays employ the use of antisera raised against individual or groups of histocompatibility antigenic determinants. The use of antibody for in vitro assays of such antigens was a major advance in the field of transplantation. Using hemagglutination with heterologous antisera, Gorer (59, 60) described four alloantigens on mouse red blood cells that subsequently proved to be antigens of the major transplantation antigen system of the mouse (H-2). Subsequently, it was found that not all of the H-2 antigens could be detected on red blood cells, and Amos (6) demonstrated that leucoagglutination was more comprehensive in defining the H-2 specificities. Hemagglutination techniques have not been applicable in primates as the red blood cells do not seem to possess the major transplantation antigens in the species studied. Leucoagglutination has been employed to detect the transplantation antigens of man (13, 43, 56).

The other important serologic techniques employ the use of fixation of complement as an assay for the presence of transplantation antigens. The direct complement fixation tests utilize lymphocytes as well as platelets as the primary target cells (35, 138). The most widely used methods of detection employ cytotoxic tests using lymphocytes as target cells. The presence of antigen is detected by the ability of the antisera to fix complement and kill the cells. Death of the cell is measured by dye uptake or the release of radioactive isotopes. The dye inclusion test is widely employed in human tissue typing, by a method or modification of a method described by Terasaki (9, 151, 158). Release of radioisotope, particularly ^{51}Cr, by the target cells is based on tests described by Sanderson (121) and Wigzell (160) to detect the H-2 alloantigens. This assay system can also be used to detect the human histocompatibility antigens (HL-A) (119, 120). This test has advantages over the dye inclusion tests in that it is more objective and quantitative.

The dye inclusion and ^{51}Cr release techniques are the most commonly employed assay systems for detecting histocompatibility antigens in solubilized materials (97, 120, 127). The principle of the assay is the inhibition of cytolysis. This test is quite sensitive and can be made quantitative, so that approximate yields of soluble antigens can be determined and comparison made of the relative recovery of antigens during purification procedures.

Some cell surface alloantigens are probably missed by the complement-dependent cytotoxicity reactions now employed in histocompatibility antigen testing. The antigens identified so far have been detected by IgG complement-fixing antibodies. Two such antibody molecules must be in close approximation on the cell surface to fix complement and cause cytolysis (25, 26, 90). Thus, only those antigens which are in clusters (90) or with at least two antigens closely associated on the cell surface are likely to be detected by present cytotoxicity systems.

Other alloantigens may very well be present on the cell surface, but spaced more widely, and may thus require other (noncomplement-dependent) immunological tests for their detection. Indeed, additional leukocyte alloantigens identified by CNAP (cytotoxicity-negative, absorption-positive) and ANAP (agglutination-negative, absorption-positive) reactions have been described (30, 153).

Studies with other immunochemical methods can be expected to detect presently unrecognized alloantigens on nucleated cells. Of course, the possible significance of such alloantigens as transplantation antigens will have to be established by appropriate in vivo tests.

There are other in vivo and in vitro tests for transplantation antigens which at present are more cumbersome and probably less sensitive than the serological assays. One test of considerable significance is the mixed lymphocyte cultures of allogeneic cells (18).

This test utilizes the principle of blast transformation and measurable increases in DNA synthesis in lymphocytes exposed to foreign antigens—in this case, other transplantation antigens. To limit the detection of the antigens represented on lymphocytes from one individual or in mice of one strain, the cells to be tested are treated with mitomycin or X ray to inhibit cell division and DNA synthesis. This then becomes a one-way stimulation test and the major antigenic determinants can be detected. Using the one-way mixed lymphocyte culture, Bach & Amos (16) demonstrated that the inheritance of the human transplantation antigen determinants could be correlated with serological analysis of HL-A alloantigens (41). They demonstrated that siblings sharing all the major HL-A or human transplantation antigenic determinants would not stimulate either way in the mixed lymphocyte reaction. Their studies as well as those of other investigators indicated that this technique is able to detect only the major transplantation antigenic determinants.

This and other test systems such as induction and elicitation of delayed hypersensitivity reaction are indirect tests for the presence of foreign antigenic materials. They would seem to have application only in conjunction with the more direct tests such as specific acceleration of graft rejection and serologic tests for specific antigenic determinants.

The minor histocompatibility antigens at present can be clearly defined only in mouse systems. These antigens can be detected by accelerated rejection of skin grafts where H-2 systems are identical in two animal strains (65), and in some instances by hemagglutination techniques (144). Recent reports by independent investigators indicate that the mixed lymphocyte culture can also detect minor transplantation antigens (91, 113). Detection of the minor histocompatibility antigens will become increasingly important as the techniques are perfected for the isolation, purification, and chemical characterization of the major transplantation antigens. This importance lies in the genetic, molecular, and biologic identity of these cell membrane components and their relationship to components bearing the major transplantation antigens.

THE GENETICS OF TRANSPLANTATION ANTIGENS

In each species studied thus far there seems to be one single strong histocompatibility locus controlling the rejection of grafted tissues. This locus has been designated *H-2* for the mouse, (129, 142), the *AgB* locus in rats (110), the *B* locus in chickens (125), the *RhL-A* locus in monkeys (22), the *DL-A* locus in dog (114), the *CL-A* locus in chimpanzees (21), and the *HL-A* locus in man (38). The most thoroughly studied genetic control of transplantation antigens has been in mouse systems. These studies have been facilitated by the development of inbred strains of mice which have been demonstrated over a period of time to bear a consistent set of transplantation antigenic determinants.

H-2 Antigen System of Mice

Information obtained from studies with inbred mice have made the mouse *H-2* system a prototype for development of knowledge about histocompatibility antigens and their genes in other species. Gorer (60) first identified an antigenic determinant on mouse red blood cells that subsequently proved to be part of the major *H-2* histocompatibility antigens. The studies of Snell (142, 143), Hoecker (72), and others (62, 132, 144), using transplantable tumors from different mouse strains, skin grafting procedures, and hemagglutination and leukoagglutination with isoantisera, identified the antigenic specificities of the *H-2* system. The major specificities are controlled by a single genetic complex (87) which is in the 9th linkage group of the mouse (3).

H–2 Genetic Map

Alleles					
H-2^d	3, 4, 13		Ss^hSlp^a	8,31	
H-2^k	"3", 32	"1, 5"	Ss^lSlp^o	1, 5	3, 8, 11, 23
	D				K

Figure 1. Genetic map of the H-2 complex based on Shreffler's (133) duplication model. Individual specificities are designated by numbers. The serum protein Ss, Slp is coded for by the region indicated. The D region of the H-2 complex is on the left, the K region is on the right.

In the *H-2* system, certain alleles are associated with different strains of mice. Each allele codes for a set of individual antigenic determinants. Some of these determinants are shared by more than one allele. As an example, animal strains bearing the *H-2^a* allele have at least 13 detectable different antigenic determinants. Nine different antigenic determinants are found in those animals carrying the *H-2^b* allele, of which six are shared by the animal strains carrying the *H-2^a* allele. Twenty different alleles have currently been recognized. The *H-2* region seems to be rather large, having a recombinant frequency of 0.5 (136). Shreffler & Klein (136) have calculated that this region may be comprised of about 500 genes.

The observation that genetic recombination (134) within the *H-2* region occurred when inbred strains were crossed has lead to a definition of the fine structure of the *H-2* complex (61, 87, 130, 132, 137). On the basis of the serologic analysis of the 15 recombinants currently available, a linear map of the *H-2* complex has been constructed. Such a map for two alleles is illustrated in Figure 1. In the midst of the *H-2* complex is genetic material which controls for the quantitative or qualitative presence of a serum protein (Ss, Slp) (111, 130, 131, 137). The Ss protein has been found in all mouse strains studied thus far, being high or low (Ss^h or Ss^l). In the qualitative sense two allotypes of this serum protein have been described (Slp^a or Slp^o).

Studies to date indicate that the product of the Ss region is a serum protein which has no anitgenic or structural similarity to the transplantation antigen. The precise mapping of this genetic material (Ss, Slp) within the *H-2* complex is of considerable importance in understanding the evolutionary development of this locus. Recently, Shreffler and colleagues have pointed out inconsistencies in the recombination data (133, 135). These inconsistencies have their genesis in the association of certain *H-2* antigenic specificities with the (Ss, Slp) serum protein. In order to explain the results of certain recombinants, an unlikely triple cross-over in the genetic material would have had to take place. In order to satisfy the observed inconsistencies, Shreffler postulated that the *H-2* locus was historically a single genetic region which has undergone gene duplication and in some instances inverted gene duplication. These genetic events resulted in the separation of the genetic regions controlling transplantation antigens and inserted genetic material which controls for a completely independent serum protein. Duplication of genetic material and subsequent mutations could account for the several distinct regions (i.e., *D* and *K*) that control different antigenic specificities.

Studies of the solubilized *H-2* antigens and characterization of the molecules controlled by the different regions of the *H-2* complex is beginning to provide new insight into the genetic control of transplantation antigens.

HL-A ANTIGEN SYSTEM OF MAN

The *HL-A* antigens or human transplantation antigens are probably the most heterogeneous groups of antigenic determinants known in man. The description of these antigens began with Dausset's demonstration of an antigen on white cells which he designated as Mac (40). Subsequently, van Rood & von Leeuwen (155) identified two leukocyte antigenic determinants, 4a and 4b, which seemed to be allelic. Payne et al (112) described an allelic pair of antigens which were designated *LA-1* and *LA-2,* the latter being found to be the same antigen as the Mac antigen described by Dausset. Other white cell antigens were described by these investigators and others (10, 32, 138, 139, 156, 157), and it appeared that these antigens belonged to the same system (29). The identification of these antigens has been through the notable work of a number of investigators and their co-workers who include Drs. Amos, Walford, Dausset, van Rood, Terasaki, Ceppellini, Kissmeyer-Nielsen, and Payne, to mention a few. These investigators cooperated in international workshops in which antisera were compared (see Histocompatibility Testing publications 1965–1970). Through these efforts it became apparent that the 15–20 different identifiable antigenic determinants belong to one single genetic system like the *H-2* system of the mouse. Dausset, Ivanyi & Ivanyi (42) proposed that this locus be called *HU-1.* Several investigators attached a variety of nomenclatures to this particular locus; however, following an international workshop, uniform notations and the term *HL-A* was adopted (2).

With the development of functionally monospecific antisera detecting only single antigens, it was demonstrated that each individual could possess only four of the major antigenic determinants. Data gathered from family studies indicated that the genetic information for two antigenic determinants (haplotype) was found to segregate independently from the corresponding haplotype, indicating that these antigens were alleles of a single locus (8).

Further investigation indicated that the *HL-A* genetic material on each chromosome could be divided into two regions or mutually exclusive alleles, each controlling the expression of a series of different antigenic determinants (45, 86, 150). These two regions termed the "*LA*" and "*4*" subloci, or

1st and 2nd regions of the *HL-A* complex are similar to the genetic regions within the *H-2* complex. The antigens determined by the two gene regions were first described as the "LA" series of antigens and the "4" series of antigens but have subsequently become known as the first and second segregant series. Since the description of the separate segregant series of antigens, all of the data currently available seem to fit into this concept. A third region and series of antigens has not been positively identified. This evidence is documented in the Proceedings of the Fourth Conference on Histocompatibility Testing held in Los Angeles, 1970 (150).

The genetic regions of the HL-A complex and the HL-A alloantigens determined by the 1st and 2nd regions are summarized in Figure 2.

HL–A SYSTEM

Regions:	1st or "LA"	2nd or "4"
Antigens:	*1st Series*	*2nd Series*
	1	5
	2	7
	3	8
	9	
	10	12
	11	13

FIGURE 2. Schematic structure of the HL-A system. The numbers designate the antigenic specificities controlled by the "LA" (1st) and "4" (2nd) regions of the HL-A complex. (Only antigenic specificities given HL-A designations are listed).

Dausset (41) has estimated that 2 percent of the genes are as yet unidentified in the first segregant series and 6 percent of the genes unidentified in the second segregant series. With the total number of antigens known at the present time, it is possible to generate 127 different HL-A haplotypes in the general population which, in various combinations, could be expressed as some 8000 different genotypes. The genetics of the HL-A system are examined in more detailed reviews by Albert (1), Terasaki (152), Walford (159), Amos (7), and Dausset (41).

The above picture is an oversimplification of a very complex picture of antigenic expression. One of the major problems in identifying specific antigenic determinants is the serologic reagents (15, 147). These reagents are obtained from multiply transfused individuals who generally have high titer antibody which lacks specificity. Some of the most specific antisera were obtained from multiparous women, it being assumed that these women were immunized by the white cells of the fetus expressing the HL-A haplotype of the father. Other methods of preparation are deliberate immunization of volunteers with white blood cells or skin grafts, or both, to produce an antiserum detecting a specific HL-A alloantigen. Some of these antiserums seem to function as monospecific reagents; however, it has been well demonstrated by a number of investigators that some "monospecific" antisera have reactivity patterns that were not completely identical to other "monospecific" antisera which seem to detect the same antigen (7, 41, 159). On the other hand, some "multispecific" antisera appear to react as monospecific sera when analyzed by absorption and quantitative analysis. Amos (7) has described a situation in which deliberate immunization produced an antiserum that detected antigens lacking or undetectable in the donor. These occurrences exemplify only a few of the inconsistencies in the serologic reagents which are intended to identify the HL-A alloantigens. These problems are discussed more extensively in several recent reviews of the genetics of HL-A alloantigens (1, 7, 152, 159).

The lack of uniformity of reactivity of the various antisera creates considerable difficulty in the analysis of soluble alloantigens. These sera are used in cytotoxic inhibition assays for the detection of alloantigens in solubilized form. We have found that the reactivity of some of these antisera changes with the solubilization of the antigens. In general, the reactivity of the antisera broadens. As an example, antisera which are not cytotoxic to or absorbed out by a particular cell (A) are inhibited in their reactivity to cell B by solubilized membranes from cell A. This implies that certain antigenic determinants are hidden as they exist on whole cell membranes and are exposed in the process of solubilization. If the configuration of a particular component of the cell membrane plays a role in antigenicity, small changes in the external environment or topography of the cells may indeed change their antigenicity.

Transplantation Antigens in Other Species

Other species (70) are now being extensively studied for the existence of a complex transplantation locus or region like that of man and mouse just described. Of particular interest are the studies in the primates closely related to man in evolution. Initial studies by Metzger & Zmijewski (105) and Shulman et al (140), and the more recently expanded studies by Balner et al (21) in the chimpanzee, demonstrated that some HL-A alloantisera reacted with chimpanzee white cells, suggesting that the chimpanzee carries transplantation antigenic determinants that are similar in some respects to the human HL-A alloantigen. Because of the obvious limitation of numbers, this animal system has not been well studied although, at the present time, some five different CL-A antigens can be identified (21).

The rhesus monkey has been an important tool in studies of bone marrow allograft survival. Balner et al (19, 20) pioneered the work in the identification of the RhL-A alloantigens. Some 12 different antigenic determinants can presently be identified (22). Rogentine and his colleagues (118) have also been developing antisera to detect specific antigenic determinants in the rhesus system. At present, they have results suggesting that there may be two subdivisions in the rhesus system similar to that found in man. This work in primates will add to information and understanding of the phylogenetic development of the transplantation antigens and the seemingly complex genetic mechanisms that control their expression.

PHYSICOCHEMICAL PROPERTIES OF TRANSPLANTATION ANTIGENS

Solubilization

In order to proceed with isolation and purification, the components of the membranes that bear transplantation antigens must be separated into soluble forms by techniques which produce a product that 1. allows assessment of specific antigenic determinants, and 2. is amenable to subsequent purification for physicochemical characterization. Most of the attempts at solubilization of transplantation antigens from various animal species have used lymphoid cells or cell sources containing a high percentage of lymphocytes. Recently, lymphoid cells in long-term tissue culture have proven to be an excellent source of large quantities of cellular starting materials (98). The expression of a set of human transplantation antigens in a given cultured cell line has been shown to be stable over a three-year period which represents many growth cycles of the cells (96).

Four approaches have been used to solubilize transplantation antigens. These are: 1. detergents and organic solvents; 2. proteolytic enzymes; 3. sonication; 4. chaotropic agents. These techniques and the properties of the materials solubilized will be discussed in this order in the next three sections. The information is summarized in Table 1.

Detergents and organic solvents.—Kandutsch and co-workers (71, 84, 85) attempted to solubilize the H-2 antigens using Triton X-100 and later Triton X-114. They obtained a material which was not removed by high-speed centrifugation but reaggregated quite easily after the removal of the detergent. More complete solubilization of this material was attempted with snake venom, of which the active component was identified as phospholipase A (84). This treatment resulted in a transplantation antigen preparation which did not reaggregate and could be further characterized by Sephadex gel filtration chromatography.

Deoxycholate and cholate have been reported by several investigators to solubilize H-2 and HL-A antigens. Metzger et al (104) successfully solubilized 4A and 4B antigens from human epithelial cells in culture as well as these same cross-reactive antigens from chimpanzee lymph node cells using deoxycholate. About 30 percent of the total antigenic activity was recovered in a soluble form. Hilgert et al (71) compared the efficiency of solubilization of Triton X-100, Triton X-114, and cholate. The latter two detergent-like materials were more efficient than the former and cholate was more convenient, being easier to remove. Recoveries were in the order of 60 percent of the antigenic activity of the starting materials. Bruning et al (28) were successful by using cholate to extract some soluble non-HL-A human cell surface antigens from placental tissue.

TABLE 1. Summary of Physicochemical Properties of Transplantation Antigens Solubilized by Different Techniques

General method	Specific technique	Physical properties Estimated molecular weight	Chemical properties
Detergent	Triton X-100, X-114	H-2, <200,000	CHO, Lipid, Protein
	Cholate	H-2, <200,000	N.D.[a]
	SDS	HL-A multiple sizes 150,000 90,000 60,000	N.D.
	SDS-Stearate	HL-A>200,000 H-2 >200,000	N.D.
Enzymes	Autolysis	H-2, HL-A (two sizes) >200,000 60,000	Protein, CHO
	Trypsin	H-2 10,000	N.D.
	Papain	H-2, two sizes Class I 57,000 Class II 34,000 HL-A two sizes Class I 57,000 Class II 34,000	 Protein, CHO Protein, CHO Protein, CHO Protein, CHO
Sonication	Sonication	H-2 (only H-2.31) 34,000 150,000 Guinea Pig 15,000 HL-A 34,000	 Protein, CHO Protein, CHO Protein Protein
Other	KCl	H-2 N.D. HL-A N.D.	N.D.
	TIS	HL-A two sizes 90,000 60,000	 Protein, CHO Protein, CHO

[a] N.D.—Not determined.

Another detergent-like compound, sodium dodecylsulfate (SDS) has been used to attempt to solubilize transplantation antigens. Manson & Palm (101) extracted soluble H-2 and non-H-2 material from mouse microsomal lipoprotein preparations that had H-2 activity in biologic test systems. The active component apparently remained in solution after removal of the SDS. Complete removal of SDS, however, is very difficult. Our own studies (94) show that extensive dialysis removes only about 95 to 97 percent of the SDS, using 35S-labeled SDS. Since SDS in high concentration destroys antigenic activity, we have successfully used SDS in concentration ratios of 0.02 mg per mg of membrane protein to obtain HL-A alloantigenic material with high specific activity. Higher concentrations of SDS solubilized more protein while antigenic activity declined. In addition to dialysis, we have attempted to remove SDS by several other means. The most effective seems to be ion exchange chromatography with DEAE Sephadex or Dowex-1. Even with these methods a small amount of SDS remains associated with the alloantigenic fractions. SDS is strongly negatively charged and therefore the physical properties of the solubilized materials are difficult to ascertain. Davies and associates (49, 66) used a combination of SDS with starch stearate to solubilize both H-2 antigenic substances and HL-A antigens. The solubilized material was not removed by high-speed centrifugation and could be characterized by gel filtration chromatography. However, small yields were obtained, amounting to about 3 to 5 percent of the total membrane activity.

There have been some attempts to solubilize cell surface antigenic components with organic solvents. Manson & Palm (101) reported recovery of approximately 75 percent of the H-2 and some non-H-2 antigenic determinants by treating microsomal lipoprotein fractions from lymphoid cell sources with butanol. Kandutsch (82) also was successful in solubilizing H-2 antigens with this solvent. Organic solvents theoretically react with the polar groups of the lipid and protein, resulting in the diassociation of lipid-protein complexes of the membrane. As butanol is sparingly soluble in water, the proteins remain in the water phase of the system and the lipid in the organic phase. However, the material derived by this technique has not been extensively characterized and is assumed to have a highly complex structure in the water phase.

Extensive attempts are needed to characterize the histocompatibility antigens solubilized with detergents. Detergents theoretically provide a good reagent in that they act by dispersing membranes in aqueous solution. One of the problems encountered is the difficulty in removing the detergent while maintaining the membrane components in solution. Anionic detergents confer a strong charge to the membrane component bearing the antigenic determinant and seriously limit the procedures that can be used for subsequent separation and purification. Further investigation of the nonionic detergents is needed.

Enzymes.—Autolysis is a method employed first by Nathenson & Davies (108) and resulted in yields of 5 to 10 percent of soluble H-2 antigens. This technique consists of suspending a preparation of spleen cell membranes in Tris-buffered saline at physiologic pH at 37°C. The removal of the cellular debris is accomplished by centrifugation at 100,000 × g for several hours. In our own hands (97), as well as in the hands of Davies et al (50), this technique is successful in solubilizing small amounts of HL-A alloantigens from cell membranes prepared from spleen tissue or from lymphoid cells in long-term tissue culture. The mechanism of release of the antigens by autolysis is presumably the enzymatic effect of proteolytic enzymes which act on cell surface components. No attempts have been made to discern the nature of this enzymatic reaction or if, indeed, this is the mechanism. This technique is limited by the lack of reproducibility and low yields of soluble antigenic materials.

Edidin (53) employed the enzyme trypsin in the presence of EDTA to solubilize H-2 alloantigens from embryonic and adult tissues. Cherry and co-workers (34) compared this enzyme with other proteolytic enzymes such as chymotrypsin, ficin, and papain. Relatively low yields of soluble antigenic substances were reported. Recovery of individual H-2 antigenic specificities was dependent on time of exposure and the enzyme employed.

The use of the proteolytic enzyme, papain, has proven to be one of the most successful methods of solubilizing the H-2 or HL-A alloantigens from cell membranes. This work was initiated by Nathenson & Shimada (109, 127), and has subsequently been used by a number of investigators (27, 34, 48, 98, 122). The technique utilizes crude cell membranes prepared from lymphoid cell sources, either spleen or lymphoid tissue culture cells. The membranes are prepared by hypotonic lysis of cells and the repeated extraction of the cell debris with normal or low tonicity saline solutions (98, 108, 127). With the mouse systems, this procedure concentrates approximately 70 to 80 percent of the H-2 antigenic activity detected on the whole cell into a crude cell membrane preparation. With human materials, we have found approximately 50–60 percent of the total antigenic activity of the cells in the crude cell membrane preparations. In both instances, estimates of recovery were based on inhibition of cytolysis. Davies and associates have used papain to solubilize H-2 (48) and HL-A alloantigens (50) from spleen cells.

The technique for papain solubilization of histocompatibility antigens is relatively standard. This procedure has been described in detail by Shimada & Nathenson (127). Maximum release of the H-2 antigens has been found to occur after 60 to 90 min exposure of the cell membranes to crystalline papain. The recovery of antigenic activity decreases with longer incubation times. Nathenson (127) has found that individual antigenic determinants are solubilized in different amounts while the recovery of solubilized protein remains constant. The differential recovery of H-2 antigenic specificities was also observed by Cherry et al (34). This evidence indicates that the antigenic determinants may be on different molecules on the cell membrane, with different susceptibilities to proteolytic action. Crude papain was found to be more efficient than crystalline papain in solubilizing the HL-A alloantigenic substances from membranes of cultured lymphoid cells (97). The kinetics of release of the HL-A alloantigens are nearly identical to that reported for H-2 antigens. We have found no essential differences in the rates of release of different transplantation antigenic determinants after 1 hr of exposure to papain. However, if the reaction is stopped with iodoacetamide at 1 hr and the membranes sedimented, washed, and reincubated for the second hour with papain, there is an increased yield of some antigens compared to others (73). The yields after 1 hr incubation are in the order of 20 to 30 percent of the HL-A activity of the original cell membrane.

Papain digestion of cell membrane has proved to be a useful and reproducible technique for the solubilization of transplantation antigens from lymphoid cells from several species. The limitations are that some antigenic specificities are lost or recovered in low amounts during the solubilization procedure. In addition, transplantation antigens solubilized from membranes with proteolytic enzymes may not truly represent the entire molecules (that carry the antigens) as the molecules exist on the cell membranes.

Sonication.—Sonication is another technique which has been successful in solubilizing the H-2 alloantigenic determinants from mice (75), transplantation antigens from two inbred strains of guinea pigs (77), and human HL-A alloantigens (80, 116, 117). Billingham et al (24) demonstrated the release of small quantities of immunogenic H-2 alloantigens with high-frequency sound. The key to successful solubilization of cell membrane antigens using sonic energy was pointed out by Haughton (68) in experiments which demonstrated that high frequencies (20 kcps) caused rapid destruction of the antigenic material. Kahan & Reisfeld (77) employed a low-frequency sound of approximately 9 to 10 kcps. Whole cells were used as sources. Exposure is for a very brief period after which the cellular debris is removed by ultracentrifugation at 130,000 × *g*. The yields of antigens are 15–20 percent of the total cell membrane protein and approximately 10–15 percent of the antigenic activity.

Other techniques.—Several other methods have recently been used to attempt to solubilize the cell membrane components bearing the transplantation antigens. Whole lymphocytes exposed to 3 *M* KCl for 16 hr at 4°C yielded 15 percent of the total antigenic activity of the crude cell particulate fraction while only 15–20 percent of the total cell membrane protein was solubilized (115). This in itself represents a considerable increase in the specific activity of unfractionated solubilized antigens compared to other reported techniques.

Chaotropic agents which have anionic groups such as SCN^-, $C10_4^-$, $SO_4^=$,

I^-, and Cl^- have been shown to disassociate membrane structures effectively (67). Employing a chemical compound TIS (Tris, 2-hydroxy-3,5-diiodobenzoic acid) possessing chaotropic properties, attempts were made to solubilize HL-A alloantigens (93). Concentrations greater than 0.2 *M* were effective in solubilizing 80 percent of the membrane protein but no antigenic activity was retained. With optimal conditions, using 0.07 *M* to 0.1 *M* TIS, 10 to 12 percent of the antigenic activity was solubilized. The low yields of antigenic activity has made characterization difficult.

The lack of a single preferred procedure to solubilize transplantation antigens speaks to the difficulty of obtaining soluble materials which can be easily characterized. In addition, evaluation of individual approaches to the solubilization of histocompatibility antigens is complicated by the lack of comparisons of techniques in the same laboratory. Hollinshead et al (73), however, compared sonication with papain to determine the relative efficiency of these two techniques. The yields of soluble HL-A2 alloantigen and HL-A3 alloantigen were comparable, being about 15 percent of the total membrane activity. The yields of the HL-A7 antigen, however, were considerably higher with papain digestion than with sonication treatment. Such evaluations are badly needed in an area in which results by all techniques are suboptimal. Comparison of the products derived by different techniques also may help in understanding the molecular representation of the transplantation antigens on cell surfaces.

Physical Characteristics of Soluble Transplantation Antigens

The physical characteristics of the solubilized transplantation antigens depend in large part on the techniques of solubilization. However, it appears that each technique extracts or solubilizes histocompatibility substances from several species that have similar physical characteristics. Much of the physical characterization has been done either with centrifugation analysis or by the use of column chromatography.

Detergents.—H-2 alloantigenic material solubilized by Triton X-100 was partially included in G-200 Sephadex and had a sedimentation coefficient of 2.65 in the presence of Triton, indicating a molecular weight somewhat smaller than 200,000 (85). The Triton X-100-solubilized H-2 antigens treated with snake venom (phospholipase A) appeared polydisperse in ultracentrifugation analysis (83). In other studies (71), the material solubilized by cholate and Triton X-114 was found for the most part to be retained within G-200 Sephadex near the excluded volume of the column. It appears that these materials are still rather complex structures, probably consisting of membrane subunits which include the molecule bearing the transplantation antigens. H-2 antigens extracted with the organic solvents have generally been found to be high molecular weight materials (101), and there have been no reported attempts to characterize such material by reducing it to smaller constituents.

Transplantation antigens from mouse or man solubilized by SDS and starch stearate appear to be highly complex forms, possibly units of membrane structures (49, 66). SDS (94) alone yields soluble HL-A alloantigens that appear heterogenous in size on G-200 Sephadex chromatography. The majority of the alloantigenically active materials are found in the first retarded fractions from G-200 columns. Smaller amounts of activity are found where proteins of a molecular weight of 120,000 and 60,000 are eluted. These results suggest that detergents solubilize a high molecular weight material that can be broken down into constituent components which retain alloantigenicity.

Enzymes.—Trypsin-solubilized H-2 alloantigens were found in the included volumes of G-25 and G-10 (53). This is the smallest material reported to show alloantigenic activity. Whether or not this activity was associated with peptides or polypeptides was not determined. Autolytically derived materials chromatograph as proteins having a molecular weight greater than 200,000 (108, 126). In some instances, a small amount of antigenic material was found included within the G-200 or G-150 Sephadex, with chromatographic properties similar to papain-derived alloantigens (50). Autolysis may well release a highly complex form of the antigen representing disperse cell membrane fragments or subunits.

Independent studies of Davies (50), Sanderson (23, 122), Mann (97), Boyle (27), and Nathenson and their co-workers (127) have demonstrated that papain solubilizes alloantigenically active material (H-2 or HL-A, or both) that is found in included fractions of G-150 or G-200 Sephadex columns. It has been generally concluded that this material has a molecular weight between 40,000 and 60,000 if these materials chromatograph as do serum proteins.

Nathenson and his colleagues (37, 127) have used spleen cell membranes from two inbred strains of mice as starting materials for the physical and chemical characterization of papain-solubilized H-2 alloantigens. With careful gel filtration chromatography they have demonstrated that papain released two sizes of antigenic fragments, class 1 and class 2. The molecular weight of the class 1 fragments is about 57,000, and of the class 2 fragments, about 34,000. The two classes of fragments have different alloantigenic determinants. The H-2.3, 4, 10, and 13 alloantigens are associated with the class 1 fragment, and the H-2.31 antigen is associated with the class 2 fragment in the histocompatibility substances derived from the H-2^d mouse strain (37). Alloantigens derived from the H-2^b allele have antigens H-2.5 and 33 associated with the class 1 fragment, and the H-2.2 antigen is associated with the lower molecular weight or class 2 fragment (37). Studies by Davies (47) have also demonstrated that molecular fragments bearing different H-2 alloantigenic determinants could be separated using DEAE cellulose chromatography. He (46) has confirmed Cullen & Nathenson's results (37) showing antigens H-2.3, 4, 10, and 13 to be on the same fragment. The significance of molecular fragments bearing different antigenic specificities will be discussed in a later section.

Summerell & Davies (146) characterized some physical properties of purified H-2 alloantigens with specificities of 5, 7, and 9. Sedimentation velocity in sucrose density gradients revealed a sedimentation coefficient of 3.6S. The buoyant density of this material was calculated to be 1.329 g/sec and the derived partial specific volume 0.733. Using these data they calculated from the Svedberg equation a molecular weight of 54,000 ± 5000. This agrees with the observations of Nathenson on the size of the class 1 fragment.

The HL-A alloantigens were shown to exist on two membrane components that differ in size after papain treatment (99, 100). These fragments can be separated on the basis of their size on G-150 or G-200 Sephadex chromatography and also can be separated by polyacrylamide gel electrophoresis. The HL-A alloantigenic determinants associated with the first genetic region (HL-A1, 2, 3, 9, 10, and 11) have been shown to be on the smaller sized (mol wt 35,000) fragments, while the antigenic specificities determined by the second region (HL-A5, 7, 8, 12, and 13) have been found on the larger fragment (mol wt 57,000). These studies indicate that the two gene regions within the HL-A complex code for molecules differing in gross structure as well as antigenic specificity.

In collaborative studies, Mann & Nathenson (95) directly compared the physical properties of the H-2 and HL-A alloantigenic materials solubilized by papain. They demonstrated that the class 1 and class 2 fragments from H-2 sources chromatographed in a nearly identical fashion with the two fragments derived from human materials. Polyacrylamide gel electrophoresis of combined H-2 and HL-A alloantigens demonstrated small differences in migration rates, probably due to the known differences in the isolectric points of the alloantigens from the two species. Colombani and associates (36) have also shown that papain-derived fragments bearing HL-A alloantigens can be separated by ion exchange chromatography similar in many respects to the separation of the H-2 alloantigenic specificities. These results indicate that the H-2 and HL-A membrane-bearing components are acted on in similar fashion by the papain, resulting in products with very similar physical properties.

Sonication.—Low-frequency sonication of whole cells yields a material which is largely excluded from G-200 Sephadex, indicating a highly complex form of the antigen (77, 79). When this material is run on acrylamide gels at pH 9.5 a single, rather fast migrating component is reported to have the alloantigenic activity in guinea pig (77) and man (79).

Studies in the guinea pig revealed that this material at this stage of purification is still rather physically complex depending on the solvent in which the material is suspended. The basic subunit of this material was reported to have a molecular weight of 15,000 or multiples thereof, using as a criterion the sedimentation velocity (77). McKenzie et al (103) solubilized H-2 alloantigens by sonication and characterized the purified materials in Sephadex columns. The material had an approximate molecular weight of 33,000 at pH 4 and a molecular weight of 150,000 at pH 7.

The human alloantigenic materials solubilized by sonication have the same general physical characteristics as those reported for the guinea pig (79). The solubilized HL-A alloantigens are excluded from G-200 Sephadex and are found in a rapidly migrating component on polyacrylamide gel electrophoresis identical to that found in the guinea pig histocompatibility substances. However, the HL-A alloantigen material was reported to have a molecular weight of 34,000, considerably different from that of the purified transplantation antigens derived from the other two species (115).

Other techniques.—The physical properties of alloantigenic materials solubilized by KCl have not been reported. TIS solubilized HL-A alloantigens chromatograph on G-200 Sephadex as does protein with a molecular weight of 150,000. On acrylamide gels this fraction has two major components (92, 93). It appears that these components bear the different alloantigens controlled by the genetic subdivisions of the HL-A region.

Chemical Characteristics of Soluble Transplantation Antigens

There is much interest in the chemical nature of the cell membrane materials bearing the various antigenic specificities. If these materials are proteins one can assume that they are primary gene products and that the antigenic determinants represent allotypic expression similar to that seen with immunoglobulins. If the antigenic specificities are determined by carbohydrates they can be considered analogous to the red blood cell antigens, i.e., secondary gene products. Thus, determination of the chemical constitutions of the substances bearing antigenic specificities will help interpret the nature of their genetic control.

Detergents.—H-2 alloantigens solubilized by Triton X-100 had high lipid content in addition to protein and carbohydrate constituents (84). The extracted lipid from these preparations had no H-2 alloantigenic activity, indicating that alloantigenicity was conferred by the protein or carbohydrate components or both (84). The butanol-solubilized material had relatively high lipid content as well as protein constituents (101). As is indicated in the previous section these materials proved to be complex, indicating aggregate forms of the cell membranes.

Enzymes—Nathenson and associates (107, 127, 128, 161) have done an extensive chemical analysis of the purified H-2 alloantigens solubilized from mouse cells with H-2^d or H-2^b alleles. Purification was achieved by ammonium sulfate precipitation, ion exchange chromatography, gel filtration chromatography, and polyacrylamide gel electrophoresis. The protein composition of this material was 80 to 90 percent of the dry weight, the remainder being carbohydrate. Little or no lipid in the form of phospholipid was detected in these purified fractions.

The carbohydrate component was found to consist of neutral sugars (mannose, galactose, fucose), glucosamine and sialic acids. These carbohydrate moieties were further studied in alloantigenic materials purified by complexing with specific antiserum. Radio-labeled sugars were used as markers. The antigen-antibody complexes were extensively digested with pronase and the physical properties of the product characterized on gel filtration chromatography. Two carbohydrate chains per glycoprotein fragment were found on both the H-2^d- and H-2^b-derived materials. The size of the carbohydrate chains was the same from both strains based on Sephadex chromatography. This study indicates that there is carbohydrate associated with the fragment carrying the H-2 activity and suggests, on the basis of overall similarity, that the carbohydrate is not responsible for the different H-2 specificities in these two mouse strains.

The amino acid composition was determined on purified, papain-solubilized H-2 alloantigens from the H-2^d and H-2^b strains. No significant difference in individual amino acid content was found when comparing the two strains or the two fragments from a single strain. The H-2^d and H-2^b alloantigens were further compared, using thin-layer peptide mapping. Purified materials were treated with cyanogen bromide, then trypsinized and chromatographed by a two-dimensional thin-layer cellulose micromethod. There were three peptides in the H-2^b alloantigenic materials which were not present in the H-2^d alloantigenic material.

An additional comparison was made by ion exchange chromatography of the H-2^d- and H-2^b-derived alloantigenic material after cyanogen bromide cleavage. Eleven peptides were resolved by this method and there were differences of two to three peptides in the material derived from the two different strains. These findings demonstrate specific differences in protein composition of alloantigenic materials with different H-2 specificity. These differences may represent the different alloantigenic specificities. Direct proof is difficult to obtain, however, since the cyanogen bromide- and trypsin-derived fragments no longer retain the ability to inhibit the cytotoxic alloantiserum. Nevertheless, these studies, taken together with the studies of the carbohydrate moiety on the glycopeptide fragments, indicate that the protein portion of the molecule is responsible for the expression of the allotypic H-2 antigenic determinants.

The amino acid composition of purified HL-A alloantigenic fragments derived from lymphoid cells by papain digestion has been reported (97). These studies were performed on materials derived from three cell lines bearing a spectrum of different antigenic determinants. No differences in individual amino acids could be related to alloantigenic determinants. The carbohydrate composition consisted of neutral sugars and sialic acid which was 6 to 8 percent of the total dry weight. Hexosamine was not detectable by colormetric techniques, but was found at a level of approximately 1 percent using gas chromatography analysis.

The finding of similar amino acid composition in different HL-A alloantigenic materials is similar to the observation made in the two mouse species studied. Mann & Nathenson (95) compared the amino acid compositions of transplantation antigens from mouse and man and found marked similarities in the general distribution of individual amino acids. These differences were of the same magnitude as those observed in amino acid comparisons of immunoglobulin light chains (k) derived from mouse and man. The evolutionary significance of the similarities in amino acid sequences of immunoglobulin light chains (k) of these two species has already been noted, and these detailed studies of the transplantation antigens from mouse and man suggest that they are derived from a common ancestral gene.

Most of the evidence to date indicates that polypeptide chains are responsible for the transplantation antigenic determinants. Recent studies by Sanderson et al (123, 124), however, have raised some question as to participation of carbohydrate in HL-A alloantigenic expression. Sanderson solubilized HL-A alloantigens with papain and found that periodate digestion of isolated HL-A alloantigenic materials destroyed the HL-A2 specificity. They further studied the participation of the carbohydrate portion of the isolated glycoprotein by extensive digestion of this material with pronase. On Sephadex G-50 gel filtration chromatography a substance that inhibited the cytotoxicity of some HL-A alloantisera was found in the included fractions. Approximately 8 to 20 percent of the inhibitory activity of the starting material was recovered. This activity was found to be coincident with the carbohydrate elution peak where approximately 90 percent of the carbohydrate was recovered. In addition, most of the polysaccharides contained in the original glycoprotein moiety was stated to be present in this fraction. About 90 percent of the amino acids present were recovered as free amino acids or low molecular weight peptides and had no inhibitory activity. It was concluded from these studies that a small molecular weight glycopeptide had alloantigenic activity inhibiting antisera detecting HL-A alloantigenic specificity. The question, however, still remains open as to whether or not the residual peptide to which the carbohydrate moiety is attached is responsible for this inhibitory activity. Thus, there is no definitive proof that the carbohydrate portion of this isolated glycopeptide is responsible for the alloantigenic specificity.

Sonication.—Histocompatibility materials from strain 2 and strain 13 guinea pigs were solubilized by sonication and analyzed for amino acid composition (76). Small but significant differences of serine, alanine, isoleucine, leucine, and valine were found. The investigators suggested that these findings could be explained on the basis of single base substitution in the genetic code, and that the differences in amino acid composition of guinea pig transplantation antigens from the two strains were responsible for allotypic specificities. As yet, however, the histocompatibility relationships (i.e., major or minor H system differences) between strain 2 and the strain 13 guinea pigs has not been defined. A definitive conclusion as to the significance of the amino acid differences awaits serologic identification and genetic and immunologic characterization of the specificities involved.

HL-A alloantigens solubilized by this technique and purified by the same procedures were analyzed for amino acid composition as well as for carbohydrate and lipid (116). No detectable amounts of lipid or carbohydrate were found. Amino acid compositions of materials having different antigenic determinants were stated to be different in aspartic acid, serine, proline, alanine, and tyrosine. Reisfeld & Kahan (116) concluded that these differences may represent a form of allotypic expression but may not necessarily be responsible for individual antigenic determinants.

McKenzie et al (103) have reported the chemical composition of H-2 alloantigenic materials isolated by sonication and purified by ammonium sulfate precipitation, gel filtration chromatography, and polyacrylamide gel electrophoresis. These materials were stated to contain 8 percent carbohydrate and 92 percent protein and no detectable phospholipid. This contrasts with the reports of no detectable carbohydrate in the sonication-solubilized alloantigenic materials from guinea pigs and humans.

Other techniques.—Only the chemical composition of histocompatibility antigen (HL-A) solubilized by TIS has been reported (92). Purified HL-A alloantigens were found to be glycoproteins and did not differ significantly from the HL-A alloantigens solubilized by papain.

Several conclusions can be drawn from the studies on the chemistry of soluble histocompatibility antigen. The lipid portion of the cell membranes confers no alloantigenic specificity. It appears that the solubilized H-2 alloantigens are glycoproteins and that the protein portion of the molecules is responsible for antigenic expression. This also seems to be true for the HL-A alloantigen, although the studies by Sanderson raise significant questions.

GENETIC-MOLECULAR CORRELATIONS OF SOLUBILIZED TRANSPLANTATION ANTIGENS

The H-*2* genetic complex can be considered in simplest fashion to be comprised of two regions, the D and the K, separated by the Ss-Slp region (87). As was previously noted, these subdivisions confer different alloantigenic specificities. A similar situation exists in man where it now seems reasonable to postulate that the HL-A antigens are controlled by a complex gene area containing (at least) two subdivisions or regions coding for a different series of HL-A alloantigenic specificities. Papain-solubilized glycoproteins from cells bearing either the H-*2* or HL-A alloantigens yielded fragments of two different sizes (95, 99, 128). In the two strains of mice studied, the two sizes of fragments carry alloantigens of either the D end of the locus or the K end (37, 128). Papain-solubilized HL-A materials from three different lymphoid cell sources (i.e., different genotypes) resulted in fragments bearing different alloantigenic specificities (99). The class I or larger fragment had the specificities of the second genetic subdivision while the smaller fragments, those of class II, had specificities that have been found to be associated with the first segregant subdivision. This information suggests that there is a high degree of homology in the genetic-molecular structures of the alloantigenic materials from these two species (100). One conclusion that can be drawn from these observations is that there are two operating genes in the HL-A or H-2 complex, each coding for a series of antigenic determinants that are on different components on the cell membranes. In the case of the mouse, the D subdivision codes for one set of molecules while the K subdivision controls the expression of the other antigenic determinants on different molecules. The analogous situation is the expression of the HL-A alloantigens of the first and second subdivision on separate cell membrane structures in man. One criticism of this interpretation has been that the products seen may be the result of proteolytic digestion of a single molecule on the cell membrane. If this is true, one would have to assume that the apparent separate genetic subdivisions code for a single polypeptide chain which is then cleaved quite specifically by papain, giving rise to fragments which bear the separate antigenic determinants. In the mouse, this does not seem likely, as the *Ss-Slp* locus is interposed between the D region and the K region of the H-2 complex.

Nonenzymatic approaches to solubilization of these materials are needed to resolve these questions. We have approached this by using a chemical compound TIS (Tris,2-hydroxy-3,5-diiodobenzoate) which acts as a chaotropic agent (92). When the solubilized material was chromatographed on G-200 Sephadex columns, the alloantigens of both series were eluted in the included volumes where 7S serum proteins were found to be eluted. On polyacrylamide gel electrophoresis the alloantigenic material appeared in several components. In preliminary investigation, it appears that the separate components bear different antigens. Studies were carried out with ^{3}H-labeled membrane components. Antisera directed toward the specificities of the first subdivision removed only one of the components, while antisera detecting antigenic determinants of the second series removed another component. This evidence strongly supports the hypothesis that there are separate genetic regions in the HL-A complex and that these regions code for different cell surface molecules.

THE IMMUNOLOGIC ACTIVITY OF SOLUBILIZED TRANSPLANTATION ANTIGENS

As indicated in the previous sections, solubilized transplantation antigens are detected by their interaction with antibody detecting the same antigenic determinants on cells. This reaction demonstrates that the integrity of the antigenic determinants is maintained in the solubilization and purification procedures. Other criteria, such as the ability of the soluble antigens to produce antibody, cause delayed hypersensitivity rections or stimulate blast transformation, only indicate that the materials are still antigens. The very nature of these reactions does not identify them as transplantation antigens. The only definitive proof that these solubilized antigens are still functioning as transplantation antigens is the demonstration that these solubilized materials will specifically accelerate or delay graft rejection.

The highly complex forms of the H-2 alloantigens solubilized by detergents and organic solvents have been demonstrated to cause specific accelerated graft rejection (63, 101). Graff and associates (64) demonstrated that papain-solubilized H-2 antigen possesses both H-2 and non-H-2 determinants in studies demonstrating specific acceleration of rejection of skin grafts. The observations were confirmed by Strober et al (145).

H-2 alloantigens (75) and guinea pig antigens (78) derived from strain 2 by the procedure of sonication have both been demonstrated to retain their activity as transplantation antigens specifically accelerating the rejection of skin grafts. Kahan & Reisfeld (78) reported that as little as 1 to 3 μ μg of the guinea pig strain 2 antigens solubilized by sonication and purified by gel chromatography and polyacrylamide gel electrophoresis could sensitize an animal and induce accelerated graft rejection.

Of considerable interest is the potential use of soluble transplantation antigens to induce immunologic tolerance, e.g., specific immune unresponsiveness to a particular antigen. The ability to suppress specifically and selectively, reactivity to transplantation antigens would greatly aid the field of clinical transplantation. A host of investigators have attempted to induce immunologic tolerance to transplantation antigens. Several comments about this particular problem seem indicated.

Prolongation of graft survival with antigen administration has usually required co-administration of a general immune suppressive agent such as cytotoxic chemicals, radiation, or antilymphocyte serum. The prolongation of graft survival that is observed is only temporary and the biologic mechanism responsible is poorly investigated and understood.

The results of successful graft prolongation after antigen administration could be due to tolerance induction or to another phenomenon which can be produced when injecting transplantation antigens in the form of cells or subcellular materials, e.g., enhancement. Enhancement is the result of specific antibody which suppresses or interferes with the immune response of the host animal to the graft in such a fashion that the graft survival is prolonged. Administration of enhancing antibodies can prolong tumor or normal tissue allograft survival across major histocompatibility barriers. Normal tissue (in contrast to tumor) will not survive for indefinite periods. Some enhancing antibodies, i.e., antibodies against transplantation antigens, act at the efferent end of the immune response, presumably by covering the histocompatibility antigens on the target cells and preventing access of cytotoxic lymphocytes and subsequent damage (81, 106). Evidence for additional mechanisms of antibody suppression of homograft rejection has been obtained (81, 106). The alternatives, tolerance and enhancement (and, possibly, different forms of enhancing antibody), need to be assessed in experiments in which antigen administration produces graft prolongation.

The specific opportunity provided by soluble antigen is that of presenting antigen to the host in a form which will not elicit the usual strong immune response. The immediate model is the induction of immune tolerance to soluble foreign serum proteins as Dresser & Mitchison (52) have done. Their experience, however, may not be directly relevant unless transplantation antigens can be prepared from membrane without damage and in a form which does not correspond to the aggregated proteins which prevented tolerance in their work with foreign serum proteins. On the other hand, availability of soluble antigen opens up possibilities for modification of antigens which are not feasible when only whole cells are used. The addition of chemical groups in order to modify the solubility or fate of the antigen in the recipient, or coupling of antigen to immunologically important carrier molecules, are possibilities that remain to be explored. Indeed, for best prolongation of graft survival, it is not yet clear whether antigen should be administered in an attempt to achieve immune tolerance or to produce optimal humoral (enhancing) antibody, and thus to minimize the cell-mediated immune response.

The successful solubilization and purification of transplantation antigens provides the starting materials needed for the investigations of these and other important fundamental and clinical problems.

MEMBRANE LOCALIZATION OF TRANSPLANTATION ANTIGENS

The transplantation antigens are cell surface components. Their presence on the external surface of the cell membrane is demonstrated by the ability of antibody to combine with viable cells bearing an antigenic determinant, fix complement, and to cause lysis of the cells. Visual identification of histocompatibility antigens has been achieved with ferritin-labeled antibody and electron microscopy. H-2 antigens were found on the cell surfaces (4, 5, 51) and HL-A antigens were more recently identified by Kourilsky and associates (88).

The distribution of histocompatibility antigens on the cell surface, i.e., whether they are evenly distributed or are grouped in restricted areas, has been investigated with electron microscopic studies. Both H-2 (5) and HL-A (88) antigens were found to be grouped in clusters on the surface of lymphocytes. Heterozygous individuals (F1 mice with $H\text{-}2^d$ and either $H\text{-}2^k$ or $H\text{-}2^b$ alleles) were studied to determine if allelic products were in the same region. Both Aoki et al (5) and Davis & Silverman (51) found that the two alleles were expressed in separate clusters on the cell surface.

Clustering of cell surface antigens may be essential for them to be significant as histocompatibility antigens, or as a major histocompatibility antigen. If histocompatibility function and complement fixation by antibody both require clustering of antigens, then cytotoxicity reactions may be particularly well suited to detecting the antigens of major histocompatibility systems. This is not to imply that complement is essential to homograft rejection. Indeed, much evidence indicates that an intact hemolytic complement system is not required for tissue rejection.

The H-2 antigens are not part of a structural membrane protein as indicated by the experiments of Swartz & Nathenson (148). They removed about 85 percent of the H-2 antigens from cell surface by enzyme (papain) treatment. The cells were still intact and viable but did not absorb antibody directed to H-2 antigenic determinants. The H-2 antigens were regenerated quite rapidly, appearing in full complement at the end of 6 hr.

With this information, the relationship of transplantation antigens to other components of cell membranes can be considered. No specific model for cell membranes is generally accepted. Hendler has suggested in a recent review (69) that the unit membrane model first advanced by Danelli & Davson (39) is still the most satisfactory model of membrane structure. The model hypothesizes that there are two basic layers of lipid components with the polar ends of these lipids extended toward the external surface and internal surface of the membrane. The apolar tails of these lipid components extend toward the middle of the membranes. There is some evidence that there are globular protein units which are dispersed through this inner portion of the membrane. The protein constituents and the glycolipid constituents are interspersed in and extend from the lipid core. According to our calculations, the material bearing the transplantation antigens comprises less than 1 percent of the total membrane protein. This would indicate that there are many protein structures within the cell membrane which serve other functions. The evidence cited above indicates that the transplantation antigens are on the external components of this membrane. Perhaps these components are attached by hydrophobic bonding to the polar ends of the lipid constituents of the membrane. This hypothesis has some support by virtue of the fact that substances which are known to compete with hydrophobic bonding are useful in solubilizing the transplantation antigens. These materials include the chaotrophic agents and high concentrations of salt such as KCl.

The availability of serologically detectable cell surface markers, such as the transplantation antigens, should aid in dissecting the cell membrane and contribute to better understanding of the molecular interrelationships of the various constituents.

BIOLOGIC SIGNIFICANCE OF TRANSPLANTATION ANTIGENS

The importance of histocompatibility antigens in tissue transplantation in animal systems is undisputed. In man, the significance of the serologically detected HL-A antigens in transplantation has been established more slowly. In a series of elegant experiments, Amos et al (12, 14) and Ceppellini et al (31, 33) transferred skin grafts between related individuals, especially siblings, as well as unrelated individuals, and demonstrated a clear-cut correlation of HL-A antigens with graft survival. When skin grafts were carried out between unrelated individuals or unmatched siblings with no parental haplotype in common, the graft survived 12–13 days. When grafts were between siblings, or between parent and child, differing at a single parental haplotype, the survival was 13–14 days. Grafts exchanged between siblings with the same combination of parental HL-A haplotypes, however, survived 20 days ($\pm$ 4.5) (12, 14). The relative significance of individual antigens and of the two loci was examined by Dausset et al (44) who compared skin allograft survival in 238 subjects. They concluded that the influence of antigens at the two loci are approximately equal in transplantation. More studies will be needed, however, to determined if some antigen combinations (haplotypes) are more effective in transplantation than others.

Studies relating HL-A antigens to tissue transplantation in man are complicated by the underlying disease and by variations in therapeutic immune suppressions, and other factors. Nevertheless, the necessity for HL-A matching for avoidance of graft-versus-host disease after bone marrow transplantation is widely recognized. Massive immune suppression of the recipient may allow successful bone marrow engraftment between unmatched individuals, but subsequent disastrous graft-versus-host disease has not been prevented.

In kidney transplantation, the data (154) support the view that graft survival is generally better if the donor and recipient differ in 0, 1, or 2 antigens than if they differ in 3 or 4 antigens.

Transplantation of tissues among individuals within a species is not a natural phenomenon, yet nature has generated and maintained extensive polymorphism in this antigen system. It would seem logical that these antigens as cell membrane surface components subserve some very important biological function. Some of these possibilities are mentioned and discussed below.

Transplantation antigens have been shown to be directly associated with the susceptibility of certain inbred strains of mice to a viral infection. Lilly (89) has shown that mice homozygous for the allele controlling the $H\text{-}2^k$ transplantation antigens were 100 percent susceptible to infection of the Gross leukemia virus. In contrast, the Tennant leukemia virus was shown to affect only those inbred strains of mice bearing the $H\text{-}2^d$ allele (149). Another inbred strain, bearing the $H\text{-}2^b$ allele, was the least susceptible to the Tennant virus infection and the subsequent development of leukemia. Strains bearing other alleles, the $H\text{-}2^a$ and the $H\text{-}2^k$, demonstrated intermediate susceptibility to infection and this viral agent. This susceptibility to viral infection may be associated with the particular cell membrane receptor for these viruses such as a specific H-2 antigen, or cell membrane components determined by genetic material linked to the *H-2* allele.

Differences in immune response in respect to high or low antibody production, at least to some antigens, may be genetically determined. McDevitt & Sela (102) demonstrated that the response to several synthetic polyamino acid antigens is controlled by a locus, designated IR-1, which is linked to or is part of the H-2 region. A similar phenomenon has been suggested by Ellman et al (55) in guinea pigs in their immunological response to poly-L-lysine. The interesting possibility that genetically determined differences in susceptibilities to viral or other infections (which seem to be associated with histocompatibility genes) may reflect differences in associated genes controlling the immune response is beginning to be investigated.

An immune-related lymphocyte function responsible for cell transformation in the mixed lymphocyte culture test apppears to be closely associated with the major histocompatibility locus. In general, mixing lymphocytes from individuals differing in major histocompatibility antigens in short-term culture leads to a positive response. By contrast, the mixture of cells from individuals with the same (HL-A) antigens does not result in synthesis of DNA or transformation. The close association of this response with the major histocompatibility system is well established. Recent studies (54, 162) indicate that this reaction may be controlled by a responder locus, responsible for the mixed lymphocyte reactions, which is closely linked to the HL-A locus.

The H-2 genes (and, presumably, major H genes in other species) are associated with genes controlling so many significant functions that, possibly, the H-2 region should be viewed in a larger context, as one member (and not necessarily the most important) of a larger gene region or super gene.

With the establishment of fairly complete systems for HL-A antigen recognition it is now possible to develop information about the relation of the HL-A system to a wide variety of cell functions by investigating correlation with disease in man. Although this work is just getting underway, already there are suggestions that associations will be found with malignant disease. Walford et al (159) have studied a population of patients with acute lymphocytic leukemia and have found an increased frequency of two of the major transplantation antigens of man. In another study of patients with Hodgkin's disease, Forbes & Morris (57) demonstrated that leukocytes from these patients possessed an increased frequency of the antigen 5w which is included in the human transplantation antigen HL-A5. Whether this represents a genetic predisposition or increased susceptibility to an oncogenic viral infection remains an open and provocative question.

The extensive and apparently stable (Dausset) polymorphism of histocompatibility antigens and the strength of the immune response against foreign histocompatibility antigens remains to be suitably explained, from the standpoint of genetic control and cellular mechanisms, as well as in the context of biologic role or need.

Snell, in a speculative essay (141), notes the survival value to a species of extensive cell surface polymorphism if the various histocompatibility alleles are related to differences in susceptibility to different infectious agents. A single type of infectious agent would infect (e.g., the more susceptible) part of a heterogeneous population of animals, thus removing some but not all of a diverse population. The danger of "molecular mimicry" by infectious agents, such as the ability to mutate so as to be compatible with one or a few alleles, would be countered by the many forms of cell surfaces (i.e., of histocompatibility antigens) available in the general population.

The maintenance of the remarkable polymorphism of the H-2 in mouse populations may be determined by its linkage to the T locus. This locus has a lethal allele, t, which is usually lethal (or viable but sterile) when homozygous. This allele serves to facilitate preservation of heterozygosity of the T locus and of the associated H-2 locus. There is approximately a 10 percent crossing over between the H-2 and *T* loci, thus assuring that individuals homozygous for some H-2 alleles will emerge in the population. Alternatively, selective fertilization may be an effective means of maintaining histocompatibility heterozygosity. The significance of these or other devices for maintaining histocompatibility diversity, however, is not established.

The fact of heterozygosity in the major H locus in the animal species studied is incontrovertible. The biologic significance has been speculated on in terms of resistance to neoplastic mutations (141). These or other reasons may underlie the functional role of the major histocompatibility cell surface components. Extensions of current information and acquisition of data from new approaches such as the HL-A antigens in human disease, and H-2 antigens in wild murine populations, are likely to provide more certain information in this area, which is of considerable significance in animal biology. Further, unraveling of the genetic and molecular mechanisms for development and maintenance of heterozygosity in the major H systems is an exciting challenge.

LITERATURE CITED

1. Albert, E. D., Terasaki, P. I. 1970. *Transplantation,* ed. J. Najarian, R. Simmons. In press
2. Allen, F. H. et al. 1968 *Bull. World Health Organ.* 39:483
3. Allen, S. L. 1955 *Genetics* 40:627–50
4. Aoki, T. et al. 1970. *Proc. Nat. Acad. Sci. USA* 65:569–76
5. Aoki, T., Haemmerling, U., de Harven, E., Boyse, E. A., Old, L .J. 1969. *J. Exp. Med.* 130–979–1002
6. Amos, D. B. 1953. *Brit. J. Exp. Pathol.* 34:464–70
7. Amos, D. B. 1970. *Fed. Proc.* 29: 2018–25
8. Amos, D. B. 1967. *Transplantation* 5:1015–23
9. Amos, D. B. 1969. *Transplantation* 7:220–23
10. Amos, D. B., Cohen, I., Nicks, J. P., MacQueen, M. M., Mladick, E. 1967. In *Histocompatibility Testing,* ed. E. S. Curtoni, P. L. Mattiuz, R. M. Tosi, 129–38. Copenhagen: Munksgaard. 458 pp.
11. Amos, D. B., Gorer, P. A., Mikulska, Z. B. 1955. *Proc. Roy. Soc. London Ser. B* 144:369–80
12. Amos, D. B., Hattler, B. G., MacQueen, M. M., Cohen, I., Seigler, H. F. 1968. In *Advance in Transplantation,* ed. J. Dausset, T. Hamburger, C. Mathe, 203–12. Copenhagen:Munksgaard. 779 pp.
13. Amos, D. B., Peacock, N. 1963. *Proc. Congr. Eur. Soc. Hematol., 9th* 1132–36
14. Amos, D. B., Seigler, H. F., Southworth, J. G., Ward, F. E. 1969. *Transplant. Proc.* 1:342–46
15. Amos, D. B., Yunis, E. 1969. *Science* 165:300–2
16. Bach, F. H., Amos, D. B. 1967. *Science* 156:1506–8
17. Bailey, D. W., Mobraaten, L. E. 1969. *Transplantation* 7:394–400
18. Bain, B., Vos, M. R., Lowenstein, L. 1964. *Blood* 23:108–16
19. Balner, H. 1966. *Vox Sang.* 11:306–14
20. Balner, H., Dersjant, H., van Rood, J. J. 1965. *Transplantation* 3: 402–22
21. Balner, H., Dersjant, H., van Vreeswijk, W. 1971. *Transplantation* In press
22. Balner, H., Dersjant, H., van Vreeswijk, W., van Rood, J. J. 1971. *Nature (London)* In press
23. Batchelor, J. R., Sanderson, A. R. 169. *Transplant. Proc.* 1 :489–90
24. Billingham, R. E., Brent, L., Medawar, P. B. 1956. *Nature (London)* 178:514–19
25. Borsos, T., Colten, H. R., Spalter, J. S., Rogentine, G. N., Rapp, H. J. 1968. *J. Immunol.* 101:392–98
26. Borsos, T., Rapp, H. J. 1965. *Science* 150:505–6
27. Boyle, W. 170. *Transplant. Proc.* 1:491–93
28. Bruning, J. W., Masurel, M., Brent, V. D., van Rood, J. J. 1967. In *Histocompatibility Testing,* ed. E. S. Curtoni, P. L. Mattiuz, R. M. Tosi, 303–6. Copenhagen: Munksgaard. 458 pp.
29. Ceppellini, R. 1968. In *Advance in transplantation,* ed. J. Dausset, J. Hamburger, G. Mathe, 195–202. Copenhagen:Munksgaard. 779 pp.
30. Ceppellini, R. 1965. In *Histocompatibility Testing Series Haematologica II,* 158:61. Copenhagen: Munksgaard.
31. Ceppellini, R. et al. 1966. *Ann. N Y Acad. Sci.* 129:421–45
32. Ceppellini, R. et al. 1967. In *Histocompatibility Testing,* ed. E. S. Curtoni, P. L. Mattiuz, R. M. Tosi, 129–238. Copenhagen:Munksgaard. 458 pp.
33. Ceppellini, R., Mattiuz, P. L., Schudeller, G., Visetti, M. 1969. *Transplant. Proc.* 2:385–89
34. Cherry, M., Hilgert, I., Kandutsch, A. A. Snell, G. D. 1970. *Transplant. Proc.* 2:48–58
35. Colombani, J., Colombani, M., Benajam, A., Dausset, J. 1967. In *Histocompatibility Testing,* ed. E. S. Curtoni, P. L. Mattiuz, R. M. Tosi, 413–19. Copenhagen:Munksgaard. 458 pp.
36. Colombani, J. et al. 1970. *Transplantation* 9:228–39
37. Cullen, S., Nathenson, S. G. 1969. *Fed. Proc.* 28:379
38. Curtoni, E. S., Mattiuz, P. L., Tosi, R. M. 1967. In *Histocompatibility Testing,* ed. E. S. Curtoni, P. L. Mattiuz, R. M. Tosi. Copenhagen:Munksgaard. 458 pp.
39. Danielli, J. F., Davson, H. 1935. *J. Cell. Comp. Physiol.* 5:495–508
40. Dausset, J. 1959. *Acta Haematol.* 20:156–66
41. Dausset, J. 1970. *Transplant. Proc.* In press

42. Dausset, J., Ivanyi, P., Ivanyi, D. 1965. In *Histocompatibility Testing,* ed. H. Balner, F. J. Cleton, J. G. Eernisse, 51–62. Copenhagen:Munksgaard. 192 pp.
43. Dausset, J., Nenna, A. 1952. *C. R. Soc. Biol.* 146:1539–41
44. Dausset, J., Rapaport, F. T., Legrand, L., Colombani, J., Marcelli-Barge, A. 1970. In *Histocompatibility Testing,* ed. P. I. Terasaki, 381–97. Copenhagen:Munksgaard. 658 pp.
45. Dausset, J. et al. 1969. *Transplant. Proc.* 1:331–38
46. Davies, D. A. L. 1970. In *Blood and Tissue Antigens,* ed. D. Aminoff, 101–15. New York:Academic. 533 pp.
47. Davies, D. A. L .1969. *Transplantation* 8:51–69
48. Davies, D. A. L., Colombani, J., Viza, D. C., Dausset, J. 1968. In *Histocompatibility Testing,* ed. E. S. Curtoni, P. L. Mattiuz, R. M. Tosi, 287–93. Copenhagen:Munksgaard. 458 pp.
49. Davies, D. A. L., Colombani, J., Viza, D. C., Hammerling, V. 1970. *Clin. Exp. Immunol.* In press
50. Davies, D. A. L., Manstone, A. J., Viza, D. C., Colombani, J., Dausset, J. 1968. *Transplantation* 6: 571–86
51. Davis, W. C., Silverman, L. 1968. *Transplantation* 6:535–43
52. Dresser, D. W., Mitchison, N. A. 1968. *Advan. Immunol.* 8:129–81
53. Edidin, M. 1967. *Proc. Nat. Acad. Sci. USA* 57:1226–31
54. Eijsvoogel, V. P., Schellekeus, P. Th. A., Breur-Vriesendorp, B., Koning, L. 1970. *Transplant. Proc.* In press
55. Ellman, L., Green, I., Martin, W. J., Benacerraf, B. 1970. *Proc. Nat. Acad. Sci. USA* 66:322–28
56. Engelfriet, C. P., Britten, A. 1965. *Vox Sang.* 10:660–74
57. Forbes, J. F., Morris, P. J. 1970. *Lancet* In press
58. Gibson, T., Medawar, P. B. 1943. *J. Anat.* 77:299–310
59. Gorer, P. A. 1937. *J. Pathol. Bacteriol.* 44:691–97
60. Gorer, P. A. 1938. *J. Pathol. Bacteriol.* 47:231–52
61. Gorer, P. A., Mikulska, Z. B. 1959. *Proc. Roy. Soc. London Ser. B* 151:57–59
62. Graff, R. J., Hildemann, W. H., Snell, G. D. 1966. *Transplantation* 4:425–37
63. Graff, R. J., Kandutsch, A. A. 1966. *Transplantation* 4:465–73
64. Graff, R. J., Mann, D. L., Nathenson, S. G. 1970. *Transplantation* 10:59–65
65. Graff, R. J., Snell, G. D. 1969. *Transplantation* 8:861–76
66. Haemmerling, U., Davies, D. A. L., Manstone, A. J. 1970. *Immunochemistry.* In press
67. Hatefi, Y., Hanstein, W. G. 1969. *Proc. Nat. Acad. Sci. USA* 62: 1129–36
68. Haughton, G. 1964. *Transplantation* 2:251–60
69. Hendler, R. W. 1971. *Physiol. Rev.* 51:66–97
70. Hildemann, W. H. 1970. In *Immunogenetics.* San Francisco:Holden-Day. 300 pp.
71. Hilgert, I., Kandutsch, A. A., Cherry, M., Snell, G. D. 1969. *Transplantation* 8:451–61
72. Hoecker, G., Counce, S., Smith, P. 1954. *Proc. Nat. Acad. Sci. USA* 40:1040–51
73. Hollinshead, A., Herberman, R. B., Mann, D. L. 1971. *Proc. Soc. Exp. Biol. Med.* In press
74. Jerne, N. K. 1970. *Proc. Nat. Acad. Sci. USA* In press
75. Kahan, B. D. 1965. *Proc. Nat. Acad. Sci. USA* 53:153–61
76. Kahan, B. D., Reisfeld, R. A. 1968. *J. Immunol.* 101–237–41
77. Kahan, B. D., Reisfeld, R. A. 1967. *Proc. Nat. Acad. Sci. USA* 58: 1430–37
78. Kahan, B. D., Reisfeld, R. A. 1969. *Proc. Soc. Exp. Biol. Med.* 130: 765–69
79. Kahan, B. D., Reisfeld, R. A. 1969. *Transplant. Proc.* 1:483–88
80. Kahan, B. D. et al. 1968. *Proc. Nat. Acad. Sci. USA* 61:897–904
81. Kaliss, N. 1966. *Ann. N Y Acad. Sci.* 129:155–63
82. Kandutsch, A. A. 1960. *Cancer Res.* 20:264–68
83. Kandutsch, A. A., Jurgeleit, H. C., Stimpfling, J. H. 1965. *Transplantation* 3:748–61
84. Kandutsch, A. A., Stimpfling, J. H. 1965. In *Int. Symp. on Immunopathology, 4th, III,* ed P. Grabar, P. Miescher, 134–44. New York: Grune & Stratton
85. Kandutsch, A. A., Stimpfling, J. H. 1963. *Transplantation* 1:201–16
86. Kissmeyer-Nielsen, F., Svejgaard,

A., Hauge, M. 1968. *Nature (London)* 219:1116–19
87. Klein, J., Shreffler, D. 1971. *Transplant. Rev.* In press
88. Kourilsky, F. M. et al. 1971. *J. Immunol.* 106:454–66
89. Lilly, F. 1966. *Genetics* 53:529–39
90. Linscott, W. D. 1970. *J. Immunol.* 104:1307–9
91. Mangi, R. J., Mardiney, M. R. 1970. *Transplant. Proc.* In press
92. Mann, D. L. 1971. *Advan. Immunol.* In press
93. Mann, D. L., Fahey, J. L. 1970. *Transplant. Proc.* In press
94. Mann, D. L., Levy, R. Unpublished observations.
95. Mann, D. L., Nathenson, S. G. 1969. *Proc. Nat. Acad. Sci. USA* 64: 1380–87
96. Mann, D. L., Rogentine, G. N. Unpublished observations
97. Mann, D. L., Rogentine, G. N., Fahey, J. L., Nathenson, S. G. 1969. *J. Immunol.* 103:282–92
98. Mann, D. L., Rogentine, G. N., Fahey, J. L., Nathenson, S. G. 1968. *Nature (London)* 217–1180–81
99. Mann, D. L., Rogentine, G. N., Fahey, J. L., Nathenson, S. G. 1969. *Science* 163:1460–62
100. Mann, D. L., Rogentine, G. N., Fahey, J. L., Nathenson, S. G. 1969. *Transplant. Proc.* 1:494–97
101. Manson, L. A., Palm, J. 1968. In *Advances in Transplantation,* ed. J. Dausset, J. Hamburger, G. Mathe., 301–4. Copenhagen: Munksgaard. 779 pp.
102. McDevitt, H. O., Sela, M. 1967. *J. Exp. Med.* 126:969–78
103. McKenzie, I. F. C. et al., 1970. *Proc. Prague Symp. on H-2 Genetics* In press
104. Metzger, R. S., Flanagan, J. F., Mendes, N. F. 1968. In *Histocompatibility Testing,* ed. E. J. Curtoni, P. L., Mattiuz, R. M. Tosi, 307–13. Copenhagen:Munksgaard. 458 pp.
105. Metzger, R. S., Zmijewski, C. M. 1966. *Transplantation* 4:84–93
106. Möller, G., Möller, I. 1966. *Ann. NY Acad. Sci.* 129:735–49
107. Muramatsu, T. Nathenson, S. G. 1970. *Biochem. Biophys. Res. Commun.* 38:1–8
107. Muramatsu, T., Nathenson, S. G. 1970. *Biochem. Biophys. Res. Commun.* 38:1–8
108. Nathenson, S. G., Davies, D. A. L. 1966. *Proc. Nat. Acad. Sci. USA* 56:476–83
109. Nathenson, S. G., Shimada, A. 1968. *Transplantation* 6:662–64
110. Palm, J. 1964. *Transplantation* 2: 603–12
111. Passmore, H. C., Shreffler, D. C. 1970. *Biochem. Genet.* 4:351–65
112. Payne, R., Tripp, M., Weigle, J., Bodmer, W., Bodmer, J. 1964. *Cold Spring Harbor Symp. Quant. Biol.* 29:285–95
113. Popp, D. M., Davis, M. L. 1970. *Transplant. Proc.* In press
114. Rapaport, F. T., Hanaoka, T., Shimada, T., Cannon, F. D., Ferrebee, J. W. 1970. *J. Exp. Med.* 131:881–93
115. Reisfeld, R. A., Kahan, B. D. 1970. *Advan. Immunol.* 12:117–200
116. Reisfeld, R. A., Kahan, B. D. 1970. *Fed. Proc.* 29:2034–40
117. Reisfeld, R. A., Pellegrino, M., Papermaster, B. W., Kahan, B. D. 1970. *J. Immunol.* 104:560–65
118. Rogentine, G. N. Personal communication
119. Rogentine, G. N., Plocinik, B. 1968. *Transplantation* 5:1323–33
120. Sanderson, A. R. 1967. *Nature (London)* 216:23–25
121. Sanderson, A. R. 1965. *Transplantation* 3:557–62
122. Sanderson, A. R., Batchelor, J. R. 1968. *Nature (London)* 219:184–86
123. Sanderson, A. R., Cresswell, P., Welsh, K. I. 1971. *Nature (London)* 230–8:12
124. Sanderson, A. R., Cresswell, P., Welsh, K. I. 1970. *Proc. Prague Symp. on H-2 Genetics* In press
125. Schierman, L. W., Nordskog, A. W. 1961. *Science* 134:1008–9
126. Shimada, A., Nathenson, S. G. 1967. *Biochem. Biophys. Res. Commun.* 29:828–33
127. Shimada, A., Nathenson, S. G. 1969. *Biochemistry* 8:4048–62
128. Shimada, A., Yamane, K., Nathenson, S. G. 1970. *Proc. Nat. Acad. Sci. USA* 65:691–96
129. Shreffler, D. C. 1967. *Ann Rev. Genet.* 1:163–84
130. Shreffler, D. C. 1970. In *Blood and Tissue Antigens,* ed. D. Aminoff, 85–99. New York:Academic. 533 pp.
131. Shreffler, D. C. 1964. *Genetics* 49: 973–78
132. Shreffler, D. C. 1967. In *Proc. Int. Congr. on Human Genetics,* ed. J.

F. Crow, J. V. Neel, 217–31. Baltimore:Johns Hopkins Press. 578 pp.

133. Shreffler, D. C. 1970. *Transplant. Proc.* In press.

134. Shreffler, D. C., Amos, D. B., Mark, R. 1966. *Transplantation* 4:300–22

135. Shreffler, D. C., David, C. S., Passmore, H. C., Klein, J. 1970. *Transplant. Proc.* In press

136. Shreffler, D. C., Klein, J. 1970. *Transplant. Proc.* 2:5–14

137. Shreffler, D. C., Owen, R. D. 1963. *Genetics* 48:9–25

138. Shulman, N. R. 1965. In *Histocompatibility Testing,* 7–16. Washington, D.C.: Nat. Acad. Sci. Nat. Res. Council.

139. Shulman, N. R., Aster, R. H., Pearson, H. A., Hiller, M. C. 1962. *J. Clin. Invest.* 41:1059–66

140. Shulman, N. R., Moor-Janowski, J., Hiller, M. C. 1965. In *Histocompatibility Testing,* ed. H. Balner, F. J. Cleton, J. G. Eernisse, 113–23. Copenhagen:Munksgaard. 192 pp.

141. Snell, G. D. 1968. *Folia Biol.* 14: 335–58

142. Snell, G. D. 1948. *J. Genet.* 49:87–103

143. Snell, G. D., Smith, P., Gabrielson, F. 1953. *J. Nat. Cancer Inst.* 14: 457–80

144. Snell, G. D., Stimpfling, J. 1966. In *Biology of the Laboratory Mouse,* ed. E. L. Green, 457–91. New York: McGraw-Hill. 706 pp.

145. Strober, S., Appella, E., Law, L. W. 1970. *Proc. Nat. Acad. Sci. USA* 67:765–72

146. Summerell, J. M., Davies, D. A. L. 1970. *Biochim. Biophys. Acta* 207:92–104

147. Svejgaard, A., Kissmeyer-Nielsen, F. 1968. *Nature (London)* 219: 868–69

148. Swartz, B., Nathenson, S. G. 1970. *Transplant. Proc.* In press

149. Tennant, J., Snell, G. D. 1968. *J. Nat. Cancer Inst.* 597–604

150. Terasaki, P. I. 1970. In *Histocompatibility Testing,* ed. P. I. Terasaki, 53–287. Copenhagen:Munksgaard. 658 pp.

151. Terasaki, P. I., McClelland, J. D. 1964. *Nature (London.)* 204: 998–1000

152. Terasaki, P. I., Singal, D. P. 1969. *Ann. Rev. Med.* 20:175–88

153. van Rood, J. J. 1962. *Leukocyte grouping: A method and its application.* Thesis, Univ. of Leiden, Netherlands

154. van Rood, J. J., Dausset, J. et al. 1971. *Eurotransplant* Unpublished observations

155. van Rood, J. J., van Leeuwen, A. 1965. In *Histocompatibility Testing,* 21–44. Washington, D. C.: Nat. Acad. Sci. Nat. Res. Counc.

156. van Rood, J. J., van Leeuwen, A., Schippers, A. M. J., Vooys, W. H., Frederick, S. E. 1965. *In Histocompatibility Testing,* ed. H. Balner, J. G. Cleton, J. G. Eernisse, 37–50. Copenhagen: Munksgaard. 192 pp.

157. Vredevoe, D. L., Mickey, M. R., Goyette, D. R., Magnuson, N. S., Terasaki, P. I. 1966. *Ann. NY Acad. Sci.* 129:521–28

158. Walford, R. L., Gallagher, R., Troup, G. M. 1965. *Transplantation* 3:387–401

159. Walford, R. L., Waters, H., Smith, G. S. 1970. *Fed. Proc.* 29:2011–17

160. Wigzell, H. 1965. *Transplantation* 3:423–31

161. Yamane, K., Nathenson, S. G. 1970. *Biochemistry* 8:4048–62

162. Yunis, E., Plate, J. M., Ward, F. E., Seigler, H. F., Amos, D. B. 1970. *Transplant Proc.* In press

OTHER REVIEWS OF MICROBIOLOGICAL INTEREST

1. Kirk, J. T. O. 1971. Chloroplast Structure and Biogenesis. *Ann. Rev. Biochem.* 40:161–96

2. Bishop, N. I. 1971. The Electron Transport System of Green Plants. *Ann. Rev. Biochem.* 40:197–226

3. Holzer, H., Duntze, W. 1971. Metabolic Regulation by Chemical Modification of Enzymes. *Ann. Rev. Biochem.* 40:345–76

4. Sussman, A. J., Gilvarg, C. 1971. Peptide Transport and Metabolism in Bacteria. *Ann. Rev. Biochem.* 40:397–406

5. Lucas-Lenard, J., Lipmann, F. 1971. Protein Biosynthesis. *Ann. Rev. Biochem.* 40:407–48

6. Perlman, D., Bodanszky, M. 1971. Biosynthesis of Peptide Antibiotics. *Ann. Rev. Biochem.* 40:449–64

7. Goldberg, I. H., Friedman, P. A. 1971. Antibiotics and Nucleic Acids. *Ann. Rev. Biochem.* 40:775–810

8. Echols, H. 1971. Lysogeny: Viral Repression and Site-Specific Recombination. *Ann. Rev. Biochem.* 40:827–54

9. Calvo, J. M., Fink, G. R. 1971. Regulation of Biosynthetic Pathways in Bacteria and Fungi. *Ann. Rev. Biochem.* 40:943–68

10. Smith, G. P., Hood, L., Fitch, W. M. 1971. Antibody Diversity. *Ann. Rev. Biochem.* 40:969–1012

11. Holloway, B. W., Krishnapillai, V., Stanisich, V. 1971. *Pseudomonas* Genetics. *Ann. Rev. Genet.* 5:Publ. Dec. 15

12. Hooker, A. L., Saxena, K. M. S. 1971. Genetics of Disease Resistance in Plants. *Ann. Rev. Genet.* 5:Publ. Dec. 15

13. Joklik, W. K., Zweerink, H. J. 1971. Morphogenesis of Animal Viruses. *Ann. Rev. Genet.* 5:Publ. Dec. 15

14. Fogel, S., Mortimer, R. K. 1971. Recombination in Yeast. *Ann. Rev. Genet.* 5:Publ. Dec. 15

15. Pink, R., An-Chuan Wang, Fudenberg, H. H. 1971. Antibody Variability. *Ann. Rev. Med.* 22:145–70

AUTHOR INDEX

A

C

D

H

L

Q

R

U

V

Y

SUBJECT INDEX

N

CUMULATIVE INDEXES

VOLUMES 21-25

INDEX OF CONTRIBUTING AUTHORS

INDEX OF CHAPTER TITLES

VOLUMES 21-25